SURFACE EFFECTS IN CRYSTAL PLASTICITY

NATO ADVANCED STUDY INSTITUTES SERIES

Proceedings of the Advanced Study Institute Programme, which aims at the dissemination of advanced knowledge and the formation of contacts among scientists from different countries.

The series is published by an international board of publishers in conjunction with NATO Scientific Affairs Division

A	Life Sciences	Plenum Publishing Corporation
B	Physics	London and New York
C	Mathematical and Physical Sciences	D. Reidel Publishing Company Dordrecht and Boston
D	Behavioural and Social Sciences	Sijthoff International Publishing Company Leyden, The Neth. and Reading, Mass., USA
E	Applied Sciences	Noordhoff International Publishing Leyden, The Neth. and Reading, Mass., USA

Series E: Applied Science - No. 17

SURFACE EFFECTS IN CRYSTAL PLASTICITY

edited by

R. M. LATANISION

associate professor of materials science and engineering
Massachusetts Institute of Technology
Cambridge, Mass., U.S.A.

and

J.T. FOURIE

head of the Electron Microscopy Division
National Physical Research Laboratory
Pretoria, South Africa

Springer-Science+Business Media, B.V. 1977

Proceedings of the NATO Advanced Study Institute on
Surface Effects in Crystal Plasticity, Hohegeiss, Germany,
September 5-14, 1975

ISBN 978-94-011-9693-2   ISBN 978-94-011-9691-8 (eBook)
DOI 10.1007/978-94-011-9691-8

Copyright © 1977 by Springer Science+Business Media Dordrecht

Originally published by Noordhoff International Publishing, division of A. W. Sijthoff International Publishing Company bv in 1977.

Softcover reprint of the hardcover 1st edition 1977

All rights reserved. No part of this publication may be reproduced, stored in a retrieval system, or transmitted, in any form or by any means, electronic, mechanical, photocopying, recording, or otherwise, without the prior permission of the copyright owner.

PREFACE

Friction, wear, adhesion, catalysis, oxidation, and corrosion are all examples of phenomena in which interactions occurring at the free surface undoubtedly play a major role. These are important phenomena which affect technology ranging from, for example, energy generation and conversion to chemical processing to biomedical prostheses. The same may be said of environmentally-induced embrittlement phenomena--stress corrosion cracking, liquid metal embrittlement, hydrogen embrittlement, etc.--and other manifestations of surface effects in crystal plasticity. Despite years of activity in this general area, however, we still do not know on an atomic scale why hydrogen embrittles iron and nickel but not copper, why Cl^- in aqueous solutions leads to the premature and often catastrophic fracture of otherwise ductile austenitic stainless steels, why various surfactants affect the mechanical behavior of some materials, what role the free surface plays in the deformation of metals, and the like. Enlightened understanding in these areas may well lead to the control and perhaps even useful application of some otherwise catastrophic embrittlement phenomena to technology--for example, in the fragmentation of materials as in machining operations. This, however, requires the concerted action of surface scientists and materials scientists alike. The surface analytical tools which are required to examine the atomic scale interactions that are of interest are by and large now available. It seemed clear to us that in order to get on with this job, surface scientists and materials scientists must meet together and must develop lines of communication as well as mutual understanding.

It was in this spirit that the Advanced Study Institute on Surface Effects in Crystal Plasticity was held from 5-14 September 1975 at the Panoramic Hotel in Hohegeiss, a small village in the

Harz Mountains near Hannover in the Federal Republic of Germany. Our roster included 71 delegates from 14 nations. This interdisciplinary Institute began with three introductory lectures which served to establish a sense of perspective. These were followed by five tutorial sessions on surface science and on crystal plasticity, the object of which was to develop for all participants a framework within which the theme of the ASI could be discussed and understood. In these sessions, the role of student and lecturer were effectively interchanged. Attention was then directed towards various manifestations of surface effects in crystal plasticity by means of a series of five sessions organized in workshop style. Each session began with a survey lecture in which critical appraisals of the phenomenology, existing models, and key issues were presented on the following topics: surface effects in uniaxial tension and fatigue, the influence of surface films on mechanical properties, chemisorption-induced variations in the plasticity and fracture of nonmetals, environmentally-induced lowering of surface energy and the mechanical behavior of solids, and surface effects in dissolution-related embrittlement phenomena. Short contributed papers were delivered subsequent to the survey lectures. The remainder of each workshop session was devoted to detailed discussions in the form of four independent workshop groups each of which was managed by a Discussion Leader, whose job it was to direct and develop the conversation, and a Recorder, who was responsible to see that an accurate record was kept of the points raised and of the essence of the discussion. The entire delegation then reconvened following the intensive discussions of the groups and each Discussion Leader presented a brief summary of his groups deliberations. Written versions of each summary were edited by the Session Chairman who then prepared a statement which appears in the Proceedings as the Workshop Summary. In total, five workshop sessions and twenty independent discussion groups were convened. The next general session considered technological applications of surface effects in crystal plasticity. The program concluded with a summary lecture and general discussions.

We made two excursions as a group during the Institute. The first was an afternoon visit following the tutorial lectures on Surface Science to Professor Bauer's surface analytical laboratories at Technische Universität Clausthal. The second was to Goslar—an ancient Imperial city which dates to 922 and is without a doubt one of the most historically interesting and fascinating cities in the F.R.G. We are grateful to Professor Bauer for his generous assistance in arranging both tours.

All of our delegates were cautioned at the outset that the Planning Committee had organized a hard working Institute, but they were also advised that we expected this to be a pleasant experience as well. It seems safe to judge from the general level of exhaustion and the convivial spirit of the entire delegation during the Institute that this initial admonition had not been made without basis. We worked hard--some days from early in the morning until late in the evening! But it is difficult if not impossible to recall another series of lectures, discussions and workshops that were on so consistently high a level. These intense and absorbing interactions are reflected in the photographs on the following pages. We are grateful to all of the delegates not only for their contributions and their endurance but also for their spirit of cooperation and good cheer which persisted throughout.

Thanks are also due to the members of the Planning Committee, each of whom contributed enthusiastically in the evolution of our program. Some people must be given a special vote of thanks: H. Mughrabi for his efforts on a number of special projects, F.R.N. Nabarro for taking on an arduous and unique fact finding excursion in the U.S. and Europe in preparation for his plenary lecture, D. J. Duquette for handling the logistics of Professor Nabarro's tour in the U.S., and R. W. Staehle who is responsible for most of the photographs shown on the following pages. We are sincere in our gratitude to all the sponsors of the ASI, without whose generosity and encouragement this meeting would not of course have occurred. Finally, we happily acknowledge the gracious assistance of the ASI secretaries, Pam Vincent and Jutta Sossong who were present in Hohegeiss, and Heather Kraemer who handled the final typing in Cambridge.

It was not anticipated that general agreement would be reached on the issues addressed during this Institute. Nor, in fact, did we achieve that situation as is reflected in the discussions and workshop summaries included herein. But, one of our objectives was to excite interdisciplinary communications and thereby to develop a lasting and productive link between surface science and materials science--a link through which significant academic and technological advances might emerge. We believe that this ASI met that objective.

R. M. Latanision
J. T. Fourie

ORGANIZATION

Chairmen:

R. M. Latanision
: Massachusetts Institute of Technology, U.S.A.

J. T. Fourie
: National Physical Research Laboratory, South Africa

Planning Committee:

C. B. Duke
: Xerox Webster Research Center, U.S.A.

D. J. Duquette
: Rensselaer Polytechnic Institute, U.S.A.

T. E. Fischer
: Exxon Corporate Research Laboratories, U.S.A.

J. P. Hirth
: Ohio State University, U.S.A.

I. R. Kramer
: Naval Ship Research and Development Center, U.S.A.

H. Mughrabi
: Max-Planck-Institut für Metallforschung, F.R.G.

W. M. Mularie
: 3M Company, U.S.A.

F.R.N. Nabarro
: University of the Witwatersrand, South Africa

P. Neumann
: Max-Planck-Institut für Eisenforschung, F.R.G.

R. A. Oriani
: United States Steel Research Laboratory, U.S.A.

R. W. Staehle
: Ohio State University, U.S.A.

A.R.C. Westwood
: Martin Marietta Laboratories, U.S.A.

Sponsors:

North Atlantic Treaty Organization
   and
Deutsche Forschungsgemeinschaft
Deutsche Physikalische Gesellschaft
U. S. Army Research Office - Durham
U. S. Office of Naval Research

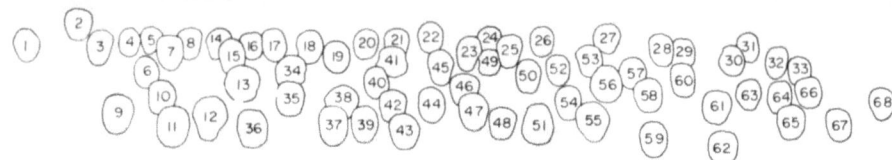

1. H. Henzler	18. H.-J. Engell	35. J. P. Hirth	52. W. R. Tyson
2. E. Bauer	19. M. Smialowski	36. R. M. Johnson	53. H.G.F. Wilsdorf
3. R. A. Oriani	20. E. D. Shchukin	37. R. E. Cuthrell	54. C. F. St. John
4. J. Lothe	21. H. Müller	38. F. C. Frank	55. J. Sossong
5. P. A. Clarkin	22. H. Viefhaus	39. H. Mughrabi	56. W. H. Weinberg
6. G. Champier	23. I. G. Greenfield	40. S. Kitajima	57. J. T. Fourie
7. F. Delamare	24. A. Broese van Groenou	41. Z. Szklarska-Smialowska	58. R. Bullough
8. A. George	25. P. Neumann	42. F.R.N. Nabarro	59. Ch. Schwink
9. S. Weissmann	26. A. Seeger	43. W. Schröter	60. N. H. Macmillan
10. T. E. Fischer	27. W. M. Mularie	44. H. C. Gatos	61. C. A. Brookes
11. A.R.C. Westwood	28. J. F. Prins	45. O. Lohne	62. M. Cetincelik
12. C. M. Preece	29. K. H. Johnson	46. M. V. Swain	63. J. A. Moskovitz
13. J. J. Mills	30. R. F. Firestone	47. D. Maugis	64. C. B. Duke
14. D. J. Duquette	31. H. P. Bonzel	48. I. R. Kramer	65. R. M. Latanision
15. A. S. Argon	32. G. Meister	49. Mrs. van Groenou	66. R. Gibala
16. N. V. Pertsov	33. J. H. van der Merwe	50. E. Kröner	67. D. L. Davidson
17. Z. S. Basinski	34. A.M.C. Oliveira	51. P. Vincent	68. R. W. Staehle

XII

XIII

CONTENTS

Page

Preface — V

INTRODUCTORY LECTURES

R. M. Latanision — 3
  Surface Effects in Crystal Plasticity: General Overview

F.R.N. Nabarro — 49
  Surface Effects in Crystal Plasticity: Overview from The Crystal Plasticity Standpoint

T. E. Fischer — 127
  Surface Effects in Crystal Plasticity: Overview from A Surface Science Point of View

TUTORIAL LECTURES ON SURFACE SCIENCE

C. B. Duke — 165
  Atomic Geometry and Electronic Structure of Solid Surfaces

H. C. Gatos and Jacek Lagowski — 221
  Space Charge Layers

W. Henry Weinberg — 267
  Surface Chemistry I: Multi-Component Systems

J. H. van der Merwe — 301
  Surface Chemistry II: Overlayers (New Phases)

TUTORIAL LECTURES ON CRYSTAL PLASTICITY

R. Bullough — 321
  Microplasticity

A. S. Argon — 383
  Inelastic Deformation and Fracture of Crystalline Solids

WORKSHOP SESSION 1: SURFACE EFFECTS IN UNIAXIAL TENSION AND FATIGUE

Z. S. Basinski
    Survey Lecture   433

Contributed Papers:

    D. J. Duquette, H. Hahn, and P. Andresen   469
        The Role of Surface-Environment Interactions on Cyclic Deformation

    H. Mughrabi   479
        A Surface Effect Specific to Cyclically Deformed BCC Metals - The Surface Roughening of Polycrystalline α-Iron During Cyclic Deformation

    O. Lohne   487
        Dislocation Mobility in the Surface Layer

    S. Kitajima   495
        Role of Free Surface in Yielding and Easy Glide Deformations of Highly Perfect Copper Crystals

    Ch. Schwink and H. Neuhäuser   505
        Microcinematographic Studies of the Development of Slip Bands

    J. T. Fourie   511
        The Plastic Flow Stress in the Core and Surface Regions of Deformed Crystals

    J. T. Fourie   527
        An Analysis of the Unloading Yield Point With Reference to the Pre-Macro-Yield Region

    H. Mughrabi   533
        New Studies of Surface Effects in Deformed Copper Single Crystals

    S. Weissmann   543
        Influence of Surface Plasticity on the Deformation Behavior and Fracture Mode of Silicon

	Page
E. Kröner 　On the Residual Stresses in Plastically 　Deformed Polycrystals and Composites	551

Workshop Summary　　　　　　　　　　　　　　　557

WORKSHOP SESSION 2: THE INFLUENCE OF SURFACE FILMS ON MECHANICAL PROPERTIES

	Page
H.G.F. Wilsdorf and G. E. Ruddle 　Survey Lecture	565

Contributed Papers:

	Page
R. M. Johnson 　Surface Damage Resulting from Film Cracking	593
V. K. Sethi and R. Gibala 　Effect of Anodic Oxide Films on Low Temperature Mechanical Behavior of Niobium Single Crystals	599
I. G. Greenfield and A. Purohit 　The Effect of Near-Surface Diffusion Layers on Fatigue of Copper	609

Workshop Summary　　　　　　　　　　　　　　　621

WORKSHOP SESSION 3: CHEMISORPTION-INDUCED VARIATIONS IN THE PLASTICITY AND FRACTURE OF NONMETALS

	Page
N. H. Macmillan 　Survey Lecture	629

Contributed Papers:

	Page
R. F. Firestone and A. H. Heuer 　Surface Effects on the Yield Point of Sapphire	663
C. A. Brookes 　Microdeformation of Hard Non-Metallic Crystals By Softer Indenters and Sliders	671
W. Schröter and P. Haasen 　The Chemomechanical Effect in Semiconductors	681

	Page
Workshop Summary	689

WORKSHOP SESSION 4: ENVIRONMENTALLY-INDUCED LOWERING OF SURFACE ENERGY AND THE MECHANICAL BEHAVIOR OF SOLIDS

E. D. Shchukin Survey Lecture	701

Contributed Papers:

E. M. Gutman Chemical-Induced Variations in the Plasticity and Fracture of Metals and Minerals - Chemomechanical Effect	737
Workshop Summary	745

WORKSHOP SESSION 5: SURFACE EFFECTS IN DISSOLUTION-RELATED EMBRITTLEMENT PHENOMENA

H.-J. Engell Survey Lecture	749

Contributed Papers:

R. E. Cuthrell Mechanical Behavior of Oxide Free Stainless Steel Surfaces in A Low Pressure Hydrogen Environment	773
W. Frank and L. Graf A Unifying Model of Intergranular Corrosion, Intergranular Penetration of Liquid Metals, Stress Corrosion Cracking and Liquid Metal Embrittlement of Substitutional Solid Solutions	781
W. R. Tyson Cleavage vs. Shear At Crack Tips in Metal Crystals	793
D. L. Davidson Nondestructive, Near-Surface Plasticity Determination by Electron Channeling	801
Z. Szklarska-Smialowska Correlation Between Pitting and Stress	811

Corrosion Cracking of Stainless Steels
in Chloride Solutions

R. W. Staehle   819
What I Know and Would Like to Know
Concerning Environmental Effects on
the Mechanical Properties of Metals

Workshop Summary   827

APPLICATIONS OF SURFACE EFFECTS IN CRYSTAL PLASTICITY TO TECHNOLOGY

A.R.C. Westwood and J. J. Mills   835
Survey Lecture: Applications of Chemomechanical
Effects to Fracture-Dependent Industrial Processes

N. V. Pertsov   863
Environmentally-Assisted Hard Materials
Machining

C. M. Preece   889
Erosion of Metals and Alloys

I. R. Kramer   911
Enhancement of Fatigue, Creep, and Stress
Corrosion Resistance By Surface Treatments

J. F. Prins   925
Microcutting: A Method to Study Surface
Plasticity at Very High Strain Rates

SUMMARY SESSION

F. C. Frank   933
Surface Effects in Crystal Plasticity:
Concluding Summary Lecture

LIST OF PARTICIPANTS   939

INTRODUCTORY LECTURES

Session Chairman: J. T. Fourie

SURFACE EFFECTS IN CRYSTAL PLASTICITY: GENERAL OVERVIEW

R. M. Latanision

Department of Materials Science and Engineering
Massachusetts Institute of Technology
Cambridge, Massachusetts 02139, U.S.A.

ABSTRACT. This paper which is intended to establish a general perspective, includes discussion of not only the phenomenology and current mechanistic understanding of manifestation of surface effects in crystal plasticity but also the application of these phenomena to technology. At the outset, certain aspects of the surface chemistry and physics of various crystalline inorganic solids will be examined in order to illustrate that the surfaces of such solids are likely to be structurally, chemically, and electronically distinct from the bulk. This is followed by brief discussion of the topics to be treated in the workshop sessions which occur later in the Institute program--namely, surface effects in uniaxial plastic deformation, the influence of surface films on mechanical properties, chemisorption-induced variations in the plasticity and fracture of metals and nonmetals, and surface effects in embrittlement phenomena. Throughout these discussions emphasis will be placed on areas of controversy and, likewise, an attempt will be made to invoke some of the principles of surface chemistry and physics in describing each topic. The remainder of the paper is intended to demonstrate that our understanding of surface effects in crystal plasticity, imperfect though it may be, has been applied by various investigators with remarkable success to such technological problems as metal cutting, the machining of ceramics, the rapid excavation of hard rock, improvement in the fatigue life of various alloys, and, potentially, even to the control of earthquakes!

1. INTRODUCTION

The fact that the mechanical properties of solids are sensitive to the environment to which these solids are exposed is, of course, not newly discovered. I am reminded by Westwood [1] that more than 100 years ago Reynolds [2] first associated certain detrimental effects on the ductility of iron with the presence of hydrogen. Not only is hydrogen embrittlement still a major industrial problem--though there is considerable disagreement, it now seems that in the case of nearly every major engineering alloy system there are some researchers who believe that hydrogen is responsible, at least in part, for serious mechanical degradation--but it seems safe to say that in a mechanistic sense, we still do not know that hydrogen (but not nitrogen or helium, etc.) does on an atomic scale to induce this degradation. Likewise, it is almost 50 years ago that the late Academician Rebinder [3] in the Soviet Union first reported that the plasticity and fracture of solids were remarkably affected by surface-active media. Here, too, years of disagreement and uncertainty in interpretation have followed that first report. More recently, an interesting and often spirited debate has developed over the role of the free surface, in the absence of specific environments, in the plasticity of crystalline solids. Despite some very clear indications that the free surface does play a role in determining bulk mechanical properties, it is nevertheless true that surface effects have by and large been ignored in modern theories of work hardening and in general discussions of crystal plasticity.

In what follows, I shall attempt to objectively establish a general sense of perspective, with due regard given to the interdisciplinary character of this meeting. I will, therefore, describe some of the phenomenology of surface effects in crystal plasticity pointing out areas of controversy. Subsequently, attention will be directed to a brief summary of applications of this to areas of technology. Where possible I will emphasize areas in which an interdisciplinary surface science/materials science approach might be useful in providing new insight into the mechanism of some manifestations of surface effects in crystal plasticity. I hasten to add that so broad a treatment cannot hope to be comprehensive. Indeed, this overview like most others may ultimately be characterized by the nature of its omissions. On that point, my detailed views on many of the issues treated here--as well as others that are not--are given in more lengthy accounts [4-7] and are left to the reader.

The logical starting place for this discussion is the surface itself. In the next section, therefore, is a brief account of some of the properties of surfaces, particularly those which may well affect mechanical behavior.

2. SOME ASPECTS OF THE PHYSICS AND CHEMISTRY OF SURFACES IN RELATION TO MECHANICAL PROPERTIES*

---

*An extended version of this discussion is given in Reference [5].

The simple view of the surface or interface between a solid and its environment as a sharp discontinuity in an otherwise infinite continuum is adequate for many purposes. In many cases, this simplification is invoked in order to arrive at solutions to otherwise intractable analytical problems. Perhaps the classic example here is the concept of an image dislocation which has an electrostatic analogue. In fact, however, it must be recognized that there is a region of finite thickness between any two media that may differ structurally, chemically, and electronically from either. (For recent general treatments, see References [8 and 9]).

Now, it is also recognized that the mechanical behavior of a crystalline solid is determined principally by the generation, motion, and interaction of dislocations. The fact that the plasticity and fracture behavior of such solids can be affected by environmental circumstances suggests that the surface and its condition must somehow affect dislocation behavior. Indeed, it has been proposed over the years that the surface or near-surface layers may serve as regions in which dislocations are easily generated or as barriers to the escape of dislocations from within the solid and that these functions may be modified by the presence of adsorbed surface-active species, charge double layers, nonequilibrium concentration of solute atoms or vacancies, solid surface films, corrosive (solvent) environments, and the like. In a simplified form, the formidable path that a dislocation must travel as it approaches the surface of a crystalline solid is suggested schematically in Fig. 1 in which structural, chemical, and electronic perturbations at the surface are indicated, for the purpose of illustration, as lattice parameters variations, solute concentration gradients, and an electrical double layer, respectively. By and large, the impact of these perturbations, particularly surface structure, on plasticity is poorly understood. In the next section, this situation is discussed briefly.

2.1 SURFACE STRUCTURE

The model that is often used for the structure of crystalline surfaces derives from the idea of Kossel [10], Stranski [11], and Burton, Cabrera, and Frank [12]. This model which has been referred to as the terrace-ledge-kink model (TLK) has been used extensively in discussions of crystal growth, evaporation, crystal dissolution, and surface diffusion. The TLK model, Fig. 2, includes low index terraces and monatomic height ledges which are occasionally displaced by atomic distances at kink sites. In considering such models, however, it should be remembered that they represent an ideal surface, no allowances having been made, for example, for the relaxation of surface atoms which seems analytically likely, particularly in the direction perpendicular to the surface and perhaps as well parallel to the surface. On this point, there is a vast literature of relatively recent origin

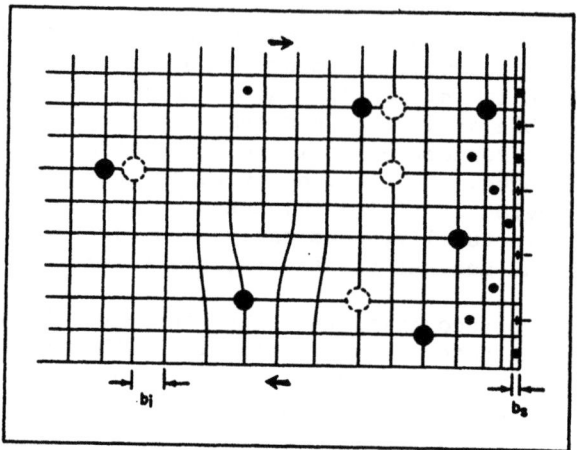

Fig. 1 Schematic illustration of some of the structural, chemical, and electronic perturbations seen by a dislocation as it approaches a crystal surface (after Latanision [5]).

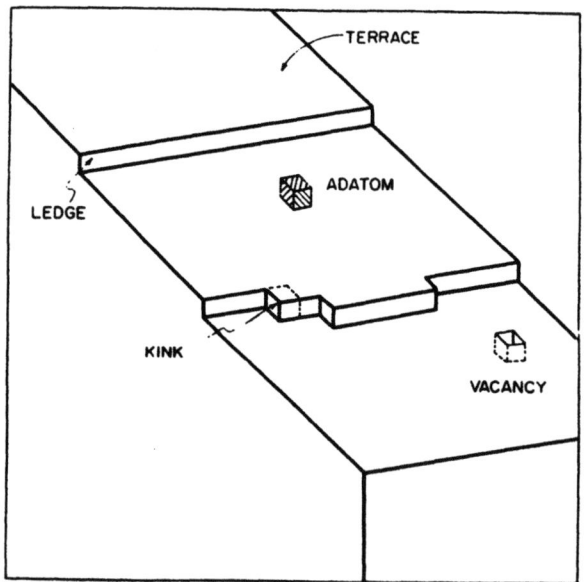

Fig. 2 Schematic of the terrace-ledge-kink (TLK) model for a crystal surface (after Latanision and Westwood [4]).

related to the structure of crystalline surfaces. There is also a considerable uncertainty in the interpretation of some information collected over the years by surface analytical tools such as LEED [13]. Some of this uncertainty appears to have been resolved by refined methodology and diffraction theory. At any rate the general consensus based upon observations seems to be that atoms in the topmost layer of many <u>clean</u> metal surfaces have the same arrangement in the plane of the surface as atoms on parallel planes in the body of the crystal, with perhaps only slight adjustment of the spacing of atoms normal to the free surface. The magnitude of this variation in lattice parameters seems uncertain and the many claims--some in excess of 10% perturbations--are too numerous to mention here. Duke's opinion [13] is that this variation (expansion or contraction) is not likely to be more than ±5% of the bulk lattice parameters for most clean single crystal surfaces--not an insignificant variation. There are exceptions, of course, and indeed crystal periodicity and/or symmetry are sometimes reported to be perturbed at the surface of presumably clean metals [8,9]. For example, phase transformations have been observed on platinum surfaces as a function of temperature though similar transformations are not observed in the bulk. Likewise a hexagonal surface structure has been reported on the cube face of fcc gold. Although there is some debate as to whether some of the surface structures reported for metals are the property of clean metal surfaces or associated with surface impurities, recent evidence tends toward the former view. In contrast, it is apparently well agreed that constructive rearrangements (reconstruction) of varying complexity do in fact occur on covalently bonded semiconductor surfaces (Si, Ge, etc.) and surface superlattices with unit cells ranging from 2-8 times the unit cell dimension of the substrate have been reported.

One of the most surprising and exciting observations to come from studies of surface structure is the large effect small amounts of adsorbed gas atoms appear to have on the structure of otherwise clean surfaces. Indeed, it seems that in many cases, adsorbed gases dislodge substrate atoms from their normal positions leading to reconstruct surfaces composed of both substrate atoms and chemisorbed gas atoms in periodic arrays. Apparent reconstructive interactions following chemisorption have been observed in many circumstances and are summarized by Somorjai [9].

Despite all that is known about the structure of crystalline surfaces, very little consideration appears to have been given to the influence of surface structure on mechanical properties. Fleischer [14] has suggested that the gradual change in lattice spacing over 4-5 atom distances from the surface may lead to the development of an obstacle to the emergence of dislocations from the crystal. Likewise, it seems not unlikely that ledges or kinks on otherwise flat surfaces may act as stress concentrators [15] and increase the probability that dislocations sources may operate preferentially in the near-surface layers. The potential catalyzing effect of vicinal surfaces on dislocation nucleation [16] may also play a role in determining the mechanical behavior of crystalline solids. While there are some other possibilities as well, the fact remains that, qualitatively and quantitatively, the influence of surface structure on mechanical properties is virtually unexplored. It is difficult to say now how significant these aspects of surface structure are relative to mechanical properties. It does seem, however, that there exist several good analytical problems worthy of attention.

2.2 CHEMICAL SEGREGATION AT FREE SURFACES

In addition to the contamination of the surface by adsorption of impurities from an external source, which was discussed briefly in the preceding section, impurities which originate in the bulk of the solid may also accumulate at the free surface. In short, the segregation of bulk impurities may occur as well at the free surface as at internal interfaces (grain boundaries, twin boundaries, etc.) or at dislocations. The driving force for solute segregation at a free surface may be thought of in terms of a relaxation of near-surface elastic distortion by equilibrium or Gibbsian adsorption of solutes of appropriate size. Alternately, in a thermodynamic sense, any solute which lowers the surface energy of the solvent might be expected to segregate to the free surface. In either case, this would lead to an enrichment or depletion of solute over a few atom layers from the surface, since lattice distortion is thought to drop off rapidly with distance into the bulk and essentially extinguishes by the fourth or fifth atom layer from the surface as indicated schematically in Fig. 1.

Aust, Westbrook, and co-workers [17,18] have also pointed out that non-equilibrium solute segregation may occur because of the development of a gradient in vacancy concentration near free surfaces, since the surface is an effective sink for vacancies. In those cases where solute-vacancy interactions are strong, it is suggested that the vacancies will diffuse toward the surface dragging impurities with them, Fig. 1. The vacancy-solute couples dissociate at the surface, vacancies are annihilated, and a net excess of solute remains at or near the free surface. Extrapolating measurements of segregation at internal interfaces to the free surface, one may predict that non-equilibrium segregation near the latter may occur to depths on the order to several tens of microns depending upon the system involved.

Solute segregation near the free surface may well exert an influence on every aspect of surface properties and behavior, not the least of which is mechanical behavior. It is perhaps more obvious, however, that thermionic emission, surface diffusion, surface energy, surface structural transformations, catalytic efficiency, etc., are more likely to be affected by perturbations in surface composition. On the other hand, evidence [17,18] for nonequilibrium segregation is considered to account for significant increases (or decreases) in microhardness near grain boundaries and, in some cases, near the free surface over distances covering about 100 µm. Hardening is presumably due to the formation of solute clusters [17] or secondary vacancy defects [18], respectively. Likewise, considerable importance has been attached to the equilibrium segregation of impurities at internal interfaces (grain boundaries) in understanding various embrittlement phenomena [19]. In most cases, embrittlement is thought to be due to solute-induced decreases in cohesion. More recent work by Latanision and Opperhauser [20], however, has indicated that segregation alone is not always sufficient to lead to embrittlement but that embrittlement may be a function of the subsequent interaction between the segregated impurities and the environment to which the solid is exposed--i.e., to impurities and microchemistry.

It should also be appreciated that near-surface concentration gradients may also affect the flow stress and tensile behavior of metals. Indeed, in view of recent indications that yielding of metal monocrystals begins in the near-surface layers (see Section 3.1.1) recognition of the potential effects of surface contamination becomes even more important. An indication of the sensitivity of tensile behavior to solutes may be seen by studying the effect of cathodically-produced hydrogen on the behavior of (impure) nickel monocrystal electrodes [21]. In Fig. 3, the behavior of a crystal deformed in air is indicated by Curve C, while B shows the effect of simultaneous hydrogenation and deformation, and Curve A, the effect of precharging with hydrogen followed by simultaneous charging and deformation. The important point to note here is that serrated yielding occurs when the specimens are deformed and cathodically charged simultaneously, even if precharging is eliminated. Serrated yielding implies interaction between mobile dislocations and solute atoms, and it is significant that this occurs even though calculations indicate that the depth of penetration of hydrogen should not have been greater than 15 microns and must have been considerably less in the early stages of deformation. The implications of this work are, therefore, that slip begins in the near-surface layers of the crystal and that the resultant stress-strain behavior is dependent on the composition of the near-surface regions of the solid. Other observations consistent with this view have been summarized recently [4,7].

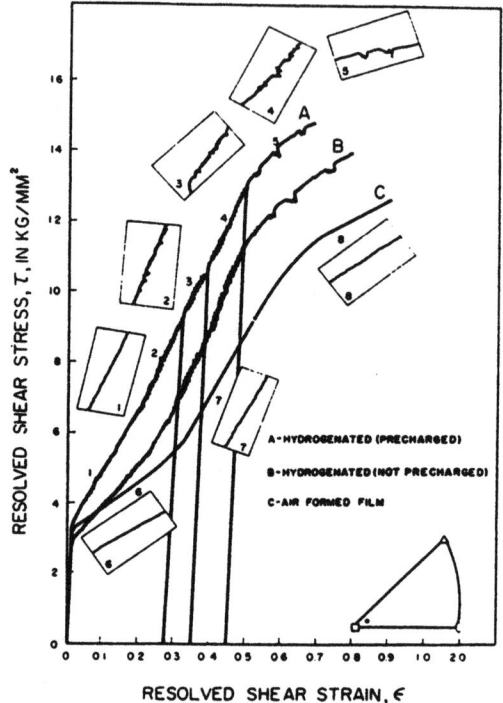

Fig. 3 Effect of hydrogen charging on the deformation on nickel monocrystals. Insets show detailed recorder tracings (load-extension) (after Latanision and Staehle [21]).

2.3 ELECTRONIC PROPERTIES OF SURFACES

In addition to structural and chemical singularities, it should be recognized that at the surface there is in general an electrostatic potential difference associated with some kind of charge double layer [22]. In the case of a metal surface in a vacuum, for example, we know that quantum mechanically there is a finite probability of finding an electron outside the metal surface. Hence, there is an electron excess just beyond the surface and a deficiency in the near-surface region of the solid. The result is an electrical double layer or dipole with its negative side outward--i.e., the doulbe layer creates an electrostatic potential that is more positive inside than outside as indicated schematically in Fig. 1.

While the above simplified illustration of a complex problem serves to indicate that electrical double layers are characteristic of the interface between almost any two phases, our major interest for the purpose of this discussion is the double layer present at the solid-electrolyte interface. I will return to this point later. It should be sufficient at this stage to point out that because of the high density and mobility of conduction electrons in a metal, interfacial electric fields cannot penetrate more than an angstrom of so into the bulk of a metal crystal. For nonmetals, however, because of the lower density and mobilities of the charge carriers involved, the influence of such fields may be sizable and much greater depth, for example, in excess of 1 μm for insulator materials. As we shall see later (Section 3.3.2), changes in the space charge distribution appear to exert a significant influence on mechanical behavior.

3. THE PHENOMENOLOGY OF SURFACE EFFECTS IN CRYSTAL PLASTICITY

In this section I will briefly describe some of the issues of importance to the theme of each workshop session, throughout attempting to invoke some of the principles of surface chemistry and physics in understanding these phenomena.

3.1 SURFACE EFFECTS IN UNIAXIAL TENSION

The role of the free surface in the plastic deformation of homogeneous and presumably clean* crystalline solids has been the subject of considerable interest, particularly recently. The details of these relatively many studies are complex and often confusing and I think need not be treated here. These details have been discussed elsewhere [4,7] and will be presented throughout the course of this Institute. In essence the centers of controversy focus on two main issues: (1) the influence of the surface on dislocations multiplication; and (2) the distribution of dislocations and consequent flow stress gradients which might be expected after large scale yielding has occurred--i.e., the hard-versus-soft surface layer controversy.

3.1.1 The Surface and Dislocation Sources

A carefully prepared and undeformed metal monocrystal typically contains about 10^6 cm of dislocation line per cm^3, although wide variations are possible. This dislocation density rises with increasing deformation, perhaps to a value as high as 10^{12} cm/cm^3 in heavily cold worked specimens. The means by which dislocation multiplication occurs is not surprisingly an important aspect of the theory of dislocations [23,24]. From the point of view of this conference, the effect of the surface on dislocation multiplication is also a central issue. In short there are at present three models which consider the role of the surface in this respect, each of which tries to answer the question of whether the surface

---

*It is now very obvious indeed that the mechanical properties of crystals with clean surfaces in a surface analytical sense have never really been examined.

in fact performs in the role of an easy source of dislocations, an obstacle to the emergency of dislocation, or perhaps both.

The differences between each model are shown in Fig. 4. In each case the mechanism of dislocation generation is essentially that proposed by Frank and Read in 1950 [25] (and which now remains the most likely multiplication mechanism) although the location of the source relative to the surface differs in each model. Frank-Read sources have in fact been observed in operation, the classic observation being that of Dash in 1956 [26].

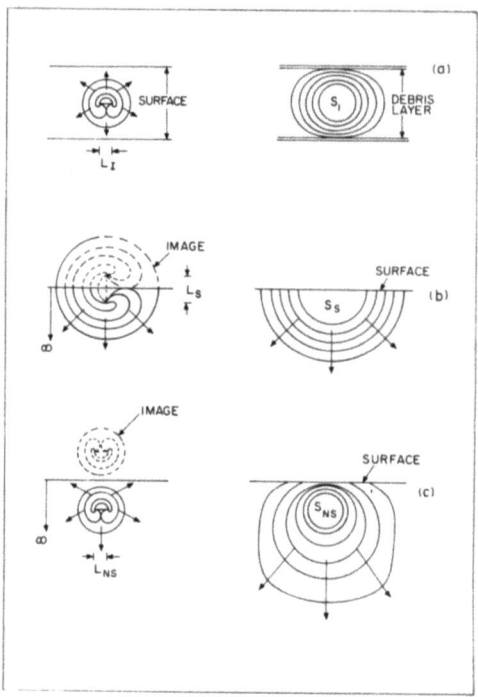

Fig. 4 Schematic illustration of (a) the "debris-layer" model, (b) a surface source of dislocations, and (c) dislocation generation by a near-surface source (after Latanision and Westwood [4]).

In the debris layer model, Fig. 4(a), first suggested by Kramer [27] in 1961, some dislocations leaving the crystal in the early stages of deformation are presumed to be trapped in the near-surface region, forming a zone of high dislocation density--the debris layer--which introduces a back stress opposing the motion of other dislocations in the surface region. The near-surface layers thus act as an obstacle to dislocation egress. Typically this debris layer is considered to be on the order of 50-100 μm in thickness. It seems implicit in this model that dislocations are generated by sources far removed from the surface-i.e., internal Frank-Read sources. Presumably these sources lie outside the debris layer and, hence, are elastically well removed from the range of influence of image forces which are considered important in Fig. 4(b) and (c). Dislocations are thus presumably generated somewhere in what might be considered the core of the crystal and move toward the surface.

Others have invoked the suggestion by Fisher [28] that the free surface provides an easy source of dislocations. The basis for this view is that a dislocation line terminating unpinned at the surface, but being pinned somewhere in the interior (at a distance L from the surface), behaves elastically as a dislocation line in a uniform medium which is anchored by the internal pinning point and its image above the surface. Thus, the effective length of this surface Frank-Read source may be twice its actual length as shown in Fig. 4(b). Since the stress required to operate such sources is inversely proportional to the length of the dislocation segment between pinning points, a surface source (2L) should operate at some stress lower than that required to operate an internal source of the same length (L) and, in fact, at half the stress if the dislocation line happens to lie normal to the surface. Half loops produced by such a source would then propagate from the surface into the interior.

The third alternative, namely that the surface acts simultaneously as both a source and obstacle was first suggested by Kuhlmann-Wilsdorf [29]. Latanision and Staehle [30] extended this argument and proposed that the most likely dislocation generator in the early stages of deformation is a near-surface source. This is pictured, Fig. 4(c), as a source of the Frank-Read type which happens to lie near enough to the surface (on the order of a micron away) that it operates in the presence of its own image, but, unlike the Fisher course, does not presume to have an arm swinging freely at the surface--i.e., both pinning points are within the crystal. The image force diminishes as the reciprocal of the distance from the surface, and, hence, the segments near the surface are strongly attracted toward the surface while the motion of segments running into the interior is more weakly resisted by a similar but proportionately smaller attractive image force. Sumino [31] had considered similar sources analytically in 1962 and showed that these sources produced dislocations more easily than otherwise equivalent internal sources. Such sources produce loops just as do internal sources, but because of the proximity of the surface, the half loops nearest the surface may either slip readily out of the crystal or be held up (i.e., the surface acts as an obstacle) depending upon environmental conditions, while the remaining half loops move into the interior. There are sensible reasons for believing that under normal atmospheric conditions the surface of a crystal should act as an obstacle.

Among others, surface films are almost certainly present. Likewise, composition gradients (segregation) are likely to occur near the surface of annealed single crystals, hence, affecting the motion of dislocation into such regions (see [4] and [7] for summary). One might therefore suspect that dislocations may accumulate at the surface, much as Kramer [27] suggests, producing a back stress which limits the operation of initially active near-surface sources. On the other hand, if the surface is slowly but actively dissolved (i.e., in the absence of surface films), potential barriers may be removed to an extent and the development of a back stress which would otherwise limit the operation of a given near-surface source would be delayed. One might expect that in the latter case a given source might generate a relatively unlimited number of loops which upon emerging from the crystal would produce slip steps with much greater than ordinary step height. Precisely this observation has been reported by Latanision and Staehle [30]. They found that in Stage II, for example, step heights on 99.8% nickel monocrystals actively dissolved during deformation were about 2000 $\underline{b}$—more than an order to magnitude larger than for crystals deformed in the absence of surface dissolution.

There is increasing evidence that in monocrystalline metals [30,32-40] as well as nonmetals [41-43] yielding does in fact begin somewhere near the free surface. Likewise, there is reason to suspect that slip processes in the surface grains of a polycrystal are relatively less restricted compared to grains in the interior [44] and etch pitting studies by Vellaikal and Washburn [40] indicate that preferential multiplication of dislocations occurs near the external surface of copper polycrystals. It appears, therefore, that one must decide whether the Fisher or near-surface models best account for these observations. My own view is that the Fisher surface source, while it is attractive in principle, is not likely to operate in practice as has been discussed elsewhere [4,7]. In the first place, the mechanical properties of atomically clean surfaces have yet to be examined. Specimens examined in the laboratory ambient are unavoidably contaminated with adsorbed gas or, in most cases, oxide films. In such situations one would expect that the free end of any potential Fisher source would be pinned. Likewise, one must wonder about the possible influence of segregates near the free surface. Moreover, even if it were possible to produce and maintain a clean free surface on tensile specimens, surface atomic relaxation and reconstruction may conceivably interfere with the operation of such sources. In short, the surface is not the simple discontinuity that it is presumed to be in the Fisher source model, and it seems to me, therefore, that the operation of classical surface sources is unlikely under ordinary circumstances.

Fourie and Dent [33] have invoked the near-surface model in

explaining flow stress gradients in Cu-Al monocrystals, and some others [38,40] have proposed explanations for preferential yielding in the surface layers by models similar to that in Fig. 4(c), but taking into account the possibility that the active segment may be inclined to this surface rather than parallel to it. In either case, the source operates in the presence of its image with both pinning points just inside the crystal surface. At any rate, it seems safe to conclude that the details of dislocation multiplication and the influence of the surface have yet to be clarified. Much of the debate on this point has been summarized elsewhere [4] and need not be treated further here. The view does emerge, however, that in the early stages of plastic deformation the net flow of dislocations is expected to be into the crystal from the surface.

3.1.2 The Hard-versus-Soft Surface Layer Controversy

The question of what role the surface plays after large-scale yielding has occurred is much debated. Again, this has been summarized in detail elsewhere [4]. In essence, Kramer [27(b)] has made the striking observation, Fig. 5, that the initial yield

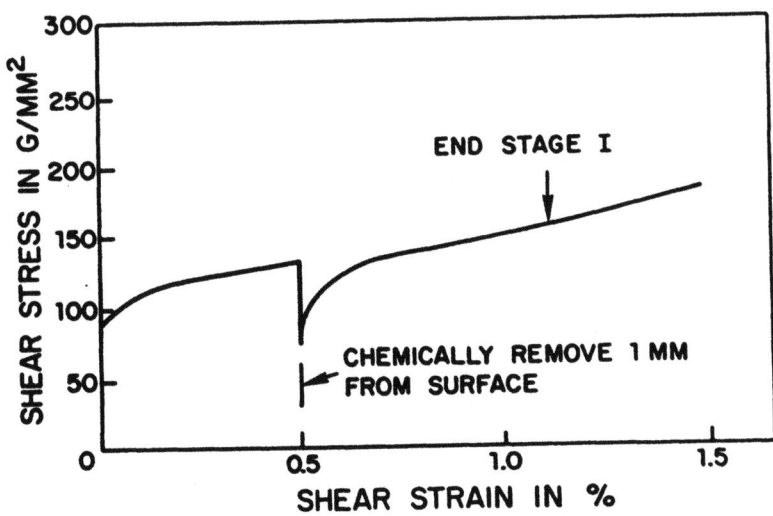

Fig. 5   Effect of electrochemically removing 1 mm from the surface of an aluminum monocrystal following 0.5% strain. Note recovery of the original yield stress upon reloading (after Kramer and Demer [27b]).

stress and work-hardening behavior of an aluminum monocrystal could be recovered by chemically removing a 1 mm envelope from the surface, the implication being that work hardening in Stage 1 is concentrated in the near-surface region. Moreover, continuous surface removal during deformation was shown to affect the work-hardening characteristics of fcc monocrystals, and similar observations have been reported by Latanision and Staehle [20]. Kramer has interpreted this in terms of the formation of a dislocation-rich (and, hence, hardened) debris layer in the surface regions of plastically deformed crystals, as explained earlier. The debris layer is considered to impede the motion of subsequent dislocations and serves as a barrier against which they pile up. The consequent debris layer (back) stress is reportedly dissipated by dissolution or by allowing specimens to relax without dissolution for various periods of time. This layer is presumed to extend to depths of about 60 µm in aluminum and 100 µm in gold and iron, for example. Kramer has presented a recent review of his work in Reference [45].

Kitajima et al. [34] also suggested that a hardened subsurface layer develops just after yielding, consistent with his etch pit observation that the dislocation density within 70 µm of the surface in lightly deformed copper crystals is higher than at a point further into the interior. By virtue of the geometry of their crystals, however, Kitajima et al. conclude that this result is not likely to derive from the accumulation at the surface of dislocations emitted from internal sources, i.e., the debris-layer model. Rather, they prefer the view that dislocations are generated by surface sources and that the advance of primary glide dislocations into the crystal is hindered by the creation of barriers developed through interacting near-surface primary and secondary dislocations, evidence for preferential activation of sources on conjugate and critical slip planes near the surface also having been observed.

It is interesting that based on the observation that work hardening in Stage I can be completely eliminated by removing the surface layers, Kramer [46] suggests that in Stage I additional internal dislocation obstacles are not formed in sufficient numbers to affect mechanical behavior--i.e., the Stage I work hardening is confined to the surface region. Moreover, it seems reasonable on this basis to suppose [46] that secondary slip systems may be operative in Stage I deformation but only in the surface region. This implies that Stage II deformation should begin first at the surface and propagate inward since the transition to Stage II, on the above basis, seems more likely in the surface than in the interior. What is interesting is that this is just opposite to the point of view developed by Fourie [47], namely, that the transition to Stage II occurs first in the core of deformed crystals. Indeed, while Kramer proposes a hard surface-soft core model, Fourie imagines just the opposite with, as is also true of Kramer's view, considerable evidence being available to support his view. In essence, Fourie has observed that flow stress gradients develop during plastic deformation and that, indeed, the flow stress near the surface is less than in the core. These dynamical flow stress measurements are made by slicing larger (14 mm wide in the direction of the primary slip vector) predeformed crystals into thin component crystals (ranging from 0.065-0.6 mm in thickness) and then restraining these components. One result is shown in Fig. 6. Notice that in initially large crystals, the soft surface layer appears to extend to distances into the bulk on the order of 2 mm--comparable to the mean free path of edge dislocations at the beginning of Stage I. Mughrabi [49], on the other hand, has observed evidence for similar gradients in crystals with diameters not much larger than the mean free path. Fourie [47] initially accounted for this flow stress gradient on the basis of a model that assumed a uniform distribution of dislocation sources throughout the crystal, but later with regard to Cu-Al monocrystals [33] preferred a model based on near-surface sources.

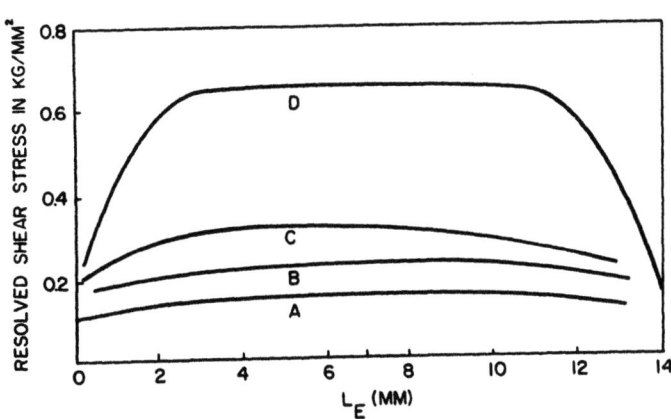

Fig. 6 The flow stress distribution in copper monocrystals plotted as a function of the length of the glide path of edge dislocations, L_E, which has the value 0 and 14 mm at the original surfaces. Curve A is for an as-grown crystal; Curves B, C, and D are for prestrains of 0.02, 0.029, and 0.058, respectively (after Fourie [47]).

In essence, it is expected that plastic deformation gives rise to a disparity of dislocation signs between the surface and interior, Fig. 7 [48]. Assuming that work hardening occurs by a mechanism in which the transition from Stage I to Stage II is initiated by the creation of dipole bundles, Fourie suggests [47] that a deficiency of dislocations of either sign near the surface would delay the onset of Stage II hardening there. In contrast, in the core or central region, the density of dislocations of both signs might be expected to be about equal. Hence, on this basis, the interior of such crystals should work-harden more rapidly than the surface, and one might envision that even in a crystal deformed macroscopically into Stage II, the surface layer might still be deforming in a Stage I mode. Indeed, by means of transmission electron microscopy Fourie [47] and later Mughrabi [49] have observed precisely that. For a crystal deformed into Stage II, the dislocation distribution near the surface is still typical of Stage I processes, Fig. 8(a)--more single primary dislocations and isolated edge dipoles--whereas the characteristics of Stage II hardening--cell formation, higher dislocation density, etc.-- are observed in the interior, Fig. 8(b).

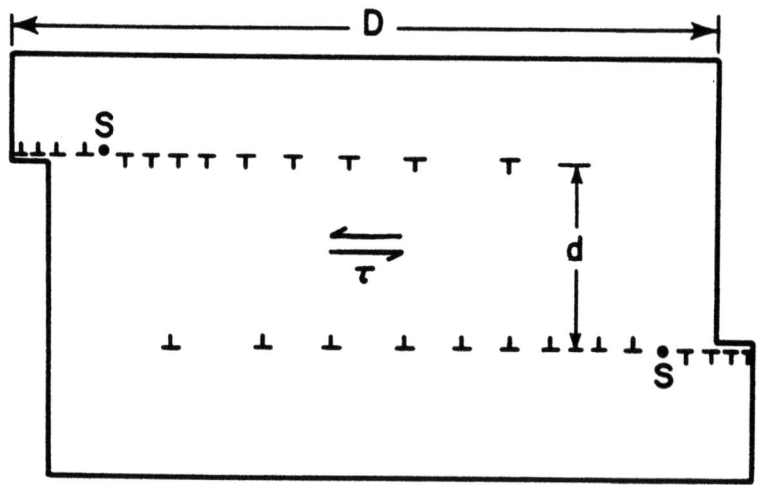

Fig. 7   Model of near-surface source operation illustrating the action of the free surface as a barrier to dislocation emergence as well as the disparity of dislocation signs developed by the glide of dislocations into the bulk from the near-surface layers (after Latanision [48]).

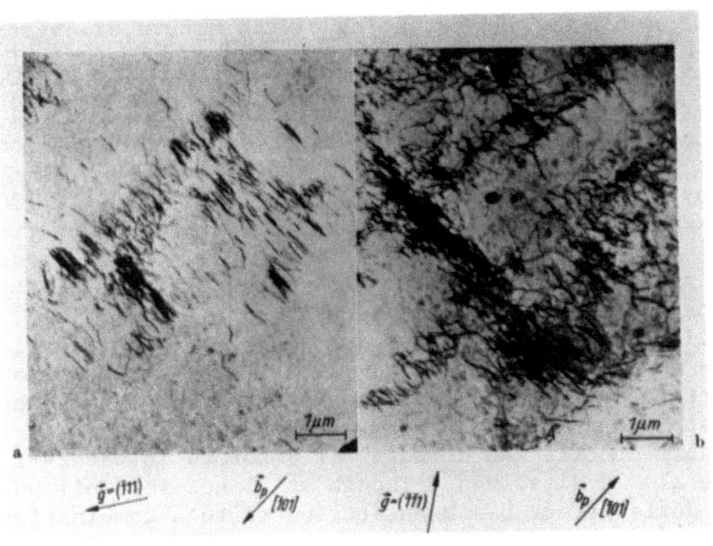

(a) (b)

Fig. 8 Dislocation arrangements in a copper crystal deformed into Stage II (a) 0.5 mm from the surface, (b) center (after Mughrabi [49]).

There is much more to be said regarding the hard-versus-soft surface layer controversy, but I believe the above represents the essence of the problem. Moreover, this will be treated in more detail by Basinski and in the workshops. This controversy has been much discussed recently (see discussion to References [45] and [50] as well as References [7], [51], and [52]). While there are clearly great differences in opinions, what is necessary now is to determine, opinions aside, if apparently divergent experimental observations might be understood from some rational perspective. More often than not, the experimental bases for the various views which have been developed have been so dissimilar and vague in the literature that direct comparisons have been quite impossible. Inquiries regarding experimental details should be tractable at this Institute. Various questions have been raised about the possible influence of (a) specimen bending in Fourie's dynamical flow stress measurements (see discussion to [50]), (b) dislocation loss and rearrangement in thin foil searches for the debris layer [51], (c) solid surface films present on the surface of electrodes during electropolishing in regard to Kramer's surface removal experiment [7], (d) the injection of vacancies by dissolution and their effect on mechanical behavior during surface removal experiments [52], (e) size effects [32], and many others that will surely be aired during these deliberations. In the light of our improving understanding of the nature of crystal surfaces, one must also now consider, for example, the role of near-surface solute and vacancy concentration gradients in crystals deformed under ambient conditions and, similarly, the removal of such gradients by dissolution prior to or during deformation. While it is true that surface removal experiments such as those performed by Kramer over the years and by others as well [30,53] are complex in the sense that they do represent a moving frame of reference [47,32], those experiments which are done under carefully controlled anodic conditions [30,53,54] probably most closely approach the deformation of a solid with a "clean" surface. It is interesting to note in passing that the increased plasticity observed in such experiments is similar in some respects to the Joffe effect [55]. In the classic demonstration of this phenomenon, a salt crystal is shown to be weak and brittle if deformed in air but remarkably ductile and stronger is deformed in water--i.e., while being slowly dissolved. Though still not yet well understood, the Joffe effect has been attributed to a combination of factors--the dissolution of pre-existing notches or cleavage cracks (Joffe's original explanation) or brittle surface films, removal or at least decrease in the active length of surface sources [42], etc. In any case, it seems to me reasonable to expect that the near-surface regions of a crystal may act as both a preferential source of dislocations as well as a barrier to the exit of dislocations at the surface. What remains in view of the conflicting observations, I think, is to decide the circumstances under which one of these competitive processes suppresses and dominates the other.

3.2 THE INFLUENCE OF SURFACE FILMS ON MECHANICAL PROPERTIES

In 1934, Roscoe [56] discovered that thin oxide films on the surface of cadmium crystals caused a significant increase in their yield stress. Similar observations have been reported many times in various solid-film combinations. Often such films have been found to increase not only the yield points but also the rate of work hardening and in many cases the three-stage work hardening behavior of fcc metals has been totally suppressed by films. For reference, Fig. 9 shows the influence of a 2 μm nickel-chromium coating on a copper monocrystal oriented for easy glide [57]. There is now no doubt that the presence of oxide layers, electrodeposited or evaporated metal films, or alloyed layers can profoundly affect the mechanical behavior of a variety of crystalline metals and nonmetals. There have been many explanations for such phenomena--the pinning of surface or near-surface sources, etc.-- any many controversies remain. Over the years it has been considered sufficient in many instances to imagine that such films act as barriers to dislocation emergence and that this accounts for some measure of strengthening. The means by which a surface film might serve as a barrier in a mechanistic sense continues to be debated. Moreover, there are more recent indications that surface films can under certain circumstances, in fact, induce softening rather than hardening of some solids (see below).

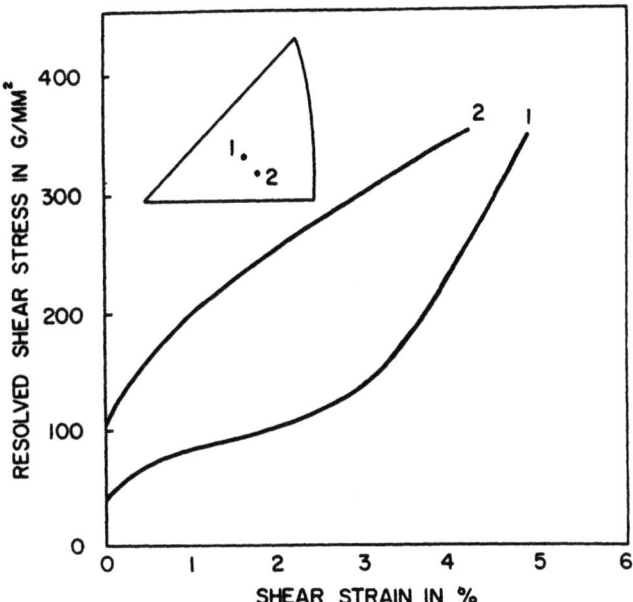

Fig. 9 Effect of nickel-chromium film on the stress-strain curve of similarly oriented copper monocrystals: (1) unplated, (2) plated (after Garstone et al. [57]).

The mechanisms by which films affect near-surface dislocation behavior is solids have been related to the elastic interaction of dislocations with a film of a shear modulus different from that of the substrate, the formation of accommodation dislocation networks at the film-substrate interface, and the existence of surface damage caused by film cracking due to residual plating stresses. In what follows, I will briefly consider some indications of the relevance of these parameters.

3.2.1 Elastic Interactions

Dislocation theory (see Reference [58]) predicts that if a dislocation approaches a surface that is covered with a film whose modulus, μ_F, is less than that of the crystal substrate, μ_S, then the dislocation will experience an image force attracting it to the surface. Conversely, if $\mu_F > \mu_S$ the dislocation will feel a long-range attraction and a short-range repulsion (over a distance comparable to the film thickness, t), the consequence of which is that a position of stable equilibrium is expected at a distance of about t from the film-substrate interface, Fig. 10. The force between the dislocation and image in the film is given by

$$F = \frac{\mu_S b^2}{4\pi r} \left(\frac{\mu_F - \mu_S}{\mu_F + \mu_S}\right)$$

where r is the distance of the dislocation from the interface and b the Burgers vector. When no film is present, the ratio of the moduli becomes unity and equation represents the attractive image force experienced by a dislocation approaching a "clean" free surface. Notice that when $\mu_F < \mu_S$ the image force is attractive but of a lesser magnitude than when no film is present. For a dislocation to emerge from the crystal in the case where $\mu_F > \mu_S$, sufficient stress must be applied in order to overcome the repulsive force discussed above (neglecting all other factors). Hence, the film acts as a barrier or resistance to dislocation emergence, and one might expect to find a dislocation accumulation just under the interface.

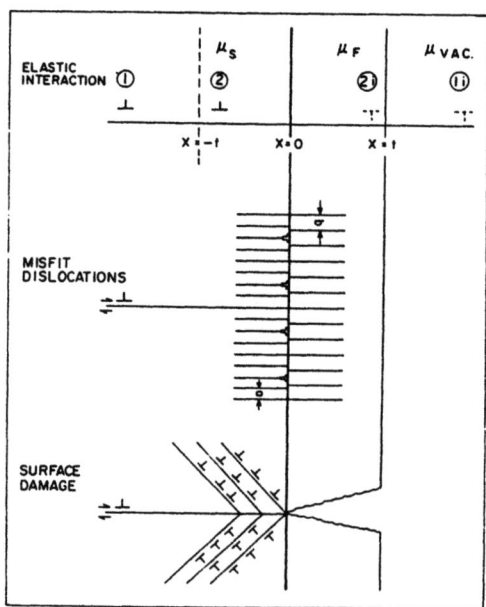

Fig. 10 Schematic illustrating the possible means by which a surface film may act as an obstacle to dislocation emergence.

An elastically harder surface film should, thus, induce a certain strengthening in the softer substrate as observed, for example, by Roscoe [56]. On this basis, the strengthening observed should be a systematic function of the modulus of the film. However, in a relatively few cases where this prediction has been examined [59,60] no such systematic correlation has been observed. Moreover, it is difficult to imagine how softening effects observed in some cases [61,62] could be accounted for on this basis. In short, some other factors must also play a role in determining the influence of surface films on mechanical properties.

3.2.2 Interfacial Mismatch or Misfit

Evans and Schwarzenburger [63] have pointed out that the transfer of dislocations from a substrate to a surface film of the same crystal structure but different orientation is difficult because of the misorientation of slip planes across the interface. This mismatch may thus serve as an obstacle to slip. If the film and substrate are of different crystal structures, or if one is a metal and the other a ceramic, the transfer of dislocations becomes even more difficult since changes in Burgers vector, stacking fault energy, etc., must be accommodated. Even in a less complex case where the film and substrate lattices are of the same structure and occur in parallel orientations, the atomic misfit in lattice spacing between the film and substrate may be accommodated by a two-dimensional grid of dislocations in the interface, the mesh size of the network decreasing as the degree of mismatch increases. The accommodation of such misfits, sometimes in part by dislocations and in part as well by homogeneous strains, is described in detail by Van der Merwe [64]. An example of such a misfit dislocation network is shown in Fig. 11 [65]. This interfacial dislocation network may serve as an effective barrier to glide dislocations approaching the surface from the substrate, Fig. 10, and one might expect that the strength of this barrier would decrease as the lattice mismatch decreases. There is some evidence to support this view [66,67]. On the other hand, there are cases [61,62] where coatings exert a distinct softening, not hardening, influence on mechanical properties. For example, Ruddle and Wilsdorf [61] report that oriented copper monocrystals coated with 600 Å electrodeposits of nickel yield in tension at approximately half the stress required for unplated crystals. In this case, it is presumed that the misfit network acts not as an obstacle to dislocations approaching the surface but rather as a source of dislocations and, importantly, a source at stresses less than the macroscopic yield point of uncoated copper crystals.

Fig. 11 Transmission electron micrograph showing long, straight misfit dislocations in a deposit of platinum on gold. The plane of the figure is (100) and its vertical and horizontal boundaries are parallel to <110> directions. Misfit, 3.9%, magnification about 250,000X (after Matthews and Jesser [65]).

3.2.3 Surface Damage Due to Residual Stresses in Films

Johnson and Block [59] report that chromium and rhodium films markedly strengthen copper monocrystals. They also conclude that strengthening on the basis of either a lattice misfit or elastic effects is incompatible with their observations. Instead, they propose that strengthening is associated with cracking of these coatings, either prior to or during deformation of the composite, and the consequent local generation of dislocation-dense regions through the stress pulses accompanying film cracking. Strengthening then occurs as a result of interaction between dislocations approaching the surface and the dislocation-dense regions near the interface associated with film cracking as suggested schematically in Fig. 10.

It is well known that electrodeposits and evaporated metal coatings as well as oxide films often contain significant residual

strains after growth. While there has been disagreement on this
point [68], it seems not improbable that such stresses and their
release could, in fact, contribute to surface film strengthening
as described above. On the other hand, it is not clear why the
injection of dislocations via the release of residual stresses in
the coating could not just as well lead to a softening effect--
i.e., to large-scale dislocation motion at stresses lower than the
yield point of the uncoated substrate. Indeed, Sethi and Gibala
[62] find substantial decreases in the flow stress and stress-
strain behavior of niobium and tantalum single crystals coated
with anodic oxides and account for this observation in just such
a manner.

In summary, the influence of solid surface films on the mechanical behavior of crystal substrates is likely to depend in a
complex and interrelated way on such factors as the elastic
properties of both components, the degree of atomic misfit at the
interface, residual stresses in the film, etc. But while all of
these factors (and probably others as well) contribute to the
effects of surface films on mechanical behavior, the important
interrelations between them remain to be established. It should
be important, for example, to determine under what conditions
during the deformation of a film-substrate composite slip is
transferred from the substrate to the film, in which case the film
may act as a barrier, or vice versa, in which case the film may
provide an easy source of dislocations.

3.3 ADSORPTION-INDUCED CHANGES IN THE PLASTICITY AND FRACTURE OF METALS AND NONMETALS

Adsorption-induced reductions in the hardness of nonmetallic
solids were first reported in 1928 by Rebinder [3] and are known
as Rebinder effects. In the period of time since Rebinder's
first announcement, many others have reported that surface-active
media (long-chain organic compounds, liquid metals, etc.) affect
the plasticity and/or fracture of a variety of solids, including
metals, covalent and ionic crystals, molecular crystals, polymers,
amorphous glasses, etc. These are far too numerous to even hope
to begin to summarize here, but there are several reviews which
might serve as useful guides [4,69-72] including those by Shchukin
[71] and Macmillan [72] in these proceedings.

While there have been a variety of explanations for Rebinder
effects, I will limit discussion in this section to the two that
are best supported by observation. The point of view developed
by Rebinder and his colleagues is that adsorption-induced softening
and strength reduction occur as a result of the lowering of the
specific free surface energy of the solid, i.e., the work of
formation of new surfaces during deformation and fracture. On the
other hand, the view of Westwood and his co-workers [73] is that

such phenomena--adsorption-induced hardening as well as softening--
may be understood in a conceptual way from consideration of the
type, concentration, mobility, and adsorption-induced redistribution
of charge carriers in the solid. In this context such phenomena
might well be called Westwood effects. While these views are at
first glance quite different, they have in common a basis for
understanding which is derived from colloid- and/or electrochemistry. In order to better understand this basis, the following
section begins with a discussion that itself began in Section 2.3,
namely, a description of the charge double layer present at a
solid/electrolyte interface.

3.3.1 Solid/Electrolyte Interface

Let us first distinguish the metal/electrolyte interface from
the semiconductor/electrolyte interface. The structure of the
double layer in the former case is likely to take the form illustrated in Fig. 12(a) for the case of a dilute aqueous electrolyte.
The compact double layer consists of an excess or deficiency of

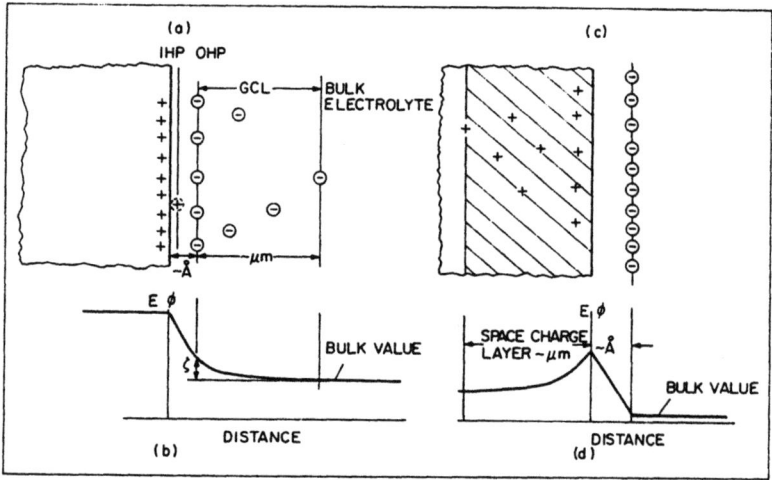

Fig. 12 Distribution of excess charge and corresponding variation
of electron energy in the solid and the potential of
ions in solution at interfaces between a metal and a
dilute electrolyte (a and b) and between a semiconductor
and a concentrated electrolyte (c and d) (after
Latanision [6]).

electrons at the metal surface and a layer of ions of charge opposite in sign to that at the surface of the metal adsorbed at the interface. This system of charges is known as the Helmholtz double layer (HDL). The locus of electrical centers of hydrated ions in contact with the electrode surface is known as the outer Helmholtz plane (OHP) while the locus of centers of unhydrated specifically adsorbed ions in contact with the electrode surface is known as the inner Helmholtz plane (IHP). The Gouy-Chapman layer (GCL) represents a distribution of charges in a space charge layer with an excess of ions similar in sign to that in the outer Helmholtz plane smeared out over a distance of up to about 1 μm. Beyond this point, the ions are present in concentrations typical of the bulk electrolyte. A schematic illustration of the potential drop expected across the double layer is shown in Fig. 12(b). There is a linear potential drop across the HDL and then a more gradual exponential decay in the diffuse double layer. The difference in potential between the OHP and the bulk electrolyte, i.e., the contribution of the diffuse double layer, is called the electrokinetic or zeta (ζ) potential to which I shall return shortly.

The relation between the charge distribution in the space charge layer near the surface and the electrostatic potential change at the surface is given by a Poisson-Boltzmann equation, the consequence of which is that the thickness of the space charge layer is essentially a function of the density of mobile charge carriers in each phase, i.e., solid or electrolyte. In essence, as the carrier concentration increases the thickness of the space charge layer decreases. In the case of a metal the density of free electrons is very high so that all of the charge behaves like a surface charge. Hence, the charge redistribution extends to only a few angstroms at the most beneath the metal surface. Likewise, the diffuse double layer becomes essentially compressed into the OHP when the electrolyte concentration is sufficiently high. In contrast, in semiconductors the density of charge carriers (electrons and holes) is orders of magnitude lower than in a metal. In pure germanium, for example, the concentration of electrons and holes at room temperature is on the same order of magnitude as the concentration of H^+ and OH^- in pure water. Consequently, the space charge layer in a typical semiconductor is on the order of a micron--10^4 times thicker than in a metal. The double layer present at the interface between a semiconductor and a concentrated electrolyte is shown in Fig. 12(c). A hybrid model showing the electron energy levels in the solid and the potential of ions in solution [74] is given in Fig. 12(d).

3.3.2 The Space Charge Layer and the Mechanical Properties of Nonmetals

It is possible to change the distribution of charge at the

electrode surface by several means--by the application of an external field, by illumination with light, or of interest in the present context by the chemisorption of charged ionic or polar molecules, for example. Now, is it possible that adsorption-induced changes in the space charge distribution can affect mechanical behavior? Well, there are many good indications of this, perhaps the best available illustration being the ζ-potential correlation studied extensively by Westwood and his colleagues at Martin Marietta Laboratories and summarized by Macmillan [72]. The ζ-potential correlation, which has been observed in a variety of inorganic solids including Al_2O_3, MgO, quartz, as well as a variety of other minerals, is shown schematically in Fig. 13 which shows the variation of ζ-potential with solution concentration (a), along with the corresponding variation in hardness (b) and dislocation mobility (c). The ζ-potential is considered to be related in sign and magnitude to the surface charge on the solids, i.e., as ζ changes sign it must pass through a zero point, and it is presumed that the surface is uncharged when $\zeta = 0$, although this is perhaps only rigorously true when specific adsorption is absent. The interesting correlation is Fig. 13 is that the hardness is a maximum when $\zeta = 0$ and decreases with increasing magnitude of the ζ-potential regardless of its sign.

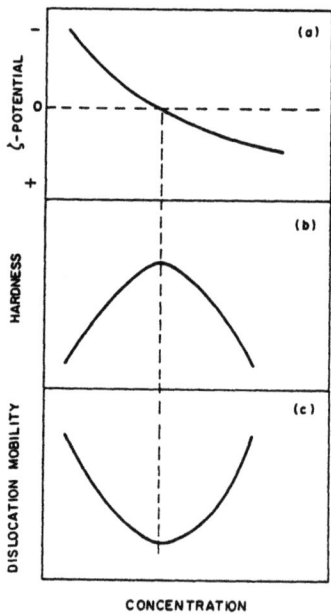

Fig. 13 Schematic of the apparently generic correlation between ζ-potential, hardness, and dislocation mobility observed in solids such as MgO, Al_2O_3, SiO_2 as well as a variety of rocks and minerals.

Just how the ζ-potential correlation may be understood is not now clear. In a general sense, the view developed by Westwood et al. [75] in order to account for adsorption-induced changes in mechanical behavior is that these phenomena are a consequence of the influence of adsorbed species on the mobility of near-surface dislocations. In short, adsorption-induced changes in a near-surface charge distribution are thought to alter the state of ionization of point and line defects in the near-surface region. Such changes induce variations in the mutual interactions between dislocations, between dislocations and point defects, between dislocations and the lattice, etc., and these variations are in turn reflected in changes in near-surface dislocation mobility or microhardness. Specifically, however, what one must account for is the symmetry that is observed in the hardness about the iso-electric point (i.e., when ζ = 0). One possibility has been suggested by Westwood and Mills in these proceedings [76]. Another which also has its origins at Martin Marietta Laboratories has been proposed by Swain and Latanision [77]. This is shown schematically in Fig. 14. In essence, one expects that the charge distribution

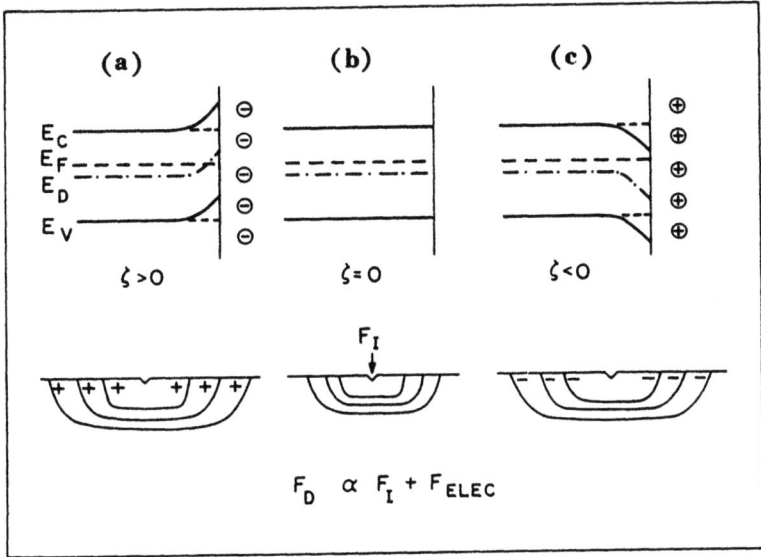

Fig. 14   Schematic illustrating a possible rationale for the symmetry observed in the ζ-potential correlation shown in Fig. 13 based upon the contribution of electrostatic interactions between near-surface dislocations.

in the space charge layer of the solid must respond to the charge (in sign and magnitude) populating the solution side of the double layer, Fig. 12. For simplicity, it is assumed that this charge is localized in the outer Helmholtz plane. Electron energy levels in the near-surface region must bend up or down depending on the nature of the adsorbed charge. An electronegative adsorbate would induce upwards band bending while the converse would be true in the presence of an electropositive adsorbate, Fig. 14(a) and (c), respectively. Now, as mentioned earlier, the ζ-potential tells us something about the surface charge. If one associates the flat band condition of Fig. 14(b) with $\zeta = 0$, one might then consider the possibility that the hardness maximum (minimum dislocation mobility) observed in relation to the ζ-potential measurements may be typical of the flat band condition in the space charge layer. Recognizing that dislocations may be assigned energy levels in the band gap [78], as shown in Fig. 14, it is tempting to suggest that dislocations in the space charge layer may also acquire an excess charge (i.e., bands bend up or down) in response to the presence of adsorbates. If this were true, then in addition to the force imposed upon near-surface dislocations mechanically by an indenter, electrostatic repulsion between adjacent and like-charged dislocations--the charge acquired by adsorption--would add a further driving force to their motion. Admittedly, this neglects electrostatic image forces, and other factors which should be considered. Nevertheless, except at a flat band condition the mechanical and electrostatic forces would combine to give greater apparent dislocation mobility--decreased hardness--as is suggested by the ζ-potential correlation.

While this explanation may not be the correct one--it is true that the relation between the ζ-potential, surface charge, and electronic band structure is not well understood--the above model is susceptible to concerted interdisciplinary examination by materials scientists and surface scientists. The critical experiment in terms of the above explanation would be to examine dislocation mobility while at the same time monitoring the distribution of charge in the space charge layer. To an extent, this has been done via the ζ-potential measurements. But, although the ζ-potential may be determined by a relatively simple electrokinetic measurement, its interpretation is not always very direct or unambiguous. On the other hand, it is possible to change the distribution of charge in the space charge layer by several means as described earlier, one of which is by illumination with light. In principle it would seem that if minimum near-surface dislocation mobility in semiconductors (and presumably insulators as well) is associated with the flat band condition, the means by which the flat band is reached should be of no consequence. In this context I hope to soon begin a study of the mobility of dislocations in cadmium sulfide monocrystals examined in such a manner that the space charge distribution may be controlled by sub-band gap illumination and monitored by means of surface photovoltage spectroscopy (SPS) which has been extensively studied by Gatos and co-workers at M.I.T. [79]. CdS is a seemingly appropriate choice since, firstly, the electronic characteristics of the basal and prismatic surfaces have been recently studied and, secondly, since the band theory for dislocation energy states in CdS has also been evaluated by Elbaum and Holmes [80]. If chemisorption-induced changes in dislocation mobility may be related to coincident changes in band structure, the above surface analytical approach is attractive since through it both elements of this problem become accessible. A crucial experiment will be to determine the dislocation mobility under flat band conditions, as identified by SPS, and to compare this with the dislocation mobility observed when band bending is known to occur.

3.2.3 The Double Layer and the Mechanical Properties of Metal Electrodes

At this point, we might return to consideration of the double layer in order to understand a possible rationale for the ζ-potential correlation based on the Rebinder school's view of adsorption-induced changes in mechanical behavior. This will then be extended to the case of metal electrodes which are also known to be sensitive to changes in the surface charge density.

One hundred years ago Lippmann [81] observed the following relation between the surface energy, γ, the surface charge density, q, and the applied potential, ϕ, for an ideal polarized electrode:

$$\left(\frac{\partial \gamma}{\partial \phi}\right)_{T,P,\mu_i} = -q$$

The Lippmann equation tells us that the surface energy of such an electrode—for example, Hg in aqueous alkali halide solutions—passes through a maximum when the surface is uncharged, i.e., when $q = 0$, and, indeed, this has been demonstrated experimentally on many occasions. The potential at which the surface becomes uncharged is known as the potential of zero charge, pzc. The above relationship may be understood physically by recognizing that because of electrostatic repulsion between the like charges in the surface charge layer, the work required to expand the interface is smaller in the presence of a net surface charge density of either sign than in the absence of electrostatic interaction. Because many of the early experiments in this field were done with a Lippmann capillary electrometer, the above is often described as the electrocapillary effect.

The Lippmann equation in the form given above is valid only for liquid electrodes (e.g., Hg). In the case of a solid electrode, it is the surface stress (the stress required to deform the surface) which passes through a maximum when the surface is uncharged [82]. In essence, because the much reduced mobility of near-surface atoms in a solid, as compared to liquid, prevents the process of creating new surface from occurring under anything like thermodynamically reversible conditions in a finite period of time, some extra work must be done and it is the sum of this extra work and the surface energy which is maximized at the pzc. Hence, even in the case of a solid electrode, one may cautiously conclude that, except when the surface is uncharged, electrostatic repulsion between like charges in the surface charge layer will reduce the total effort required to produce unit area of new surface.

It is possible to argue, therefore, that the symmetry of the dependence of hardness on ζ-potential in Fig. 13 could conceivably be attributed to the electrocapillary effect as has been acknowledged by Westwood et al. [83] and Shchukin [71]. Indeed, Shchukin et al. [84] suggest that dislocation mobility (hardness) becomes sensitive to adsorbed environments only when two conditions are met: (1) surface steps are formed during dislocation motion and (2) the environment causes large reductions in the energy of surface slip steps formed during deformation. In short, hardness would be greatest and dislocation mobility least when the surface step energy is a maximum.

While there are clearly many more facets to this difference of opinion regarding the basis for understanding adsorption-induced changes in mechanical behavior--i.e., electronic effects versus surface energy reduction--the above represents the essence of this difference. Recently Macmillan, Huntington, and Westwood [83] have shown that in MgO and LiF the mobilities of both edge dislocations (which did not produce surface slip steps in the particular hardness test performed) and screws (which did) were environment sensitive. This would seem to argue strongly away from the surface energy concept presented above at least in the case of nonmetallic solids. On the other hand, it seems possible that the contribution of the surface step energy may be more significant in some other instances, particularly in the case of metal electrodes. In contrast to semiconductors and insulators in which the distribution of charge in a space charge layer that extends to depths on the order of microns may well influence the mobility of near-surface dislocations, the same argument cannot be invoked in the case of metal electrodes. As explained earlier, because of the high density of mobile charge carriers (electrons) in a metal, the charge in the solid consists of an excess or deficiency of electrons within at most a few angstroms of the surface. Correspondingly, any effect of charge density on mechanical behavior must result from events occurring right at the metal surface.

It has in fact been demonstrated that the mechanical behavior of metal electrodes is a function of charge density on the metal surface [85-88]. The first reported studies of electrocapillary effects on mechanical behavior were reported by Rebinder and co-workers [85]. They reported that the pendulum hardness of metal electrodes (determined from the amplitude damping of an oscillating pendulum) passed through a maximum at the pzc. However, Bockris and Parry-Jones [89] later observed precisely the opposite effect in pendulum experiments, using a smooth fulcrum rather than the ground glass fulcrum used by Rebinder, and suggested that friction between the fulcrum of the pendulum and the metal, rather than hardness, may be the underlying physical property that is most affected by the charge double layer in such experiments. While there is evidence to show that friction in such cases is potential sensitive [90], others have shown [86,87] that the creep rate of metals passes through a minimum at the pzc, and it is clear that this could not be related to frictional effects. The latter investigators have considered a contribution to the activation energy for creep arising from the energy required to create surface steps. Recent work by Latanision et al. [88] on zinc monocrystal surfaces has confirmed the notion that electrocapillary changes in the charge density can indeed affect the motion of dislocations producing slip steps. In these experiments the diamond pyramid hardness and (etch pit) dislocation distribution about the indentation were monitored as a function of applied electrode potential on the basal $\{0001\}$ and prism $\{10\bar{1}0\}$ planes of zinc monocrystal electrodes. As shown in Fig. 15, observable slip about indentations on the basal plane occurs on second order pyramidal planes $\{11\bar{2}2\}$ and in $<11\bar{2}3>$ directions. Hence, screw dislocations on the pyramidal planes that intersect the basal plane have components of their Burgers vector normal to the surface and slip around the hardness impression exposes $\{11\bar{2}2\}$ steps. The motion of such dislocations is resisted in part by the work required to produce the slip steps left in their trails, and one expects that at the pzc, where more effort is required to create surfaces, the extent of glide of dislocations about a hardness impression would be a minimum. Conversely, the hardness would be a maximum as observed, Fig. 16. It is therefore seemingly possible to harden or soften zinc surfaces by changing charge density. Note that the hardness maximum at -1200 mV on Fig. 16 corresponds well with the anticipated value of the pzc for the $\{11\bar{2}2\}$ slip plane as should be expected since the new area being formed by slip is that of the $\{11\bar{2}2\}$ plane. On the other hand, if one indents the prism surface, in which observable pyramidal dislocations have their Burgers vector in the plane of the surface, then such dislocations do not produce steps and are expected to show no sensitivity to potential variation. This, too, has been observed, Fig. 16.

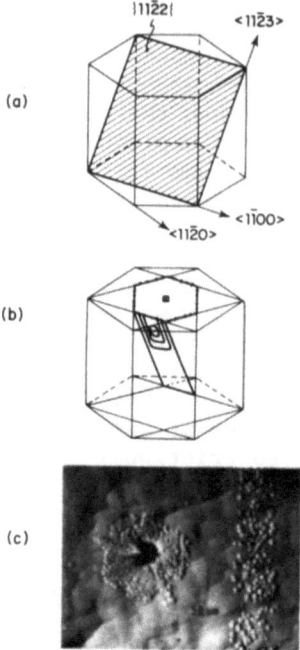

Fig. 15 Nonbasal slip about a microhardness indentation on (0001) zinc surfaces showing second-order pyramidal slip systems and a schematic of the dislocation distribution at an indentation (after Latanision et al. [881]).

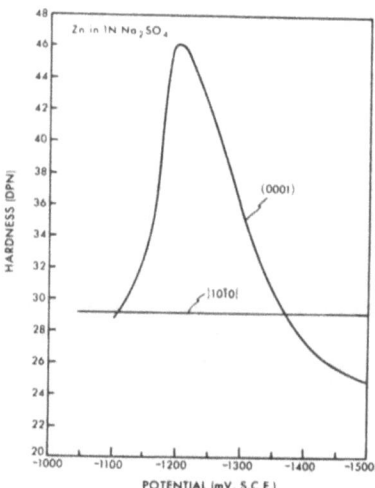

Fig. 16 Potential dependence of hardness on (0001) and {10$\bar{1}$0} surfaces of zinc (after Latanision et al. [881]).

We now have the situation that in the case of ceramics like
MgO the mobilities of both edge and screw dislocations are environment sensitive whereas in metals like zinc [88], gold [86],
and lead [87] only the mobilities of those dislocations which
trail slip steps appear to be sensitive. One explanation for this
difference between such metals and nonmetals may arise because
dislocation mobility in the latter is principally a function of
electrostatic interaction between dislocations and point defects--
which vary similarly with environment for both edge and screw
dislocations--and the additional resistance, if present, due to
slip step formation is so small as to be negligible. Alternately,
one might compare the energy consumed in creating slip steps with
the energy consumed in sweeping out the area traveled by the dislocations loops which produce those steps as a consequence of
plastic deformation beneath an indenter. In this case one would
find [91] that only in materials with relatively high surface
energies and low yield stresses (some pure metals) would the
surface energy contribution be significant and, hence, reflect
changes in surface energy (however they are induced) in mechanical
testing. Conversely, in materials of high flow stress (ceramics,
metallic alloys, etc.) or relatively low surface energy (LiF,
NaCl, MgO, etc.) variations in surface step energy might not be
reflected in dislocation mobility. In short, only if the surface
energy and plastic work terms were of the same magnitude could a
change in one significantly affect the other. This seemingly can
only happen in the case of relatively pure metals. At any rate,
while it is true that the correlation between hardness (dislocation
mobility) and surface energy is poor for most solids, it may be
that this is more a matter of degree than a failure in principle.

3.4 SURFACE EFFECTS IN THE ENVIRONMENT-SENSITIVE FRACTURE OF SOLIDS

The Joffe effect, mentioned earlier (see Section 3.1.2) is
one example of the remarkable influence that environments--some-
times very innocuous one at that--may have on the fracture behavior
of solids. In the case of the Joffe effect, the presence of
ordinary water has the staggering effect of turning a typically
weak and brittle material (e.g., KCl), into one with increased
strength and almost unbelievable ductility. This is a relatively
pleasant and useful surprise. Often the surprises are not so
happy: small amounts of Hg on the surface of otherwise (relatively)
ductile zinc polycrystals lead to a catastrophic loss of strength
and ductility; hydrogen has a similar effect on alloys of iron,
nickel, etc.; certain mildly corrosive electrolytes can lead to
the premature cracking of certain (often normally ductile) alloys
(304 stainless steel in Cl^- environment, etc.) at stresses far
below their normal fracture stresses, and so on. All of these
embrittlement phenomena--liquid metal embrittlement [92], hydrogen
embrittlement [93], stress corrosion cracking [94]--and others

[95] have been the subject of recent reviews and conferences and will not be discussed in detail here. Some of the issues of concern, particularly in relation to dissolution-related embrittlement, will be treated by Engell [96] at this Institute.

In all of the above, the action begins at the interface between the solid and the environment which surrounds it, and it should not be too surprising to find that many of the mechanistic models for these phenomena involve film formation, adsorption of critical species from the environment, interfacial solute and vacancy concentration gradients, etc. However, we know little about the possible influence of yielding initiation in the surface layer on the embrittlement of polycrystals [20,97]. Nor, for example, do we seem to know very much about the influence of surface charge on crack propagation in metal electrodes, about the detailed atomic-order interactions between embrittling species and strained crystal surfaces, about the atomistics of decohesion or about the electron distribution at a crack tip (which has, however, received some attention [98]). Some of these will be discussed briefly in this section.

In the previous section, evidence was presented to show that changes in the surface charge density on metal electrodes appear to affect their plasticity--the electrocapillary effect. It would also seem possible that changes in the charge density on a metal electrode may affect its fracture behavior. For example, an equilibrium crack in a solid subject to an increasing force, Fig. 17, will either propagate by cleavage or grow slowly by shear,

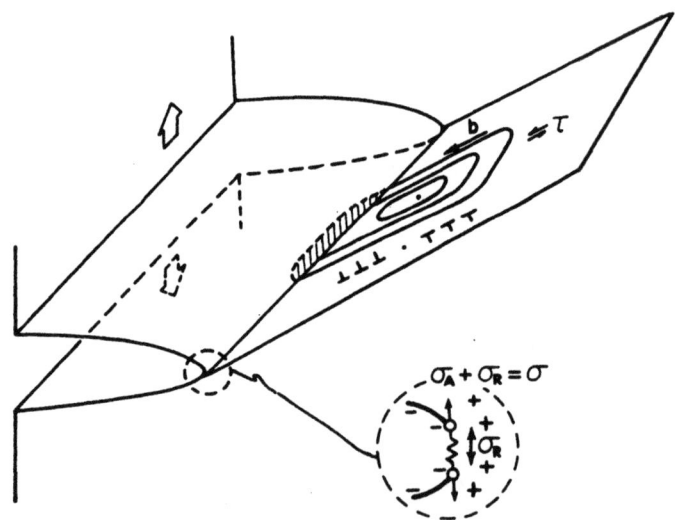

Fig. 17 Crack tip showing the possible influence of the double layer on σ and τ (after Latanision [6]).

depending on whether the tensile fracture stress, σ, for the atom-atom bond or the shear stress, τ, to cause dislocation motion on a favorable slip system is achieved first. Hence, as the ratio of σ/τ decreases, cleavage failure becomes more likely and, conversely, shear failure becomes more probable as the ratio increases [99]. As discussed earlier, the microhardness of metals (and, hence, crystal plasticity, τ) may be affected by changes in the charge density, provided that glide dislocations produce surface steps. On this basis, one anticipates that the motion of such dislocations, Fig. 17, away from sources near the crack tip may be inhibited (i.e., increasing τ) at the pzc encouraging cleavage. Likewise, because of the repulsion between like charges, the bond strength or cohesion between atoms in the surface layer of the crack tip, Fig. 17, may also be affected by the charge density in the double layer. In this case one expects that at the pzc, where the surface charge is extinguished, cohesion of surface atoms would be maximized--i.e., cleavage would be discouraged.

Although it is difficult to predict whether the expected influence of charge density on τ or on σ should be most significant, recent straining electrode experiments [100] suggest that in the case of zinc monocrystal electrodes the former may be quite important. In these experiments, cylindrical electrodes with their tensile axis normal to the basal plane were deformed at (a) a constant strain rate (10^{-4}sec^{-1}) or (b) a fixed load of 90% of the yield stress and their fracture strain and time-to-fracture, respectively, were measured at a function of applied potential over the same range of potentials as shown in Fig. 16. In this orientation, the limited slip which does occur takes place on second-order pyramidal planes allowing correlation of fracture behavior with the known [88] plastic response of pyramidal slip to electrode potential. Both the fracture strain and time-to-failure were observed to pass through sharp minima at about -1200 mV, SCE, i.e., at the pzc for the $\{11\bar{2}2\}$ plane. The latter is shown in Fig. 18. This coincides with the hardness maximum--i.e., decreased mobility of dislocations (plasticity)--which was observed in Fig. 16. This suggests that once a crack is initiated it may be expected to propagate most rapidly at the pzc since crack blunting, via dislocation emission from the crack tip, would be most restricted (i.e., the ratio of σ/τ is decreased), provided dislocations produce surface steps as shown in Fig. 17. This increases the tendency for cleavage and should be reflected in reduced time-to-fracture and fracture strain.

It should be noted in passing that we really cannot say whether the variation in σ had any effect on the observed results. While one would argue that a maximum in τ would occur at the pzc for the $\{11\bar{2}2\}$ slip plane, the maximum in σ would occur at the pzc for the plane of the free surface which in this case cannot be specified. Since the pzc is anisotropic, there is no reason to

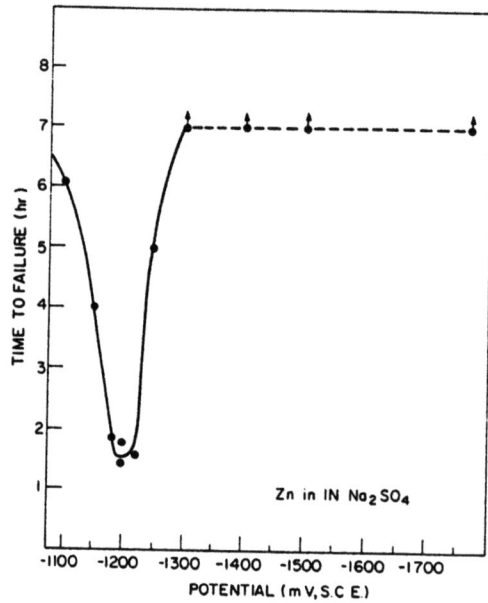

Fig. 18 The potential dependence of the time-to-failure of zinc monocrystal electrodes.

expect singularities in τ and σ to occur at the same potential. Hence, by inference, we presume to have seen a dependence of the initiation and propagation of cleavage cracks in zinc by a shear and not a bond strength mechanism.

It may be worth noting as well that the model proposed in Fig. 17 would allow for the possibility that adsorbates may significantly affect the fracture of metals since the propagation of a surface-initiated crack involves the consecutive rupture of surface bonds, and chemisorption, which is likely to involve some change in the distribution of charges in the double layer, may thus affect the bond strength or cohesion (perhaps as described above) between the atoms constituting the crack tip. Liquid metal embrittlement is a classic example of this [92]. Note that in liquid metal embrittlement, the embrittling species is considered to be adsorbed directly on the solid surface and, hence, strongly affects the strength of bonds between surface atoms. Thus, the consecutive rupture of surface atomic bonds, induced by reduction in the cohesive strength of surface atoms, leads to catastrophic fracture. In contrast, in aqueous electrolytes and in the absence of specific adsorption, ions present in the outer Helmholtz plane, Fig. 11, are effectively shielded from the surface by water molecules, and are less likely to significantly influence cohesive strength (i.e., fracture). If it occurs, specific adsorption on the other hand (for example, of Cl^-) may lead to a situation in an aqueous electrolyte approximating that of liquid metal embrittlement. Adsorption models based on this premise have been proposed for stress corrosion cracking (see [94]). At any rate, it should be appreciated that σ and in some cases τ may both be affected by variations in the charge density in the electrical double layer, regardless of whether the variations are due, for example, to the application of external potential or to the chemisorption of surface-active species. It should be mentioned that Gilman [101] has also recently considered some aspects of surface effects in embrittlement phenomena.

While it is possible to describe liquid metal embrittlement, hydrogen embrittlement, etc., on the basis of environmentally induced decohesion, the atomic scale interactions between the adsorbate (Hg, H, etc.) and strained metal surfaces are not well understood. In this context, one interesting electronic model of hydrogen embrittlement proposed in 1960 by Troiano [102] but never really examined experimentally, may now be tractable through the use of available surface analytical tools. In this model, Troiano presumes that hydrogen behaves in an electropositive sense and donates its electron to the unfilled d-bands of the metallic cores as shown schematically in Fig. 19. The increase in electron density

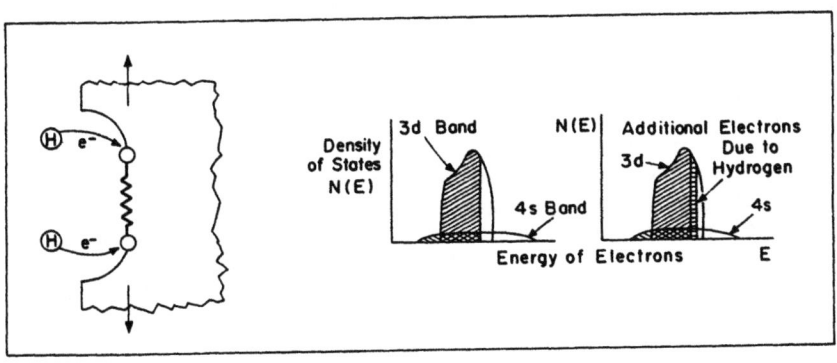

Fig. 19 An electronic model for hydrogen embrittlement (after Troiano [102]).

leads to an increase in the repulsive force between adjacent metal cores or, in other words, a decrease in the cohesive strength of the lattice, i.e., in not unlike the manner described in Fig. 17. This may be studied by means of chemical shifts in the Auger spectrum for iron, for example, due to the presence of adsorbed hydrogen. If hydrogen does in fact behave in an electropositive sense, iron's Auger peaks should undergo positive chemical shifts. The converse is true if hydrogen behaves in an electronegative sense, as is suggested in fact by work function measurements [103]. Of course, there may be other ways of understanding the atomistics of hydrogen embrittlement based on knowledge of the charge transfer process which occurs when hydrogen is adsorbed on a suitable metal surface. For example, recognizing that the great cohesive strength of the transition metals may be associated with their band structure, particularly the filling of d-orbitals, it is tempting to suggest [104] that an electropositive adsorbate on iron, for example, donates its electrons to iron surface atoms allowing them to behave, at least insofar as cohesion is concerned, rather like cobalt atoms. Conversely, in the presence of an electronegative adsorbate, iron surface atoms might behave as their neighbor in the periodic chart with one less d-electron, namely, manganese. It is interesting that cobalt has a reportedly [105] lower cohesive strength than iron suggesting in this context that the former rather than the latter is more likely to occur, i.e., hydrogen behaves as an electropositive adsorbate. At any rate, the nature of the charge transfer process which occurs should be accessible and would provide valuable insight into the mechanism of hydrogen embrittlement as well as other presumably adsorption-induced embrittlement phenomena such as liquid metal embrittlement. In short, we are now in a position to use surface analytical tools to study the atom-scale interactions between strained surfaces and embrittling species (adsorbates), and we should get on with it.

In this brief commentary, I have not attempted to discuss the specific and often exciting areas of controversy that have arisen in the studies of environmentally induced embrittlement--for example, the role of hydrogen (embrittlement), if any, in stress corrosion cracking [94]. This will surely be aired in the workshop sessions. What I have attempted to do, however, is to suggest, firstly, some relatively unexplored means of viewing environment-sensitive fracture, and, secondly, an example of a surface analytical approach to studying the atomic-order processes which occur when embrittling species interact with crystal surfaces. Perhaps by means of such experiments we may better understand the age-old question of why normally ductile metals become embrittled when exposed to certain environments.

4. APPLICATIONS OF SURFACE EFFECTS IN CRYSTAL PLASTICITY TO TECHNOLOGY

Generally the effects of environments on mechanical behavior are considered to be adverse, and to be sure often that is the case. Stress corrosion cracking, corrosion fatigue, hydrogen embrittlement, liquid metal embrittlement, etc., are all examples of failure phenomena which take on catastrophic consequences. Sometimes, however, the effects of environments on the mechanical behavior of solids are more beneficial and can be used to advantage. For example, we have already seen that through variations in the surface or near-surface charge density on various metals and nonmetals, the hardness of such solids may be controlled--if, however, for different reasons. Moreover, we also know that the coefficient of friction may be affected by changes in the charge density. Hence it should not be surprising to find that much attention has recently been directed toward making application of research on the surface- and environment-sensitive mechanical behavior of solids in metal cutting, the machining of ceramics, rapid excavation of hard rock, comminution, even the control of earthquakes, all of which are related to such parameters as friction, flow, and the fracture behavior of solids. A detailed account of these efforts in both the U.S. and in the USSR will be presented by Westwood [76], Pertsov [106], and others in this Institute and a superb review in this context has been recently prepared by Westwood [1]. In any case, a brief commentary on some of these follows. What is important to recognize now, however, is that not only are there many exciting academic problems to tackle but, despite an admitted lack of detailed understanding, we are able to apply this understanding with remarkable success to technological problems.

Let's first consider metal cutting. In conventional metal cutting, the removal of material is considered to be governed principally by two parameters: friction at the tool-chip and tool-workpiece interfaces and the strength or workhardening capacity of the workpiece. Significantly, both the frictional and mechanical behavior of metals may be made controllable in the presence of electrolytes, as demonstrated earlier. Not only does the presence of a charge double layer affect mechanical behavior, but the presence of anodic surface films, surface dissolution, and the generation of the various gases on the electrode surfaces are also known to affect mechanical and frictional behavior. Hence, one expects that the machining of metals should be significantly affected by making the workpiece an electrode in a suitable electrolyte and then applying an electrode potential which imparts appropriate physical and mechanical properties to the electrode. Thus, in this approach metal removal still occurs by contact of a tool with the workpiece; however, frictional and mechanical behavior of the tool-workpiece system are controlled electrochemically. This technique, developed particularly for use in machining hard-to-cut alloys and patented at Martin Marietta Laboratories [107] under the name of electromechanical machining (EMM), is already being examined in pilot-scale production operations.

The use of surface active environments in the machining of ceramic materials has its origins in the book "Hardness Reducers in Rock Drilling", published in the USSR [108]. Today surfactants are being examined for various uses in the cutting, grinding, and drilling of ceramics, minerals, and rocks--both on the very large scale, as when driving a railroad tunnel into the side of a mountain, and on the very small scale, as when manufacturing the Al_2O_3 substrates for modern microelectronic components. By maximizing the surface hardness through ζ-potential control (i.e., by changing the composition of the surface active environments surrounding the solid), researchers at Martin Marietta Laboratories have been able to increase the rate of drilling by factors of 6-10 in the case of alumina, up to 14 in the case of quartz, and 3-4 for various types of granites. From a practical viewpoint, the importance of such improvements in drilling rate--which potentially represent an enormous reduction in drilling costs--is that they may be achieved merely by adding a very small amount of a relatively cheap and innocuous chemical to the drilling water normally employed for cooling, lubrication, and flushing away of drilling debris.

A very different application for environmental control of the coefficient of friction may be as a tool for reducing earthquake hazards in [109]. In simple terms, an earthquake occurs when the shear forces resulting from movements in the earth's crust exceed the frictional resistance along a place of crustal weakness (a fault), and this frictional resistance varies directly with the coefficient of friction between the fault surfaces and inversely with the pressure of any liquid present in the fault. One can therefore imagine influencing an earthquake either by pumping natural pore water (usually brine) out of the fault, or by pumping in some manmade fluids of suitable chemistry, thereby changing the pressure tending to force the fault surfaces apart and/or varying the coefficient of friction between them. In this context it might be possible to "defuse" a large and potentially destructive earthquake by environmentally releasing its stored strength and energy as a succession of small and harmless tremors rather than as a less frequent but large-scale and catastrophic release of strain energy, i.e., a major earthquake.

The above are just a few of several exciting and potentially useful applications of the phenomenology of surface effects in crystal plasticity to real technological problems. All of the above topics and others will be treated in detail in the session devoted to applications.

5. CONCLUDING REMARKS

My objective in this paper was to provide a general overview of the science and technological applications of surface effects in crystal plasticity. I have, therefore, emphasized topics which

I think best describe some of the areas of interest to the workshop sessions in this Institute. I have not discussed adhesion, oxidation, catalysis, etc., all of which are dependent as well on the surface and its condition. Likewise, this effort has been confined to inorganic crystalline solids, thereby excluding vitreous and polymeric solids whose mechanical properties are also environment-sensitive.

I believe that materials scientists have made considerable progress in understanding surface effects in crystal plasticity, but I also believe that the detailed, atom-scale understanding that will lead to definitive treatments of these phenomena and to application on a broad scale demands the concerted action of materials scientists and surface scientists alike. Regrettably, there has been a historical lack of communication between these fields. It is encouraging to think, however, that the interdisciplinary efforts being developed here may lead to unprecedented progress in this area in the reasonably near future--an exciting prospect indeed!

6. ACKNOWLEDGEMENTS

This review was begun during a visit to the Max-Planck-Institut für Eisenforschung in Düsseldorf as a recipient of a Senior U.S. Scientist Award administered by the Alexander von Humboldt Foundation. I am grateful to the Foundation and to Professor H.-J. Engell, Director of MPI, for providing me with the facilities and resources to pursue this activity. I am also particularly pleased to acknowledge many helpful and informative discussions over a number of years with my friends and colleagues, A.R.C. Westwood, N. H. Macmillan, and R. W. Staehle.

REFERENCES

1. A.R.C. Westwood, J. Matl. Sci. 9, 1871, 1974.
2. O. Reynolds, Manchester Lit. Phil. Soc., 13, 93, 1874.
3. P. A. Rebinder, Reports to the VI Congress of Physicists, Moscow, 29, 1928.
4. R. M. Latanision and A.R.C. Westwood, in Advances in Corrosion Science and Technology, Plenum Press, N.Y., 51, 1970.
5. R. M. Latanision, in Corrosion Fatigue, N.A.C.E., Houston, 185, 1972.
6. R. M. Latanision, in Proc. Intl. Conf. on Surface Technology, S.M.E., Dearborn, 1, 1973.
7. R. M. Latanision, A. J. Sedriks and A.R.C. Westwood, in Structure and Properties of Metal Surfaces (Honda Memorial Series on Materials Science), Maruzen Co., Tokyo, 500, 1973.

8. J. M. Blakely, *Introduction to the Properties of Crystal Surfaces*, Pergamon, N.Y., 1973.
9. G. A. Somorjai, *Principles of Surface Chemistry*, Prentice-Hall, Englewood Cliffs, 1972.
10. W. Kossel, *Nach. Ges. Wiss. Göttingen*, 135, 1927.
11. I. N. Stranski, *Z. Phys. Chem.*, 136, 259, 1928.
12. W. K. Burton, N. Cabrera, and F. C. Frank, *Phil. Trans. Roy. Soc.*, London, 243A, 299, 1950.
13. C. B. Duke, *Adv. Chem. Phys.*, 27, 1, 1974.
14. R. L. Fleischer, *Acta Met.*, 8, 598, 1960.
15. D. M. Marsh, in *Fracture of Solids*, Interscience, N.Y., 119, 1963.
16. J. P. Hirth, in *Relation Between Structure and Strength of Metals and Alloys*, H.M.S.O., London, 218, 1963.
17. K. T. Aust, P. Niessen, R. E. Hanneman, and J. H. Westbrook, *Acta Met.*, 16, 291, 1968.
18. K. T. Aust and J. H. Westbrook, *Acta Met.*, 19, 521, 1971.
19. C. J. McMahon, in *Proc. 4th Bolton Landing Conf.: Grain Boundaries in Engineering Materials*, June 1974.
20. R. M. Latanision and H. Opperhauser, Jr., *Met. Trans.*, 5, 483, 1974; 6A, 233, 1975.
21. R. M. Latanision and R. W. Staehle, *Scripta. Met.*, 2, 667, 1968.
22. C. Herring, in *Metal Interfaces*, ASM, 1, 1952.
23. F.R.N. Nabarro, *Theory of Crystal Dislocations*, Clarendon Press, Oxford, 1967.
24. J. P. Hirth and J. Lothe, *Theory of Dislocations*, McGraw-Hill, N.Y., 1968.
25. F. C. Frank and W. T. Read, in *Symposium on Plastic Deformation of Crystalline Solids*, Carnegie Institute of Technology, 44, 1950.
26. W. C. Dash, *J. Appl. Phys.*, 27, 1193, 1956.
27. (a) I. R. Kramer and L. J. Demer, *Prog. Matl. Sci.*, 9, 133, 1961; (b) *Trans. A.I.M.E.*, 221, 780, 1961.
28. J. C. Fisher, *Trans. A.I.M.E.*, 194, 531, 1952.
29. D. Kuhlmann-Wilsdorf, in *Environment-Sensitive Mechanical Behavior*, Gordon and Breach, N.Y., 681, 1966.
30. R. M. Latanision and R. W. Staehle, *Acta Met.*, 17, 307, 1969.
31. K. Sumino, *J. Phys. Soc. Japan*, 17, 454, 1962.
32. H. Mughrabi, *phys. stat. sol.*, 44, 391, 1971.
33. J. T. Fourie and N.C.G. Dent, *Acta Met.*, 20, 1291, 1972.
34. S. Kitajima, H. Tanaka, and H. Kaieda, *Trans. Japanese Inst. Metals*, 10, 12, 1968.
35. V. I. Vol'shakov and L. G. Orlov, *Soviet Phys.-Solid State*, 12 [3], 576, 1970.
36. D. Vesely, *phys. stat. sol.*, 29, 685, 1968.
37. F. W. Young and F. A. Sherrill, *Can. J. Phys.*, 45, 747, 1967.
38. O. Lohne and O. Rustad, *Phil. Mag.*, 25, 529, 1972.
39. O. Lohne, *phys. stat. sol. (a)*, 18, 473, 1973.

40. G. Vellaikal and J. Washburn, J. Appl. Phys., 40, 2280, 1969.
41. Y. Tsunekawa and S. Weissmann, Matl. Sci. Eng., 17, 51, 1975.
42. S. Mendelson, J. Appl. Phys., 33, 2175, 2182, 1962.
43. For review, see: A.R.C. Westwood, in Environment-Sensitive Mechanical Behavior, Gordon and Breach, N.Y., 1, 1966.
44. K. Kolb and E. Macherauch, Phil. Mag., 7, 415, 1962.
45. I. R. Kramer and A. Kumar, in Corrosion Fatigue, N.A.C.E., Houston, 146, 1972.
46. I. R. Kramer, Trans. A.I.M.E., 233, 1462, 1965.
47. J. T. Fourie, Phil. Mag., 17, 735, 1968.
48. R. M. Latanision, in Proc. Intl. Conf. on Strength of Metals and Alloys -2, ASM, 446, 1970.
49. H. Mughrabi, phys. stat. sol., 39, 317, 1970.
50. J. T. Fourie, in Corrosion Fatigue, N.A.C.E., Houston, 164, 1972.
51. I. R. Kramer, Scripta Met., 8, 1231, 1974.
52. R. W. Revie and H. H. Uhlig, Scripta Met., 8, 1235, 1974.
53. D. J. Duquette and H. H. Uhlig, Trans. ASM, 61, 445, 1968; 62, 839, 1969.
54. D. J. Duquette, H. Hahn, and P. Andresen, These Proceedings.
55. A. Joffe, M. W. Kirpitschewa, and M. A. Lewitsky, Z. Physik, 22, 286, 1924.
56. R. Roscoe, Nature, 133, 912, 1934.
57. J. Garstone, R.W.K. Honeycombe, and G. Greetham, Acta Met., 4, 485, 1956.
58. R. Bullough, These Proceedings.
59. R. M. Johnson and R. J. Block, Acta Met., 16, 831, 1968.
60. W. A. Jemian and C. C. Law, Acta. Met., 15, 143, 1967.
61. G. E. Ruddle and H.G.F. Wilsdorf, Appl. Phys. Letters, 12, 271, 1968.
62. V. K. Sethi and R. Gibala, Scripta Met., 9, 527, 1975.
63. T. Evans and D. R. Schwarzenburger, Phil. Mag., 4, 889, 1959.
64. J. H. van der Merwe, These Proceedings.
65. J. W. Matthews and W. A. Jesser, Acta Met., 15, 595, 1967.
66. D. R. Brame and T. Evans, Phil. Mag., 3, 971, 1958.
67. L. C. DeJonghe and I. G. Greenfield, Acta Met., 17, 1411, 1969.
68. J. Pridans, B. Berkowitz, and J. C. Billelo, Scripta Met., 5, 701, 1971.
69. P. A. Rebinder and E. D. Shchukin, in Prog. Surface Science, 3[2], 97, 1972.
70. A.R.C. Westwood and N. H. Macmillan, in Science of Hardness Testing and its Research Applications, ASM, 372, 1973.
71. E. D. Shchukin, These Proceedings.
72. N. H. Macmillan, These Proceedings.
73. A.R.C. Westwood, C. M. Preece, and D. L. Goldheim, in Molecular Processes on Solid Surfaces, McGraw-Hill, N.Y., 591, 1969.

74. P. J. Boddy, J. Electroanal. Chem., 10, 199, 1965.
75. A.R.C. Westwood, D. L. Goldheim, and R. G. Lye, Phil. Mag., 16, 505, 1967; 17, 951, 1968.
76. A.R.C. Westwood and J. J. Mills, These Proceedings.
77. M. V. Swain and R. M. Latanision, unpublished work, Martin Marietta Laboratories, 1974.
78. P. Haasen and W. Schröter, in Fundamental Aspects of Dislocation Theory, N.B.S. Spec. Pub. 317, 1231, 1970.
79. J. Lagowski and H. C. Gatos, Surface Science, 30, 491, 1972.
80. C. Elbaum and R. R. Holmes, Ref. 78, p. 1293.
81. G. Lippmann, Ann. Chim. Phys., 5, 494, 1875.
82. R. M. Latanision, N. H. Macmillan, and R. G. Lye, Corrosion Science, 13, 387, 1973.
83. N. H. Macmillan, R. D. Huntington, and A.R.C. Westwood, Phil. Mag., 28, 923, 1973.
84. E. D. Shchukin, V. I. Savenko, L. A. Kochanova, and P. A. Rebinder, Dokl. Akad. Nauk SSSR, 200[2], 406, 1971.
85. P. A. Rebinder and E. K. Venstrem, Acta Phys. Chem-USSR, 19, 36, 1944.
86. A. Pfutzenreuter and G. Mazing, Z. Metallk., 42, 361, 1951.
87. V. I. Likhtmann, L. A. Kochanova, D. T. Leykis, and E. D. Shchukin, Elektrokhimiya, 5, 729, 1969.
88. R. M. Latanision, H. Opperhauser, Jr., and A.R.C. Westwood, in Ref. 70, p. 432; Proc. 5th Intl. Congress on Metallic Corrosion, NACE (Houston, 1974) p. 111.
89. J. O'M. Bockris and R. Parry-Jones, Nature, 171, 930, 1953.
90. J. O'M. Bockris and R. K. Sen, Surface Science, 30, 237, 1972.
91. R. M. Latanision, unpublished work.
92. A.R.C. Westwood, C. M. Preece, and M. H. Kamdar, in Fracture, Vol. 3, Plenum Press, N.Y., 589, 1970.
93. I. M. Bernstein and A. W. Thompson, eds., Hydrogen in Metals, ASM, 1970.
94. R. W. Staehle, in Theory of Stress Corrosion Cracking, NATO Scientific Affairs Division, Brussels, 1971.
95. R. W. Staehle, A. J. McEvily, and O. F. Devereux, eds., Corrosion Fatigue, N.A.C.E., Houston, 1973.
96. H. -J. Engell, These Proceedings.
97. H. W. Liu, in Proc. 1st Intl. Conf. on Fracture, 191, 1965.
98. W. A. Tiller, Scripta Met., 8, 487, 1974.
99. A. Kelly, W. R. Tyson, and A. H. Cottrell, Phil. Mag., 15, 567, 1967.
100. R. M. Latanision, H. Opperhauser, Jr., and A.R.C. Westwood, unpublished work, Martin Marietta Laboratories, 1974.
101. J.J. Gilman, Phil. Mag., 26, 801, 1972.
102. A. R. Troiano, Trans. ASM, 52, 54, 1960.
103. F. C. Tompkins, in Solid-Gas Interface, Marcel Dekker, N.Y., 765, 1967.
104. F.R.N. Nabarro, private communication, April, 1975.
105. R. E. Watson and H. Ehrenreich, Comments on Solid State Physics, 4, 109, 1970.

106. N. V. Pertsov, These Proceedings.
107. R. M. Latanision, *Electromechanical Machining Method*, U.S. Patent No. 3,873,512, March 25, 1975.
108. P. A. Rebinder, L. A. Schreiner, and K. F. Zhigach, *Hardness Reducers in Rock Drilling*, English Translation CSIRO, Melbourne, 1948.
109. A.R.C. Westwood, Martin Marietta Laboratories, TP-386, 1971.

SURFACE EFFECTS IN CRYSTAL PLASTICITY - OVERVIEW FROM
THE CRYSTAL PLASTICITY STANDPOINT

F.R.N. Nabarro

Department of Physics, University of the Witwatersrand,
Johannesburg, South Africa

ABSTRACT. The influence of the surface on the plasticity and fracture properties of crystals is surveyed, and possible applications of the phenomena are discussed.

1. INTRODUCTION
2. THE INFLUENCE OF A CLEAN SURFACE ON PLASTICITY
    2.1 Summary
    2.2 Surface Sources
    2.3 Dislocation Sources near the Surface
    2.4 Evidence for Soft or Hard Surface Layers
    2.5 The Unloading Yield Point
    2.6 Annealed Surfaces
    2.7 Oxide Layers, Alloy Layers and After-Effects
3. PLASTICITY IN AN ENVIRONMENT OF WATER OR SURFACE-
    ACTIVE AGENT
    3.1 Water
    3.2 Surface-active Agents
4. APPLICATIONS OF SURFACE EFFECTS IN CRYSTAL PLASTICITY
5. FRACTURE
    5.1 Fracture of a Crystal with a Clean Surface
    5.2 Fracture in a Liquid Medium : General Considerations
    5.3 Stress Corrosion and Hydrogen Embrittlement
    5.4 Reduction of the Surface Energy : The Rebinder
        Effect and Liquid-Metal Embrittlement
    5.5 Decreased Dislocation Mobility near the Surface
    5.6 Increased Dislocation Mobility near the Surface
    5.7 Fatigue and Erosion
    5.8 Surface Effects in Glasses
6. APPLICATIONS OF SURFACE EFFECTS
    REFERENCES

1. INTRODUCTION

John Keats, writing to Fanny Brawne [1], said "I long to believe in immortality.... . I wish to believe in immortality - I wish to live with you forever". So much of this talk will be concerned with the ductile behaviour of crystals, plasticity in its narrower sense. We shall consider a crystal which is deforming by slip, and shall expose a surface in this crystal. We first think of the surface as a simple mathematical cut along a low-index plane. Then we allow for the relaxation of the newly-exposed atoms, and for surface irregularities, and we consider the effect of lattice vacancies which can enter at the surface. We consider the effect of dissolving off the surface layers, either intermittently or continuously. Then the effects of adsorption or oxidation by normal constituents of the atmosphere must be considered, the effects of surface alloying, and finally those of special surface-active agents.

But "All/Life death does end and each day dies with sleep" [2], and plasticity in its broader sense includes the fracture which terminates flow. Here there is a bewildering array of effects. The medium in which the crystal flows may enhance its ductility enormously, or it may cause it to break almost without plastic deformation, or under a load which it has already supported. Our problem is often to decide whether the medium has made the crystal more or less brittle, or whether it has made it less or more ductile. It will become, as W.T. Read wrote in 1953 [3], "apparent that dislocations could explain not only any actual result but virtually any conceivable result, usually in several different ways".

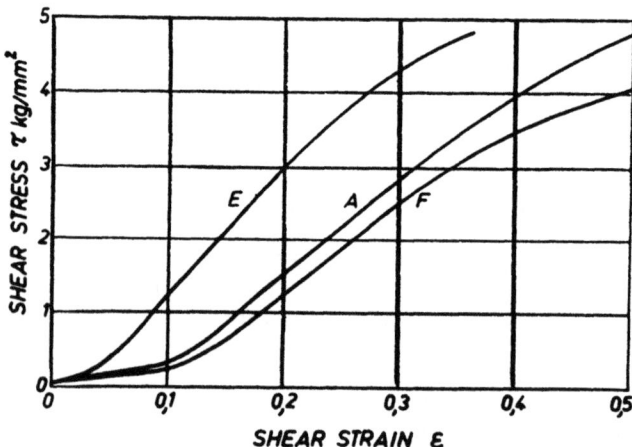

Fig. 1. Stress-strain curves of three copper crystals of different initial orientations. After Seeger et al. [15].

There have been many extensive reviews (e.g. [4] - [11]), as well as conference reports on more applied aspects (e.g. [12] - [14]).

We have spoken of "crystals" as if all crystals were alike. When we consider surface effects, at least two types of distinction may be important. Metallic crystals with a simple face-centred cubic structure show monotonically increasing stress-strain curves (Fig. 1). The speed of a dislocation increases very rapidly with the applied stress and a small density of dislocations can sustain the imposed rate of flow. Additional dislocations produced by plastic strain merely impede each others' motion. Crystals such as corundum and germanium (Fig. 2a and b) show a rounded yield point in the stress-strain curve. The speed of a dislocation increases only slowly with the applied stress. The few dislocations present initially require a high stress if they are to move fast enough to accommodate the imposed rate of strain. As the dislocations multiply, the stress falls. A surface effect which promotes the multiplication of dislocations will affect the plastic properties of these two types of crystal in quite different ways.

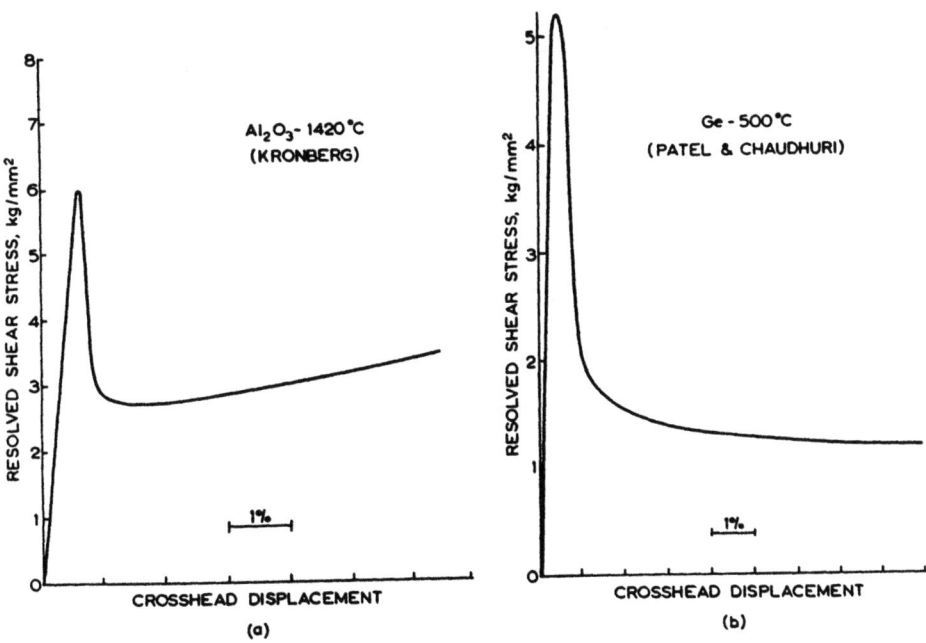

Fig. 2. (a) Stress-strain curve of a single crystal of Al_2O_3. From Johnston [16], after Kronberg. (b) Stress-strain curve of a single crystal of germanium. From [16], after Patel and Chaudhuri.

If molecules or ions are adsorbed to a surface, they may change the electrostatic potential of the surface layer. The thickness of this layer is $(\epsilon kT/2ne^2)^{\frac{1}{2}}$, where ϵ is the dielectric constant of the solid, kT the Boltzmann factor, n the density of charge carriers, and e their charge. For an insulator or a semiconductor with small n, this depth may exceed the length Λ of the dislocation loops which act as sources; in a metal, the layer of surface charge is only one atom thick [6].

2. THE INFLUENCE OF A CLEAN SURFACE ON PLASTICITY

2.1 Summary

We first give some general theoretical considerations of the effect of a clean surface, and then outline the evidence that a clean surface is a source of new dislocations. There follows an outline of the often contradictory evidence that the surface has no unusual properties during plastic flow, that it acts as a soft region, and that it acts as a hard region. A brief discussion is given of the influence of the surface on the yield point which may occur if a crystal is unloaded and then reloaded during a tensile test.

<u>General Theoretical Considerations.</u> We think of a crystal deforming in single glide, with loops of dislocation spreading from sources and moving predominantly on a single set of parallel glide planes. In Stage I of the deformation, represented by the regions of low slope in the stress-strain curves of Fig. 1, we need consider only this primary glide system. In Stage II, the linear region which has a larger slope of $d\sigma/d\epsilon \simeq \mu/300$, where μ is the shear modulus, dislocations on other glide planes also move. These "forest dislocations" contribute little to the total strain, but they tangle with the primary dislocations and cause the observed rapid hardening.

Now we fix our attention on a mathematical cut in the crystal (ABCD in Fig. 3), parallel to a low-index plane which intersects the primary glide plane LMNO in the line EF. The slip vector $\overline{b}$ is parallel to LM. Loops of dislocation spread from sources $\overline{P}$ and Q above and below EF. When edge components of loops from P and Q meet on their glide plane, they are of opposite sign, and annihilate. The same is true of screw components. If P and Q are not on the same glide plane, but on neighbouring planes, the screw components of loops spreading from the two sources may often cross-slip and annihilate as they pass, but the edge components form stable dipoles or locked counter-processions of edge dislocations of opposite sign [17]. This seems to be the principal source of work hardening in Stage I. If the deformation has reached Stage II, the tangling of dislocations moving on other glide planes is compounded by the fact [18] that the intersection of dislocations produces jogs, and the dragging of these jogs produces vacancies.

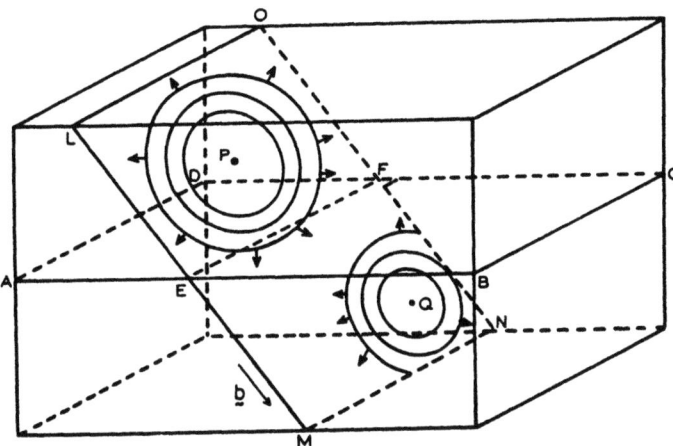

Fig. 3. Dislocation loops spread from two sources P and Q on the same glide plane LMNO, which is intersected in EF by a mathematical cut ABCD parallel to a low-index plane in the crystal.

The vacancies diffuse to dislocations and cause them to climb out of their glide planes, thereby distorting the dislocation configuration still further and causing further hardening.

Now let the plane ABCD be a real physical cut in the crystal, so that the half crystal containing the source P is completely removed. Though some complications will appear later, we may think of ABCD as the clean surface of a finite crystal. The first effect of the removal of sources such as P is clearly that the dislocation loops growing from Q can now travel freely to the surface without meeting dislocations moving in the opposite direction. Work hardening is suppressed in a surface layer of the crystal to a depth which, in Stage I, is of the order of D_e, the mean free path of an edge dislocation. We should expect this to be the dominant effect in pure clean ductile metals where the flow stress is controlled essentially by the interactions between dislocations. It has been extensively studied by Fourie [19] - [22] and Mughrabi [23,24]. (This "soft-surface" effect in single crystals is quite different from the "soft-surface" effect in polycrystals (Fig. 4) which arises because [25] grains which meet the surface are less constrained than are those in the interior.) Since work hardening in Stage I is produced essentially by edge dislocations, the effect is observed [19] only under an "edge" surface such as ABCD, on which edge dislocations emerge, and not on a "screw" surface such as LBMA on which screw dislocations emerge. Fourie [21] took single crystals of

54

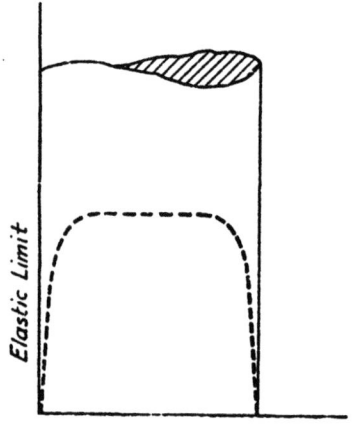

Fig. 4. The "soft-surface" effect in polycrystals. After Thompson and Millington [25].

copper, strained them in tension, and then (Fig. 5) cut them with a jet of acid into slices parallel to their "edge" faces. He then measured the flow stresses of the slices, and found (Fig. 6) that the slices near the original "edge" faces were weaker than the

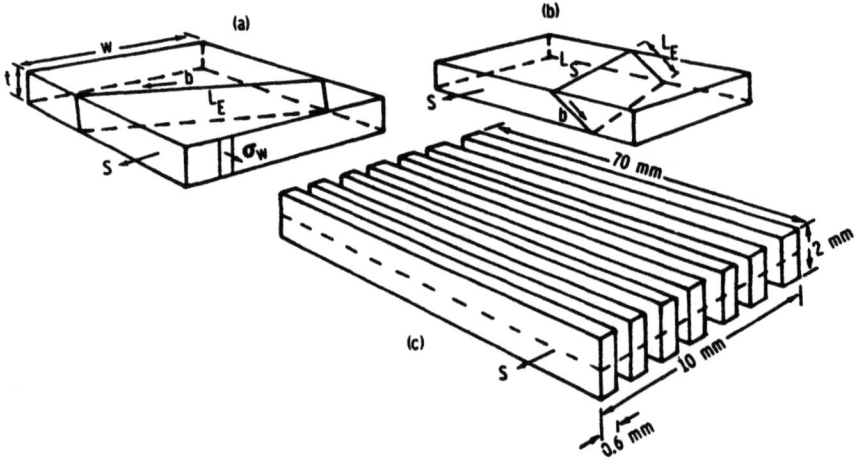

Fig. 5. (a) A copper crystal oriented for cutting into slices with "edge" faces in Fourie's experiments. (b) A similar crystal for producing slices with "screw" faces. (c) The method of slicing. After Fourie [21].

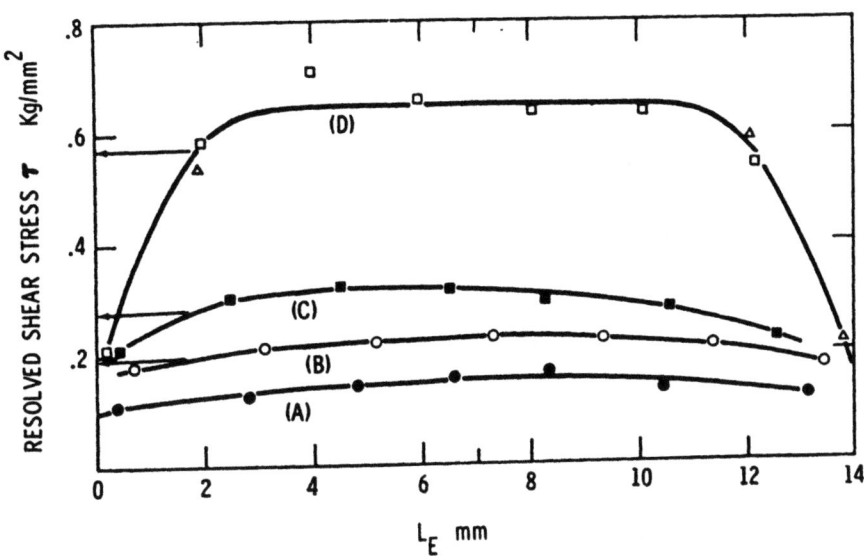

Fig. 6. The flow stresses of "edge" slices. The final stress to which the whole crystal had been pre-strained is indicated by an arrow for each of the samples (A) - (D), which were pre-strained to shear strains of 0, 0.020, 0.029 and 0.058 respectively. After Fourie [21].

original crystal, while those from the middle were stronger. When we come to compare this result with others, it will be important to remember that the soft layer extends to a depth of 1-2 mm.

This very direct and compelling observation may be criticized on the following argument, due essentially to Kramer ([14], p.173). Suppose a slice of thickness w, length ℓ and unit depth (Fig. 7) has a mean flow stress σ, and a flow stress gradient of either sign, so that the flow stresses on its surface are $\sigma \pm \frac{1}{2}\Delta\sigma$. Then, when the slice is cut free, the residual elastic stresses will bend it into a cylinder of radius $R = wE/\Delta\sigma$, where E is Young's modulus. If the slice is now tested in tension to a load $w\sigma$, at which plastic flow would begin if the slice were straight, the bending moment acting across the central section of the slice is $w\sigma\ell^2/8R = \sigma\Delta\sigma\ell^2/8E$. This moment requires the stress across the middle section to vary from $\sigma - \frac{1}{2}\delta\sigma$ to $\sigma + \frac{1}{2}\delta\sigma$, where $\delta\sigma$ is given by

$$\frac{w^2 \delta\sigma}{12} = \frac{\sigma \Delta\sigma \ell^2}{8E} \tag{1}$$

or

$$\frac{\delta\sigma}{\Delta\sigma} = \frac{3}{2} \frac{\sigma}{E} \left(\frac{\ell}{w}\right)^2 . \tag{2}$$

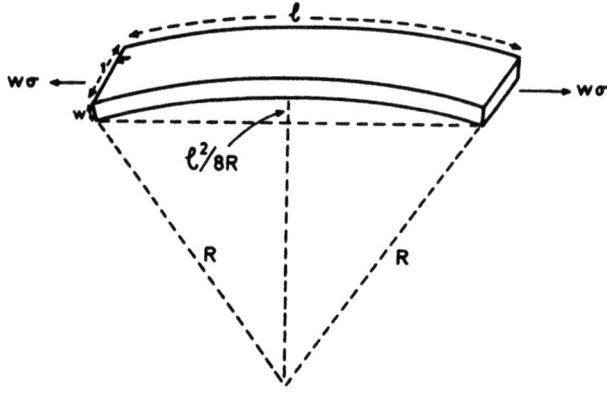

Fig. 7. If the flow stresses on the two sides of a slice of width w differ by $\Delta\sigma$, the residual stresses in the slice when it is cut free from the parent crystal will bend it elastically into a cylinder of radius $R = wE/\Delta\sigma$.

For Fourie's experiments, σ had a maximum value of 6×10^7 dyn cm^{-2}, $E \simeq 12 \times 10^{11}$ dyn cm^{-2}, $\ell = 70$ mm, $w = 0.6$ mm. Eqn. (2) then leads to $\delta\sigma/\Delta\sigma = 1$. In this case the additional stress $\frac{1}{2}\delta\sigma$ on the concave surface just compensates for the higher flow stress of the material on this surface, and the whole central cross section yields at once. But if ℓ/w had been larger, $\delta\sigma$ would have exceeded $\Delta\sigma$, and plastic deformation would have begun below the stress σ as a result of the initial gradient of flow stress, whether the gradient represented a soft surface or a hard surface layer in the original crystal.

The correctness of the direct interpretation of Fourie's experiments is supported by other observations. Fourie [22], using crystals in which the dislocations were allowed to relax when the load was removed, and Mughrabi [23], using crystals which were neutron-irradiated under stress at low temperatures to pin the dislocations, both showed that the arrangement of dislocations near the surface was characteristic of an earlier stage of deformation than was the arrangement in the middle of the crystal. In fact the deformation near the surface can still be in Stage I while that in the middle is in Stage II (Fig. 8). However, on the present model, the regions of crystal just below an "edge" surface should show an excess of edge dislocations of the sign which travel towards the surface during tensile deformation. This should bend

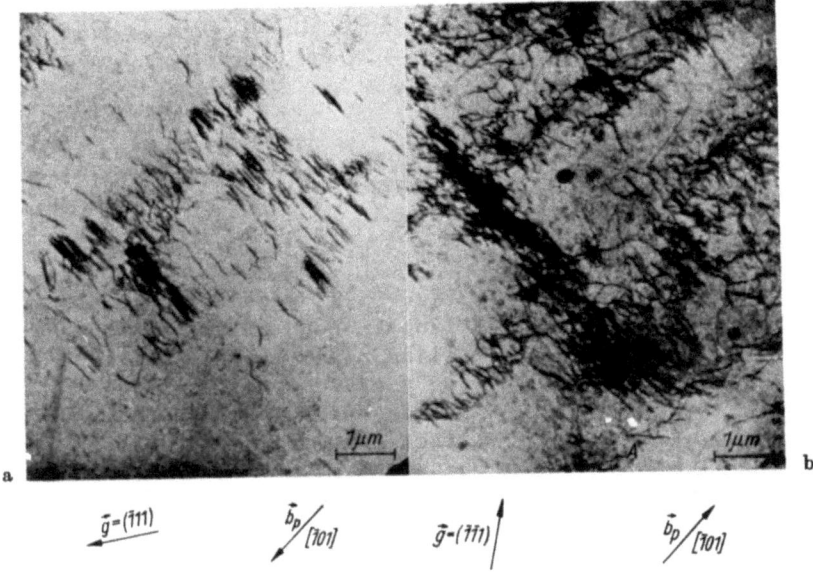

Fig. 8. A copper crystal strained to ε = 0.15, showing (a) Stage I arrangement of dislocations 0.5 mm from the surface (b) Stage II arrangement in the middle. After Mughrabi [23].

the glide planes so that their normals near the surface make smaller angles with the tensile axis than do the normals in the middle (Fig. 9). This is not what is observed [24] in the early stages of deformation. We thus have to consider other phenomena which may occur at the surface of a clean crystal.

Energy is required to form a surface step when an edge dislocation emerges. In a rough calculation, Nabarro [26] suggested that the energy of the dislocation which was annihilated would probably just fail to provide the necessary energy. Even though more recent estimates of the surface energy of a pure metal [27] are higher than those used by Nabarro and would strengthen his conclusion in the case of a single dislocation, the argument of Shchukin [28] that a succession of dislocations usually propagates on a single glide plane shows that the formation of the first surface ledge on a clean metal will not usually be an effective obstacle. If it were, the lattice curvature falsely predicted in Fig. 9 would be enhanced.

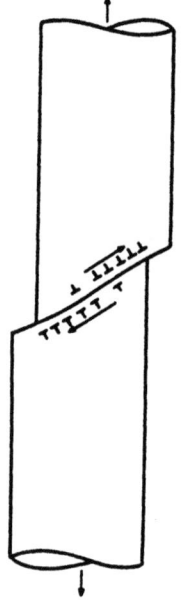

Fig. 9. If edge dislocations flow in both directions on a glide plane in the interior of a crystal, but only towards the surface in the surface regions, a characteristic curvature of the glide planes is predicted. The opposite curvature is observed. After Mughrabi [24].

2.2 Surface Sources

If the surface is not exactly parallel to a low-index crystal plane, but is vicinal (Fig. 10), we may reverse the previous argument [29,30]. There are already steps at the surface. If dislocations are nucleated at these steps, and move inwards so as to remove these steps, the process of nucleation occurs at a lower stress than it does in the bulk crystal. The steps act as surface sources.

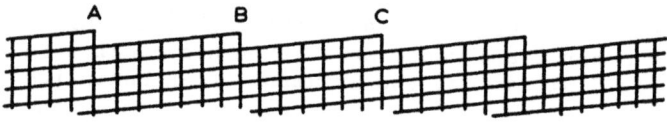

Fig. 10. A vicinal surface. There are steps of monatomic height at A, B, C... .

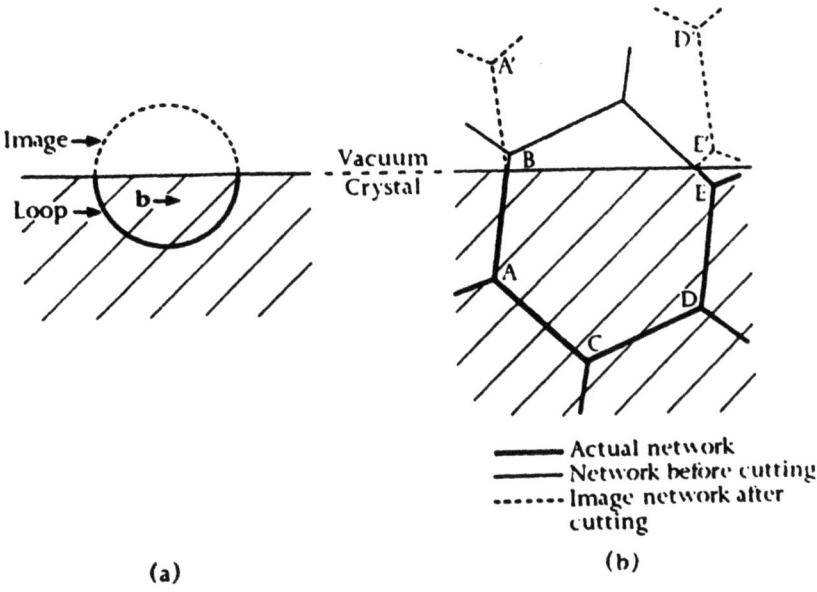

Fig. 11. A dislocation network has a typical link length Λ = AC. Close to a surface, the network, together with its image, has some links of length AA' ≃ 2Λ.

Even an atomically smooth surface can produce surface sources which operate under about half the stress required to operate sources in the interior. Fisher [31] pointed out that if the common longest length of a mobile dislocation segment in a network was Λ, segments meeting a freshly-exposed surface could, together with their images in the surface (Fig. 11) form links of effective length 2Λ which would form Frank-Read sources able to operate at a reduced stress. The importance of this simple mechanism has often been questioned, largely [32] because a dislocation meeting a free surface will try to swing normal to the surface by cross-slip of its screw segments [33,34]. (In fact [35] the equilibrium configuration represents a compromise between the forces tending to shorten the dislocation and those tending to turn it into a screw orientation of low energy per unit length.) While such a jogged dislocation may lead to the multiplication of dislocations, it will not act as a primary dislocation source under low stresses. In addition, it is sometimes suggested (e.g. [7]) that the change in lattice parameter near the surface of the crystal may act as an obstacle to the egress of edge dislocations [36]. Since this change is only of the order of 10 per cent of the lattice parameter, the residual "surface dislocation" has only 1 per cent of the energy of a lattice dislocation.

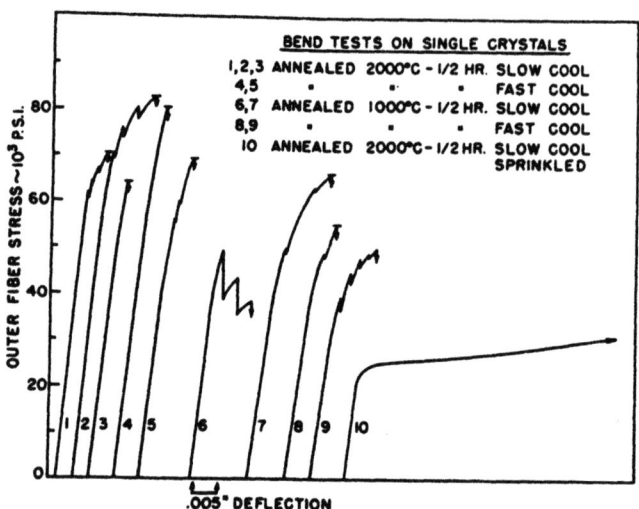

Fig. 12. Reduction of flow stress and large increase of ductility of a single crystal of MgO by sprinkling the surface with silicon carbide. After Stokes [37].

This may well cause diffuse slip bands to be preferred to sharp deep slip lines, but it is unlikely to provide a serious obstacle to the first few dislocations which emerge.

Whatever the mechanism of their action, the evidence that effective surface sources exist is overwhelming. The effects are clearest in crystals which show a falling section in the stress-strain curves as in Fig. 2. Thus a LiF crystal which was chemically polished and then sprinkled with carborundum departed from elasticity at a stress of 830 g wt mm^{-2} in a bending test, while a crystal which was just chemically polished yielded only at 1580 g wt mm^{-2} [33]. The results of similar tests in MgO single crystals are shown in Fig. 12. The introduction of surface sources reduces the yield stress by a factor of three, and greatly increases the ductility [37]. Photoelastic studies of KCl crystals [38] show deformation spreading inwards from the surface, and the flow stresses of pre-deformed crystals could be halved by polishing off less than 100 microns from the surface. In Fe 3% Si single crystals, dislocation etch pits on the sides of the crystal fan out diffusely from sharp slip bands on the "edge" faces of a bent single crystal [39] (Fig. 13), making it clear that slip originated from sources at or close to the surface. This is confirmed by the electron microscopic observations of Vol'shakov and Orlov [40]. There is much indirect evidence of the action of surface sources in very perfect single crystals of Cu [41] and Sn [42]. An alternative approach [43] is to diffuse Zn

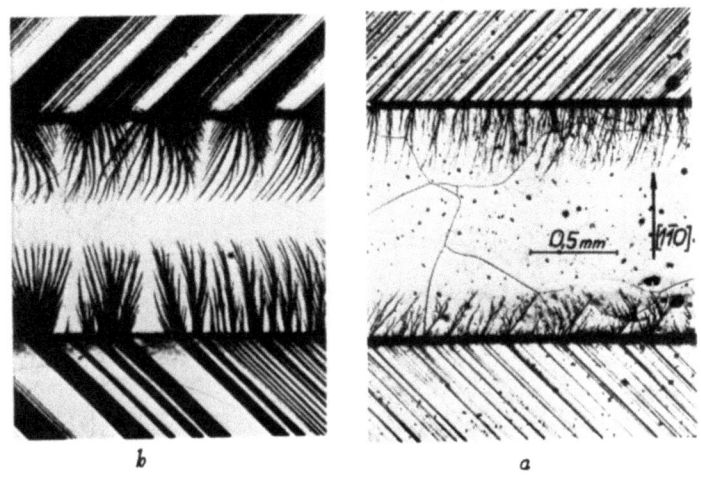

Fig. 13. Arrays of dislocation etch pits fanning out from slip lines in bent single crystals of Fe 3% Si. After Šesták and Libovický [39].

into the surface of a single crystal of Cu. Surface layers were formed containing 0, 0.5, 1, 8 and 12 per cent of Zn. The corresponding critical resolved shear stresses were 56.5, 75, 101, 149 and 136 g wt mm^{-2}. The hardening of the surface layer roughly doubles the critical shear stress. Further hardening of the surface layer causes internal dislocation sources to operate, with only a moderate increase in the critical shear stress for a large increase in the Zn content. The operation of internal sources was confirmed by the absence of a yield point at high Zn concentrations, whereas at low zinc concentrations the sources in the alloyed surface layer initiated a yield point.

The importance of surface sources even in soft single crystals is brought out very clearly by the experiments of Worzala and Robinson [44,45]. The easy-glide strain of single crystals of silver was increased from about 2 per cent to about 6 per cent by polishing off the surface after each increment of about 2 per cent strain. This could have happened either because the surface had become filled with debris which impeded the egress of dislocations, or because the newly exposed surface (either as a result of its greater perfection or because it was slightly deformed by abrasion during the process of polishing) contained active dislocation sources which could continue the process of easy glide. The second explanation was shown to be correct by replacing the procedure of intermittent polishing by one of intermittently sprinkling with silicon carbide grit. An increased range of easy glide was again found (Fig. 14).

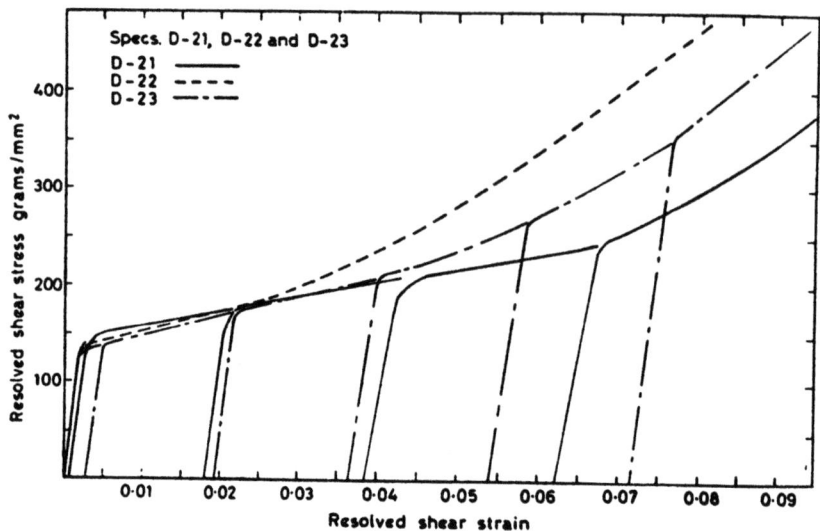

Fig. 14. Stress-strain curves of single crystals of silver: D22 strained directly, D21 strained with incremental polishing, D23 strained incrementally, and sprinkled with silicon carbide grit between increments. After Worzala and Robinson [45].

These observations may be compared with those of Nakada and Chalmers [46], who sprinkled single crystals of gold with silicon carbide grit and found a marked increase in the flow stress, with a complete suppression of Stage I hardening. Repeatedly etching the surface produced only a moderate reduction in the flow stress, but extended the region of moderately low hardening rate to a greater strain than that of Stage I in an undamaged and unetched crystal. The difference between the two sets of experiments is that Worzala and Robinson "impinged lightly with 3 micron diameter silicon carbide particles", while Nakada and Chalmers dropped one hundred grams of 200 micron silicon carbide grit from a height of four feet. This produced a heavily work-hardened surface layer, in which the single crystal Laue X-ray pattern was replaced by a few diffuse rings.

The model in which removal of the surface of the crystal extends the range of Stage I by continually exposing fresh surface sources of dislocations is compatible with the lattice curvature observations of Mughrabi [24] and the sprinkling experiments of Worzala and Robinson [45]. However, it predicts that the slip lines formed under conditions of continual surface removal would be fine and closely spaced. The observations of Latanision and Staehle [47] on nickel crystals which are continuously electropolished during deformation show exactly the opposite; the slip lines are deep and widely separated (Fig. 15), as if surface removal allowed the same

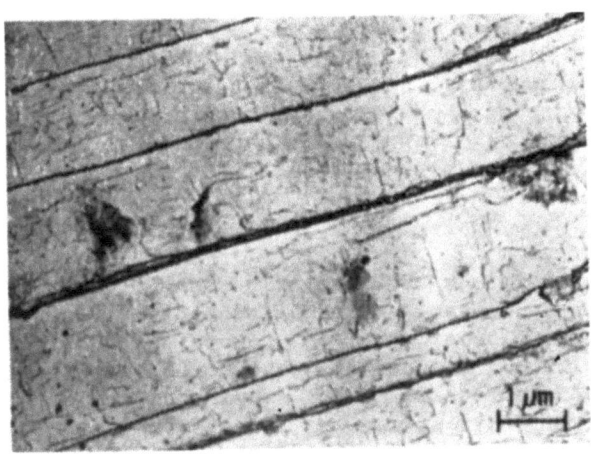

Fig. 15. Deep, widely-separated slip lines in a nickel single crystal deformed in Stage I under active electrolytic dissolution. After Latanision and Staehle [47].

dislocation sources to remain in operation. We do not understand this discrepancy, and would welcome experiments designed to examine it. The complexity of the situation is illustrated by Mughrabi's observation that the lattice curvature which indicates a predominance of surface sources begins to regress when the bulk of the crystal is in early Stage II, and ultimately changes sign to indicate that dislocations formed at sources in the interior are held up by barriers at or near the surfaces.

2.3 Dislocation Sources near the Surface

Sumino [48] pointed out that the action of a Frank-Read dislocation source was arrested by the back stress from dislocation loops which become stuck in the glide plane. If the source is near to a free surface, a large portion of each loop escapes, the back stress is reduced, and the source can emit more loops under a given applied stress. We note that this effect increases the total glide contributed by a single source, and the height of the surrounding slip line, but does not reduce the stress at which the source first operates. Vellaikal and Washburn [49] provided a mechanism whereby sources near the surface would be more effective than other sources as soon as a single loop had been emitted. Sources in annealed crystals are usually jogged. If this is the case, segments of dislocation such as X and Y in Fig. 16e which mutually annihilate during the operation of a perfect Frank-Read source approach on different glide planes. If they have a large edge component at these points, they cannot cross-slip to annihilate, but form a stable dipole which blocks further operation of the source. If one

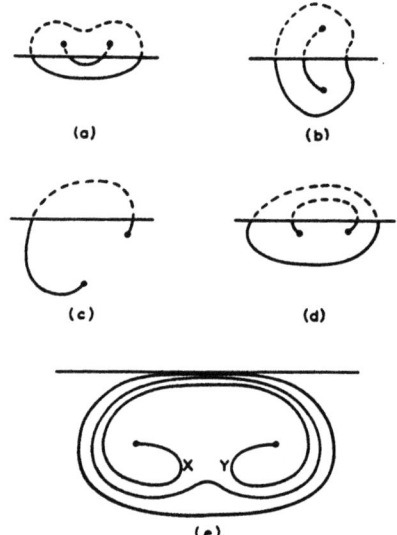

Fig. 16. Dislocation sources near a surface: (a) A surface step, represented by a Frank-Read source with both nodes outside the surface. (b) A Fisher source, with one pole inside the surface and the image pole outside. (c) A near-surface source according to Vellaikal and Washburn [49]. (d) A simple near-surface source where dislocations readily penetrate the surface. (e) a near-surface source where the surface is a barrier to dislocation motion.

of the poles of the source is close to a free surface, this dipole is formed in the hypothetical extension of the incomplete dislocation loop into the space outside the crystal, and does not impede the further operation of the source (Fig. 16c).

It has been suggested that such dislocation sources near the surface may have a special role in surface effects. We do not believe this to be the case. A source usually operates between two fixed nodes. These may both lie outside the surface (Fig. 16a), one inside and one outside (Fig. 16b), or both inside a surface which is readily penetrated by dislocations (Fig. 16c,d). All of these sources. operate more readily than do sources in the interior, and all of them lead to the inward flux of dislocations required by the observations of Fourie and Mughrabi. They cannot explain the coarse slip lines found on Latanision and Staehle's actively dissolved nickel specimens, or the hard layer some 50-100 microns deep required to explain the observations of Kramer which are discussed in §2.4. A model which provides both a soft layer which may be 1-2 mm deep and a thin hard skin is that of Latanision and Staehle

[47], illustrated in Fig. 16e. Here, a dislocation source lies close to a surface which acts as a barrier to the egress of dislocations. The weakness of the model is that the back stress of the dislocations pressed against the surface is large, and such a source will cease to operate while internal sources continue to produce dislocations which move towards the surface and accumulate to produce in the "thick" surface layer observed by X-rays a lattice curvature opposite to that which is actually seen in the early stages of deformation.

The study of etch pits produced where dislocations meet a low-index surface provides an additional method of investigation. As Latanision [50] has shown, the interpretation of the observations is not always simple. The glide plane in f.c.c. metals is close-packed. Etch pits can only be formed on another close-packed plane, and in the most convenient experimental arrangement this surface also includes the glide direction, so that it is a "screw" surface. Specific surface effects occur on "edge" faces, so that the surface which is observed is not one at which the effects of interest are occurring. It would be of interest to expose a close-packed surface which does not contain the active glide direction.

2.4 Evidence for Soft or Hard Surface Layers

We have till now concentrated on a few very direct studies, which have led to the idea of a soft surface layer which penetrates 1-2 mm into a single crystal of copper. We now balance these observations against part of the great mass of observations which suggest that there is no special surface layer, that there is a soft layer, or that there is a hard layer.

Evidence that there is no special surface layer comes from the work of Swann [51], who extended polycrystalline copper by 15 per cent, and found no change in the dislocation arrangement between depths of 0.5 micron and 1 mm. Krejčí and Lukáš [52] found a similar result for single crystals deformed 11 per cent in single glide. In single crystals of copper sheared by 0.2 - 2.5 per cent, Block and Johnson [53] found no systematic variation of dislocation density with depth over the range from 3 to 500 microns (Fig. 17). They pointed out that their crystals had glide paths for edge dislocations of only 4 mm, so that they would be "thin" on the criterion of Fourie [21], and such a uniform distribution would be predicted by Fourie's analysis. Essmann et al. [54] made similar observations on copper polycrystals strained about 1 per cent and irradiated after deformation and unloading. Observations of the kind described in this paragraph have been criticized by Kramer and Haehner [55] on the ground that the "hard" surface layer which their own experiments (to be described later) show to exist recovers rapidly - for gold in 2 hours at $150°C$ and for copper in 2 weeks at room temperature - and may well have escaped observation. Theoretically, such a rapid

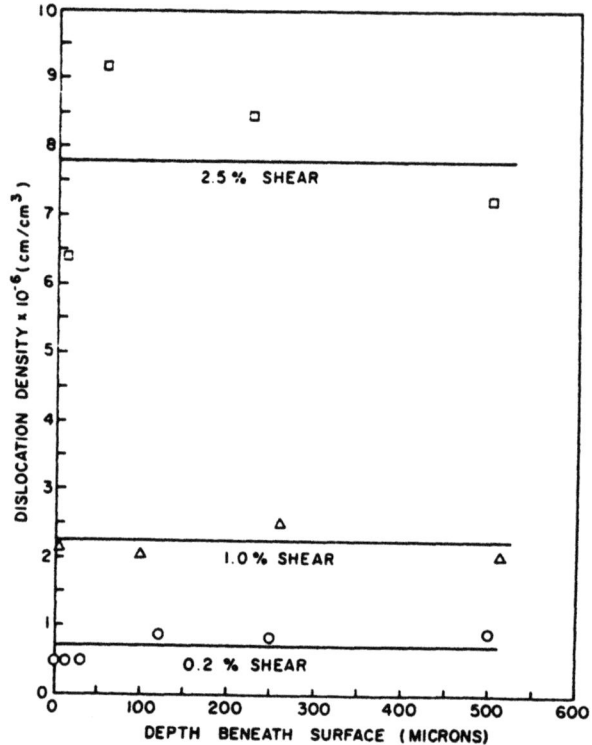

Fig. 17. Variation of dislocation density with distance from the surface of "thin" single crystals of copper at various small strains. After Block and Johnson [53].

recovery of surface stress is surprising, since the force between a dislocation and its image in the surface is no greater than that between two dislocations in the interior. One could suggest that vacancies in the interior migrate to dislocations, jog them, and inhibit recovery by glide, while vacancies near the surface are annihilated at the surface, leaving dislocations near the surface free to recover by gliding. It may even happen (see [56]) that when a surface which is being moved into the crystal by electropolishing intersects a jogged dislocation, the vacancies responsible for the jogs drain to the surface by short-circuit diffusion even at room temperature, so that the dislocation becomes able to glide. This idea should be checked against the known mobility for self-diffusion in dislocation cores. But this is clearly a case where more experiments are needed.

Observations of a "soft" surface are more numerous, particularly if one remembers that Block and Johnson found their observations to be compatible with the idea that a thick crystal has a soft surface. The early observations were indirect. If the surface is soft, the interior bears a higher stress than the exterior during a tensile test. When the load is removed, the elastically extended interior contracts, driving the surface layers into compression. Such compressed surface layers were found by X-ray methods by Kolb and Macherauch [57] in polycrystalline nickel, aluminium and copper. Nye et al. [58] made similar observations photoelastically in single crystals of silver chloride. Suzuki et al. [59] showed that the strain in Stage I of single crystals of copper all having the same orientation (near [011]) increased from 20 per cent to 43 per cent as the radius decreased from 0.890 to 0.105 mm (Fig. 18). The rate of hardening in Stage I was greatest for the thickest crystal. In gold [60] the thicker crystals showed a greater rate of hardening in Stage I, with an indication of a reduced total strain in this stage. The results of Fourie [21] were rather directly confirmed by Murphy [61], who made Vickers diamond indentations across the face of a crystal. His indentations were 30-40 microns across, and necessarily deformed the crystal severely over several times that distance. They could therefore detect a soft surface layer 1-2 mm in depth, but could not detect the hard layer 50-100 microns in depth which according to Kramer is formed near the surface.

Finally, we turn to the evidence for "hard" surface layers. Much

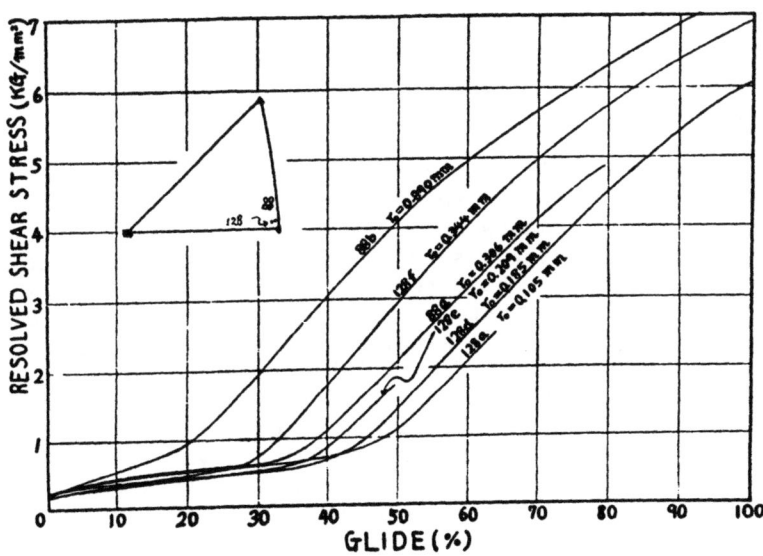

Fig. 18. Shear stress as a function of shear strain for a number of copper crystals having the same orientation but different radii. After Suzuki et al. [59].

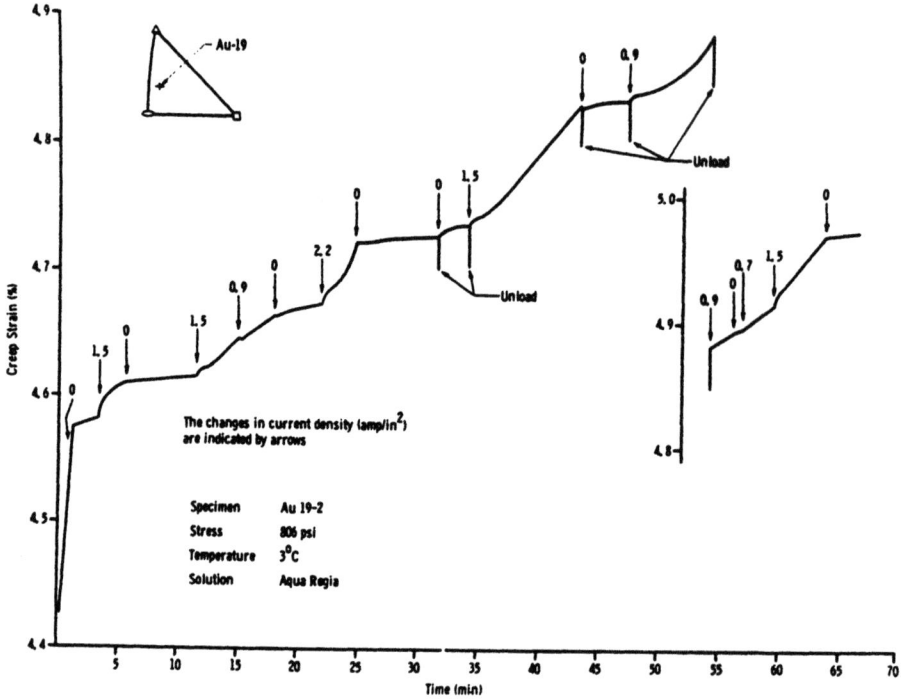

Fig. 19. The creep of a single crystal of gold subjected to varying rates of electrolytic dissolution in a bath of aqua regia. After Kramer [62].

of this evidence is due to Kramer, and most of his work has been done on aluminium, which readily forms a strong oxide coating. We think it likely that his results on aluminium are to be attributed to this oxide layer, and defer discussion of them until we treat the effects of such layers. There remains a body of evidence of a "hard" surface layer on other metals. The most direct is that of Kramer [62]. The creep rate of a single crystal of gold could be markedly increased or decreased (Fig. 19) by increasing or decreasing the rate of electrolytic dissolution in a bath of aqua regia. Since this is a crucial experiment which has hardly been repeated by other workers, it requires careful analysis. One problem, recognized by Kramer, is that the current densities were high enough to alter the temperature of the crystal appreciably. The discontinuities seen in Fig. 19 correspond to thermal expansion produced by a temperature change of 3°C. One should look for the effect on the creep of a similar temperature change produced without electrolysis. Moreover, Pfützenreuter and Masing [63] showed as long ago as 1951 that the potential difference between a gold

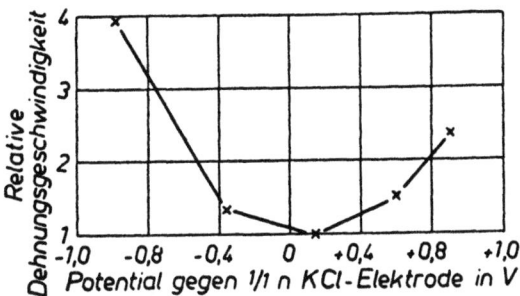

Fig. 20. Relative creep rate of a gold wire in a solution of potassium nitrate as a function of the potential of the wire against a normal potassium chloride electrode. After Pfützenreuter and Masing [63].

wire and a solution of KNO_3 would influence the rate of flow, which was a minimum when the wire was at a potential of about 0.15V positive with respect to the electrolyte, and rose symmetrically as the potential changed in either direction (Fig. 20). Since solution in this electrolyte is unlikely with one polarization and hardly possible with the other, the effect is probably purely electrostatic. The authors show that if the nucleation of flow requires the formation of a surface step, and if the step is 1 atom deep and 10 atoms wide, while the surface double layer has a capacity of the order of 10^{-5} Farad cm^{-2}, a potential difference of the order of 0.5 - 1 volt can increase the nucleation rate by a factor which may be about 10 but may be very much larger. A thorough analysis of Kramer's experiment must compare it with experiments in which potentials are present but the specimen does not dissolve. Observations similar to Kramer's were made by Kitajima et al. [64] on single crystals of copper and of alpha brass, while Kitajima et al. [65] observed high dislocation densities in a surface layer 40-70 microns thick in single crystals of copper deformed in Stage I.

In a further series of experiments, Kramer [66] determined the activation energy for creep as a function of the rate of electrolytic removal of the surfaces of single crystals of aluminium, copper and gold (Fig. 21). If the rate of strain at temperature T under a stress σ is $\varepsilon(\sigma,T)$, the activation energy $U^{\dagger}$ is defined by

$$U^{\dagger} = kT^2 (d(\ln\dot{\varepsilon})/dT)_{\sigma}.$$

In all cases, $U^{\dagger}$ decreased as the rate of surface removal increased. The effects of temperature changes associated with changes in the electrolytic current can be complicated. We shall accept the observations as they stand, and observe that the activation energies for processes such as the cutting of two dislocations or cross-slip are independent of the density of dislocations. If the activation energy for such a process in the absence of an applied stress is $U_o^\dagger$, then the activation energy in the presence of a stress σ with an activation volume $V^\dagger$ is given by

$$U^\dagger = U_o^\dagger - V^\dagger \sigma.$$

A decrease of $U^\dagger$ at constant σ implies an increase in $V^\dagger$, suggesting that increasingly rapid dissolution exposes regions with longer loops of dislocation, and thus with a lower dislocation density, contrary to the results of Fourie and Mughrabi. These results are supported by Kramer's measurements of $V^\dagger$ in single crystals of aluminium, using the formula

$$V^\dagger = kT/d(\ln\dot\varepsilon)/d\sigma)_T.$$

They showed the expected increase of $V^\dagger$ with increase in the rate of solution. To confirm or remove the conflict of these observations with those of Fourie and Mughrabi, one should try the effect of a reverse potential or of an electrolyte which does not dissolve gold. The two sets of results can still be reconciled if one assumes the existence of a thin hard layer and a thick soft

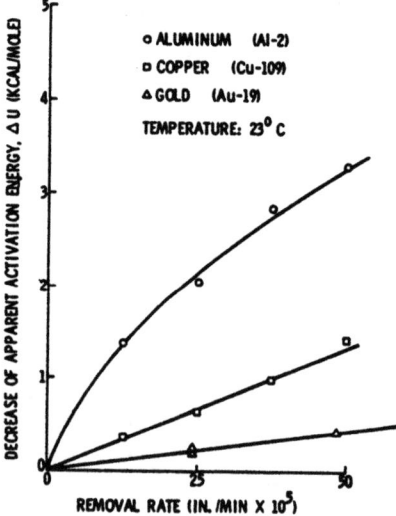

Fig. 21. Decrease in the activation energy for creep of single crystals of aluminium, copper and gold as a function of the rate of electrolytic surface removal. After Kramer [66].

layer as illustrated in Fig. 16e; the problem is then to decide why such sources should act in preference to sources in the interior. Finally, one may accept Kramer's results and re-interpret them in terms of the soft-surface model. Near to the soft surface, single dislocations are pressed against obstacles. In the interior, the dislocation loops are shorter, but processions of dislocations are likely to be piled up against a single obstacle. If there are n dislocations in a procession, the activation volume is multiplied by n.

Another series of measurements by Kramer [67] on the body-centred cubic metals iron and molybdenum, is important because it shows (Fig. 22) that there was a change $\Delta\bar{\sigma}$ in the flow stress $\bar{\sigma}$ (corrected for change in cross-sectional area) when the surfaces of polycrystalline samples were removed electrochemically and the samples were retested at the end of the electrochemical process. The problem of an applied potential no longer arises, and the depth of the hard surface layer, about 5×10^{-3} inches or 125 microns in iron, can be determined. These experiments should be repeated on copper, for comparison with those of Fourie and Mughrabi, and on gold. Single crystals should be used, because of the complication introduced by intergranular constraints in polycrystals.

Rather direct evidence was provided by Stokes et al. [68], who showed (Fig. 23) that there was a sudden increase in the rate of work hardening when an MgO crystal was transferred from orthophosphoric acid at 120°C to mineral oil at the same temperature.

While Murphy [61] made Vickers diamond indentations across the "screw" face of a copper crystal, and found soft layers about 2mm

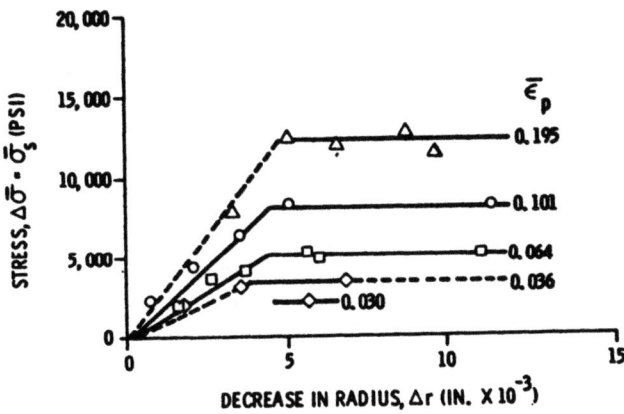

Fig. 22. Decrease in flow stress after electrochemical reduction of the radius of polycrystalline samples of Armco iron. After Kramer [67].

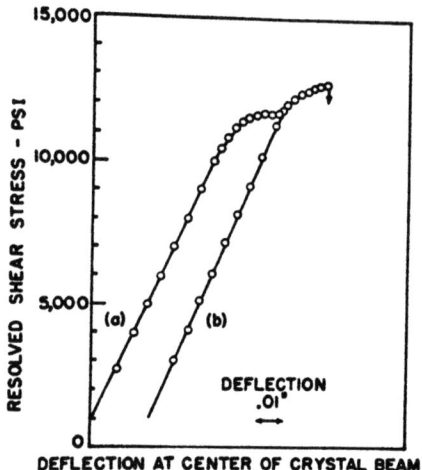

Fig. 23. Stress-strain curves for the bending of a single "flaw-free" crystal of MgO; (a) in orthophosphoric acid at 120°C, (b) in mineral oil at the same temperature. After Stokes et al. [68].

deep at the "edge" faces, Kramer and Balasubramanian [69] made similar tests on a surface which was gradually removed by electro-chemical polishing, and found a steep fall in hardness from the surface to a depth of 70 microns, with no change in hardness below this depth.

As Duquette [70] has pointed out, much of the evidence for a "soft" layer 2mm thick is compatible with the existence of an unobserved "hard" layer 50 - 100 microns thick, and vice versa. Moreover, the extent of each layer may vary not only from metal to metal and from condition to condition, but during the course of straining a single sample. For example, surface sources on the primary glide system may make the surface initially soft; similar sources on a secondary system could make the surface harden into Stage II while the interior is still in Stage I - though it is the opposite behaviour which seems to be observed. Volunteers are needed to carry out on the same, or very similar, samples both the measurements which they themselves have made previously and the exact types of measurements which have led to contradictory results in the hands of other workers.

2.5 The Unloading Yield Point

It is well known that if an f.c.c. crystal is extended into Stage II, and the load is released and re-applied, the crystal begins to flow only at a stress higher than the maximum previously applied, and may then show a yield point, the flow stress decreasing with increasing strain. In some cases this is caused by the locking of dislocations by impurities or vacancies, but in other cases it seems to result purely from rearrangements of the dislocations. In such cases [62,71,72], the yield point disappears if a layer is polished off the surface of the crystal while it is free from external stress. Feng and Kramer [71] found that the thickness of the layer which must be removed in gold was 0.0067 inches = 170 microns, while in copper Fourie [72] found about 180 microns, in remarkable agreement, and lying closer to the thickness of Kramer's "hard" layer than to that of Fourie and Mughrabi's "soft" layer. Brydges [73] has proposed an explanation of these results in terms of a soft surface layer, and we present an analysis of this argument. Suppose first (Fig. 24a) that the flow stress AD at the surface of the crystal under load is somewhat less than the flow stress BC in the middle. Then when the external load is removed the internal tensile stress distribution is given by the curve EFGHIJ. Since surface effects are important only on the "edge" surfaces, a one-dimensional model is appropriate, and the absence of external load requires the shaded area FGHI to be equal to the sum of the two shaded triangular areas. The flow stress for reverse flow is given, if we neglect Bauschinger effects which are usually small in single crystals, by KLMN. This is not attained, and, on reloading, the crystal begins to flow simultaneously across the whole cross-section, and the stress-strain curve continues as if there were no interruption. If the surface layer is removed while the crystal is loaded, this simple model would show an increase in the flow stress, followed by a period of low and possibly negative work hardening while the soft layer was re-established. This is contrary to experience.

If, however, the surface layer is much softer than the interior as in Fig. 24b, the internal tensile stress at the instant of unloading would follow the curve EFGHIJ, while the flow stress would be KLMN. These curves cut at X and Y, and the surface layers yield. If we neglect work hardening during this process, the internal stress curve becomes KXLMYN (Fig. 24c), while, in the absence of work hardening, the yield stress for flow on reloading remains OPQR. The line GH has moved downwards to maintain the equality of the shaded areas. On reloading, yielding begins at the surface, and may progress until the regions AB and GH have reached the yield stress (Fig. 24d), which remains constant if we continue to neglect work hardening during the small deformation of the surface layer. The region CDEF has not yielded plastically during this early stage of the reloading process. Finally, the whole region corresponding to XFGHIY in Fig. 24c reaches the yield point, and the crystal again deforms along the prolongation of the original stress-strain curve. The effect of removing the surface layers will be as in the first case, again contrary to experience.

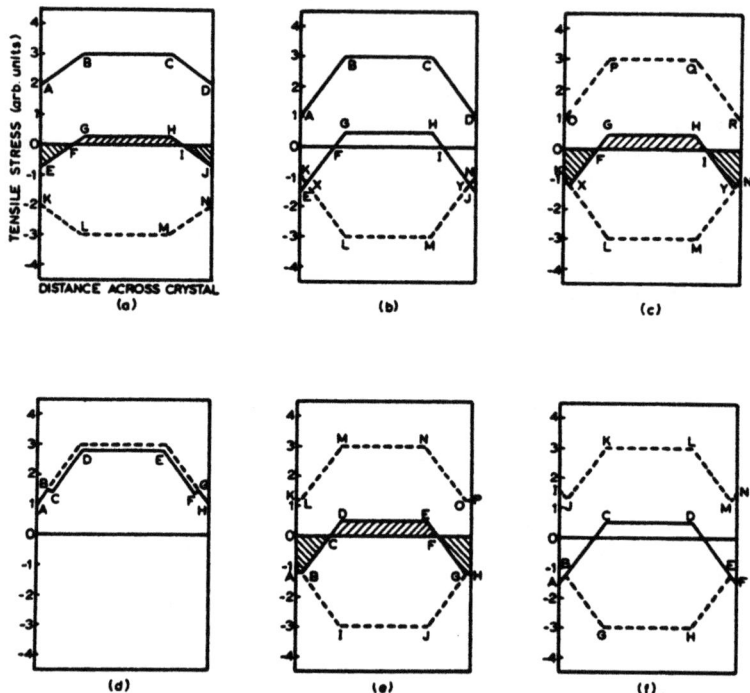

Fig. 24. The unloading and reloading of a crystal having a soft surface: (a) The flow stress at the surface A,D is slightly less than that in the interior BC. The tensile stress profile is EFGHIJ, the yield stress for reverse flow on unloading is KLMN, and both unloading and reloading are elastic. (b) The flow stress in the surface is considerably less than that in the interior, and the compressive stress in the surface exceeds the local yield stress for reverse flow. (c) If the surface flows without further work hardening, the unloaded stress profile is KXFGHIYN, while the flow stress for reloading remains OPQR. (d) The stress profile during reloading then passes through the slope ABCDEFGH. (e) If the surface work hardens rather rapidly during plastic compression on unloading, the unloaded stress profile may take the form ABCDEFGH, while the yield stress for reloading is KLMNOP. (f) If the surface hardens without appreciable plastic deformation, the unloaded stress profile is ABCDEF, while the reloading yield profile is IJKLMN.

Now suppose that the surface layers work harden during their small deformation. For simplicity, consider the case of Fig. 24e, in which the rate of work hardening is such that the whole surface region which has deformed plastically during unloading has attained the same flow stress ABGH. Again, equality of the shaded areas requires the line DE to fall during the plastic deformation of the surface which takes place on unloading. The yield stress for reloading is given by the curve KLMNOP. The whole cross-section reaches this curve under the same load, and this load is greater than that at which the crystal originally deformed. Moreover, since the usual soft surface layer is not fully developed, we may expect deformation to proceed with a low or even negative work hardening. On this model, the observed unloading yield point depends essentially on the work hardening of the soft surface layer when it is plastically compressed. Although the anomalously hardened region is at the surface, it is still the softest part of the crystal, and removing the surface should reduce the yield stress.

Finally, suppose that the rate of work hardening in the surface layer is for some reason so great that it is able to withstand after only infinitesimal deformation the whole reverse stress indicated by the curve EXFGHIYJ in Fig. 24b. Then (Fig. 24f) the tensile stress distribution in the unloaded state is still given by ABCDEF, the yield stress for reverse flow by ABGHEF, and that for flow on reloading by IJKLMN. Yielding begins in the interior along JKLM at the original stress, but the whole cross-section yields only at a higher stress, which again may be expected to fall as the soft surface is established. The surface regions IJK, LMN are still softer than the interior, and their removal should reduce the flow stress.

It is clear that we would do well to examine the phenomenon of the unloading yield point in terms of the model of a hard surface layer. The same arguments show, mutatis mutandis, that the unloading yield point is a consequence of the work hardening of the soft core during unloading, with a fall of stress on further loading while the postulated normal soft core is re-established. Removing the hard surface while the crystal is unstressed necessarily reduces the flow stress on reloading, as is observed. The argument is consistent with the observation that the depth of surface which has to be removed to remove the yield point is only about 160 microns.

2.6 Annealed Surfaces

The surfaces we have considered up to now have been exposed from the interior of the crystal. The natural surface of an as-grown crystal may be very different. We shall neglect the influence of impurities adsorbed from the surroundings. We shall also neglect the surface adsorption of impurities dissolved in the crystal, although there is ample experimental evidence that they are almost always present. We cannot even in principle neglect the fact that the crystal is usually grown at a high temperature where it contains a high concentration of vacant lattice sites. At the temperature at which the crystal is grown, dislocations climb readily, and dislocations near the surface will become heavily jogged as they climb in order to meet the surface at the equilibrium angle (see §2.2). In the terminology of Latanision and Staehle [47], the surface layer, while initially not having an unusually high concentration of dislocations, is "disarrayed" in comparison with the interior. However, we have seen in §2.4 that Fabiniak and Kuhlmann-Wilsdorf [56] have predicted exactly the opposite behaviour, in which jogged dislocations are able to discharge their jogs to a free surface which they intersect. The vacancies become supersaturated as the crystal cools. In the interior of the crystal, these excess vacancies nucleate dislocation loops or voids; they can escape at the surface, and the surface layer therefore contains a lower concentration of these small dislocation loops or voids. The effects [56] are striking and complicated. The first result is the striking difference (Fig. 25) between the stress-strain curves of crystals grown within 3K of the melting point ("zone-melting furnace") and slowly cooled, and crystals grown in a normal travelling furnace and also slowly cooled. Then the critical resolved shear stresses were determined (Fig. 26) for these two types of crystal, which were reduced in diameter either by electropolishing before straining (circles) or by repeated straining and electropolishing (crosses and triangles). The coincidence of the circles with the crosses and triangles shows that the work hardening in the latter cases, which amounted to as much as 300 g wt mm^{-2}, was entirely removed by electropolishing. The

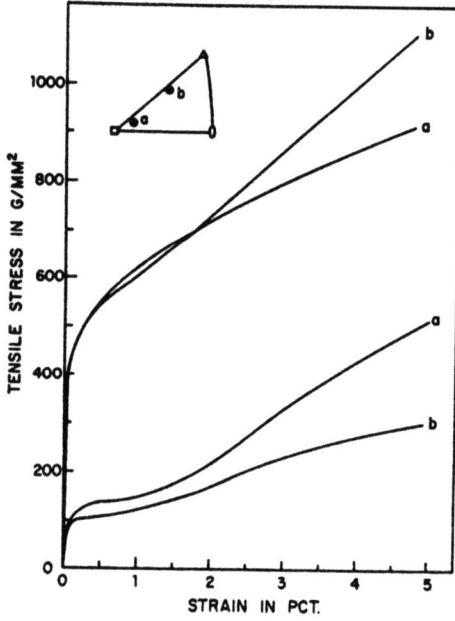

Fig. 25. Stress-strain curves of slowly cooled aluminium crystals, the lower curves referring to crystals grown within 3K of the melting point, and the two upper curves referring to crystals grown in a normal travelling furnace. After Fabiniak and Kuhlmann-Wilsdorf [56].

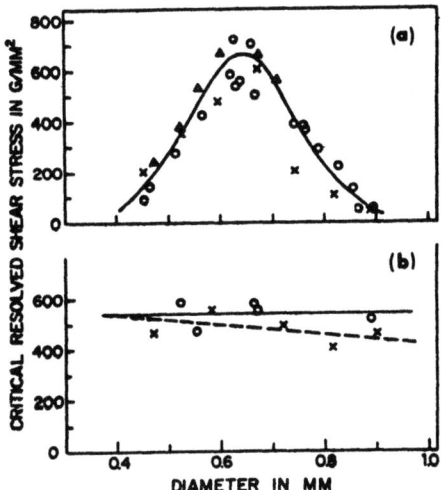

Fig. 26. Critical resolved shear stresses of (a) aluminium crystals grown within 3K of the melting point and quenched into iced brine and (b) similar crystals, but grown in an ordinary travelling furnace, then transferred to the same furnace as (a) and quenched. The crystals were originally all of the same diameter, and were reduced in diameter by electropolishing before straining (circles) or by repeated straining and electropolishing (crosses and triangles). After Fabiniak and Kuhlmann-Wilsdorf [56].

crystals were of aluminium, where the oxide film induces a very rapid surface hardening, and it would be most valuable to have these experiments repeated in gold and copper. The flow stress is controlled by the dislocation structure at the surface, and the variation of flow stress with radius correlates very closely with the observed dislocation densities at the same radii. So strong is the control of the surface that (Fig. 27) a crystal of diameter 0.75 mm increases its load at yielding when its diameter is reduced to 0.67 mm, despite the reduction in cross-sectional area.

2.7 Oxide Layers, Alloy Layers and After-Effects

Oxide layers increase both the flow stress and the rate of work hardening of metal crystals. This was clearly demonstrated by Andrade and Henderson [74], using a silver crystal which was oxidized by heating to 430°C, and then de-oxidized by heating to 800°C (Fig. 28). In magnesium single crystals [75] the extent of Stage I is greater if the tests are conducted in a vacuum than if they are conducted in air, while Roscoe [76,77] showed that both the initial flow stress and the rate of work hardening of single crystals of cadmium were increased by the presence of an oxide film, thicker

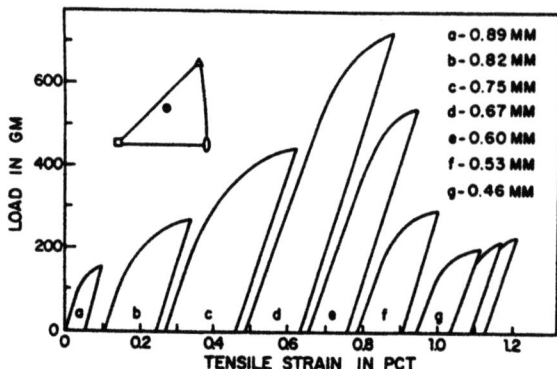

Fig. 27. Load-elongation curves of an aluminium crystal repeatedly strained and repolished. Note the increase of load associated with a decrease of diameter in going from (c) to (d). After Fabiniak and Kuhlmann-Wilsdorf [56].

films being more effective. The great body of experimental evidence is on aluminium. Nakada and Chalmers [78] showed that at strains less than 5 per cent the flow stress and the rate of work hardening were reduced by polishing the "edge" surfaces but not the "screw" surfaces; above 5 per cent strain, polishing either pair of faces reduces the flow stress but not the rate of hardening. They also [46] found that abrading the edge faces at strains of 0.5 and 2.5 per cent produced an immediate hardening. Sumino and Yamamoto [79] also found substantial drops of flow stress on polishing off 190 or 150 microns after strains respectively of 3.4 and 24.6 per cent. By far the most extensive series of observations on aluminium is due to Kramer and his colleagues. Our approach to all of this work is that the effects observed are dominated by the difficulty that dislocations, and especially edge dislocations, find in breaking through the oxide layer of a normal crystal, and that this leads to pile-ups of dislocations just below the surface. These in their turn produce slip on secondary systems, and lead to a hardened surface layer. If there are effects which are not due to the oxide film, it is hardly possible to disentangle them from the effects of the oxide. The effects are striking. The removal of 0.041 inches from the surface sharply reduces the flow stress in Stage I, and continuous surface removal increases the extent and decreases the slope of Stage I and of Stage II [80]. There is a size effect in the critical resolved shear stress, and unloading yield points are observed which are removed by electropolishing 0.006 inches off the surface [62]. Single crystals harden less rapidly in a vacuum than they do in air [81]. The contribution of the hardened surface layer to the flow stress can

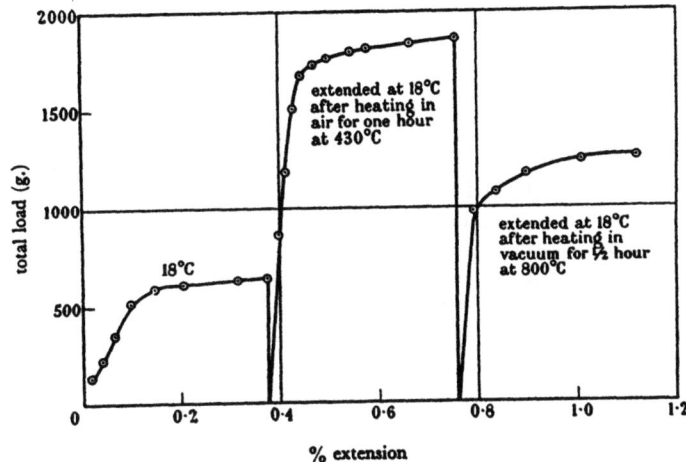

Fig. 28. Load-extension curve of a single crystal of silver oxidized by heating to 430°C and de-oxidized by heating to 800°C. After Andrade and Henderson [74].

be estimated quantitatively by etching off the surface layer, which has a depth of 0.0025 inches. Stage I ends when the difference between the resolved shear stress on the secondary glide system and the surface hardening stress is equal to the critical resolved shear stress [82]. The unloading yield point is suppressed by removing the surface [71]. Both the rate of work hardening and the activation energy for creep are reduced if the elongation is conducted in a high vacuum [83]. Assuming that the internal stress τ_i and the surface stress τ_s are state variables depending only on the present state of the sample and independent of the instantaneous rate of straining, they may both be determined [84]. First τ_s is determined by removing the surface. Then, if the total applied stress is τ_a, the effective stress acting on the dislocations near the surface which are believed to control the flow is given by

$$\tau^* = \tau_a - \tau_i - \tau_s.$$

The effective stress τ^* may be determined by observing the change in the rate of strain produced by small changes in τ_a, since the relation between the strain rate and τ^* is reasonably well known. The result, which is supported by the simultaneous work of Fabiniak and Kuhlmann-Wilsdorf [56], is that τ_s increases from the beginning of the elongation, while τ_i is negligible below 0.5 per cent resolved shear strain, and equals τ_s only at 1.4 per cent strain. There is an important ambiguity in this ingenious analysis. The internal stress τ_i is generally assumed to be a frictional stress which opposes equally deformation either in the direction of the previous strain or in the opposite direction. Any asymmetry, which would give rise to a Bauschinger effect, is assumed to be small, especially in single crystals. But in the case of τ_s we know that there is a large excess of dislocations of the same sign in the surface layer, and the studies of after-effects soon to be discussed give evidence of this. A proper analysis of τ_s must resolve it into a frictional component which opposes equally deformation in either direction, and an internal stress component which opposes forward flow but assists reverse flow. Experiments are needed in which crystals which have been extended are then tested in compression, with or without removal of the surface layers. The fatigue life is enhanced by testing in a vacuum. Unexpectedly, the effect of increasing the frequency in this region is to decrease the number of cycles to failure, presumably because of the greater rate of work hardening in the surface layer [85]. The expected interaction between strain rate and surface oxidation was shown in the experiments of Nelson and Williams [86], who pulled single crystals of magnesium at various strain rates in air and in a vacuum of 5×10^{-9} torr. In air, the hardening was greater at the higher strain rates, but the inverse effect was observed at low pressures (Fig. 29), because new slip steps could grow to considerable depths before becoming oxidized.

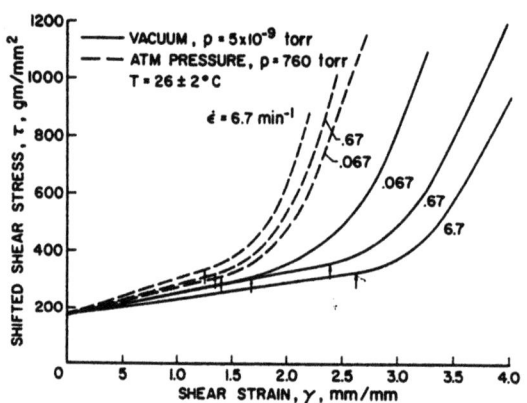

Fig. 29. Effect of strain rate on the stress-strain curves of magnesium crystals tested in air or under a pressure of 5×10^{-9} torr. After Nelson and Williams [86].

The effect of sample diameter on the flow of polycrystalline aluminium has been studied [87]. Kramer and Haehner [88] compared the recovery in the flow stress of pure polycrystalline aluminium after relaxing at 3°C for 24 hours with that produced by removing 0.01 inches from the surface. For each initial pre-strain, the recoveries produced by the two treatments were identical (Fig. 30). Moreover, the recovery in each case was "metarecovery" in the sense of Cherian et al. [89], the stress-strain curve joining the extrapolation of the uninterrupted curve after a further strain of about

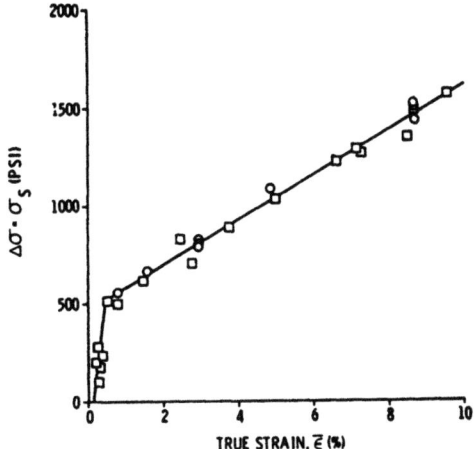

Fig. 30. Comparison of the reduction in the flow stress of pure polycrystalline aluminium produced by annealing at 3°C for 24 hours (squares) with that produced by removing 0.01 inches from the surface (circles), after various strains. After Kramer and Haehner [88].

0.5 per cent. Annealing at 50°C produced "orthorecovery", the entire curve remaining depressed after the anneal. It is clear that these very important experiments should be repeated on aluminium, copper and gold. Low-temperature creep in polycrystalline aluminium can be satisfactorily analyzed in terms of a model in which the flow stress in the interior increases with strain, while that in the surface layer, though increasing with strain, decays with time in the manner already observed. The number of mobile dislocations decreases as transient creep progresses [90]. If the stress is suddenly reduced and creep stops, the delay time before it recommences can be reduced in a predictable way by removing a thin surface layer [91].

We now turn from oxide layers, which are normally present on the surface of a metal, to alloy layers, which are more controllable. The work of Adams [43] has already been discussed. Even earlier, Pickus and Parker [92] alternately electroplated a thin layer of copper on to a single crystal of zinc creeping under constant load, and dissolved the layer off. Large changes in the creep rate and in the activation energy for creep were observed (Figs. 31,32), although no changes were found in polycrystalline samples.
Rosi [93] measured the stress-strain curves of single crystals of copper (A) with clean surfaces, (B) with a surface coating of 4×10^{-3} mm of silver, and (C) with this silver diffused to form an alloy layer 0.3 mm thick. Treatment (B) slightly reduced the

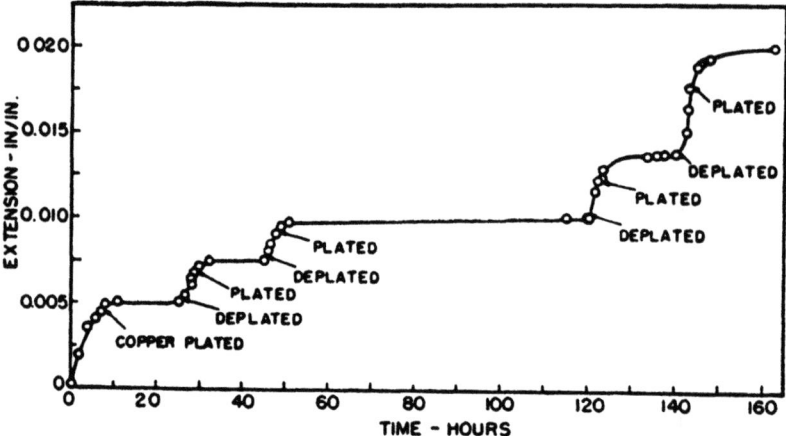

Fig. 31. Changes in the creep rate of a single crystal of zinc on electroplating and removing a thin layer of copper. After Pickus and Parker [92].

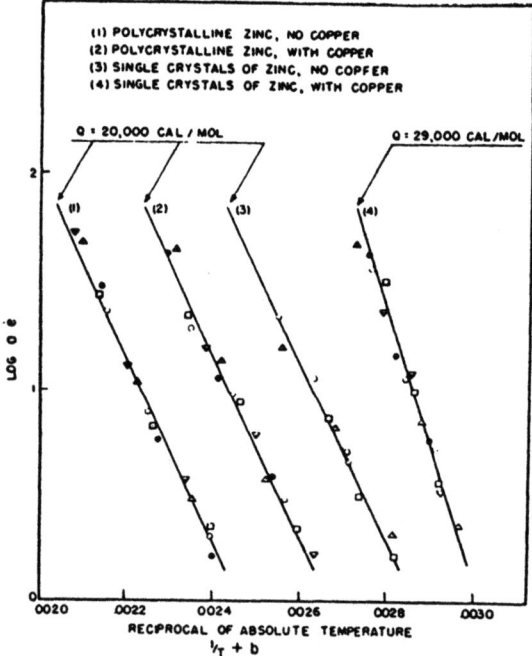

Fig. 32. The activation energy for creep of single crystals and polycrystals of zinc, with and without a thin layer of copper. After Pickus and Parker [92].

extent of easy glide, while treatment (C) greatly increased the
critical resolved shear stress and the extent of Stage I, and
slightly reduced the rate of hardening in Stage II (Fig. 33). Since
the crystals were ⅜ inches ≃ 9 mm in diameter, the disproportionate
influence of the surface on the plastic properties is very apparent. Anomalies such as serrated stress-strain curves are also
controlled by the surface layer [94]. Single crystals of nickel
2 mm in diameter exhibited serrated curves when charged electrolytically with hydrogen under conditions in which the hydrogen is
not likely to have penetrated more than about 15 microns from the
surface. In an ingenious experiment, Ruddle and Wilsdorf [95]
found that a thin epitaxial layer of nickel (about 0.06 microns)
deposited on either the "edge" or the "screw" face of a copper
crystal 2 mm thick roughly halved its critical resolved shear
stress, the slip-line structure being very fine. The epitaxial
layer is subjected to a tensile strain of 2.5 per cent. A small
additional tension initiates slip in the epitaxial layer, and, since
this is coherent with the substrate, dislocations penetrate into
the copper and cause it to deform plastically. Effects of this
kind are not confined to metals. Doping the surface of lithium
fluoride crystals with magnesium to a depth of less than 10 microns
increases the yield stress in compression by about 50 per cent, and
reduces the rate of work hardening [96]. The latter effect seems
to arise because four glide systems are equally stressed. If surface dislocation sources are eliminated by doping, the rapid multi-

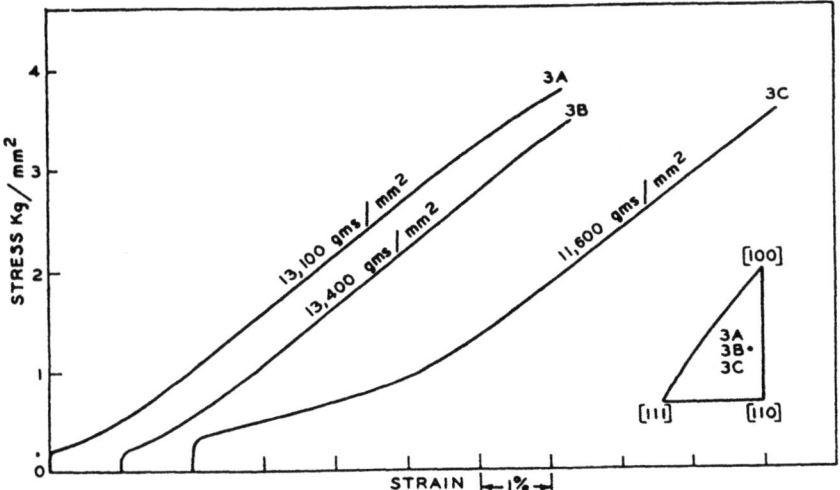

Fig. 33. Stress-strain curve of a single crystal of copper, with
(A) a clean surface, (B) a surface layer of silver 4 x 10^{-3} mm
thick, and (C) the same layer diffused to form an alloy layer
0.3 mm thick. After Rosi [93].

plication of dislocations from the first source to operate which occurs at high stresses inhibits glide on other systems, and it is the intersection of glide on different planes which is responsible for work hardening.

We conclude this section with a discussion of after-effects. If a sample is strained and unloaded, the elastic recovery which reduces the strain is often followed by an after-effect in which part or even the whole of the remaining strain is recovered. This is a "normal" after-effect. An "abnormal" after-effect was predicted by Cottrell in 1957 and observed by Barrett [97,98]. A twisted wire of single crystal or polycrystalline iron or zinc [97] or polycrystalline cadmium [98] is surrounded by water, and untwists so that the angle of twist depends linearly on the logarithm of the time. If an etchant which dissolves the oxide film is applied, the direction of twisting reverses, and after some delay a logarithmic untwisting curve with a reduced slope establishes itself (Fig. 34). The obvious explanation is that dislocations piled up against the oxide film during the initial twisting are gliding back into the interior, where they annihilate with other dislocations. This produces the "normal" untwisting. When the oxide film is removed, some of the piled-up dislocations near the surface escape, producing the "abnormal" forward twist. Barriers still exist at the surface, and those dislocations in the surface pile-ups which are farthest from the surface again move towards the interior,

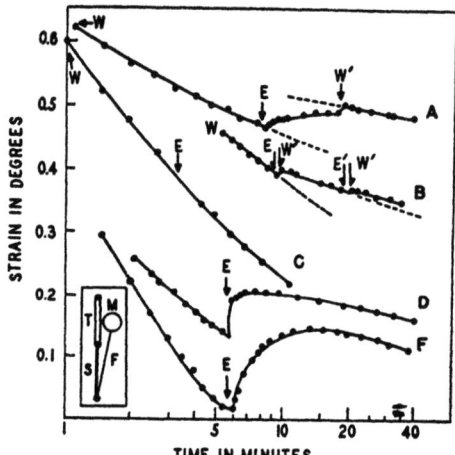

Fig. 34. The "abnormal after-effect" in twisted polycrystalline cadmium wires: (A), (B) oxidized, (C) etched, (D), (F) anodized. Etchant was applied at the points E, E', and water at the points W and W'. After Barrett [98].

producing a "normal" after-effect again, but at a reduced rate, because they are being driven inwards by a reduced stress. The effect has been studied in more detail in aluminium [99] and cadmium [100]. The effect occurs not only in the presence of an oxide layer, but also if the wire has an electrodeposited metallic covering, whether the elastic constant of the film is higher or lower than that of the substrate [101]. Holt's detailed study in polycrystalline aluminium [102] illustrates the complexity of surface effects. He repeated Barrett's experiments, twisting the wires by different amounts. At twists exceeding about 12 degrees cm^{-1} the usual "abnormal" after-effect is observed, but between about 1 degree cm^{-1} (a surface strain of 9 x 10^{-4}) and 2 degrees cm^{-1} the transient twist on removing the oxide film from an anodized specimen is in the same direction as the normal relaxation (Fig. 35). The explana-

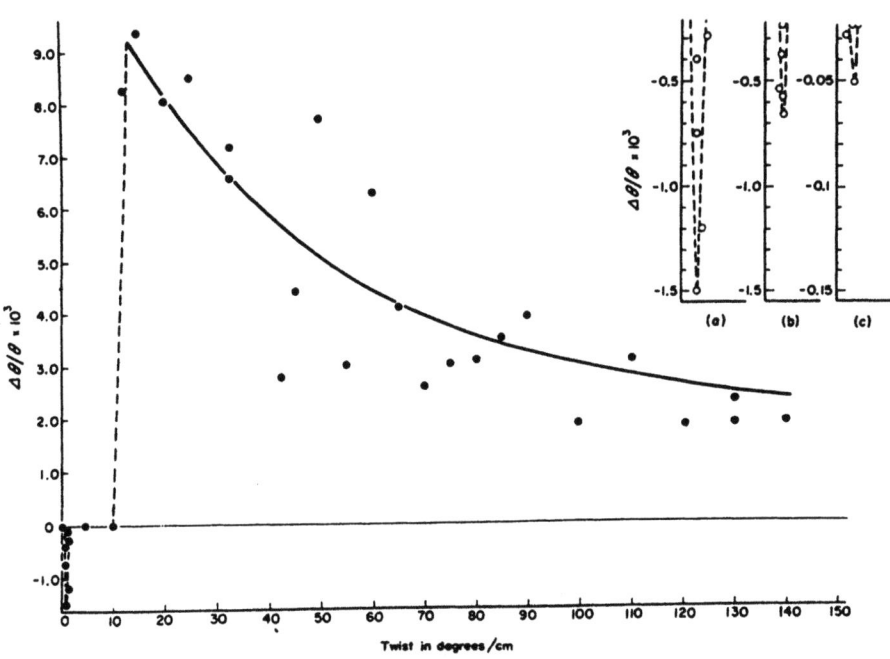

Fig. 35. The ratio of the initial twist $\Delta\theta$ produced by pouring phosphoric acid on to anodized polycrystalline aluminium wires to the critical angle of plastic twist θ. A twist of 1 degree cm^{-1} is a surface strain of about 8.7 x 10^{-4}. Inset (a) is the small-angle negative peak on a larger scale; insets (b) and (c) are similar observations under modified conditions. After Holt [102].

tion is probably that at these small twists only the relatively soft anodized oxide layer and the surface regions of the metal have deformed plastically, while the bulk of the metal is strained elastically. Removing the soft outer layer of oxide allows the elastic core region to untwist the plastically deformed outer layers of the metal.

3. PLASTICITY IN AN ENVIRONMENT OF WATER OR SURFACE-ACTIVE AGENT

3.1 Water

According to Ewald and Polanyi [103], the first observation of the enhanced plasticity of rock salt when immersed in warm water was made by Markscheider Engelhardt in 1867. Using this knowledge, he proceeded to the "Herstellung von kleinen Kunstwerken". A study by Joffé, Kirpitschewa and Lewitzky [104] followed on the demonstration by Coblentz [105] that prisms of rock salt could be bent between the tongue and the teeth after being sucked in the mouth (Fig. 36). Coblentz noted that after the salt had been wet for

NOTE ON THE BENDING OF ROCK SALT.¹

BY WM. W. COBLENTZ.

WHILE splitting rock salt plates it was noticed that occasionally some plates were bent (not from splitting), showing that the whole mass must have been subjected to strain.

The conditions under which such a bending will occur is of interest, since at room temperature rock salt is very brittle.

It was found that plates of rock salt .5 mm. thick would bend after being held in the mouth for a few minutes. The bending is done by pushing the plate against the teeth with the tongue — temperature of the mouth is about 36° C. The rock salt plates can also be bent by holding them in a warm (50° to 70° C.), saturated salt solution by means of a forceps, and pressing lightly against the side of the vessel. The same is true when held in a bunsen flame, temperature 1400° C. The accompanying mounted specimens have been bent by the three different methods, the thicker ones by the second and third.

In all cases the rock salt remains brittle for a short time, after which bending begins, and (in the methods employed) then suddenly occurs more readily. It can then be bent back and forth like a piece of whalebone. The plates can also be twisted after they have become pliable.

A small plate 3 x .5 x .1 cm. held in a vise, with a weight of about 75 grams suspended from the free end, did not show a visible bending after leaving it at room temperature for 25 days.

Whether this is a case of slipping due to gliding planes, such as occurs in certain other minerals, or whether it is simply a case of viscosity like that of pitch, has not been determined. The thick plates show but slight double refraction.

PHYSICAL LABORATORY OF CORNELL UNIVERSITY,
April, 1903.

¹ Presented at the meeting of the Physical Society held on April 25, 1903.

Fig. 36. Apparatus for studying the effects of aqueous solutions on the plasticity of rock salt, and the observation of a reduction in the flow stress. After Coblentz [105].

some time "bending ... suddenly occurs more readily". Ritzel [106] had already explained the phenomenon by assuming that the crystal had a hard surface which was removed by holding in warm water. Joffé's explanation was different. The water did not affect the plasticity, which was a property of the bulk material, but dissolved away surface cracks which initiated brittle fracture. Having safely passed the stress required to fracture a dry specimen, the material reached the plastic flow stress, flowed, and work hardened. This increased both the flow stress and the stress for brittle fracture, which occurred at a stress of 30-160 kg wt mm^{-2} instead of the usual 0.45 kg wt mm^{-2}. While we shall find in §5.1 evidence in favour of Joffé's explanation, it was vigorously attacked by Ewald and Polanyi [103]. They showed (Fig. 37) that a dry crystal which had not yielded under the applied load (so that no hardened surface layer could have been formed by deformation) would immediately yield when wetted and repeatedly ceased to flow when dried and flowed again when wetted. A remarkable, and unexplained, observation, which is directly opposed to any model in which water enters surface cracks was that water is more effective in plasticizing a bent crystal if it is applied to the compressed side than if it is applied to the extended side. This observation does not seem to have been repeated during the past 50 years. Joffé [107] found that Laue asterism set in at the same stress whether the crystal was wet or dry. We might think nowadays that Laue asterisms measure the dislocations remaining in the crystal, while plastic strain depends mainly on those that

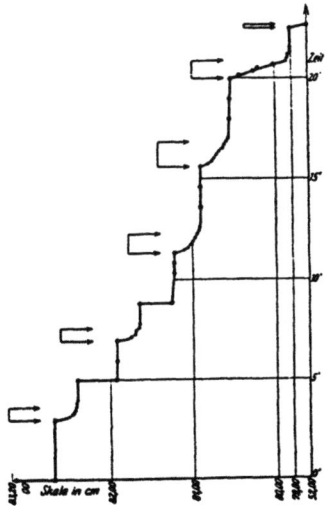

Fig. 37. Bending of a prism of rock salt which is alternately dry and wet. The crystal is under water at the times indicated by the pairs of arrows. Note the changes of scale for the strain. After Ewald and Polanyi [103].

have escaped from the surface. Joffé found, contrary to the arguments of Ewald and Polanyi, that a crystal stressed under water to 5 kg wt mm^{-2} would afterwards break at 0.4 kg wt mm^{-2} when dried. One is forced to the conclusion that water must both reduce the fragility and increase the ductility.

Moisture also increases the plasticity of many other crystals if they are studied by observing the time dependence of the indentation produced by a constant load. The effect occurs not only in salts and oxides such as Al_2O_3 which are hardly soluble in water, but also in semiconductors such as Ge. On the other hand, there is no observable effect in metals or alloys, whether they show no creep under either wet or dry conditions (Cu, NiAl) or whether they creep equally under both conditions (Sn) [108,109] (Fig. 38). We shall discuss later the significance of the absence of the effect in good conductors. As is shown in the data for Al_2O_3 in Fig. 39, the softening extends only to a small depth, about 1.5 microns in this material. Opposite faces of non-centrosymmetric III - V semiconductors behave differently [110]. The conditions under which the photomechanical effect (reduction under

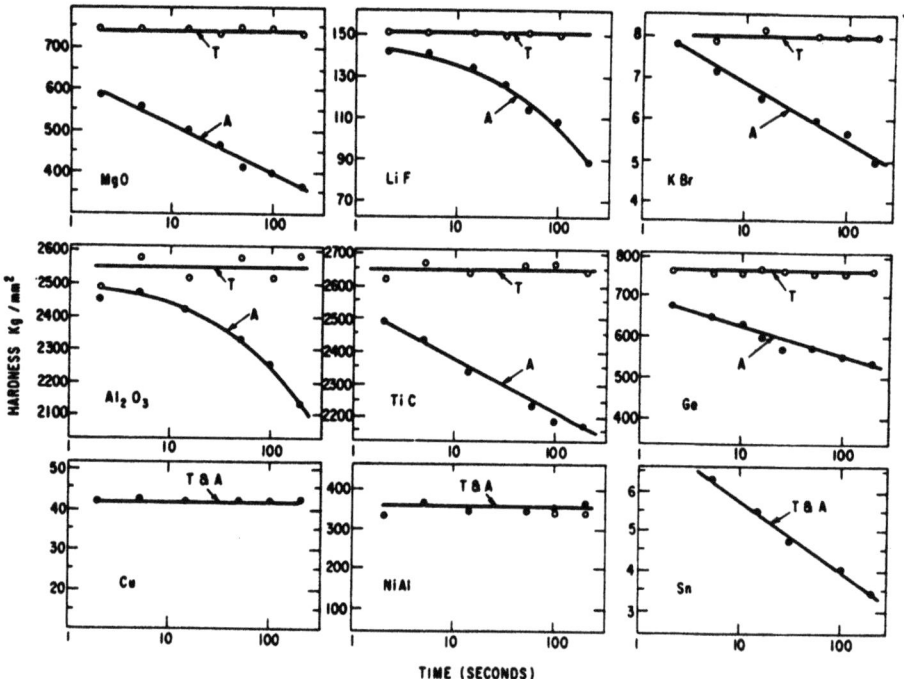

Fig. 38. Indentation creep of insulators, semiconductors, metals and alloys under dry and moist conditions. After Westbrook and Jorgensen [108].

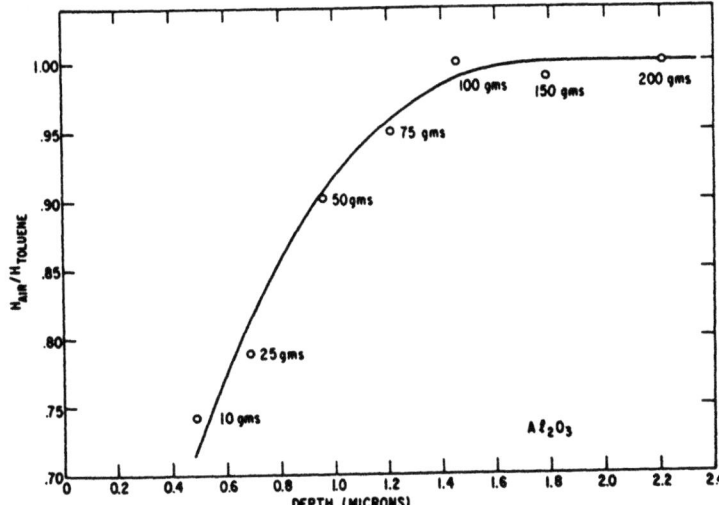

Fig. 39. Ratio of the hardness of Al_2O_3 in air to its hardness in dry toluene, as a function of the calculated depth of the indentation. The load on the indenter is marked against each point. After Westbrook and Jorgensen [108].

illumination of the indentation hardness of a semiconductor) and the electromechanical effect (reduction in hardness when a potential is applied between the sample and the indenter) occur are similar to those under which indentation creep occurs [109]. The phenomenon is complicated, because Holt [111] found a surface hardening under illumination, sometimes on materials for which Westbrook and Jorgensen [108,109] had found softening.

3.2 Surface-active Agents

The enhancement of plasticity by surface-active agents has been a special study of the school of Rebinder, and has been thoroughly reviewed [9,10,11]. Re(h)binder and Wenström [112] found that the flow stress of single crystals of tin decreased by about 40 per cent when the crystals were immersed in 1N KCl + C_8H_{17}COOH rather than simple 1N KCl, while Rebinder and Likhtman found reductions both in the critical resolved shear stress and in the rate of hardening in Stage I on immersing tin single crystals in "vaseline oil" containing various concentrations of oleic acid [10,113] (Fig. 40). The reduction in hardness is often accompanied by a refinement of the slip bands (ref. [11], p.173). We note two remarkable effects. (1) Water, itself a highly polar medium, has no effect on the plastic properties of metals. (2) The effect of a surface-active agent rises to a maxi-

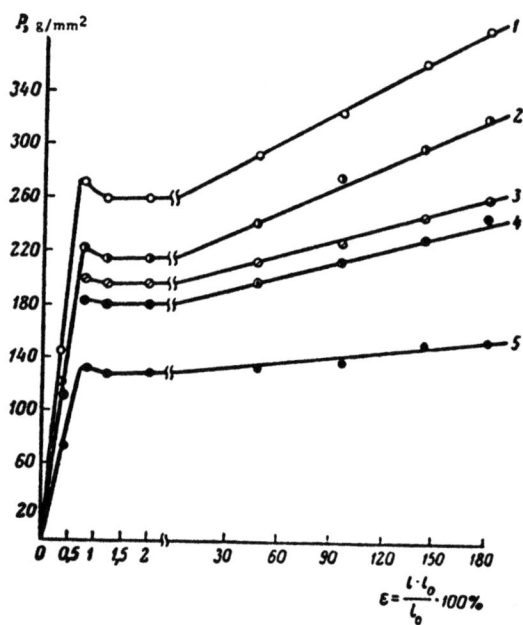

Fig. 40. Load-elongation curves for single crystals of tin of identical orientation immersed in "vaseline oil" containing various concentrations of oleic acid: (1) zero, (2) 0.1 per cent, (3) 1 per cent, (4) 0.5 per cent, (5) 0.2 per cent. After Likhtman et al. [10].

mum and decreases again as the concentration is increased. The evidence is that the effect arises because the agent dissolves oxide layers which hinder the egress of dislocations. This was very apparent in the experiments of Harper and Cottrell [114] on the creep of zinc crystals (Fig. 41). Hydrochloric acid was more effective than oleic acid (Figs. 41/1 and 41/4) when applied to an oxidized crystal. Neither was effective when applied to a polished crystal (Figs. 41/2 and 41/3). The influence of paraffin alone observed in Fig. 41/4 was not confirmed by Klinkenberg et al. [115], and was probably caused by the presence of some active impurity. Kramer [116] showed that the critical resolved shear stress of aluminium single crystals was independent of the concentration of stearic acid in the paraffin in which they were immersed, while (Fig. 42) the total strains at the ends of Stage I and of Stage II reached a maximum at a concentration of about 0.002 mol litre^{-1} of stearic acid. If, however, this most effective solution was pre-saturated with aluminium stearate, the glide in Stages I and II had the same value as in very dilute or rather concentrated stearic acid solutions. This strongly suggests that the stearic acid acts

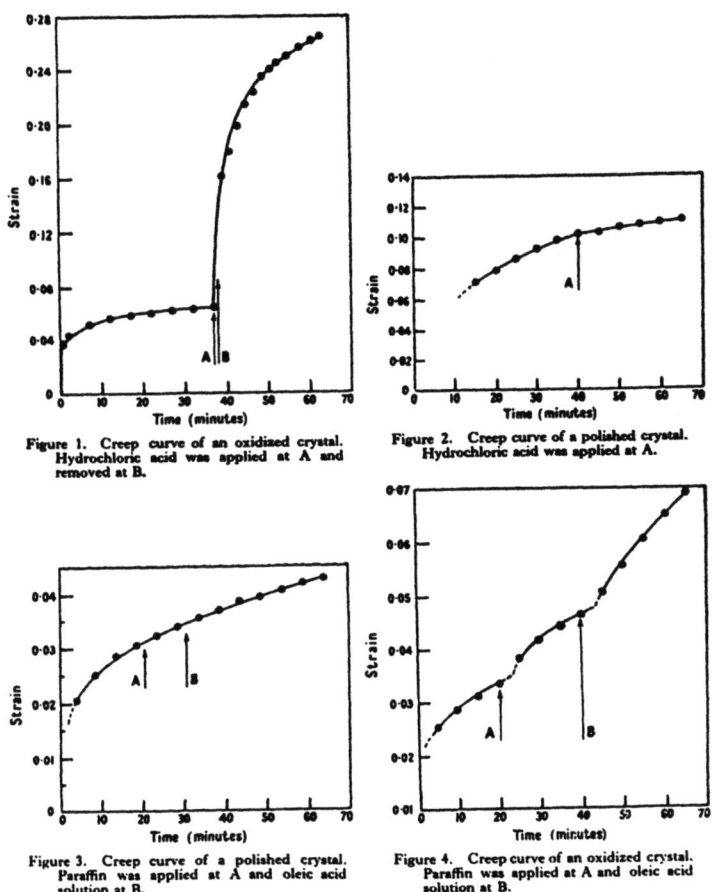

Fig. 41. The influence of hydrochloric acid and oleic acid on the creep of oxidized and polished zinc crystals. After Harper and Cottrell [114].

by dissolving aluminium oxide to form aluminium stearate. This action is prevented if the solution is already saturated with aluminium stearate, or if the solubility of aluminium stearate is suppressed by a high concentration of stearic acid. Kramer [117] confirmed this interpretation by experiments on copper, which showed directly (Fig. 43) that the softening in Stage I correlated directly with the amount of copper taken into solution on immersing a copper sample into a solution of the same concentration of stearic acid in benzene. If this explanation is correct, surface-active agents should have no effect on the creep of gold. Klinkenberg et al. [115] did observe a small effect in poly-

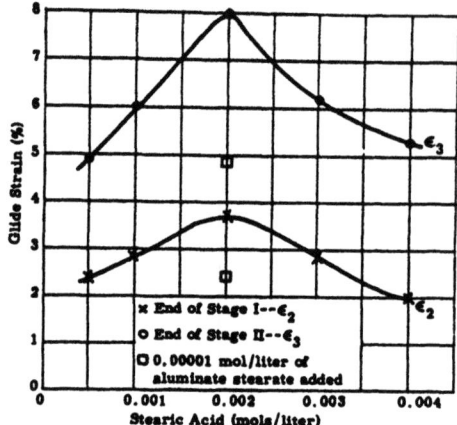

Fig. 42. Effect of concentration of stearic acid in paraffin on the extent of glide of aluminium single crystals in Stages I and II. After Kramer [116].

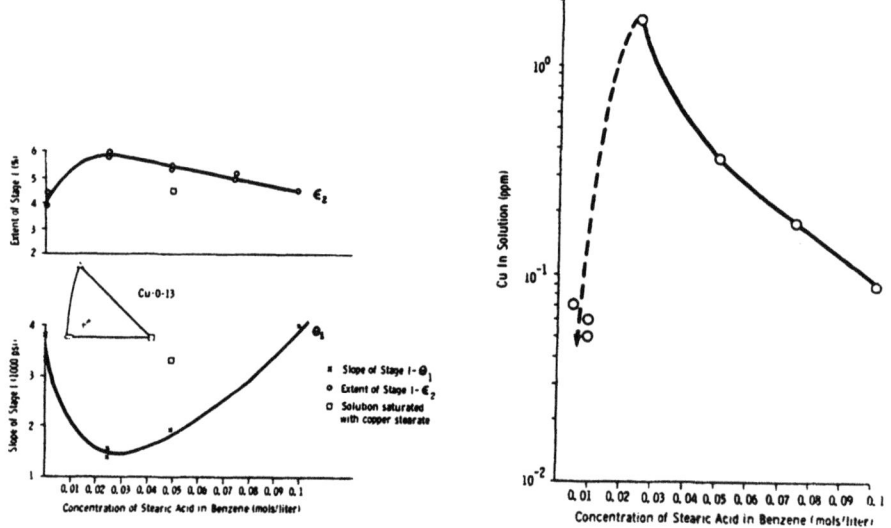

Fig. 43. On the left, the change in the extent and slope of Stage I in the deformation of copper single crystals as a function of the concentration of stearic acid in the benzene in which they were immersed. On the right, the amount of copper present in these solutions after samples had been immersed in them for 3 hours. After Kramer [117].

crystals, but Kramer [117] found no effect in gold single crystals, and no trace of gold in the solution after the experiment. This contradiction should be resolved by further experiments.

The effect of surface-active agents on the mobility of dislocations near the surface of hard ionic crystals can be studied in detail, because these crystals readily cleave along low-index planes, and the points of emergence of dislocations on these planes can be revealed by etch-pitting. Shchukin et al. [118,119] found that in NaCl, but not in LiF, the distance travelled by screw dislocations from an indentation was greater in moist air than in dry heptane, while the distance travelled by edge dislocations was the same in air as in heptane for both crystals. This suggests that the motion of screw dislocations, which trail a surface step as they move, is aided by the strong adsorption of water to the step which occurs in NaCl (surface energy reduced by a factor of 3) but not in LiF (surface energy reduced by 30 per cent). The interpretation of these results was vigorously criticized by Westwood et al. [120]. They studied NaCl, LiF and MgO, and used much longer loading times than did Shchukin et al., in order to approach more closely the conditions under which dislocations are just unable to glide under the local stress. While they confirmed the results of Shchukin et al., they found that in almost every system other than LiF in heptane or water tested for 2 seconds a surface-active environment affects the mobility of edge dislocations more than that of screws, in accordance with the earlier observations of Westwood et al. [121,122] which we shall discuss shortly.

While simple theory suggests that surface-active agents should enhance the mobility of screw dislocations which meet the surface, Westwood [123] has shown just the opposite effect in LiF. Surface-active agents increase the flow stress and reduce the rate of work hardening of as-cleaved crystals or of crystals sprinkled with 100 mesh carborundum powder, but have no effect on the deformation of polished crystals (Fig. 44). The interpretation is that the abundant dislocations meeting the surface in the first two types of crystal were locked by adsorption of the active agent. The reduced rate of hardening could arise either from the rapid rate of primary dislocation multiplication under the increased stress, or from the locking of surface dislocation sources on secondary systems. Later observations [124] on CaF_2 showed that the surface-active agent was as effective in controlling the shrinkage of dislocation loops after the indenter had been removed as it was in controlling their growth while the indenter was under load, so that it does not act by influencing the friction between the indenter and the sample. They also showed the remarkably complicated dependence of the effect on the concentration of the active agent which is suggested in Fig. 44 and will soon be discussed more fully.

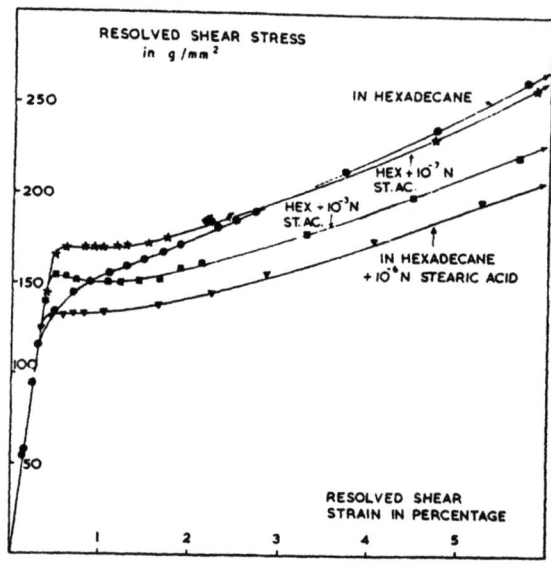

Fig. 44. The effect of the concentration of stearic acid on the stress-strain curves of LiF crystals chemically polished and sprinkled with carborundum, and tested in hexadecane. After Westwood [123].

We complete this section by a discussion of the electrocapillary effects which occur when the potential between the sample and its surroundings is varied either by direct imposition of an electric field or as the result of the formation of an adsorbed double layer. Two mechanisms can be operative. In the case of a metal, any internal electric charges are screened out within one atomic spacing, and the only effects which are possible are those which occur strictly at the surface. The argument of Pflützenreuter and Masing [63] shows that if there is an electric double layer in the liquid just outside the solid, the surface free energy in the presence of an applied potential passes through a maximum when the potential across this double layer vanishes, and decreases quadratically if the applied potential is varied in either direction from this condition. The flow stress thus passes through a maximum, and the rate of creep under constant load through a minimum. This is widely observed ([112], pp. 33-39 of ref. [10], [125], [126]). However, recent observations [127] have complicated the picture. Anodic and cathodic polarizations of a polycrystalline copper wire creeping in an aqueous solution of electrolyte increase the rate of creep comparably, as is required by our model, which depends on changes in the surface free energy. But the effect of applying cathodic polarization is immediate, and it disappears within a few minutes of the removal of the polarization, while the response to either applying or removing anodic polarization takes ten minutes. Revie and Uhlig [127] suggest that anodic polarization injects vacancies into the metal; these combine to form divacancies which can diffuse to a depth of 0.25 microns in 4 minutes. It is assumed that these vacancies soften the surface layers by allowing dislocation relaxation by climb. We might expect this effect, if it occurs, to be superposed on the instantaneous surface-energy effect which is observed with cathodic polarization. The acceleration of creep produced by anodic polarization increases fifty-fold with increase in the time of creep before the anodic polarization was applied. This seems to support Kramer's view that the anodic polarization dissolves off a hardened surface layer formed during the previous creep, yet if the anodic current is passed while the wire is free from load and the wire is then loaded without polarization, creep proceeds no faster than it does if the surface layer is not removed. This is another set of observations which deserves independent confirmation.

If we now turn to non-metals, we must remember that the surface-free-energy effect which is predominant in metals will also be present in non-metals, except in experiments in which only one slip system operates and the surface under observation contains the Burgers vector. Such experiments might be particularly valuable. It seems likely that the electromechanical effect which is observed in many semiconductors, but not in metals or in ionic crystals [128], is associated with some change in the electronic state of the dislocations as the energy bands are shifted by the applied potential, even though the surface-free-energy model directly predicts a softening for either sign of the applied potential, while this observed result (Fig. 45) is not an obvious consequence of

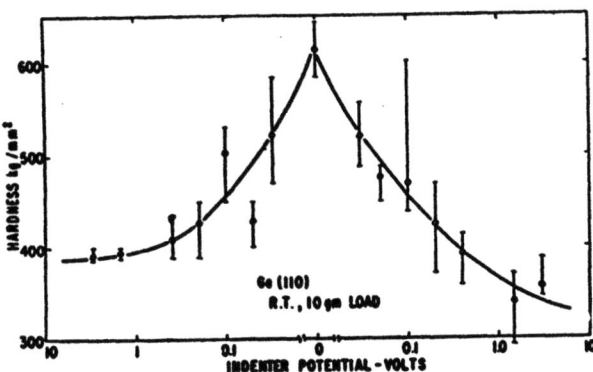

Fig. 45. Effect of the potential difference between the indenter and the specimen on the hardness of germanium. After Westbrook and Gilman [128].

the band-bending model. The present explanation is supported by the work of Patel and others which is summarized in ref. [129], and shows, for example, that n-type and p-type doping of Ge have no effect on the dislocation mobility until the impurity concentration becomes equal to the intrinsic carrier concentration. Above this concentration, p-type doping reduces the mobility, but n-type increases it.

We come finally to the influence of surface-active agents on dislocation mobility in ionic crystals. As in the case of semiconductors, the effect is generally that a change in the potential of the surface layer in either direction increases dislocation mobility and weakens the crystal. A partial explanation has been suggested by Mills [130]. In an ionic or a covalent crystal, the bonding orbitals are filled and the anti-bonding orbitals are empty. If polarization requires the removal of electrons, these come from bonding orbitals, and weaken the crystal; if polarization requires the addition of electrons, these must be added in anti-bonding orbitals, and weaken the crystal. Superposed on this effect is the adsorption-locking of the points of emergence of dislocations. The experiments of Westwood et al. on MgO [121,122] are illustrated by Fig. 46. The mobility of edge dislocations is measured by the distance $\Delta L(t)$ that the slip band formed by edge dislocations spreads in the time between 2 sec and t sec from the instant of application of a diamond indenter. The results of Fig. 47 show that dislocation mobility is enhanced by the presence of highly-charged ions of either sign, such as Ag_3Cl^{2+} or $AgCl_4^{3-}$. A remarkably similar pattern is presented if the parameter ΔL (1000) is plotted against the logarithm of the concentrations of dimethyl sulphoxide (DMSO) and dimethyl formamide (DMF) in the solution.

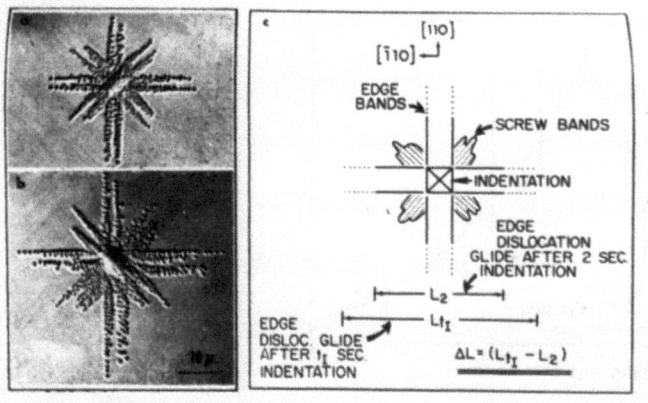

Fig. 46. To illustrate the experiments of Westwood et al. [121].

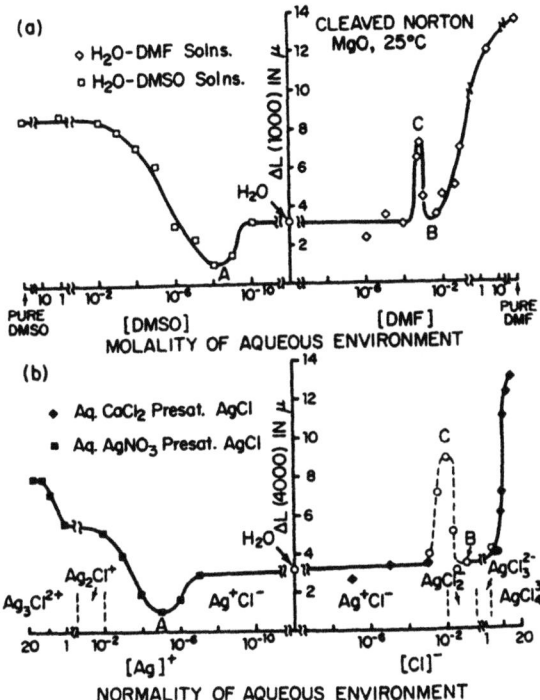

Fig. 47. (a) The mobility of edge dislocations in the surface of a crystal of MgO, measured by the parameter ΔL (1000 sec), in aqueous solutions with various concentrations of dimethyl sulphoxide (DMSO) and dimethyl formamide (DMF). (b) A similar plot of ΔL (4000) in solutions with various concentrations of Ag^+ and Cl^- ions. The predominant complex ions present are indicated. After Westwood et al. [122].

In simple mixtures of DMSO and DMF, the form of the curve is unchanged but the concentrations at which various features appear is somewhat shifted if an MgO crystal of different purity is employed (Fig. 48). The results in CaF_2 [124] show that in the same crystal the features of the curves for the spreading of dislocation loops under load and for their shrinkage on recovery appear at identical concentrations. Westwood, Macmillan and others were able to show that in a number of systems the pendulum hardness in a liquid environment was greatest when the ζ-potential was zero, and Macmillan et al. [131], using the simple systems of MgO in buffered aqueous NaCl and of LiF in water, toluene and dimethyl formamide, were able to show that the mobilities of both edge and screw dislocations were least when ζ = 0 (Fig. 49). This

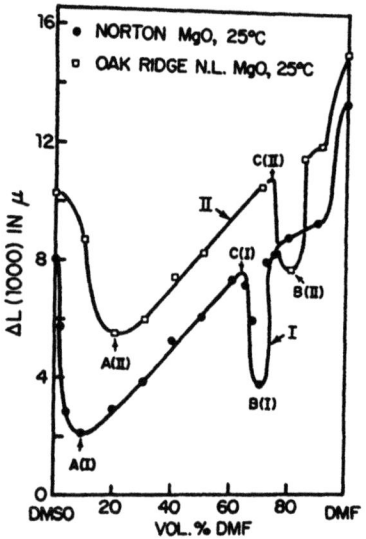

Fig. 48. The dislocation mobility parameter ΔL (1000) for crystals of MgO of high purity ("Oak Ridge") and of lower purity ("Norton"), as a function of the composition of a DMSO-DMF environment. After Westwood et al. [122].

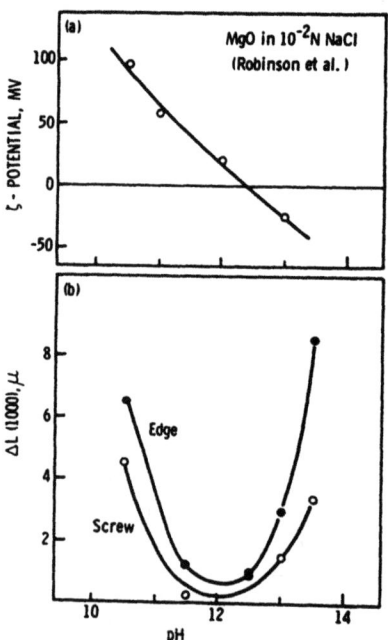

Fig. 49. (a) Variation with pH of the ζ-potential of MgO in a buffered NaCl solution. (b) The mobilities of edge and screw dislocations near the surface of a crystal immersed in this solution. After Macmillan et al. [131].

result is perhaps surprising, since the ζ-potential relates to a layer in the liquid appreciably outside the interface, while the dislocations lie within the crystal. We must recognize [132]

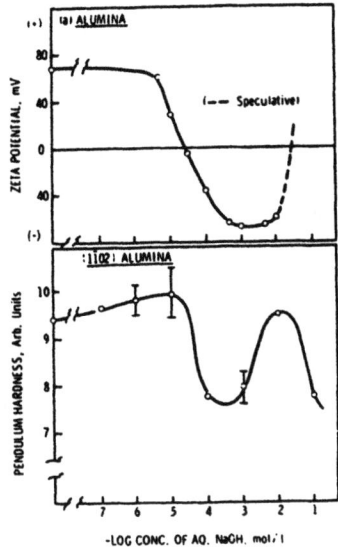 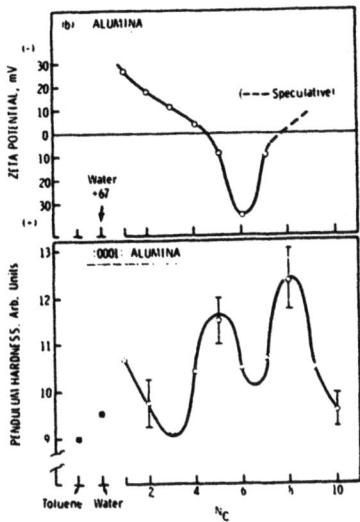

Fig. 50. (a) Variations of ζ-potential and of pendulum hardness for Al_2O_3 with the concentration of NaOH in aqueous solution. (b) Similar variations with the chain-length of n-alcohols. After Westwood et al. [133].

that the behaviour of dislocations can be influenced to a depth greater than that to which the electronic energy bands are shifted, because dislocation kinks generated near to the surface can travel along the dislocation to a deep-lying dislocation node. The correlation between ζ-potential and plasticity extends to quite complicated changes in the ζ-potential, which may pass through two zeros as the pH of the surrounding aqueous medium or the chain length of a surrounding normal-alcohol medium is varied. The hardness shows a clear peak at each zero of the ζ-potential [133] (Fig. 50). The experiments of Macmillan et al. [131] also show (1) that the mobilities of edge dislocations can be increased by surface-active agents, even though their motion does not produce surface steps and (2) that DMF greatly increases the mobility of edge dislocations, while it has no significant influence on the surface energy (as determined by fracture measurements). Thus the surface-energy model cannot account for all of the surface-sensitive flow phenomena in ionic crystals.

4. APPLICATIONS OF SURFACE EFFECTS IN CRYSTAL PLASTICITY

Many practical forming operations depend more on fracture than on ductile deformation, and we shall discuss these later. However, a good lubricant for hydrostatic extrusion will be one which facilitates the plastic deformation of the metal as well as reducing its friction against the die (which itself is controlled by the plastic properties of the surface layer of the metal) [9, 10, 11, 134]. The creep resistance of polycrystalline alloys at relatively low stresses can be substantially increased by pre-stressing to just below the proportional limit, when plastic deformation can be observed in the surface grains, and then chemically removing the surface layers [135]. It is not clear if this result depends on single-crystal surface effects or on the polycrystal surface effect [25].

5. FRACTURE

In considering fracture, we shall first consider the fracture of a crystal with a clean surface, and then consider in turn the effects of water, of surface-active agents, and of liquid metals. Finally we discuss failure under fatigue conditions, and applications of the principles involved.

5.1 Fracture of a Crystal with a Clean Surface

Surface effects in the fracture of crystals must be seen in the context of a general theory of crystal fracture [136,137]. If a crystal has surface energy γ, lattice spacing a and Young's modulus E, its ideal fracture strength σ_m is given by

$$\sigma_m \simeq (E\gamma/a)^{\frac{1}{2}} \simeq E/5.5.$$

If the crystal contains a crack of length c, the stress σ_G required to spread the crack is given by Griffith's theory as

$$\sigma_G \simeq (E\gamma/c)^{\frac{1}{2}}.$$

These criteria approach one another if the crack is of atomic dimensions, $c \simeq a$. Crystals and glass with very perfect surfaces can approach the strength given by the first formula, while there is much experimental evidence for the second formula. But the arguments leading to this second formula hold only if there is no plastic flow at the tip of the crack. Plastic flow can have two effects. It amy simply increase the work which has to be done to produce unit area of crack surface, so that γ must be replaced by an effective surface energy γ' which may [138] be as high as $10^5\gamma$. It may, however, allow the crack to spread slowly by plastic deformation until the Griffith criterion is satisfied, when the crack spreads catastrophically [139,140]. A quantitative analysis [141] supports the simple idea that the fracture will be accompanied by some plastic flow if the ratio of the largest tensile stress to the largest shear stress at the tip of an equilibrium crack is less than the ratio of the ideal cleavage strength to the ideal shear strength, while in the converse case the crystals break their bonds asunder [142].

We have been concerned with the propagation of an existing crack. We must also consider the nucleation of cracks. These may occur on the surface as the result of abrasion, corrosion, or shrinkage of the surface layer, or as a result of plastic flow near the surface. They may arise in the interior from the shrinkage of inclusions or by plastic deformation round inclusions, through the plastic extension of voids under creep conditions, or as the result of the accumulation of tensile stresses where edge dislocations pile up (Fig. 51).

The surface can influence the formation of interior cracks by influencing the nucleation either of intersecting slip bands or of parallel non-intersecting bands. This last effect was demonstrated very clearly in MgO by Stokes et al. [143] (Fig. 52). It is likely that many surface cracks are initiated from surface steps formed either as cleavage steps or by the presence of surface deposits either of the crystal itself or of its oxidation or decomposition products [144]. Marsh [145] has shown that the stress concentration near the tip of a step with a radius of curvature ρ is

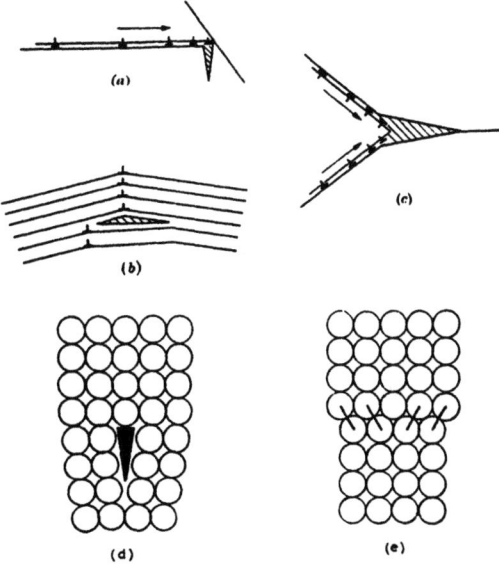

Fig. 51. Nucleation of a crack by edge dislocations. (a) A pile-up at a barrier. (b) A jog in a tilt boundary. (c) The intersection of two slip planes. (d) The incipient crack in a dislocation in a brittle material. (e) The glide dissociation of a dislocation in a ductile material. After Cottrell [136,137].

102

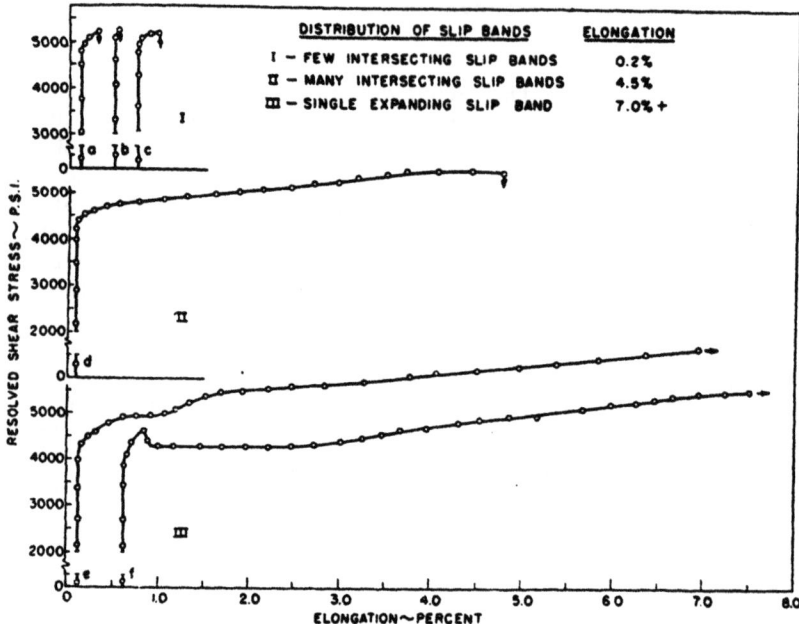

Fig. 52. The stress-strain curves of chemically polished MgO crystals with: I few intersecting slip bands, II many intersecting slip bands, and III a single extending slip band. After Stokes et al. [143].

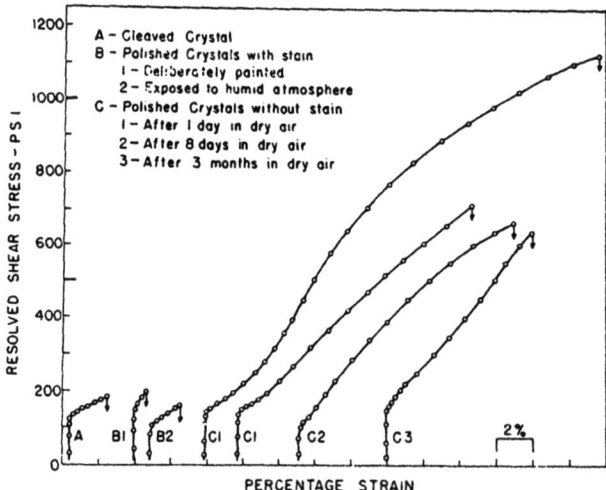

Fig. 53. The effect of: A cleavage steps, B surface stains, and C a clean surface on the stress-strain curves of single crystals of NaCℓ. After Stokes et al. [68].

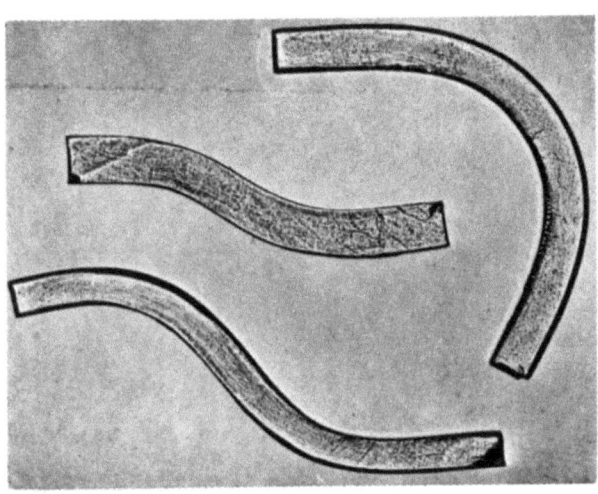

Fig. 54. Crystals of KCℓ bent by hand immediately after cleaving. After Gorum et al. [146].

similar to that near the tip of a crack with tip radius ρ, and Stokes et al. [68] clearly demonstrated the initiation of fracture by such surface deposits (Fig. 53). There is an interesting and unresolved discrepancy between these results and those of Gorum et al. [146]. Stokes et al. [68] found cleaved crystals of NaCℓ to be almost brittle, while Gorum et al. found that freshly cleaved crystals of KCℓ were very ductile (Fig. 54), becoming brittle after less than a minute's exposure to the air. A resolution of this disagreement, and an explanation of the ductility of a specimen having a roughly cleaved surface, would be valuable.

5.2 Fracture in a Liquid Medium : General Considerations

Many of the factors affecting the fracture of a crystal may be affected if the crystal is placed in a liquid medium. The liquid may penetrate the crystal to dissolve out thin layers, making cracks which initiate fracture. This may occur even in the absence of an external stress if there are sheets of high energy density such as occur in subgrain boundaries or pile-ups of dislocations [147] (Fig. 55). We shall discuss this effect more fully under the heading of Stress-corrosion Cracking (§5.3). The liquid may reduce the surface energy of the crystal, so enabling a crack to spread more readily and rendering the crystal more brittle. This will be discussed under the headings of the Rebinder Effect and Liquid-Metal Embrittlement (§5.4). The liquid may bond to the atoms at the tip of a crack, thus preventing the mechanical spread of the crack, making the material less brittle. All of

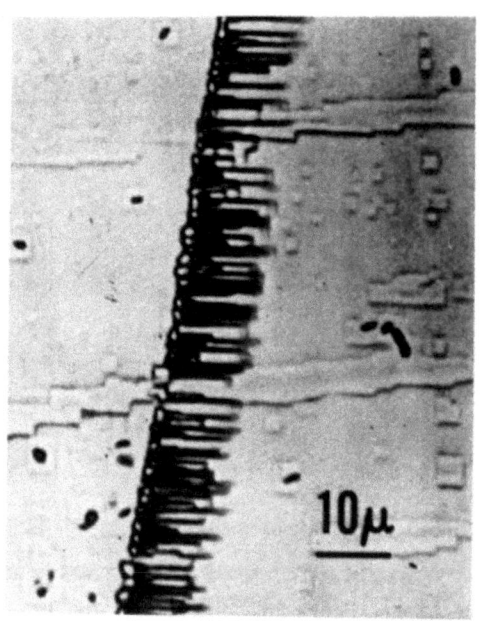

Fig. 55. Etch tunnels in a subgrain boundary in a crystal of LiF in water containing stearic acid. After Westwood and Rubin [147].

these are true surface effects. The liquid may also affect the plasticity of the surface layers of the crystal. If it decreases the mobility of dislocations, it will render the crystal more brittle. This we may call the Westwood Effect (§5.5). Conversely, if the decrease of dislocation mobility is most marked in the highly stressed regions near the crack tip, the slow growth of a sub-critical crack may be hindered, and the creep life of the specimen increased. Or the liquid may increase the mobility of dislocations near the surface, thereby blunting cracks and increasing the ductility of the crystal, or aiding the growth of sub-critical cracks and reducing the creep life. The liquid may influence the pattern of slip bands, increasing or reducing the generation of internal cracks by the intersection of slip bands. It is not always easy to argue back from the observation to the mechanism.

5.3 Stress Corrosion and Hydrogen Embrittlement

One mechanism of stress corrosion, the etching of laminar cracks along subgrain boundaries, has already been mentioned. The second mechanism mentioned, etching along dislocation pile-ups, occurs most readily in alloys of low stacking-fault energy in which cross-slip cannot easily occur. It is probably for this reason that the addition to copper of up to 20 per cent of gold actually reduces its resistance to stress corrosion [148] (Fig. 56). Another method

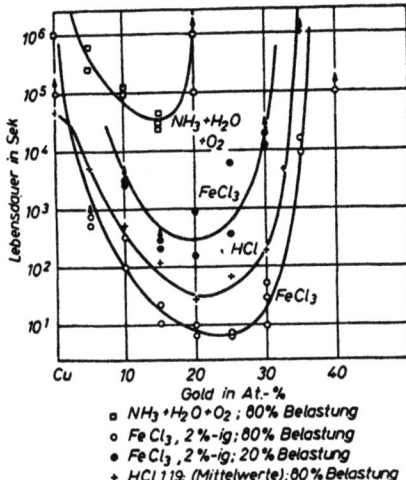

Fig. 56. Lifetime of Cu Au alloys under conditions of stress corrosion. After Graf and Budke [148].

by which the attack of the medium can be concentrated into planar cracks is the alternate formation and cracking of a brittle tarnish layer. The importance of this mechanism has been demonstrated by Edeleanu [149], Forty [150], Forty and Humble [151] and Pugh and Westwood [152]. As long ago as 1953, Rozhanskii and Rebinder [153] explained the observation that vapours such as ethyl alcohol and ethylamine greatly increased the creep rate of tin single crystals having oxide films but had hardly any effect on the creep rate of crystals free from oxide films by saying that cracks in the oxide film produced regions of high local stress on the surface, and that the surface-active agents were most effective in these highly stressed regions.

A special type of stress corrosion is hydrogen embrittlement. Gilman [154] has shown that hydrogen atoms bond strongly to single metal atoms, less strongly to metal surfaces, and least strongly in interstitial solution in the bulk. Dissolved hydrogen therefore tends to migrate to the surface of an extending crack, reducing the surface energy and leading to embrittlement, especially in cases in which the criterion for the ductile-brittle transition is marginally satisfied. One would naturally expect both the concentration and the mobility of hydrogen in a metal such as nickel to be greatest in the grain boundaries. The mechanism is probably more complicated, since impurities such as Sb and Sn also segregate into the grain boundaries, and may enhance the absorption of hydrogen where the boundaries meet the surface [155].

There are several recent reviews of stress-corrosion cracking (e.g. [13],[7]), which discuss the relative importance of the mechanisms we have considered, and also of mechanisms such as the injection of lattice vacancies during the process of corrosion [156].

5.4 Reduction of the Surface Energy : The Rebinder Effect and Liquid-Metal Embrittlement

The basic model of Rebinder and his school (e.g. [157]) is that a surface-active agent reduces the surface energy and hence assists the propagation of cracks. The method of testing employed in much of their early work was the pendulum sclerometer, in which the hardness of a sample is measured by the decay time of the swinging of a pendulum suspended from two hard steel spikes resting on the surface of the sample. It is not clear how far the damping in such an experiment is controlled by the generation of new surface and how far by plastic deformation. Even if most of the energy is dissipated by plastic deformation below the surface, the extent of this plastic deformation might be controlled by the surface energy of slip steps, but such steps can hardly be called cracks. The modern development of this approach is fully reviewed in ref. [11]. Most important from a theoretical viewpoint is the analysis of liquid-metal embrittlement taken from the work of Pines and Geguzin. This is based on the "quasichemical" model in which only bonds between AA, AB and BB atoms are considered, and the important energetic quantities are the entropies of melting of A and B and the ordering energy $V = V_{AB} - \tfrac{1}{2}(V_{AA} + V_{BB})$, where V_{AB}, etc. are the energies of bonds between A and B atoms. It is assumed that V may have different values in the solid and the liquid phases. Similar, but less detailed, analyses have been given by Westwood et al. [158] in terms of the relative electronegativities of the solid and liquid metals, and by Kelley and Stoloff [159] in terms of electronic bond energies and promotion energies derived from spectroscopic data. A particularly interesting development of this approach has been the computer modelling of the initiation of fracture in a two-dimensional crystal. In a "pure" crystal initially containing a cavity [160], loading in tension produces thermoelastic cooling, followed by plastic deformation which was accompanied by heating and apparent local melting. The crystal tore away from the confining walls, and a crack initiated in the surface of the void. Under rapid loading, the thermoelastic cooling amounted to as much as 20 per cent of the initial kinetic energy. The computer experiments were then repeated [161] with "surface-active" atoms in the cavity. For conditions in which plastic deformation takes place in a pure system, brittle fracture was usually observed in the presence of the foreign atoms. When the foreign atoms were highly mobile, they penetrated the cracks (Fig. 57/2), but they could be effective in initiating fracture without penetrating the crack when it opened (Fig. 57/3).

Since, because of electronic screening, the motion of dislocations

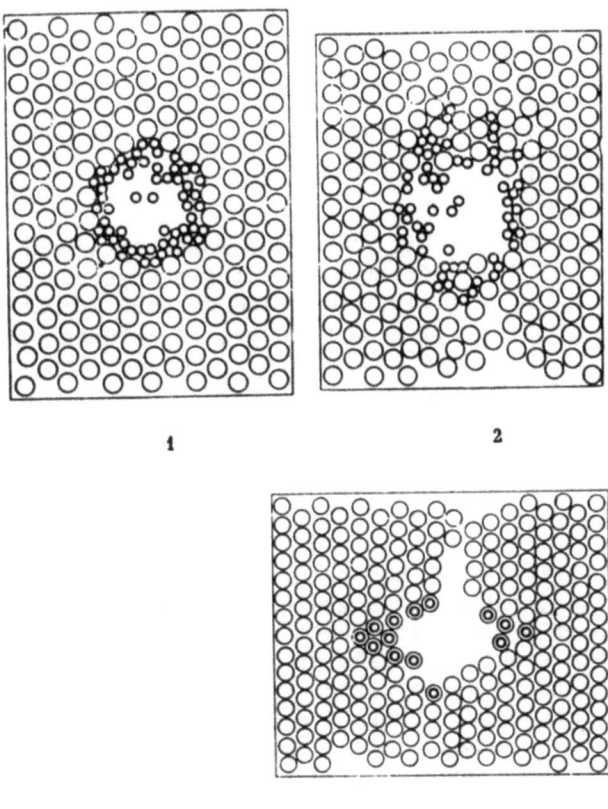

Fig. 57. Computer simulation of tensile test of a two-dimensional close-packed crystal containing a cavity in which surface-active atoms are present: (1) Initial distribution of atoms. (2) Penetration of atoms of the surface-active medium into incipient cracks in the crystal and formation of a crack filled with atoms from the medium. (3) Formation of a crack not filled with atoms from the medium (upper part), and plastic deformation of the crystal (lower part). After Yushchenko et al. [161].

in a metal can hardly be influenced at any appreciable depth by the presence of a liquid metal on the surface, the widespread phenomenon of liquid-metal embrittlement can well be regarded as a pure Rebinder Effect, in which the liquid metal reduces the surface energy of the solid. The subject has been thoroughly reviewed by Westwood, Preece and Kamdar [158], and the following outline follows their arguments rather closely. Surface cracks do not normally exist on a ductile metal crystal, nor do cleavage steps, which can act as stress concentrations in a brittle ionic

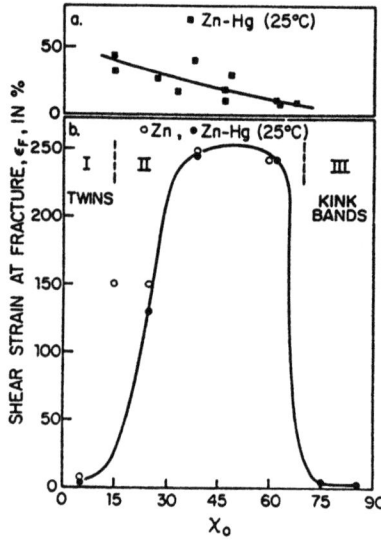

Fig. 58. The shear strain at fracture of single crystals of zinc, with and without a coating of mercury, as a function of the angle χ_o between the basal plane and the tensile axis at the beginning of the test. (a) After Shchukin et al. (b) After Westwood et al. [158].

crystal. Liquid-metal embrittlement can occur only if a crack can be initiated at the surface, either at a grain boundary or along the line in which a twin or a kink band meets the surface. Thus mercury does not embrittle single crystals of zinc which have been handled very carefully if their glide planes are oriented at about 45° to the tensile axis (Fig. 58). There are many suggested criteria for the initiation of a crack, depending on the details of the model assumed, but they are nearly all derived by an argument of the following kind: Let the shear stress on the glide plane be τ, and the shear modulus μ. Then if dislocations pile up on the glide plane at a distance L from their source, the number piled up is roughly $\tau L/\mu b$, b being the Burgers vector. The normal stress σ on the cleavage plane of surface energy γ which is required to initiate a cleavage is about γ/nb. Substituting the value of n, we obtain

$$\sigma \tau L \simeq \gamma \mu.$$

As in the case of stress corrosion, we expect the nucleation of cracks by pile-ups of dislocations to be most effective in alloys of low stacking-fault energy, and this is observed (Fig. 59).

In considering the propagation of a crack which exists in an embrittled metal, we recognize that it is moving in a metal that is normally ductile, so that considerable plastic flow occurs at the crack tip. The work γ' required to create unit area of new surface is much greater than the surface energy γ. Can we expect the presence of a liquid medium, which reduces γ, to have an appreciable influence on γ'? If the radius R of the tip of the crack

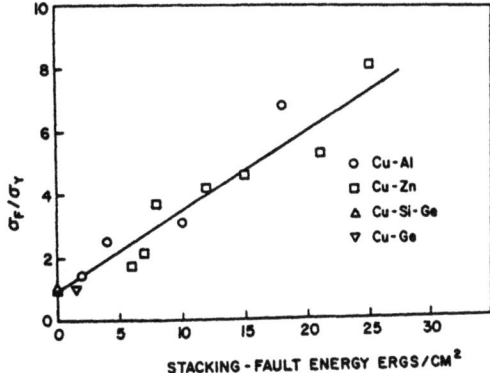

Fig. 59. The ratio of the fracture stress σ_F in mercury to the flow stress σ_Y in mercury or in air for a series of copper-base alloys of different stacking-fault energies. After Johnston et al. [162].

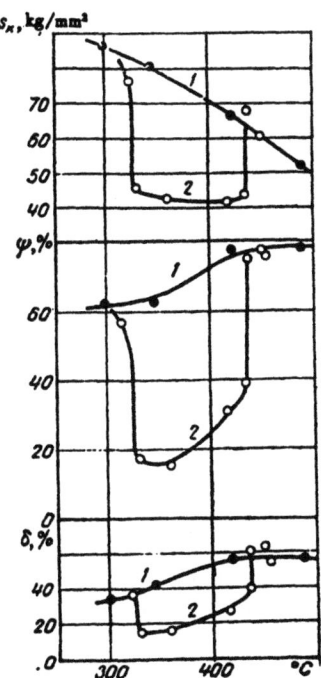

Fig. 60. Cohesive strength, reduction of area and elongation of polycrystalline Armco iron with and without a 15 micron coating of cadmium, as a function of temperature. After Dityatkovskii et al. [163].

is independent of the presence of the embrittling medium, and we write $R/a = \rho$, then the stress concentration at the tip of the crack is less than that at the tip of an atomically sharp crack by a factor $\rho^{-\frac{1}{2}}$. The stress which must be applied in order to break an interatomic bond at the crack tip is therefore larger by a factor $\rho^{\frac{1}{2}}$, and the Griffith criterion $\sigma_G \simeq (E\gamma/c)^{\frac{1}{2}}$ is replaced by

$$\sigma_G \simeq (E\gamma\rho/c)^{\frac{1}{2}}.$$

The blunting of the crack has multiplied the effective surface energy by a factor ρ. If now the addition of an embrittling medium reduces γ to $\eta\gamma$ ($\eta < 1$), the fracture criterion becomes

$$\sigma_G \simeq (E\eta\gamma\rho/c)^{\frac{1}{2}}.$$

If, then, we assume that neither the crack length c nor the radius R of its tip is altered by the presence of the medium (or that R/c is constant), the fracture stress is reduced by a factor $\eta^{\frac{1}{2}}$ when the medium is applied.

The effect of temperature on liquid-metal embrittlement is complicated, but seems to be well understood. There is no effect below a temperature close to that at which the environment freezes [163,164]. The effect again vanishes at higher temperatures (Fig. 60). The explanation seems to be that the atoms of the medium have insufficient mobility to reach the crack tip at low temperatures, while their adsorption to the surface of the metal at the crack tip becomes weak at high temperatures [165].

5.5 Decreased Dislocation Mobility near the Surface

In insulators and materials with a low charge-carrier density, a layer of adsorbed polar molecules can bend the electronic energy

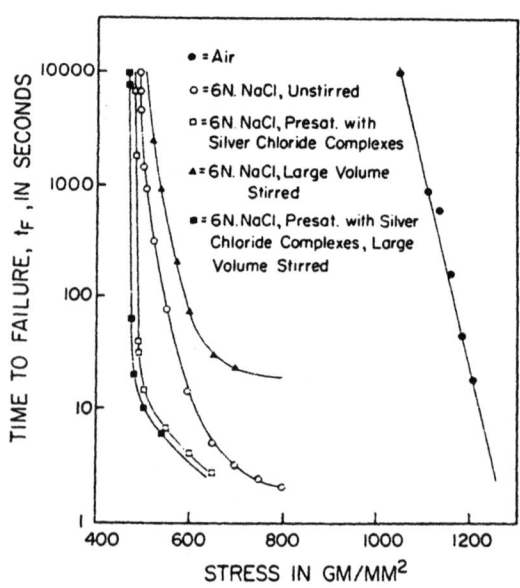

Fig. 61. Time to failure of polycrystalline AgCl under tensile test in aqueous NaCl solutions without or with previous saturation with silver chloride complexes. After Westwood et al. [166].

bands, resulting in changes in the state of ionization of defects and impurities, and possibly in changes in the concentration of mobile charged impurities. If these changes reduce the mobility of dislocations near the surface, the material is embrittled - not through a direct reduction in the stress required for brittle cleavage, but through a reduction in the plastic work associated with the spread of a crack. Westwood, Pugh and Goldheim ([166], see also [167]) showed that polycrystalline AgCl was embrittled by aqueous NaCl much more effectively if the solution was presaturated with silver chloride complexes than if it was not, that the embrittlement in NaCl solutions free of complexes was much reduced if the liquid had a large volume and was kept stirred, but that stirring in a large volume did not reduce the embrittlement in the presence of silver chloride complexes (Fig. 61). It is clear that it is the presence of the complexes on the surface and not the dissolution of the AgCl which causes the embrittlement. Copper complex ions have the same effect [168]. Since [169] the characteristic defects of AgCl are Ag^+ interstitials and Ag^- vacancies, and the activation energy for the failure process in the temperature range 30 - 100°C is in good agreement with that for the motion of the appropriate defect, depending on whether the sign of the charge on the adsorbing complexes is negative or positive respectively, there is little doubt that the embrittlement is produced by the migration of these vacancies in the electric fields of highly-charged adsorbed complex ions.

The same mechanism could also increase the stress required to produce failure after long periods of loading by preventing the spread of sub-critical cracks.

An interesting possibility (Shand [170]) is that the classical agreement between the value of the surface energy of glass derived from Griffith's fracture analysis and that obtained by other means may arise from the effects of atmospheric moisture in preventing plastic distortion at the crack tip. The surface energy derived from fracture analysis can be as much as twenty times the expected value if the glass is fractured in a vacuum.

5.6 Increased Dislocation Mobility near the Surface

In crystals which have no mobile charged defects other than electrons or holes, the effect of charged complex ions usually is to increase the dislocation mobility near the surface [121,122]. This usually blunts cracks and increases the work of fracture (e.g. [171]). On the other hand, at least in glass [172], there is some evidence that increased surface plasticity induced by water vapour may increase the length of sub-critical cracks.

5.7 Fatigue and Erosion

The phenomena of fatigue are too complicated to be discussed in detail in this review, even though it is generally accepted that fatigue failure nucleates at the surface of a sample [173], and that changes in surface conditions may drastically affect the fatigue behaviour. Much of the recent work is summarized in ref. [14], and we shall draw attention to some articles in this report. Laird and Duquette [174] show that the initiation of a fatigue crack occurs differently in metals which show planar slip, where deep slip steps act as stress raisers, and in metals which show much cross-slip, where fracture is initiated by the formation of intrusions and extrusions. Lukáš and Klesnil ([175], see also [176]) describe in detail the dislocation structures near the surface of fatigued f.c.c. metals. Kramer and Kumar ([177], see also [178]) draw attention to the influence of the heavily work-hardened surface layer which occurs, for example, in aluminium, while Fourie [179] relates fatigue to the unloading yield-point phenomenon, and Argon [180] emphasizes the likely importance of the formation of lattice vacancies which can aggregate to form regions of porosity. An interesting observation [181] is that polycrystalline copper, which hardens during fatigue testing if it has not been pre-strained and softens during fatigue testing if it has been pre-strained, hardens if it is pre-strained and then has 0.005 inches removed electrochemically from the surface. Unfortunately, there does not seem to be a direct comparison in this work between two sets of samples which are tested identically except for the surface removal.

Shevelya and Kostetskii [182] have demonstrated by direct electron microscopy the increase in dislocation density near the surface of fatigued steel if the test is carried out in paraffin containing 0.2 per cent of oleic acid.

It seems that the process of erosion by cavitation is, at least in part, one of fatigue. A surface-active agent which reduces the surface tension of water can reduce the violence of the cavitation. At the same time, it may reduce the fatigue strength of the metal, and the technical application of surface-active agents to the control of erosion therefore faces serious problems [183].

5.8 Surface Effects in Glasses

Surface effects analogous to those observed in crystals are common in glasses. It is normal to cut glass by making a scratch and then moistening it before applying stress. Glass which shows no time-dependent creep in water will creep in an aqueous solution of iso-amyl alcohol [184] (Fig. 62). Westwood et al. [185,186] have explained this behaviour in terms of the migration of sodium and hydroxyl ions, and have correlated the effects with changes in the ζ-potential.

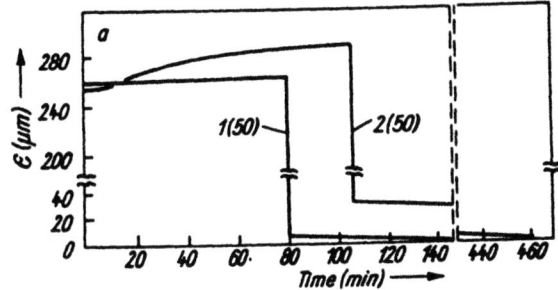

Fig. 62. Time-deflection curve for samples of aluminium borosilicate glass tested under the same load of 50 kp. mm^{-2} (1) in water, (2) in a 2.5 per cent aqueous solution of iso-amyl alcohol. After Aslanova and Rebinder [184].

The analysis in terms of the pinning of dislocations is clearly not applicable to glasses. However, Ashby and Logan [187] have shown that rows of dangling bonds could well exist in glasses, and would have many of the mechanical and electrical properties of dislocations.

6. APPLICATION OF SURFACE EFFECTS

The early applications of surface effects were on a small scale

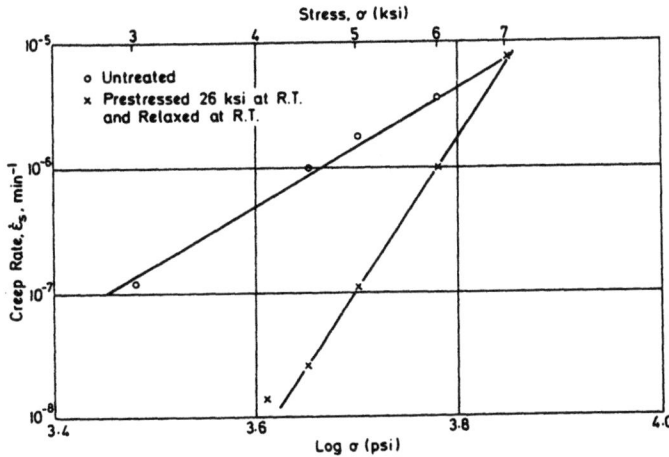

Fig. 63. Stress dependence of the creep rate at 1400°F of 321 steel untreated and after pre-stressing to 26 ksi and chemically removing 0.005 inches from the surface. Redrawn from Kramer and Balasubramanian [135].

(e.g. §3.1, and the application of surface-active agents to ease the polishing of porcelain tubes [188]). Technological applications are of two types, those in which the aim is to preserve the form of a component, and those in which the aim is to produce a new form.

Surface control as a means of lengthening the life of a component seems to have been principally exploited by Kramer and his collaborators. For example, the creep rate of steel can be reduced by pre-stressing and chemically removing 0.005 inches from the surface [135] (Fig. 63). Similar treatments can increase the limiting fatigue strength of titanium by 7-8 per cent [177]. A heavy anodized film on aluminium alloy sheet, permeated with palmitic acid, can give a fatigue strength 25 per cent greater than that of a similar sheet with normal surface treatments (Fig. 64) [189].

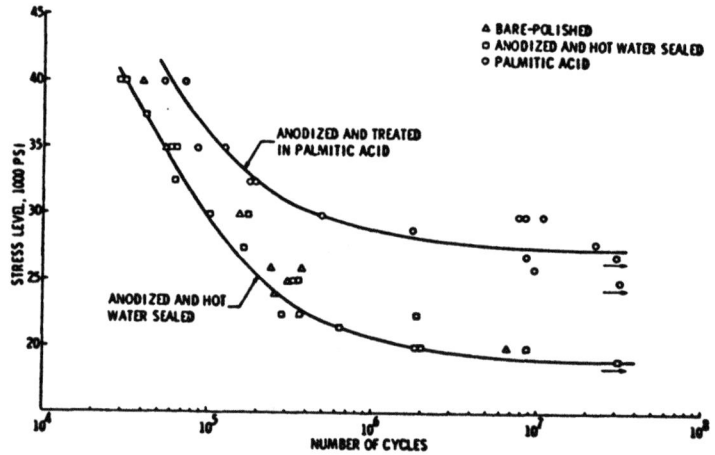

Fig. 64. Flexure fatigue tests of 7075-T6 aluminium alloy sheets with normal surface treatment and with surfaces anodized and permeated with palmitic acid. After Kramer [189].

The classical account of Russian work on hardness reduction in drilling was given by Re(h)binder et al. in 1944 [190]. Russian work has also been concerned with the reduction of stresses in metal drawing [10] (Fig. 65).

The work of Westwood and his school has been summarized recently [191]. There have been two major advances. The first is the extension, over a very wide range of variables, of the correlation already established between the zeros of the ζ-potential and the minima of dislocation mobility to their correlation both with the

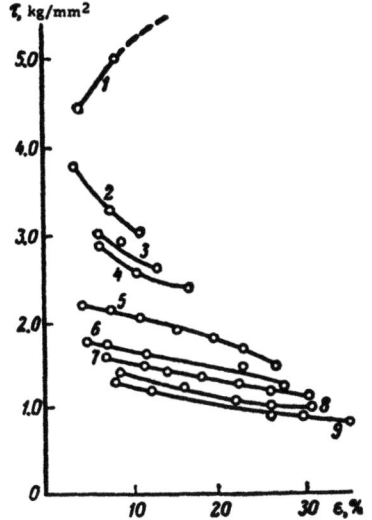

Fig. 65. The effective shearing stress τ for drawing aluminium in octane containing: (1) 0 per cent, (2) 0.25 per cent, (3) 0.5 per cent, (4) 3 per cent, (5) 5 per cent, (6) 10 per cent, (7) 18 per cent, (8) 25 per cent and (9) 100 per cent octyl alcohol. After Likhtman et al. [10].

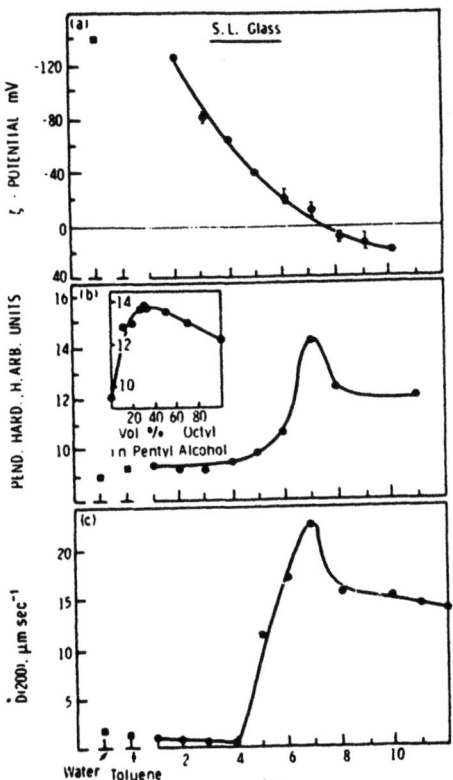

Fig. 66. (a) ζ-potential, (b) pendulum hardness, and (c) rate of penetration with a diamond-studded bit of soda-lime glass in toluene, water, and n-alcohols of various chain lengths. After Westwood [191].

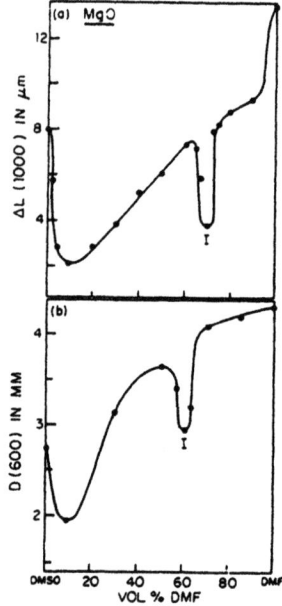

Fig. 67. (a) Dislocation mobility ΔL(1000), (b) penetration in 600 sec of a carbide spade drill, for MgO in DMSO-DMF mixtures. After Westwood and Goldheim [192].

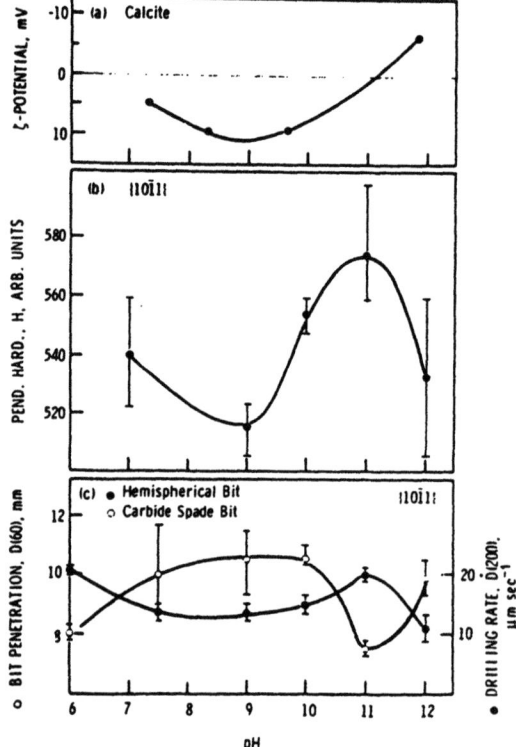

Fig. 68. (a) ζ-potential, (b) pendulum hardness and (c) penetration of a diamond-studded hemispherical bit and of a carbide spade bit into single crystals of calcite as a function of pH. After Westwood [191].

maxima of pendulum hardness and with the rate at which the material is drilled. This is illustrated for soda-lime glass in Fig. 66 [191], and for MgO in Fig. 67 [192]. The second, and perhaps more general, advance has been the explicit demonstration of the intuitively obvious idea that increases in plasticity aid the penetration of a sharp tool, but hinder the penetration of a blunt tool [191] (Fig. 68).

Finally, we note the connection between these quantities and the coefficient of friction [193,194].

ACKNOWLEDGEMENTS

It is a pleasure to thank many colleagues in the U.S.A. and Germany for their advice, hospitality, and provision of offprints and preprints, and Professor E.D. Shchukin for the provision of reprints. Dr. R.D. Zwicker has helped with Russian translations, and Mrs. A.B. Alexander has spared no effort in helping me prepare the manuscript.

REFERENCES

1. J. Keats, (1820). Letters, ed. M.B. Forman, Oxford Univ. Press, 1935, No. 223.
2. G.M. Hopkins, No Worst, There is None.
3. W.T. Read, Dislocations in Crystals, McGraw-Hill, New York, 1953, p. ix.
4. E.S. Machlin, Surface Sensitivity of Strength Properties, in Strengthening Mechanisms in Solids, Amer. Soc. Metals, Metals Park, Ohio, 1962, pp. 375-404.
5. Environment - Sensitive Mechanical Behavior, Metallurgical Society Conferences 35, ed. A.R.C. Westwood and N.S. Stoloff, Gordon and Breach, New York, 1966.
6. R.M. Latanision and A.R.C. Westwood, Surface-and Environment-Sensitive Mechanical Behavior, in Advances in Corrosion Sciences and Technology 1, Plenum Press, New York, 1970, pp. 51-145.
7. R.M. Latanision, A.J. Sedriks and A.R.C. Westwood, Surface-Sensitive Mechanical Behavior of Metals, in Structure and Properties of Metal Surfaces, Honda Memorial Series on Materials Science 1, Maruzen, Tokyo, 1973, pp. 500-538.
8. I.R. Kramer and L.J. Demer, Effects of Environment on Mechanical Properties of Metals, in Progress in Materials Science, ed. B. Chalmers, Pergamon, New York, 1961, pp. 131-199.
9. V.I. Likhtman, P.A. Rebinder and G.V. Karpenko, Effect of a Surface-Active Medium on the Deformation of Metals, Academy of Sciences, Moscow, 1954; English Translation : Her Majesty's Stationery Office, London, 1958.
10. V.I. Likhtman, E.D. Shchukin and P.A. Rebinder, Physicochemical Mechanics of Metals, Moscow, 1962; English Translation :

Israel Program for Scientific Translations, Jerusalem, 1964.
11. P.A. Rehbinder and E.D. Shchukin, Surface Phenomena in Solids During Deformation and Fracture Processes, in *Progress in Surface Science* 3, ed. S.G. Davison, Pergamon, Oxford, 1972, pp. 97-188.
12. *Physical Metallurgy of Stress Corrosion Fracture*, Metallurgical Society Conferences 4, ed. T.N. Rhodin, Interscience, New York, 1959.
13. *Fundamental Aspects of Stress Corrosion Cracking*, ed. R.W. Staehle, A.J. Forty and D. van Rooyen, National Association of Corrosion Engineers, Houston, Texas, 1969.
14. *Corrosion Fatigue*, ed. O.F. Devereux, A.J. McEvily and R.W. Staehle, National Association of Corrosion Engineers, Houston, Texas, 1972.
15. A. Seeger, J. Diehl, S. Mader and H. Rebstock, *Phil. Mag.* 2 323-350 (1957).
16. W.G. Johnston, *J. Appl. Phys.* 33 2716-2730 (1962).
17. J.W. Mitchell, J.C. Chevrier, B.J. Hockey and J.P. Monaghan, Jr., *Canad. J. Physics* 45 453-479 (1967).
18. D. Kuhlmann-Wilsdorf and H.G.F. Wilsdorf, in *Electron Microscopy and Strength of Crystals*, ed. G. Thomas and J. Washburn, Interscience, New York, 1963, pp. 575-604.
19. J.T. Fourie, *Phil. Mag.* 15 187-198 (1967).
20. J.T. Fourie, *Canad. J. Physics* 45 777-786 (1967).
21. J.T. Fourie, *Phil. Mag.* 17 735-756 (1968).
22. ibid. 21 977-985 (1970).
23. H. Mughrabi, *phys. stat. sol.* 39 317-327 (1970).
24. H. Mughrabi, *phys. stat. sol.* (b) 44 391-402 (1971).
25. F.C. Thompson and W.E.W. Millington, *J. Iron Steel Inst.* 110 61-84 (1924).
26. F.R.N. Nabarro, *Adv. Physics* 1 269-394 (1952) (pp. 332-333).
27. M.C. Inman and H.R. Tipler, *Metall. Revs.* 8 105-166 (1963).
28. E.D. Shchukin, *Dokl. Akad. Nauk SSSR* 118 1105-1108 (1958) : *Soviet Physics - Doklady* 3 143-146 (1958).
29. J.P. Hirth, in *The Relation between the Structure and Mechanical Properties of Metals*, Her Majesty's Stationery Office, London, 1963, pp. 217-228.
30. L.G. Orlov, *Fiz. Tverd. Tela* 9 2345-2349 (1967) : *Soviet Physics - Solid State* 9 1836-1839 (1968).
31. J.C. Fisher, *J. Metals* 4 (*Trans. A.I.M.E.* 194) 531-532 (1952).
32. J. Friedel, ref. [18], pp. 605-649.
33. J.J. Gilman and W.G. Johnston, in *Dislocations and Mechanical Properties of Crystals*, ed. J.C. Fisher et al., John Wiley, New York, 1957, pp. 116-163.
34. J.J. Gilman, *Phil. Mag.* 6 159-161 (1961).
35. R. Gevers, S. Amelinckx and P. Delavignette, *Phil. Mag.* 6 1515-1526 (1961).
36. R.L. Fleischer, *Acta Met.* 8 598-604 (1960).

37. R.J. Stokes, Trans. Met. Soc. A.I.M.E. 224 1227-1237 (1962).
38. T. Suzuki, ref. [33], pp. 215-231.
39. B. Šesták and S. Libovický, Acta Met. 11 1190-1191 (1963).
40. V.I. Vol'shakov and L.G. Orlov, Fiz. Tverd. Tela 12 745-751 (1970) : Soviet Physics - Solid State 12 576-581 (1970).
41. F.W. Young, Jr. and F.A. Sherrill, Canad. J. Physics 45 757-763 (1967).
42. R. Fiedler and A.R. Lang, J. Mater. Sci. 7 531-542 (1972).
43. M.A. Adams, Acta Met. 6 327-338 (1958).
44. F.J. Worzala and W.H. Robinson, ref. [5], pp. 183-212.
45. F.J. Worzala and W.H. Robinson, Phil. Mag. 15 939-957 (1967).
46. Y. Nakada and B. Chalmers, Trans. Met. Soc. A.I.M.E. 230 1339-1344 (1964).
47. R.M. Latanision and R.W. Staehle, Acta Met. 17 307-319 (1969).
48. K. Sumino, J. Phys. Soc. Japan 17 454-462 (1962).
49. G. Vellaikal and J. Washburn, J. Appl. Phys. 40 2280-2286 (1969).
50. R.M. Latanision, Scripta Met. 3 465-469 (1969).
51. P.R. Swann, Acta Met. 14 900-903 (1966).
52. J. Krejčí and P. Lukáš, Czech. J. Phys. B 18 954-955 (1968).
53. R.J. Block and R.M. Johnson, Acta Met. 17 299-306 (1969).
54. U. Essmann, M. Rapp and M. Wilkens, Acta Met. 16 1275-1287 (1968).
55. I.R. Kramer and C.L. Haehner, Acta Met. 15 678 (1967).
56. R.C. Fabiniak and Doris Kuhlmann-Wilsdorf, ref. [5], pp. 147-182.
57. K. Kolb and E. Macherauch, Phil. Mag. 7 415-426 (1962); Z. Metallkde. 58 238-242 (1967).
58. J.F. Nye, R.D. Spence and M.T. Sprackling, Phil. Mag. 2 772-776 (1957).
59. H. Suzuki, S. Ikeda and S. Takeuchi, J. Phys. Soc. Japan 11 382-393 (1956).
60. Y. Nakada, U.F. Kocks and B. Chalmers, Trans. Met. Soc. A.I.M.E. 230 1273-1278 (1964).
61. R.J. Murphy, Scripta Met. 3 905-910 (1969).
62. I.R. Kramer, Trans. Met. Soc. A.I.M.E. 227 1003-1010 (1963).
63. A. Pfützenreuter and G. Masing, Z. Metallkde. 42 361-370 (1951).
64. S. Kitajima, H. Oasa and H. Kaieda, Trans. Japan Inst. Metals 8 185-189 (1967).
65. S. Kitajima, H. Tanaka and H. Kaieda, Trans. Japan Inst. Metals 10 10-14 (1969).
66. I.R. Kramer, Trans. Met. Soc. A.I.M.E. 230 991-1000 (1964).
67. I.R. Kramer, Trans. Met. Soc. A.I.M.E. 239 520-528 (1967).
68. R.J. Stokes, T.L. Johnston and C.H. Li, Trans. Met. Soc. A.I.M.E. 218 655-662 (1960).
69. I.R. Kramer and N. Balasubramanian, Acta Met. 21 695-699 (1973).
70. D.J. Duquette, Scripta Met. 3 513-516 (1969).

71. C. Feng and I.R. Kramer, Trans. Met. Soc. A.I.M.E. 233 1467-1473 (1965).
72. J.T. Fourie, Phil. Mag. 22 923-929 (1970).
73. W.T. Brydges, Scripta Met. 2 557-560 (1968).
74. E.N. da C. Andrade and C. Henderson, Phil. Trans. Roy. Soc. A244 177-203 (1951).
75. D.P. Williams and H.G. Nelson, Trans. Met. Soc. A.I.M.E. 233 1339-1345 (1965).
76. R. Roscoe, Nature 133 912 (1934).
77. R. Roscoe, Phil. Mag. 21 399-406 (1936).
78. Y. Nakada and B. Chalmers, J. Appl. Phys. 33 3307-3308 (1962).
79. K. Sumino and M. Yamamoto, J. Phys. Soc. Japan 16 131 (1961).
80. I.R. Kramer and L.J. Demer, Trans. Met. Soc. A.I.M.E. 221 780-786 (1961).
81. I.R. Kramer and S. Podlaseck, Acta Met. 11 70-71 (1963).
82. I.R. Kramer, Trans. Met. Soc. A.I.M.E. 233 1462-1467 (1965).
83. H. Shen, S.E. Podlaseck and I.R. Kramer, Trans. Met. Soc. A.I.M.E. 233 1933-1938 (1965).
84. I.R. Kramer, ref. [5], pp. 127-146.
85. H. Shen, S.E. Podlaseck and I.R. Kramer, Acta Met. 14 341-346 (1966).
86. H.G. Nelson and D.P. Williams, ref. [5], pp. 107-125.
87. I.R. Kramer, Trans. Met. Soc. A.I.M.E. 239 1754-1758 (1967).
88. I.R. Kramer and C.L. Haehner, Acta Met. 15 199-202 (1967).
89. T.V. Cherian, P. Pietrokowsky and J.E. Dorn, J. Metals 1 (Trans. A.I.M.E. 185) 948-956 (1949).
90. I.R. Kramer, Trans. Quart. Amer. Soc. Metals 60 310-317 (1967).
91. I.R. Kramer, Scripta Met. 6 601-606 (1972).
92. M.R. Pickus and E.R. Parker, J. Metals 3 (Trans. A.I.M.E. 191) 792-796 (1951).
93. F.D. Rosi, Acta Met. 5 348-350 (1957).
94. R.M. Latanision and R.W. Staehle, Scripta Met. 2 667-672 (1968).
95. G.E. Ruddle and H.G.F. Wilsdorf, Appl. Phys. Lett. 12 271-273 (1968).
96. A.R.C. Westwood, Phil. Mag. 5 981-990 (1960).
97. C.S. Barrett, Acta Met. 1 2-7 (1953).
98. C.S. Barrett, J. Metals 5 (Trans. A.I.M.E. 197) 1652-1654 (1953).
99. C.S. Barrett, P.M. Aziz and I. Markson, J. Metals 5 (Trans. A.I.M.E. 197) 1655-1661 (1953).
100. B.I. Edelson and W.D. Robertson, Acta Met. 2 583-590 (1954).
101. W.A. Jemian and C.C. Law, Acta Met. 15 143-144 (1967).
102. D.B. Holt, Acta Met. 10 1021-1035 (1962).
103. W. Ewald and M. Polanyi, Z. Physik 28 29-50 (1924).
104. A. Joffé, M.W. Kirpitschewa and M.A. Lewitzky, Z. Physik 22 286-302 (1924).
105. W.W. Coblentz, Phys. Rev. 16 389 (1903).

106. A. Ritzel, Z. Krist. (Z. Kryst. Min.) 53 97-148 (1913).
107. A.F. Joffé, The Physics of Crystals, McGraw-Hill, New York, 1928, pp. 62-63.
108. J.H. Westbrook and P.J. Jorgensen, Trans. Met. Soc. A.I.M.E. 233 425-428 (1965).
109. J.H. Westbrook, ref. [5], pp. 247-268.
110. R.E. Hanneman and J.H. Westbrook, Phil. Mag. 18 73-88 (1968).
111. D.B. Holt, ref. [5], pp. 269-292.
112. P. Rehbinder and E. Wenström, Acta Physicochimica URSS 19 36-50 (1944).
113. P.A. Rebinder and V.I. Likhtman, Dokl. Akad. Nauk SSSR 56 723-726 (1947).
114. S. Harper and A.H. Cottrell, Proc. Phys. Soc. B63 331-338 (1950).
115. W. Klinkenberg, K. Lücke and G. Masing, Z. Metallkde. 44 362-369 (1953).
116. I.R. Kramer, Trans. Met. Soc. A.I.M.E. 221 989-993 (1961).
117. I.R. Kramer, Trans. Met. Soc. A.I.M.E. 227 529-533 (1963).
118. Ye. (E.) D. Shchukin, V.I. Savenko, L.A. Kochanova and P.A. Rebinder, Dokl. Akad. Nauk SSSR 200 406-409 (1971).
119. V.I. Savenko, L.A. Kochanova and E.D. Shchukin, Kristallografiya 17 995-999 (1972) : Soviet Physics - Crystallography 17 874-877 (1973).
120. A.R.C. Westwood, R.D. Huntington and N.H. Macmillan, J. Appl. Phys. 44 5194-5195 (1973).
121. A.R.C. Westwood, D.L. Goldheim and R.G. Lye, Phil. Mag. 16 505-519 (1967).
122. ibid. 17 951-959 (1968).
123. A.R.C. Westwood, Phil. Mag. 7 633-649 (1962).
124. A.R.C. Westwood and D.L. Goldheim, J. Appl. Phys. 39 3401-3405 (1968).
125. V.I. Likhtman, L.A. Kochanova, D.I. Leikis and E.D. Shchukin, Elektrokhimiya 5 729-733 (1969) : Soviet Electrochemistry 5 679-682 (1969).
126. R.M. Latanision, H. Opperhauser, Jr. and A.R.C. Westwood, in The Science of Hardness Testing and its Research Applications, Amer. Soc. Metals, Metals Park, Ohio, 1973, pp. 432-9.
127. R.W. Revie and H.H. Uhlig, Acta Met. 22 619-627 (1974).
128. J.H. Westbrook and J.J. Gilman, J. Appl. Phys. 33 2360-2369 (1962).
129. H. Alexander and P. Haasen, in Solid State Physics 22, ed. F. Seitz et al., Academic Press, New York, 1968, pp. 75-81 and 117-119.
130. J.J. Mills, private communication (1975).
131. N.H. Macmillan, R.D. Huntington and A.R.C. Westwood, Phil. Mag. 28 923-931 (1973).
132. N.H. Macmillan and A.R.C. Westwood, Mat. Sci. Res. 7 493-513 (1974).
133. A.R.C. Westwood, N.H. Macmillan and R.S. Kalyoncu, J. Amer. Ceram. Soc. 56 258-262 (1973).

134. R.M. Latanision, in *Proc. Internat. Conf. Surface Technology*, Soc. Manufact. Eng., Dearborn, Mich., 1973, pp. 1-22.
135. I.R. Kramer and N. Balasubramanian, *Metall. Trans.* **4** 431-436 (1973).
136. A.H. Cottrell, *Proc. Roy. Soc.* A**276** 1-18 (1963).
137. A.H. Cottrell, The 1963 Tewksbury Lecture, in *Tewksbury Symposium on Fracture*, ed. C.J. Osborn, Univ. of Melbourne, 1965, pp. 1-27.
138. G.R. Irwin, J.A. Kies and H.L. Smith, *Proc. Amer. Soc. Test. Mater.* **58** 640-660 (1958).
139. F.J.P. Clarke and R.A.J. Sambell, *Phil. Mag.* **5** 697-707 (1960).
140. F.J.P. Clarke, R.A.J. Sambell and H.G. Tattersall, *Phil. Mag.* **7** 393-413 (1962).
141. A. Kelly, W.R. Tyson and A.H. Cottrell, *Phil. Mag.* **15** 567-586 (1967).
142. Psalms ii, 3.
143. R.J. Stokes, T.L. Johnston and C.H. Li, *Phil. Mag.* **6** 9-24 (1961).
144. W.H. Class, E.S. Machlin and G.T. Murray, *Trans. Met. Soc. A.I.M.E.* **221** 769-775 (1961).
145. D.M. Marsh, in *Fracture of Solids*, Metallurgical Society Conferences **20**, ed. D.C. Drucker and J.J. Gilman, Interscience, New York, 1963, pp. 119-142.
146. A.E. Gorum, E.R. Parker and J.A. Pask, *J. Amer. Ceram. Soc.* **41** 161-164 (1958).
147. A.R.C. Westwood and H. Rubin, *J. Appl. Phys.* **33** 2001-2007 (1962).
148. L. Graf and J. Budke, *Z. Metallkde.* **46** 378-385 (1955).
149. C. Edeleanu, ref. [12], pp. 79-98.
150. A.J. Forty, *ibid.* pp. 99-120.
151. A.J. Forty and P. Humble, *Phil. Mag.* **8** 247-264 (1963).
152. E.N. Pugh and A.R.C. Westwood, *Phil. Mag.* **13** 167-183 (1966).
153. V.I. Rozhanskii and P.A. Rebinder, *Dokl. Akad. Nauk SSSR* **91** 129-131 (1953).
154. J.J. Gilman, *Phil. Mag.* **26** 801-812 (1972).
155. R.M. Latanision and H. Opperhauser, *Metall. Trans.* **5** 483-492 (1974).
156. A.J. Forty, ref. [13], pp. 64-71.
157. P. Re(h)binder, *Z. Physik* **72** 191-205 (1931).
158. A.R.C. Westwood, C.M. Preece and M.H. Kamdar, in *Fracture, An Advanced Treatise* **3**, ed. H. Liebowitz, Academic Press, New York, 1971, pp. 589-644.
159. M.J. Kelley and N.S. Stoloff, *Metall. Trans.* **6A** 159-166 (1975).
160. V.S. Yushchenko, A.G. Grivtsov and E.D. Shchukin, *Dokl. Akad. Nauk SSSR* **215** 148-151 (1974).
161. *ibid.* **219** 162-165 (1974).
162. T.L. Johnston, R.G. Davies and N.S. Stoloff, *Phil. Mag.* **12** 305-317 (1965).
163. Ya. M. Dityatkovskii, I.V. Andreyev and V.F. Gorshkov, *Fiz. Metal. Metalloved.* **15** 435-438 (1963) : *Phys. Met. Metallogr.*

15 [3] 94-97 (1963).
164. E.D. Shchukin, L.A. Kochanova and A.V. Pertsov, Kristallografiya 8 69-74 (1963) : Soviet Physics - Crystallography 8 49-52 (1963).
165. C.M. Preece and A.R.C. Westwood, Trans. Amer. Soc. Metals 62 418-425 (1969).
166. A.R.C. Westwood, E.N. Pugh and D.L. Goldheim, Phil. Mag. 10 345-347 (1964).
167. A.R.C. Westwood, D.L. Goldheim and E.N. Pugh, Discuss. Faraday Soc. 38 147-156 (1964).
168. A.R.C. Westwood and D.L. Goldheim, First Internat. Conf. Fracture 3, Japan Soc. Strength Fracture Mater., 1966, pp. 1999-2014.
169. A.R.C. Westwood, D.L. Goldheim and E.N. Pugh, Phil. Mag. 15 105-120 (1967).
170. E.B. Shand, J. Amer. Ceram. Soc. 44 21-26 (1961).
171. D.A. Shockey and G.W. Groves, J. Amer. Ceram. Soc. 52 82-85 (1969).
172. F.R.L. Schoening, J. Appl. Phys. 31 1779-84 (1960).
173. N. Thompson and N.J. Wadsworth, Adv. Physics 7 72-169 (1958).
174. C. Laird and D.J. Duquette, ref. [14], pp. 88-117.
175. P. Lukáš and M. Klesnil, ibid. pp. 118-132.
176. P. Lukáš, M. Klesnil, J. Krejčí and P. Ryš, phys. stat. sol. 15 71-82 (1966).
177. I.R. Kramer and A. Kumar, ref. [14], pp. 146-163.
178. I.R. Kramer, Metall. Trans. 5 1735-1742 (1974).
179. J.T. Fourie, ref. [14], pp. 164-175.
180. A.S. Argon, ibid. pp. 176-182.
181. I.R. Kramer, Trans. Amer. Soc. Metals 62 521-536 (1969).
182. V.V. Shevelya and B.I. Kostetskii, Dokl. Akad. Nauk SSSR 175 1270-1272 (1967) : Soviet Physics - Doklady 12 836-838 (1968).
183. S.P. Kozyrev and E.D. Shchukin, Dokl. Akad. Nauk SSSR 179 825-828 (1968) : Soviet Physics - Doklady 13 332-335 (1968).
184. M.S. Aslanova and P.A. Rebinder, Dokl. Akad. Nauk SSSR 96 299-302 (1954).
185. A.R.C. Westwood, R.M. Latanision and R.G. Lye, phys. stat. sol. (a) 3 K17-20 (1970).
186. A.R.C. Westwood and R.D. Huntington, in Internat. Conf. on Mechanical Behavior of Materials 4, Soc. Mater. Sci., Japan, 1972, pp. 383-393.
187. M.F. Ashby and J. Logan, Scripta Met. 7 513-522 (1973).
188. W. v. Engelhardt, Naturwiss. 33 195-203 (1946).
189. I.R. Kramer, in Fatigue - Interdisciplinary Approach, Proc. 10th Sagamore Army Mater. Res. Conf., ed. J.J. Burke et al., Syracuse Univ. Press, New York, 1964, pp. 245-259.
190. P.A. Re(h)binder, L.A. Shreiner and K.F. Zhigach, Hardness Reducers in Drilling, Akad. Nauk SSSR, Moscow, 1944 : Council Sci. Industr. Res., Melbourne, 1948.
191. A.R.C. Westwood, J. Mater. Sci. 9 1871-1895 (1974).

192. A.R.C. Westwood and D.L. Goldheim, J. Amer. Ceram. Soc. 53 142-147 (1970).
193. E.D. Shchukin, V.I. Savenko, L.A. Kochanova and P.A. Rebinder, Dokl. Aka. Nauk SSSR 200 1329-1332 (1971): Soviet Physics - Doklady 16 913-915 (1972).
194. N.H. Macmillan, R.D. Huntington and A.R.C. Westwood, J. Mater. Sci. 9 697-706 (1974).

DISCUSSION

Comment by H. Mughrabi:

I wish to make two comments-

1) The etch pit technique is not suitable to detect differences in dislocation density between near-surface and interior regions. Even in weakly deformed single crystals, the dislocation pattern is inhomogeneously distributed, e.g., in dense dipole or multi-pole configurations. These closely spaced dislocations are not all resolved by the etch pit technique.

2) Thinking of Fourie's distribution of the flow stress in the virgin crystal, I feel that not enough effort has been made to ascertain that the crystals have a constant initial defect density throughout the cross section. Even a small initial gradient in flow stress can lead to misleading results. If one assumes homogeneous hardening in stage I, then the stress level at which the transition to stage II work-hardening occurs is first obtained in the regions of highest initial flow stress. Because of the associated substantial local increase in work-hardening rate, initially small differences in local flow stress are rapidly magnified.

Reply:

1) We do know the trend of the errors - the density will be under-estimated when it is either high or very non-uniform.

2) Agreed,

Comment by J. T. Fourie:

The point raised by Dr. Mughrabi, namely, that a small prior flow stress gradient in the crystal could be magnified when the crystal is pulled through the transition region of the stress strain curve, was considered by Fourie (J.T. Fourie, Phil. Mag. 17, 735 (1968)). To test this possibility, he sliced a virgin crystal which was known to be slightly harder in the core than near the surface, lengthwise down the centre. One of the half crystals was then pulled into stage II, whereafter it was unloaded and sliced into component crystals. It was found, after testing the flow stress in the components, that near both surfaces where edge dislocations escaped, the usual flow stress distribution was to be found, in spite of the fact that the initial flow stress befor straining was highest near the one edge. The point raised by Dr. Mughrabi was thus shown to be of minor importance regarding its influence on the development of the flow stress distribution.

Comment by F.R.N. Nabarro:

Does Dr. Fourie have evidence that the flow stress can actually decrease if the interior of a strained crystal is exposed and the crystal is again strained?

Reply by Fourie:

There is some indirect evidence for an actual decrease in the flow stress, based on the measurement of dislocation density in a newly exposed surface before and after pulling. This is contained in the paper: J.T. Fourie, Proc. 2nd Int. Conf. on Strength of Metals and Alloys, Asilomar, USA, 1970, pp. 441-445.

Comment by J. T. Fourie:

With reference to Professor Nabarro's interpretation of my results on the effect of electropolishing on the unloading yield point in deformed copper crystals, I should like to point out the following. In the paper which Professor Nabarro refers to, it is suggested that the process of unloading causes a shift in the flow stress distribution such that the surface region becomes harder than it was during the dynamic deformation process before unloading. However, that it was still considerably softer than in the core is consistent with experimental measurements.

SURFACE EFFECTS IN CRYSTAL PLASTICITY: OVERVIEW FROM A SURFACE
SCIENCE POINT OF VIEW

T. E. Fischer

Exxon Research and Engineering Company, Linden, New
Jersey 07036 U.S.A.

The purpose of this Nato Institute is to explore what progress in the knowledge and understanding of environmental effects on crystal plasticity can be achieved by increased interaction of materials scientists and metallurgists with surface physicists. Many phenomena of importance to the strength of structural materials involve physical and chemical processes at the surface of these materials. Since surface physics is presently very active and has vastly increased the diversity and sophistication of its experimental and theoretical techniques of investigation, one expects that a cooperative effort of the two disciplines will lead to insights into the phenomena of interest here that neither discipline would provide alone. In order to facilitate the dialogue during this institute, it is useful to have a broad look at the present state of surface physics, its methods of investigation, the type of information that can be obtained, the knowledge already acquired and the limitations of the techniques. We shall do this in the first section of this paper. In the second part, we shall address ourselves to the interaction between surface physics and materials science with the help of a few examples, namely the chemisorption-hardening of nonmetals (Rebinder effect), the adsorption model of stress corrosion cracking and liquid metal embrittlement and a chemical model of fatigue wear in lubricated bearings.

1. SURFACE PHYSICS

Certain areas of surface science, most notably electrochemistry, have played an important role in metallurgy for a long time; they are well known and have already made the contributions that we can presently think of. We shall not discuss them further but concentrate our attention on the new methods of investigation which have appeared in the recent past.[1]

We are now capable of preparing surfaces that are free of all but a small fraction (usually less than 1%) of a monolayer of contaminants; we can determine the elemental composition of the surface with a sensitivity better than 1% and follow its change within about a second; as of recently, the exact positions of the atoms on a surface can be determined, although at considerable expense, but some useful information on the surface structure of clean or adsorbate-covered surfaces is obtained readily; several techniques of electron spectroscopy are providing information on the electronic structure of the bulk, its modification at the surface and the behavior of the valence electrons of metal and gas molecules in chemisorption. Simultaneously, the theoretical description of surface effects is progressing rapidly. We are now capable of a unified description of surface phenomena which combines the extended-wave and many-electron concepts essential to solid state physics with the quantum chemical treatment of electrons in terms of localized molecular orbitals and charge distributions which prevail in chemical bonds and reactions.

The experiments just mentioned are usually performed in ultrahigh vacuum, under pressures of a few 10^{-10} torr to insure that a surface, once clean, will stay so for the duration of a measurement. Multiple measurements in one vacuum chamber are the rule; a typical set up would include facilities for sample preparation, composition and structure determination, measurements of work function, controlled gas handling and a mass spectrometer capable of detecting partial pressures as low as 10^{-13} torr and perhaps an electron spectroscopy experiment. It is important to recognize that most experiments measure average properties over an area of about 10^{-4} cm^2 and are unsuitable for the study of defects which are so important in crystal plasticity.

A closer look at the techniques used will provide us with an appreciation of their capabilities and limits.

Rather than describing the physics underlying the various techniques, which will be treated by C. B. Duke and H. Gatos in this volume, we shall adopt the standpoint of the materials scientist and ask ourselves the following questions concerning their usefulness.

- What kind of information does this technique provide?
- What is the sensitivity of the method?
- How deep a surface layer does it sample?
- What is its lateral resolution on the surface?

- Is the information quantitative?
- To what extent does the measurement alter the properties of the surface?
- Is the method well developed, so that it can be used conveniently by others than surface physicists?
- Is the interpretation straightforward or does it require extensive work in theoretical surface physics?

The experimental methods used in surface physics involve various combinations of the absorption, scattering or emission of photons (u.v. light and x-rays) of electrons and of ions. For the later analysis of the techniques, in terms of the depth of the surface they sample, it is useful to review briefly how far the particles in question can travel in the solid before they become useless for the experiment (by absorption or inelastic scattering).

The penetration depth of light depends on the photon energy. It ranges from a minimum of about 200 Å (where $h\nu \approx 6$ eV) to several thousand Å in the soft x-ray region (where $h\nu \approx 1000$ eV) and centimeters at the high photon energies. Light in itself is thus not a surface sensitive tool. Changes in optical reflectivity caused by surface effects are extremely small and are not generally used for the study of surfaces.

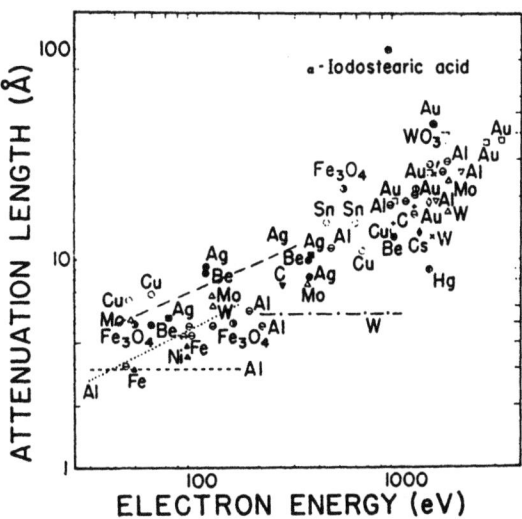

Fig. 1. Escape depth of electrons from metals as a function of their energy. This represents the average distance travelled by an electron before it suffers a collision that makes it lose enough energy (several eV) so that it does not carry any useful information and becomes part of the background (After Powell, Reference 2).

The mean free path of electrons before loss of substantial amounts of energy (1 eV or more) has an interesting dependence on the kinetic energy of the electrons that is very useful to surface physicists. Figure 1 shows a compilation[2] of this mean free path as a function of the kinetic energy of electrons escaping from the solid. We notice that this "escape depth" becomes as small as 1 to 3 monolayers for electrons whose kinetic energies are between 50 and 100 eV. The escape depth increases to 20 Å as the kinetic energy reaches 1000 Å.

As the energy is lowered from 50 eV, the escape depth increases again; it is approximately 10 Å at 20 eV and about 30 Å near the zero of kinetic energy in vacuo. To the extent to which the measuring technique allows to vary the energy of the probing electrons, it permits to control the depth of the layer on samples. Most techniques of surface physics use electrons with energies below 1000 eV and therefore study layers that are less than 30 Å deep.

The surface layer probed by ions is even more shallow: sputtering methods probe a few atomic layers and elastic ion scattering gathers information on the topmost layer only. We shall come back to this point in the discussion of the methods of surface analysis.

It is important to be aware of the contrast between the very shallow layers measured with the methods of surface physics and the depth of several tenths of a millimeter over which a surface influences mechanical properties of solids.

1.1 Sample Preparation

Surfaces under study must be cleaned in the same vacuum chamber in which the measurements are performed. The sample, most often a single crystal, having been cut to expose a surface of desired orientation and polished is then cleaned by alternating cycles of bombardment with ions (usually Argon) of a few hundred volt energy and annealing. Most impurities can be removed in this fashion, although chemical reaction at elevated temperature with a gas at about 10^{-6} torr is often used for selected impurities. Cleavage of single crystals is a convenient way of exposing clean surfaces of brittle materials; it is widely used, for example, in the study of semiconductor surfaces. Obviously, also, fracture surfaces that are to be studied by methods of surface physics must be obtained by breaking the sample in the experimental chamber itself under ultrahigh vacuum to avoid the contamination and reaction of the surface under study by the ambient gases. A useful rule of thumb is that a monlayer can be deposited (or modified by reaction) in one second at 10^{-6} torr partial pressure

of reactive gas. A pressure of 10^{-10} torr will thus maintain a
surface with less than 1% modification for about 20 minutes.
Polycrystalline samples of high purity prepared by evaporation,
which is of course done in the vacuum chamber, can be used for
the measurements. Multiple evaporations provide a convenient means
to prepare alloy samples. Other thin film preparation methods,
such as sputtering or chemical vapor deposition, are seldom used
because they are less compatible with multiple measurements in
ultrahigh vacuum.

1.2 Surface Composition

The measurement of the surface composition of a sample is of
the highest importance. It will tell us whether our cleaning procedure has been successful and will identify the surface impurities.
In the case of an alloy or a compound, it will determine the extent of surface segregation. With adsorption or chemical reactions
it will identify the adsorbed species or the reaction products.
There are several methods of nondestructive instantaneous surface
chemical analysis, namely Auger electron spectroscopy (AES), ESCA
(electron spectroscopy for chemical analysis), also called x-ray
photoemission spectroscopy (XPS), ion scattering spectroscopy (ISS),
secondary mass spectroscopy (SIMS) and appearance potential spectroscopy. Of these methods, Auger spectroscopy and ESCA are the
most widely used for their flexibility and relative ease of operation. The equipment is now commercially available for all methods
except appearance potential spectroscopy.

1.2.1 Auger Electron Spectroscopy

This method permits the identification of the elements present
in a thin layer near the surface. Since it uses electrons only, its
probing depths varies from 5 to 30 Å. New developments allow the
scanning of a surface for a given element with a lateral resolution
of 3μm.

In this technique, one bombards the surface with electrons of
approximately 3 kV energy. These electrons eject core electrons
from the atoms of the sample. The emptied orbital of a given atom
can be refilled by several mechanisms, of which the Auger process
is particularly efficient. This process is illustrated in Figure
2. Orbital A has just been emptied as described above. Electron
B can transfer to orbital A if it can get rid of the energy $\Delta E = E_A - E_B$ where E_B and E_A are the binding energies of orbitals A and
B. Since electrons carry a charge, they interact with each other;
and electron B gets rid of the Energy ΔE by transmitting it to
electron C via Coulomb repulsion. If ΔE is larger than the binding
energy of electron C, the latter is ejected with the kinetic energy

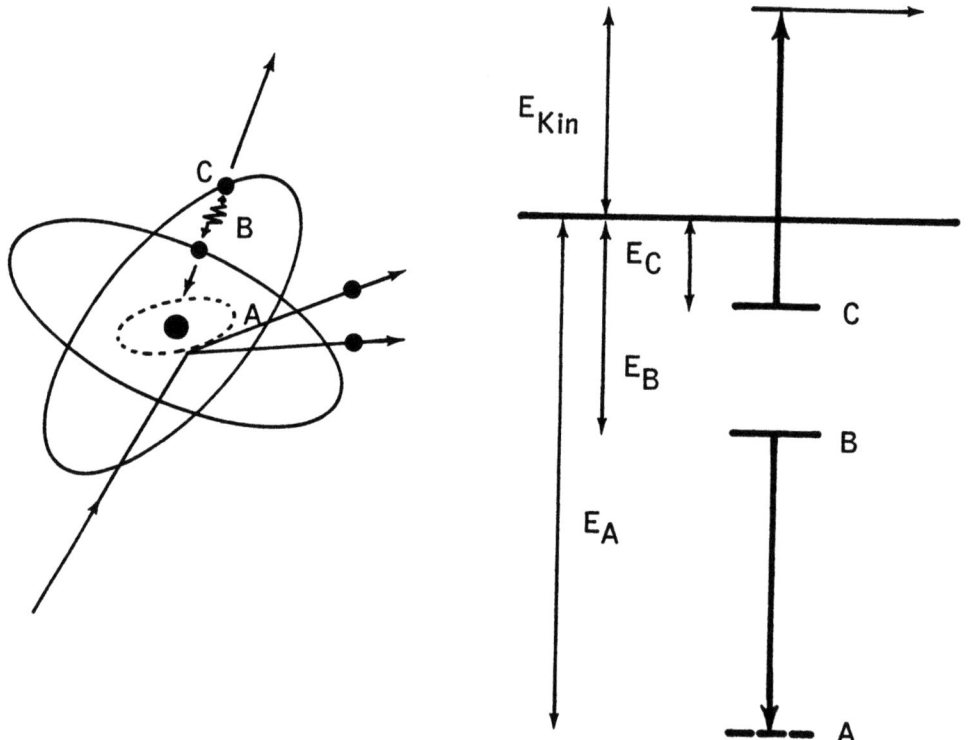

Fig. 2. The principle of Auger electron spectroscopy. Electron A has been ejected by impact from primary electron. Electron B transfers to orbital A by transmitting its excess energy through Coulamb repulsion to electron C which is ejected.

$$E_{kin} = \Delta E - E_C = E_A - E_B - E_C \qquad (1)$$

This kinetic energy depends only on the binding energies of the three orbitals involved. There are of course, quite a number of kinetic energies at which Auger electrons from a given atom will be emitted since electrons A, B, and C are chosen at random among many possible candidates. The Auger process is one of many mechanisms by which the primary electron dissipates its energy and causes the emission of electrons. Auger electron spectroscopy is part of the phenomenon called secondary emission: experimentally, one bombards a surface with electrons of relatively high energy and measures the energy distribution of secondary electrons re-emitted by the sample. Figure 3 a is a portion, measured from 0 to 600 eV of such an energy distribution. The Auger electrons represent small peaks superimposed on the background of secondary

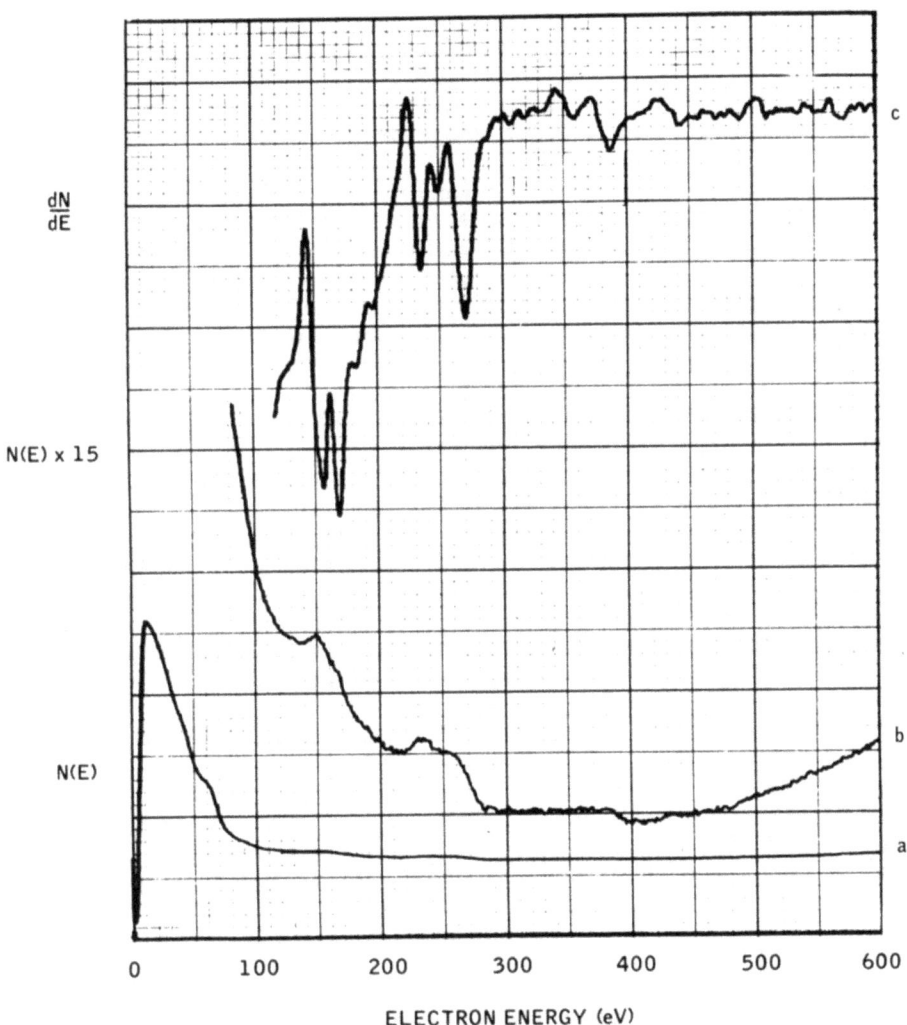

Fig. 3. An example of Auger electron spectrum: Platinum (Primary electron energy 2300 eV)

3a. The energy distribution of secondary electrons as emitted. Notice the small contributions from the Auger peaks.

3b. The same energy distribution at 15 times larger sensitivity.

3c. An electronic differentiation of the spectrum 3a.

All peaks are due to platinum, except the peak at 270 eV which reveals contamination by carbon. The curves shown are direct reproduction of traces from the x-y recorder.

electrons emitted by other mechanisms,[3] as shown in Figure 3b, which is the curve of Figure 3 a redrawn with 15 times higher vertical sensitivity. A convenient way of extracting the Auger signal from the background is an electronic differentiation[4] dN/dE of the energy distribution, as shown in Figure 3c.

Thus, every element emits a fingerprint of Auger peaks. The element is identified simply by comparing the Auger spectrum with those available in published compilations.[5] The number of electrons emitted at the characteristic energies is proportional to the number of atoms present within the escape depth of the electrons and provides us with a semiquantitative chemical analysis of the surface region.

It is customary to measure the amount of an element by the amplitude of the derivation of the Auger peak. This procedure is convenient but dangerous: any change in the experimental line shape will introduce errors. A careful measurement of the amount present requires that one determines the area of the Auger peak which amounts a double integration of the differentiated Auger peak (Fig. 3c). This is especially important when chemical reactions with the element under study can occur. The reason is that chemical reactions change the energies of valence electrons and change the shape of the corresponding Auger peaks.

An interesting new development is Scanning Auger Microscopy (SAM) where the primary beam (the one producing the vacancies in orbital A, Fig. 2) is carefully focused to a spot of about 3 micron diameter and scans the surface. The energy analyzer of secondary electrons is tuned to one of the Auger energies (eq. 1); the amplitude of the Auger peak is displayed as intensity on a cathode ray tube, as in scanning microscopy and gives an image of the spatial distribution of the element, with a resolution of 3 microns (Fig. 4).

The sensitivity of Auger spectroscopy is such that an impurity representing less than 1% of the top atomic layer can easily be detected. Our discussion of Fig. 1 shows that the depth probed by the method depends on the energy of the Auger electrons. Many elements have Auger lines at quite different energies. An example is shown in Figure 5 where the segregation of copper to the surface of a Cu-Ni alloy was studied. Copper and Nickel lines exist near 100 eV, probing a depth of 5 Å, and at 870 eV which probe 30 Å. The high energy line shows no marked variation in the average composition over 30 Å, the low energy line shows a total enrichment in copper within 5 Å: surface segregation of copper was thus shown to be very shallow.[6]

Information on the variation of concentration of various elements with depth can be obtained by the simultaneous etching

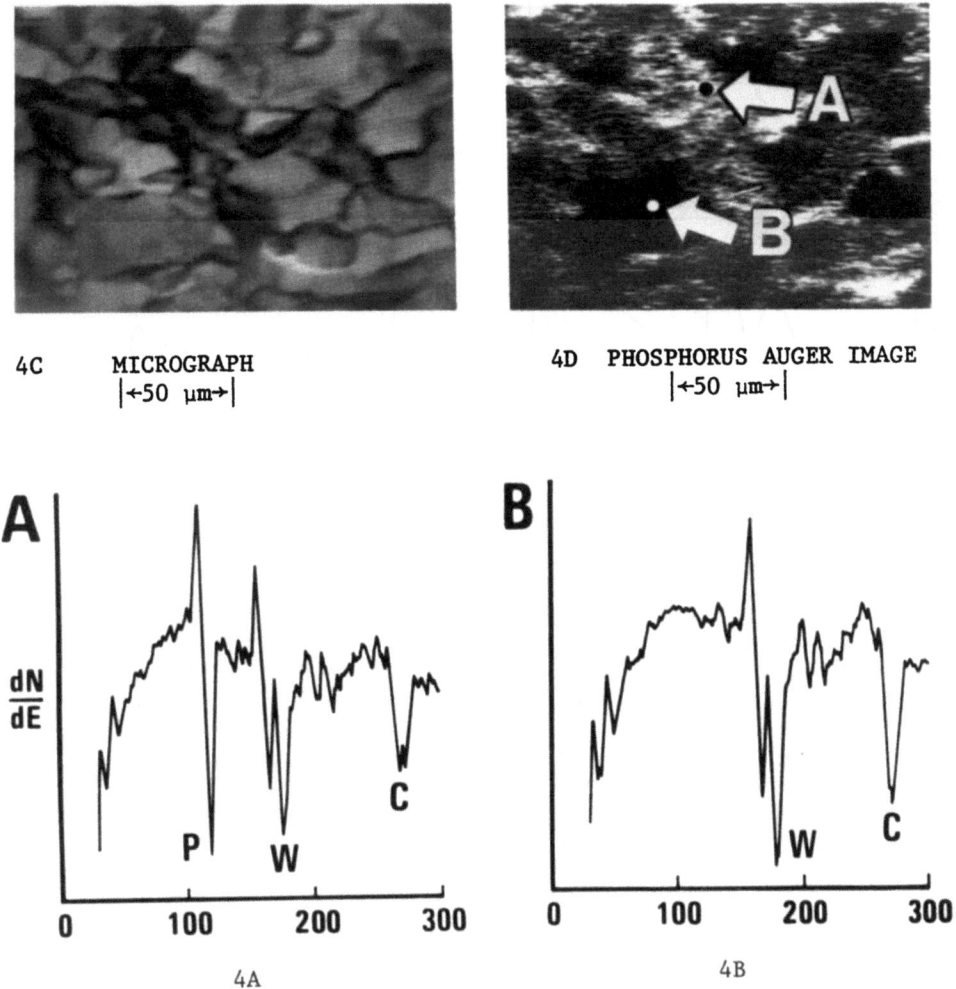

Auger Electron Spectra of Points A and B

Fig. 4. Study of a fracture surface of embrittled tungsten in the Scanning Auger Spectrometer. 4C is a micrograph obtained by measuring total secondary current as a function of scanning primary beam position, as in scanning electron microscopy. 4D is obtained by displaying the phosphorus signal (as brightness) as a function of primary beam position. Fig. 4a and 4b are Auger spectra obtained at points A and B of Fig. 4D. (Courtesy of P. W. Palmberg of Physical Electronics Industries).

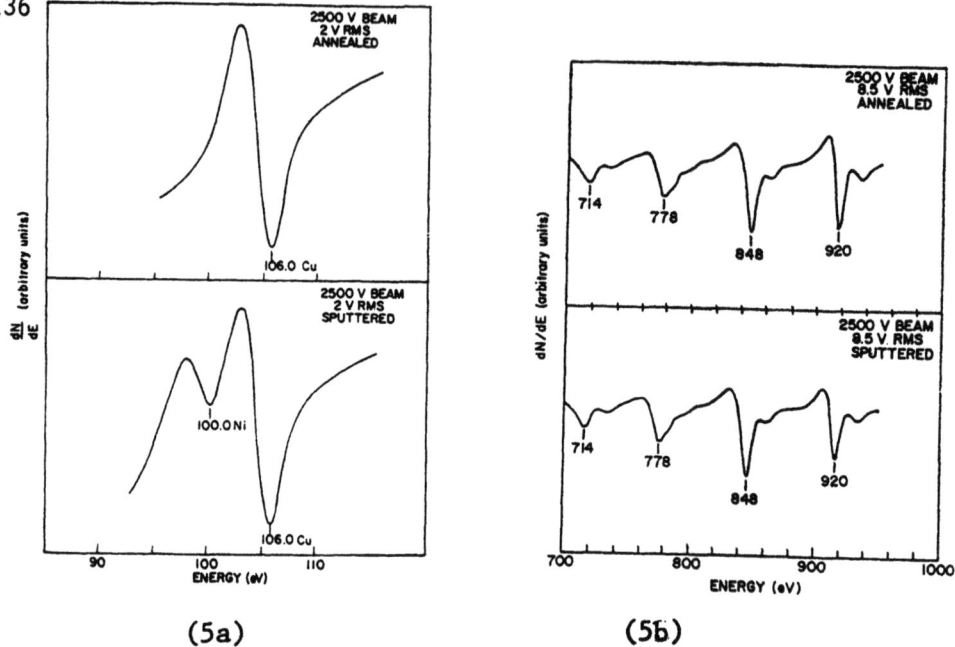

(5a)   (5b)

Fig. 5. Auger spectra of a 50% Cu 50% Ni alloy with a monolayer enriched in copper by surface segregation. Notice that the low energy nickel peak (Fig. 5a) has disappeared and the high energy Ni signal (5b) is virtually unchanged. (After Helms, Reference 6).

of the surface by ion bombardment and Auger measurements. Fig. 6 shows such a study on nichrome steel.[7] The precision of this depth profile decreases with increasing depth of the removed layer since one must expect a certain roughing of the sputtered surface.

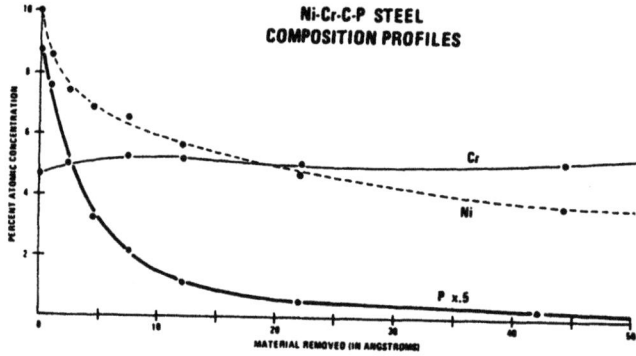

Fig. 6. Composition profile of fracture surface from de-embrittled Ni-Cr-C-P Steel. The ordinate shows the concentration of elements from the amplitude of Auger peaks after a surface layer of thickness shown in the abscissa was removed by ion bombardment. (After Palmberg and Marcus, Reference 7).

A further, although usually not severe, limitation of depth profiling results from forward scattering of surface atoms. Incoming energetic ions will cause the ejection of surface atoms; they will also push a few surface atoms deeper into the bulk, so that a small fraction of the surface layer travels with the new surface.[8] In general, this forward scattered layer represents less than 1% of the surface composition.

Auger electron spectroscopy is a destructive method for certain materials (mostly insulators) and for many adsorbates. The incoming electron beam interacts with the surface molecules and may induce chemical reactions on the surface. It is well known, for instance, that CO adsorbed on metals is decomposed by the Auger beam into oxygen which desorbs and carbon which covers the surface.[9] Electron beam irradiation also causes decomposition of the SiO_2 surface and creates an elemental Si overlayer.[10] Many organic molecules are likewise modified during measurement. The electron beam can also cause the desorption of gases. No case of modification of surface concentration of metal atoms is known to this author.

It is useful to compare Auger spectroscopy with the electron probe and the non-dispersive x-ray analysis of the Scanning Electron Microscope. In both techniques one bombards the sample with a well focused beam of high energy electrons ($\approx$ 50 keV). These electrons again eject core electrons of the atoms in the sample. These vacancies can be filled by the Auger process, as described above, but this effect is not measured. An alternate way of filling the vacancy in orbital A is for the B electron to emit a photon (i.e. an x-ray) of energy.

$$h\nu = E_A - E_B \qquad (2)$$

These x-rays are then collected and their photon energy $h\nu$ reveals the identity of the atom. In the scanning microscope the photon energy is measured with a non-dispersive solid state detector; in the electron probe, the wavelength of the x-rays is measured by diffraction, which provides a better resolution and elemental separation. Solids are quite transparent to x-rays, so the technique samples a region limited by the distance travelled by the primary electron. This is a plume of approximately 1 micron diameter and about 5 micron depth, in contrast with the extreme surface sensitivity of Auger spectroscopy.

1.2.2 ESCA

ESCA is a chemical analysis technique that uses the phenomenon of photoelectric emission. One shines monochromatic x-rays (photon energy $h\nu \approx 1$ keV) onto the samples. These photons are absorbed by exciting core electrons of binding energy E_B which are emitted with kinetic energy (Fig. 7)

$$E_{kin} = h_\nu - E_B \qquad (3)$$

One measures the energy distribution of the emitted electrons and obtains lines at energies characteristic of the elements with amplitudes that are proportional to the number of atoms present in the volume sampled (Fig. 7). The latter is again determined by the escape depth of the electrons shown in Figure 1. We see that electrons emitted at different energies sample different depths raning from 2 to 20 atomic layers. The signal is usually averaged over several mm^2 of surface.

The interpretation of the measurements proceeds as in Auger spectroscopy: comparison with published spectra[11] identifies the element, and the amplitude of the lines gives a semiquantitative measure of the amount present. Lateral resolution is 1 mm at best and generally poorer. ESCA has the great advantage that the interaction of the x-rays with the surface atoms is much weaker; this

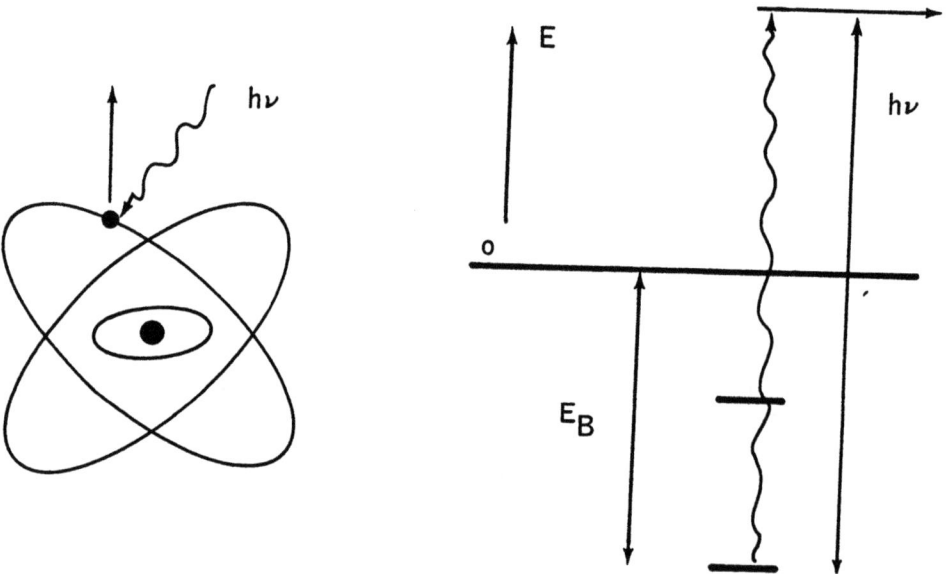

Fig. 7. The principle of photoemission. Monochromatic light shining on the sample is absorbed by the core or valence electrons which acquire enough energy to be emitted. Ultraviolet photoemmision (UPS) and x-ray photoemission (XPS or ESCA) differ in the photon energy of the exciting light.

technique is far less destructive than Auger. No scanning facility exists and the large sampled surface makes depth profiling less convenient, although not impossible.

Helium Ion Scattering[12] is conceptually simple and sensitive strictly to the top layer of atoms (Fig. 8). One bombards the surface with He ions impinging at 45° angle and measures the energy of the ions that are specularly reflected. Simple collision mechanics relate the energy E_S of the ion to the mass of the particular atom responsible for the scattering:

$$E_S = \frac{M^2 - m^2}{(M+m)^2} E$$

where E is the primary energy of the ion, m its mass and M is the mass of the surface atom. An interesting property of this technique is its sensitivity to the top atomic layer only; its use is somewhat restricted because rather severe alignment requirements limit the possibility of incorporating it into a multiple technique chamber.

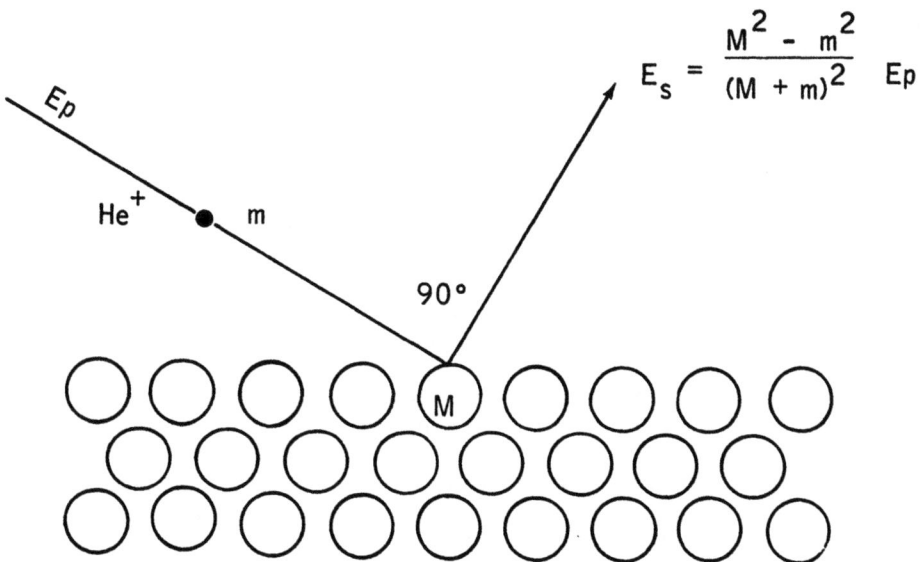

Fig. 8. Schematic of ion scattering spectroscopy. Helium ion with primary energy E_p suffers elastic collision with surface atom of mass M. The energy of ions scattered by 90° is subsequently measured and reveals the mass of the scattering atom.

Secondary Ion Mass Spectroscopy (SIMS)[13] consists of bombarding the sample with ions and collecting the ions ejected from the surface in a mass analyzer. This technique is extremely sensitive to certain elements, being capable of detecting 10^{-6} monolayer of certain elements. Unfortunately, the sensitivity varies from one element to another by several orders of magnitude.

SIMS measures the composition of the top atomic layer; scanning the surface as in Auger spectroscopy is possible in principle but has not been realized to date. Identification of the element is straight-forward since one measures directly the mass of ions. It is at first sight surprising that SIMS is not very destructive. In general, the sputtering of 1% of a monolayer is sufficient to measure its composition. Depth profiling is obviously part of the method itself: it suffices to sputter more extensively.

Most of these techniques are fast enough to allow to measure changes of the surface composition within a few seconds.

We have seen that various measuring techniques provide at least a semiquantitative measure (the absolute amount being difficult to obtain with a better accuracy than a factor of two) of the elemental composition of the surface. Can we do more and determine what molecules are adsorbed? Ultraviolet photoemission spectroscopy (UPS) is particularly well suited for this purpose. This technique differs from ESCA only in the wavelength of the exciting radiation. Instead of X-rays, one irradiates the sample with monochromatic ultraviolet light and measures the energy distributions of emitted electrons. The ultraviolet light is usually of short wavelength, below 1000 Å; most often one uses the resonance lines of helium: He I at $h\nu = 20.8$ eV; He II at $h\nu = 40.8$ eV; Ne at $h\nu = 16.8$ and Ar at $h\nu = 13.5$ eV, unless one has access to the continuous, polarized, and intense ultraviolet emission of a synchrotron. At these relatively low photon energies, the electrons are emitted into vacuum by excitation out of the valence band of the substrate and the molecular orbitals of adsorbed molecules.[14] (Figure 9) Thus ultraviolet photoemission is capable of identifying the adsorbed species which may result from dissociative chemisorption[16] or surface reactions. Because of the lower electron energy (Fig. 1), ultraviolet photoemission is more surface sensitive than ESCA but it also samples an area of several square millimeters.
A more widespread use of this technique for chemical surface analysis can be expected in the near future. ESCA, besides identifying an element, also determines its state of oxidation by means of the chemical shift: If the bonding of an atom to its surrounding involves charge transfer, the binding energies of all electrons are modified by a few electron volts. (Figure 10) At this point, the correlation between chemical shift and formal state of ionization is entirely empirical but sufficient for analytical purposes.

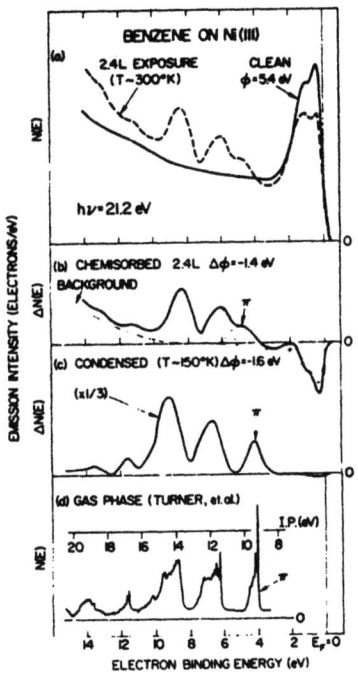

Fig. 9. Ultraviolet photoemission spectra of a clean nickel surface (9a) of the benzene molecule in gas phase (9d), of physisorbed (9c) and chemisorbed (9b) benzene on nickel. Figures 9b and c plot the change in photoemission from a nickel surface due to adsorption. (i.e., 9b is the difference between the two curves of 9a). Note the shift of the peak attributed to the π - electrons in fig. 9b. (After Demuth and Eastman, reference 15).

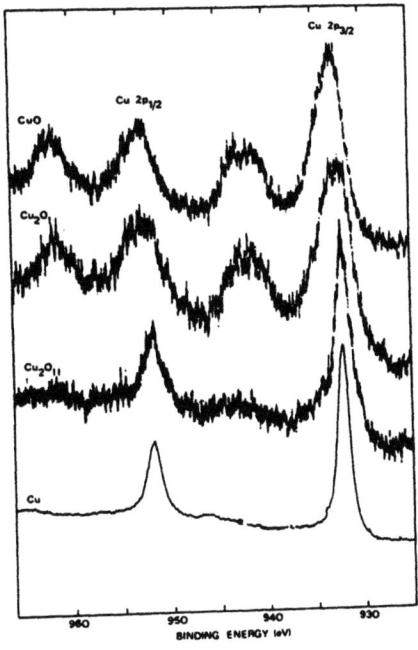

Fig. 10. X-ray photoemission (ESCA) of the Cu 2p levels from metallic copper, CuO and CuO_2, (after Schön, reference 17).

Chemical shifts are not generally used in Auger spectroscopy, because the process and its interpretation are more complicated than for ESCA (compare eqs. (1) and (3) and Figures 2 and 7). However, when electrons B and/or C are valence electrons, the shape of the Auger line will depend on the chemical state of the atom. See Figure 11. Auger line shapes have not been as well studied as ESCA chemical shifts; like ultraviolet photoemission they are not yet a well developed chemical tool, but have considerable potential.

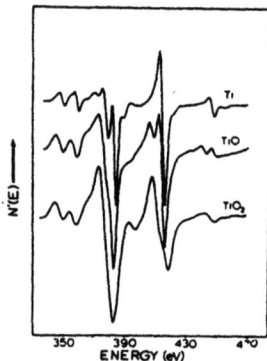

Fig. 11. Titanium Auger derivative spectra of the pure metal, TiO and TiO_2 (After Solomon and Baun, reference 18).

1.3 Surface Structure

The determination of the surface structure is less advanced than the chemical analysis. Two techniques are used for this purpose, Low Energy Electron Diffraction (LEED) and Field Ion Microscopy. The latter is capable of displaying the positions of single atoms but requires samples of a very restrictive geometry, namely fine points with a radius of curvature well below one micron. LEED is practiced on flat samples which must be single crystals or polycrystalline solids with grain sizes larger than 0.1 mm.

A typical LEED apparatus (Fig. 12) consists of an electron gun capable of producing a reasonably monochromatic and well focused beam of electrons with energy varying between zero and about 1000 volts. This beam impinges on the surface and is diffracted by the periodic array of equivalent scatterers. The backscattered electrons travel in a field free space defined by a hemispherical grid and are accelerated onto a phosphor screen where they produce a diffraction pattern. The pattern reveals the symmetry and size of the two dimensional period in the surface structure. By itself it does not reveal the positions of the atoms

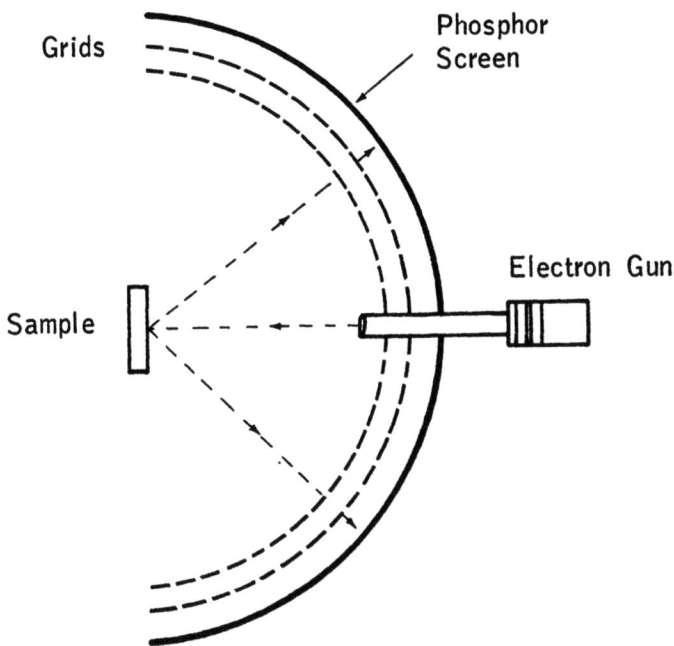

Fig. 12. Schematic of a LEED apparatus.

within this period. More information can be obtained by measuring intensity profiles, i.e. the variation of the intensities of the diffracted beams as a function of the electron energy. Simple kinematic theory familiar from X-ray diffraction, cannot be used to interpret the intensity profiles because of the same phenomenon that accounts for the surface sensitivity of LEED. The scattering cross section for electrons of 10 to 200 eV is so large that multiple scattering and inelastic scattering of electrons play an important role. The theory of LEED is complex and the determination of surface structures has only recently been accomplished with great expense of computer time.[19,20] (See Fig. 13) Most often, especially in the study of the different surface structures obtained with chemisorption, one limits oneself to obtaining the diffraction pattern and proposes a plausible structure with the help of additional (e.g. chemical) information. We have stated that the crystallites of the material under study cannot be much smaller than 0.1 mm in diameter. The reason for this limitation is that LEED is a backscattering technique where the signal is not intense enough for the observation of Debye-Seberrer rings. The sample must therefore be monocrystalline over the size of the electron beam. The latter cannot be focused to a spot smaller than 0.1 mm with the electron energies of interest in LEED (usually

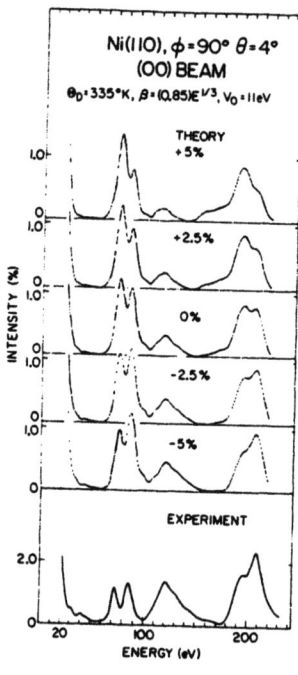

Fig. 13. Intensity of specular (0,0) beam in Low Energy Electron Diffraction from Ni as a function of electron energy. Comparison with calculated intensity curves points to a small (<2%) modification of the interatomic spacing at the surface.

below 200 eV). Since LEED depends on the periodicity of the surface structure it cannot be used to study the structure of defects.

Field Ion Microscopy[21,22] is capable of revealing the positions of individual atoms on a surface; as such then, it can be used to visualize surface defects. Unfortunately, the samples must have the very specialized shape of a fine tip with radius of curvature of less than 1000 Å. Despite this, it has been applied for a variety of metallurgical problems. Surface diffusion,[23] preferred sites of adsorption of single metal atoms, forces between atoms on a surface, radiation damage, grain boundaries are but a few of the phenomena that were investigated by this method.[22] The apparatus is of great simplicity. One places the tip in a vacuum chamber equipped with a large counter electrode covered with a phosphor screen. One admits approximately 10^{-3} torr of an imaging gas (usually involving He, H_2 or Ar) and applies a potential large enough to produce a positive field in the order of 2×10^8 v/cm at the surface of a tip. The gas molecules are ionized in the electric field and the positive ions are accelerated towards the phosphor screen, where they produce the emission of light. The ionizing field varies at the surface of the tip, being stronger above substrate atoms than between them, therefore the ions impinging on the phosphor screen produce in image of the positions of the atoms.

If the applied field is increased further, it removes metal atoms from the surface. This field evaporation is used to clean and shape the tip and also to peel off atom layers from the surface in a controlled fashion to explore the three dimensional structure of features of interest. A fascinating development is the atom probe whereby chosen individual atoms are field-evaporated from the tip, escape through a small hole in the phosphor screen and are collected in a time-of-flight mass spectrometer to reveal their identity. The latter technique however seems too difficult to enjoy widespread use.

Field ion microscopy, besides its obvious limitation of a very specialized sample geometry, is also restricted to metals that possess enough tensile strength to withstand the enormous stresses applied by the field.

1.4 Electron Spectroscopy

It is well known that the cohesive energy of molecules and solids resides in the lowering of the energy of an atom's valence electrons as their shape is changed from atomic to molecular orbital or to itinerant crystal wave function. Therefore, the determination of the energy distribution of valence electrons in the bulk and at the surface of a solid is an important input into the theory of interatomic forces. The discovery[14] that one could measure the energies of valence orbitals of molecules as they are absorbed on surfaces by means of photoemission is thus an exciting new development in surface science. We shall briefly review the different measuring techniques in use and the information they provide. We shall treat, in turn, the electronic spectrum of solids and their surfaces, the spectroscopy of chemisorption and the determination of band bending near the surface of semiconductors and insulators.

Photoemission, as we have already seen, is the emission of electrons due to their excitation by absorption of the photon energy $h\nu$. The experiment consists of shining monochromatic light onto the surface and measuring the energy distributions of the emitted electrons. These energy distributions give some information on the spectrum of the electrons in the solid or on its surface. The interpretation of these measurements is however complicated by optical selection rules, relaxation of the electrons remaining in the solid around the newly created hole and diffraction of the emitted electron by the lattice. At this point, a quantitative interpretation of photoemission data is not possible, but theoretical work is progressing rapidly in this direction. The relative contribution of the surface to the emitted current is determined by the escape depth of the emitted electrons (Fig. 1) since the light penetrates deep into the solid. Low (hν < 10 eV) and very high (hν > 500 eV) energy photons excite electrons probing the "bulk" density of states over a depth of about 10 - 30 Å. The far ultraviolet range hν = 20 to 100 eV is very sensitive to the surface. It has been used to measure the spectrum of surface states on semiconductors and the variation of the electron energy spectrum of solid surfaces (e.g. platinum) with changes in the surface structure.[24] Their most important application is the measurement of the energies of valence electrons of molecules adsorbed on the surface (Fig. 9). In view of the difficulties of interpretation, one is mostly limited to determining the identity of the adsorbed molecules. The coupling of electron spectroscopy measurements on surfaces with the development of the corresponding theories represents perhaps the most exciting frontier of surface physics since it holds the promise of relating chemical properties and interatomic forces to the properties of valence electrons.

Photoemission measures the spectrum of electrons in their ground state. It would be interesting, from theoretical point of view, to determine their excitation energies (or, in other words, the spectrum of their excited, unoccupied, states). Optical measurements are being done on surfaces, but they are difficult because of the large penetration depth of the light. An interesting new development is low energy electron spectroscopy[25] where one measures the energy lost by a beam of incoming electrons in the energy range for which their penetration is small (E ≈ 50–200 eV). These energy losses occur by excitation of the electrons in the solid and can be described in terms of the optical properties of the surface layer.[26]

Two further techniques for surface spectroscopy merit mention because they rely on the tunneling of wavefunctions into vacuum and are therefore most specifically surface sensitive.

In ion neutralization spectroscopy,[27] one bombards the surface with helium ions. These ions are neutralized by electrons extending out of the surface and overlapping the incoming ion. The neutralization of the ion then occurs via an Auger process as shown in Fig. 2. The technique has not seen the widespread use it deserves because the experimental apparatus is somewhat intimidating. Nevertheless, this technique was far ahead of photoemission in observing the electronic energy levels of chemisorbed molecules[28] and it continues to provide important information on electronic properties of surfaces and adsorbed species that lies outside the capabilities of photoemission.[29]

Field emission spectroscopy measures the energy distribution of electrons emitted by the quantum mechanical tunnel effect when a field of 1 to $3 \cdot 10^7$ V/cm is applied to a sharp tip. Since the emission process depends on tunneling of wave functions into vacuum, the technique is exclusively surface sensitive.[30,31] Unfortunately, it only allows the measurement of molecular orbitals lying not more than 2 eV below the Fermi level.

It is quite clear from the above discussion that electron spectroscopy techniques are not developed to a point where they can be used profitably for the study of crystal plasticity. However, they represent the frontier of surface science, and when further developed, could be a very fruitful area of cooperation between surface physicists and materials scientists towards an understanding of surface phenomena in plasticity and corrosion from the most fundamental point of view.

Semiconductors and insulators present a further problem arising from the existence of space charges and the accompanying macroscopic electric fields and potentials. Figure 14 shows the electronic energy structure at the surface of a semiconductor. The bulk of this material is characterized by a completely filled valence band of energies separated from a normally empty conduction band by a bandgap in which no extended electron wave functions can exist. Impurities produce localized shallow donors and acceptors (which are used in semiconductor devices to produce the desired type of conduction) and deep traps. Shallow impurities are close enough to the bands so their occupation by electrons is in thermodynamic equilibrium at room temperature. By contrast, the thermal repopulation of deep traps is extremely slow so that their occupancy is not adequately described by equilibrium (Fermi-Dirac) statistics. This is especially important in insulators. The

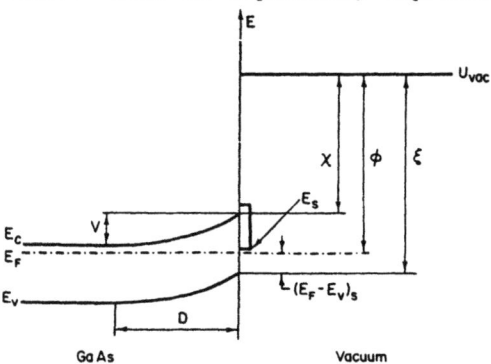

Fig. 14. Energy diagram for the surface and space charge region of an n-type semiconductor. E_F is the Fermi level, E_c the conduction band edge, E_s the lower edge of a band of surface states, U_{vac} the potential energy of electrons in vacuo, V the band bending, D the depth of the space charge layer, χ the electron affinity, ϕ the work function and ξ the ionization energy.

position of the Fermi level (or electrochemical potential) is determined, in the bulk of the semiconductor, by the concentration of impurities. The surface of the solid is often the seat of surface charges that originate in the occupation of intrinsic surface states of the solid or in adsorbed molecules. From electrostatics we know that these charges induce an electric field which penetrates into the solid and is progressively screened off by a space charge near the surface. The potential due to these charges is added to the periodic potential of the crystal atoms and shifts all energy levels and causes the band bending shown in the figure. This macroscopic potential thus changes the position of the band edges with respect to the Fermi level as one approaches the surface. Photoemission provides a convenient measure of the energy splitting between the Fermi level and the top of the valence band on the surface of semiconductors.[32] If a wavelength of light is chosen so that a shallow layer is sampled, one compares the energy distributions of electrons emitted from the semiconductor to that from a metal reference. The highest energy at which the metal can emit is due to excitation from the Fermi level, E_F so that

$$E_{max} = E_F + h\nu$$

From the semiconductor, by contrast, electrons are emitted from the top of the valence band or from surface states. The two can usually be distinguished with the help of adsorption experiments.

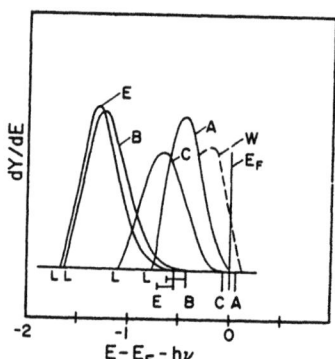

Fig. 15. Energy distributions of photoelectrons from GaAs. Sample A is highly doped p-type, C is moderate p-type, B and E are n-type. The dashed line (W) is the high-energy extremity of an energy distribution from tungsten and serves to fit the Fermi level. The marks below the base line indicate the position of the top of the valence band. Thus, the values of $(E_F - E_V)_s$ of Fig. 13 are -0.07 eV for sample A, 0.07 eV for sample C, .45 eV for B and .55 eV for E. (After Dinan, Galbraith and Fischer, Reference 33).

Comparison of the energy distributions from a metal and a semiconductor usually give a direct measure of E_F-E_V, namely the position of the valence band with respect to the Fermi level. If the corresponding quantity in the bulk is known (from conductivity measurements), one has a measure of band bending. Changes in the band bending due to adsorption are easier to measure than the absolute value: one merely observes the shift of the electron energy distributions on the energy scale.

A different experiment, surface photo voltage, measures the changes in band bending and the time required to reach the equilibrium value (which is especially important in the wide band gap materials represented by insulators). It consists of measuring the contact potential between the semiconductor and a metal reference and shining light onto the former. The absorption of light will cause a change in the population of the valence band, impurity levels and conduction band near the surface. These changes will modify the space and surface charges and the band bending. Application of thermodynamic statistics for equilibrium and of electron-hole recombination theory will usually provide a measure of the band bending. In the case of semiconductors, the latter can also be measured with the help of surface conductivity measurements. More details on these techniques will be provided in the paper of H. Gatos. The determination of electric potentials and fields near the surface is important in view of their influence and the motion of dislocations and defects.

The application of measuring techniques involving electrons and ions to insulators requires special experimental precautions to prevent charging of the material. Otherwise, the energy of emitted electrons is shifted or smeared by the electrostatic potential associated with the charge. The problem is thus more severe where high energy resolution is required and less important for chemical analysis by Auger spectroscopy or ESCA. Some of the corrective techniques include the use of very thin samples or of a compensating, low-energy, electron gas.

Great care must be exercised in interpreting surface effects in the plasticity of insulators with the concepts of solid state physics. The large activation energies for the movement of electrons (3 to 12 eV for interband excitations, several eV for the change of charging of point defects) usually prevents thermodynamic equilibrium from ever being obtained, except at high temperatures. Consequently, phenomena and properties depending on equilibrium statistics, such as band bending and its change by chemisorption, cannot be invoked. In fact, the Fermi level in an insulator most often cannot be defined.

1.5 Theory

The theory of surface and chemisorption phenomena represents a formidable challenge because the simplifying symmetries of the bulk solid and of the molecule are not available. In fact, a successful theory of surfaces and chemisorption must combine the points of view of the interactions between many extended itinerant electrons of the solid with the descriptions of local changes in potential and charge distributions that occur when a single molecule is adsorbed on the surface. In view of these difficulties a comprehensive theory of the electronic structure, interatomic binding energies and equilibrium positions of atoms and their modifications by chemisorption is some way off and most rapid progress will be made through the interplay of theory and experiment.

Three main areas of theoretical activity can be identified, namely the theory of the various surface measuring techniques described above, the calculation of the electronic properties of free surfaces and the theory of chemisorption.

The complex situation described above is being approached from the points of view of the solid,[1] where the adsorbed molecule is a local perturbation, and from the standpoint of molecular theory, where the solid is approximated by a cluster.[34] The latter approach seems to us particularly well suited for exploring the role of the local configuration of atoms with a view towards including the treatment of defects and dislocations. It also seems logical to expect that theory will represent the most efficient interface between the science of crystal plasticity and surface science.

This, then, is a rapid overview of the present state of surface physics which will be presented in more detail by other authors in this conference. The recent activity of surface physics has mostly consisted of developing new methods of investigation and that quantitative information will be forthcoming in the fundamental processes of importance to environmental effects in crystal plasticity.

In the remainder of this paper we shall chose a few examples from the literature of crystal plasticity and analyze how the concepts and methods of surface science could lead to a better understanding of these phenomena.

2. APPLICATIONS OF SURFACE SCIENCE TO ENVIRONMENTAL EFFECTS IN CRYSTAL PLASTICITY

The purpose of this section is to explore the mode of interaction between surface physics and metallurgy. We shall choose three examples of known environmental effects on the strength of materials, try to identify the frontier of knowledge and see how the tools of surface physics described in the first section could be brought to bear on the problem in order to advance our knowledge. This essay, as it predates the Nato Institute, will of necessity be tentative and will merely show that there is ample opportunity for a fruitful collaboration between the two disciplines.

2.1 The Rebinder Effect in Non-Metals

Rebinder[35] discovered in 1928 that the adsorption of appropriate liquids modifies the microhardness of certain minerals. A correlation was first established by Heins and Street[36] and confirmed by Westwood[37] between the surface hardness and the zeta potential. (Fig. 16). When a solid is immersed in a polar liquid, a surface charge can be collected on the solid by the preferential adsorption of ions of a certain sign. Beyond this surface charge there exists, in the liquid, a space charge (which may be of the same or opposite sign as the surface charge depending on the properties of the solid liquid interface). If one displaces the solid, the ions adsorbed as surface charge move as integral part of the solid while the ions in the space charge are sheared away. The zeta potential is the difference between the electrostatic potential at this shear plane and its asymptotic value at large distance from the solid. It is the liquid analogue of the band bending V in the semiconductor (Fig. 14). The correlation measured by Westwood and collaborators, shown in Fig. 17, establishes that those liquids in which the zeta potential is zero induce maximum hardness and cause the most rapid drilling of the solid.

The mechanisms of embrittlement of non-metals by chemisorption have not been established yet, nor do we have an understanding of the correlation between ξ potential and hardness of non-metals. The ξ-correlation itself tends to encourage explanations in terms of electric fields applied to the surface, and penetrating into the solid. One can invoke forces the electric field exerts on charges associated with the dislocations or, as in the case of semiconductors, a change in band bending V (Fig. 14) associated with changes in surface charge due to adsorbates. This change in band-bending could result in transfer of charge to and from dislocations and other defects. In this context two remarks are in order. First, it is important to give due consideration to the difference between semiconductors and insulators. As their names imply, the two classes of solids differ in their ability to allow

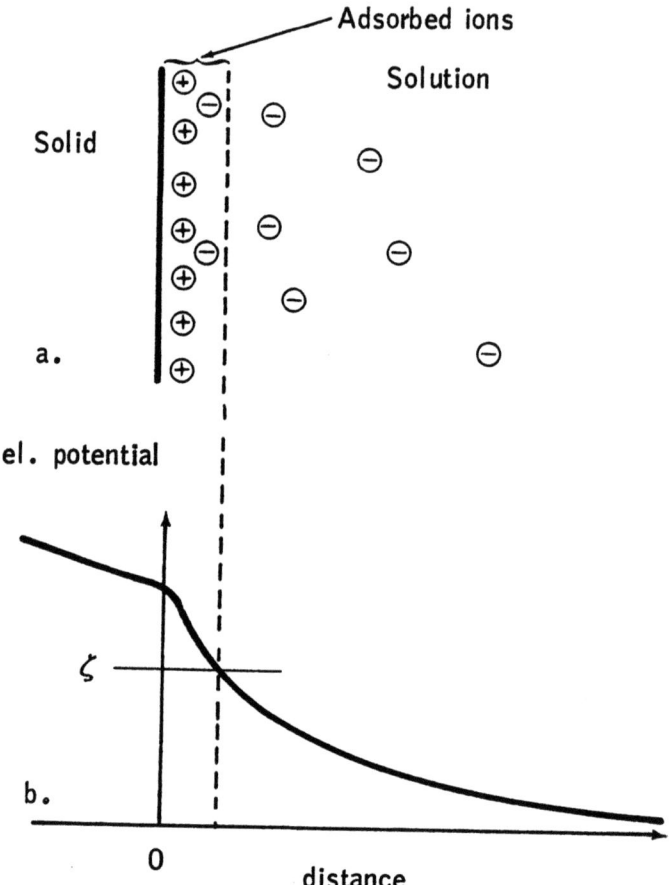

Fig. 16. Definition of the ξ-potential. It is the electrostatic potential due to the mobile space charge in the liquid surrounding the immersed solid.

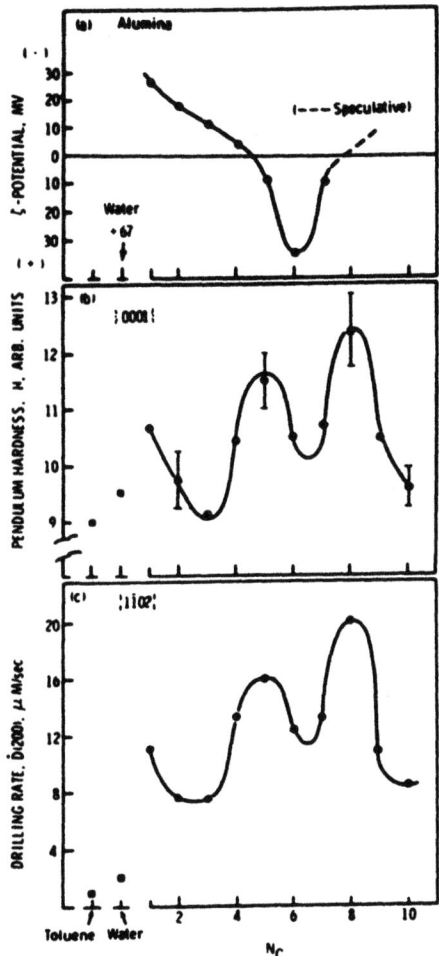

Fig. 17. (a) ξ-potential, (b) pendulum hardness and (c) rate of core bit drilling for alumina in toluene, water and n-alcohol environments. N_c is the number of carbon atoms in the alcohol molecule. (After Westwood, Macmillan and Kalyoncu, Reference 38).

the movement of charges. In the semiconductor, an electric field penetrating into the solid can be screened off within distances usually shorter than 1μ by a space charge which results from the movement of electrons or holes. Still by electrical conduction, the space charge can vary in response to changes in surface charge and localized charges are bound by potential wells shallow enough to permit charge transfers that restore thermodynamic equilibrium, once disturbed, within a few seconds. In an insulator, such movements of charge are impossible, space charge layers near the surface do not exist, or at least do not change in response to variations of the surface charge. Localized charges are usually bound so tightly that, as a rule, they cannot be described in terms of thermodynamic equilibrium (i.e. Fermi-Dirac or Maxwell-Boltzmann) statistics.

The second remark pertains to possible contributions of surface physics to the investigation of the Rebinder effect and the verification of models. Any model that explains the Rebinder effect in terms of changes in electric field in the solid produced by the adsorption of the appropriate liquid can be tested in vacuo. Electric fields can be applied to a solid by alternate methods routinely used in field effect measurements and in semiconductor technology. Modifications of the band bending of semiconductors can also be caused by the absorption of light and so can the recharging of defects. In short, surface physics, especially as described later in this volume by Gatos, provides the means for isolating particular aspects of a model and testing their validity.

2.2 Chemisorption Embrittlement

Of the many explanations of environmental effects on the plasticity of metals, such as stress corrosion cracking, liquid metal embrittlement and the Rebinder effect in metals, the model that describes these mechanical changes in terms of a modified interatomic bond strength by adsorbed molecules[39,40] is particularly attractive from the point of view of surface science. Consider a crack in a metal. If a stress is applied such as to open the crack, two things can occur (Fig. 18), either the material will slip along a slip plane, which will result in blunting of the crack, or the bond between atoms A and A_O will be broken and the crack propagates. The former behavior will result in ductile deformation of the metal and the latter will result in cleavage or catastrophic failure.

The adsorption of molecule or (liquid metal) atom B can result in a decrease of the bond strength A-A_O and can thereby change the response of the solid from ductile deformation to brittle fracture. If we adopt a chemical point of view of this phenomenon, it is easy to see that chemisorption embrittlement, corrosion and catalysis are connected by a common fundamental origin which we shall discuss with the help of Figure 19. The figure depicts schematically the adsorption of a molecule AB from the gas phase onto a metal surface. When the molecule is chemisorbed, the electron orbitals which were responsible for its cohesion and chemical reactivity in the gas phase are modified by the bonding of the molecule to the metal. This same bonding modifies the shape and energy of the valence electrons around metal atoms C and D. As a result, the intramolecular bond strength AB,

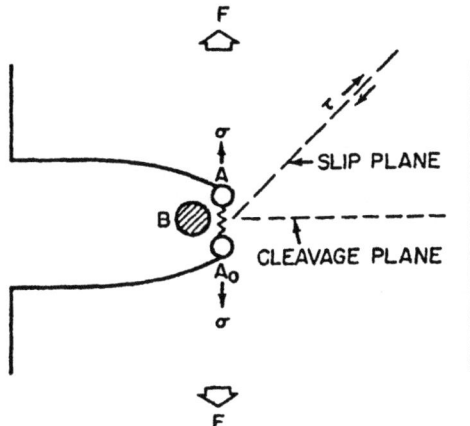

Fig. 18. Schematic of a crack in a solid, subjected to an increasing force F. The bond $A-A_0$ constitutes the crack tip, and B is a surface-active adsorbed atom or molecule. (After Westwood et. al.[41])

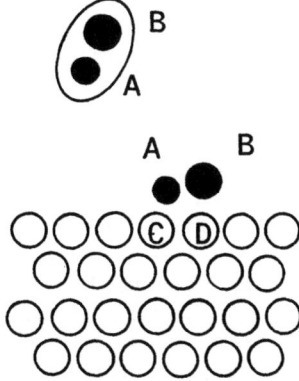

Fig. 19. Schematic representation of the adsorption of a molecule AB onto a metal surface (see text).

the metallic bond strength CD and the chemical reactivity of the molecule AB are modified by the adsorption of the latter on the metal surface. We can distinguish three cases which differ only in degree of these changes.

If the properties of the metal are little affected, but the

reactivity of the adsorbed molecule towards another gas molecule is increased, we have a case of heterogeneous catalysis.[42]

If in the other extreme, the chemical interaction with the metal is so large that molecules AC and BD are formed, with diffusion of these molecules into the solid, or their dissolution in a surrounding fluid, we are in presence of corrosion. Obviously, the formation of a strong inert film of the form AB represents passivation.

The intermediate case where the adsorption of the molecule causes a weakening of the bond CD, represents chemisorption embrittlement (Fig. 18).

From this fundamental standpoint, chemisorption embrittlement is an integral part of the current frontier of surface physics which is otherwise approached with the tools of electron spectroscopy and quantum theory. We therefore expect embrittlement studies to contribute as much to the advancement of surface physics as the latter can contribute to metallurgy.

2.3 Surface Fatigue Wear in Lubricated Bearings

Lubricating oils represent an important but complex application of surface phenomena. Their purpose is to reduce the friction between the moving surfaces in contact and also to minimize their corrosion and wear. Lubricating oils consist of a hydrocarbon base stock and various additives. The formulation of the base stock itself is adapted to the particular use. It is known, for instance that polynuclear aromatics lubricate well in oxidizing atmospheres and lead to severe wear in the absence of oxygen and that, on the contrary, paraffins lead to wear in air but lubricate well in slightly reducing ambients. Mixtures of the two kinds of hydrocarbons perform well in both atmospheres. The addition of fatty acids to the oil presents an interesting case with ramifications in surface science and metallurgy. It is well known that the addition of fatty acids to the base stock dramatically improves the lubrication properties of the oil (i.e. reduces friction) but also shortens the lifetime of the moving parts by promoting surface fatigue. (Acids are thus added to oils where there is little movement and where the reduction of friction is paramount, as in key ways and low speed reciprocating motion; they are carefully removed with the help of basic additives in motor oil, for example). A model of surface fatigue initiated by fatty acids has been proposed by Goldblatt[43] (Fig. 20). Accordingly, aromatic molecules adsorb on freshly abraded surfaces; the adsorbtion is assumed to be accompanied by a transfer of electrons from the metal and the formation of radical anions (Fig. 20a). When a fatty acid is coadsorbed, it reacts with the aromatic radicals, carboxylate anions and a "cationic surface". The

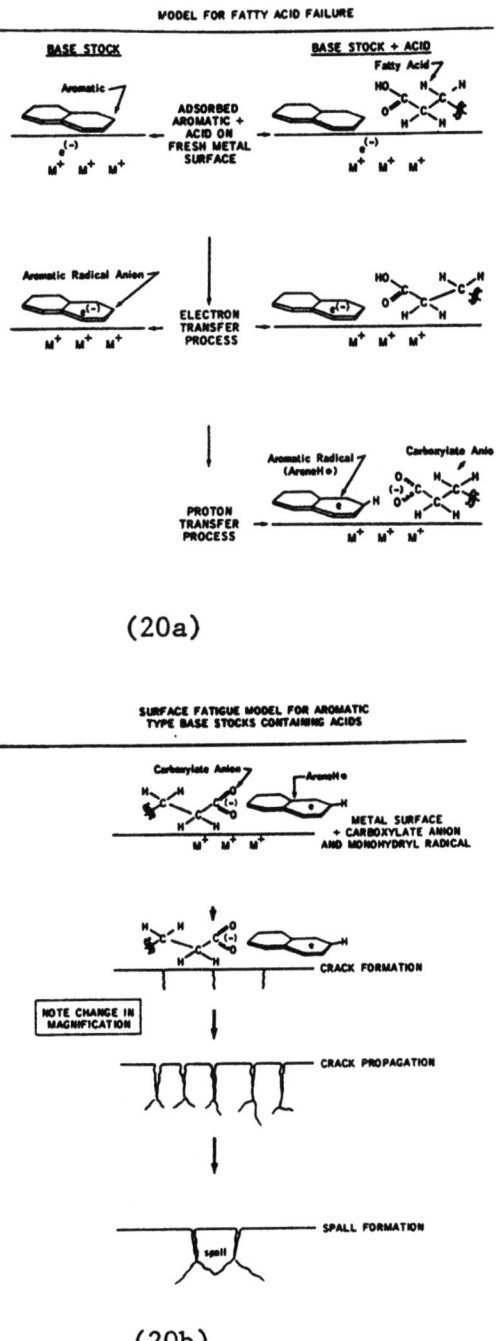

Fig. 20. Model for surface fatigue wear due to fatty acids in lubricant. (After Goldblatt, Reference 43).

carboxylate anion is then assumed to promote the rapid growth of cracks (Fig. 20b) which are present in the metal due to other processes and which are propagated by the cyclic stressing of the surface. Coalescence of the cracks can lead to the formation of large spalls or to wear particles.

This particular model has been selected for discussion because it is accessible to experimental investigation by the methods of surface physics and of metallurgy. For instance, the transfer of electronic charge to the aromatic, which is an essential step in Goldblatt's model and accounts for the fact that fatty acids do not promote wear in purely aliphatic oils, can be checked by measurements of the change in work function and of the electronic spectrum by photoemission (as in Fig. 9). Such measurements were performed for the adsorption of benzene on platinum,[44] for instance, where the decrease of the work function pointed to a charge transfer in the opposite direction from that postulated by Goldblatt. (No corresponding measurements with iron are known to this author). The embrittlement of the surface by the adsorbed species can also be measured by conventional metallurgical methods. The clean surface on which to adsorb the aromatic can be provided by surface science techniques and/or by the propagating crack itself.

3. CONCLUSION

Some of the methods of surface physics described above are already finding widespread use in metallurgy. The determination of the elemental composition of surfaces by Auger Electron Spectroscopy can perhaps already be considered a standard research method for the study of corrosion and fracture. In fact, the electronic differentiation of the spectrum of secondary electrons, which is universally employed to measure the Auger electrons, has been introduced by Harris[4] in order to solve a metallurgical problem.

We have presented an overview of the capabilities of surface physics and have chosen some examples of crystal plasticity to show that the areas of overlap between the two disciplines are not restricted to the measurement of surface compositions. It is true that the latter has the advantage of being relatively simple to perform, that the necessary equipment is commercially available and that the results can be immediately used by the materials scientist. The other methods yield information that requires much more familiarity with surface physics and the electronic properties of solids and are therefore better suited for a true interdisciplinary approach.

REFERENCES

1. Very readable reviews have been published in the April 1975 and October 1975 issues of Physics Today: J. R. Schrieffer and P. Soven, Theory of the Electronic Structure, Physics Today, April 1975, p. 24; J. Estrup, The Geometry of Surface Layers, Physics Today, April 1975, p. 33; D.E. Eastman and M. I. Nathan, Photoelectron Spectroscopy, Physics Today, April 1975, p. 44; R. L. Park, Inner-shell Spectroscopy, Physics Today, April 1975, p. 52; E. W. Plummer, J. W. Gadzuk and D. R. Penn, Vacuum Technology Spectroscopy, Physics Today, April 1975, p. 63 and T. N. Rhodin and D. S. Tong, Structure Analysis of Solid Surfaces, Physics Today, October 1975, p. 23.
2. C. J. Powell, Surface Science, 44, 29 (1974).
3. J. J. Lander, Phys. Rev., 91, 1382 (1953).
4. L. A. Harris, J. Appl. Phys., 39, 1419 (1968).
5. C. C. Chang, Surface Science, 25, 53 (1971); P. W. Palmberg, G. E. Riach, R. E. Weber and N. C. McDonald, Handbook of Auger Electron Spectroscopy, (Physical Electronics Inc., Edina, Minnesota, 1972); Y. E. Stausser and J. J. Uebbing, Varian Chart of Auger Electron Energies, (Varian Vacuum Division, Palo Alto, California).
6. C. R. Helms, J. Catalysis, 36, 114 (1975).
7. P. W. Palmberg and H. L. Marcus, Trans. ASM, 62, 1016 (1969).
8. M. L. Tarng and G. K. Wehner, J. Appl. Phys., 43, 2268 (1972).
9. H. H. Madden and G. Ertl, Surface Science, 35, 211 (1973).
10. S. Thomas, J. Appl. Phys., 45, 161 (1974).
11. J. A. Bearden, Rev. Mod. Phys., 39, 78 (1967); J. A. Bearden and A. F. Burr, Rev. Mod. Phys. 39, 125 (1967).
12. R. F. Goff and D. P. Smith, J. Vac. Sci. Technol., 7, 72 (1970).
13. A. Benninghoven, Z. Physik, 230, 403 (1970).
14. D. E. Eastman and J. K. Cashion, Phys. Rev. Letters, 27, 1520 (1971).
15. J. E. Demuth and D. E. Eastman, Phys. Rev. Lett., 32, 1123 (1974).
16. H. P. Bonzel and T. E. Fischer, Surf. Sci., 51, 213 (1975).
17. G. Schön, Surf. Sci., 35, 96 (1973).
18. J. S. Solomon and W. K. Baun, Surf. Sci., 51, 228 (1975).
19. J. E. Demuth, D. W. Jepsen and P. M. Marcus, Phys. Rev. Lett., 31, 540 (1973); S. Anderson, B. Kasemo, J. B. Pendry and M. A. VanHove, Phys. Rev. Lett., 31, 595 (1973); and G. E. Laramore, Phys. Rev. 9B, 515 (1974).
20. J. E. Demuth, D. W. Jepsen and P. M. Marcus, Phys. Rev., B11, 1460 (1975).
21. E. W. Müller, Z. Physik, 131, 136 (1951).
22. E. W. Müller and T. T. Tsong, Field Ion Microscopy, Principles and Applications (Elsevier, New York, 1969).
23. G. Ehrlich and F. G. Hudda, J. Chem. Phys. 36, 3233 (1962).

24. H. P. Bonzel, C. R. Helms, and S. Kelemen, Phys. Rev. Lett., 35, 1237 (1975).
25. J. E. Rowe, H. Ibach and H. Froitzheim, Surface Science, 48, 44 (1975).
26. D. L. Mills, Surface Science, 48, 59 (1975).
27. H. D. Hagstrum, J. Vac. Sci. Techn., 12, 7 (1975).
28. H. D. Hagstrum, Phys. Rev., 150, 495 (1966).
29. H. D. Hagstrum and G. E. Becker, Phys. Rev., B8, 1592 (1973).
30. J. W. Gadzuk and E. W. Plummer, Rev. Mod. Physics, 45, 487 (1973).
31. W. B. Shepherd, W. T. Peria, Surf. Sci., 38, 461 (1973) and B. F. Lewis and T. E. Fischer, Surf. Sci., 41, 371 (1974).
32. T. E. Fischer, Surf. Sci., 13, 30 (1959).
33. J. H. Dinan, L. K. Galbraith and T. E. Fischer, Surf. Sci., 26, 587 (1971).
34. J. C. Slater and K. H. Johnson, Physics Today, 27, October 1974.
35. P. A. Rebinder, Proc. 6th Physics Conference, Moscow, 29, (1928).
36. R. W. Heins and N. Street, Soc. Pet. Eng. J., 5, 177 (1965).
37. A. R. C. Westwood, This volume.
38. A. R. C. Westwood, N. H. Macmillan and R. S. Kalyoncu, J. Amer. Ceramic Soc., 56, (5) 258-62 (1973).
39. A. R. C. Westwood and M. H. Kamdar, Phil. Mag., 8, 787 (1963).
40. N. S. Stoloff and T. L. Johnston, Acta Met., 11, 251 (1953).
41. A. R. C. Westwood, C. M. Preece and D. L. Goldheim, Molecular Processes on Solid Surfaces, (McGraw-Hill, New York 1971).
42. T. E. Fischer, Physics Today, May 1974.
43. I. L. Goldblatt, ASLE Trans., 16, 150 (1972).
44. J. L. Gland, and G. A. Somorjai, Surface Science 28, 157 (1973).

DISCUSSION

Comment by N. H. Macmillan:

Why should (non-equilibrium) defects such as dislocations not produce additional energy levels in the band gap of an insulator that are close enough to the valence or conduction band to produce thermally stimulated motion of electrons or holes at room temperature?

Reply:

Dr. Schröter will discuss this point. The defects do, indeed, produce energy levels inside the band gap. My contention is that in insulators these levels are too deep in the gap (i.e., energetically too far removed from the bands) to permit charge exchange with the bands in reasonable times (i.e., less than centuries), otherwise these materials would not be insulators in the first place.

It is also useful to consider that not every level close to a band will cause electronic conduction. For instance, a level that is close to the conduction band must be a donor: its chemical nature must be such that when the defect is neutral, the donor level is occupied by an electron which can then be thermally excited into the conduction band. Small perturbations of the lattice would cause an acceptor level near the conduction band and a donor level near the valence band with no increase in conductivity.

Comment by F. C. Frank:

Could I envisage an insulator with semiconductive dislocations in it?

Reply:

One can think of two types of conductivity along a dislocation: ionic and electronic conduction. Ionic conduction is akin to pipe diffusion except that the diffusing particles must be ions and that, of course, the asymmetry of diffusion is provided by the electric field rather than a concentration gradient. Electronic conduction is possible in principle. It is well known that local perturbations of the periodic potential cause the apparition of electronic levels inside the band-gap. Since a dislocation is a one-dimensional, periodic perturbation of the potential, it is well possible that the corresponding electronic states overlap enough to form a one-dimensional energy bands along the dislocation. This would produce a one-dimensional metal, or semiconductor or, again, insulator, depending on the material. Dr. Schröter tells me that electronic conduction along dislocations has been observed is cadmium sulfide. The concept of a conducting dislocation in an insulator leads to a model of chemisorption-induced modification of the hardness on non-metals to be described by Drs. Shchukin and Westwood. I shall propose this model in the Workshop Session.

Comment by R. W. Staehle:

I wish to support Dr. Fischer's suggestion of the applicability of Auger analyses with three examples from work by Dr. Lumsden in our laboratory together with other co-workers. The important effects observed are:

1) Identification of valence states of iron for passive films found in aqueous electrolytes. We have seen distinction amoung Fe^{+3}, Fe^{+2}, and Fe.

2) Identification of stoichiometric regions. In particular, we see the Fe_2O_3 and Fe_3O_4 layers on iron in aqueous electrolytes.

3) Identification of the role of alloy additions. In particular we have shown that the Cr and Mo in iron dissolve from the passive film in aqueous electrolytes according to the predictions of solubility of their ions as a function of the pH of the environment.

TUTORIAL LECTURES ON SURFACE SCIENCE

Session Chairmen: R. A. Oriani and C. B. Duke

ATOMIC GEOMETRY AND ELECTRONIC STRUCTURE OF SOLID SURFACES

C. B. Duke

Webster Research Center, Xerox Corporation
800 Phillips Road, Webster, New York 14580 U.S.A.

ABSTRACT. This article is an extended outline of the text of three lectures on the characterization of solid surfaces. In the first lecture the quantities which specify the state of a surface are defined; their relevance to the mechanical properties of solids is indicated; and the techniques for their measurement are reviewed. The second lecture deals with the crystallography of clean and adsorbate-covered surfaces of crystalline solids. Following a review of the atomic geometry of periodic surface structures and of surface defects, the application of elastic low-energy electron deffraction to determine these structures is described. Our discussion of this topic concludes with a synopsis of the structures which have been determined to date. The final lecture is concerned with the electronic structure of surfaces. We consider the qualitative features of surface potentials and charge densities, and review current calculations and measurements of surface energies, adsorbate binding energies, work functions, localized electronic surface states and surface plasma oscillations.

1. THE CHARACTERIZATION OF SOLID SURFACES

This section is devoted to an exposition of three topics. First, we inquire into the definition of the quantities which characterize a surface. Second, we review some proposed tentative relationships between these quantities and the mechanical properties of thin slabs. Third, we indicate the currently used techniques for surface characterization.

1.1 Definition of a Surface

The analytical definition of a "surface" is given by $f(\vec{r}) = c$ in which $\vec{r}$ is the three-dimensional coordinate vector, c is an arbitrary constant, and f is an arbitrary, nonsingular function. This equation describes the relationship between the three components of the coordinate vectors $\vec{r}$ which terminate on a two-dimensional "surface" of area A embedded in the three-dimensional space. Such a definition is appropriate for the description of macroscopic objects of given shapes and sizes because it is an abrupt change in some physical quantity, usually the density of matter, which leads us to perceive such objects as distinct entities. To describe surfaces on a microscopic scale, however, we must extend our concepts to encompass the consequences of the atomistic nature of matter.

From an atomistic point of view, surfaces are spatial boundary layers between two different phases of matter (for example, solid-gas, liquid-gas). As such, they exhibit a finite thickness which depends on the thermodynamic variables characterizing the two-phase system, the history of the system, and the phenomenom under consideration. In our present discussion, we limit our consideration to "solid-vacuum" interfaces characterized by the removal of the ambient gas in contact with a solid until the residual pressure is $p \sim 10^{-10}$ torr. At these pressures, if every gas atom which strikes the solid sticks to it, the solid would become coated with a monolayer of gas in roughly an hour.

It is useful to characterize in several different ways the quantities which describe a surface [1,2]. First, since essentially all mathematical models of surface properties are based on the adiabatic hypothesis [1] we distinguish between electronic properties (e.g. the work function, ϕ; surface plasmon energy, ($\hbar\omega_s$) and atomistic properties (e.g. atomic positions and vibrations). Second, because models of the various properties exhibit different limitations, we differentiate between ground-state properties (e.g. surface energies), which are difficult either to measure or to calculate reliably, and excitation spectra (e.g. surface phonon [3] or plasmon [2,4] energies) which usually are accessible to direct spectroscopic measurement and which can be evaluated using simpler models [1]. Moreover, these excitation spectra are ultimately responsible for the thermodynamic properties [5], (e.g. the specific surface stress tensor, g_{ij}) of solids [6,7]. Finally, we must contrast [1] those models and phenomena in which the surface simply plays the role of a boundary condition (e.g. space-charge effects in semiconductors and insulators [8]) and those which require a microscopic description of the structure of the upper few atomic layers of the solid (e.g. low-energy electron diffraction [2,9,10] or atomic surface diffusion [11]). Our interest in these lectures is focused upon the experimental techniques and mathematical models required to describe the microscope structure and properties of the uppermost few atomic layers of a solid in the vicinity of one of its interfaces.

A complete microscopic description of a surface involves answering the following questions [2,12]:

1. What atomic species are present on the surface?

2. How are they arranged?

3. How are their valence electrons distributed?

4. What are their atomic motions?

Different experimental measurements are appropriate for responding to each of these questions. Elastic low-energy electron diffraction (ELEED) provides the most satisfactory probe of the atomic geometry [2,9,13-15] and surface vibrational properties [9,16] of macroscopic crystalline samples. Inelastic low-energy electron diffraction (ILEED) is currently the only high-precision, surface-sensitive technique for measuring the energy-momentum relation of surface branches of the electronic excitation spectra of solids [2,17]. Core-level excitation and emission spectroscopies are appropriate for identifying the atomic species present on a surface [18-20]. The results obtained by using all of the various techniques in concert must be fitted together in a consistent fashion in order to characterize adequately a solid surface. We arrive, therefore, at the popular multiple-technique experimental strategy [2,12,19] whereby multiple experimental measurements are utilized in a single ultra-high-vacuum ("UHV", $p \sim 10^{-10}$ torr) chamber to characterize the composition, structure and properties of a solid surface.

A final important concept in the definition of "surface" properties is the recognition that such definitions must be stated in terms of observable quantities associated with specific experimental measurements. Yet, the distance (or "depth") resolution (i.e., "surface sensitivity") of any given experimental probe depends primarily on the nature of the interaction of the probe with the constituents of the solid rather than on externally controllable parameters characteristic of the experiment itself. Electromagnetic radiation in and above the range of visible frequencies penetrates a solid for distances $\lambda_{em} \sim 10^4 \text{Å}$. Static electric fields penetrate metals for $\lambda_m \sim 1$ to 10Å and semiconductors and insulators by $\lambda_{sc} \sim 10^3$ to 10^6 Å. Low energy (10 eV $\lesssim$ E $\lesssim 10^3$ eV) electrons exhibit inelastic collision mean free paths $\lambda_{ee} \sim 2$ to 10Å. Low-energy (E $\lesssim$ 10 eV) atoms (obtained, e.g., from atomic beams) and kiloelectron-volt ions do not penetrate the surface, although mega-electron-volt ions can if the incident beam is aligned along a crystallographic axis of high symmetry. Thus, we see that scattering experiments using the various probes will measure properties of the solid characteristic of different depths below its surface: Another reason why the utilization of multiple techniques is necessary in order to achieve a reasonably complete description of a surface. Moreover, the selection of experimental methods and their corresponding theoretical methodologies for data analysis proceeds according to a series of well-defined principles as described, for example, by Duke [2].

1.2 Relevance of Surface Properties to Mechanical Properties

Having indicated the nature of the quantities which characterize a solid surface, it seems germane to the thrust of this institute to consider which of these quantities are currently believed to bear some relationship to mechanical properties. In this context, two distinctions should be introduced immediately: that between metals and non-metals (glasses, insulators, semiconductors); and that between mechanical deformations the primary effect of which is the generation of new surface area (e.g. the elongation (creep) or fracture of thin rods [21]) and those whose primary consequence is the generation of damage within the sample (e.g. microhardness measurement via indentation [22]).

The most obvious influence of UHV surface characterization on studies of mechanical properties occurs via the determination of the composition and atomic geometry of adsorbed overlayers known to influence mechanical properties [21-23]. Some existing general results already are of direct interest, such as the fact that the low-index ("singular") surfaces of most clean metals do not exhibit expanded layer spacings [2,13-15] in contrast to both early expectations and the hypothesis that surface stresses could be relieved by the accumulation of near-surface dislocations [23]. Moreover, it seem plausible that the atomic composition and geometry of the initial few adsorbed overlayers, in both chemical attack (e.g. oxidation) and epitaxy, will exert appreciable influence on the density and motion of dislocations in the immediate vicinity of the interface. In metals the atomic composition of grain-boundary layers already is well-known to exert a marked influence on mechanical properties [24].

In addition to direct correlations between surface structure and mechanical properties, however, various models have been proposed linking these properties to the thermodynamic variables characterizing surfaces. Specifically, for metals the specific surface work, γ, is thought to be a suitable intermediate variable [21] whereas for insulators the surface charge density, σ, has been proposed for this purpose [22]. While surface thermodynamic parameters are notoriously difficult to measure reliably, models used to predict such quantities as surface energies also may be applied to evaluate surface structure, work functions, thermal atomic displacements and other similar quantities amenable to more direct and reliable experimental test. Thus, surface analytical techniques can provide independent verifications of models commonly used to calculate the thermodynamic variables entering into empirically proposed relations between these variables and observed mechanical properties.

Although modern surface analytical and theoretical methods have much to offer, a few words of caution seem appropriate. First, fatigue and hardness, like catalytic reactions, are defect-dominated phenomena. Moreover, the dislocations of interest in analyses of mechanical properties are extended defects, persisting over distances $d \sim 10^4 - 10^5 \text{Å}$. The conventional technique for determining the atomic geometry at planar surfaces, ELEED, is notoriously insensitive to defects and "yields" the incoherent superposition of (coherent) diffraction signals from extremely small (i.e. $A \sim 10^4 \text{cm}^2$) areas of a surface. Consequently, for studies of surface effects on mechanical properties, its use to determine surface atomic geometry should be supplemented with a high-resolution version of scanning electron microscopy [25] to determine surface topography. Second, the surface analytical methods are, by construction, sensitive to the properties of the uppermost few atomic layers. This fact coupled with their insensitivity to defects, does not render them directly suitable for such such required studies as determining the dislocation density as a function of depth. Therefore in applying these techniques to study surface effects on mechanical properties, some ingenuity (e.g. decorating the dislocations with foreign atoms and ion etching to obtain a depth profile) will have to be employed to design experiments suitable for the resolution of the central conceptual issues in this field. Finally, we should recognize that rather different techniques should be utilized to examine metals (for which the electric fields vanish a few angstroms inside the surface) and insulators (for which electric fields usually penetrate for 10^5-10^7Å inside the surface). Conventional techniques [2,12,18-20] have been designed primarily for metals. Insulator surface characterization, especially to obtain σ, requires other measurements as, e.g., surface photovoltage spectroscopy [26].

1.3 Characterization Techniques: A Survey

In this section we indicate briefly the techniques used to determine the atomic composition, atomic geometry, atomic vibrational motion, and electronic structure of the uppermost few (1-10) atomic layers of a crystalline solid. Different ones are used to determine the properties of the space-charge region in semiconductors and insulators [8,27] as discussed by Prof. Gatos in his lectures [28].

Two entirely different sample geometries have been employed, historically, for surface characterization. In the first, to which we refer as the "tip geometry", the sample is a small, more-or-less hemispherical tip of radius approximately 1000Å. Such tips consist of many atomically flat regions containing up to 100 atoms. If a positive voltage of several thousand volts is applied between a hemispherical screen (a few cm away) and this tip, electrons are field emitted to the screen leading to an image of the tip on the screen: a technique known as field-emission microscopy (FEM) [29,20]. If the energy distribution of these electrons is measured, the method is called field-emission-energy-distribution spectroscopy (FEED) [30]. Alternatively, if the region between the screen and tip is filled with a "imaging" gas (usually H_2, He, or Ne) and a large negative voltage is applied between the screen and tip, the gas atoms are ionized in the vicinity of the tip and yield an image of the tip on the screen: a technique known as Field Ion Microscopy (FIM) [11,29-32]. Individual atoms and complexes can be identified by inserting a drift tube behind a hole in the screen, an instrument known as the atom probe (AP) [32,33]. In all of these techniques based on the tip geometry, the surface is actually imaged on the screen, so that we refer to them as <u>imaging techniques</u>.

The second prominent sample geometry is that of a planar surface of a single crystal solid. The structure of these surfaces is determined by scattering and emission experiments, reviewed in detail elsewhere [2,34]. We call such experiments <u>statistical techniques</u> (as opposed to imaging techniques) because they describe the average properties of the surface area examined. This area usually is determined by the spot size of the incident beam, sometimes reduced by the focusing characteristics of the exit-particle analyzer. It is typically $A \sim 10^{-8} cm^2$ for electron and ion scattering spectroscopies and $A \sim 10^{-2} cm^2$ for photon-based spectroscopies. As in the case of imaging techniques, a wide variety of statistical techniques have been developed for the characterization of planar surfaces. A synopsis of the common examples of both types of techniques is given in Table I.

Planar-surface scattering spectroscopies may be based on the use of either charged (electrons, ions) or neutral (electromagnetic radiation, neutrons, atoms, molecules) particles as the incident or exit species. Although almost all possible incident-exit combinations have been tried [2,12,34,35], only charged-particle spectroscopies are popular both because of the ease in

Table 1: Commonly Used Techniques for Measuring the Chemical, Geometric, Vibrational, and Electronic Properties of Surfaces in Ultrahigh Vacua as reviewed, e.g., by Duke [2,34], Gomer [29], Van Oostrom [32], and Hobson [35].

Physical property	Tip geometry (imaging spectroscopies)	Planar geometry (statistical spectroscopies)
Atomic identity of surface species	Atom probe (AP)	o Auger-electron spectroscopy (AES) o X-ray photoelectron spectroscopy (XPS, ESCA) o Characteristic-loss spectroscopy (CLS) o Ion-scattering spectroscopy (ISS) o Secondary-ion mass spectroscopy (SIMS) o Appearance-potential spectroscopy (APS)
Atomic geometry of surface species	Field ion microscope (FIM)	Elastic low-energy electron diffraction (ELEED)
Surface-atom vibrations, surface diffusion	Field ion microscope	ELEED, CLS, Infrared spectroscopy
Surface electronic excitation spectra	Field emission energy distributions (FEED)	o Ultraviolet photoelectron spectroscopy (UPS) o Inelastic low-energy electron diffraction (ILEED) o Ion neutralization spectroscopy (INS) o AES o APS

focusing and energy-analyzing such projectiles and because of the strong interactions of charged entities with the constituents of solids.

The surface sensitivity of a given scattering experiment is a direct consequence of strong particle-solid interactions experienced by either the incident or exit projectile. Indeed, the strength of such interactions is the distinguishing characteristic between surface (e.g. Table I) and bulk (e.g. x-ray diffraction) spectroscopies [2,34]. Charged particle spectroscopies are based on the observation that the projectile loses appreciable ($\Delta \gtrsim 10$ eV) quantities of energy upon traversing a distance equal to an inelastic-collision mean free path, λ_e. For either electrons or ions, λ_e depends upon the energy of the particle. Typical results for electrons are shown in Fig. 1 [9,36]. If an energy analyzer is utilized to detect only particles scattered elastically (or emitted without experiencing energy loss) from a surface, then these particles must have emanated from within a distance $d \sim \lambda_e$ from that surface. The increase in λ_e at higher energies, evident in Fig. 1, is typical of all charged particle

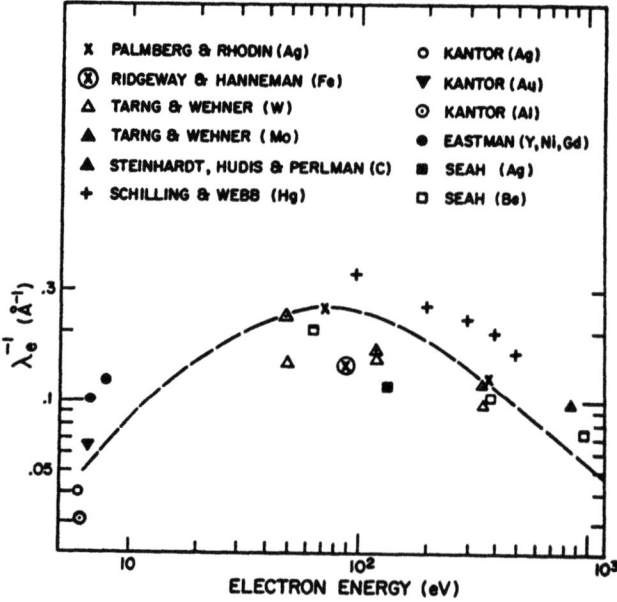

Figure 1. Measured inelastic collision mean free paths for electrons of varying energy. The data points and authors responsible therefor are indicated. The dashed curve is a roughly "universal" extrapolation formula.

spectroscopies because it is a consequence of the decrease as E^{-2} of their scattering cross sections for a given inelastic process [2,34]. The increase in λ_e at low energies ($E \lesssim 25$ eV) is peculiar to electrons, however, since it results from the incident electron being identical to those in the solid. Consequantly, the occupied electronic states in the solid are not available for it to scatter into [34].

The transition from a surface to a bulk measurement with increasing projectile energy may be illustrated by ion scattering pectroscopies [2,34,37]. The stopping power, dE/dx, of a fast heavy charged particle in matter often is described by the Bethe-Bloch equation [38], which, for incident particles of charge ze, mass M, and energy $E \ll Mc^2$ becomes approximately

$$\frac{dE}{dx} = - \frac{1.44 \times 10^6 z^2}{E(eV)} \frac{Z\rho}{A} \frac{M}{M_p} \ln \frac{E}{I} \cdot \frac{m}{M} \frac{eV}{\text{Å}} . \quad (1)$$

For typical targets, $Z\rho/A \sim 1$. Therefore, setting the logarithm equal to unity we obtain the estimate that for proton of energy $E \sim 1$ eV loses its full energy in about 1 Å of typical solids. (For electrons, this would occur when $E \sim 10$ eV if the exclusion principle could be neglected.) Therefore, the elastic backscattering of low-energy protons $E \sim 1$ KeV, is employed to determine the properties of the surface of a solid, whereas higher energy protons, normally with $E \sim 1$ MeV, provide penetrating radiation useful in the determination of the "bulk" composition of thin (1000Å) films [37].

Incident ions of energy $E \lesssim 500$ eV become neutralized in the vicinity of a surface by electrons which tunnel from the filled electronic states in the solid into the empty electronic states ("holes") on the ion. Usually this process is of the Auger type, with the energy lost by the tunneling electron being given to a second electron which may be emitted from the solid. The measurement of the energy distribution of such low-energy ejected electrons is called Ion Neutralization Spectroscopy (INS) and is utilized by Hagstrum to examine the electronic structure of clean surfaces and of adsorbed molecules [39]. Unlike the conventional scattering experiments, however, INS relies on the phenomenon of resonance-tunneling [40] for its sensitivity to the electronic structure of surface species. The electrons in the solid preferentially tunnel into the vacuum outside the surface at the energies associated with the nearly localized orbitals of the surface species. Since the ionization probability is much higher for these initial electrons, the Auger spectrum of emitted electrons exhibits a resonance maximum at energies associated with the eigenstates of the adsorbate-substrate system [39]. Indeed, the dual concepts of resonance tunneling [40] and strong inelastic collision damping [41], alluded to in Fig. 1, form the theoretical bases for the use of electron scattering and emission experiments as surface-structure spectroscopies (See, e.g., [12], [34], and Sec. 3.1 below).

Having articulated the conceptual foundations underlying the selection of scattering experiments suitable for surface (as opposed to bulk) characterization, we conclude this section with a brief dictionary of the terms utilized in Table I. In principle, all of these experiments are extremely simple, embodying a photon, electron or ion source in conjunction with an energy analyzer for the emitted particles. In order to achieve surface sensitivity, either the incident or exit projectile (or both) must be an atom (molecule) or a charged particle (electron or ion), with the use of the latter being considerably more convenient, reliable, and common for surface characterization.

Elements are characterized by the mass and charge of their nuclei. The mass of surface species is determined most easily by the elastic backscattering of "low-energy", $0.5 \text{ KeV} \lesssim E \lesssim 3 \text{ KeV}$, ions: a technique called ion-scattering spectrometry (ISS) [42] or Noble-Gas-Ion Reflection Mass Spectrometry [43]. Alternatively, the incident ion beam may be utilized to sputter away the surface species, the mass of which is determined subsequently in a mass spectrometer. This technique is known as secondary-ion mass spectrometry (SIMS). It has been evaluated relative to other surface analysis techniques several times [44,45] in the past two years, leading to the conclusions that sensitivity (10^{-5} monolayer) and good spatial resolution (1-5μm lateral and 1 monolayer depth) are its outstanding virtues. Incident ions also generate x-rays [46] and optical transitions [47], but these processes are not commonly used for surface analysis. Finally, in the tip geometry a particular atom may be selected in the field ion microscope, stripped off the surface by a pulsed electric field, and identified directly in a special mass spectrometer. The instrument utilized to perform these measurements is called an atom probe (AP) [32,33].

The charge of a surface species is determined via the energy of its tightly-bound ("core") electrons. The most direct measurement of this energy is obtained by observing core electrons emitted from the solid without energy loss under the impact of X-ray photons: a technique referred to as X-ray photoelectron spectroscopy (XPS) or electron-spectroscopy for chemical analysis (ESCA) [19]. Alternatively, the unscattered emitted Auger electrons (Auger-electron spectroscopy (AES) [18]) or X-ray photons (soft-X-ray appearance potential spectroscopy (SXAPS) [20]) created when an electron or photon generated core hole is filled can be detected. Finally, if electrons are utilized to generate the core hole, the energy of this hole may be determined by measuring the energy lost by the incident electron [48,49]; a technique referred to as characteristic loss spectroscopy (CLS)[49] or ionization spectroscopy [48].

Determinations of the geometrical structure of planar crystalline solid surfaces are performed using electron diffraction because the wavelengths for electrons $E \lesssim 1$ KeV are suitable for diffraction studies and the strong electron-solid interaction renders the elastic scattering of electrons by solids a surface-sensitive spectroscopy [2]. The most common experimental configuration [2,34] involves the use of low-energy, $50 \text{ eV} \lesssim E \lesssim 500 \text{ eV}$, electrons incident nearly normal to the crystal and the detection of those electrons reflected elastically back from the surface: a technique known as elastic low-energy electron diffraction (ELEED). An alternative measurement, called reflection-high-energy electron diffraction (RHEED) [12,34] embodies KeV electrons both incident and elastically diffracted at glancing incidence to a surface. This technique is well suited for industrial process control, but unlike the ELEED case, the diffracted RHEED intensities cannot yet be analyzed quantitatively to determine the surface atomic geometry. The most direct determination of surface atomic structure, however, is achieved by use of the field-ion microscope with tip-geometry samples [32,33]. Substantial progress in the analysis of such data has been reported recently [33].

Examination of the temperature dependence of LEED intensities is the primary technque for determining the vibration amplitudes of surface species [2]. In the cases of light adsorbates on heavy substrates [50] and intra-molecular vibrations of either bulk materials or adsorbed species [51] high resolution characteristic loss spectroscopy may be utilized to determine vibration energies. For molecules adsorbed on powders or films, however, infra-red absorption spectroscopy is customarily employed for this purpose [52,53].

If atomic diffusion rather than vibrations at surfaces is the topic of interest, then field-ion microscopy on tip geometry specimens is the useful technique. Direct observation of atomic movement is achieved by repeated imaging of a cooled tip upon which additional atoms have been adsorbed. The tip is heated between observations so that the atoms move about. By virtue of the measurement of the changes in atomic position caused by the atomic motion during the heat cycle, the diffusion constants for surface species as well as the interactions between adatoms can be measured directly [11].

In discussing the electronic structure of solid surfaces, we must delineate precisely the nature of the desired information. As noted earlier, considerations of the mechanical properties of solids tend to invoke <u>ground state</u> surface electronic properties (e.g., surface energies, adsorbate binding energies) as appropriate intermediate variables. Yet neither the techniques for measuring such quantities [5,53] nor the models for their computation [1,54,55] have been developed to the point of providing quantitative as opposed to qualitative information. Rather, recent attention has been devoted primarily to spectroscopic techniques which probe the excitation spectra of the valence electrons of a solid in the vicinity of its surfaces.

Two separate types of spectroscopic measurement are common: those that measure the <u>local</u> electronic structure in the vicinity of a particular kind of atom and those that measure the <u>global</u> electronic excitations of the valence electrons in the vicinity of the surface. Many of the former are based on electronic transitions involving core electrons of the type of atom in question [12,56]. Not only are the energies of these core-electron states influenced by the local valence-electron charge density ("chemical shifts"), but also Auger transitions occur in which one or both Auger electrons emanate from the valence band, and radiative recombination of a valence electron and a core hole may be detected. Thus, the lineshapes in Auger-electron spectroscopy (AES) and appearance-potential spectroscopy (APS) reflect the <u>electronic</u> structure of surface species when the associated electronic transitions involve one or more valence electrons in the solid. The resonance-tunneling principle [40] discussed earlier provides, however, an alternative vehicle for the achievement of sensitivity to the local properties of a particular surface species. This principle was first proposed [40] and verified [30] within the context of field emission energy distribution (FEED) spectroscopy from tip-geometry samples. Nevertheless, it also underlies the use of ion-neutralization spectroscopy (INS)[39] and ultraviolet-photoemission spectroscopy (UPS) [57] to measure the energies of the orbitals of surface chemical complexes.

Global surface excitations of the valence electrons are those characteristic of the surface as a whole rather than of specific elements or chemical species adsorbed thereon. Typical among these are the collective electronic excitations, i.e., "surface plasmons" [2], localized near the surface of a solid. While UPS and CLS can yield a general measure of the energies of such excitations, their energy-momentum relationship (i.e., "dispersion relation") has been measured precisely only by inelastic low-energy electron diffraction (ILEED) [2,17].

In summary, therefore, we have indicated in this section how the two concepts of strong inelastic-collision damping and resonance tunneling conspire to render ion and electron spectroscopies convenient measures of the surface properties of solids. The commonly utilized surface-sensitive characterization techniques based on such charged-particle scattering and emission processes are listed in Table I, and described in more detail in the references cited.

2. SURFACE CRYSTALLOGRAPHY

This section is devoted to a description of the specification of the atomic geometry of the surfaces of clean and adsorbate covered crystalline solids, and to a discussion of the determination of this geometry via elastic low-energy electron diffraction (ELEED). We proceed by first examining, in turn, the structure of ideal clean single-crystal surfaces and of certain defect structures (e.g. ordered adsorbed overlayers) thereon. Then we consider the determination of the symmetry and atomic geometry of periodic surface structures via analyses of ELEED spot-patterns and intensities, respectively.

2.1 The Atomic Geometry of Ideal Surfaces

An <u>ideal</u> (or "atomically flat") surface is defined to be that obtained by cutting a single crystal by a plane and subsequently removing all atoms whose centers lie on one side of this plane. Ideal surfaces are designated by the chemical symbol of the material and Miller indices (hkℓ) of the dividing plane, e.g., Aℓ(100). A detailed discussion of the mathematical specification of such surfaces is given by Nicholas [58]. A thorough tabulation of the atomic geometry of the low-index surfaces of monatomic and diatomic solids crystallizing in the fcc, bcc, hcp, NaCl, and diamond (zinc-blende) lattices is provided in his book [58].

All ideal surfaces are characterized by both translational and point-group symmetry in the plane of the surface [59]. The combined (space-group) symmetry operations restrict the possible surface geometries to five two-dimensional (Bravais) <u>nets</u> of points. The association with each point in this net of a <u>basis</u> of atoms defines the actual crystal surface. These bases extend throughout the crystal in the direction normal to the chosen surface. The unit areas in the five Bravais nets are called <u>unit meshes</u>. A tabulation of the five possible meshes is given by Wood [59]. The specification of both the unit mesh and the elements of the space group results in a unique characterization of the symmetry of an ideal surface. In order to describe its atomic geometry, however, the basis also must be prescribed.

2.2 The Atomic Geometry of Imperfect Surfaces

In this subsection we consider three topics. First, we introduce a convenient classification scheme [2] for cataloging imperfect surfaces. Then, we examine in more detail the simpler types of overlayer and multi-step geometries.

In classifying defect structures, it is convenient to visualize a surface as comprised of close-packed two-dimensional atomic diffraction gratings stacked on top of each other. Therefore these two-dimensional layers constitute the building blocks of a surface just as the atomic scatterers are the units out of which the layers themselves are constructed. An example of this concept is shown in Fig. 2 which illustrates the construction of the (755) face of a monatomic fcc solid out of close-packed (111) layers of atomic scatterers.

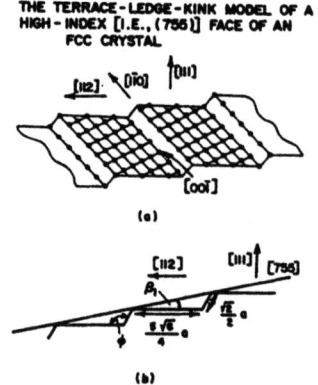

Figure 2. Schematic diagrams of the construction of the (755) surface of a monatomic fcc solid in terms of layers of close-packed (111) planes. Various crystallographic directions are indicated in the figure. The quantity $\underline{a}$ designates the simple-cubic lattice distance of the fcc crystal. Panel (a) indicates the atomic structure of the (755) face whereas in panel (b) the relevant structural parameters are indicated on a cross-sectional view of this face. (after Duke [2].)

This view of surfaces being composed of arrays of close-packed atomic planes is often referred to as the terrace-ledge kink (TLK) model [60]. The flat expanses of the close-packed planes are the "terraces"; the abrupt atomic steps from one terrace to another are the "ledges" (often referred to as "steps"); and both intersections of two ledges and atomic vacancies in the individual ledges form the "kinks". The close-packed faces (i.e., (100) and (111) in fcc and (110) in bcc structures) evidently exhibit a preferred status in this model because all other atomically flat faces are considered to be constructed from individual close-packed atomic layers. The model is most useful for the description of high-index faces (i.e., those inclined at a small angle relative to the (100), (110), and (111) faces) and surface defects. Low-index faces are more easily described by considering them to be composed of two-dimensional atomic layers parallel to the surface.

The TLK model permits us to classify surface structures according to the number and spatial extent of their low-index terraces. Surfaces which consist of such terraces whose spatial extent is large relative to the ELEED coherence zone [$(\Delta x)^2 \sim 10^4 \text{Å}^2$ [2]] are considered to exhibit single-step topography. Those for which the areas of the terraces are small relative to the coherence zone are said to exemplify multi-step topography. By basing this definition on the area of the ELEED coherence zone, we recognize explicitly the instrumental restrictions on ELEED as a technique for surface defect crystallography. Although optical and electron microscope techniques are capable of detecting considerably larger terraces, their resolution horizontal to the surface is less than or comparable to the size of the ELEED coherence zone [2].

Models of single-step topography usually are utilized in ELEED structure analyses. Within this class of models, however, several different types of surface imperfections (i.e., deviations from the ideal surface geometry) can occur. Two of these exist even for atomically-flat, chemically-clean surfaces: imperfections caused by a non-uniform spacing of the upper few atomic layers parallel to the surface and by a non-uniform stacking sequence of these layers. Both of these types of defect structure manifest themselves as alterations in the atomic basis associated with the surface Bravais net. Imperfections caused by alterations in the structure of the surface atomic layers themselves can be classified in one of two ways. Uniform overlayer structures preserve the periodic long-range order parallel to the surface of an atomically-flat single-crystal, although they may reduce the space-group symmetry of this surface (i.e., increase the size of the unit mesh and possibly alter its shape). Such structures can occur via the adsorption of new surface species, the spontaneous rearrangement of surface atoms, or the diffusion of impurities to the surface. Non-uniform overlayer structures arise via the same physical mechanisms, but fail to preserve long-range order parallel to the surface. They can be characterized by the degree of order in the overlayer structure itself.

The model which provides the foundation for our visualization of single-step surface structures consists of an atomically-flat, low-index face of a crystal on which a fraction of a monolayer coverage of atomic or molecular adsorbates has been deposited to form an "overlayer". Typically, the underlying atomically-flat crystal is referred to as the substrate and the overlayer is called the selvedge [59]. Although more general models of planar surfaces have been proposed, few if any of our considerations will depend on either the chemical nature of the overlayer or its thickness. Therefore our discussion describes equally well rearranged low-index surfaces of clean solids, simple monatomic overlayers on such surfaces, and reconstruction of the underlying substrate by the adsorbate.

The most general description of the periodic space-group symmetry of a uniform (ordered) overlayer-substrate system is given in terms of a matrix notation [2]. Consider a substrate whose primitive (i.e., smallest-possible) unit mesh is described by the translation vectors $\underline{a}$ and $\underline{b}$. On this substrate there is an overlayer associated with the unit mesh vectors $\underline{a}_s$ and $\underline{b}_s$. If we fix a set of cartesian unit vectors $\hat{x}$ and $\hat{y}$, then the unit meshes of the substrate and the overlayer may be described by matrices $\underset{\sim}{A}$ and $\underset{\sim}{B}$, respectively, whose matrix elements are determined by

$$\underline{a} = A_{11} \hat{x} + A_{12} \hat{y} \qquad (2a)$$

$$\underline{b} = A_{21} \hat{x} + A_{22} \hat{y}$$

and similarly for $\underset{\sim}{B}$. The relationship between the unit meshes of the overlayer and substrate may be expressed in terms of the matrix transformation $\underset{\sim}{G}$ between them, i.e.,

$$\underset{\sim}{G} = \underset{\sim}{B} \, \underset{\sim}{A}^{-1} . \qquad (3)$$

The determinants of these matrices are designated according to $G \equiv \det | \underset{\sim}{G} |$.

The matrix $\underset{\sim}{G}$ permits us to classify the different possible types of relationship between the substrate and overlayer Bravais nets [2]. If G is an integer the nets are said to be simply related and the combined structure is referred to as simple. If G is a rational number the nets are rationally related and the composite is called a coincidence-site structure. Finally, if G is an irrational number the nets are irrationally related and the structure is designated as <u>incoherent</u>. In the first two of these three cases the combined system is itself characterized by a Bravais net and the overlayer structure is said to be <u>in register</u> with that of the substrate. In all three cases, however, the composite system can be described by a symbol of the form M(hkℓ)-($\underset{\approx}{G}$)-S in which M(hkℓ) designates the (hkℓ) face of a substrate of substance M; $\underset{\approx}{G}$ is defined by Eqs. (2) and (3); and S is the chemical symbol of the overlayer.

Wood [59] uses a more common although less general notation which we designate as the <u>net-quotient</u> overlayer labelling procedure. In the case that the unit mesh vectors of the overlayer can be expressed as rational linear combinations of those of the substrate, we can identify the combined substrate-overlayer system by the symbol

$$M(hkℓ) - \frac{p}{c} \left(\frac{a_s}{a} \times \frac{b_s}{b} \right) \zeta - S \qquad (4)$$

In this symbol M(hkℓ) and S are defined as in the preceding paragraph, p(c) designate primitive (p) or centered (c) unit meshes of the composite system, ((a_s/a) x (b_s/b)) denotes the ratios of the sides of the unit surface mesh vectors to those of the substrate mesh, and ζ is the angle between the overlayer and substrate unit vectors $\underset{\sim}{a_s}$ and $\underset{\sim}{a}$, respectively (which is omitted if it is zero).

Fig. 3 provides an illustration of the use of both notations. If a Na overlayer is deposited on an Ni(100) substrate, then the structure shown in the top panel of this figure is designated as

$$Ni(100) - \begin{pmatrix} 1 & 1 \\ -1 & 1 \end{pmatrix} - Na, \qquad (5a)$$

$$Ni(100) - p(\sqrt{2} \times \sqrt{2})45° - Na, \qquad (5b)$$

$$Ni(100) - c(2 \times 2) - Na. \qquad (5c)$$

Eqs. (5b) and (5c) reveal that the overlayer structure shown in Fig. 3 can be regarded as either a primitive or centered unit mesh.

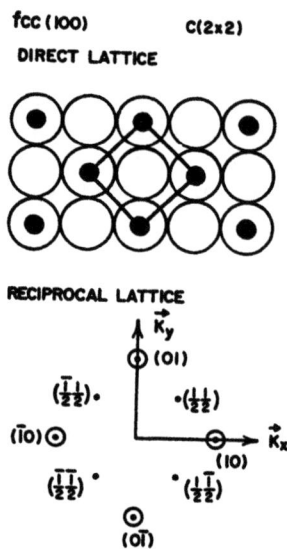

Fig. 3. Schematic indication of the space group symmetry (top panel) and its associated elastic-low-energy electron diffraction intensity pattern (lower panel) for the adsorbate structure commonly referred to as "c(2x2)" on the (100) face of an f.c.c. crystal. These structures typically appear at half-monolayer coverage when, as indicated in the figure, one adsorbate species (small heavy circles) is associated with two atoms (large open circles) in the substrate. (After Duke [2].)

Turning to our discussion of non-uniform single-step imperfections, we identify the major cause of such structures as the occurence of disorder in the overlayer atomic positions for an arbitrary temperature and coverage. Only at low temperatures and for coverages which are rational fractions of a monolayer are uniform overlayer structures to be expected. We proceed by giving examples of the common types of non-uniform single-step structures.

The overlayer and substrate structures are said to be in
<u>translational register</u> when the Bravais net of the composite
system is invariant under translations by prescribed multiples
of the unit mesh vectors of both the overlayer and the substrate.
Often, however, the species in the overlayer may coalesce into
islands or <u>domains</u>, each of which is in register with the substrate,
but which are not separated from each other by the translation
vectors of the overlayer unit mesh. In this case, examples of
which are illustrated in Fig. 4, we refer to them as <u>out-of-phase</u>
<u>domains</u>.

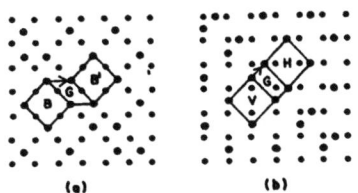

Fig. 4. Illustration of out-of-phase c(2x2) domains on the
(100) face of an fcc or bcc crystal. The direct nets for
the c(2x2) structure are indicated. Small and large dots
represent substrate and overlayer sites respectively. In
panel (a) the overlayer atoms are located in positions of
maximum 4-fold coordination. Unit meshes B and B', representative of two overlayer domains, are connected by the
unit mesh G. In panel (b) the overlayer atoms are located
in nearest neighbor 2-fold bridged positions: vertical
bridged positions (V) and horizontal ones (H). The two
overlayer domains are connected by a unit mesh G. (After
Duke [2].)

Evidently the vector connecting the domains B and B' as well as
that linking V and H does not belong to the overlayer translational symmetry group. Park and Houston [61] have formalized
the description of such domains by introducing the concept of
equivalence. Two domains are said to be related by <u>equivalent</u>
tranformations if the atomic positions in one can be mapped into
those in the other by an element of the overlayer symmetry group.
All other transformations relating two domains are called <u>in-</u>
<u>equivalent</u>. The <u>registry degeneracy</u> of a given overlayer structure
is defined to be the minimal number of non-equivalent transformations required to relate all possible domains consisting of this
structure on a perfect substrate. Thus, the registry degeneracy
is a measure of the number of different composite overlayer-
substrate structures which can be constructed from a given overlayer structure, substrate structure, and set of symmetry-

equivalent positions of the overlayer species in the substrate unit cell.

In addition to translational registry degeneracy, similar degeneracies occur because domains occur which are not related to each other by the elements of the point group of the overlayer. The examples of rotation and reflection registry degeneracy have been discussed by Duke [2].

We now turn to our final topic in this subsection: multistep topographies. The simplest of these are the high index crystal faces whose mathematical description has been developed by Nicholas [58]. They are uniquely and completely described by the Miller indices (hkℓ) of the surface plane. An example of this description is illustrated in Fig. 2. Lang et.al. [62], introduced a rather cumbersome notation for high-index surfaces based on (h k ℓ) terraces in which they are designated by M(S) -[m($h_1k_1\ell_1$)xn($h_2k_2\ell_2$)] for a crystal designated by the chemical symbol M. The index m is the number of atomic rows in the ($h_1k_1\ell_1$) terraces and n is the number of atomic layers in the ($h_2k_2\ell_2$) ledges. Thus a (755) surface of Ni such as that shown in panel (a) of Fig. 2 would be designated by Ni(S) - [2(111) x (100)]. Unfortunately, the absorption of foreign species on high-index faces is a complicated topic. If the area of the low-index terraces is large and the unit mesh of the overlayer is small, then the overlayer structures may be those characteristic of the low index terraces. If, in addition, out-of-phase domains are common, the notation of Lang et.al. may prove appropriate. In the case that size of the overlayer mesh is comparable to or larger than that of the terraces, however, we recover the situation described in the preceding discussion, i.e., the structure of the overlayer is described best in terms of atomic planes parallel to the substrate surface.

Since high-index surfaces are energetically unstable[60], they are likely to be imperfect in which case it is necessary to specify statistically the parameters describing the terraces and the ledges. Such a description also is required for disordered low-index crystal faces. In these cases the surface topography is prescribed by the Miller indices of the close-packed terrace faces and the combined distribution of terrace widths, lengths, and orientations and of ledge heights. Several simplified models in which the terraces are taken to exhibit a random distribution of sizes and orientations or, more generally, this distribution is taken to be separable from that of their heights have been considered in the literature [2]. Determinations of average step heights and widths have been based on kinematical analyses of elastic low-energy electron diffraction intensities from high-index surfaces. No serious attempt has yet been made, however, to utilize ELEED as a technique for the quantitative determination of statistical multi-step topographies.

2.3 Determination of Surface Space Group Symmetry via ELEED: Spot Pattern Analysis

Any consideration of surface structure determination via ELEED must begin with a discussion of the type of exeriment required to extract the various types of structural information. Specifically, two types of measurement are common in the literature [2]: those of intensity ("spot") patterns and those of intensity profiles as indicated schematically in Fig. 5.

EXPERIMENTALLY MEASURED QUANTITIES IN
ELASTIC LOW-ENERGY ELECTRON DIFFRACTION
(ELEED)

1. INTENSITY PATTERN: FIXED INCIDENT BEAM ENERGY

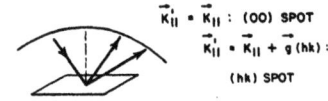

$\vec{K}'_{\|} \cdot \vec{K}_{\|} : (00)$ SPOT

$\vec{K}'_{\|} \cdot \vec{K}_{\|} + \vec{g}(hk) :$

(hk) SPOT

2. INTENSITY PROFILE: ENERGY DEPENDENCE OF THE INTENSITY OF A GIVEN SPOT

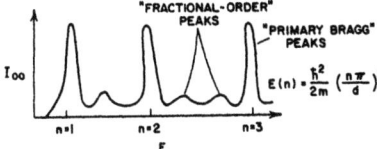

$E(n) = \frac{\hbar^2}{2m} \left(\frac{n\pi}{d}\right)^2$

Fig. 5. Illustration of the distinction between an intensity pattern and an intensity profile. In both cases, the detector is biased relative to the gun in such a fashion that only the electrons quasi-elastically scattered from the solid (i.e., $E' \cong E$) are collected. The diffracted beams are labeled by Miller indices of the reciprocal lattice vectors of the Bravais net for translational symmetry parallel to the surface as defined in Fig. 6 (After Duke [2].)

In this subsection we describe the study of the configuration of the diffracted beams (i.e. the intensity pattern) to extract the translational symmetry of the surface. In the following subsection we discuss the analysis of the diffracted intensities, usually but not necessarily obtained in the form of intensity profiles, to determine the atomic geometry of surfaces (i.e. the structure of the basis associated with the unit mesh deduced from the spot pattern).

The nature of electron diffraction from crystals can be understood most simply by regarding a crystal as comprised of geometrically equivalent layers of atoms parallel to a given surface. An incident electron of wavevector k (i.e. momentum $p = \hbar k$) is diffracted from the array of these two-dimensional gratings stacked together to form the crystal [2]. The periodic translational symmetry of these diffraction gratings is manifested in the electron-solid scattering cross sections by virtue of the reflected electrons emerging from the solid in a series of diffracted beams as indicated in Fig. 6.

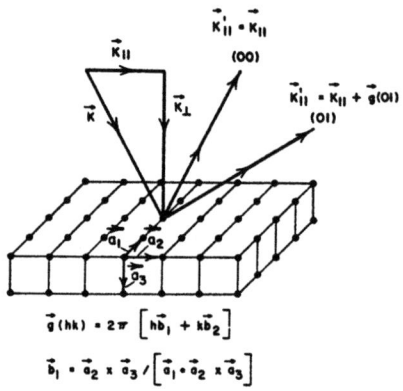

Fig. 6. Schematic illustration of an incident electron beam of wavevector $\vec{k} = \vec{k}_\perp + \vec{k}_\parallel$ diffracted from a single crystal surface into the state $\vec{k}' = \vec{k}'_\perp + \vec{k}'_\parallel$. The construction of the reciprocal lattice associated with the single-crystal surface also is shown. The vectors g(hk) designate the reciprocal lattice vectors associated with the lowest-symmetry Bravais net parallel to the surface.

Low-energy electron diffraction (LEED) is the study of the electrons coherently scattered from the solid into these beams. Both elastically and inelastically scattered electrons are diffracted by the periodic potential of the solid [2], but we confine our attention here to elastic low-energy electron diffraction (ELEED). In this case, the configuration of the diffracted beams relative to the beam of incident electrons may be predicted from the translational symmetry of the system which leads to the momentum conversation law

$$\underline{k}'_\parallel = \underline{k}_\parallel + \underline{g}\ (hk) \tag{6a}$$

$$k_\parallel = (2mE/\hbar^2)^{1/2} \sin\theta \tag{6b}$$

for electrons of energy E incident on a planar surface at an angle θ relative to its exterior normal. The vectors g are the reciprocal lattice vectors associated with one of the five possible two-dimensional Bravais nets with their associated unit meshes.

Examination of Eqs. (6) and Fig. 6 reveals that the only information conveyed by the configuration of the diffracted beams is the translational symmetry parallel to the surface of the two-dimensional atomic diffraction gratings which comprise the solid. Determination of the unit-cell structure within each of these gratings and of the packing sequence of the gratings relative to each other requires analysis of the diffracted intensities. The deduction of possible translational symmetries of adsorbate-solid systems from measurements of the configuration of the diffracted electron beams has been a standard surface-characterization experiment since Davisson and Germer's initiation of the field of "LEED" in 1927 [63]. Only in the past few years, however, has the theory of the electron-solid diffraction process been developed to the point that an analysis of the <u>intensities</u> of the diffracted beams is feasible. It is the construction of this theory and its application to extract surface structural parameters from measured intensities that is the major new development (1968-1973) in the field of ELEED which has permitted this technique to assume its present place as the routine surface structure-determination spectroscopy when the atomic geometry as well as symmetry of the surface is desired [2,15].

Each of the five possible unit meshes associated with an allowed surface space-group symmetry is associated with a reciprocal lattice (see, e.g. Fig. 6) which may be constructed using the matrix notation introduced earlier. If we use Eq. (2) to define the matrix $\underline{A}$ which describes a given surface unit mesh, then the associated reciprocal lattice is comprised of translations by multiples of the vectors

$$\underline{a}^* = A^*_{11}\ \hat{x} + A^*_{12}\ \hat{y} \tag{7a}$$

$$\underline{b}^* = A^*_{21}\ \hat{x} + A^*_{22}\ \hat{y}\ . \tag{7b}$$

The matrix $\underset{\sim}{A}^*$ is defined by

$$\underset{\sim}{A}(\underset{\sim}{A}^*)_t = \underset{\sim}{1} \qquad (8)$$

in which a subscript t indicates the transpose. The vectors $\underset{\sim}{g}$ in the reciprocal lattice are specified by

$$\underset{\sim}{g}_{hk} \equiv 2\pi(h\underset{\sim}{a}^* + k\underset{\sim}{b}^*). \qquad (9)$$

Using this notation, the construction of the reciprocal lattice associated with a given Bravais net is elementary. The net vectors can be written as

$$\underset{\sim}{a} = a_o (s\hat{x} - t\hat{y}) \qquad (10a)$$

$$\underset{\sim}{b} = a_o r\hat{y} \qquad (10b)$$

in which a_o is a unit of length. We find from Eqs. (7)-(10) that

$$\underset{\approx}{A} = a_o \begin{pmatrix} s & -t \\ o & r \end{pmatrix} \qquad (11a)$$

$$\underset{\approx}{A}^* = (rsa_o)^{-1} \begin{pmatrix} r & o \\ t & s \end{pmatrix}. \qquad (11b)$$

Eqs. (11) constitute the relationships which permit us to infer the lattice unit mesh (the $\underset{\approx}{A}$ tensor) from the observed reciprocal lattice vectors (the $\underset{\approx}{A}^*$ tensor). For example, in the case of normally-incident electrons, the symmetry of the intensity pattern (Fig. 6) is identical to that of the reciprocal lattice of a uniform substrate-overlayer target system. Moreover, since in this case the sine of the exit polar angle is given by

$$\sin \Theta'_{\underset{\sim}{g}} = |\underset{\sim}{g}|/(2mE/\hbar^2)^{1/2}, \qquad (12)$$

the slopes of plots of $\sin^{-2}\theta'_g$ versus energy yield a direct determination of the lengths of $\underline{a}$ and $\underline{b}$ via Eqs.(10)-(11). Therefore for all surface structures exhibiting long range order parallel to the surface the unit mesh vectors in principle can be constructed by inspection from the intensity pattern provided domain degeneracies can be neglected. The practice of this art is described in several recent reviews [64,65]. Unfortunately, in the absence of an analysis of the magnitudes of the scattered intensities we can say no more. In general many atomic geometries are consistent with a given Bravais net.

2.4 Determination of the Surface Atomic Geometry via ELEED: Intensity Analysis.

On the basis of the considerations set forth in Section 1, we might anticipate serious difficulties in constructing models of the electron-solid diffraction process which are adequate for structure analysis. Specifically, since a _strong_ electron-solid interaction is required to establish the surface sensitivity of ELEED, we expect theoretical difficulties on two fronts. From the technical perspective, strong interactions imply the failure of perturbation theory, in particular linear response theory. Thus, to construct a proper description of the diffraction process, the full apparatus of renormalized quantum field theory must be invoked [66-70], although for practical structural analysis suitable simplified computer algorithms are available [2]. Second, from a physical point of view, the strong electron-solid interactions imply difficulties in separating the geometrical structure of a surface from its electronic and vibrational structure [71]. As an analogy, consider trying to determine the crystal structure of Si and GaAs from their energy-band structures as determined from optical and transport measurements. Consequently, some labor is required to establish that any given model of electron-solid interactions is adequate to determine atomic geometries in the face of uncertainties in the vibrational and electronic structure of solid surfaces.

Two general hypotheses underlie almost [70] all models of ELEED from solids. The first of these is the neglect of the presence of surface defects. In particular, the surface is regarded as being a periodic planar array of scatterers consisting of a selvedge of specified symmetry on an otherwise ideal single-crystal substrate. We refer to such a visualization of the surface structure as the _uniform-overlayer_ _model_. In this model the determination of the contents of the unit cells is reduced to that of the structure of the selvedge layers and the registry of these layers relative to the substrate. Moreover, we regard the selvedge itself as comprised of individual monatomic planes

of identical symmetry parallel to the surface. The unit-cell
structure of the selvedge, therefore, is expressed by using
multiple "subplanes" of identical scatterers, displaced relative
to each other but all exhibiting the space-group symmetry of the
selvedge as a whole.

The second general hypothesis is that fairly simple models
of electron-solid interactions, taken over from bulk solid-state
physics, suffice for surface structure determination [2,15].
To establish the validity of such models, extensive sensitivity
analyses of ELEED from aluminum, nickel and copper have been
performed [2,15]. These analyses suggest three conclusions of
major importance in studies of surface crystallography. First,
simple overlapping-atomic-potential models seem adequate for the
purpose of surface-structure determination to within $\Delta a \sim 0.1 \text{\AA}$.
Consequently, the electronic structure of surface scatterers
need not be known in detail in order to determine their geometrical structure: an extremely significant result. Second, crude
empirical models of the electron-solid optical potential suffice
for applications to surface crystallography provided that only
the lineshapes of intensity profiles (plots of scattered intensity
versus incident beam energy) are examined and that this examination is carried out for angles of incidence $\theta \lesssim 20°$. Third, for
the reliable separation of vibrational and geometrical effects
in a structure determination, low-temperature ($T \ll 150°K$) intensity data should be used in the analysis. Thus, studies of ELEED
from clean metals have both established the validity of a renormalized quantum-field-theoretic multiple-scattering formalism
for the description of the ELEED process and identified three
important procedural results permitting the use of simple model
force laws within the context of this formalism for the determination of surface atomic geometries.

Having demonstrated [2] the general adequacy of simple models
based on the planar surface hypothesis, the use of overlapping
electron-ion-core potentials to describe the scattering of the
incident electron from the constituents of the solid, and the
utilization of uniform optical potentials to describe its interaction with the valence electrons in the solid, we can turn to
the issue of devising methodologies for the use of such models
for surface structure determination. Two fundamentally different
approaches have been proposed to determine the atomic geometry
of the surfaces of crystalline solids via the analysis of ELEED
intensities. The objective of <u>microscopic-model</u> <u>approaches</u> is
the development of a model of electron-ELEED intensities from
solid surfaces [2,13-16]. This approach utilizes a complete
specification of the electron-solid interaction subject to the
relaxations of this requirement permitted by the sensitivity
analyses discussed above [2]. The alternatives to microscopic-model approaches, referred to as <u>data-reduction</u> <u>approaches</u>, are

based on exploitations of the fact that despite the occurrence of strong dynamical phenomena, both model calculations and ELEED intensity data exhibit unmistakable and universal residual manifestations of the purely geometrical conditions for intensity maxima which characterize the single-scattering Born-approximation ("kinematical") analyses. By manipulating observed ELEED intensities to enhance these geometrical effects while smoothing or ignoring the dynamical ones, analytical procedures based on kinematical concepts can be used for structure determination. The general features of both methods have been reviewed in detail elsewhere [2,9,13-16].

The selection of the method to be used in a particular structure determination is based on both technical and physical considerations. At the present time, computational limitations restrict microscopic-model structure analyses to cases characterized by small unit cells in each layer parallel to the surface (i.e., those containing four atoms or less). The use of data-reduction methods, however, is limited by physical constraints. Whereas these methods clearly are applicable in the high-energy region (150eV $\lesssim$ E), the surface sensitivity of ELEED diminishes in this region due to increases in the inelastic-collision damping length (λ_{ee}) and decreases in the atomic elastic scattering cross sections caused by thermal vibrations of the ion-cores. Consequently, the desire that the kinematical model be applicable for the analysis of measured ELEED intensities tends to be mutually exclusive with the requirement that the ELEED technique be surface sensitive: an observation which is evidently more restrictive for adsorbed overlayers than for undistorted clean surfaces. Also, the data-reduction methods require a considerably larger body of intensity data than the microscopic-model techniques. Given these considerations, the atomic geometries of small (e.g., low-coverage) overlayer structures currently are inferred from dynamical, multiple-scattering analyses of the intensities of the diffracted beams [2,13-16]. Nevertheless, the geometry of more complex (high-coverage) coincidence-lattice structures can be determined by suitable data-averaging procedures [9,72]. Indeed, the first example of the analysis of the atomic geometry of an adsorbed overlayer was that of such a coincidence lattice: Rh(100)-c(2x8)-O [72]. Consequently, by a judicious choice of analytical techniques it is possible to attempt an analysis of even fairly complex overlayer structures.

For the purposes of these lectures, it seems appropriate to conclude our discussion of surface atomic geometry with an indication of what has been learned to date about this subject. Specifically, we consider three topics. Since the vast majority of structural analyses have been performed for the low index faces of metals, we begin by summarizing our current picture of clean metal surfaces. We then proceed to consider the situation for clean surfaces of insulators and semiconductors. Finally, we conclude with a discussion of adsorbed overlayer structures.

The first important result to emerge from ELEED spot-pattern studies of the low-index surfaces of clean metals is that their translational symmetry parallel to the surface is almost always identical to that in the bulk [1,2,15]. The only known (possible) exceptions are the (100) faces of Au, Pt, and Ir [1] and the (110) faces of Pt [73] and Ir [74]. The former exhibit close-packed hexagonal uppermost layers whereas the latter are characterized by (2x1) structures. Intensity analyses revealed, however, a greater surprise: The uppermost layer spacings are either nearly equal to their bulk value (to within 0.1Å) or else contracted on the (110) face of fcc metals. These effects were first obtained by Duke and coworkers [75,76] for aluminum, and later confirmed on aluminum, copper, and nickel by several groups [2,15,77,78] and extended to (bcc) tungsten [79] and (hcp) Zn [80]. Thus, in sharp contrast to numerous early predictions of <u>expanded</u> upper layer spacings based on static two-body force law models [1,81] the layer spacings as well as translational symmetries of the high-index faces of clean metal surfaces are found to correspond surprisingly closely to their bulk values for close-packed faces and contracted rather than expanded for the more open faces. Both of these results are believed to be consistent with more accurate analyses of the effect of the electronic structure of a metal surface on its atomic geometry [82,83].

In contrast to observation of nearly "bulk" atomic surface geometries for the low-index faces of metals, analyses of the temperature dependence of the ELEED intensities leads to the conclusion that the rms surface vibrational amplitudes, $<u_s^2>_T$, are two to four times their bulk values [16]. Such values are predicted by two-body force law models for expanded surfaces [1,16,84]. We see, therefore, that such models cannot seem to describe the observations both of large increases in $<u_s^2>_T$ relative to the bulk values and of either undistorted or contracted upper layer spacings. Consequently, as discussed elsewhere [1], the use of two-body force law models to describe surface atomic dynamics is a procedure of dubious validity.

Turning our attention from metals to semiconductors and insulators, an important pattern emerges. Early work [64,85] indicated that the space-group symmetry of most "clean" surfaces of non-metallic solids was lower that than of corresponding bulk atomic layers. Indeed, the literature of 1958-70 is filled with attempts to construct models of the atomic geometry of these "reconstructed" surfaces which are consistent with the observed (complicated) ELEED spot patterns [64]. As vacuum technology improved, however, Auger electron spectroscopy increasingly came to be used together with ELEED to characterize surfaces. This combination of techniques revealed immediately that many of the reconstructed surface structures are impurity stabilized [86], and hence associated with adsorbed overlayers rather than clean surfaces. Therefore reports of "clean" reconstructed surface structures prior to 1970-71 should be regarded with caution.

More recent studies [86-89], suggest that whereas the low-index clean surfaces of mon-atomic semiconductors (diamond, Si, Ge) exhibit reconstruction, those of insulating solids with any non-zero degree of ionicity (including, e.g., InSb, GaAs) tend to exhibit the same space-group symmetry as their bulk counterparts. The only exceptions to this general result are the clean polar surfaces of zincblende and wurtzite binary semiconductors, which can exhibit reconstruction under appropriate thermal treatments [87-89]. We conclude, therefore, that the Madelung energy of the surface layers stabilizes the bulk atomic geometry at the surfaces of polar semiconductors and insulators. Atoms at the surfaces of purely covalent semiconductors rearrange themselves in order to maximize the energy of their local chemical bonding. Because of long range electrostatic forces, however, the atoms at the non-polar surfaces of polar semiconductors are stabilized in their bulk configuration in spite of the failure of their chemical bonds to be saturated. This stabilization is more subtle at the polar faces, and favors domain formation or possibly reconstruction of these faces.

Only two intensity analyses of ELEED from non-metallic solids have been reported to date. In both cases (LiF(100) [90] and ZnO($000\bar{1}$), (0001) [91]) distortions of the layer spacings from their values in the bulk appear likely, but the intensity data are inadequate to draw a definitive conclusion. Activity in this area seems likely to grow rapidly.

The development of surface-structure analysis for adsorbed overlayers exhibits interesting parallels to that for clean surfaces. Early studies of alterations in the spot patterns of metal surfaces upon the adsorption of reactive gases (especially O on Ni) usually were interpreted in terms of extensive surface atomic reconstruction, e.g. the formation of surface oxides, carbides, and sulfides [64]. Perhaps surprisingly, the recent intensive burst of activity in ELEED intensity analyses [2,13-15,79, 92-94] of adsorbed overlayers of O, S, Se and Te on the low-index faces of metals has revealed that in the UHV environment reconstruction almost never occurs except possibly under extreme thermal treatments. Instead, even reactive gases sit in the hollows of the metal surface at distances above it determined by the nature of the substrate-overlayer chemical bonds. The only case to date in which a reconstructed overlayer seems favored is Cu(100)-c(2x2)-O [14,95], although Ni(100)-c(2x2)-O is a possible candidate [13,14]. Moreover, analogous results are obtained for metallic overlayers in those cases for which intensity data have been taken and analyzed [2,15].

The picture of chemisorbed structures which emerges from these ELEED studies can best be described in terms of three generic regions in a coverage-temperature phase diagram. At low (i.e. less than or equal to about a half a monolayer) coverages and temperatures (T $\lesssim$ 300°K), ordered overlayer structures occur for those coverages (e.g., 1/4 or 1/2 monolayer coverage on (100) faces) for which overlayer-substrate registry is possible when the adsorbate atoms sit in the hollows of the clean surface in 2-dimensional periodic arrays. As the coverage is increased at modest temperatures, all of the hollows can fill up (giving a (1x1) structure) or the overlayer can form its own structure "on top" of the substrate (coincidence lattices). Alternatively, heating an adsorbate-covered surface or depositing reactive species at high temperatures often leads to compound-formation and surface reconstruction. This process can occur at very low coverages (e.g. less than 1/4 monolayer) but the resulting structures are incoherent just as in the case of adsorption "on top". If compounds form, however, the adsorbate will not be mobile. Evidently, just as in the solid state quite complex coverage-temperature phase diagrams can occur. The present state-of-art in ELEED intensity analysis permits the systematic determination of the atomic geometries associated with the various regions in such diagrams. Yet even for the most extensively-studied system, Ni(100)-S, the phase diagram is not yet available because of the lack of suitable intensity data taken under a wide variety of preparatory conditions. Clearly, the taking and analyzing of such data is emerging as the major frontier in surface chemistry. The resulting phase diagrams are necessary input data to quantitative analyses of the influence of chemisorption on the mechanical properties of solids.

3. THE ELECTRONIC STRUCTURE OF SURFACES

In this section we examine three aspects of the electronic structure of surfaces. First, we discuss the qualitative features of the ground-state valence electron charge density and the effective one-electron potential in the vicinity of a clean surface. Second, we consider the stability of surfaces, i.e. the calculation of surface energies and adsorbate binding energies. Finally, we describe the measurement and interpretation of surface features of the electronic excitation spectra. Particularly interesting single-electron excitations include those involving one or more electronic states localized near the surface ("surface states"). New features of the excitation spectrum appear when we consider particle-hole excitations. Of these we focus our attention on the collective surface electronic excitations ("surface plasmons"). We conclude with a brief discussion of extrinsic (i.e. adsorbate induced) surface states.

3.1 Ground-State Charge Density and One-Electron Potential

Perhaps the most important concept underlying an understanding of the ground-state electronic properties near a surface is that the bulk solid is a stable, self-bound system. Thus, the valence electrons would be localized inside the solid even in the _absence_ of surface effects. For example, the work functions of most metals are about 4.5 ± 0.5eV with fluctuations of at most about ±0.5eV from one crystal face to another. Consequently, we must develop models of surface electronic structure in which specifically surface effects play a minor role relative to the bulk cohesive energy. Since, in "simple" (e.g. alkali) metals the electron-electron interactions among nearly planewave valence electrons dominate the bulk cohesive energy, it is convenient to discuss models suitable in this limiting case [96]. The extension of these ideas to more realistic models has been described by Lang [54].

The consequences of the electron-electron interactions in simple metals may be examined using a "jellium" model in which the lattice potential is replaced by the potential due to a uniform positive charge in each unit cell of the metal. The magnitude of this charge is taken to make the metallic system electrically neutral. The Schrodinger Equation for bulk jellium is given by

$$(H_{ee} - \bar{V} - E_T)\Psi = 0 \tag{13a}$$

$$\bar{V} \equiv \frac{1}{2} \int d^3r \, d^3r' \, \frac{e^2}{|\underline{r}-\underline{r}'|} \tag{13b}$$

For a uniform electron fluid, Eqs. (13) can be solved in the Hartree-Fock (HF) approximation. The ground-state energy at zero temperature is given by [96].

$$E_T = E_{HF} + E_{corr} \tag{14a}$$

$$E_{HF} = \sum_{\underline{k}} \left[\frac{\hbar^2 k^2}{2m} + \frac{V(k)}{2} \right] \theta(k_F - k) \tag{14b}$$

$$V(k) = -\frac{e^2 k_F}{\pi}\left[1 + \frac{k^2-k_F^2}{2kk_F} \cdot \ln\left(\frac{k_F-k}{k_F+k}\right)\right] \qquad (14c)$$

$$k_F = (3\pi^2 n)^{1/3} \qquad (14d)$$

In Eqs. (14) n denotes the number of electrons per cm^3 and E_{corr} denotes the correlation energy of the uniform electron fluid [96]. An important feature of Eqs. (14) is that the potential energy of an electron with wave number k depends explicitly on k. In particular, the "inner-potential", $V(k)$, becomes less negative as k (i.e., the energy of the electron) increases.

Conceptually, the most important aspect of bulk jellium is that it forms a self-bound system. The separation energy, S_n, is defined to be the energy released by the removal of an electron [together with its associated uniform positive background] from the system:

$$S_n \equiv \partial E_T/\partial N \qquad (15)$$

in which E_T is the ground-state energy of the N-electron system. In the Hartree-Fock approximation, the separation energy is given by

$$S_n = \frac{\hbar^2 k_F^2}{2m} + V(k_F) \qquad (16)$$

When $S_n < 0$, energy is required to remove an electron from the system.

A schematic illustration of the consequences of the self-binding character of the electron fluid on the potential distribution at a metal-vacuum interface is shown in Fig. 7.

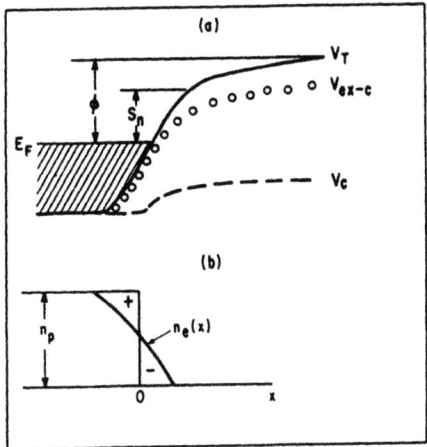

Fig. 7. (a) Schematic diagram of the self-consistent potential energy of an electron at the Fermi surface in a semi-infinite half-plane of electron fluid (jellium). The total potential is denoted by V_T, the exchange-correlation potential V_{ex-c}, and the Hartree potential by V_c.
(b) The self-consistent electronic charge density, $n_e(x)$, at a jellium-vacuum interface. The interface is specified by the step discontinuity in $n_p(x)$ at $x = 0$. (After Duke[96]).

The interface is specified by a discontinuous drop of the positive background from $n_p(x)=n$ to $n_p(x)=0$ along the [infinite] plane x=0. If the electron fluid at the surface were constrained to equal its bulk value, i.e. $n_e(x) \equiv n_p(x)$, then the work function ϕ shown in Fig. 7 would be equal to $-S_n$. At a "real" interface, however, the electronic charge density at the surface is not frozen at its uniform-fluid value. The electronic states in the semi-infinite fluid respond to the potential drop shown in Fig. 7a, and exhibit evanescent components for $x > 0$ due to tunneling into the vacuum. Therefore the observed work function has a contribution to the potential, $V_c(x)$, shown in Fig. 7a. $V_c(x)$ is determined by Poisson's equation

$$\frac{d^2 V_c}{dx^2} = 4\pi e^2 [n_p(x) - n_e(x)] . \tag{17}$$

The electron density, $n_e(x)$ in Eq. (17) is obtained by solving the one-electron Schrödinger equation

$$\left[\frac{-\hbar^2}{2m}\frac{d^2}{dx^2} + V_T(x) - E_x\right]\Psi_{E_x}(x) = 0 \tag{18a}$$

$$\Psi(\underline{r}) = \frac{e^{i\underline{k}_\parallel \cdot \underline{\rho}}}{2\pi}\Psi_{E_x}(x) \tag{18b}$$

$$E = \hbar^2 k_\parallel^2/2m + E_x \tag{18c}$$

in which $\underline{\rho}$ is a vector parallel to the plane of the interface. The one-electron potential is evaluated in a self-consistent fashion as the sum of two terms:

$$V_T(x) = V_c(x) + V_{ex-c}[n_e(x)]. \tag{19}$$

The coulomb term, $V_c(x)$, is calculated from Eq. (17) as a functional of the electron density, $n_e(x)$. In the "local-density approximation" [54,96], the exchange-correlation potential, $V_{ex-c}(x)$, also is written as the value it would assume in a <u>uniform</u> electron fluid of density $n_e(x)$, and then computing $n_e(x)$, self-consistently via

$$n_e(x) = \frac{(2s+1)}{2\pi}\frac{m}{\hbar^2}\sum_i \int_0^{E_F} \left|\Psi_{E_x}^{(i)}(x)\right|^2 (E_F - E_x)dE_x. \tag{20}$$

The "i" labels the two degenerate solutions to Eqs. (18) in a doubly infinite continuum, $s = 1/2$ is the electron spin quantum number, and E_F is the Fermi energy. Fig. 7 illustrates the first solution to equations (17)-(20), as obtained by Bennett and Duke [96]. These calculations have been extended to greater numerical accuracy and to include the electron-ion-core interactions by Lang and Kohn [54]. Evidently, the valence electron interactions with the ion-cores in the surface region differ from one crystal face to another, giving rise to the explicit dependence of the electronic surface potential (and hence work function) on the surface atomic geometry. A critical review of the experimental determinations of work functions has been given by Riviére [97].

Two additional concepts are required to understand electron spectroscopic measurements from surfaces. First, the effective one-electron "optical" potential acting upon an electron near a surface depends upon the electron's energy and angle of incidence [98]. Moreover, it is non-local and becomes complex at electron energies $E > E_F$, reflecting the fact that an incident electron can lose energy by exciting lattice vibrations or electron-hole excitations of the solid [16,34,41,98]. Second, the potential of an adsorbed species can dramatically influence the one-electron potential in its vicinity, leading to resonant emission of electrons through adsorbed complexes [30,40,99]. A schematic indication of the modified one-electron potential in the vicinity of an adsorbed complex is shown in panel (a) in Fig. 8.

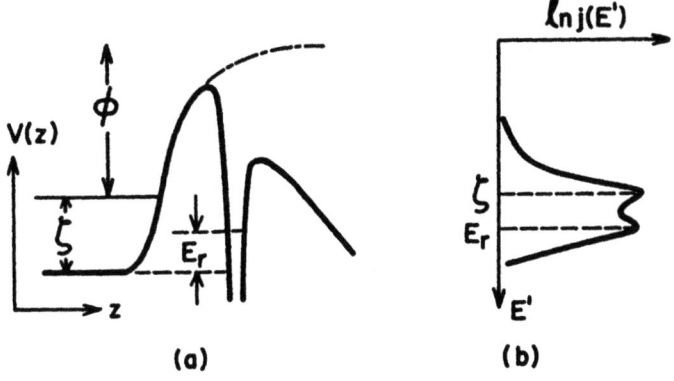

Fig. 8. Schematic indication of the potential energy [panel (a)] and isothermal field emission energy distribution (panel (b)] of an adsorbed species on a metal surface. The adsorbate exhibits a resonant energy level at E_r in the presence of the metal and the external field. Tunneling from the metal occurs preferentially at this energy because the barrier penetration probability is greatly enhanced for $E \cong E_r$.

Panel (b) of this figure displays the additional tunneling resonance in the field-emission energy distribution predicted by Duke and Alferieff [40] to occur at the energy, E_r, of a quasi-stationary resonance state of the surface species. These resonances subsequently were observed [30] and now constitute a routine probe of the electronic structure of adsorbed complexes [99]. Their existence reveals, quite directly, that the electronic orbitals of surface chemical complexes do not, in general, lose their identity because of interactions with the valence electrons of the solid. Rather, they simply acquire a lifetime, $\tau \sim 10^{-13}$-10^{-14} sec., associated with the possibility of the exchange of an orbital electron in the complex with a valence electron in the solid.

3.2 Surface Energies

The surface energies of condensed phases are difficult either to measure or to calculate reliably. In fact, even the definition of surface "energy" is itself a source of confusion in the literature [5], our convention being to refer to the partial derivative of the Helmholz free energy with respect to area at constant volume, temperature, and particle number as the "surface energy", γ, a quantity otherwise known as the specific surface work [5] or surface tension [100]. This rather loose notation is correct for clean surfaces at zero temperature, in which case the surface tension and Helmholtz free energy per unit area are equal. Our discussion is confined to models of γ appropriate in this limiting case. Reviews of experimental techniques from which this quantity may be inferred are given elsewhere [5,100]. One should be cautioned, however, that few measurements have been performed under suitable ultra-high vacuum conditions, and the values of γ are extremely sensitive to adsorbed species.

Calculations of surface energies have been carried out using at least four qualitatively distinct types of models. We consider the results obtained with each in turn.

The simplest description of the surface energy is achieved using the broken-bond model [101] in which the energy required to form a surface is the sum of the energies in the pariwise bonds broken when the surface is generated. This model is convenient for examining the anisotropy in γ from one crystal face to another, although reasonable interpretations of existing data for γ have not yet been achieved [102].

The pair-potential model is an obvious extension of the broken-bond model in which the "atoms" in a solid are taken to interact by pairwise interactions. Using such a model, one can calculate both the equilibrium geometry and the energy of solid surfaces [5,81,100,102]. As noted earlier, these models predict surface geometries in poor agreement with ELEED surface structures, and there is no reliable data on surface energies with which to

compare their predictions [5,100]. One does not, however, expect
these models to be useful for metals and covalent solids because
of the explicit dependence of the effective interatomic interaction on the electron density in the vicinity of the interacting
pair [1,82,83].

Recognizing the necessity of an electronic theory of surface
energies, the simplest approach is the extension of the local-density formalism of Eqs. (17)-(20) to encompass this property.
Lang and Kohn [103] carried out this calculation in a self-evident fashion by evaluating the difference in ground-state
energies of a slab of jellium after and before it is divided into
two equal parts. The lattice structure is treated as a perturbation on the results obtained using the jellium model. Some insight can be achieved by noting that the surface energy for
jellium can be written as the sum of three terms

$$\gamma = \gamma_s + \gamma_{ex-c} + \gamma_c . \qquad (21)$$

The γ_s term is that due to the change in the kinetic energy of
the electrons when the slab is divided, γ_{ex-c} is that associated
with their change in potential energy (exchange and correlation),
and γ_c is that caused by the classical dipole at the surface.
γ_s is large and negative because after splitting, the valence
electrons tunnel into the vacuum at each new interface thereby
expanding the effective volume of the self-consistent potential
constraining these electrons and hence lowering their kinetic
energies. γ_{ex-c} is positive and usually larger than γ_s because
the electrons in the vicinity of one of the newly created surfaces
lose their attractive exchange-correlation interactions with those
in the other half of the initial slab. (Indeed, $\gamma_{ex-c} + \gamma_s > 0$
is a stability criterion for a self-bound system.) Finally, γ_c
is always positive because the electrons in both slabs do work
in creating the dipole layer at the surface. Thus, $\gamma_{ex-c} + \gamma_s > 0$
automatically for internally consistent models, and γ_c merely
adds some extra surface energy

$$\gamma_c = \frac{1}{2} \int_{-\infty}^{\infty} dx \, V_c(x) [n_e(x) - n_p(x)] \qquad (22)$$

which increases rapidly with increasing metallic density (i.e.,
more charge in the dipole layer). The quantitative details of
these calculations are reviewed by Lang [54]. Typically,

for metals,

$$\gamma \cong [56(r_s-6)^2 + 100] \text{erg/cm}^2 \qquad (23a)$$

$$r_s = (3\pi a_B^3 n/4)^{1/3} \qquad (23b)$$

in which $a_B = 5.27 \times 10^{-9}$ cm is the Bohr radius, and n is the valence electron density.

A fourth (and final) model of surface energies is obtained by recognizing that in a rough sense ("rough" because some extra integrals and other formula modifications occur in the complete expressions [104]) the ground state energy can be written as an appropriate sum of the energies of the various particle-hole excitation energies (i.e., electron "normal modes") of the system. Since when a new surface is introduced all of these excitation energies change, the surface energy is simply a sum over the resulting shifts in excitation energy.

As an elementary example, consider two dielectric continua, described by the dynamic dielectric functions $\varepsilon_1(\omega)$ and $\varepsilon_2(\omega)$, respectively. A simple model of $\varepsilon_i(\omega)$ is

$$\varepsilon_i(\omega) = 1 + \omega_{pi}^2 / [\Delta_i^2 - \omega^2] \qquad (24a)$$

$$\omega_{pi}^2 = 4\pi n_i e^2 / m \qquad (24b)$$

$$E_{gi} = \hbar \Delta_i \qquad (24c)$$

in which ω_{pi} is the bulk plasma frequency of the medium labeled by i and E_{gi} is the energy-gap associated with optical transitions in this medium (i.e., $E_g = 0$ signifies a metal in this model). The surface energy of the interface between these media is [105]:

$$\gamma = \int \frac{d^2 k_{11}}{(2\pi)^2} \int_0^\infty \frac{\hbar d\omega}{4\pi} \ln \left\{ \frac{[\varepsilon_1(i\omega) + \varepsilon_2(i\omega)]^2}{4\varepsilon_1(i\omega)\varepsilon_2(i\omega)} \right\}. \qquad (25)$$

This formula admits an elementary interpretation. Surface plasmons, i.e., collective electrodynamic excitation localized at the interface, occur at frequencies Ω_s defined by

$$\varepsilon_1(\Omega_s) + \varepsilon_2(\Omega_s) = 0 . \tag{26a}$$

Bulk plasma oscillations exist if

$$\varepsilon_i(\omega_{bi}) = 0 . \tag{26b}$$

The transverse optical excitations occur for frequencies

$$\varepsilon_i^{-1}(\omega_{ti}) = 0 . \tag{26c}$$

Thus, $\omega_{bi} = \omega_{pi}$ and $\omega_{ti} = \Delta_i$ for the model given by Eqs. (24). A formula about meromorphic functions [105] indicates that the integral over ω in Eq. (25) yields

$$\gamma = \frac{1}{4} \int \frac{d^2 k_{11}}{(2\pi)^2} \left\{ 2\sum_s \hbar\Omega_s - \hbar\omega_{b1} - \hbar\omega_{b2} - \hbar\omega_{t1} - \hbar\omega_{t2} \right\} \tag{27}$$

in which the first term is the total energy of the surface plasmons, the second that of the bulk plasmons of the components, and the third that of the transverse excitations of the components. We interpret Eq. (27) as indicating that when an interface between two dielectric media is formed, we get extra surface energy from the (new) surface plasmons (Ω_s) while losing the contributions to the cohesive energy from bulk plasmons at $k_\perp = 0$ (ω_{bi}) and transverse excitations at $k_\perp = 0(\omega_{ti})$. Early estimates [106] led to the recognition that for a metal-vacuum surface, Eq. (27) can lead to

$$\gamma = cr_s^{-5/2} \tag{28}$$

in tolerable agreement with Eq. (23a) over the range of metallic densities $2 \leq r_s \leq 5$.

This sort of analysis precipitated a heated controversy in the literature, because collective-mode contributions to the surface-energy are not obviously included in the local-density model. The resolution of this controversy is that neither model is quite correct. The local density model is numerically inaccurate [105] and the non-locality of more accurate dielectric functions [108] conspire to render the collective plasmon contributions to the self energy smaller than those of single-particle excitations. The most complete models to date, however, are built on extensions of Eq. (25) to encompass the non-local response of a quantum dielectric medium [107].

The calculation of the binding energy of adsorbed species is even more challenging than that of obtaining the surface energy of a "clean" surface. In essence, this calculation requires the evaluation of the <u>change</u> in the cohesive energy of the adsorbate-substrate system as a function of the adsorbate's position and orientation above the surface [109],(i.e., the adsorption potential-energy diagram). It is, of course, a hopeless task to attempt quantitative evaluation of such potential energy diagrams, but numerous semi-empirical schemes have been proposed, ranging from simple pairwise interaction models [110] to modern techniques in quantum chemistry [111-113]. While qualitative, the semi-empirical potential-energy diagrams often are helpful in examining the nature (e.g., molecular versus disassociative chemisorption) of an adsorption reaction. The field is in a state of rapid flux, so that several modern reviews [111-114] are available for consultation.

3.3 Valence Electron Excitation Spectra

A direct measure of the electronic structure of surfaces is the eigenvalue spectrum of valence electron states which have appreciable wave functions in the vicinity of a surface. A common combination of spatial and energetic information is embodied in the concept of a <u>local density of states</u> (LDOS):

$$n(\vec{r},E) = \sum_i | \psi_{E_i}(\vec{r}) | \delta(E-E_i) \qquad (29)$$

in which $\psi_E(\vec{r})$ designates a one-electron valence electron eigenfunction. In the case that $\vec{r}$ lies in the surface region $n(\vec{r},E)$ is a measure of the surface valence electron charge density associated with eigenstates having eigenvalues $E_i \cong E$. Thus the evaluation of $n(\vec{r},E)$ is a standard model computation in studies of the electronic structure of surfaces [55].

Experimentally, it is possible only to measure the energy spectrum of valence electrons emitted from a surface in response to various external perturbations; external electric fields (field emission energy distributions [30,40]), incident photons (photoemission spectroscopy [34,55]), incident electrons (secondary emission or characteristic loss spectroscopy [2,34], inelastic low-energy electron diffraction [2,17]), or incident ions (ion-neutralization spectroscopy [34,39]). The spatial origin of emitted (or scattered) electrons is determined by the electron-solid interaction. Consequently, the local density of states is <u>not</u> an observable quantity, and spatial information can be extracted from valence electron emission spectroscopies only by inference via a model of the emission process.

A final important concept is that of a <u>surface state</u>. This is a solution, $\psi_E(\vec{r})$, to the effective one-electron Schrodinger equation which is localized in the vicinity of a surface. The most obvious example of such a state is a olution to Schrodinger's equation for an insulator-vacuum interface which yields a bound-state wave function whose energy eigenvalue lies within the insulator's bulk energy band gap so that the wave function decays away from the surface both within the insulator and with the vacuum. The possible existence of such <u>intrinsic</u> surface states on clean insulator and semiconductor crystals is an old idea [115,116], but only within the past few years has photoemission spectroscopy [117-120], characteristic loss spectroscopy [121,122] and surface photovoltage spectroscopy [26] provided direct evidence for their existence. This subject has been reviewed often [123,124]. The concept of surface states also has been extended to metallic surfaces as well [125], although direct evidence for these surface states seems less firm [126-129].

Thus far we have considered surface states only in the one-electron ("single-particle") exictation spectrum. They also occur, however, in the "particle-hole" spectrum associated with excited electrons which crudely speaking may be regarded as "bound" to the holes which they vacated. The best-known examples of such surface states are the surface plasma oscillations [See, e.g., Eq. (26a)] associated with a charge density which is localized on the surface and electric fields which decay as $\exp(-k_\parallel |x|)$ for plasma oscillations of wave-vector $\vec{k}_\parallel$ and energy $\hbar\Omega_s(k_\parallel)$. These oscillations are consequences of the rapid change in dielectric response across an interface. The existence criteria, electrodynamic fields, and consequences of these "collective" surface excitations recently have been reviewed by several authors [3,4]. The major contribution of electron spectroscopy in this area has been the determination via inelastic low-energy diffraction of the energy-momentum relationship, i.e., Ω_S vs. $k_\parallel$ or "dispersion relation", of surface plasma oscillations on Aℓ(111) [17] and Aℓ(100) [130]. The results of these measurements suggest that tht electronic charge density on Aℓ(100) drops off less rapidly than that on Aℓ(111), as one would expect because Aℓ(100) is a more open face. Other types of collective particle-hole excitations, e.g., literally bound electron-hole pairs called "excitons", also exhibit surface branches in their excitation spectra which, however, have not yet been studied extensively using modern surface spectroscopies.

Although we have confined our discussion to intrinsic (i.e., clean-surface) surface excitations, it is evident that the adsorption of different chemical species on a surface also can lead to electronic states localized in the vicinity of the adsorbates. It is convenient in this context to introduce the concept of a <u>surface resonance</u> [40] which consists of a localized electronic state of the adsorbate (or of a complex which it forms with surface atoms) weakly interacting with the valence electrons of the bulk solid. An indication of the potential appropriate in such a case already has been illustrated in Fig. 8. The wave functions of these resonance states are not strictly localized at the surface, but $n(\vec{r},E)$ exhibits a sharp maximum in the vicinity of the adsorbate orbital [30,40,57]. Similarly, collective particle-hole excitations occur which are characteristic of the adsorbate rather than substrate [131]. Electron scattering and emission experiments are particularly sensitive to these surface resonances both because of inelastic collision damping and because the associated transitions have a high probability (oscillator strength) in energy regions for which few bulk transitions occur. This fact makes the observation of adsorbate resonance emission (either field-emission [30], photoemission [57], or ion-induced Auger emission [39]) a convenient direct probe of the electronic structure of surface complexes, although a detailed theory of the photoemission and ion-induced Auger emission has not yet been given.

4. SYNOPSIS

In these three lectures we surveyed the characterization in ultra-high vacuum of the surfaces of solids. In particular, we focused our attention on the specfication of the atomic geometry and electronic structure of crystalline solid surfaces. Such studies are germane to investigations of the mechanical properties of solids because surface energies and charge densities are strongly influenced by the geometrical and electrochemical structure of the surface in question. Moreover, such studies permit a direct examination of the nature of the epitaxial growth of one material on another [132], the formation of surface alloys [133], and the oxidation and corrosion of surfaces [13,14,95,134-136].

ACKNOWLEDGEMENTS: The author is indebted to J. Graham for invaluable secretarial assistance, to Professor P. Mark and Drs. A. R. Lubinsky and R. G. Barerra for helpful discussions, and to Dr. R. A. Oriani and Professor W. H. Weinberg for many helpful comments on the manuscript.

REFERENCES

1. Duke, C. B., Annu. Rev. Mater. Sci. 1, 165, 1971.

2. Duke, C. B., Adv. Chem. Phys. 27, 1, 1974.

3. Kliewer, K. L. and Fuchs, R., Adv. Chem. Phys. 27, 355, 1974.

4. Economou, E. N. and Ngai, K. L., Adv. Chem. Phys. 27, 355, 1974.

5. Linford, R. G., Solid State Surface Sci., 2, 1, 1973.

6. Maradudin, A. A., Montroll, E. W., Weiss, G. H., and Ipatova, I. P., Theory of Lattice Vibrations in the Harmonic Approximation, 2nd Edition, Academic, New York, 1971, chapt. 9.

7. Harris, J. and Jones, R. O., J. Phys. F: Metal Phys., 4, 1170, 1974.

8. Greene, R. F., Crit. Revs. Solid State Sci. 4, 477, 1974.

9. Webb, M. B. and Lagally, M. G., Solid State Phys. 28, 301, 1973.

10. Duke, C. B., Crit. Revs. Solid State Sci. 4, 371, 1971.

11. Ehrlich, G., Crit. Revs. Solid State Sci. 4, 205, 1974.

12. Duke, C. B. and Park, R. L., Physics Today 25, 23, August, 1972.

13. Duke, C. B., Lipari, N. O. and Laramore, G. E., J. Vac. Sci. Technol. 11, 180, 1974.

14. Duke, C. B., Lipari, N. O. and Laramore, G. E., Nuovo Cimento, 23B, 241, 1974.

15. Duke, C. B., Japan Jour. Appl. Phys. Suppl. 2, part 2, 13, 641, 1974.

16. Duke, C. B., In *LEED: Surface Structures of Solids*, Vol. II, M. Laznicka, Ed., Czechoslovak Union of Mathematicians and Physicist, Prague, 1972, 1.

17. Duke, C. B. and Landman, U., *Phys. Rev. B* 8, 505, 1973.

18. Sickafus, E. N., *Jour. Vac. Sci. Technol.* 11, 299, 1974.

19. Brundle, C. R., *Jour. Vac. Sci. Technol.*, 11, 212, 1974.

20. Park, R. L. and Houston, J. E., *Jour. Vac. Sci. Technol.* 11, 1, 1974.

21. Rebinder, P. A. and Schchukin, E. O., *Usp. Fiz. Nauk*, 108, 3, 1972: [trans. *Sov. Phys. Usp.* 15, 533, 1973].

22. Westwood, A. R. C. and Macmillan, N. H., *Science of Hardness Testing*, ASM, Cleveland, 1973, p. 377.

23. Latanision, R. M., *Corrosion Fatigue*, National Assoc. of Corrosion Eng., 1972, p. 185.

24. Stein, D. F., *Jour. Vac. Sci. Technol.* 12, 268, 1975.

25. Broers, A. N., *Crit. Rev. Solid State Sci.* 4, 333, 1974.

26. Brillson, L. J., *Jour. Vac. Sci. Technol.* 12, 76, 1975.

27. Many, A., *Crit. Rev. Solid State Sci.* 4, 515, 1974.

28. Gatos, H. C., This volume, p.

29. Gomer, R., *Field Emission and Field Ionization*, Harvard University Press, Cambridge, 1961.

30. Gadzuk, J. W. and Plummer, E. W., *Rev. Mod. Phys.* 45, 487, 1973.

31. Müller, E. W., and Tsong, T. T., *Field Ion Microscopy*, American Elsevier, New York, 1969.

32. Van Oostrom, A., *Crit. Rev. Solid State Sci.* 4, 353, 1974.

33. Graham, W. R. and Ehrlich, G. *Surface Sci.*, 45, 530, 1974.

34. Duke, C. B., in *Proceedings of the International School of Physics "Enrico Fermi", Course LVIII*, Editrice Compositori, Bologna, 1975, pp. 52-249.

35. Hobson, J. P., *Japan Jour. Appl. Phys. Suppl.* **2**, part 1, 13, 317, 1974.

36. Powell, C. J., *Surface Sci.* 44, 29, 1974.

37. Chu, W. K., Mayer, J. W., Nicolet, M-A., Buck, T. M., Amsel, G., and Eisen, F., *Thin Solid Films*, 17, 1, 1973.

38. Evans, R. D., *The Atomic Nucleus*, McGraw Hill, N. Y., 1955, pp. 637-646.

39. Hagstrum, H. D., *J. Vac. Sci. Technol.* 12, 7, 1975.

40. Duke, C. B. and Alferieff, M. E., *J. Chem. Phys.* 46, 923, 1967.

41. Duke, C. B. and Tucker, C. W., Jr., *Surface Sci.* 15, 231, 1969.

42. Goff, R. F., *J. Vac. Sci. Technol.* 10, 355, 1973.

43. Brongersma, H. H. and Mul, P. M., *Surface Sci.* 35, 393, 1973.

44. Benninghoven, A., *Appl. Phys.* 1, 3, 1973.

45. Evans, C. A., Jr., *J. Vac. Sci. Technol.* 12, 144, 1975.

46. G. Carter, *J. Vac. Sci. Technol.* 10, 95, 1973.

47. White, C. W., Simms, D. L., and Tok, N. H. *Science*, 177, 481, 1972.

48. Gerlach, R. L., *J. Vac. Sci. Technol.* 8, 599, 1971.

49. Coad, J. P. and Riviere, J. C., *Phys. Stat. Sol.* (a), 7, 571, 1971.

50. Propst, F. M. and Piper, T. C., *J. Vac. Sci. Technol.*, 4, 53, 1967.

51. Ibach, H., *J. Vac. Sci. Technol.*, 9, 713, 1972.

52. Little, L. H., *Infrared Spectra of Adsorbed Species*, Academic, N. Y., 1966.

53. Greenler, R., *Crit. Rev. Solid State Sci.*, 4, 415, 1974.

54. Lang, N. D., *Solid State Phys.*, 28, 225, 1973.

55. Schrieffer, J. R. and Soven. P., Physics Today, 28, 24, April 1975.

56. Park, R. L., Physics Today, 28, 52, April 1975.

57. Eastman, D. E. and Nathan, M. I., Physics Today, 28, 44, April 1975.

58. Nicholas, J. F., An Atlas of Models of Crystal Surfaces, Gordon and Breech, N. Y., 1965.

59. Wood, E. A., J. Appl. Phys., 35, 1306, 1964.

60. Dunning, W. J., in the Solid Gas Interface, 1, E. A. Wood, Ed., Marcel Dekker, N. Y., 1967, pp. 271-306.

61. Park, R. L. and Houston, J. E., Surface Sci. 18, 213, 1969.

62. Lang, B., Joyner, R. W., and Somorjai, G. A., Surface Sci. 18, 213, 1969.

63. Davisson, C. J. and Germer, L. H., Phys. Rev. 30, 705, 1927.

64. May, J. W., Adv. Catal., 21, 151, 1970.

65. Estrup, P. J. and McRae, E. G., Surface Sci. 25, 1, 1971.

66. Duke, C. B. and Laramore, G. E., Phys. Rev. B, 2, 4765, 1970.

67. Duke, C. B. and Laramore, G. E., Phys. Rev. B, 3, 3183, 1971.

68. Duke, C. B., Smith, D. L., and Holland, B. W., Phys. Rev. B, 5, 3358, 1972.

69. Duke, C. B. and Landman, U., Phys. Rev. B, 6, 2956, 1972.

70. Duke, C. B. and Liebsch, A., Phys. Rev. B, 9, 1126, 1974.

71. Duke, C. B. and Tucker, C. W., Jr., Phys. Rev. B, 3, 3561, 1971.

72. Tucker, C. W., Jr. and Duke, C. B., Surface Sci., 29, 237, 1972.

73. Bonzel, H. P. and Ku, R., J. Vac. Sci. Technol. 9, 663, 1972.

74. Christman, K., and Ertl, G., Z. Naturforsch, 28a, 1144, 1973.

75. Duke, C. B., Laramore, G. E., Holland, B. W., and Gibbons, A. M., Surface Sci. 27, 523, 1971.

76. Laramore, G. E. and Duke, C. B., Phys. Rev. B. 5, 267, 1972.

77. Demuth, J. E., Marcus, P. M., and Jepsen, D. W., Phys. Rev. B, 11, 1460, 1975.

78. Burkstrand, J. M., Kleiman, G. G., and Arlinghaus, F. J., Surface Sci. 46, 43, 1974.

79. Legally, M. G., Bucholz, J. C., and Wang., G. C., J. Vac. Sci. Technol. 12, 213, 1975.

80. Unertl, W. N. and Thapliyal, H. V., J. Vac. Sci. Technol. 12, 263, 1975.

81. Jackson, D. P., Canad. J. Phys. 49, 2093, 1971.

82. Finnis, M. W. and Heine, V., J. Phys. F: Metal Phys. 4, L37, 1974.

83. Alldredge, G. P. and Kleiman, L., J. Phys. F: Metal Phys. 4, L207, 1974.

84. Jackson, D. P. Surface Sci. 43, 431, 1974.

85. Somorjai, G. A. and Farrell, H. H., Adv. Chem. Phys. 20, 215, 1971.

86. Sickafus, E. N., and Bonzel, H. P., Prog. in Surface and Membrane Sci. 4, 115, 1971.

87. Chang, S. C. and Mark, P., Surface Sci. 45, 721, 1974; 46, 293, 1974.

88. Chang, S. C. and Mark, P., J. Vac. Sci. Technol. 12, 624, 1975; 12, 629, 1975.

89. Mark, P., Chang, S. C., Creighton, W. F., and Lee, B. W., Crit. Revs. Sol. State Sci. (to be published).

90. Laramore, G. E. and Switendick, A. C., Phys. Rev. B, 7, 3615, 1973.

91. Duke, C. B. and Lubinsky, A. R., Surface Sci. (to be published).

92. Duke, C. B., Laramore, G. E., and Lipari, N. O., J. Vac. Sci. Technol. 12, 222, 1975.

93. Ignatiev, I., Jona, F., Jepsen, D. W., and Marcus, P. M., J. Vac. Sci. Technol. 12, 226, 1975.

94. Van Hove, M. and Tong, S. Y., J. Vac. Sci. Technol. 12, 230, 1975.

95. McDonnell, L., Woodruff, D. P., and Mitchell, K.A.R., Surface Sci. 45, 1, 1974.

96. Duke, C. B., J. Vac. Sci. Technol. 6, 152, 1969.

97. Riviere, J. C., in Solid State Surface Sci., M. Green, Ed., Marcel Dekker, N. Y., 1969, 1, 179.

98. Feibelman, P. J., Duke, C. B., and Bagchi, A., Phys. Rev. B. 5, 2436, 1972.

99. Gomer, R., Adv. Chem. Phys. 27, 211, 1974.

100. Benson, G. C. and Yun, K. S. in The Solid Gas Interface, E. A. Flood, Ed., Dekker, N. Y., 1967, 1, 203-269.

101. Mackenzie, J. K., Moore, A. J. W., and Nicholas, J. F., J. Phys. Chem. Solids 23, 185, 1962.

102. Nicholas, J. F., Aust. J. Phys. 21, 21, 1968.

103. Lang, N. D. and Kohn, W. Phys. Rev. B 1, 4555, 1970.

104. Pines, D., Elementary Excitations in Solids, Benjamin, N. Y. 1963, 132-163.

105. Barerra, R. G. and Duke, C. B. (to be published).

106. Schmit, J. and Lucas, A. A., Solid State Commun. 11, 415, 1972.

107. Harris, J. and Jones, R. O., J. Phys. F 4, 1170, 1974.

108. Griffin, A., Kranz, H., and Harris, J., J. Phys. F 4, 1744, 1974.

109. Bond, G. C., Catalysis by Metals, Academic, N. Y., 1962, Chapt. 4.

110. Horiuti, J. and Muller, K., J. Res. Inst. Catalysis, Hokkaido Univ., 16, 605, 1968.

111. Grimley, T. B., *J. Vac. Sci. Technol.* 8, 31, 1971.

112. Schrieffer, J. R., *J. Vac. Sci. Technol.* 9, 561, 1972.

113. Blyholder, G., in *Modern Aspects of Electrochemistry*, J. O. M. Bockris and B. E. Conway, Eds., Plenum, N. Y. 8, 1, 1972.

114. Horiuti, J. and Toya, T., in *Solid State Surface Science*, M. Green, Ed., 1, 1, 1969.

115. Tamm, I., *Z Physik* 76, 849, 1932.

116. Schockley, W., *Phys. Rev.* 56, 317, 1939.

117. Eastman, D. E. and Grobman, W. D., *Phys. Rev. Lett.*, 28, 1378, 1972.

118. Wagner, L. F. and Spicer, W. E., *Phys. Rev. Lett.* 28, 1381, 1972.

119. Eastman, D. E. and Freeouf, J. L., *Phys. Rev. Lett.* 33, 1601, 1974.

120. Rowe, J. E., Christman, S. B., and Ibach, H., *Phys. Rev. Lett.* 34, 874, 1975.

121. Rowe, J. E. and Ibach, H., *Phys. Rev. Lett.* 31, 102, 1973.

122. Ludeke, R. and Esaki, L., *Phys. Rev. Lett.* 33, 653, 1974.

123. Mark, P., *Catal. Rev.* 1, 165, 1967.

124. Davison, S. G. and Levine, J. D., *Solid State Phys.* 25, 1, 1970.

125. Gurman, S. J. and Pendry, J. B., *Phys. Rev. Lett.* 31, 637, 1973.

126. Plummer, E. W. and Gadzuk, J. W., *Phys. Rev. Lett.*, 25, 1493, 1970.

127. Waclawski, B. J. and Plummer, E. W., *Phys. Rev. Lett.* 29, 783, 1972.

128. Feurbacher, B., and Fitton, B., *Phys. Rev. Lett.* 29, 786, 1972.

129. Eastman, D. E., *Phys. Rev. B* 3, 1769, 1971.

130. Duke, C. B., Pietronero, L., Porteus, J. O., and Wendelken, J. F., Phys. Rev., (to be published).

131. MacRae, A. U., Muller, K., Lander, J. J., Morrison, J., and Phillips, J. C., Phys. Rev. Lett. 22, 1048, 1969.

132. Matthews, J. W., J. Vac. Sci. Technol., 12, 126, 1975.

133. Perdereau, J., Biberian, J. P., and Rhead, G. E., J. Phys. F 4, 798, 1974.

134. Eastman, D. E. and Cashion, J. K., Phys. Rev. Lett. 27, 1520, 1971.

135. Helms, C. R. and Spicer, W. E., Phys. Rev. Lett. 28, 565, 1972.

136. Baker, J. M. and Eastman, D. E., J. Vac. Sci. Technol. 10, 223, 1973.

GLOSSARY OF SYMBOLS

A : surface area

p : pressure

$p_{\parallel}$: component of momentum parallel to a (planar) surface

ϕ : work function

Θ : polar angle for electrons incident on a planar surface

ψ : one electron wave function

$\hbar\omega_s$, $\hbar\Omega_s$: surface plasmon energy

$\hbar\Omega_i$: bulk collective excitation energy

$\hbar\omega_p$: bulk plasmon energy of a free-electron metal

γ : specific surface work

g_{ij} : surface stress tensor

σ : surface charge density

λ_e : inelastic collision mean free path

E	:	energy (in eV)
ρ	:	density
$\vec{\rho}$	:	vector within a planar surface
A	:	atomic weight
M_p	=	proton mass = 1.67×10^{-24} gm
M	=	mass
I	=	ionization potential
Z	=	atomic number
c	=	speed of light = 3×10^{10} cm/sec.
$\underset{\sim}{k}$	=	wave vector
$k_\parallel$	=	component of wave vector parallel to a planar surface
$k_\perp$	=	component of vector perpendicular to a planar surface
$\underset{\sim}{g}$	=	reciprocal lattice vector of a surface mesh
$\underset{\sim}{a}, \underset{\sim}{b}$	=	surface mesh unit vectors
e	=	electronic charge, 1.6×10^{-19} coul.
n	=	particle density in cm^{-3}
k_F	:	Fermi wave number
E_F	:	$\hbar^2 k_F^2/2m$: Fermi energy
m	:	electronic mass, 9.1×10^{-28} gm.
S_n	:	separation energy
V_c	:	coulomb "dipole" potential
V_T	:	total one-electron potential
s	:	electron spin quantum number, s = 1/2
$\varepsilon_i(\omega)$	:	frequency dependent local dielectric function
ω	:	angular frequency
E_g	:	optical energy gap.

DISCUSSION

Comment by F. C. Frank:

I would very much like to know the state of the water molecules on the surface of a single crystal of ice at -5°C and -15°C. I think that none of the things you have told us about would help me in the slightest for that question. How would you go about it?

Reply:

I would prepare thick water layers for insertion into photoelectron spectrometers run with both x-ray and ultraviolet sources at the highest possible pressures. Then I would increase the XPS core-level spectrum and UPS valence electron spectrum as I pumped down, thus subliming the H_2O. A reasonable signal giving the dynamic composition and electronic structure should be achievable prior to the complete disappearance of the sample. An estimate of required sample thickness could be obtained from the (known) vapor pressure of ice at the temperatures cited.

Comment by R. M. Latanision:

While I, of course, accept your report that diffraction measurements show no significant change (contraction or expansion) in near-surface lattice spacing, from perhaps a naive materials science point of view, I would expect intuitively that such changes should occur. It seems to me that when one separates a solid into two parts, atoms at the newly created surfaces should (because of changes in nearest neighbor interactions, etc.) like to relax to lower energy configurations and should not be expected to maintain the bulk lattice spacing. Why do you suppose this is not observed? Is it possible that diffraction theory is now still not adequate to handle this--i.e., would present theory allow you to detect relaxation if it occurred?

Reply:

The limits of accuracy of diffraction theory are determined primarily by uncertainties in the electron-solid force law. A detailed discussion of the derivation of the customarily accepted uncertainty, i.e., $\Delta a \sim 0.1 Å$, has been given, e.g., by C. B. Duke, Adv. Chem. Phys. 27, 1 (1974). Different groups, analyzing comparable data, consistently reach structures to within this uncertainty despite the use of widely different analytical procedures. Therefore, I see little reason to doubt that the intrinsic resolution of the diffraction-theory analyses are in the range $0.1 - 0.05$ Å at the present time for surfaces that are primarily "flat" on a distance scale of $\Delta X \sim 100$ Å (in practice, well annealed low index surfaces of simple (mostly fcc) metals, III-V and II-V semiconductors).

Concerning "why" surface layer expansion is not pervasive, it seems cognent to emphasize that this expectation was based on the predictions of two-body force-law models which neglected the (very substantial) electronic charge redistributions which occur in the outermost cells of a solid. We know, however, that these charge rearrangements exert a sizeable influence on the position of ion-cores in the surface cells (see e.g., M.W. Finnis and V. Heine, J. Phys. F. $\underline{4}$, L37 (1974), G. P. Alldredge and L. Kleinman, ibid., L207 (1974)). Thus, the absence of surface expansion should not be surprising once one realizes that the reduction of the effective number of harmonic "springs" in the surface layer is counterbalanced by the surface ions' larger attraction to the higher electronic charge density on the "solid" side of the surface unit cell as opposed to the rarified charge density on the vacuum side.

Comment by J. J. Mills:

Is it not possible that the averaging which the beams make could make surface expansions difficult to detect in LEED measurements, on, e.g., Al(100).

Reply:

George Laramore and I examined this possibility in Phys. Rev. $\underline{B5}$, 267 (1972). We came to the conclusion that, if there had been a relaxation there, it should have been seen. Several subsequent independent analyses (for a review see C.B. Duke, Adv. Chem. Phys. $\underline{27}$, 1 (1974)) have confirmed our findings.

Comment by J. P. Hirth:

Presumably ledges and terraces on surfaces would produce both pseudo domain boundaries and "normal to the surface" intensity shifts. Could this not produce important effects in interpreting overlayer patterns? What is the state of understanding of the relative effect of ledges versus effects arising from differences between reconstruction - no reconstruction and between a single lattice and superposed domains?

Reply:

The consequence of surface typography can, in principle, be substantial and our present understanding of them is virtually non-existent. For a review with references see C.B. Duke, Japan Jour. Appl. Phys. Suppl. 2, part 2, 641 (1975) or C.B. Duke and A. Liebsch, Phys. Rev. $\underline{B9}$, 1126 (1974).

Comment by R. Bullough:

With regard to John Hirth's question, I understand that you

get observations that imply surface relaxations of the ions and that you need to invoke a certain 'step' density to remove this relaxation interpretation. If so, then can you really be certain that such relaxations are absent?

Reply:

Yes, because only in the case of the larger relaxations on Al(110) is the agreement between the observed and calculated LEED intensities poor. Because of this poor quality of the description of the experimental data both surface steps and surface relaxations can provide comparable interpretations of the measurements.

Comment by A. Seeger:

Are the present theories of adsorption of hydrogen on otherwise clean metal surfaces capable of saying whether individual hydrogen atoms are absorbed as atoms or as protons screened by the conduction electrons? Is there any evidence on this question?

Reply:

Although self consistent potentials for hydrogen adsorption on metals have been computed by Smith, Ying and Kohn (Phys. Rev. spring of 1975) no structure analyses have been performed for specific systems because of the lack of adequate LEED intensity data.

Comment by A. Seeger:

In calculations of simple defects in crystals (e.g., monovacancies), the bond-breaking model (with the energy per bond determined empirically from the cohesive energy) gives an upper limit to the correct defect formation energy. As one would expect, the deviations from this upper limit are smallest for covalent crystals such as Si and Ge; they are very large for metal and ionic crystals. Can a similar statement be made for surface energies?

Reply:

Presumably so, but one must keep in mind that even the cleavage surfaces of the group IV semiconductors reconstruct in such a fashion as to redistribute much of the charge in the broken (i.e., "dangling") bonds into other bonds between the surface and substrate atoms. Thus, on the basis of bond-energy-bond-order chemical considerations, one might expect the surface to be more stable (i.e., require less energy for formation) than predicted by a simple bond-breaking model which ignores the consequences of the subsequent reconstruction. These considerations are consistent with the predictions of physical models of the surface reconstruction which regard this reconstruction as a Peierls distortion to an insulating state of the electrons in a surface state band built out of the "dangling-bond" orbitals on the individual group IV atoms.

Comment by R. M. Latanision:

What effect would you expect elastic strain to have on Auger or ESCA spectra, i.e., would you expect shifts in the spectra of a strained surface compared to an unstrained surface.

Reply:

In principle, the core electron energy levels will undergo a strain induced shift. I suspect, however, this shift will be small and difficult to observe (it should be at least $\Delta E \sim 0.1$ ev or larger to be readily detectable).

Comment by R. Firestone:

Are there any data on the surface structure of ceramic materials such as Al_2O_3, SiC, etc.?

Reply:

Electron intensity profiles for Al_2O_3 have been measured but the data has not been analyzed. However, there is no inherent difficulty in performing the experiments. If you will make the measurements, I will analyze the data.

Comment by W. R. Tyson:

Regarding the "broken bond" models of surface energy, in spite of their doubtful theoretical justification, they at least have the virtue of explaining the correlation between surface energy and cohesive energy which seems to be an established experimental fact. This is a very useful empirical correlation, and it is to be hoped that more sophisticated treatments of the surface energy problem will also address the cohesive energy calculation within the same theoretical frame-work, particularly for the technologically important metals like iron.

Reply:

Surface energies have proven notoriously difficult to measure reliably (see, e.g., C.B. Duke, Annu. Rev. Mater. Sci., $\underline{1}$, 165 (1971) and R.G. Linford, Solid State Surface Sci., $\underline{2}$, 1 (1973) for reviews)). Therefore, their correlation with bulk cohesive energies is, in my judgement, not well established at the present time.

SPACE CHARGE LAYERS

Harry C. Gatos and Jacek Lagowski*

Massachusetts Institute of Technology
Cambridge, Massachusetts 02139

ABSTRACT. A brief general account is first given of the types and origin of space charge layers in solids and in liquids; their common characteristics are pointed out. Then the space charge layer at semiconductor surfaces is considered under equilibrium conditions. The depletion space charge layer in semiconductors is singled out because of its importance in high-energy gap materials and in piezoelectric phenomena. Methods for the study of space charge layers and effects on the properites of materials are briefly discussed. Chemisorption on semiconductor surfaces is presented in conjunction with its effects on the structure of the surface states and the associated space charge layer. Surface piezoelectric phenomena and their importance in the study of semiconductor surfaces and space charge layers are given special emphasis. Space charge layers in insulators are briefly considered. Finally, the main features of space charge layers in semiconductor-electrolyte interfaces are outlined in conjunction with the associated chemisorption processes and electrode reactions.

INTRODUCTION

Space charge layers occur in materials systems (solids, liquids or gases) containing mobile carriers, i.e., electrons, ions, or charged defects which distribute themselves under the influence of diffusion and/or potential gradients. Regardless of its specific nature a space charge layer can be generally

---

* On leave from the Institute of Physics, Polish Academy of Sciences, Warsaw, Poland

defined as a micro- or macro-region within which the density of electrically charged species varies significantly resulting in a potential drop across the region. The formation of a space charge region is the result of entropy and energy changes that a system undergoes towards minimizing its free energy.

The principal characteristic of space charge layers is that they produce potential barriers; such barriers may have a pronounced influence on the motion of charged species across their boundaries, on adsorption equilibria, on photo-electric effects, on mechanical properties and on the nature of the electrical contact between dissimilar materials.

Phenomena directly related to space charge barriers in solids were reported as early as 1874, [1] when it was found that the current-voltage characteristics of a system consisting of a metal in contact with a metal compound (e.g., PbS, galena) were not symmetric with respect to the voltage polarity. Two years later the photovoltaic effect was observed,[1] i.e., the appearance of a voltage across a selenium-metal contact. Although selenium rectifiers and photocells were introduced into practical applications at about the turn of this century, the origin of their characteristics was not understood until the early 1930's[1].

In the case of metal-electrolyte interfaces the concept of space charge (electrical double layer) was introduced by Helmholtz in 1879 and refined by Gouy (diffused double layer) in 1909[2]. The origin, nature and properties of space charge layers and related phenomena were developed much earlier in electrolyte systems than in semiconductors; the subsequent detailed understanding of space charge layers in semiconductors served as the basis for the rapid development of solid state electronics. In this context it is interesting to quote from an article by Brattain (1953), the co-discoverer of the transistor, in which the similarities between electrolytes and semiconductors are discussed:[3] "In view of these similarities, it is not surprising that the mathematical formalism which has been developed for handling problems connected with the electrical properties of semiconductors bears a considerable resemblance to that developed long ago for solutions [reference is made to H. J. Reiss, J. Chem. Phys. $\underline{21}$, 1209 (1953)]. The physicist, you may say, is a Johnny-come-lately in this respect. He is trying to impress chemists with one comparatively new subject of semiconductors using concepts which he has had to develop for himself for the first time; although if he had taken the trouble to consult the chemists, he would have been told that the necessary formalism has been available for many years."

In this paper an attempt will be made to give an overview of the origin of space charge layers encountered in solids (semiconductors and insulators) and in electrolytes in contact with solid electrodes. Their characteristics are best understood in equilibrium, i.e., under conditions where the net flow of their charges is essentially zero.

Methods of characterizing space charge layers will be outlined. Properties affected by space charge layers will be discussed, and in particular surface properties in the presence of a surface space charge region. Space charge regions have been most extensively studied, best understood and have led to greatest technological applications in semiconductor materials.

There are numerous rigorous treatments in the literature of the various theoretical and applied aspects of space charge layers[4-9]. The present text is not intended to be rigorous, but rather is designed to project the origin, the general nature and broad implications of space charge layers to physical and chemical phenomena.

1. PHENOMENOLOGY AND MICROSCOPIC MODELS

1.1 General discussion

To understand the phenomena associated with space charge layers (SCLs) and their effects on materials properties it is essential to know the origin of the SCLs; i.e., it is most useful to develop models which, at least on a phenomenological basis, provide an insight to the mechanisms leading to the formation of the SCLs. There can be as many models of mechanisms as there are types of SCLs in various materials systems. Some generalizations, however, can be made. To provide some necessary background representative types of SCLs will be briefly discussed.

Metals will be excluded since their free-electron density (of the order of $10^{22}/cm^3$) is too high to permit the existence of a macroscopic SCL. In contrast, semiconductors and insulators are characterized by an energy band gap (ranging from a fraction of an eV to several eV) which is forbidden to mobile carriers. Consequently, the intrinsic density of mobile charges in these materials is many orders of magnitude smaller than in metals and an SCL can be readily formed (typically the density range of mobile charges in semiconductors at room temperature is 10^{15} to $10^{18}/cm^3$). For the same reason SCLs can be formed in electrolyte solutions.

As an introduction to the discussion of SCLs the surface of a solid will be considered (as in Fig. 1). On the surface,

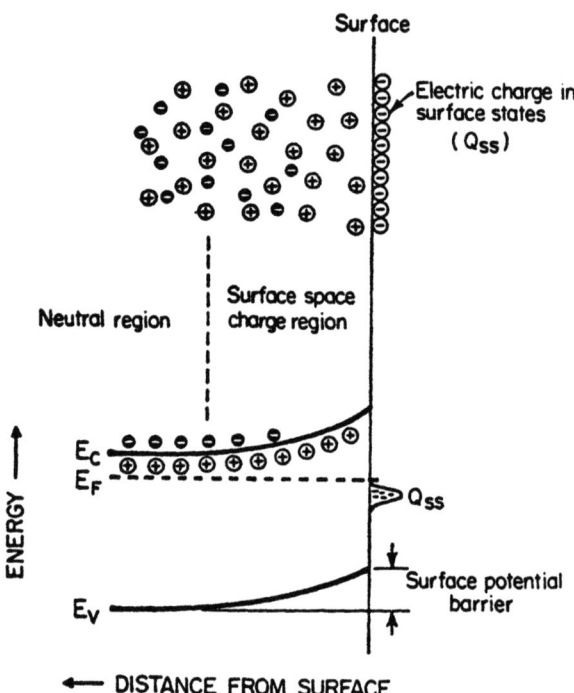

Figure 1. Upper Part: space charge layer in a solid with mobile negative carriers, ⊖, and fixed positive charges; the negative charges in the surface states are not mobile. Lower part: energy band diagram of the solid; E_c and E_v is the electron energy at the bottom of the conduction band and the top of the valence band, respectively; E_f is the Fermi energy.

defined as the termination of the solid (right-hand side of upper part of figure), there exist sites, referred to as surface states, which can capture the mobile negative charges. The origin of the surface states may be associated with the termination of the crystalline lattice or with the presence of adsorbed molecules, impurities located at the surface and structural defects. A discussion of the nature of the surface states is beyond the scope of this paper[10-12]. It will be accepted here, however, that surface states do exist and are charged. Accordingly, the surface is charged with a charge density Q_{ss} (in q/cm^2). The mobile charges in the solid underneath the surface will be redistributed due to electrostatic interactions, but electrical neutrality will be maintained in the system. In Fig. 1 the negative charges removed from the region underneath the surface and captured at the surface are compensated by the fixed positive charges in that region. Thus, underneath the surface a space charge layer is formed characterized by a nonvanishing net electrical charge per unit volume (charge density, $\rho \neq 0$).

It is important to note that the SCL separates the surface from the neutral region in which the positive and negative charges are present at equal densities and the net charge density, $\rho = 0$.

The lower part of Fig. 1 is a schematic energy band diagram corresponding to the solid in the upper part of the figure. The electron energy varies, leading to a surface potential barrier.

It is important to note that the characteristics of SCLs are strictly related to the nature of the electric charges present in the material. Thus, in ionic materials the contribution to the charge density is made by positive and negative ions. Charged structural defects, mobile and nonmobile,[8] (for example, in isolators) or impurities can also contribute to space charge. In semiconductors the charge density is determined primarily by free electrons and holes and shallow ionized donors and acceptors, although in some cases other electric charges, pointed out above, must also be considered[4]. In the energy band scheme these species (charge sources or charges) are often described in terms of a characteristic energy; whether or not they are of importance depends on their concentration and their energy position with respect to the Fermi energy.

The surface SCL in semiconductors is best understood among SCLs. A schematic representation of two types of surface SCLs is shown in Fig. 2 for an n-type semiconductor. They result from the capture of mobile charge carriers by the surface state, E_t. In (b) electrons (majority carriers in this case) are captured at the surface and a depletion layer is formed which has a free charge carrier density, and, thus an electrical conductivity smaller than the bulk; a pronounced depletion layer can essentially isolate the surface from the underlying bulk[13,14]. In (d) an accumulation layer is formed which has a carrier density and, thus an electrical conductivity greater than the bulk. In Fig. 3 an accumulation, a depletion, and an inversion SCL is shown in a p-type semiconductor; in the inversion layers the minority carrier concentration is greater than the majority carrier concentration in the bulk; inversion layers, like accumulation layers, exhibit high electrical conductance (for a detailed discussion see reference [4]).

The type of a surface SCL is determined by the sign and the value of the net electric charge in the surface states (Q_{ss}) and by the type and value of the electrical conductivity in the bulk

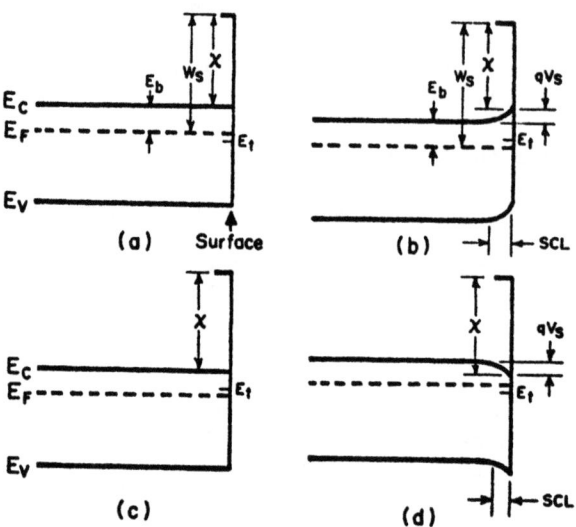

Figure 2. Schematic energy band representation of an n-type semiconductor. W_s is the work function and χ the electron affinity. (a) Surface state E_t is unoccupied; (b) Surface state has captured an electron; qV_s is the surface potential barrier; the thickness of the resulting depletion SCL is indicated; (c) Same as (a); (d) Surface state has captured a positive charge leading to an accumulation SCL.

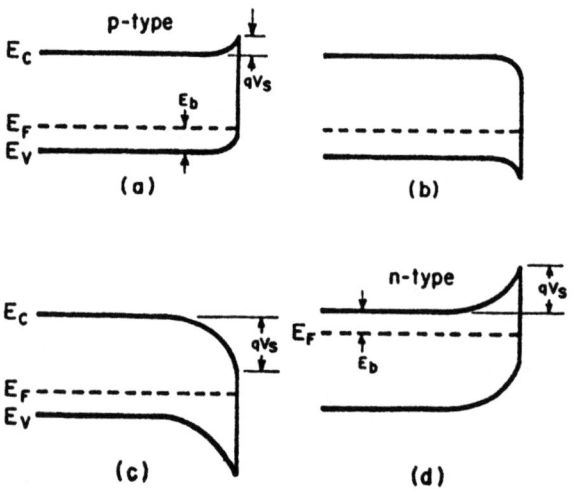

Figure 3. Schematic energy band representation of (a) an accumulation, (b) a depletion, and (c) an inversion layer in a p-type semiconductor; (d) represents an inversion layer in an n-type semiconductor (see also Fig. 2)

region of the semiconductor. The surface SCLs can be treated in fact as one-dimensional cases in which the quantities involved vary only in the direction perpendicular to the surface. An example of a different type of an SCL with cylindrical symmetry formed by an edge dislocation is shown in Fig. 4. In this case a space charge compensates the electric charge along the dislocation

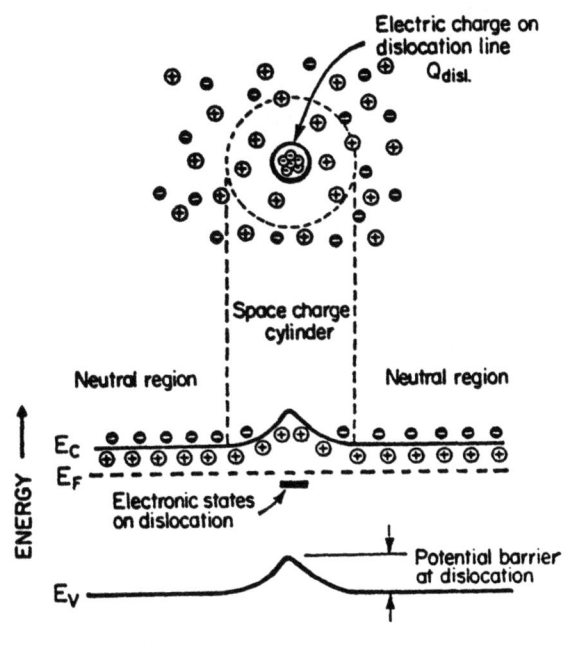

Figure 4. Schematic representation of an SCL and the corresponding energy band diagram about an edge dislocation in an n-type semiconductor;
⊖ mobile negative charges
⊖⊕ fixed negative and positive charges.

line[7,15]; The charges are located at electronic states in the dislocation. In Fig. 5 the space charge associated with a p-n junction is schematically illustrated[5]; the potential barrier due to the SCL is responsible for the rectifying current-voltage characteristics of these structures[5,6]. Space charge layers at metal-semiconductor contacts and solid-electrolyte interfaces are shown in Figures 6 and 7, respectively.

In the above pictorial discussion it was indicated that SCLs are formed without applying external forces. Their formation results from the tendency of the system to minimize its free energy by redistribution of the charges to accommodate energy states at surfaces and dislocations (or other defects), abrupt changes in impurity concentrations, or differences in the Fermi energy (electrochemical potential) between two phases brought into contact.

In addition to the "spontaneous" SCLs discussed above, SCLs can be induced by external forces, e.g., electric or magnetic

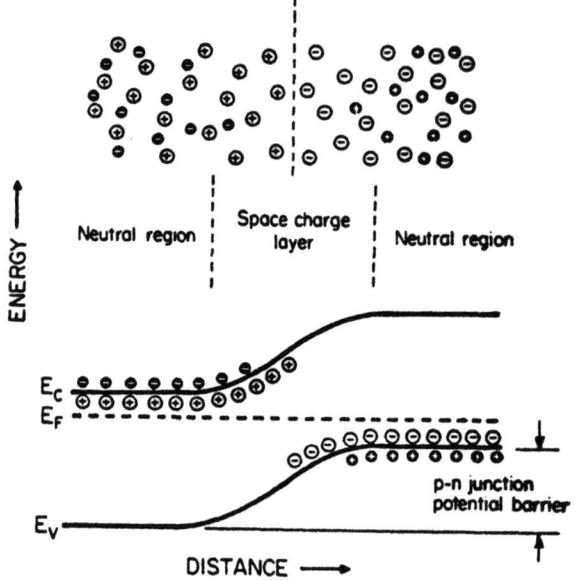

Figure 5. Schematic representation of an SCL at a p-n junction; the corresponding energy band diagram is also shown. ⊖⊕ mobile negative and positive charges. ⊖⊕ fixed negative and positive charges.

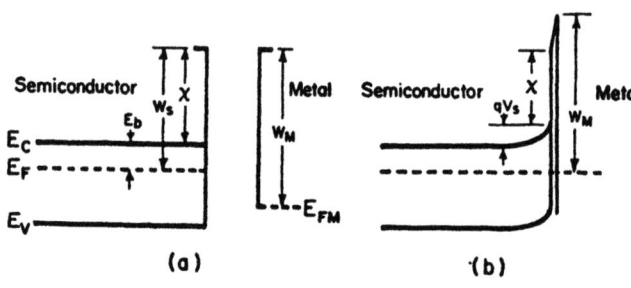

Figure 6. Space charge layer formed when a semiconductor and a metal, (a), are brought into contact, (b), as a result of electron transfer from the semiconductor into the metal; electron transfer takes place because the work function of the semiconductor is smaller than the work function of the metal; in the absence of surface states $-|qV_s| = W_M - W_S$. Actually $-|qV_s| < W_M - W_S$ because of screening by surface states.

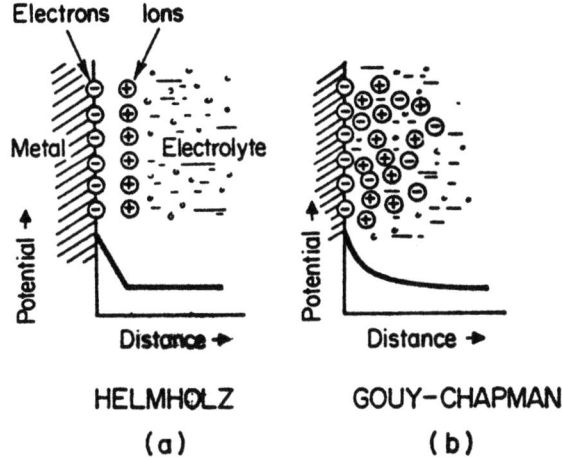

HELMHOLZ   GOUY-CHAPMAN
  (a)   (b)

Figure 7. Space charge layers at metal-electrolyte interface; ⊕ and ⊖ in the electrolyte represent ionic charges (see text in part 3).

fields, temperature gradients, nonuniform illumination, mechanical stress, etc. We will not attempt to discuss further the various types of SCLs or the factors causing their formation and affecting their properties. We will rather outline some features of the SCL in general which may help to understand the detailed treatment of any specific case.

1.2 SCL under equilibrium

When the cause is neglected, the problem of the space charge region reduces to the discussion of a region in which the net-electric charge density, $\rho \neq 0$; positive and negative charges, mobile and nonmobile, contribute to ρ. Thus, $\rho = \Sigma_k (n_k^+ - n_k^-)$ where n_k^+ and n_k^- is the density of the positive and negative charges, respectively. A simple illustration of a positively charged region is seen in Fig. 8. It is evident that in the SCL the concentration of mobile charges, negative in this case, varies; this concentration gradient produces a diffusion current. At equilibrium (dn/dt = 0) this current must be compensated by an equal flow of charges in the opposite direction caused by the electric field in the SCL[16]. Another and very important contribution to the current in the SCL may arise from charge generation recombination-processes. Under equilibrium it is commonly assumed, however, that the generation recombination-currents are low enough to be neglected.

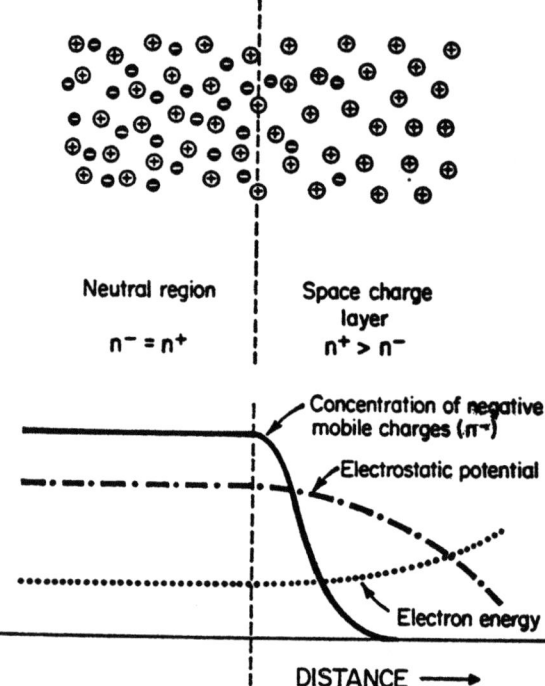

Figure 8. Schematic representation of a positively charged SCL.

Equilibrium is approximated by zero current conditions. Accordingly, no current equations need to be involved and the description of the SCL can be obtained on the basis of Poisson's equation:

$$\nabla^2 V = - \frac{\rho}{\varepsilon \varepsilon_o}$$

or for planar geometry:

$$\frac{d^2 V}{dx^2} = - \frac{\rho}{\varepsilon \varepsilon_o} \qquad (1)$$

where V is the electrostatic potential, ε is the dielectric constant and ε_o is the permittivity of free space.

Often, instead of V, the dimensionless potential is used, defined as $u = qV/kT$. An electrostatic description of SCL is provided by a function $V(x)$ which satisfied eq. 1 with appropriate boundary conditions. Unfortunately, in most instances, $V(x)$ cannot be obtained in an analytical form. Accordingly, it is more useful in practice to describe the phenomena related to the SCL by introducing other quantitites, namely: the electric field in the SCR, $\mathcal{E}(V)$; the maximum absolute value of the potential in the SCL, V_{max} (i.e., the potential barrier) and the net electric charge in the SCL, $Q_{ss}(V_{max})$.

In this procedure, all quantitities which vary with x are expressed as functions of the electrostatic potential V. It should be noted that, with the exception of very simple cases, the charge density, ρ, cannot be given, a priori, as a function of x. On the other hand, even in the most complex cases, ρ can be expressed as a function of V. This very essential simplification is possible due to the interrelationship between the electrostatic potential and the electron energy in solids. From the microscopic point of view, it is very important to point out that most of the treatments of the SCL in solids is based on the approximation that the electrostatic potential varies slowly in comparison with the periodic lattice potential[11] (see Fig. 9).

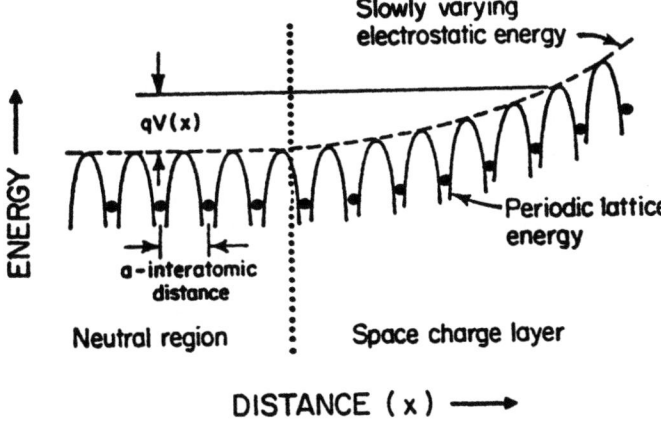

Figure 9. One-dimensional representation of the approximation of slowly varying electrostatic potential in crystals.

In this approximation (i.e., $\mathcal{E}$ = -grad V, for relatively low electric field), the only effect of the electrostatic potential change is a corresponding shift of the energy bands and localized levels in the energy scale (qV)[14]. A schematic representation of this effect is shown in Fig. 10.

For high electric fields (i.e., when the electrostatic potential varies noticeably on an atomic scale) the above approximation is no longer valid and quantum effects in SCL are encountered,

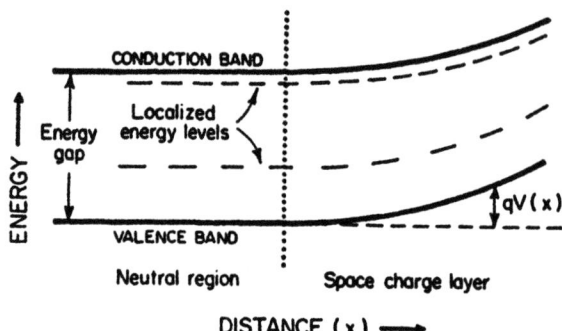

Figure 10. Schematic representation of slowly varying electrostatic energy of the energy bands and the localized levels in a semiconductor crystal.

[12] e.g., splitting of the quasicontinuous energy bands into a set of discrete levels, change in a spectrum of localized levels, etć.

Within the validity of the approximation discussed here, the relation between the electrostatic potential and the charge density is quite simple: The electrostatic potential causes a shift of any characteristic energy E in the neutral region to an energy E - qV in the space charge region. Thus, if n_k describes the charge associated with some energy level E_k in the neutral region of the solid, this energy is shifted in the space charge layer to E_k-qV. Accordingly, the number of charges at any energy level (or in any energy band) contributing to ρ can be related to the electrostatic potential V.

$$n_k^- = N_k^- f_k^- (\frac{E_k - qV}{kT}) \qquad (2)$$

where N_k^- is the density of the contributing states and f_k is the fractional occupancy (f_k can be obtained from Fermi statistics).

Considering eq. 2, the space charge density can be expressed as a function of the electrostatic potential V:

$$\rho = q\Sigma_k[N_k^+ f_k^+(\frac{E_k+qV}{kT}) - N_k^- f_k^-(\frac{E_k-qV}{kT})] \quad (3)*$$

in the neutral region $\rho = 0$; $V = 0$.

The more specific form of the charge density function $\rho(V)$ can be obtained when the nature and the parameters of the sources of charges are specified together with the parameters of the material. To solve Poisson's equation, the knowledge of the boundary conditions and of the characteristic geometry of the problem is required. The boundary conditions are often given by the requirement of electrical neutrality, or are formulated on the basis of electrostatic reasoning (e.g., a drop of the electric displacement vector on the charged plane)[4].

It is evident that when $\rho(V)$ is formulated and the boundary conditions are given it is only a question of mathematics to find the quantities $\mathcal{E}(V)$, $Q_{sc}(V_{max})$ or any other pertinent quantities related to the SCL.

1.3 Energetics of SCL

The formation of an SCL must be treated as a process affecting the free energy of the system. In such treatment, the cause of formation of the SCL needs to be taken into account. As an illustration, the case of the surface space region will be considered. The space charge is formed due to the transition of charges from the bulk region of the solid into a two-dimensional set of energy levels at the surface (surface states). For simplicity acceptor-like surface states will be considered in an n-type material (see Fig. 11).

The transfer of charges represents the transition of electrons from the bulk into the energy levels at the surface, E_t. The free energy in the bulk (neutral region) per one electron equals the Fermi energy E_F (electrochemical potential)[17]. Thus, the change of the free energy of the system per electron captured by the surface state is

$$\Delta F = E_F - E_t + E_S - T\Delta S \quad (4)$$

---

* Note the sign (+) in the term describing the density of positive charges ((E_k+qV)/kT); the energy of the positive charges in an energy band diagram increases in a direction opposite to that of the negative charges.

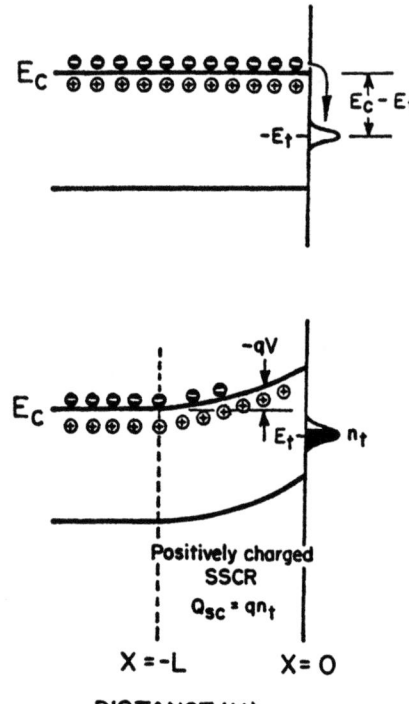

Figure 11. Schematic representation of the formation of a positively charged surface space charge layer.

where E_s is the electrostatic energy in the system: space charge region minus charged surface (per electron), T is the temperature and ΔS is the change in the entropy (for a uniform distribution of electrons in surface states $\Delta S = 0$).

Transition of electrons into the surface states and the corresponding formation of the surface space charge layer proceeds until the total change of the free energy of the system ($n_t \Delta F$) reaches its minimum value. Thus, at equilibrium:

$$\frac{\partial n_t \Delta F}{\partial n_t} = 0 \text{ or } \frac{\partial f_t \Delta F}{\partial f_t} = 0 \qquad (5)$$

where $f_t \equiv n_t/N_t$ defines the fractional occupancy of the surface states.

For uniform electron distribution, from eqs. (4) and (5) one obtains:

$$E_F - E_t + \frac{\partial f_t E_s}{\partial f_t} = 0 \qquad (6)$$

It can be shown that when the interaction between electrons in the surface states is negligible, $\partial f_t E_s/\partial f_t = qV_s$, where V_s is the surface potential. For strong interactions, i.e., when each electron added to the surface states causes a spatial redistribution of the charges already captured by surface states, $\partial f_t E_s/\partial f_t \neq qV_s$ and the fractional occupancy, f_t, deviates significantly from Fermi statistics.* In a view of the small density of surface states (usually about $10^{12}/cm^2$, i.e., three orders of magnitude smaller than the surface density of atoms) the interaction between charges in the surface states can be neglected.

1.4 Depletion surface space charge layer

The specific case of SCL employed above (Fig. 11) to illustrate the energetics involved in the formation of the SCL will now be considered in more detail from the electrostatic point of view. This case is representative of a depletion surface space charge layer, and is very commonly encountered in "real" surfaces of a number of semiconducting materials. The present treatment will provide a basis for considering further the SCL in piezoelectric semiconductors and the role of the surface SCL in chemisorption.

As shown in Fig. 11 and also in Fig. 8, in the neutral region, the negative charge of the mobile electrons is compensated by the positive charges of the ionized shallow donors. In fact, the thermal ionization of these donors (dopant atoms) is responsible for the presence of mobile electrons. Thus, in the bulk $\rho = 0$, and $n_b = N_D$ where n_b is the concentration of free electrons and N_D is the concentration of the shallow donors. In the space charge layer the electrons are practically removed and the net positive charge of the ionized donors, $Q_{sc} = qN_D L$ (where L is the thickness of surface SCL), compensates the electric charge in the surface states Q_{ss}. From the requirement of electrical neutrality, $Q_{sc} = -Q_{ss}$, the thickness of the surface depletion layer is simply obtained $L = -(Q_{ss}/qN_D)$. Typical values of Q_{ss} fall in a region of 10^{11} to $10^{12} q/cm^2$ [4]. In high energy gap materials N_D varies from 10^{14} to $10^{18}/cm^3$ and thus the thickness of the surface depletion layer varies, in practice, from 10^{-7} to $10^{-2} cm$.

---

* For a treatment of the energetics of the space charge region which takes into account the interaction between electric charges in the space charge region at edge dislocations see ref. 15.

On the basis of the above eq. (1) reduces to

$$\frac{d^2u}{dx^2} = -\frac{q^2}{\varepsilon\varepsilon_o kT} n_b \quad \text{where } u = \frac{qV}{kT} \quad (7)$$

With the boundary conditions:
   in the bulk ($x \geqslant -L$) : $u = 0$ and $du/dx = 0$
   at the surface ($x=0$) : $u = u_s$ and $\varepsilon\varepsilon_o \frac{kT}{q} \frac{du}{dx} = Q_{ss}$
and since $\frac{d}{du}(\frac{du}{dx})^2 = 2\frac{d^2u}{dx^2} = -\frac{2q^2}{\varepsilon\varepsilon_o kT} n_b$ (8)

one obtains after integration

$$\frac{du}{dx} = -\sqrt{2}\left(\frac{q^2 n_b}{\varepsilon\varepsilon_o kT}\right)^{1/2} |u|^{1/2} \quad (9)$$

Integration of eq. 9, taking into account the boundary conditions, leads to the following basic relationships describing the surface depletion layer:

(a) $\quad u(x) = -\dfrac{q^2 n_b}{2\varepsilon\varepsilon_o kT}(x+L)^2 \quad (10)$

i.e., the dimensionless potential inside the depletion layer varies parabolically with distance.

(b) $\quad \mathcal{E}(x) = -\dfrac{kT}{q}\dfrac{du}{dx} = \dfrac{qn_b}{\varepsilon\varepsilon_o}(x+L) \quad (11)$

i.e., the electric field is a linear function of distance.

(c) $\quad u_s = -\dfrac{q^2_{ss}}{2\varepsilon\varepsilon_o kT n_b} \quad (12)$

where u_s is the dimensionless potential at the surface.

At room temperature, in materials with a dielectric constant $\varepsilon \simeq 10$ and $n_b \simeq 10^{16}$, a surface potential u_s of about $-20 (V_s \simeq -0.5 \text{ volt})$ is obtained for a net electric charge at the surface less than $10^{12} q/cm^2$. Such a high surface barrier is practically nontransparent to thermal electrons (the thermal penetration factor: $\exp(qVs/kT) \simeq 10^{-8}$); thus, it would be expected that the surface states are in poor communication with the bulk (quasi-isolated surface states)[13,14]. Such behavior was, in fact, observed in high energy gap semiconductors and leads to the very significant conclusion that, in these materials, the equilibrium occupation of surface states (Fermi statistics) cannot be reached within a reasonable period of time[18,19]. Under these conditions, an analysis of the SCL - surface states configuration must be based on nonequilibrium dynamic considerations, not involving the equilibrium Fermi statistics, but rather the charge transfer equations (generation-recombination rate equations) [13,14,20]. This type of analysis will be employed in conjunction with charge transfer chemisorption on high-gap semiconductor surfaces.

1.5 Experimental methods for studying SCLs

The methods for studying SCLs are based on phenomena occurring in or influenced by the presence of these layers. Space charge layers may affect a broad spectrum of phenomena in solids, e.g., carrier transport, recombination-generation processes, optical and photo-electric phenomena, the interaction of solids with the ambient, elastic properties and mechanical properties. Thus, it is not surprising that a large number of methods can be employed for the direct or indirect investigation of SCLs. In the present general discussion these methods will be considered in the light of the SCL characteristics and their interrelation with the sources of the SCL (e.g., in the light of the characteristics of the surface SCL and their interrelation with the surface states). On this basis, three types of methods can be distinguished:
(1) methods based on phenomena related directly to the SCL itself;
(2) methods sensitive to the SCL through its interrelation with the charge source; and (3) methods which cannot isolate the contribution of the SCL from the contribution of the source.

It is apparent that an SCL can best be characterized through type (1) methods. On the other hand, type (3) methods can be of importance for studying various phenomena in solids rather than the SCL itself.

Some of the methods generally employed are summarized in Table I. Three of these methods, representative of types (1), (2) and (3) will be discussed briefly below.

<u>Saturation photovoltage at high generation level: Type (1)</u>. It is well known that the potential barrier in solids decreases upon bandgap illumination [21,22] due to the generation of excess carriers (electrons and holes) which directly affect the charge density, ρ. For a high excitation level, (i.e., when Δn, Δp, become much greater than the bulk values of n and p in the dark) the potential barrier tends to vanish completely. The value of photovoltage generated at the barrier (by illumination of intensity, I), $\lim \Delta V(I)$, for $I \to \infty$, provides a simple measure of the initial value of the potential barrier V_{max}, the important quantity characterizing SCLs.

Table I

Common Methods for Studying Space Charge Layers [4, 20]

Method	Type of SCL	Effect on SCL	Effect on Source	Comments
Field Effect	SCL at surfaces & interfaces	Direct	Indirect	Suitable for studying the properties of SCL, electronic transport in SCL & charge exchange between surface states & the bulk
Photovoltaic effect				
a) Saturation photovoltage	Any SCL	Direct	None or indirect	Standard method for evaluation of the potential barrier height
b) Subbandgap spectroscopy	Surface SCL	Indirect	Direct	Applicable to high gap semiconductors for surface state studies & charge transfer
c) Spectroscopy in the intrinsic region	Any SCL	Indirect	Indirect	Suitable for studying recombination-generation processes
I-V characteristics C-V characteristics	p-n junctions, metal-semiconductor contacts MIS structures	Indirect or direct (C-V)	Indirect or direct (C-V)	Characterization of SCL and/or of the source
Ambient cycling	SCL at surfaces & interfaces	Indirect	Direct	Characterization of SCL & of the source. Often applied in conjunction with Field Effect
Bulk doping	Any SCL	Direct	Indirect	Characterization of SCL & of the source
Piezoelectric effects	Surface SCL	Direct	Indirect	See Table II

Subbandgap photovoltage spectroscopy: Type (2). If the charges in the source are in poor communication with the bulk (isolating SCL) it is possible to modulate the net electric charge in the source employing subbandgap illumination of certain wavelengths. The resulting photovoltage is due to the change of the potential barrier across the space charge region neutralizing the change of charge in the source. In this case the charge density, ρ, is essentially not affected by the illumination and the change of the potential barrier occurs indirectly because of the interrelation of the source with the SCL itself.

For example, in high-gap semiconductors characterized by a depletion surface SCL (e.g., ZnO, CdS and GaAs) it was found that the surface photovoltage under subbandgap illumination is associated with discrete electron transitions from the surface states into the conduction band and from the valence band into the surface states. The analysis of this photovoltage as a function of the energy of the incident light, together with the analysis of the transients of the associated transitions, leads to the direct determination of the energy position and the dynamic parameters of the surface states. This method for studying the electron configuration of semiconductor surfaces is known as "Surface Photovoltage Spectroscopy"[18,19,23].

Spectroscopy in the intrinsic region: Type (3). Photovoltage near intrinsic excitation arises due to the generation of excess carriers, [21,22,24] exciton transitions, [25] and trapping of photo-generated carriers[26,27]; it should be noted that all of these contributing mechanisms are influenced by surface recombination[28]. Since the measured photovoltage depends on a large number of parameters, the contribution from the SCL itself and from the source cannot be isolated unless complementary studies are carried out. Nevertheless, the photovoltage spectroscopy near intrinsic excitation is of considerable interest for practical applications in many devices (e.g., solar cells) whose performance is related to SCLs.

1.6 Surface piezoelectric effects in polar semiconductors

Crystallographically polar semiconductors represent a unique group of materials in which a strong interaction may occur between electrical and mechanical properties[29]. This interaction has significant consequences in the presence of SCLs[30].

Thus, it has been shown that in depleted surface layers of polar semiconductors piezoelectric coupling of electrical and mechanical properties leads to various macroscopic effects. Table II summarizes these effects and their application to the study of semiconductor surfaces. So far, the piezoelectric effects have been observed on the polar surfaces of semiconductor compounds and have been treated as surface phenomena. There is no reason, however, to restrict these effects only to surface layers. Similar effects should occur at p-n junctions or depleted regions in the bulk of polar semiconductors and should serve as the basis for new types of light sensoring and other piezoelectric devices.

In crystallographically polar crystals a mechanical stress, S, induces an electric displacement, $\mathscr{D}$, (piezoelectric effect). On the other hand, an electric field, $\mathscr{E}$, gives rise to a mechanical strain, T. Thus, in a convenient matrix notation the set of equations describing the space charge region involves the Poisson equation and the elasticity and the piezoelectric equations:

$$\text{Div}\,\mathscr{D} = \rho$$
$$\mathscr{D}_i = d_{ij}S_j + \varepsilon\varepsilon_o\mathscr{E}_i$$
$$T_j = s_{jk}S_k + d_{ij}\mathscr{E}_i$$
$$\mathscr{E}_i = -\frac{\partial V}{\partial x_i}$$
$$j,k = 1,2\ldots 6 \quad (13)$$
$$i = 1,2,3$$

where s and d are the compliance and the piezoelectric constants, respectively, ε is the dielectric constant (which is taken as a scalar). ε_o is the permittivity of free space, ρ is the charge density, and V is the electrostatic potential.

On the basis of the treatment of a depletion-type surface SCL discussed in section 1.4 it is possible to solve the above set of equations and to obtain a quantitative description of the various phenomena presented in Table II. The problem is considerably simplified by the fact that the macroscopic piezoelectric effects are confined in a practically isolating layer (in the bulk a redistribution of mobile charges effectively screens out the piezoelectric polarization).

An illustration of the effect of the surface barrier on the mechanical properties of a thin crystal of GaAs is shown in Fig. 12. The electric fields in the surface depletion layers on the parallel polar surfaces induce surface stresses (indicated by arrows) of opposite sign. In effect the thin wafer is bent, the radius of curvature, R, being inversely proportional to the surface barrier height $1/R \sim |V_s|$.

When the surface barrier height (i.e., the electric field in the SCL) is modified by incident illumination, the piezoelectric contribution to the surface stresses changes leading to a corresponding change of the radius of curvature (surface photopiezoelectric effect). In fact it was found that thin (00.1) CdS wafers and (111) GaAs wafers undergo pronounced bending upon

Table II

Surface-piezoelectric and Photo-piezoelectric Effects Associated with Surface Depletion Layers in Polar Semiconductors

Phenomenon	Conditions	Measured Quantity	Resulting Information	Observed in
Surface piezoelectric effect	$S \neq 0, I = 0$	cpd vs. mechanical strain	Surface barrier height, identification of A and B surfaces	CdS, ZnO
Piezoelectric modulation of surface charges	$S \neq 0, I = 0$	cpd transients	Dynamic parameters of surface states	CdS, GaAs, ZnO
Piezochemisorption effect	$S \neq 0, I = 0$	Ambient induced cpd transient or resistivity transient	Processes associated with charge transfer between the solid & chemisorbed species	ZnO
Stress-induced amplification of photo-voltage	$S \neq 0, I \neq 0$	Photovoltage vs. mechanical strain	Processes associated with photovoltage identification of light-induced electronic transitions	CdS
Illumination-induced bending	$S = 0, I \neq 0$	Light-induced changes in radius of curvature	Piezoelectric contribution to surface stress	CdS, GaAs
Photomechanical vibration	$S = 0, I \neq 0$ chopped illumination	Amplitude of vibrations vs. $h\nu$ & illumination intensity Frequency of vibration	Electrical & mechanical properties of surfaces Surface stress, adsorption induced changes of surface stress	CdS, GaAs GaAs

S = external stress applied to the crystal; I = incident illumination.

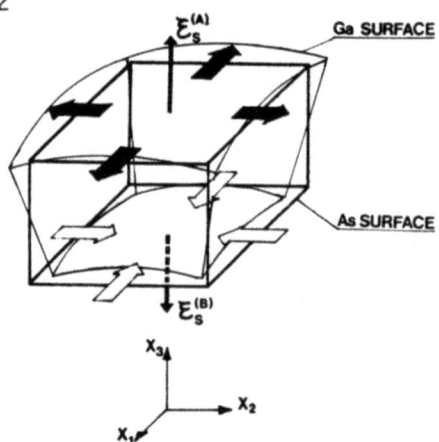

Figure 12. Schematic representation of mechanical bending of a GaAs wafer due to the converse piezoelectric effect at the surface; X_3 is the <111>, crystallographically polar direction.

strong illumination with white light. A striking manifestation of this effect is provided by chopped-light: when the frequency of light-chopping approaches the frequency of the wafer's natural vibration, the wafer is excited to this vibration (Fig. 13). A vibration amplitude as large as 5 mm is observed for wafers 10 mm long and 10 μm thick. As indicated in Table II, the amplitude of the vibration vs. the incident photon energy provides information

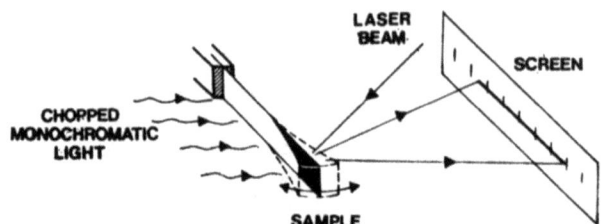

Figure 13. Schematic representation of the experimental arrangement employed in the study of photomechanical vibration. The very low-power laser beam.

on the electronic processes taking place at the surface during illumination; while the frequency of vibration serves as a basis for evaluation of the surface stress, as will be discussed in part 2.

The interrelation between the electrical parameters characterizing the surface SCL and mechanical stress applied to a polar crystal (surface piezoelectric effect) can be clearly seen from the experimental data on CdS shown in Fig. 14. A stress applied to a crystal through mechanical bending results in a change of the surface barrier ΔV_s (measured as change in the contact potential difference) as high as 2 volts. ΔV_s satisfies

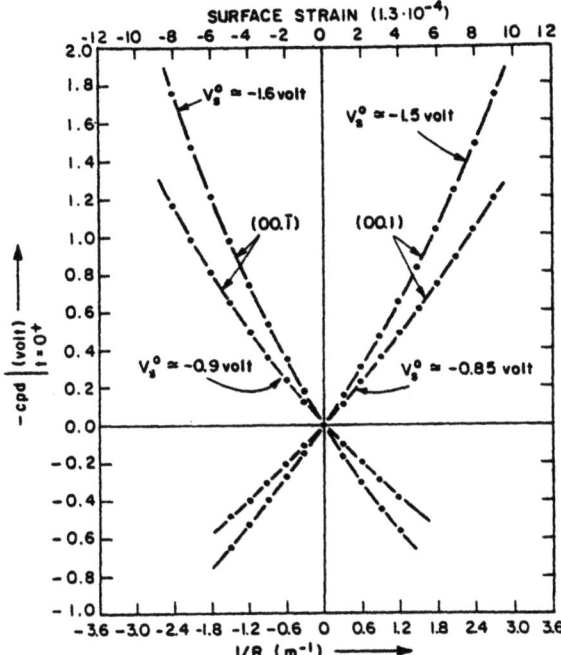

Figure 14. Changes in cpd (surface barrier height) as a function of the radius of bending and the corresponding surface strain. The zero point on the cpd axis corresponds to the steady state contact potential in the unbent condition.

a simple relation derived from eqn. 1:

$$|V_{so} - \Delta V_s|^{1/2} - |V_{so}|^{1/2} \simeq C \frac{1}{R} \text{ for } \left|\frac{q(V_{so}-\Delta V_s)}{kT}\right| \gg 1 \quad (14)$$

where V_{so} is the initial value of the surface barrier height and C is a constant which depends on material parameters and on crystallographic orientation.

It is important to note that the surface piezoelectric effect provides a simple means for changing at will the surface potential barrier in polar crystals. Accordingly, it can be employed for studying surface phenomena related to the surface barrier as will be discussed in part 2 in conjunction with charge-transfer chemisorption.

2. CHEMISORPTION ON SEMICONDUCTORS

2.1 General discussion

Chemisorption is characterized by short-range chemical bonding associated with the evolution of an amount of heat often comparable to that evolved in chemical reactions. In contrast, physical adsorption is characterized by long range Van der Waal-type bonding; the heat of adsorption is usually less than 0.2 eV.

In considering chemisorption on solid surfaces, such as metal surfaces, it is generally important to formulate the nature of the bonding forces and the structural or chemical changes undergone by the surface atoms. Changes in the electrical properties of the metal surfaces are manifested essentially only through changes in the work function. In the case of semiconductors, however, chemical changes associated with chemisorption may affect profoundly the surface state structure and thus the concentration of the mobile carriers in the SCL; conversely, the surface state structure and the carrier concentration at the surface may control the rate and extent of chemisorption[37].

As pointed out earlier, trapping of electrical charges on the semiconductor surface leads to the formation of a space charge layer (Fig. 1). Chemisorbed species, in forming a chemical bond at the surface, may shift the energy levels of the surface states changing their density and thus, changing (increasing or decreasing) the surface barrier, or they may introduce new surface states, again causing changes in the surface barrier, qV_s. Physically adsorbed species do not affect the space charge barrier (as they do not lead to charge transfer at the surface), but they may cause a change in the electron affinity χ (see Fig. 2) when they have a dipole component perpendicular to the surface. Changes in qV_s or χ are reflected as changes in the work function of the semiconductor, W_s. Changes in χ can be distinguished from those in qV_s as they do not affect the surface conductance or surface photovoltage.

Direct evidence of changes in the surface barrier through adsorption or chemisorption was established by exposing germanium surfaces to different gaseous ambients and carrying out contact potential and surface photovoltage measurements (Brattain-Bardeen cycle)[38]. Changes in the work function of about 0.5 eV were observed; the highest value of work function was obtained in ozone and the lowest in wet oxygen. It was estimated that about 20% of the change in W_s was due to the change in qV_s, i.e., charge transfer at the semiconductor surface.

Theoretical models on chemisorption based on charge transfer of semiconductor surfaces were independently advanced in 1952 by three research groups[39]. In these models (presented here in their simplest form) it is assumed that no surface states are present (i.e., no space charge layer exists) prior to chemisorption and that the binding energy of the chemisorbed species is derived from charge exchange with the semiconductor. If the electron affinity E_A of an electronegative molecule, e.g., oxygen, is larger than the work function, W_s, of the semiconductor (consider the semiconductor to be n-type) then the molecule will acquire an electron from the semiconductor and become chemisorbed. The energy of chemisorption is E_A-W_s and determines the position of the energy level introduced at the semiconductor surface. The process can be illustrated by Fig. 2 (a) and (b); the level E_t is introduced in this case by the adsorbed molecule. As chemisorption proceeds the surface barrier increases (the trapped negative charge at the surface increases) until no further electron transfer is possible and equilibrium is established with a depletion space charge layer being formed. Considering an electropositive species, e.g., an alkali atom, if its ionization energy E_I is less than W_s, then an electron is transferred from the alkali atom to the semiconductor. In this case the work function is decreased (positive charge is trapped at the surface) and an accumulation space charge is established as in Fig. 2 (c) and (d).

It is apparent from the above elementary presentation of the charge transfer model that the barrier at the semiconductor surface determines the chemisorption equilibrium (extent of chemisorption). Similarly, since the height of the surface barrier controls the rate of charge transfer from the bulk of the semiconductor to the surface (and vice versa), it must also control the kinetics of chemisorption.

There are extensive experimental studies which provide sound--although by and large qualitative--evidence of the validity of the charge transfer model of chemisorption[37,40]. In these studies measurements of changes in contact potential (work function), photoemission, surface conductance and surface photovoltage were carried out which permitted the determination of the sign of the surface charge and its rate of change under various gaseous ambients. Conversely, changes in the space charge region induced by illumination and by externally or internally applied electric fields have been shown to cause changes in the rates and extent of chemisorption (or desorption).

Since there are extensive accounts in the literature of the theoretical and experimental work on the charge transfer model of chemisorption, [37,40,41-44] particularly on the ZnO and CdS surfaces, this subject matter will not be pursued further here. Instead the discussion will be limited to recent work in which the charge transfer model was quantitatively studied by means of the surface piezoelectric effect (discussed earlier) where the surface barrier is varied, at will, by mechanical stress (bending) without altering the surface state structure.

2.2 The role of surface barrier in charge transfer

As pointed out earlier, mechanical stress induces a pronounced change in the surface barrier at the polar surfaces of semiconductors, provided a depletion surface space charge layer is present, i.e., provided that free carriers in the space charge region do not screen out the piezoelectric polarization. Some recent results on the role of the surface barrier in the chemisorption of oxygen on the ZnO polar surfaces will be summarized here.

Mechanical strain is introduced by bending thin wafers (approximately 60 μm thick) mounted on an apparatus which permits the controlled change and measurement of the radius of curvature[31]. During chemisorption, contact potential and resistivity measurements are carried out simultaneously for determining, as a function of time, changes in the surface barrier and in the carrier concentration within the space charge layer. For reproducible results the etched surfaces must be exposed, in vacuum, to bandgap illumination ($h\nu > 3.2$ eV); this exposure is known to cause the desorption of oxygen adsorbed on the surface[40]. It is also known that after this treatment an accumulation space charge layer is present at the ZnO surfaces[37]. In the presence of the accumulation layer the surface barrier does not change by applying mechanical stress, as the piezoelectric polarization is screened out by the mobile carriers. At an oxygen pressure of several torr, the accumulation space charge layer is inverted to an essentially insulating depletion layer, as electrons from the conduction band are trapped at the oxygen-induced surface states. Upon formation of the depletion layer the transients of the cpd (qV_s) and resistivity follow the well known logarithmic dependence on time[45]. Upon slight mechanical bending (leading to a decrease in the surface barrier) the slopes of the cpd and resistivity relaxations changed abruptly as shown in Fig. 15[33].

The chemisorption process will now be discussed in the presence of a depletion layer.

If the change in the number of negatively charged adsorbed species per unit surface area is dn_t, the corresponding change

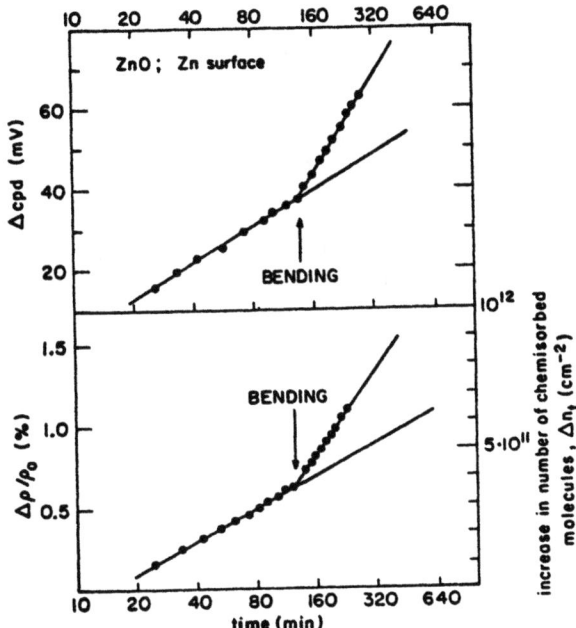

Figure 15. Typical changes in cpd and resistivity under an oxygen pressure of 40 torr. The sample was bent at a radius of curvature R = -30 cm at the time indicated by arrows.

in the surface barrier dV_s is

$$dV_s = -\alpha |V_s|^{1/2} dn_t \qquad (15)$$

where V_s is the surface potential barrier ($-\frac{qV_s}{kT} \gg 1$), and $\alpha = (8q^3 n_b \varepsilon \varepsilon_o)^{1/2}$.

Neglecting the change in electron affinity[46]

$$dV_s \simeq -d(\Delta cpd) \text{ and } \frac{d(\Delta cpd)}{dt} = \alpha |V_s|^{1/2} \frac{dn_t}{dt} \qquad (16)$$

Equation 16 shows that the rate of cpd change reflects the chemisorption rate, dn_t/dt, and is in fact proportional to dn_t/dt for small changes in V_s. The resistivity changes are also related to the chemisorption rate: $d\rho/dt \sim dn_t/dt$. Accordingly, the bending-induced changes of the transient slopes shown in Fig. 15 represent in essence the strain-induced changes in the chemisorption rates.

Since the observed trasients are logarithmic, the bending effect on chemisorption rate can be conveniently examined on the basis of the quantity $dn_t/d\ln t$ and the related quantity $d\Delta cpd/d\ln t$ (slopes of curve in Fig. 15). The changes of $d(\Delta cpd)/d\ln t$ induced by bending are shown in Fig. 16: at $t = 0$ the oxygen pressure was rapidly increased to 40 torr and chemisorption was initiated; at time $t = t_1$ the sample was bent to a radius of curvature $R = +30$ cm (corresponding to a strain of $+10^{-3}$); the bending accelerated the chemisorption process manifested as an increase in $d(\Delta cpd)/d\ln t$. At time $t = t_2$ the sample was bent in the opposite direction $R = -30$ cm (strain -10^{-3}) and the chemisorption rate was decreased below its initial value under zero strain. At times $t = t_3$ and $t = t_4$ the sample was bent to a radius of curvature $R = +45$ cm and $R = +15$ cm, respectively.

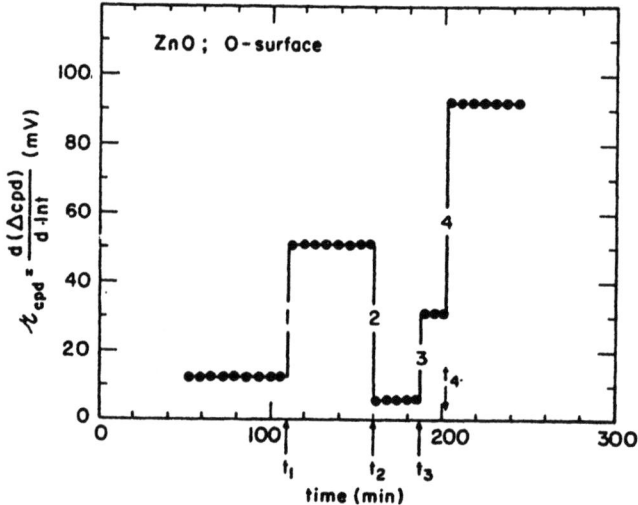

Figure 16. Bending induced-changes in the rate of oxygen adsorption on the oxygen surface of ZnO, in terms of the quantity $d(\Delta cpd)/d\ln t$ (see text) under a pressure of 100 torr. At t_1, t_2, t_3 and t_4 the sample was bent to a radius of curvature $R_1 \simeq 30$ cm; $R_2 \simeq -30$ cm, $R_3 \simeq 45$ cm, and $R_4 \simeq 15$ cm, respectively.

It is thus seen from the data of Fig. 16 that mechanical bending accelerates or decelerates the adsorption process without affecting its logarithmic dependence on time ($\Delta cpd \sim \ln t$). This piezo-chemisorption effect can be understood if one takes into account that: (1) oxygen chemisorption on ZnO is associated with a transfer of electrons from the bulk of the crystal into the adsorbed species; and (2) the charge transfer is controlled by the surface potential barrier, V_s, which can be modulated through mechanical bending (surface piezoelectric effect).

The thermal desorption of oxygen is negligible at room temperature and the rate of oxygen chemisorption on ZnO can be related to charge transfer from the semiconductor bulk into the acceptor-like surface states introduced by the chemisorbed oxygen species as follows[37]:

$$\frac{dn_t}{dt} = K_n n_b \exp\left(\frac{qV_s}{kT}\right) M_p \qquad (17)$$

where K_n is the surface capture cross-section for electrons and M_p is the density of oxygen species available for electron capture. For a depletion layer $-qV_s/>>1$ and thus the $\exp(qV_s/kT$ affects the adsorption rate significantly. If R is the radius of curvature, the bending-induced change of the surface barrier, ΔV_s, is given by eq. 14. It follows from eqs. 14 and 17 that even a small variation of V_s caused by slight mechanical bending does lead to a pronounced change in the chemisorption rate.

The model of the piezo-chemisorption effect indicates that a compression parallel to the basal planes will lead to an acceleration of chemisorption on the zinc surfaces due to the resulting decrease in the surface barrier [30] and to a deceleration of chemisorption on the oxygen surface due to the resulting increase in the surface barrier. Reversal in strain should lead to reverse behavior, e.g., an increase of the chemisorption rate on the oxygen surfaces and a decrease on the zinc surface. Such reversal of the chemisorption behavior was experimentally observed[46].

The piezo-chemisorption effect is in some respects analogous to the field-effect-induced changes of chemisorption[47]. However, in the present case the charge transfer process is modulated not by an external field applied capacitively to the sample across the gaseous phase, but by internal polarization.

As pointed out above, eq. 16, for $V_s \simeq$ constant the rate $d\Delta\dot{c}pd/dt$ is proportional to the rate of electron-capture at the surface. In Fig. 17 the measured dcpd/dt rates are plotted for various values of $cpd(qV_s)$ against the oxygen pressure; these "iso-cpd" lines were obtained from a family of cpd transients of a ZnO unbent wafer under various pressures of oxygen. It is seen that over a broad range of oxygen pressure the rate of electron capture is nearly proportional to the pressure. It is thus apparent that the number of states, M_p, available for capturing electrons from the conduction band can be proportional to the oxygen pressure; accordingly, the rate of electron capture dn_t/dt can indeed be expressed as in Eq. 17.

For a constant pressure, the dependence of the rate of electron-capture as a function of the surface barrier (as varied by controlled mechanical bending) is shown in Fig. 18. It is seen

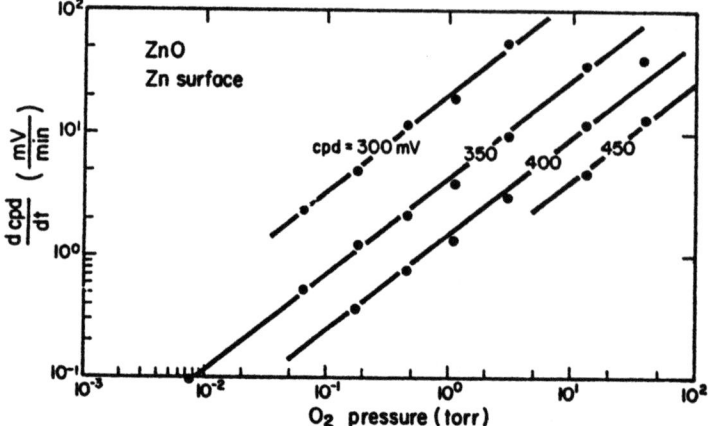

Figure 17. Rate of change of cpd (qV_s) as a function of oxygen pressure for various values of cpd.

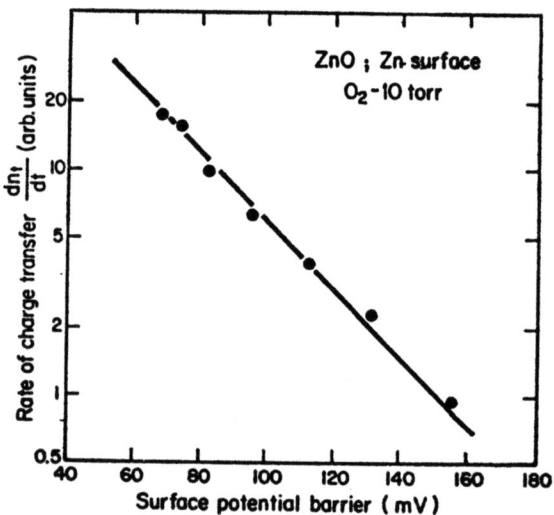

Figure 18. Rate of change transfer during oxygen chemisorption (under constant oxygen pressure) as a function of the surface barrier height, varied by controlled bending.

that the capture rate of electrons is proportional to the Boltzmann factor $\exp(qV_s/kT)$ which shows that thermal penetration of electrons over the surface barrier is involved in charge transfer chemisorption.

In summary and generalizing on the basis of the results available in the literature[37] and those discussed above:
(a) charge transfer between adsorbent and adsorbate characterizes chemisorption on semiconductors; (b) at constant pressure the charge transfer rate is proportional to $\exp(qV_s/kT)$ i.e., it is an activated process with an activation energy equal to the surface barrier; (c) at a constant surface barrier the charge transfer rate is proportional to the ambient pressure (in the present case proportional to the oxygen pressure for pressures < 20 torr); (d) charge transfer is an irreversible process (with respect to the ambient pressure).

2.3 Chemisorption and surface stress

In solids the surface forces acting within the surface are referred to as surface stresses[48]. While the surface tension measures the work required to create a new surface, the surface stresses measure the work required to deform the surface; thus the surface stress can be related to surface tension:

$$f_{ij} = \delta_{ij}\gamma + \partial\gamma/\partial T_{ij} \qquad (i, j = 1, 2) \qquad (18)$$

where f_{ij} is the surface stress tensor, γ is the surface tension, and $\delta_{ij} = 0$, if $i \neq j$, $\delta_{ij} = 1$ if $i = j$; and T_{ij} is the strain tensor; the quantity $\partial\gamma/\partial T_{ij}$ can be positive or negative corresponding to compressive or tensile stress. In the case of liquids $\partial\gamma/\partial T_{ij} = 0$ and the surface stress reduces to the surface tension, i.e., a tensile force, normal to the surface having a constant magnitude γ per unit length. The surface tension is equal to the surface free energy per unit area dF/dA only in single component systems where the Gibbs surface excess (change in the composition at the surface) is zero.

Although it is well established that chemisorption affects significantly (decreases) the surface tension (surface stress) of solids, experimental methods for determining surface stress are quite limited (either in scope, and/or reliability). A method developed recently[36] for the determination of surface stress and the effects of chemisorption, will be outlined here because of its relative simplicity and its sensitivity in determining chemisorption effects through changes in "mechanical properties". It is based on the dependence of the normal mode of vibration of thin wafers on surface stress.

As pointed out earlier, thin wafers of semiconductor compounds (mounted on one end) can be excited to their normal mode of vibration by modulating the space charge region in the two parallel polar surfaces[30,35]. The frequency of this vibration is constant and characteristic of the physical constants of the material and the dimension of the wafer; the amplitude of the vibration on the other hand reflects the structure of the surface states and their interaction with gaseous ambients[35].

When the thickness of the wafer, however, is below a certain value (e.g., below approximately 15 μm in the case of GaAs) its cohesive energy becomes much smaller than its surface stress and, thus, its surface stress controls the natural frequency of vibration. For example, as seen in Fig. 19, the natural frequency of

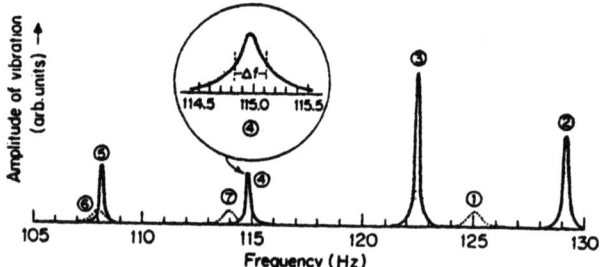

Figure 19. Normal mode of vibration of a 12-μm-thick GaAs wafer for different ambient conditions: (1) room atmosphere just after etching; (2) 30 min in 10^{-4} torr; (3) 48 h in 10^{-5} torr; (4) 10^{-6} torr after corona discharge; (5) 10^{-6} torr; outgassing at ≃200°C for 3 h; (6) immediately after exposure to room atmosphere; (7) 2 h after exposure to room atmosphere.

vibration of a 12 μm thick GaAs wafer varies by more than 25% depending on the gaseous ambient.

On the basis of the analysis of the normal mode of vibration of a cantilever beam[49] it can be shown that the natural frequency of vibration, f, of a thin wafer is:

$$f = f_o \{1-\alpha(\ell^2/h^3)\sigma_s\}^{1/2} \qquad (19)$$

where f_o is the natural frequency of vibration, when there is no contribution from surface stress, $\alpha = 8.6(1-\nu^2)E$, γ is Poisson's ratio, E is young modules, ℓ and h are the length and thickness of the wafer, respectively, and σ_s is the surface stress.

In solids, $(1-\nu^2)/E$ is of the order of $10^{-2} cm^2/dyne$ and thus, for a wafer with a thickness of the order of several microns, the natural frequency of vibration becomes a function of surface stress; in this case the natural frequency decreases, i.e., $f<f_o$, as confirmed experimentally[36]. From eq. 19, the surface stress, σ_s, can be readily determined from the slope of the plot of $(f/f_o)^2$ vs. ℓ^2/h^3. In the case of etched (111) GaAs wafers, σ_s, was found to be 325 dyne/cm in room atmosphere.

For a small contribution of surface stress to f and for small changes of σ_s, eq. 19 can be reduced to:

$$\Delta f/f_o = - \alpha \ell^2 \Delta \sigma_s / 2h^3 \qquad (20)$$

i.e., the change in the natural frequency is proportional to the change in surface stress.

The above relationship can be conveniently employed for the study of adsorption-desorption processes. Thus, as seen in Fig. 20,

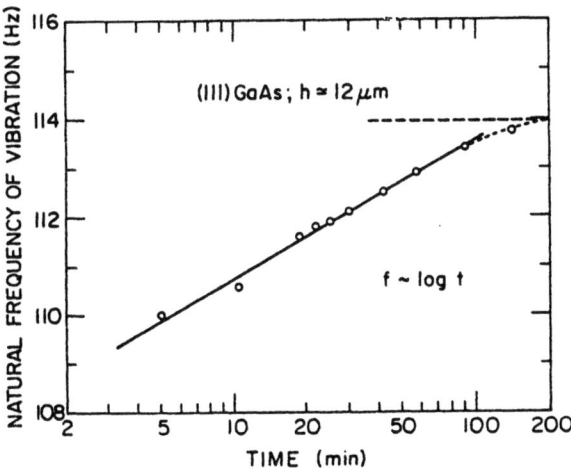

Figure 20. Transient of the natural frequency of a 12-μm-thick (111) GaAs wafer upon exposure to room atmosphere.

the natural frequency of vibration or $\Delta \sigma_s$ is a logarithmic function of time during the adsorption process; since the number of adsorbed molecules on semiconductor surfaces is proportional to the logarith of time,[45] then the adsorption-induced decrease of surface stress is proportional to the number of adsorbed molecules. It should be pointed out that the excitation of a wafer to its natural frequency of vibration can be obtained not only by optical means (modulation of the surface barrier) but also by mechanical and acoustic means.

Thus, employing the natural frequency of vibration, it is possible to determine the surface stress and establish quantitative relationships between adsorption processes, surface stress and the electrical characteristics of semiconductor surfaces. It is believed that this approach should be conveniently applicable to solids other than semiconductors.

3. SPACE CHARGE LAYERS AT SOLID ELECTROLYTE INTERFACES

3.1 Metal-electrolyte interface

As pointed out in the Introduction, space charge layers in metal-electrolyte interfaces have been recognized since before the turn of this century and their characteristics have been extensively studied, primarily because of their important role in electrode reactions. The simplest concept of metal-electrolyte charged interface was advanced by Helmholtz (1879) who proposed that the potential difference lay across two charges of opposite sign. In the Helmholtz "electrical double layer" illustrated in Fig. 7(a) the electrostatic potential varies linearly and the charge distribution is analogous to a parallel-plate condenser. Thus, the capacity, C, per unit area of the double layer should be

$$C = \varepsilon \varepsilon_o / d \qquad (21)$$

where ε is the dielectric constant and d is the thickness of the double layer. Taking an effective radius for the ions of 2×10^{-8} cm and a dielectric constant of 8, a value of 35 $\mu F/cm^2$ is found for the capacity. Values of this order of magnitude have been experimentally obtained. However, the variation of the capacity with the nature and concentration of the electrolyte as well as with the potential difference across the metal-electrolyte interface are more consistent with the presence of a "diffused double layer" (Gouy, 1909) as illustrated in Fig. 7(b) and even more consistent with a double layer consisting of a combination of the Helmholtz and the diffused layers[50]. The capacity, C_S, of this composite double layer (Stern-Chapman) is treated as two capacitors in series:

$$1/C_S = 1/C_H + 1/C_D \qquad (22)$$

where C_H and C_D are the Helmholtz and diffuse components. In the case of mercury in contact with aqueous solution alkali halides (NaF) it was shown that C_H is the major component of C_S (of the order of 25 $\mu F/cm^2$) and essentially independent of variations in C_D[51].

Poisson's equation has been solved for several types of metal-electrolyte interfaces. The metal, in the one limit, is considered as an ideally reversible electrode always in equilibrium with its ions ($M \rightleftarrows M^{2+}$); in this case either the metal or the electrolyte constitute a controllable source of charge. In the opposite limit the metal is an ideally polarizable electrode (blocking electrode), i.e., the interface is impermeable to charges.

Although no current flow is possible across the interface, however, charges may be brought to it, leading to the formation of an SCL, by adsorption or by an external potential. Although the conditions in the two extreme cases are different, the SCL phenomenology is the same. Actually, the characteristics of most metal-electrolyte interfaces lie between these two extremes and their quantitative treatment presents major difficulties.

The composite space charge layer, pointed out above, describes best the metal-electrolyte interface at equilibrium, in the light of extensive experimental evidence[52].

In solving Poisson's equation,* the electrolyte is considered bounded by the electrode plane, $x = 0$, and occupies the space $x > 0$; the mean electric field vanishes for large values of x and the concentration of the positive and negative ions, $C_\pm$, are given in terms of bulk concentration of the electrolyte, C_o and the electrostatic potential, V, across the SCL.

$$C = C_o \exp(\pm zqV/kT) \qquad (23)$$

where z is the valency of the ion; the charge density $\rho = (C_+ - C_-)$ since only positive and negative ions contribute to the charge density. Then the potential and concentrations of the positive and negative ions are determined as a function of distance, x, when a given electric field acts at $x = 0$ normal to the plane. The plane $x = 0$ might correspond to the "outside Helmholtz plane" but the field at this plane would again depend on the electrical state of the electrode. At $x = 0$, $dV/dx - Q/\varepsilon\varepsilon_o$, i.e., the maximum value of the electric field. It is thus seen that the approach to the solution of Poisson's equation in metal electrolyte SCLs is the same as that outlined in part 1 for the SCLs in semiconductors. Actually, as was pointed out in the Introduction, the formal treatment of SCLs developed for metal-semiconductor interfaces (1913) was rediscovered about 40 years later for semiconductor SCLs.

There is extensive literature on the energetics and kinetics of the metal-electrode interface and the subject will not be pursued further here. Similarly, effects of ion adsorption on the surface-free energy of metals are extensively discussed in the literature from the theoretical and experimental point of view[53].

The work on mercury-electrolyte interfaces is of particular interest as the surface tension of mercury, σ, can be determined directly as a function of the nature and concentration of the electrolyte and the potential difference, V, between the mercury and the electrolyte. Thus, the charge density, Q, in the double

---

* For a comprehensive review of the electrical double layer at metal-electrolyte interfaces see ref. [6].

layer and its capacity can be readily obtained:

$$d\sigma/dV = Q \text{ and } d^2\sigma/dV^2 = C \qquad (24)$$

It is well known that the surface tension of mercury exhibits a maximum when measured as a function of V, referred to as zero point of charge, where $d\sigma/dV = 0$, $Q = 0$ and C is a minimum.

Ion-adsorption or surface charges are known to have pronounced effects on the mechanical properties of solids. These effects, particularly with regard to ionic solids, are discussed elsewhere in this volume[54]. No single mechanism or theory among those proposed, thus far, seems to account satisfactorily for these effects. Unfortunately, the direct study of SCLs, which might help identify the origin or nature of such effects, is not as yet possible in ionic solids.

3.2 Semiconductor-electrolyte interface

<u>General Discussion</u>. Under equilibrium conditions the space charge layer in the electrolyte phase at a semiconductor-electrolyte interface can be treated in the identical manner as that at the metal-electrolyte interface. Charge transfer processes at the semiconductor-electrolyte interface, however, may differ considerably from those at the metal-electrolyte interface[55].

A basic difference between a metal and a semiconductor is that the concentration of mobile carriers in the latter is (typically 10^{14} to $10^{18}/cm^3$) several orders of magnitude smaller than in the former (of the order of $10^{22}/cm^3$). The metal electrode is treated as an ideal conductor, with its interior un-- charged when polarized by an externally applied voltage; in the semiconductor electrode, however, electric charge is distributed well in its interior, i.e., within its surface space charge region. Thus, the potential drop at a semiconductor-electrolyte interface is much smaller than in the metal-electrolyte interface. It will be recalled[4] that the width of an SCL is proportional to the inverse square root of the carrier concentration; since the carrier concentration in electrolytes (e.g., $\sim 10^{19}$ carriers in a 0.01 M solution) is typically greater than in semiconductors, the space charge region in the interior of the semiconductor extends much further than in the interior of the electrolyte in a semiconductor-electrolyte system.

A semiconductor-electrolyte interface is schematically illustrated in Fig. 21[55]. Assuming that no dissolution of the semiconductor in the electrolyte takes place, charge flows from one phase into the other until the electrostatic potential gradient (indicated in the lower part of the figure as E_I) does not permit further charge flow, i.e., until the electrochemical potential of electrons becomes the same in the two sides of the interface. In Fig. 21 the semiconductor is n-type and the bands are bent up

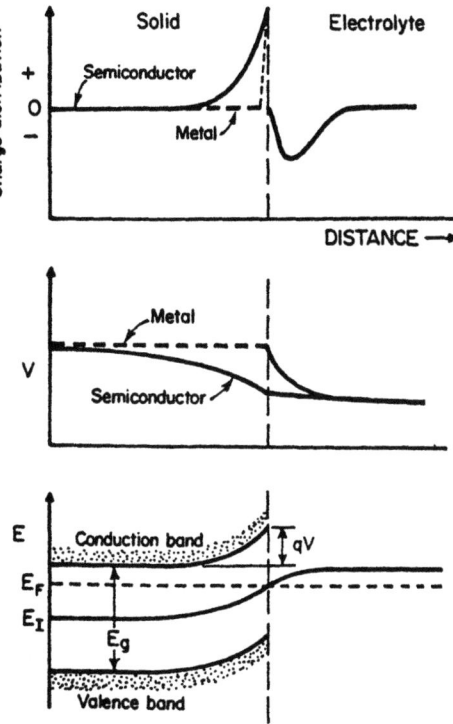

Figure 21. Schematic representation of the charge distribution, the electrostatic potential, V, and the energy of electrons E across a solid-electrolyte interface.

indicating the presence of oxidizing ions in the electrolyte (compare with chemisorption of gases on semiconductors, 2.1). The potential drop from the bulk of the semiconductor to the bulk of the electrolyte is Galvani or "inner" potential of the electrode. Obviously, the Galvani potential is determined by the position of the Fermi level in the semiconductor and the oxidation potential of the electrode reaction.(e.g., $M^{++} \rightleftarrows M^{+++}$).

As indicated in Fig. 21, there are four types of charges contributing to the semiconductor-electrolyte double layer: The semiconductor space-charge, Q_{sc}, surface state charge, Q_{ss}, the charge due to adsorbed ions, Q_I, and the charge in the diffused double layer of the electrolyte, Q_E. Electrical neutrality requires that:

$$Q_{sc} + Q_{ss} + Q_I + Q_E = 0 \tag{25}$$

Non-Equilibrium Phenomena. Under equilibrium conditions the metal-electrolyte and semiconductor-electrolyte interfaces are described by the same space-charge layer formalism. Under non-equilibrium conditions, however, there are basic differences between the two types of interfaces: Under an externally applied field, the field strength at the semiconductor surface is much smaller than at the metal surface. Accordingly, the influence of such field on electrode reactions is significantly different in the two cases.

The participation of two types of mobile carriers, i.e., electrons and holes, at the semiconductor-electrolyte interface is a fundamental aspect of the semiconductor-electrode reactions. Furthermore, excitation of electrons and holes at semiconductor-electrolyte interfaces, by illumination or other means, is not encountered in metal-electrolyte interfaces.

In a reaction at a semiconductor electrode (where no dissolution takes place) charge transfer can take place via electrons or holes:

$$M^+ \rightarrow M^{++} + e^- \tag{26}$$

$$M^+ + h^+ \rightarrow M^{++} \tag{27}$$

Since charge transfer must take place between charge states of a similar energy level (conservation of energy), the type of charge transfer will depend on the position of the energy levels of M^+ and M^{++} relative to the bottom of the conduction band and the top of the valence band of the semiconductor. The electron energy levels at M^+ and M^{++} can be defined in the same terms as the donor and acceptor levels in the semiconductor, i.e., by the energy of "ionization" of donor ($M_1^+ \rightarrow M_1^{++} + e^-$) or acceptor ($M_2^{++} + e^- \rightarrow M_2^+$) states. The M_1^+ and M_2^+ correspond to filled ion-states and the M_1^{++} and M_2^{++} to empty states. It should be pointed out in this context that the energy position of an occupied level of an ion may be significantly different from the energy position of the same level when empty, because of strong interactions of charged ions with the polar solvent.

Strongly oxidizing systems (e.g., $Fe^{+++} \rightleftarrows Fe^{++}$) tend to bend the semiconductor bands up (making the semiconductor surface p-type) and exchange electrons with the valence band (corresponding to acceptor states in the semiconductor); strongly reducing systems ($V^{+++} \rightleftarrows V^{++}$) bend the bands down (making the semiconductor surface n-type) and exchange electrons with the conduction band.

The adsorption of electron-acceptor or electron-donor ions on semiconductor surfaces is expected to be in many respects similar to charge transfer chemisorption discussed in part 2.

Regarding anodic and cathodic reactions at an n-type semiconductor under an externally applied emf, minority carriers are involved at the cathodic reaction (negative electrode) and majority carriers at the anodic reaction. Accordingly, the I-V curves exhibit saturation when the n-type semiconductor is made the cathode of a cell and no saturation when the n-type semiconductor is made the anode. The reverse situation prevails in p-type semiconductors. It is apparent that the saturation level (controlled by the availability of minority carriers) is photosensitive whereas the reactions involving majority carriers are not.

The equilibrium potential at the semiconductor-electrolyte interface does not depend on the concentration ratio of the holes and electrons in the semiconductor but on the ion ratio in the electrolyte. Thus, for a given electrolyte the surface potential of the semiconductor is fixed; as the Fermi level moves up (transition from p- to n-type material), the potential in the semiconductor rises or falls to adjust to the steady state with respect to the electrolyte. Thus, a convenient reference potential for characterizing the space charge behavior is the flat band potential (zero point of charge). This point is obtained by differential capacity measurements and corresponds to the minimum of the capacity curve (Fig. 22)[56]. Thus, the surface potential $-|\Delta V_s|$ can be determined from the externally applied voltage, V_E (measured against a reference electrode) and the flat band potential V_{fb}:

$$-|\Delta V_s| = V_E - V_{fb} \qquad (28)$$

Differential capacity data have contributed significantly to the understanding of the semiconductor electrolyte interface.

The flat-band potential can also be obtained by photovoltage measurements as a function of externally applied potential. At the flat band potential (Fig. 23)[57] the photovoltage changes sign for a small change in the applied potential.

3.3 Insulator-electrolyte interface

An insulator-electrolyte interface differs from a semiconductor-electrolyte interface in that the energy gap of the insulator is much greater than that of the semiconductor. Thus, only highly oxidizing systems can exchange electrons with the valence band

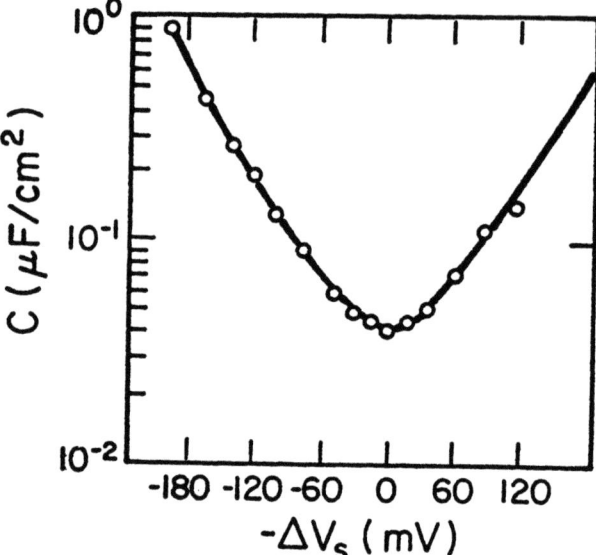

Figure 22. Differential capacity of a germanium electrode in an aqueous electrolyte.

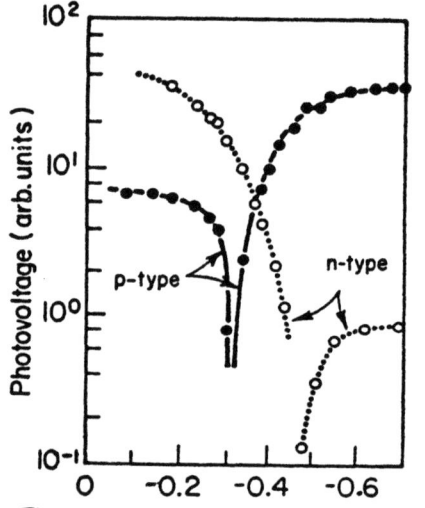

Figure 23. Instantaneous photovoltaic response as a function of electrode potential for a germanium electrode.

(inject holes into the insulator) and high reducing systems can exchange electrons with the conduction band (inject electrons). Due to the very small concentration of mobile carriers the space charge layer can extend throughout the insulator, whereas a semiconductor space charge layer is confined near the surface. It is apparent that high electric fields are necessary for the collection of injected carriers in an insulator. Experimental studies of insulator-electrolyte interfaces have been rather limited because of the difficulties associated with the application of high electrical fields to the electrode[58]. In fact, the work on insulator-interfaces has been essentially limited to organic materials and particularly anthracene single crystals.

A typical current voltage behavior of an insulator is shown in Fig. 24. At low voltages the injected carriers are not

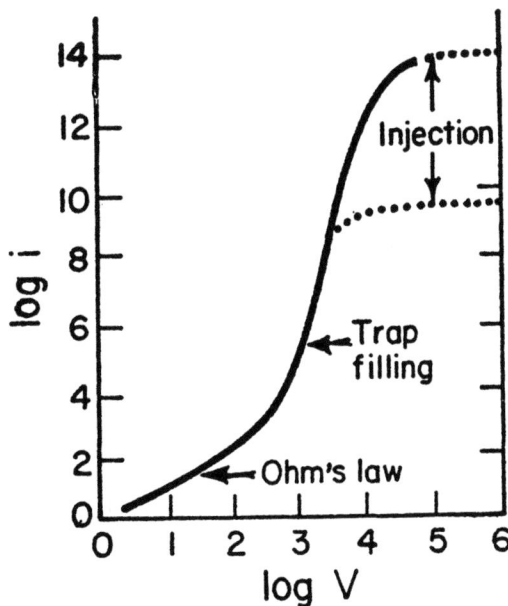

Figure 24. Current-voltage characteristics of insulator electrodes.

transferred from the surface into the bulk because of the high resistance of the electrode (Ohm's law region). As the applied voltage is increased the carriers move into the bulk and increase its conductivity. Thus, the current increases more steeply than linearly with applied voltage; the increase in current, however, is controlled by the pressure of localized energy states in the bulk (traps) which capture mobile carriers. As the applied voltage is further increased, the current reaches a saturation value determined by the rate of carrier injection.

SYNOPSIS

Space charge layers are spontaneously formed in solids and in liquids (SCLs in gases were not considered in this paper) when separation of charges (mobile or nonmobile) lower their free energy. The width of an SCL at a given temperature is in general proportional to the square root of the dielectric constant and inversely proportional to the square root of the mobile charge carrier concentration. Accordingly, in metals SCLs do not extend beyond a very few angstroms, whereas in semiconductors, liquids (electrolytes) and insulators they can extend to significant distances (in insulators they can extend through the entire bulk).

In semiconductors SCLs result primarily from trapping of mobile carriers at lattice discontinuities and lattice defects (surfaces, grain boundaries, dislocations, etc.) from abrupt changes in chemical composition (as in p-n junctions) and from strains (in piezoelectric semiconductors). SCLs play a very important role in the electronic processes encountered in semiconductors. The theoretical and experimental study of SCLs is far more feasible in semiconductors than in any other materials and has contributed significantly to the development of solid state science and technology.

In electrolytes SCLs result primarily from adsorption of ions on solids they are in contact with. SCLs in electrolytes play a fundamental role in electrochemical reactions, in colloid systems and biological processes. The experimental study of SCLs in electrolytes is extremely difficult and progress in their fundamental understanding continues at a slow rate.

In insulators SCLs result primarily from accumulation of charged vacancies or interstitials at lattice discontinuities or defects and from variations in chemical composition. The importance of SCLs in the various properties of insulators has been recognized for many years and detailed theoretical treatments on their origin and nature have been advanced[59]. However, their direct experimental study has met with as yet insurmountable difficulties. The essential requirements of the experimental methods successfully employed for the study of SCLs in semiconductors are not satisfactorily met in the case of insulators, i.e., moderate electrical conductivity, controlled generation and transport of charge carriers, controlled chemical composition and high crystalline perfection.

ACKNOWLEDGEMENT

The authors are grateful to the National Aeronautics and Space Administration and the National Science Foundation for their continued financial support and encouragement.

REFERENCES

1. For a historical review see Many, A., Goldstein, Y. and Grover, N.B., Semiconductor Surfaces, North-Holland, Amsterdam, 1965, Chap. I.
2. See for example, Potter, E.C., Electrochemistry, Cleaver-Hume, London, 1956, Chap. VII.
3. Brattain, W.G., and Garrett, C.G.B. N.Y. Acad. of Sciences, 58, 951 (1953).
4. Many, A., Goldstein, Y., and Grover, N.B., Semiconductor Surfaces, North-Holland, Amsterdam, 1965; Frankl, D.R., Electrical Properties of Semiconductor Surfaces, Pergamon, N.Y., 1967.
5. Kahng, D. and Nicollian, E.H., Applied Solid State Science, Wolfe, R., Ed., Academic, N.Y., 3, 2 (1972); Goetzberge, A. and Sze, S.M., ibid, 1, 153 (1969); Sze, S.M., Physics of Semiconductor Devices, Interscience, N.Y., 1969; Grove, A.G., Physics and Technology of Semiconductor Devices, John Wiley, N.Y., 1967.
6. Barlow, C.A., Jr., Physical Chemistry, Eyring, H., Ed., 9A, 167 (1970).
7. Matare, H.F., Defect Electronics in Semiconductors, Wiley-Interscience, N.Y., 1971, Chap. 8.
8. Kliewer, K.L. and Koehler, J.S., Phys. Rev. 104, A1226 (1965); Blakely, J.M. and Danyluk, S., Surface Sci. 40, 37 (1973).
9. Stern, F., Critical Reviews in Solid State Sciences, CRL, Cleveland 4, 499 (1974).
10. Davison, S.G. and Levine, J.D., Solid State Physics, Ehrenreich, W., Seitz, F. and Turnbull, D., Eds., Academic 25, 1 (1970).
11. Frankl, D.R., Critical Reviews in Solid State Physics, CRL, Cleveland 4, 455 (1974).
12. Heine, V., Proc. Tenth Conf. on the Physics of Semiconductors, Cambridge, Mass. (1970) 228.
13. Lagowski, J., Balestra, C.R. and Gatos, H.C., Surface Sci. 27, 547 (1971).
14. Many, A., Shappir, J., and Shaked, U., Molecular Procedures in Solid Surfaces, Drauglis, E., Gretz, R.D., and Jaffee, R.I., Eds., McGraw-Hill, N.Y. 1969, 199.
15. Read, W.T., Jr., Phil. Mag., 45, 775 (1954); ibid 45, 1119 (1954); ibid 46, 111 (1955).
16. Ref. 1, 79.

17. Hannay, N.B., *Semiconductors*, Hannay, N.B., Ed., Reinhold, N.Y., 1959, 24.
18. Lagowski, J., Balestra, C.L., and Gatos, H.C., *Surface Sci.* 29, 203 (1972); *ibid*, 29, 213 (1972).
19. Lagowski, J., Sproles, E.S., Jr., and Gatos, H.C., *ibid*, 30, 653 (1972).
20. Gatos, H.C., and Lagowski, J., *J. Vac. Sci. Technol.* 10, 130 (1973).
21. Garrett, C.G.B., and Brattain, W.H., *Phys. Rev.*, 99, 76 (1955).
22. Johnson, E.D., *Phys. Rev.*, 111, 153 (1958).
23. Lagowski, J., Baltov, I., and Gatos, H.C., *Surface Sci.* 40, 216 (1973).
24. Williams, R.J., *J. Phys. Chem. Solids*, 23, 1057 (1962).
25. Lashkarev, V.E., Salkov, E.A., and Khvostov, V.A., *Ukr. Fiz. Zh.* 14, 363 (1969).
26. Frankl, D.R., and Ulmer, E.A., *Surface Sci.* 6, 115 (1967).
27. Buimostrov, V.M., Gorban, A.P., and Litovchenko, V.G., *Surface Sci.* 3, 445 (1965).
28. Dimitruk, N.L., Lyashenko, V.I., Tereshenko, A.K. and Spektor, S.A., *Phys. Status Solidi* (a) 20, 53 (1973); Dimitruk N.L., Zuev, V.Z., Lyashenko, N.L., and Tereshenko, A.F., *Fiz. Tekh. Poluprovodnikov*, 4, 663 (1970).
29. See for example, Nye, I.F., *Physical Properties of Crystals*, Oxford University, 1969.
30. Lagowski, J., and Gatos, H.C., *Surface Sci.* 45, 353 (1974).
31. Lagowski, J., and Gatos, H.C., *Surface Sci.* 30, 49 (1972).
32. Lagowski, J., and Gatos, H.C., *Proc. Conf. on the Physics of Semiconductors*, Warsaw, 1972, 1462.
33. Lagowski, J., Gatos, H.C., and Sproles, E.S., Jr., *Appl. Phys. Lett.*, 27, 420 (1975).
34. Lagowski, J., Morawski, A., and Gatos, H.C., *Surface Sci.* 40, 216 (1973).
35. Lagowski, J., and Gatos, H.C., *Appl. Phys. Lett.*, 20, 14 (1972).
36. Lagowski, J., Gatos, H.C., and Sproles, E.S., Jr., *Appl. Phys. Lett.*, 26, 493 (1975).
37. For a recent review see Many, A., *Critical Reviews in Solid State Sciences*, CRC, Cleveland, 4, 515 (1974).
38. Brattain, W.H., and Bardeen, J., *Bell System Tech. J.*, 32, 1 (1953); Morrison, S.R., *J. Phys. Chem.* 71, 717 (1953).
39. Aigrain, P., and Dugas, C., *Z. Elektrochem.* 56, 363 (1952); Hauffe, K. and Engell, H.J., *ibid*, 56, 366 (1952); Engell, H.J., and Hauffe, K., *ibid*, 57, 762 (1953); Weisz, P.B., *J. Chem. Phys.* 20, 1483 (1952); *ibid*, 21, 1531 (1953).
40. For a recent review see Volkenstein, F.F., *Advances in Catalysis*, Academic, N.Y., 23, 157 (1973).
41. Krysemeyer, H.J., and Thomas, D.G., *J. Phys. Chem. Solids* 4, 78 (1958).
42. Law, J.T., *Semiconductors*, Hannay, N.B., Ed., Reinhold, 1959, Chap. 16.

43. Garrett, C.G.B., *J. Chem. Phys.* 33, 966 (1960).
44. Heiland, G., Mollwo, E., and Stöckmann, F., *Advances in Solid State Physics*, Seitz, F. and Turnbull, D., Eds., Academic, N.Y., 8, 193 (1959).
45. Ref. 1, Chap. 3.
46. Lagowski, J., Gatos, H.C., and Sproles, E.S., Jr., to be published.
47. Volkenstein, F.F., *The Electronic Theory of Catalysis on Semiconductors*, McMillan, N.Y., 1963; Hoenig, S.A. and Lane, J.R., *Surface Sci.* 11, 163 (1968).
48. Herring, C., *Structure and Properties of Solid Surfaces*, Gomer, R., and Smith, C.S., Eds., The University of Chicago, 1953, Chap. 1; Mullins, W.W., *Metal Surfaces*, ASM, Cleveland 1963, Chap. 2.
49. Timoshenko, S., *Vibration Problems in Engineering*, 2nd ed. Van Nostrand, N.Y., 1937, 364.
50. Stern, O., *Z. Elektrochem.* 30, 508 (1974); Chapman, D.L., *Phil. Mag.*, 25, 475 (1913).
51. Grahame, D.C., *Chem. Rev.*, 44, 441 (1947).
52. Delahay, P., *Double Layer and Electrode Kinetics*, Wiley, N.Y., 1965.
53. See for example review articles and references therein *Physical Chemistry, An Advanced Treatise*, Eyring, H., Ed., Academic, N.Y., 94 (1970).
54. Macmillan, N.H., this volume.
55. For comprehensive reviews on semiconductor-electrolyte interfaces see Dewald, J.F., *Semiconductors*, Hannay, N.B., Ed., Reinhold, N.Y., 1959, 727; Gerischer, H., *Advan. Electrochem. Electrochem. Eng.*, Interscience, N.Y., 1, 139 (1961); Gerischer, H., *Physical Chemistry*, Eyring, H., Ed., Academic, N.Y., 9A, 463 (1970).
56. Gerischer, H., *Physik. Chem.*, 69, 578 (1965).
57. Boddy, P.J., and Brattain, W.H., *Ann. N.Y. Acad. Sci.*, 101, 683 (1963).
58. For a review see, Mehl, W., and Hale, J.M., *Advan. Electrochem. Electrochem. Eng.*, Delahay, P., Ed., Interscience, N.Y., 6, 399 (1967).
59. See for example: Kliewer, K.L. and Koehler, J.S., *Phys. Rev.* 140, A1226 (1965); Kliewer, K.L., *Phys. Rev.* 140, A1241 (1965); Poeppel, R.B. and Blakely, J.M., *Surface Sci.* 15, 507 (1969: Blakely, J.M. and Danyluk, S., *Surface Sci.* 40, 37 (1973) and references therein.

SURFACE CHEMISTRY I: MULTI-COMPONENT SYSTEMS*

W. Henry Weinberg

Division of Chemistry and Chemical Engineering
California Institute of Technology, Pasadena,
California U.S.A.

ABSTRACT. In this review of the surface chemistry of multi-component systems, we discuss a classical description of adsorption with special emphasis on the gas-solid interface. We derive the classical Gibbs adsorption equation, and we then use this equation to deduce adsorption isotherms from various assumed two-dimensional equations of state of the adphase. We then discuss both two-dimensional condensation and two-dimensional critical phenomena and show that the latter is useful in determining values of the parameters in the equations of state of the adphase. We include comparisons between our derived results and experimental data.

1. INTRODUCTION

A description of adsorption, the binding of a molecule (the adsorbate) at a surface (the adsorbent or substrate), represents one of the most important problems in an understanding of the surface chemistry of multi-component systems. For example, from the point of view of mechanical and metallurgical properties of solids, adsorption at the gas-solid interface is important in such varied disciplines as lubrication, friction and wear, corrosion, and the adhesion of two solids. Nucleation and crystal growth are also affected fundamentally by surface composition, i.e., adsorption. It should also be noted that preferential segregation, or "adsorption", at the solid-solid interface of the new generation of microelectronic devices may have a profound influence on both their lifetime and stability. Finally, adsorption represents one of the

---

* The preparation of this chapter was facilitated by the generous support of the National Science Foundation.

elementary steps in all heterogeneously catalyzed surface reactions, and from this point of view a detailed understanding of the phenomenon of adsorption is mandatory.

Adsorption on surfaces may be considered at any of several levels of sophistication or generality. First, it may be considered from what is perhaps the most general point of view, namely that of classical thermodynamics. Second, adsorption may be considered from the point of view of statistical mechanics. This view is less general than the first and requires the formulation of microscopic models which contain physically measurable parameters, e.g., the number of adsorption sites on the surface, the binding energy of the adsorbate-substrate bond, the adsorbate-adsorbate interaction energy, and the phonon spectrum of the adsorbate overlayer. Third, the most detailed approach aimed at understanding adsorption is a quantum mechanical calculation both of the nature of the chemisorption bond as well as the perturbation of the substrate as a result of chemisorption, e.g., the change in the density of electronic states upon adsorption as a function both of the geometry of the adsorption site (adsite) and the coupling strength of the adsorbate to the substrate. In this review, we will consider adsorption both from a classical thermodynamic and a statistical mechanical point of view, but we will have essentially nothing to say about the quantum mechanical calculations. It should be intuitively obvious that this latter approach is a much less mature arena than are the former two.

When beginning to consider adsorption, especially on solid surfaces, it is useful to pose the following two questions: (1) Is the adsorbate localized at specific sites on the surface, or is it mobile in the two-dimensional plane of the surface? and (2) What is the nature of the interaction both between the adsorbate and the substrate and between the adsorbate and other adsorbate molecules on the surface? We will now consider these two questions in turn.

Another way of posing the first question above is to consider both the mean residence time of an adparticle at a specific adsite (the reciprocal of the mean "hopping" frequency between adsites) as well as the mean total residence time of the adparticle on the surface prior to desorption. These two characteristic times depend on the activation energy to surface diffusion and the activation energy to desorption, respectively. Almost all real surfaces are energetically heterogeneous, i.e., the binding energy is a function of the fractional coverage of adsorbate, but, as a rule of thumb, the activation energy to surface diffusion is approximately one-quarter to one-half the binding energy at the coverage of interest. Making use of the expression $\tau = \tau_0 \exp(+|q|/kT)$, we may calculate either the mean residence time on the surface or the mean residence time at an individual adsite according to our interpretation of the energy barrier q. Thus, we view the problems of surface diffusion

TABLE I

Mean Adsorbate Lifetime Either at a Particular Adsite or Total Lifetime on the Surface (Assuming $\tau_o = 10^{-13}$ sec).

q, kcal/mole	τ, sec at 300°K	τ, sec at 500°K
0.1	1.2×10^{-13}	1.1×10^{-13}
1.0	5.4×10^{-13}	2.7×10^{-13}
10	1.9×10^{-6}	2.4×10^{-9}
30	$7.2 \times 10^{+8}$ (22.8 yrs)	1.3
100	$7.2 \times 10^{+59}$	$5.2 \times 10^{+30}$

and desorption in terms of a particle surmounting a classical energy barrier, i.e., $1/\tau$ is the rate of success in such an endeavor, $1/\tau_o$ is the rate at which an attempt is made, and $\exp(-|q|/kT)$ is the probability of a successful venture. Realizing that $1/\tau_o$ is on the order of a vibrational frequency, approximately 10^{+13} sec^{-1}, we may calculate τ at various temperatures for various values of the activation energy. Results are presented in Table I for 300°K and 500°K and for activation energies ranging from 0.1 kcal/mole to 100 kcal/mole. At the temperatures used in the construction of Table I, we see that the measurement of the surface residence time of an adsorbate whose binding energy is 0.1 kcal/mole is beyond the capabilities of the most sophisticated spectroscopic techniques, whereas the measurement of the residence time of an adsorbate whose binding energy is 100 kcal/mole would also be a futile enterprise.

The results presented in Table I help us to begin to understand an answer to a part of the second question posed above concerning the nature of the adsorbate-substrate interaction. We define, rather arbitrarily, three classes of adsorption, namely, physical adsorption, weak chemisorption and strong chemisorption, and we delineate several properties of these types of adsorption in Table II. In this review, we will be more concerned with both physical and weak chemical adsorption as compared to strong chemical adsorption due to our interest in elucidating the competing effects of adsorbate-adsorbate and adsorbate-adsorbent interactions. In the case of strong chemical bonding, the interadsorbate interactions are generally much weaker than the adsorbate-adsorbent interactions, and thus this case is of less direct concern to us.

This chapter is divided into seven sections. In Section 2, we present a brief discussion of the classical thermodynamics of adsorption and in so doing gain useful insight into the macroscopic chemical phenomena which occur upon adsorption at a surface or interface. Moreover, we present a derivation of the Gibbs equation of adsorption, and we shall see that this equation is quite important in that

TABLE II

Three Types of Adsorption on a Solid Surface

Type	Heat of Adsorption, kcal/mole	Example	Comments
Physical	$q \leq 10$	Rare gas adsorption	No specific chemical bond formed. Induction forces generally important.
Weak chemical adsorption	$10 \leq q \leq 30$	CO on Pt and H_2 on most transition metals	Formation of a (generally) weak chemical bond between adsorbate and substrate. Often associative adsorption.
Strong chemical adsorption	$q \geq 30$	O_2 on most transition metals	Formation of a strong chemical bond between adsorbate and substrate. Usually strong perturbation of substrate as in oxidation of metals. Usually dissociative adsorption.

it allows us to derive adsorption isotherms which correspond to various two-dimensional equations of state of the adlayer. In Section 3, we outline the derivation of the isotherms from the two-dimensional equations of state via the Gibbs equation both for mobile as well as localized adsorbates. We present a microscopic (statistical mechanical) derivation of the two-dimensional van der Waals equation of state. The equations of state are of importance since they give information concerning the interadsorbate interaction strength. The isotherms are important both because they allow us to calculate the surface coverage as a function of gas phase pressure and, in addition, because they contain information concerning the adsorbate-substrate and the interadsorbate interaction strengths. In Section 4, we consider two-dimensional condensation of an adsorbate which obeys the van der Waals equation of state. We note the strong similarities with familiar three-dimensional phase transitions, and we derive values of the two parameters of the two-dimensional van der Waals equation of state in terms of (tabulated) ones of the three-dimensional van der Waals equation for a number of adsorbates. We then consider, in Section 5, various aspects of two-dimensional critical phenomena. In particular, we calculate the *critical value* of a number of relevant properties of the adphase (e.g., the two-dimensional spreading pressure, the molecular area, the temperature, the compressibility, and the surface coverage) in terms of the two parameters of several equations

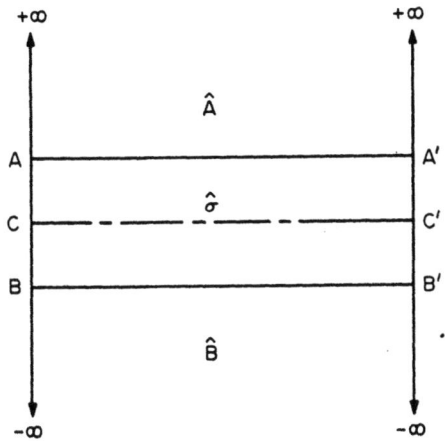

Fig. 1. Schematic representation of bulk phases $\hat{A}$ and $\hat{B}$ and the surface phase $\hat{\sigma}$.

of state. In this way, we compare directly the interadsorbate interaction strength with the adsorbate-adsorbent interaction strength. In Section 6, we compare our derived results with experimental data. In particular, we show it is possible to distinguish experimentally between an assumed mobile and immobile adlayer, that two-dimensional phase transitions (adsorbate condensation) have been observed experimentally, and that the experimental reality of an energetically heterogeneous surface makes a clear distinction between various assumed isotherms quite difficult. In Section 7, we conclude this chapter by giving a synopsis of our major results.

2. THE GIBBS ADSORPTION EQUATION

Although we are primarily concerned with the adsorption of a vapor at a solid surface, it is instructive for us to consider a general derivation of the surface tension in terms of the chemical potential. This relationship can then be written as the more familiar expression relating the two-dimensional spreading pressure to the gas phase pressure, but the detailed derivation reminds us of the important assumptions which are inherent in this so-called *Gibbs adsorption equation*.

We consider first two phases ($\hat{A}$ and $\hat{B}$) separated by a "surface phase" whose properties we denote by $\hat{\sigma}$. As may be seen in Fig. 1, we suppose that phase A exists from $+\infty$ to the plane AA', whereas phase B exists from $-\infty$ to the plane BB'. The two AA' and BB' planes are separated by the $\hat{\sigma}$ phase, and we place a third plane (CC') between and parallel to the AA' and BB' planes. For convenience, this hypothetical CC' plane is shown midway between the AA' and BB' planes in Fig. 1. Physically, we understand that above plane AA' our system possesses the properties of pure phase $\hat{A}$, and below plane BB' the system possesses the properties of pure phase $\hat{B}$. It is also clear that the surface phase $\hat{\sigma}$ will possess a character of its own, e.g., the preferential accumulation of one of the components may occur in this interfacial region. The plane we denote by CC' is a mathematical rather than physical plane, but it is nonetheless useful when considering adsorption from a classical thermodynamical point of view. This treatment is quite general and can be applied to a vapor-liquid, liquid-solid or vapor-solid interface as well as the interface between two immiscible liquids.

We may apply classical thermodynamics to the surface phase $\hat{\sigma}$, and we obtain for the total internal energy $E_{\hat{\sigma}}$

$$E_{\hat{\sigma}} = TS_{\hat{\sigma}} - PV_{\hat{\sigma}} + \gamma a + \sum_i N_{\hat{\sigma} i} \mu_i \qquad (1)$$

where all terms have their usual significance, i.e., T is temperature, S entropy, P pressure, V volume, γ surface tension, a interfacial surface area, N_i the number of molecules of component i, and μ_i the chemical potential of component i; the subscript $\hat{\sigma}$ refers to the surface phase. It is easy to obtain an expression similar to Eq. (1) for the *excess* properties of the *actual* surface, i.e., we simply subtract from Eq. (1) equivalent expressions had phase $\hat{A}$ existed from plane AA' to plane CC' and had phase $\hat{B}$ existed from plane BB' to plane CC'. If we denote the excess or surface properties by subscript s, we find

$$E_s = TS_s + \gamma a + \sum_i N_{si} \mu_i \qquad (2)$$

or expressed in differential form

$$dE_s = TdS_s + S_s dT + \gamma da + a d\gamma + \sum_i N_{si} d\mu_i + \sum_i \mu_i dN_{si} \qquad (3)$$

The similarity between Eqs. (1) and (2) should be noted as well as the constancy of the intensive thermodynamic variables and the fact that $V_{\hat{\sigma}}$ represents the volume enclosed by planes AA' and BB'.

We may now apply the first law of thermodynamics to a differential perturbation of the surface phase (an infinitesimal reversible change), and we obtain for the differential of the total internal energy of the surface phase the following expression:

$$dE_{\hat{\sigma}} = TdS_{\hat{\sigma}} - PdV_{\hat{\sigma}} + \gamma da + \sum_i \mu_i dN_{\hat{\sigma} i} \qquad (4)$$

As in the derivation of (2), we subtract the bulk properties of phases $\hat{A}$ and $\hat{B}$, had they extended to the dividing plane CC', from (4) obtaining

$$dE_\delta = TdS_\delta + \gamma d\alpha + \sum_i \mu_i dN_{\delta i} \tag{5}$$

Comparing Eqs. (3) and (5), we find that

$$S_\delta dT + \alpha d\gamma = -\sum_i N_{\delta i} d\mu_i \tag{6}$$

If we impose the restriction of constant temperature, and if we define $\alpha/N_{\delta i} \equiv \sigma_i$, then we deduce the usual form of the Gibbs equation [1,2,3], namely,

$$d\gamma\Big]_T = -\sum_i \left(\frac{1}{\sigma_i}\right) d\mu_i\Big]_T \tag{7}$$

We may proceed further by recalling that for an *ideal gas*, the chemical potential may be written as

$$\mu_i(p,T) = \mu_{oi}(T) + kT \ln p_i \tag{8}$$

where $\mu_{oi}(T)$ is the standard state chemical potential which for an ideal monatomic gas of mass m_i is given by

$$\mu_{oi}(T) = -kT \ln \left[\left(\frac{2\pi m_i kT}{h^2}\right)^{3/2} kT\right] \tag{9}$$

From (8), the differential of the chemical potential at constant temperature is found to be

$$d\mu_i\Big]_T = kT \, d\ln p_i\Big]_T \tag{10}$$

so that Eq. (7) becomes

$$d\gamma\Big]_T = -kT \sum_i \left(\frac{1}{\sigma_i}\right) d \ln p_i \tag{11}$$

We may rewrite (11) in more familiar form by (1) recognizing that $d\gamma]_T = -d\pi]_T$ where π is the two-dimensional spreading pressure with dimensions of energy divided by area, (2) considering only a single component adsorbate, thus dropping the i subscript, and (3) suppressing the dependence on the specific adsorbate surface area σ by introducing the specific surface area of an adsite β and a *fractional* surface coverage θ. For nonlocal adsorption β may be considered to be the minimum adsorbate area, i.e., that value of σ which applies to saturation coverage. Thus, Eq. (11) may be written as [4]

$$d\pi = \frac{kT}{\beta} \theta(p) d \ln p \qquad (12)$$

Eq. (12) is often known as the Gibbs equation (but one of many "Gibbs equations", one of the liabilities of being a brilliant scientist) or the "ideal gas" expression. Recalling that there are N_δ molecules of adsorbate on the surface the total area of which is a, and if we define the total number of adsites as M, then it is clear that

$$\theta(p) = N_\delta/M \qquad (13)$$

and

$$\beta = a/M \qquad (14)$$

so that $\beta/\sigma = \theta$. We note that the integrated form of Eq. (12) may be written as

$$\pi = \frac{kT}{\beta} \int \theta(p) \, d \ln p + \ln K(T) \qquad (15)$$

where K(T) is a constant of integration (which may be a function of temperature).

Prior to considering the way in which adsorption isotherms may be derived from the Gibbs equation, it is useful to compare the gas phase pressure p and molecular volume v with the corresponding two-dimensional quantities, namely, the spreading pressure π and the molecular area σ. Considering an ideal three-dimensional gas at 0°C and 1 atmosphere, we find that the molecular volume is 37,216 $Å^3$. This corresponds to a sphere, the radius of which is 20.7 Å. Let us now compare this result with the molecular area of a two-dimensional ideal gas at 0°C and a spreading pressure of 1 dyne/cm. Under these conditions, we find [using Eq. (40) of the next section] that the molecular area is 377.1 $Å^2$, i.e., a circle, the radius of which is 11.0 Å. Thus, based on this average separation distance in the three- and two-dimensional phases, we conclude that 1 dyne/cm is a rather large value for the spreading pressure. In fact, we find that the value of the spreading pressure which corresponds to an average separation distance of 20.7 Å, the three-dimensional result, at 0°C is 0.28 dynes/cm. Conversely, the three-dimensional pressure which corresponds to an average molecular separation distance of 11.0 Å (that value applying in two dimensions with a spreading pressure of 1 dyne/cm), at 0°C is found to be 6.68 atmospheres.

3. TWO-DIMENSIONAL EQUATIONS OF STATE AND ADSORPTION ISOTHERMS

Eq. (12), or equivalently Eq. (15), is quite useful since it may be used in conjunction with various assumed two-dimensional equations of state in order to derive the *adsorption isotherms* [p(θ)] corresponding to the respective equations of state. We will describe in some detail the procedure for the two-dimensional van der Waals equation of state and then summarize our results for a variety of other equations of state.

The two-dimensional van der Waals equation of state [4], by analogy with its three-dimensional counterpart [e.g., 5], may be written as

$$(\pi + \alpha/\sigma^2)(\sigma - \beta) = kT \tag{16}$$

where α and β are the analogs of the three-dimensional van der Waals constants a and b, i.e., they account for an attractive potential between adsorbates on the surface and an "excluded area" correction on the surface, respectively. The relationship between α and a and between β and b is quantified below, but we now return to the problem of deriving the adsorption isotherm corresponding to (16). In order to do this, we rewrite (16) in terms of θ rather than σ (recall $\sigma = \beta/\theta$), differentiate the resulting equation holding temperature constant, and equate the resulting expression for $d\pi$ with that given by the Gibbs equation (12). We find that

$$\frac{kT}{\beta} \theta \, d \ln p = \left[\frac{kT}{\beta}\left(\frac{1}{1-\theta}\right)^2 - \frac{2\alpha\theta}{\beta^2}\right] d\theta \tag{17}$$

Integration of (17) yields the adsorption isotherm corresponding to the van der Waals equation of state, namely

$$p = \left(\frac{K_1 \theta}{1-\theta}\right) \exp\left[\frac{\theta}{1-\theta} - \frac{2\alpha\theta}{kT\beta}\right] \tag{18}$$

where K_1 is a constant of integration which may be a function of temperature.

We now return to the question of the microscopic significance of α and β vis-à-vis their three-dimensional counterparts a and b in the van der Waals equation of state [6]. The most straightforward way to understand the relationship between the two- and three-dimensional van der Waals constants is to derive both the two-dimensional and three-dimensional equations of state using statistical mechanics. We will outline such a derivation for the two-dimensional case and simply quote the well-known result for the three-dimensional case [7].

Let us assume that our system consists of N identical and indistinguishable admolecules which are adsorbed as a *mobile film* on an adsorbent, the purpose of which is simply to provide a potential field for the admolecules. Henceforth, for convenience, we suppress the subscript δ on N. Furthermore, we allow an adsorbate-adsorbate interaction which we model using a Sutherland potential with an induced dipole-induced dipole attraction (dispersion force) and a hard sphere repulsion, i.e., we assume the interadsorbate potential is given by

$$u(r) = \begin{cases} +\infty &, r \leq R \\ -|\varepsilon|\left(\frac{R}{r}\right)^6 &, r > R \end{cases} \qquad (19)$$

where $|\varepsilon|$ is the minimum in the potential function (proportional to the square of the polarizability of the admolecule), and R is the "hard disk" radius (the collision radius which is equal, for example, to the atomic diameter of two identical colliding adatoms). With this pair interaction function we may now derive the *total* potential energy of interaction between an adsorbate at an arbitrary origin and *all other* adsorbate molecules by using Eq. (19) and performing the following integration

$$\phi = -\int_R^\infty |\varepsilon|\left(\frac{R}{r}\right)^6 \left(\frac{N}{a}\right) 2\pi r \, dr \qquad (20)$$

where $(N/a)(2\pi r)dr$ is the number of adsorbate molecules between r and r + dr. We find that

$$\phi = -\frac{|\varepsilon|\pi R^2}{2}\left(\frac{N}{a}\right) = -\frac{|\varepsilon|\pi R^2}{2\sigma} \,. \qquad (21)$$

In order to derive the two-dimensional van der Waals equation of state, we need an expression for the canonical ensemble partition function for our system (the independent thermodynamic variables are N, a and T). Since the admolecules are assumed to be indistinguishable, the canonical ensemble partition function, Q, may be written in terms of the single admolecule canonical ensemble partition function, q, as follows

$$Q(N,a,T) = q(a,T)^N/N! \qquad (22)$$

For our assumed mobile film with adsorbate-adsorbate interactions, $q(a,T)$ is composed of the product of several terms, viz., the partition function corresponding to two translational degrees of freedom, the partition function of one vibrational degree of freedom orthogonal to the surface of the adsorbent, the partition functions of the internal (vibrational and rotational) degrees of freedom of the adsorbate, and two Boltzmann factors describing two different interaction potentials. One Boltzmann factor takes account of the adsorbate-adsorbent interaction [U_0, in Eq. (23) below, is defined as the minimum in the adsorbate-adsorbent potential function] and assures us of a consistent zero of energy, namely the adsorbate infinitely removed from the adsorbent and possessing zero velocity relative to the adsorbent. The other Boltzmann factor arises from the adsorbate-adsorbate interaction and is given by an exponential of the potential function of Eq. (21)

It should be noted in passing that Eq. (22) strictly applies to a system of independent particles. This implies that we must assume that each admolecule translates independently in an *isotropic potential field* generated by all the other admolecules which in turn translate randomly on the surface. Thus, our model is in the spirit of the Bragg-Williams lattice gas theory [8]. This also implies that the two-dimensional translational partition function must reflect the excluded volume (really excluded area) of one admolecule relative to another due to the impulsive nature of the repulsive part of the interadsorbate potential. If we define a free adsorbate area, $\sigma_f \equiv \sigma - (1/2)\pi R^2$, (the factor of 1/2 in the excluded area arises from the *pair* interaction with half the effect assigned to each member of the adsorbate-adsorbate pair), then we can write the single adsorbate molecular canonical ensemble partition function as

$$q(a,T) = \left(\frac{\sigma_f}{\Lambda^2}\right)\left(\frac{e^{-\hbar\omega_z/2kT}}{1-e^{-\hbar\omega_z/kT}}\right) e^{|U_0|/kT} \prod_j\left(\frac{e^{-\hbar\omega_j/2kT}}{1-e^{-\hbar\omega_j/kT}}\right)$$
$$\times q_r q_{e\ell} e^{-\phi/2kT} \qquad (23)$$

where Λ is the effective de Broglie wavelength of the admolecule $(\Lambda = h/(2\pi mkT)^{1/2})$, ω_z is the vibrational frequency orthogonal to the surface, q_r accounts for any rotational degrees of freedom that might be present in the adsorbate, $q_{e\ell}$ is the electronic partition function of the adsorbate, and ω_j are the internal vibrational frequencies of the (in general) polyatomic adsorbate molecules. The term σ_f/Λ^2 of Eq. (23) is the two-dimensional translational partition function (analogous to the three-dimensional function, v_f/Λ^3). The form of the vibrational partition functions, i.e., $e^{-\hbar\omega_j/2kT}/(1-e^{-\hbar\omega_j/kT})$, assumes that the vibrations undergo harmonic motion (springs which obey Hooke's Law), and, as implied above, the factors $e^{|U_0|/kT}$ and $q_{e\ell}$ assure a consistent energy zero when counting vibrational energy upward from the vibrational ground state. It should be recalled that the vibrational energy of a one-dimensional harmonic oscillator is given by

$$E_n = (n + 1/2)\hbar\omega, \quad n = 0,1,2\ldots \qquad (24)$$

where n is the vibrational quantum number which is equal to zero when the oscillator is in its ground state, and the frequency ω

is given (classically) by

$$\omega = (\kappa/m^*)^{1/2} \tag{25}$$

where κ is the force constant of the oscillator and m^* is the reduced mass. The electronic levels of an adsorbate are usually unexcited, and in that case $q_{e\ell}$ is given by the degeneracy of the electronic ground state multiplied by a Boltzmann factor involving the ground state electronic energy, i.e.,

$$q_{e\ell} = w_{e\ell} \exp(|D_e|/kT) \tag{26}$$

where $w_{e\ell}$ is the ground state degeneracy and $|D_e|$ is the electronic ground state energy. In the event that excited electronic levels are occupied, Eq. (26) must be replaced by

$$q_{e\ell} = \sum_j w_{ej} \exp(-E_{ej}/kT) \tag{27}$$

where the summation is over electronic levels, and E_{ej} is the energy of the jth electronic level.

Although Eq. (23) may appear somewhat forbidding, it is in fact quite easy to derive the two-dimensional van der Waals equation of state from it. We proceed by recalling that the Helmholtz energy, A, is given by

$$A = -kT \ln Q \tag{28}$$

and the two-dimensional spreading pressure, π, is given by

$$\frac{\pi}{kT} = - \left[\frac{\partial (A/kT)}{\partial a} \right]_{N,T} \tag{29}$$

where a is the adsorbent surface area. Eq. (29) is the two-dimensional analog of the following familiar equation which applies in three dimensions:

$$\frac{p}{kT} = - \left[\frac{\partial (A/kT)}{\partial V} \right]_{N,T} \tag{30}$$

Thus, anticipating taking a partial derivative with both N and T held constant, we may combine Eqs. (22), (23) and (28) to obtain

$$-\frac{A}{kT} = N \ln \left(\frac{\sigma f}{\Lambda^2} \right) - \frac{N\phi}{2kT} + \text{terms which are not explicit functions of } a \tag{31}$$

Recalling that $\sigma_f = \sigma - (1/2)\pi R^2$ and $\sigma = a/N$, and using the expression for ϕ from (21), we make use of (29) to obtain the following expression for π:

$$\frac{\pi}{kT} = \left\{ \frac{1}{\sigma - \frac{1}{2}\pi R^2} \right\} - \frac{|\varepsilon|\pi R^2}{4kT\sigma^2} \tag{32}$$

If we recognize that the excluded area, $(1/2)\pi R^2$, is just the parameter β of Eq. (14), i.e., $\beta \to \sigma$ as monolayer coverage is approached, and if we define

$$\alpha \equiv (1/4)|\varepsilon|\pi R^2 \tag{33}$$

then Eq. (32) becomes

$$(\pi + \alpha/\sigma^2)(\sigma - \beta) = kT \tag{16}$$

which is the two-dimensional van der Waals equation of state that we had written down earlier by analogy with the three-dimensional equation. The advantage of the above derivation is that now the microscopic meaning of α and β is clarified. Moreover, if we recognize that the three-dimensional van der Waals equation of state may be written as [5,7,8]

$$(p + a/v^2)(v - b) = kT \tag{34}$$

where

$$a = (2/3)|\varepsilon|\pi R^3 \tag{35}$$

and

$$b = (2/3)\pi R^3 \tag{36}$$

then we can relate the two-dimensional van der Waals constants α and β to the respective three-dimensional ones a and b. We find that

$$b/\beta = (4/3)R \tag{37}$$

and

$$a/\alpha = (8/3)R \tag{38}$$

or, combining Eqs. (37) and (38),

$$a\beta/\alpha b = 2 . \tag{39}$$

We expect that Eqs. (37) through (39) should be most applicable to admolecules which retain best their gas phase properties upon ad-

TABLE III

EQUATION OF STATE		ISOTHERM		COMMENTS
$\left[\pi + \dfrac{a}{\sigma^2}\right](\sigma - \beta) = kT$	(16)	$p = \dfrac{K_1 \theta}{1 - \theta} \exp\left(\dfrac{\theta}{1 - \theta} - \dfrac{2a\theta}{kT\beta}\right)$	(18)	Van der Waals equation. Mobile adsorbate film with both attractive and repulsive (free area or impulsive) interactions.
$\pi\sigma = kT$	(40)	$p = K_2 \theta$	(41)	Henry's Law or ideal gas equation. Mobile adsorbate film with no adsorbate-adsorbate interaction
$\pi(\sigma - \beta) = kT$	(42)	$p = \dfrac{K_3 \theta}{1 - \theta} \exp\left(\dfrac{\theta}{1 - \theta}\right)$	(43)	Volmer or hard disk equation. Mobile adsorbate film with impulsive (free area) repulsive interaction.
$\pi\sigma = \nu kT$	(44)	$p = K_4 \theta^\nu$	(45)	Küster or Freundlich equation. Empirical equation which has been shown to apply theoretically if the adsorbate-adsorbent binding energy decreases exponentially with increasing surface coverage.
$\left[\pi + \dfrac{a}{T\sigma^2}\right](\sigma - \beta) = kT$	(46)	$p = \dfrac{K_5 \theta}{1 - \theta} \exp\left(\dfrac{\theta}{1 - \theta} - \dfrac{2a\theta}{k\beta T^2}\right)$	(47)	Berthelot equation similar to van der Waals. Mobile adsorbate film with both attractive and repulsive (free area) interactions.
$\pi\left[\exp\left(\dfrac{a}{\sigma kT}\right)\right](\sigma - \beta) = kT$	(48)	$p = \dfrac{K_6 \theta}{(1-\theta)^{1-\xi+\xi^2/2}} \exp\left[\dfrac{\theta[1-\xi+(\theta-1/2)\xi^2]}{1-\theta}\right]$	(49)	Dieterici equation. Empirical equation similar to van der Waals. Mobile adsorbate film with both attractive and repulsive (free area) interactions.
$\left[\pi + \dfrac{a}{\sigma^2}\right]\left[\sigma - \bar{c}(\beta\sigma)^{1/2}\right] = kT$	(50)	$p = K_7 \theta \exp\left[-\dfrac{2a\theta}{\beta kT} + f(\theta)\right]$	(51)	Eyring equation. Mobile adsorbate film with both attractive and repulsive (free area) interactions.
$\pi\sigma = kT\left[1 + \dfrac{B'(T)}{\sigma} + \dfrac{C'(T)}{\sigma^2} + \cdots\right]$	(52)	$p = K_8 \theta \exp\left[\dfrac{2\theta}{\beta}B'(T) + \dfrac{3}{2}\left(\dfrac{\theta}{\beta}\right)^2 C'(T) + \cdots\right]$	(53)	Virial equation. Mobile adsorbate film with both attractive and repulsive interactions.
$\pi\sigma = \dfrac{kT\sigma}{\beta}\ln\left(\dfrac{\sigma}{\sigma - \beta}\right)$	(54)	$p = \dfrac{K_9 \theta}{1 - \theta}$	(55)	Langmuir equation. Immobile adsorbate film with no adsorbate-adsorbate interactions.
$\pi\sigma = kT\dfrac{\sigma}{\beta}\ln\left(\dfrac{\sigma}{\sigma - \beta}\right) - \dfrac{C\omega\beta}{2\sigma}$	(56)	$p = \dfrac{K_{10} \theta}{1 - \theta} \exp\left(-\dfrac{C\omega\theta}{kT}\right)$	(57)	Fowler and Guggenheim equation. Immobile adsorbate film with an attractive interaction.

sorption. For example, the adsorbent should neither polarize nor orient the adsorbate, and we would thus hesitate to apply Eqs. (37) through (39) to adsorbates other than those which are spherically symmetric.

We have given a very detailed discussion of the van der Waals equation of state of an adsorbate and its associated isotherm. In so doing the assumptions inherent in the van der Waals treatment were quite obvious. For example, the adsorbate forms a *mobile* film on the adsorbent and the adsorbate-adsorbate interaction is of the Sutherland type with an attractive potential proportional to r^{-6}. In Table III, we present several other equations of state and the isotherms corresponding to each as derived using the Gibbs equation (12) in the same way the isotherm for the van der Waals equation was derived previously in (18).

Eq. (40) is the two-dimensional analog of the ideal gas equation and results in a linear adsorption isotherm commonly known as Henry's Law [Eq. (41)]. This equation applies to a mobile two-dimensional gas of independent point adsorbates. As will be noted subsequently, this isotherm is ineffectual at high surface coverages.

Eq. (42) is the Volmer equation [9] and applies in the case of a mobile two-dimensional adsorbate film with no attractive adsorbate-adsorbate interaction, but including an impulsive repulsive (free area) interaction. The parameter β is a constant for a given system and represents the excluded area of an admolecule, i.e., the specific adsorbate area at maximum surface coverage. When considering an adsorbent with structure (i.e., a lattice) β is given by Eq. (14). The isotherm, derived from the Volmer equation of state by assuming ideality in the gas phase [i.e., the Gibbs equation (12)] is given by Eq. (43). The Kuster [10] or Freundlich [11] equation of state is given by Eq. (44). The corresponding isotherm is shown in Eq. (45). This empirical isotherm is quite successful in describing the adsorption of ionic solutes at a liquid-liquid interface, and it is also used frequently in interpreting kinetic data from heterogeneously catalyzed surface reactions. It has been shown that this isotherm applies rigorously for the case of heats of adsorption which decrease exponentially with increasing surface coverage, and, in particular, it may be shown [12,13]

$$|U_a| = - |U_m| \ln \theta \qquad (58)$$

and

$$\nu = \frac{|U_m|}{kT} \qquad (59)$$

where $|U_a|$ is the heat of adsorption, $|U_m|$ is a constant, and the parameter ν, which also appears in the equation of state and the isotherm, is an empirical constant characteristic of the particular adsorbate-adsorbent system. It has been observed that ν usually lies between approximately two and ten.

Eqs. (46), (48) and (50) correspond respectively to the two-dimensional Berthelot, Dieterici and Eyring equations of state. The corresponding isotherms derived from these equations of state using the Gibbs equation are shown in Eqs. (47), (49) and (51). All these equations of state are two-parameter equations (α and β) and apply to mobile two-dimensional adsorbate gases. They are thus similar in spirit to the van der Waals equation. The parameter ξ in the Dieterici isotherm is defined by

$$\xi \equiv \alpha/\beta kT \tag{60}$$

and, due to the nonanalytical nature of the integral encountered in deriving the isotherm, the latter is an approximation which is accurate through terms of order ξ^2. We require in addition that $\xi\theta < 1$ in order for the expansion used in deriving (49) to be valid. A deeper discussion of the physical meaning of the various interaction parameters (α and β) for these equations of state will be deferred until later in this chapter. In three dimensions, the Dieterici equation of state has been found to be very accurate in so far as reproducing the critical compressibility coefficient ($Z_c \equiv pV_c/RT_c$, where the subscript c denotes the critical point) for twenty-five nonpolar gases, and it is probably the best, on balance, of all analytic two-parameter equations of state [5,14].

The Eyring equation of state, Eq. (50), has as its basis the concept of a cell model and thus in three dimensions has been successfully applied to both liquids as well as dense gases [5,15]. We would expect the two-dimensional Eyring equations [(50) and (51)] to be especially appropriate for relatively high surface coverages. The derivation of the Eyring equation of state is quite similar to that of the van der Waals equation presented earlier, but we will sketch it briefly here since it is rather interesting to understand the origin of the excluded area $\bar{c}(\beta\sigma)^{1/2}$. We see schematically in Fig. 2 that in one dimension, a particular adsorbate molecule (the so-called "wanderer" which is cross-hatched in the figure) is constrained to a cell length equal to $2\sigma^{1/2} - 2R$ where in our standard nomenclature σ is the molecular area of an adsorbate (i.e., the surface area divided by the total number of adsorbate molecules), and R is a core diameter of the molecular adsorbate (equal to the collision radius considering bimolecular encounters of two admolecules). For a potential with an impulsive repulsive part, R is just the diameter of the "hard core" of the potential as we discussed earlier in the case of a Sutherland potential, Eq. (19). If we assume simple cubic packing of our adsorbate, then we can mentally rotate the line of atoms in Fig. 2

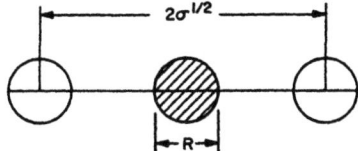

Fig. 2. Eyring unit cell in one dimension. The "wandering adatom" of diameter R is cross-hatched.

by 90° in the plane of the page, and we will have generated the two-dimensional model which we use to formulate the Eyring equation. In both orthogonal directions of our simple cubic cell the "free length" available to our wandering admolecule is given by

$$\sigma_f^{1/2} = 2(\sigma^{1/2} - R) \tag{61}$$

so that the free area available is just

$$\sigma_f = 4(\sigma^{1/2} - R)^2 . \tag{62}$$

We may now derive an expression for the spreading pressure (the equation of state) within the context of this model just as we did previously for the van der Waals model [see Eqs. (28) through (32)]. In particular, we assume the adsorbate-adsorbate interaction potential is of the Sutherland kind, so we obtain a parameter α corresponding to adsorbate-adsorbate attraction which is identical to the van der Waals α. In order to make use of Eq. (29) to derive the equation of state, it is simply necessary to rewrite our expression for σ_f in terms of a. If we associate the excluded area β with $\pi R^2/2$ then we retain the identical β parameter that we used in the van der Waals equation. We find that

$$\sigma_f = 4\left[(a/N)^{1/2} - (2\beta/\pi)^{1/2}\right]^2 , \tag{63}$$

and the equation of state is found to be that given by (50) where

$$\bar{c} \equiv (2/\pi)^{1/2} = 0.7979 \tag{64}$$

for our assumed adsorbate packing.

In deriving the isotherm from the Eyring equation (50), we again encounter a nonanalytic integral, but we obtain the isotherm given by Eq. (51) if we expand the integrand in a power series and integrate term-by-term. The function $f(\theta)$ in (51) is found to be given by

$$f(\theta) \equiv 2.3937\theta^{1/2} + 1.2732\theta + 0.8466\theta^{3/2} + 0.6079\theta^2 +$$
$$0.4527\theta^{5/2} + 0.3440\theta^3 + 0.2647\theta^{7/2} + 0.2053\theta^4 +$$
$$0.1602\theta^{9/2} + 0.1255\theta^5 + 0.0986\theta^{11/2} + 0.0777\theta^6 +$$
$$0.0613\theta^{13/2} + 0.0484\theta^7 + 0.0383\theta^{15/2} + 0.0304\theta^8 +$$
$$0.0241\theta^{17/2} + 0.0191\theta^9 + 0.0151\theta^{19/2} +$$
$$0.0120\theta^{10} + \ldots \tag{65}$$

One final equation of state worthy of mention which also applies to a mobile adsorbate film is the virial equation given by (52) with $B'(T)$, $C'(T)$ ... the second, third ... two-dimensional virial coefficients. The isotherm corresponding to the virial equation of state is given by (53). For a full discussion of the two-dimensional virial equation, reference should be made to the comprehensive review of Steele [16].

Finally, in Table III we present two equations of state and the corresponding isotherms which are applicable to *immobile* adsorbate films. Eqs. (54) and (55) are the Langmuir equation of state and isotherm, respectively. These equations apply to an immobile adsorbate film with no adsorbate-adsorbate interactions. The Langmuir isotherm is especially important since it is often used to make the connection between gas phase pressure and surface composition when interpreting observed heterogeneously catalyzed reaction rates. Eqs. (56) and (57) are the Fowler and Guggenheim equation of state and isotherm, respectively [8]. The model on which the Fowler and Guggenheim equation is based is similar to the Langmuir model, namely, an immobile adsorbate film. However, the Fowler and Guggenheim model takes into account attractive interactions between admolecules. In particular, the constants c and ω of Eqs. (56) and (57) specify the number of near neighbor "bonds" and the interaction energy per "bond". For example, an adlayer with packing corresponding to the fcc(111) surface has $c = 6$, the sc(100), fcc(100) or bcc(110) have $c = 4$, the fcc(110) has $c = 2$, etc. Since only the product $c\omega$ enters into Eqs. (56) and (57), the Fowler and Guggenheim equation of state is a two-parameter equation of state, with the parameters being $c\omega$ and β. The parameter β in both the Langmuir and the Fowler and Guggenheim formulations is the specific area of an adsite and is given by Eq. (14). Indeed, Eq. (14) is now more sensible since we have a lattice onto which admolecules may bind in a specific way.

4. TWO-DIMENSIONAL CONDENSATION

In those equations of state which contain a term describing

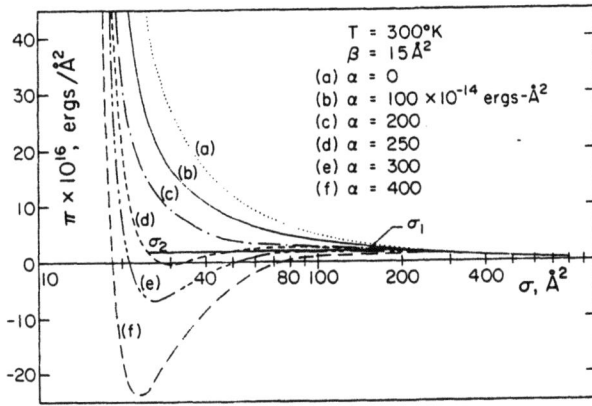

Fig. 3. The variation of the two-dimensional spreading pressure π with specific adsorbate molecular area σ for the two-dimensional van der Waals equation of state for the case of $T = 300°K$ and $\beta = 15 \text{ Å}^2$. The strength of the adsorbate-adsorbate attractive interaction is given by the parameter α and ranges from zero to 400×10^{-14} erg - Å^2 for curves (a) through (f). Two-dimensional condensation is evident in curves (d), (e) and (f).

interadsorbate attraction, it is possible to predict two-dimensional condensation under certain conditions. We will now consider adsorbate condensation in some detail, and we will discover that most of the familiar concepts which apply to three-dimensional condensation phenomena carry over in a logical way to two dimensions, i.e., the attractive interadsorbate forces cause deviations from ideal gas behavior, which eventually lead to two-dimensional condensation if the attractive forces are large enough and if the two-dimensional critical temperature is greater than the surface temperature. In order to understand two-dimensional condensation, consider the van der Waals equation of state

$$(\pi + \alpha/\sigma^2)(\sigma - \beta) = kT .\tag{16}$$

Let us now consider the variation of π with σ for various parametric values of α and assuming the (not atypical) values of β and T are 15 Å^2 and 300°K, respectively. Such a representation is given in Fig. 3 for six different values of α, namely, zero, 100, 200, 250, 300 and 400 all in units of 10^{-14} erg - Å^2. There are a number of features to notice in Fig. 3. First, note that the case of $\alpha = 0$ does *not* represent an ideal gas since we still include an impulsive repulsion term in the interadsorbate potential via the excluded area β. In fact, curve (a) of Fig. 3 is the Volmer equation of state (42), and the absence of an attractive part of the potential naturally precludes condensation. Likewise, no conden-

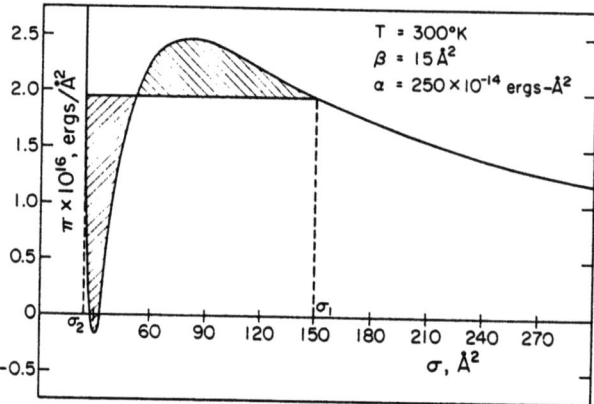

Fig. 4. The variation of spreading pressure with adsorbate molecular area for the two-dimensional van der Waals equation of state for the case $T = 300°K$, $\beta = 15 \text{ Å}^2$ and $\alpha = 250 \times 10^{-14}$ erg $-$ Å^2. Two-dimensional condensation is evident, and the horizontal tie-line connects a two-dimensional gas with $\sigma_1 \approx 150$ Å^2 to a two-dimensional liquid with $\sigma_2 \approx 25$ Å^2.

sation is observed for curves (b) and (c) corresponding to values of α of 100×10^{-14} erg $-$ Å^2 and 200×10^{-14} erg $-$ Å^2. However, curves (d), (e) and (f) of Fig. 3 are not monotonic, but rather display both a minimum at rather small values of σ and a (weak) maximum at larger values of σ. Thus, there are of necessity *three* values of σ which correspond to a single value of the spreading pressure π when the relationship between π and σ is considered at a constant temperature. This phenomenon may be seen more clearly in Fig. 4 where curve (d) of Fig. 3 ($\alpha = 250 \times 10^{-14}$ erg $-$ Å^2) is shown in more detail. The prediction of a maximum and a minimum in the $\pi - \sigma$ curves only occurs if the temperature is below the two-dimensional critical temperature T_c'. We will examine two-dimensional critical phenomena in more detail subsequently.

The maxima and minima in curves (d), (e) and (f) of Fig. 3 are manifestations of two-dimensional condensation. The value of σ of intermediate magnitude (e.g., approximately 53 Å^2 in Fig. 4) is unstable physically, whereas the smaller and larger values of σ (e.g., approximately 25 and 150 Å^2, respectively) have a definite physical meaning. For very dilute surface coverages ($\sigma > 150$ Å^2), we have a two-dimensional gas if we consider the case shown in Fig. 4. When the surface coverage reaches 150 Å^2, then condensation occurs, and we obtain a two-dimensional *liquid* with $\sigma \approx 25$ Å^2. Further adsorption then occurs at constant π as shown by the horizontal *tie-line* connecting σ_1 and σ_2 in Figs. 3 and 4. This tie-line, just as in the case of the three-dimensional condensation, is positioned such that the area between it and the $\pi - \sigma$ curve is

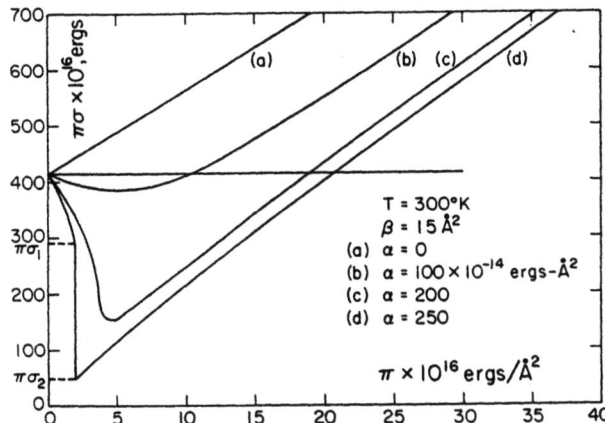

Fig. 5. The variation of the product of the spreading pressure and the specific adsorbate molecular area with spreading pressure for the two-dimensional van der Waals equation of state for the case of $T = 300°K$ and $\beta = 15$ Å^2. The strength of the adsorbate-adsorbate attractive interaction is given by the parameter α and ranges from zero to 250×10^{-14} erg - Å^2 for curves (a) through (d). The vertical tie-line in curve (d) is evidence of two-dimensional condensation.

identical both below and above the tie-line, i.e., the two cross-hatched regions in Fig. 4 above and below the tie-line corresponding to $\pi > 1.95 \times 10^{-16}$ erg/Å^2 and $\pi < 1.95 \times 10^{-16}$ erg/Å^2, respectively, are equal in area.

Yet another way in which we can consider two-dimensional condensation is by constructing a plot of the product of spreading pressure and specific adsorbate area as a function of the spreading pressure, i.e., a plot of $\pi\sigma$ versus π as shown in Fig. 5. In this figure we again assume $T = 300°K$ and $\beta = 15$ Å^2, and we present four curves corresponding to values of α (in units of 10^{-14} erg - Å^2) of zero, 100, 200 and 250. We found in Fig. 3 that of these values of α, only $\alpha = 250 \times 10^{-14}$ erg - Å^2 results in two-dimensional condensation. This view is confirmed in Fig. 5, and as we would expect, there is a vertical tie-line in curve (d) connecting the two-dimensional gas with $\pi\sigma_1 \approx 292 \times 10^{-16}$ erg to the two-dimensional liquid with $\pi\sigma_2 \approx 49 \times 10^{-16}$ erg at a constant spreading pressure of $\pi \approx 1.95 \times 10^{-16}$ erg/Å^2. In Fig. 5, the horizontal line corresponding to $\pi\sigma = 414.2 \times 10^{-16}$ erg represents ideal gas behavior, i.e., $\pi\sigma = kT$ at $300°K$. We note that curve (a), the Volmer or hard disk equation of state with no interadsorbate attraction, deviates immediately and rather seriously from ideal gas behavior. The inclusion of an attractive part of the interadsorbate potential, for example with $\alpha = 100 \times 10^{-14}$ erg - Å^2 as shown in curve (b) of Fig. 5, results in a more nearly ideal gas like behavior especially at smaller values of π. This is due to a fortuitous cancellation of opposing forces and should not be construed to imply a two-dimensional gas of particles which are independent over a wide range of π, σ and T. We also note in passing that Fig. 5 quantifies the notion that repulsive interactions result in positive deviations from ideality ($\pi\sigma > kT$), whereas attractive interactions result in negative deviations from ideality ($\pi\sigma < kT$).

Later in this chapter, we will present expressions for α and β for several equations of state in terms of two-dimensional critical point parameters, but we will now present values of the three-dimensional constants a and b of the van der Waals equation of state [17]. We can calculate the adsorbate diameter R if we know the value of b by making use of Eq. (36) written in the following form

$$R = (3b/2\pi)^{1/3} \qquad (66)$$

Values of the two-dimensional van der Waals equation of state constants α and β may then be calculated from Eqs. (57) and (58), i.e.,

$$\alpha = 3a/8R \qquad (67)$$

and

$$\beta = 3b/4R \qquad (68)$$

We summarize the reported values of a and b [17], as well as the calculated values of R, α and β in Table IV. The values of α and β in Table IV, when combined with the results presented in Figs. 3, 4 and 5, allow us to understand better when to expect the occurrence of two-dimensional condensation.

It should be noted that our entire discussion of condensation including the calculated van der Waals constants which apply to a two-dimensional system shown in Table IV has assumed the adsorbate is spherically symmetric. Unfortunately, this assumption is likely to be far less useful in two dimensions than in three dimensions. In three dimensions, the free translation and rotation of the molecules tends to average out asymmetries even in highly polar molecules. For example, the dipolar-dipolar attractive potential is proportional to the inverse cube of the separation distance, whereas the spatial averaged attractive potential is proportional to the inverse sixth power of the separation distance. This angular averaged interaction is a very useful concept in gas phase molecular physics [5]. However, the potential field of the adsorbent tends to accentuate asymmetries in the electronic distribution of the adsorbate, and this effect tends to cloud any results which

TABLE IV

Various Parameters in van der Waals Theory

Adsorbate	$a \times 10^{12}$, erg - $Å^3$	$b, Å^3$	$R, Å$	$\alpha \times 10^{14}$, erg - $Å^2$	$B, Å^2$	$k_1 (300°K)$
He	0.0953	39.35	2.66	1.34	11.09	0.058
Ne	0.588	28.37	2.38	9.26	8.94	0.50
Ar	3.757	53.45	2.94	47.92	13.64	1.70
Kr	6.474	66.05	3.16	76.83	15.68	2.37
Xe	11.71	84.76	3.43	128.02	18.53	3.34
H_2	0.683	44.18	2.76	9.28	12.00	0.37
O_2	3.798	52.85	2.93	48.61	13.53	1.73
N_2	3.882	64.97	3.14	46.36	15.52	1.44
CO	4.072	66.16	3.16	48.32	15.70	1.49
HCl	10.24	67.76	3.19	120.38	15.93	3.65
H_2O	15.26	50.62	2.89	198.01	13.14	7.28
CO_2	10.03	70.85	3.23	116.45	16.45	3.42
NH_3	11.65	61.55	3.09	141.38	14.94	4.57
CH_4	6.293	71.03	3.24	72.84	16.44	2.14
CCl_4	56.95	229.62	4.79	445.85	35.95	5.99
Cl_2	18.13	93.34	3.55	191.51	19.72	4.69
HBr	12.43	73.57	3.27	142.55	16.87	4.08
Hg	22.60	28.16	2.38	356.09	8.87	19.4
NO	3.743	46.31	2.81	49.95	12.36	1.95
H_2S	12.38	71.18	3.24	143.29	16.48	4.20
CS_2	32.45	127.59	3.93	309.64	24.35	6.14
SO_2	18.75	93.57	3.55	198.06	19.77	4.84
NO_2	14.76	73.45	3.27	169.27	16.85	4.85
N_2O	10.56	73.30	3.27	121.10	16.81	3.48
C_2H_2	12.26	85.27	3.44	133.65	18.59	3.47
C_2H_4	12.49	94.87	3.56	131.57	19.99	3.18
C_2H_6	15.33	105.93	3.70	155.37	21.47	3.49
C_3H_8	24.20	140.21	4.06	223.52	25.90	4.17
iso - C_4H_{10}	35.95	189.61	4.49	300.25	31.67	4.58
iso - C_5H_{12}	50.41	235.26	4.83	391.38	36.53	5.17

are based on a two-dimensional angular average. It is possible to account for these effects in a rather *ad hoc* way, and reference should be made to a review article by de Boer for details [18]. After having given this warning, we will nevertheless continue to make use of two-dimensional spherically averaged quantities.

5. TWO-DIMENSIONAL CRITICAL PHENOMENA

The occurrence of two-dimensional condensation that we explored in detail in the previous section implies that the two-dimensional analog of the three-dimensional critical point also exists. We will examine the implications of two-dimensional critical phenomena as well as in several cases derive expressions for the parameters α and β which appear in the two-dimensional equations of state (see Table III). At the critical point

$$\left(\frac{\partial \pi}{\partial \sigma}\right)_T = \left(\frac{\partial^2 \pi}{\partial \sigma^2}\right)_T = 0 \ . \tag{69}$$

The two equations indicated by (69) together with the actual

TABLE V

Critical Parameters of Several Two-Dimensional Equations of State
and the van der Waals Isotherm

Equation of State or Isotherm	π_c	σ_c	T'_c	Z'_c	θ_c	α	β	$\left(\frac{p}{p_o}\right)_c$	k_{1c}
Van der Waals Equation of State	$\frac{\alpha}{27\beta^2}$	3β	$\frac{8\alpha}{27k\beta}$	$\frac{3}{8}$	$\frac{1}{3}$	$\frac{9}{8}kT'_c\sigma_c$	$\frac{1}{3}\sigma_c$		
Dieterici Equation of State	$\frac{\alpha e^{-2}}{4\beta^2}$	2β	$\frac{\alpha}{4k\beta}$	$2e^{-2}$	$\frac{1}{2}$	$2kT'_c\sigma_c$	$\frac{1}{2}\sigma_c$		
Berthelot Equation of State	$\frac{(6k\alpha)^{1/2}}{36\beta^{3/2}}$	3β	$\frac{2}{9}\left(\frac{6\alpha}{k\beta}\right)^{1/2}$	$\frac{3}{8}$	$\frac{1}{3}$	$\frac{9}{8}kT'^2_c\sigma_c$	$\frac{1}{3}\sigma_c$		
Van der Waals Isotherm		3β	$\frac{8\alpha}{27k\beta}$		$\frac{1}{3}$	$\frac{9}{8}kT'_c\sigma_c$	$\frac{1}{3}\sigma_c$	$\frac{k_2}{2}e^{-7/4}$	6.75

equation of state represent three equations which may be solved simultaneously in order to evaluate α, β and $Z'_c \equiv \pi_c\sigma_c/kT'_c$. We will denote all parameters and physical quantities which apply at the critical point by a subscript c, and we use the standard notation of Z' to represent the compressibility of our two-dimensional gas. A prime is placed on both the two-dimensional critical compressibility as well as the two-dimensional critical temperature in order to distinguish them from their three-dimensional counterparts. Results are shown in Table V for π_c, σ_c, T'_c, Z'_c, $\theta_c = \beta/\sigma_c$, α and β for three different equations of state, namely, the van der Waals equation (16), Dieterici equation (48) and the Berthelot equation (46).

We might also note that the three-dimensional analog of Eq. (69), namely,

$$\left(\frac{\partial p}{\partial V}\right)_T = \left(\frac{\partial^2 p}{\partial V^2}\right)_T = 0 \tag{70}$$

may be applied to the three-dimensional van der Waals equation of state (34) just as we did above for the two-dimensional case. We find that $a = 9V_ckT_c/8$, $b = V_c/3$ and $Z_c = 3/8$. Thus, it is evident from these results coupled with Eq. (39) that

$$a\beta/\alpha b = T_c/T'_c = 2 \quad \text{or} \quad T'_c = (1/2)T_c , \tag{71}$$

i.e., the two-dimensional critical temperature is expected to be just half the value which applies to three dimensions (within the context of the van der Waals theory). Serious discrepancies are found between Eq. (71) and experimental data [6], due presumably to a preferential orientation of the adspecies on the surface.

We will next consider the characteristics which two-dimensional condensation phenomena impart to the adsorption isotherms. We will continue to concentrate on the van der Waals equation although similar ideas can be applied to other equations of state and their associated isotherms (see Table III). Our first step in the analysis of the van der Waals isotherm is to write the isotherm, given by Eq. (18), in a reduced form by dividing both sides of (18) by the vapor pressure p_o of the adsorbate at the temperature of interest. We obtain the following expression [6]

$$\frac{p}{p_o} = k_2 \left(\frac{\theta}{1-\theta}\right) \exp\left(\frac{\theta}{1-\theta} - k_1\theta\right) \quad (72)$$

where

$$k_2 \equiv K_1/p_o \, , \quad (73)$$

and

$$k_1 \equiv 2\alpha/kT\beta \, . \quad (74)$$

The constant k_2 in the reduced isotherm Eq. (72) is a measure of the adsorbate-substrate interaction energy, and k_1 is a measure of the adsorbate-adsorbate interaction energy. It is clear from Eq. (74) that k_1 is *directly* proportional to the interadsorbate interaction energy, and the form of the isotherm (72) shows that the adsorbate-substrate interaction is *inversely* proportional to k_2. In order to obtain a more quantitative understanding of realistic magnitudes for k_1, we present values in Table IV which were calculated assuming a temperature of 300°K. The propensity for condensation, as judged by the value of k_1, is 335 times greater for Hg as compared with He.

We can clearly see the inverse relationship between the strength of the adsorbate-substrate interaction and the magnitude of k_2 from (72), but the question arises as to what a typical value of k_2 actually is. In fact, k_2 can be shown to be a ratio of partition functions which under certain conditions can be shown to reduce to [19]

$$k_2 = (M/a)(\nu_o/p_o)(2\pi mkT)^{1/2} \exp(-|U_o|/kT) \quad (75)$$

where M/a is the surface site concentration (i.e., the reciprocal of the excluded area, or simply β^{-1}), ν_o is the frequency of vibration of the adsorbate along its desorption coordinate, m is the adsorbate mass, and U_o is the heat of adsorption. If we assume that $M/a = 10^{14}$ cm^{-2}, $\nu_o = 2 \times 10^{12}$ sec^{-1}, m = 30 amu, T = 300°K and p_o = 1000 torr, then we find that k_2 = 19 if U_o = 2 kcal/mole, k_2 = 0.7 if U_o = 4 kcal/mole and k_2 = 8 × 10^{-4} if U_o = 8 kcal/mole. This order of magnitude calculation is sufficient to illustrate typical values of k_2, and, in addition, it serves to show the

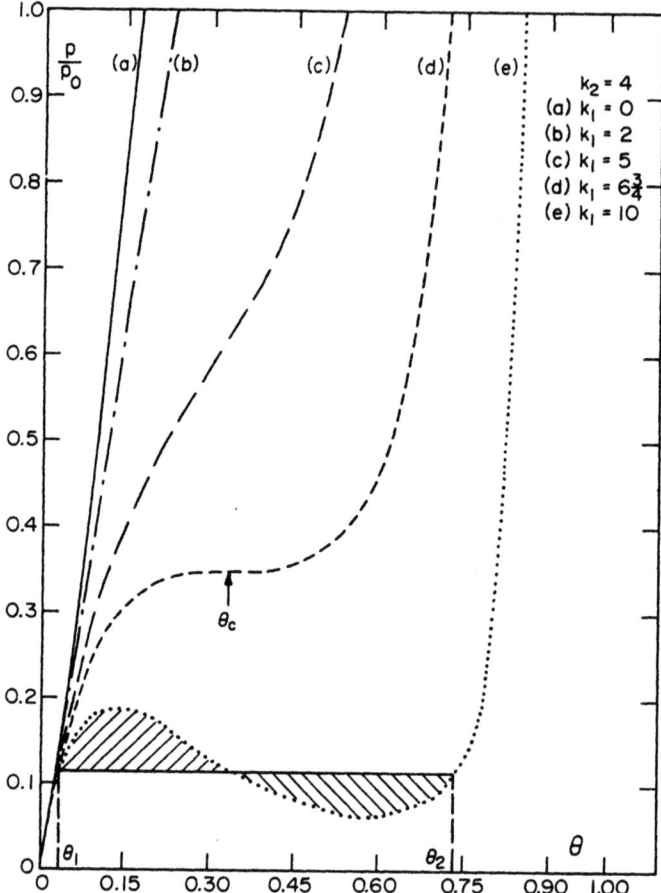

Fig. 6. The variation of the reduced gas phase pressure (the pressure divided by the vapor pressure) with surface coverage for the van der Waals isotherm with $k_2 = 4$ and $0 \leq k_1 \leq 10$ [see Eqs. (72), (73) and (74)]. The critical isotherm is given by curve (d) and corresponds to $k_1 = 6.75$. Note that the critical surface coverage corresponds to $\theta_c = 1/3$. Two-dimensional condensation is evident in curve (e) which corresponds to $k_1 = 10$.

strong dependence of the magnitude of k_2 on the adsorbate-substrate interaction strength; k_2 is an exponential function of the binding energy as seen in (75).

We present in Fig. 6 a graphical representation of Eq. (72), namely, the variation of p/p_0 with θ for the case $k_2 = 4$ and $0 \leq k_1 \leq 10$. We choose the case of relatively weak adsorption (the relatively large value of k_2) in order to illustrate the variation of p/p_0 with θ over a rather wide range of p/p_0, that is we avoid the somewhat common, but less pedagogic case of surface saturation at a very small reduced pressure. Referring to Fig. 6, curve (a) applies to the case of no adsorbate-adsorbate interaction ($k_1 = 0$) and is thus representative of the Volmer or hard disk isotherm given by Eq. (43). This curve is slightly convex with respect to the abscissa, typical of a purely repulsive interadsorbate potential. Curve (b) of Fig. 6 corresponds to a value of $k_1 = 2$ and thus represents weak interadsorbate attraction. The curve is now concave toward the abscissa. Curve (c) with $k_1 = 5$ is concave toward the abscissa at small values of θ, but it becomes convex at higher coverages passing through an inflection point. This curve is reminiscent of the case of a two-dimensional gas above its critical temperature which we treated earlier. Curve (d) of Fig. 6 with $k_1 = 6.75$ corresponds to the critical van der Waals isotherm. The critical point is described by the equation

$$\left(\frac{\partial p}{\partial \theta}\right)_T = \left(\frac{\partial^2 p}{\partial \theta^2}\right)_T = 0 \qquad (76)$$

If we apply Eqs. (76) together with the isotherm (72), then we can derive expressions for the van der Waals constants α and β as well as the critical parameters σ_c, T_c', θ_c, $(p/p_0)_c$ and k_{1c}. These quantities are shown in Table V. In the case of σ_c, T_c', θ_c, α and β, the results obtained from the isotherm are in agreement with those obtained earlier using the two-dimensional van der Waals equation of state. In addition, however, we have now shown that the reduced pressure at the critical point $(p/p_0)_c$, is equal to $0.5\ k_2 \exp(-1.75)$; and the value of k_1 corresponding to the critical isotherm is equal to 6.75.

Returning to Fig. 6, we note finally that curve (e) corresponds to $k_1 = 10$ and is representative of the physical case of attractive forces being sufficiently large that condensation occurs. A two-dimensional gas is adsorbed as p/p_0 is increased from zero until the coverage $\theta_1 \approx 0.035$ is reached, at which point additional adsorption results in the condensation of a two-dimensional liquid. At the discontinuity in coverage when θ increases from $\theta \approx 0.035$ to $\theta \approx 0.73$, there is co-existence of a two-dimensional gas and a two-dimensional liquid on the surface. The tie-line connecting these two coverages is positioned such that the cross-hatched area above the tie-line is equal to the cross-hatched area beneath it. This entire analysis is obviously similar to our earlier treatment of condensation using two-dimensional equations of state. We are now considering $p - \theta$ rather than $\pi - \sigma$ data, and we have made connection between the two by using the Gibbs adsorption equation (12).

We have now obtained a methodology for obtaining the parameters of the two-dimensional equations of state (e.g., α and β of the van der Waals equation) by using either the microscopic relations given by Eqs. (35) through (38) or by using critical point properties summarized in Table V. Alternatively, we can evaluate α and β by relating them to the two-dimensional second virial coefficient B'(T) and its first derivative with respect to temperature dB'(T)/dT. We will not elaborate further on this method other than to mention that it is quite analogous to the procedure in three dimensions [5]. It is now left for us to suggest how to determine the integration constants $K_i(T)$ for i = 1, 2 . . . 10 of the ten isotherms presented in Table III [4,20].

In order to evaluate the $K_i(T)$, it is convenient to consider the molecular change in Gibbs energy, $\Delta\mu_{ad}$, upon adsorption relative to a gas phase standard state with the adsorption occurring under isothermal conditions

$$\Delta\mu_{ad} = \Delta h_{ad} - T\Delta s_{ad} \quad . \tag{77}$$

In Eq. (77), Δh_{ad} and Δs_{ad} are the molecular changes in enthalpy and entropy upon adsorption, respectively. We can also express the change in molecular Gibbs energy upon adsorption as

$$\Delta\mu_{ad} = kT \ln(p/p^*) \tag{78}$$

where p is the three-dimensional gas phase pressure and p* is the three-dimensional standard state gas phase pressure, usually one atmosphere (1.01325 x 10^6 dynes/cm^2). Furthermore, we recall that Δh_{ad} may be written as [4,21]†

$$\Delta h_{ad} = -q_{diff} - kT + \frac{\beta T}{\theta} \left(\frac{\partial \pi}{\partial T}\right)_\theta \tag{79}$$

and we note that all the isotherm equations in Table III may be written as

$$p = K_i(T) f_i(T,\theta) \quad , \quad i = 1, 2, 3 \ldots 10 \tag{80}$$

In Eq. (79), q_{diff} is the differential heat of adsorption, i.e., the differential of the integral heat of adsorption with respect to surface coverage at constant temperature and surface area, divided by the number of adsorbate molecules at saturation (monolayer) coverage.

If we combine Eqs. (77) through (80), we find that

$$\ln K_i(T) = -\frac{q_{diff}}{kT} - 1 + \frac{\beta}{k\theta}\left(\frac{\partial \pi}{\partial T}\right)_\theta - \frac{\Delta s_{ad}}{k} +$$

---

† By convention Δh_{ad} is positive if heat is absorbed upon adsorption, whereas q_{diff} is positive if heat is liberated.

$$+ \ln(1.01325 \times 10^6) - \ln f_i(T,\theta) \quad (81)$$

if we employ cgs units. It is important to recall that the derivation of the isotherm equations in Table III, all of which may be written in the form of Eq. (80), insures that the integration constants $K_i(T)$ are not explicit functions of surface coverage. Rather, we can evaluate $K_i(T)$ at *any* coverage, and this coverage will then correspond to the two-dimensional standard state. Although we can define the standard state in an arbitrary way, it is most common to use the following convention: (1) For a mobile two-dimensional gas, the average separation distance between admolecules is equated to the average separation distance of an ideal three-dimensional gas at one atmosphere and 273°K; and (2) For an immobile adlayer, the standard state coverage is taken to be 1/2. Thus, in Eq. (81), q_{diff}, θ, Δs_{ad} and f_i are all evaluated at the standard state. This result is in agreement with our earlier conclusion that K_i is connected with the adsorbate-substrate interaction, and the adsorbate-adsorbate interactions are only considered via the parameters appearing in the two-dimensional equations of state.

6. COMPARISON WITH EXPERIMENT

Although an exceedingly large amount of experimental work has been carried out in the areas discussed in this chapter, i.e., physical and weak chemical adsorption, we will restrict our comments to a consideration of the following three questions: (1) Is it possible to distinguish experimentally between a localized and a mobile adsorbate phase? (2) Have two-dimensional phase transitions been observed, and, if so, what isotherm do the data suggest? and (3) How does the (omnipresent) energetically heterogeneous nature of real surfaces affect our conclusions, e.g., the choice of an adsorption isotherm? We will now consider these three questions in turn.

Everett [22] has attempted to distinguish between a localized and a mobile adlayer by comparing both calculated and experimental partial molar entropies of adsorption. In his calculations, he assumed the adlayer was either an ideal two-dimensional gas, a non-ideal two-dimensional gas [using an equation of state which accounts for interadsorbate repulsion, e.g., the Volmer equation (42)], or a localized layer. Accordingly, he calculated either the translational or configurational entropy and compared his results to experimental data for the adsorption of Ar [23], NH_3 [24] and CO_2 [25] on charcoal. Only in the case of CO_2 adsorption on charcoal could a choice be made between a non-ideal mobile adlayer and a local adlayer, and the latter was preferred. The choice could be made in the case of CO_2 adsorption on charcoal because the data covered a very broad surface coverage range. We conclude that although it is difficult to distinguish between mobile and immobile adlayers by comparing experimental and calculated partial molar entropies, such distinctions may be made if the data are of a high quality and cover a broad range of surface coverages.

The existence of two-dimensional phase transitions (i.e., adsorbate condensation) has been well-documented experimentally [26-28]. The data of Fisher and McMillan [28] are particularly noteworthy. These authors have conclusively shown that both Kr and CH_4 adsorbed on NaBr undergo a two-dimensional phase transition, whereas adsorbed N_2 does not. Moreover, their NaBr surfaces were sufficiently energetically homogeneous (see below) that they were able to conclude their measured isotherms for all three adsorbates were consistent only with localized adsorbates. The data followed very closely an isotherm similar to the one of Fowler and Guggenheim given in Eq. (56).

We should also note that the fact all real surfaces have a distribution of adsorbate-substrate binding energies, i.e., they are energetically heterogeneous, makes it difficult in practice to choose one particular adsorption isotherm as being correct physically. If we could make such a choice of a "correct" isotherm then we would understand better the molecular physics of adsorbate-adsorbate and adsorbate-substrate interactions. However, it is rather seldom that the energy distribution function of the substrate is known with sufficient accuracy. What is needed is extremely precise isotherm measurements over a wide surface coverage range coupled with an independent measurement of the distribution function. In that case, a straightforward inversion scheme may be invoked to determine the true physical isotherm, and thereby to understand the molecular chemistry. This detailed knowledge could then be applied to many different kinds of practical problems involving surface chemistry.

7. SYNOPSIS

We will close by reminding the reader that the derivation of the adsorption isotherms shown in Table III makes use of the Gibbs adsorption equation which in turn assumes that the three-dimensional gas phase is ideal. The two-dimensional adphase, of course, need not be ideal, and Table III is a collection of various isotherms corresponding to several different two-dimensional equations of state. The assumption of ideality in the three-dimensional gas phase is not a serious restriction since the reduced pressure, p/p_o, required for the submonolayer adsorption discussed in this chapter is sufficiently small ordinarily to preclude any effects of non-ideality in the gas phase.

The equations of state which we derived are useful in

predicting the thermodynamic properties of the two-dimensional ad-phase in a similar way in which three-dimensional equations of state are employed. The adsorption isotherms which we derived are useful in calculating adsorbate concentrations on the surface from gas phase pressures, the latter a readily measurable quantity. This knowledge of surface composition is vital in the interpretation of a myriad of different physical and chemical phenomena.

REFERENCES

1. Gibbs, J. W., *Collected Works, Volume I, Thermodynamics*, Longmans, Green and Co., New York, 1928.

2. Davies, J. T., and Rideal, E. K., *Interfacial Phenomena*, Academic Press, New York, 1961.

3. Flood, E. A., in *The Solid-Gas Interface, Volume 1*, Flood, E. A., Ed., Marcel Dekker, New York, 1967, chap. 2.

4. Ross, S., and Olivier, J. P., *On Physical Adsorption*, Interscience, New York, 1964.

5. Hirschfelder, J. O., Curtiss, C. F., and Bird, R. B., *The Molecular Theory of Gases and Liquids*, John Wiley, New York, 1954.

6. de Boer, J. H., *The Dynamical Character of Adsorption*, 2nd Ed., Clarendon Press, Oxford, 1968.

7. Hill, T. L., *Introduction to Statistical Thermodynamics*, Addison-Wesley, Reading, Massachusetts, 1960.

8. Fowler, R. H., and Guggenheim, E. A., *Statistical Thermodynamics*, Cambridge University Press, 1939.

9. Volmer, M., *Z. physik. Chem.*, 115, 253, 1925.

10. Küster, F. W. A., *Liebigs Ann.*, 283, 360, 1894.

11. McBain, J. W., *The Sorption of Gases and Vapors by Solids*, Routledge, London, 1932.

12. Halsey, G., and Taylor, H. S., *J. Chem. Phys.*, 15, 624, 1947.

13. Halsey, G., *Advan. Catal.*, 4, 259, 1952.

14. Beattie, J. A., and Stockmayer, W. H., *Reports on Progress in Physics*, 7, 195, 1940.

15. Eyring, H., and Hirschfelder, J. O., *J. Phys. Chem.*, 41, 249, 1937.

16. Steele, W. A., in *The Solid-Gas Interface, Volume 1*, Flood, E. A., Ed., Marcel Dekker, New York, 1967, chap. 10.

17. *Handbook of Chemistry and Physics*, 49th Ed., 1968, D-107.

18. de Boer, J. H., *Advan. Colloid Science, Volume 3*, Interscience, New York, 1950, 1.

19. Weinberg, W. H., *J. Catal.*, 28, 459, 1973.

20. Everett, D. H., *Trans. Faraday Soc.*, 46, 942, 1950.

21. Clark, A., *The Theory of Adsorption and Catalysis*, Academic Press, New York, 1970.

22. Everett, D. H., *Proc. Chem. Soc.*, 38, 1957, a reprint of the Tilden Lecture, delivered before the Chemical Society in London on November 17, 1955..

23. Homfray, R., *Z. phys. Chem.*, 74, 129, 1910.

24. Richardson, L. B., *J. Am. Chem. Soc.*, 39, 1828, 1917.

25. Smith, F. W., Ph.D. Thesis, St. Andrews University, Scotland, 1952.

26. Jura, G., Loeser, E. H., Basford, P. R., and Harkins, W. D., *J. Chem. Phys.*, 14, 117, 1946.
    Jura, G., Harkins, W. D., and Loeser, E. H., *J. Chem. Phys.*, 14, 344, 1946.

27. Clark, H., and Ross, S., *J. Am. Chem. Soc.*, 75, 6081, 1953.
    Ross, S., and Winkler, W., *J. Am. Chem. Soc.*, 76, 2637, 1954.
    Ross, S., and Clark, H., *J. Am. Chem. Soc.*, 76, 4291, 1954.
    Ross, S., and Clark, H., *J. Am. Chem. Soc.*, 76, 4297, 1954.

28. Fisher, B. B., and McMillan, W. G., *J. Am. Chem. Soc.*, 79, 2969, 1957.
    Fisher, B. B., and McMillan, W. G., *J. Chem. Phys.*, 28, 549, 1958.
    Fisher, B. B., and McMillan, W. G., *J. Chem. Phys.*, 28, 555, 1958.
    Fisher, B. B., and McMillan, W. G., *J. Chem. Phys.*, 28, 562, 1958.

DISCUSSION

Comment by E. Kröner:

In your discussion of the van der Waals law you assume that the two-dimensional van der Waals potential decays like $1/r^6$. Considering that the power 6 has to do with the three-dimensionality of the space and that the interactions between two absorbed atoms occur via the substrate, how obvious is it that the power 6 is the most natural choice in the two-dimensional van der Waals potential? Does the power 6 influence the reported results, for instance, the values of k and if yes, how?

Reply:

The $(6-\infty)$ potential energy function used in this chapter to model interadsorbate interactions assumes an angular averaged induced dipolar-induced dipolar interaction, a <u>through-space</u> interaction. As pointed out in the text, such a procedure lies on a weaker foundation than the equivalent Keesom average in three dimensions, and it should apply best to spherically symmetric admolecules. If one chose to use, for example, an $(n-\infty)$ interaction potential, then the averaged potential would become [Eq. (21)]

$$\phi = - \frac{2\pi R^2 |\varepsilon|}{(n-2)\sigma}$$

(for $n > 2$), and with an appropriate reinterpretation of the well depth ε. We retain the form of the van der Waals equation of state, namely,

$$(\pi + \frac{\alpha}{\sigma^2})\ (\sigma - \beta) = kT$$

except that we now associate the parameter α with $\alpha \equiv \pi R^2 |\varepsilon|/(n-2)$. The reported values of k_1 would be affected insofar as both $|\varepsilon|$ and $n-2$ are perturbed within the parameter α.

If the admolecules are highly asymmetric or if the adlayer is rather dense, perhaps the more appropriate ways to modify the analysis based on the $(6-\infty)$ potential would be to take one of the three following approaches: (1) Correct the free area concept of the van der Waals theory, e.g., by using a cell model (especially useful for dense overlayers); (2) Use a different potential function which introduces the asymmetry of the admolecule, e.g., the Kihara potential; or (3) Introduce a coverage dependence on the partition functions of the internal modes of the admolecule, e.g., by using an effective number of translational partition functions (which is greater than two).

Indirect interadsorbate interactions which occur through the substrate are a more difficult question. Model treatments both for simple cubic metallic substrates [1] as well as body-centered cubic metallic substrates [2] and insulators or semiconductors with the CsCl structure [2] have recently been carried out. For further details concerning this matter, these references should be consulted.

[1] T.L. Einstein and J.R. Schrieffer, Phys. Rev. $\underline{B7}$, 3629 (1973).

[2] W.Ho., S.L. Cunningham and W.H. Weinberg, to be published (1975).

Comment by F. C. Frank:

I guess when Douglas Everett was collecting those adsorption data you showed us he would first make a determination of the area of his charcoal by an adsorption experiment and application of the adsorption isotherm of Brunauer, Emmett and Teller. It would have been unkind to you and everyone else to ask for more than you gave in an hour, but shouldn't you tell us briefly how the BET isotherm stands in relation to all the rest?

Reply:

In order to distinguish between a local and a non-local adlayer, the essential measurement to make is that of the differential entropy of adsorption. In order for that entropy measurement to be useful, however, it is necessary to know the substrate surface area. One convenient way to judge the relevant surface area is to note experimentally when the differential heat of adsorption drops precipitously. The BET surface area is not relevant in this context since it is based on a specific (and rather naive at that) model of multi-layer adsorption, i.e., the BET surface area, while useful in accessing relative substrate areas, is not helpful in the present context where an absolute surface area measurement is required.

SURFACE CHEMISTRY II: OVERLAYERS (NEW PHASES)

J.H. van der Merwe

Department of Physics, University of Pretoria, Pretoria,
South Africa

ABSTRACT. The contents of this lecture are limited to those aspects of overlayers which are of importance to crystal plasticity. Substrate surface features and overlayer classification are briefly described. Overlayer formation by deposition from the vapour, a physical process, is considered in some detail. Certain features of overlayer formation by oxidation, a chemical process, are indicated. The accommodation of misfit between overlayer and substrate by misfit dislocations and strain is regarded as very important for crystal plasticity. The relevant aspects of a monolayer are described in fair detail. The results for a thick overlayer and the intermediate region are briefly presented. The relevance of interdiffusion and differential thermal expansion is indicated.

Notation

$a_i: a_x, a_y$ — [L] lattice spacings in substrate surface

$b_i: b_x, b_y$ — [L] interfacial lattice spacings of monolayer

b_z — [L] thickness of monolayer

$(a,b) \equiv (a_x, b_x)$ when $a_y = b_y$ i.e. one-dimensional misfit

$\bar{b}_i: \bar{b}_x, \bar{b}_y$ — [L] average interfacial lattice spacings of monolayer

A, B — symbol designating substrate/overgrowth

c — [L] lattice spacing of reference lattice

c_i — [L^{-2}] density factor for critical nuclei

$\bar{d}$ — [L] slip vector of imperfect misfit dislocation

e_x, e_y, e_{xy} — strains in monolayer

$\bar{e}_x, \bar{e}_y, \bar{e}_{xy}$ — average strains in monolayer
$\bar{E}$ — $[MT^{-2}]$ average energy per unit area
E_i — $[ML^2T^{-2}]$ binding energy of critical nucleus
E_{des} — $[ML^2T^{-2}]$ desorption energy per atom
E_o — $[MT^{-2}]$ energy factor
E_s, E_v — $[ML^2T^{-2}]$ activation energy for surface/volume diffusion
$E(\ )$ — complete elliptic integral of Second Kind
E_T — $[ML^2T^{-2}]$ total energy of system
$f_i: f_x, f_y$ — natural misfits at monolayer-substrate interface
$\bar{f}_i: \bar{f}_x, \bar{f}_y$ — average mismatches at monolayer-substrate interface
$f_x \equiv f$ — when $f_y=0$; one-dimensional mismatch
$F[\ ]$ — incomplete elliptic integral of the First Kind
g_A, g_B, g_{AB} — $[MT^{-2}]$ specific surface Gibbs free energy; surface A,B, interface AB
G_A, G_B, G_{AB} — $[ML^2T^{-2}]$ Gibbs free energy; A,B, mixture or solid solution AB
H — factor related to dislocation core energy
i — $[L]$ subscript equal to x or y
$\underline{i}, \underline{j}$ — Number of atoms in critical nucleus/$\underline{j}$-cluster
k — $[ML^2T^{-2}/\text{Degree K}]$ Boltzmann's constant
$k_i: k_x, k_y$ — Integration constants
$K(\ )$ — complete elliptic integral of First Kind
$\ell_i: \ell_x, \ell_y$ — parameters defined by eq. (18)
m — $[L^{-1}]$ parameter in eq. (29)
n — integer
$\bar{n}_i$ — $[L^{-2}]$ density of critical nuclei
$n_i: n_x, n_y$ — numbers enumerating both atoms and corresponding troughs in rectangular interface mesh
p — $[L]$ dislocation spacing
$r_i: r_x, r_y$ — ratios E_{des}/W_i in eq. (11)
t — $[L]$ overlayer thickness
T — temperature in Kelvin
V_s — $[ML^2T^{-2}]$ periodic interfacial interaction potential energy per atom

$W_i: W_x, W_y$ — $[ML^2T^{-2}]$ amplitudes of V_s, also activation energy for surface migration

x, y — $[L]$ Cartesian coordinates in interface

β — parameter in eqs. (26)

ε — $[ML^2T^{-2}]$ average energy per atom

ε_s — $[ML^2T^{-2}]$ strain energy per atom

θ — contact angle

λ — $[ML^{-1}T^{-2}]$ effective substrate-overlayer elastic modulus

λ_a, λ_b — $[ML^{-1}T^{-2}]$ elastic modulus; substrate, overlayer

μ, μ_a, μ_b — $[ML^{-1}T^{-2}]$ shear modulus; interface, substrate, overlayer

$\nu_n, \nu_{so}, \nu_{vo}$ — $[T^{-1}]$ appropriate atomic vibration frequency; normal to substrate, parallel to substrate, within crystal

ν_s, ν_v — $[T^{-1}]$ atomic migration jump frequency; on substrate surface, within crystal

$\xi_x(n_x)a_x, \xi_y(n_y)a_y$ — $[L]$ displacement of atom (n_x, n_y) from trough (n_x, n_y)

σ_a, σ_b — Poisson's ratio; substrate, overlayer

τ — $[T]$ adatom stay time

ϕ — angle between interface and slip vector

χ — parameter defined in eq. (26)

Ω — $[L^3]$ volume per atom in overlayer

1. INTRODUCTION

This tutorial lecture is limited to those aspects of overlayers which are of relevance to the plastic deformation of crystalline overlayers and the substrates on which they have grown. Since dislocations are usually the vehicles of plastic deformation, phenomena which involve dislocations will receive special consideration.

The following aspects of overlayers will be dealt with, some very briefly and others in greater detail: Substrate to overgrowth compositional and structural transition, overlayer formation by vapour deposition and oxidation and overlayer-substrate misfit accommodation.

The present considerations will be limited to phenomena for which equilibrium or quasi-equilibrium prevails. The thermodynamic equilibrium condition of minimum free energy or simply minimum energy, is usually applied as a governing principle.

2. SUBSTRATE SURFACE

The condition and structure of the substrate and its interaction with the overlayer, influence the growth of the layer. Of interest are [1]: the crystal face exposed; atomic steps on the crystal face, emerging dislocations, surface impurities, reactivity, bonding, surface strain and energy. Steps usually become part of the interface while emerging dislocations are usually continued in the overlayer. Impurities may enhance but usually hinder growth and may also be built into the overlayer. The reaction of different crystal faces with a given environment may be completely different.

3. CLASSIFICATION OF OVERLAYERS

For the present purpose overlayers may conveniently be classified according to their crystallinity, compositional and morphological uniformity and coherency.

The term *coherent* implies sufficient similarity in symmetry and dimensions of the interfacial atomic meshes of the overlayer and substrate crystals allowing the overlayer mesh to be brought into coincidence with a part or whole of the substrate mesh by a relatively small homogeneous strain. No direct consideration will be given to amorphous as well as completely incoherent overlayers. Compositional and morphological aspects will be dealt with briefly.

3.1. Interdiffusion

A composition gradient is important in crystal plasticity because it interacts with dislocations. The concentration profile depends on miscibility, time and temperature. The thermodynamic condition for miscibility is defined by (Fig. 1)

$$G_{AB} - G_A - G_B < 0 \qquad (1)$$

Because of the entropy of mixing, miscibility is enhanced at high temperatures. Interdiffusion which is a prerequisite for mixing depends on thermal activation through the migration jump frequency

$$\nu_v = \nu_{vo} \exp[-E_v/kT] \qquad (2)$$

An important implication of eq. (2) is that at low enough temperatures the rates of interdiffusion and mixing are so low that mixing may be disregarded.

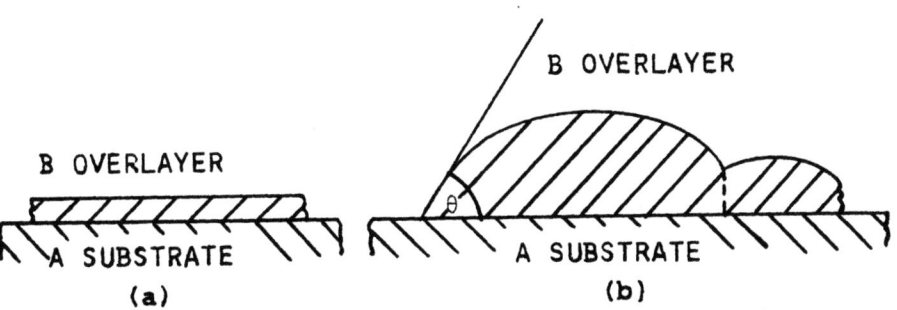

Fig. 1. Formation of overlayers (a) Two-dimensional (b) Three-dimensional.

3.2. Morphology

The term morphology here refers to two- and three-dimensional growth as shown in Fig. 1. The term "two-dimensional" indicates a growth whose lateral extent is large compared to its height. It is called "three-dimensional" if not. Admittedly these are rather imprecise definitions which is also evident from the figure and eq. (4) below.

The two-dimensional equilibrium growth of a substance B onto A has been discussed by Bauer [2] and is characterized by the requirement that

$$g_A \geq g_B + g_{AB} \qquad (3)$$

When eq. (3) is not satisfied the growth is three-dimensional. The true equality in (3) applies to a substance growing on itself and then $g_{AB} = 0$. Whether the next atomic layer also grows two-dimensionally when the inequality holds, depends on the way in which BB bonding between the topmost layers is influenced by the presence of the crystal A below the first layer of B atoms. This is a problem which still requires attention. Strong interfacial bonding is conducive to two-dimensional growth and weak bonding to three-dimensional growth. The degree of three-dimensionality is perhaps best expressed in terms of the contact angle θ (See Fig. 1) in Young's equation

$$g_A = g_B \cos\theta + g_{AB} \qquad (4)$$

4. FORMATION OF OVERLAYERS

In this lecture overlayer formation by deposition, will be discussed in some detail. Reference will be made to growth by oxidation. Neither of these phenomena is perfectly understood.

4.1. Overlayer formation from the vapour phase

Of primary interest to workers in this field is the dependence of cluster density on substrate temperature, atom incident rate, atomic bonding and misfit, defect tight-binding sites, cluster mobility, coalescence, etc. Once this dependence is known the growth rate can be calculated [3,4].

When an atom from the vapour impinges on the substrate surface it may be reflected, or adsorbed. An adsorbed atom moves along the surface either in a two-dimensional gaslike manner or by surface diffusion with a jump frequency

$$\nu_s = \nu_{so} \exp(-E_s/kT) \tag{5}$$

where T is the substrate temperature. The realization of one or the other of these processes depends primarily on T, E_s and E_{des}. It may re-evaporate after an average stay time

$$\tau = \nu_n^{-1} \exp(E_{des}/kT) \tag{6}$$

or become part of an existing adatom cluster. Adatom stay time and surface behaviour may also be greatly influenced by surface contaminants and surface defects which often constitute tight-binding sites for cluster nucleation.

Most theories now use the atomistic concepts introduced by Walton [5] and exploited by Zinsmeister [6] and others. The introduction of the concept of a critical nucleus containing i atoms and satisfying the assumption that all clusters $j \leq i$ are in quasi-equilibrium with the adatom population and all clusters $j > i$ grow and do so by the catchment of adatoms, allows one to express the density $\bar{n}_i$ of critical clusters, by

$$\bar{n}_i = C_i \exp(E_i/kT) \tag{7}$$

where C_i is independent of E_i.

A cluster which has leaked through the critical stage grows. The conditions under which it grows two- or three-dimensionally are expressed by the relation (3) when macroscopic concepts are applicable. In two-dimensional growth a monolayer sweeps across the surface at a relatively high speed incorporating adatoms which reach the perimeter from both above and below the terrace. Because the edge is an efficient sink for adatoms a surface adatom concentration gradient normal to the step edge will exist. This complicates significantly theoretical analysis. Three-dimensional growth involves the simultaneous advance of many steps.

The expressions for ν_s, τ and $\bar{n}_i$ have specifically been given to stress their energy dependence. In particular, it should be noted that E_i not only includes the binding energy between the adatoms in the cluster but also the binding energy onto the substrate [7]. If the conditions for coherency are met, E_i depends strongly on the relative orientation of the two superimposed interfacial atomic meshes. This predicts according to (7) a distribution of orientations about some mean orientation and may contribute to the realization of epitaxy. If in addition there is strong interfacial bonding, the relatively high values of E_i in favourable orientations yields a sharp distribution with more effective orientating potency towards the ideal epitaxial orientation(s). When coherency is absent, the interfacial orientating potency about the substrate normal is small and the resulting overlayer is usually polycrystalline and nonepitaxial.

It has been explained with reference to eq. (3) that strong bonding of the first adatom layer does not necessarily imply strong bonding of subsequent layers. At this stage there may be a transition from two- to three-dimensional growth as has been explained by Bauer [2].

When the conditions discussed with regard to eq. (1) are favourable interdiffusion will generate a transition zone in which the composition and lattice dimensions change continuously from that of pure overlayer B to that of pure substrate crystal A [8]. Small values of the activation energy E_v for volume diffusion, together with high substrate temperatures, are according to relation (2), conducive to interdiffusion. When the conditions for interdiffusion are not satisfied an abrupt transition from one crystal to the other at the interface, is obtained.

Of great interest in overlayers is the accommodation of misfit between overlayer and substrate by misfit dislocations and or strain, e.g. by strain as in pseudomorphism. This will be dealt with in par. 5.

Consider the three-dimensional or so-called island growth which is favoured by "weak" bonding. It follows from eq. (4) that a sharp division between two- and three-dimensional growth does not exist, e.g. if "weak"-designates all bonds whose strength varies from A-A downwards eq. (4) predicts contact angles which vary from zero upwards. Both abrupt and gradual transitions from one crystal to the other may be obtained. The phenomenon of misfit accommodation by dislocations and strain is likewise observed. Initially when the islands are small they seem to be mobile.

A phenomenon characteristic of three-dimensional growth is the liquidlike coalescence of islands which grow into contact with each other [9]. This phenomenon can be understood in terms of the rapid transport of adatoms by surface diffusion from one region to another, the driving force being the minimum energy principle. The fact that the lattices of two neighbouring islands may be displaced relative to each other by a vector different from a multiple of a lattice vector may account in part for some of the planar defects such as stacking faults. Since by (7) the lattice orientations of these islands are distributed about an ideal value, subboundaries with a low dislocation content are usually found along the surface where they meet. When two islands coalesce with-

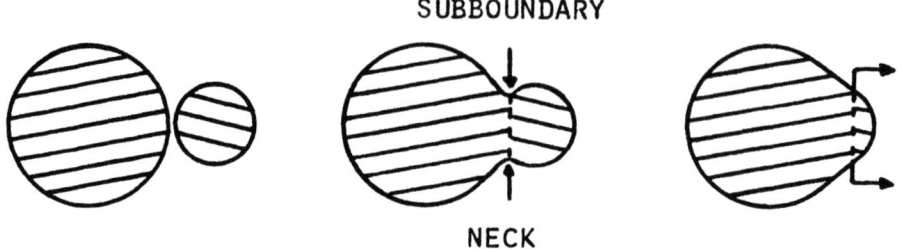

Fig. 2. Illustration of coalescence process (After Stowell [9]).

out the interference of others, as illustrated in Fig. 2, the subboundary will escape at the free surface with the disappearance of the neck between the two regions. The total energy is reduced in doing so. The effect is usually to improve the overall orientation.

After three or more islands have been interlinked a change of orientation can only be accomplished by recrystallisation. The stage when most of the islands are interlinked are usually referred to as the "hole and channel" stage [10]. This marks the final stage before the overlayer becomes completely continuous. These channels and holes may contain a relatively high density of surface impurities which have collected on the advancing island boundary. They may then be incorporated in the subboundaries which form when the holes and channels close up completely. If they agglomerate they may act as centres for stress concentration and generation of dislocations.

4.2. Overlayer formation by oxidation

Overlayer formation by oxidation is an important process, e.g. in corrosion. Many of the previous concepts apply here too, but additional ones are needed. In low temperature oxidation [11, 12, 13] three stages namely, rapid chemisorption, initial fast oxidation to form a continuous film and slow thickening of the oxide often form important steps in the process.

5. MISFIT ACCOMMODATION

The accommodation of misfit at the interface between the overlayer and substrate is important for crystal plasticity, since it involves both dislocations and/or strains which may exceed the critical values for macroscopic crystals. Again only coherent overlayers will be discussed. Even then the problem is extremely complicated and somewhat oversimplified models have to be used. These models nevertheless yield useful results: When the misfit is small below about 10% and the atomic bonding within the overlayer and onto the substrate are of average strength, the initial atomic layer is homogeneously strained to fit the substrate exactly, i.e. the misfit is completely accommodated by homogeneous strain [14]. This is referred to as the pseudomorphic configuration. As the number of atomic layers increases during the growth process a critical thickness is reached at which dislocations are generated to take up part of the misfit; hence the name misfit dislocations. The part accommodated by dislocations increases with increasing thickness. When the conditions of misfit and bonding, needed for pseudomorphism are not met, misfit dislocations are present, even for a monolayer, almost as soon as its lateral extent exceeds the vernier period of natural misfit.

The dislocations which are present in the case of large misfit are usually pure misfit dislocations, i.e. edge type dislocations with Burger's vector in the interface plane. When the misfit is very small and the misfit dislocations only appear after the overlayer is many atomic layers thick, the dislocations are usually imperfect; [15, 16] i.e. they are glide dislocations on glide planes which are inclined to the interface in contrast to the pure misfit dislocations which are generated at the free perimeter of the overlayer and fed in along the overlayer-substrate interface. It is believed that many imperfect misfit dislocations, have their origin in threading dislocations, i.e. substrate dislocations which have been continued into the overlayer during the growth process. In contrast to perfect misfit dislocations which can glide along the interface to achieve regular spacing, climb motion has to be invoked in addition to glide for the rearrangement of imperfect misfit dislocations.

At present there does not exist adequate agreement between theory and experiment on the variation of homogeneous strain with overlayer thickness. Neither has the transition from perfect to imperfect misfit dislocations been placed on an acceptable quantitative footing. The former discrepancy is believed to be caused by the fact that the experimental conditions do not conform exactly to the equilibrium assumption on which the theory is based [17]. Although the theory is imperfect it has yielded sufficient significant results to merit a more detailed description. Most theories are based on the assumption that equilibrium prevails throughout the growth process. The equilibrium assumption is normally expressed in a simplified form, [14] namely

$$E_T = \text{minimum} \tag{8}$$

This presents somewhat of a dilemma because (8) is only exactly applicable at absolute zero and there the thermally activated processes needed to achieve equilibrium, are frozen.

The existence of an attractive or adhesive bond is a prerequisite for the formation of an overlayer on a substrate. The bonding of adatoms onto the substrate is an exceedingly complicated and largely unsolved quantum mechanical problem. Qualitatively speaking the adatoms induce new surface electronic states which alter the surface electronic configuration and bond energy. The states change as the adatoms form monatomic layers and also when consecutive layers are stacked to form thick overlayers. The states reach saturation at a film thickness which mainly depends on the substances concerned. The presence of the substrate adjacent to the overlayer similarly influences the bonding within the overlayer, as mentioned previously.

<u>Monolayers</u> Various simplifications have been introduced in the description of the complexity of atomic interaction, each with its own regime of applicability and some with phenomenal success. These simplifications which include the concepts of covalent, ionic, van der Waals' and metallic bonding, pairwise potentials, nearest neighbour bonds, etc. will not be dealt with here. An important feature in the crystalline overlayer-substrate interaction potential V_s is its periodicity along the substrate. For overlayers one atomic layer thick, the approach which invokes a rigid substrate and expresses the periodic substrate-overgrowth interaction in terms of a "rigid" periodic potential energy density function V_s, seems to be the most appropriate. V_s is usually approximated by

$$V_s(x,y) = \tfrac{1}{2}W_x[1-\cos(2\pi x/a_x)] + \tfrac{1}{2}W_y[1-\cos(2\pi y/a_y)] \qquad (9)$$

This represents the case where the natural atomic configurations of the crystals have rectangular symmetry with principal axes along the x and y directions. The accuracy of relation (9) has not been established yet. Cases with different symmetry are more complicated [18]. The natural misfits are expressed by

$$f_i = (b_i - a_i)/a_i; \qquad i=x,y \qquad (10)$$

The equation

$$E_{des} = r_i W_i \qquad (11)$$

provides a measure of the bonding or desorption energy E_{des} in terms of W_i which turns out to be an important parameter in the considerations. The factor r_i depends on the participating species and their crystal structures. In the nearest neighbour approximation, for an adatom on a close-packed surface, one out of three bonds is broken when an adatom migrates. Thus $r = 3$ in this case. In this model E_{des} is half of the sublimation energy. For long range van der Waals' type forces though, $r \sim 30$.

The interaction between the overlayer atoms is assumed to be the forces in an elastic continuum taking a modulus, the shear modulus say, as a measure of the bonding. The bulk values of the moduli are usually employed. This approximation does not take into account the influence of the substrate neither the dependence of μ_b and σ_b on overlayer thickness. It was shown using pair potentials [19] that elastic and lattice constants may deviate from bulk values by as much as 60% and 2% respectively. This is significant.

The elastic representation of the overlayer is achieved by imagining the atoms to be embedded in an elastic sheet. In isotropic elasticity theory the energy per atom in the two-dimensional system is given by [14]

$$\varepsilon_s = \tfrac{1}{2}\mu_b \Omega [2(e_x^2 + e_y^2 + 2\sigma_b e_x e_y)/(1-\sigma_b) + e_{xy}^2] \quad (12)$$

$$\Omega = b_x b_y b_z \quad (13)$$

The model neglects strain variation normal to the overlayer plane.

An effective mismatch

$$\overline{f}_i = (\overline{b}_i - a_i)/a_i; \qquad i=x,y \quad (14)$$

obtains when the overlayer is subjected to overall lateral strains

$$\overline{e}_i = (\overline{b}_i - b_i)/b_i = (\overline{f}_i - f_i)a_i/b_i \quad (15)$$

The analysis yields the relations [17]

$$\pi n_i / k_i \ell_i = F[k_i, \pi(\xi_i - \tfrac{1}{2})]; \quad \xi_i(0) = \tfrac{1}{2}; \quad i=x,y \quad (16)$$

for the displacement $[a_x \xi_x(n_x), a_y \xi_y(n_y)]$ of atom (n_x, n_y) from the potential trough (n_x, n_y), both being enumerated from a position of coincidence and

$$\varepsilon = \sum_i W_i [4E(k_i)\ell_i \overline{f}_i/(\pi k_i) - 2\ell_i^2 f_i \overline{f}_i + \ell_i^2 f_i^2 + 1 - k_i^{-2}]$$

$$+ 2\sigma_b \Pi \sum_i W_i^{\tfrac{1}{2}} (\ell_i \overline{f}_i - \ell_i f_i); \qquad i=x,y \quad (17)$$

for the average energy per interfacial atom. In the above

$$\ell_i^2 = \mu_b \Omega a_i^2 / [W_i (1-\sigma_b) b_i^2] \quad (18)$$

and k_i is related to $\overline{f}_i$ by

$$\overline{f}_i = \pi / [2\ell_i k_i K(k_i)] \quad (19)$$

The parameters ℓ_i play an important role because they are simple functions of the pertinent properties of the model, in particular, W_i and μ_b are measures of the substrate-overlayer and overlayer-overlayer bonding strengths respectively - ℓ_i increases with increasing $\mu_b a_i^2/W_i$ - secondly the quantitative aspects of the predictions below are significantly dependent on the ℓ_i.

The solution (16) does not allow exactly for free boundary effects. Exact calculations have shown however that such effects are negligible [18].

The following conclusions have emanated from these considerations:

1. The relation (16) when solved for $\xi_i(n_i)$ represents a natural resolution of the interface into two orthogonal sequences of dislocations of spacings $\bar{b}_i/\bar{f}_i$ and have become known as misfit or interfacial dislocations of which the ℓ_i are the effective widths when widely spaced.

2. The stable minimum energy configuration is one in which misfit is accommodated jointly by misfit dislocations and overall strain according to the relation

$$f_x = 2E(k_x)/\pi k_x \ell_x + \sigma_b \bar{e}_y a_y / b_y \quad (20)$$

where the second term on the right accounts for the Poisson contraction.

3. The pseudomorphic configuration in which the overlayer is homogeneously deformed to fit the substrate exactly, is the stable one provided the natural misfit does not exceed the value

$$f_x = 2/\pi\ell_x - \sigma_b f_y a_y b_x / b_y a_x \quad (21)$$

along the x-direction. The case of quadratic symmetry ($a_x = a_y$, $b_x = b_y$) $f_x = f_y = 2/[\pi\ell_x(1+\sigma)]$, shows a simple though important dependence on ℓ. When the atomic forces in and between the crystals are equal a simple force model yields a value $f_x = 7\%$ [21].

4. For a misfit exceeding f_x in (21) but below a value given by

$$f_x = \ell_x^{-1} - \sigma_b f_y a_y b_x / a_x b_y \quad (22)$$

the pseudomorphic configuration is metastable. An estimate for f_x applicable under similar conditions as in 3. is 11%. We may call the contribution $\bar{f}$ of the misfit dislocations to misfit accommodation the mismatch or misfit dislocation density. The relation between f and $\bar{f}$ in (19) and (20) is shown in Fig. 3.

5. The generation of dislocations needed to maintain equilibrium in a growing monolayer, or to accomplish a transition from a metastable state, requires an energy of activation except in the case for which f_x exceeds the value defined in (22), when dislocations generate spontaneously.

The calculation of activation energies is very complicated even in this simple two-dimensional model and the theory is still

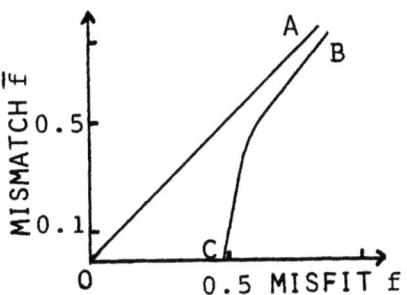

Fig. 3. Dependence of mismatch $\bar{f}$ on misfit f for the case of quadratic symmetry ($a_x = a_y$, $b_x = b_y$, $\ell_x = \ell_y = \ell$). Both $\bar{f}$ and f are expressed in units of ℓ^{-1}. Along OA $f = \bar{f}$ and misfit accommodated by dislocations only. For stable configuration the curve OCB is followed; along OC misfit is completely accommodated by homogeneous strain and along CB by strain and dislocations jointly.

very imperfect in this respect. Theory implicitly assumes that equilibrium is attained in times short compared to times of observation. This assumption is certainly not good enough in many cases. Even the motion of a dislocation may be impeded by large frictional forces such as presumably exist in semiconductors.

Thick overlayers An overlayer can be designated "thick" when its thickness exceeds half the dislocation spacing. There are two approaches to the problem of the thick overlayer. One is a generalized Peierls-Nabarro model. It presupposes the existence of a periodic interaction potential at the interface between the two crystals treated as elastic continua. The periodicity is approximated by a sinusoidal dependence which in regions where the mismatch is small, reduces to Hooke's law for shear with an elastic constant µ referred to as the "interfacial shear modulus". The other invokes dislocations on geometrical grounds and uses a Volterra dislocation description treating the entire system as an elastic continuum. While the former has less undetermined parameters the latter is more versatile and lends itself more naturally to the description of imperfect misfit dislocations. However, in the case of thin overlayers with widely spaced dislocations it is rather inaccurate.

The analysis is usually carried out for misfit in one interfacial direction only, the energies with misfit in two directions being approximately equal. The crystals are treated on the same footing by defining the misfit as [17]

$$f = c/p = (b-a)/[\tfrac{1}{2}(a+b)] \qquad (23)$$

where

$$p = ab/(b-a) \text{ and } c = ab/[\tfrac{1}{2}(a+b)] \tag{24}$$

In the analysis with the Peierls-Nabarro model the interface resolves into a sequence of dislocations. The average energy per unit area associated with them is

$$\underline{E} = (\mu c/4\pi^2)[1-\chi-\beta\ln(1-\chi^2)] \tag{25}$$

where

$$\chi = (1+\beta^2)^{\tfrac{1}{2}}-\beta, \quad \beta = 2\pi\lambda c/(p\mu)$$

$$1/\lambda = (1-\sigma_a)/\mu_a + (1-\sigma_b)/\mu_b = 1/\lambda_a + 1/\lambda_b \tag{26}$$

It is notable that β, like ℓ_i, depends here on the bonding parameters μ_a, μ_b and μ, and on the misfit c/p. Through β and $\mu c/4\pi^2$ the energy thus depends in a simple manner on the important physical parameters of the system. Of interest is also that 98% or more of the total energy associated with the interface is stored within a distance $\tfrac{1}{2}p$ from the interface. This serves partly as justification for the condition that an overlayer may be treated as "thick" when its thickness exceeds $\tfrac{1}{2}p$.

The Volterra approach was introduced in the theory of interfaces by Brooks [22] and applied extensively by Matthews [16] and others. In this model the average energy $\underline{E}$ per unit area of interface for a row of perfect misfit dislocations is given by

$$\underline{E} = E_o f(H-\ell n f) \tag{27}$$

where the factor E_o depends on elastic and lattice parameters and H is a constant associated with the core energy of the dislocation. When the misfit dislocations are imperfect, the dependence on the angle of inclination of the Burger's vector to the interface have to be included. The variation of $\underline{E}$ versus f is shown in Fig. 4.

<u>Thickening overlayer</u>: Because of the complexity of the problem the existing theories are restricted to uniform overlayers. Extrapolation of the monolayer approach upwards breaks down when the film exceeds about two atomic layers because it ignores the substrate deformation and the strain gradient in the overlayer. Likewise the extrapolation of the thick film analysis down to thicknesses below half the dislocation spacing becomes inaccurate as well. This is serious particularly for widely spaced dislocations in thin overlayers.

A model which to some extent overcomes the shortcomings of the abovementioned extrapolations is the so-called First Approximation [23]. It is essentially an interpolation between the two extremes and uses the exact solution for the displacement function of the thick overlayer as a first Approximation to the real solution. It yields for the average energy per unit area the result

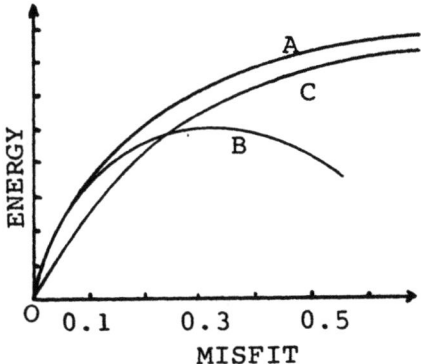

Fig. 4. Dependence of mean energy per unit area of misfitting, interface (specified in arbitrary units) on misfit c/p Curves A and B for a thick overlayer; eqs. (26) and (27) and C for a monolayer; eq. (28).

$$E \cong \frac{\mu c}{4\pi^2}[1-X-\beta \sum_{n=1}^{\infty} \frac{(m^2t^2-\bar{s}^2)X^{2n}}{n\lambda[(\bar{s}^2-m^2t^2)/\lambda_a+(\bar{s}\bar{c}-mt)/\lambda_b]}] \quad (28)$$

where

$$\bar{c} = \cosh mt, \quad \bar{s} = \sinh mt \quad \text{and} \quad m = 2\pi n/p \quad (29)$$

Its variation with f is illustrated in Fig. 4 for $t = b_z$.

The foregoing has been limited to the energy associated with the dislocations only. The energy due to the homogeneous strain, needed to minimize the total energy with respect to strain, must still be added to give the total energy. The energy of homogeneous strain can be obtained from simple elasticity theory.

This and other calculations failed to give good agreement with experiment. The measured strain exceeds that predicted by equilibrium considerations. Matthews [15, 24] has ascribed this descrepancy mainly to the difficulty of acquiring dislocations. Dislocations either originate from threading dislocations or are created at free surfaces [25] or crystal defects and have moved towards the interface by glide and climb. Only the component d cos ϕ will be effective in accommodating misfit. Hence, in order to accommodate the same misfit as perfect misfit dislocations their density must be more by a factor of sec ϕ, $0 \leqslant \phi < \frac{1}{2}\pi$.

One important aspect of overlayers which warrants a few additional remarks here is that of overlayer-to-substrate adhesion or bonding, which is not yet well understood. While W apparently plays an important role in misfit dislocation formation, E_{des} is of great significance in adhesion. Also, in a peel mechanism it is the force rather than the energy which is important in adhesion. Adhesion is generally enhanced by interface cleanliness and by the existence of a concentration gradient from one crystal to the other. When the thicknesses of overlayer and substrate are of the same order and misfit is partly accommodated by elastic strain, adhesion is manifested by the curling up of the bicrystal slab.

In retrospect it may be said that a brief review has been given in this section of stable coherent and semi-coherent overlayer-substrate systems. The concepts of misfit accommodation, misfit strain, misfit dislocations and misfit dislocation structure and its formation in a growing film have been dealt with. Important aspects not dealt with are the interdiffusion induced redistribution [26] of misfit dislocations and the differential thermal expansion component of misfit accommodation.

REFERENCES

1. Green, A., Bauer, E. and Dancy, J., in Molecular processes on solid surfaces, Grauglis, E., Gretz, R.G. and Jaffee, R.I., Eds., McGraw-Hill Book Co., New York, p. 479, 1969.
2. Bauer, E., Thin Solid Films, 12, 167, 1972.
3. Venables, J.A., Phil. Mag., 27, 697, 1973.
4. Stowell, M.J., J. Cryst. Growth, 24/25, 45, 1974.
5. Walton, D., J. Chem. Phys., 37, 2182, 1962.
6. Zinsmeister, G., Jap. J. Appl. Phys., Suppl. 2, Pt 1, 545, 1974.
7. Sato, H., Annual Rev. Mat. Sci., 2, 217, 1972.
8. Matthews, J.W., Phil. Mag., 8, 711, 1963.
9. Stowell, M.J., Thin Films, 1, 55, 1968.
10. Jacobs, M.H., Pashley, D.W. and Stowell, M.J., Phil. Mag., 13, 129, 1966.
11. Fehlner, F.P. and Mott, N.F., Oxidation of Metals, 2, 52, 1970.
12. Holloway, P.H. and Hudson, J.B., Surface Sci., 43, 123, 1974.
13. Cabrera, N. and Mott, N.F., Rept. Progress Phys. 2, 163, 1948-49.
14. Frank, F.C. and van der Merwe, J.H., Proc. Roy. Soc., A 198, 205 and 216, 1949.
15. Matthews, J.W., Thin Solid Films, 5, 369, 1970.
16. Matthews, J.W., in Epitaxial Growth, Ed. Matthews, J.W., Academic Press, New York, p. 559, 1975.
17. van der Merwe, J.H., in Treatise on materials science and technology, Ed., Herman, H., Academic Press, New York, p. 1, 1969.
18. Snyman, J.A. and van der Merwe, J.H., Surface Sci., 45, 619, 1974.
19. Auret, F.D. and van der Merwe, J.H., Thin Solid Films, 23, 257, 1974.
20. Jesser, W.A. and Kuhlmann-Wilsdorf, D., Phys. Status Solidi, 19, 95, 1967.
21. van der Merwe, J.H., J. Appl. Phys., 41, 4725, 1970.
22. Brooks, H., Metal interfaces, Amer. Soc. Metals, Cleveland, Ohio 1952.

23. van der Merwe, J.H. and van der Berg, N.G., Surface Sci. 32, 1, 1972.
24. Matthews, J.W., J. Vac. Sci. Technol. 12, 126, 1975.
25. Shinohara, K. and Hirth, J.P., Japn. J. Appl. Phys., Suppl. 2, Pt. 1, 629, 1974.
26. Vermaak, J.S. and van der Merwe, J.H., Phil. Mag. 10, 785, 1964; 12, 453, 1965.

DISCUSSION

Comment by R. A. Oriani:

Would you explain further how cracking, which relives strain, can itself produce misfit dislocations?

Reply:

An epitaxial layer of PbS grown pseudomorphically in parallel epitaxy on the (001) surface of PbSe is in a state of tension. Matthews (1971) observed such overlayers to crack along the (010) and (100) planes. When the wedges thus formed shrink to relieve the tension, misfit dislocations are needed at the substrate-overlayer interface to accommodate the misfit which previously had been taken up by strain alone. Only one parallel sequence of dislocations with Burgers vector along the [110] direction is observed. The fact that they were more closely spaced than one would expect on the basis of the natural misfit alone suggests that misfit has been increased along the [110] direction. This can be explained in terms of Poisson's contraction and the absence of relaxation along the perpendicular [$\bar{1}$10] direction.

TUTORIAL LECTURES ON CRYSTAL PLASTICITY

Session Chairman: J. P. Hirth

MICROPLASTICITY

R. Bullough

Theoretical Physics Division, A.E.R.E. Harwell,
Nr. Didcot, Oxon, U.K.

1. INTRODUCTION

2. THE DEFECT CRYSTALLINE STATE
2.1 Point Defects
2.2 Dislocations
2.3 Interfaces

3. ENERGY AND DISTORTION FIELDS OF DISLOCATIONS
3.1 The Continuum Model of a Dislocation: Elastic Fields
    3.1.1 Linear Fields In Anisotropic and Isotropic Continua
    3.1.2 Non-Linear Fields
3.2 The Partially Discrete Models of Dislocations
    3.2.1 The Peierls-Nabarro Model
    3.2.2 The Van der Merwe Model
3.3 Lattice Models of Dislocation

4. FORCES ON DISLOCATIONS
4.1 The External Stress
4.2 Interactions with other Defects
    4.2.1 Dislocation - dislocation interactions
    4.2.2 Point Defect Interactions
    4.2.3 Second Phase Particles
4.3 Interaction with Surfaces
    4.3.1 The Free Surface
    4.3.2 The Internal Surface

1. INTRODUCTION

A fundamental understanding of the plastic deformation of crystalline solids requires an understanding of the various microscopic (atomic) processes that can and do prevail during such deformation. These microscopic processes involve the formation and movement of dislocations and microplasticity is essentially the study of such dislocation phenomena as they are affected by mutual dislocation interactions, by various obstacles in the crystalline host and by interactions with the external surface. It is obvious that a comprehensive discussion of all the pertinent dislocation properties is not feasible in the limited space available and many important topics have had to be either omitted or simply referred to in passing; fortunately several excellent texts on dislocations are available and can be usefully used to supplement the present discussion [1,2,3,4].

We begin the discussion in section 2 with a qualitative description of the nature of the defect crystal. These basic geometrical and thermodynamic properties of point, line and surface defects provide the essential microscopic conceptual foundation for the forthcoming discussions and also serve to indicate the various hierarchical inter-relations between the point and extended defects in crystals. In addition to these discrete defect properties we shall also refer to the continuously dislocated crystalline state [5,6,7]; this last topic is included because it offers the framework for the eventual building of a bridge between the discrete dislocation concept as the essential element of plasticity and classical field plasticity where the basic atomic crystallinity is discarded.

More quantitative aspects of discrete dislocations are taken up in section 3 where we outline the various models that have been developed to discuss their energies and distortion fields. The importance of the elastic fields is demonstrated with reference to the isotropic continuum model; the modifications arising from anisotropy and non-linear elastic effects are also included. Such continuum models are only appropriate for a discussion of the long range properties of dislocations. Dislocation properties that arise from the detailed atomic configuration of their cores cannot be discussed with such models and require either partially discrete models or full lattice models. We thus conclude this section with a discussion of these non-continuum models of dislocations including the semi discrete Peierls-Nabarro and Van der Merwe models and the full lattice models.

Microplasticity may be defined as the response of an assembly of dislocations to an applied stress. In the final section we thus discuss the factors that influence this response beginning with the basic concept of force on a dislocation due to an external stress and concluding with the many defect interactions, both short and long range, that can constrain the response. The aim of the lectures is to lend support to the proposition that a study of microplasticity involving, as it does, a basic understanding of the properties of isolated and distributions of dislocations, does provide a useful foundation for an understanding of the plastic deformation of crystalline solids and is of worthwhile complementary value to the purely phenomenological approach of field plasticity.

2. THE DEFECT CRYSTALLINE STATE

Real crystalline materials will contain defects which may be conveniently divided into:

Point Defects - such as intrinsic interstitials, vacancies and impurity atoms which can be located either in interstitial or substitutional sites.

Line Defects - dislocations which can be edge, screw or mixed and in turn can all either be perfect or imperfect.

Surface Defects - such as grain boundaries, low angle boundaries, twin or phase boundaries and, of course, the external surface of the crystal.

The intrinsic properties of these defects coupled with their various interactions with each other provide the fundamental microscopic basis for understanding the plastic response of the crystalline body to applied external forces. The study of these defects and their interactions with each other is what we mean by microplasticity. Of course, from the point of view of microplasticity the intrinsic properties of dislocations and their interactions with the other defects are of pre-eminent importance and will therefore be emphasized in this and subsequent sections; the point defect and surface defect properties that we refer to will therefore be only those that pertain to their effect on dislocation mobility.

2.1 Point Defects (see figures 1 and 2)

The intrinsic point defects are the interstitial and vacancy, both of which can exist in thermal equilibrium in a crystal. If E_f is the increase in internal energy of the crystal due to the formation of such a defect and S_f is the change in vibrational entropy when the defect is formed, the equilibrium fractional concentration of the defects is

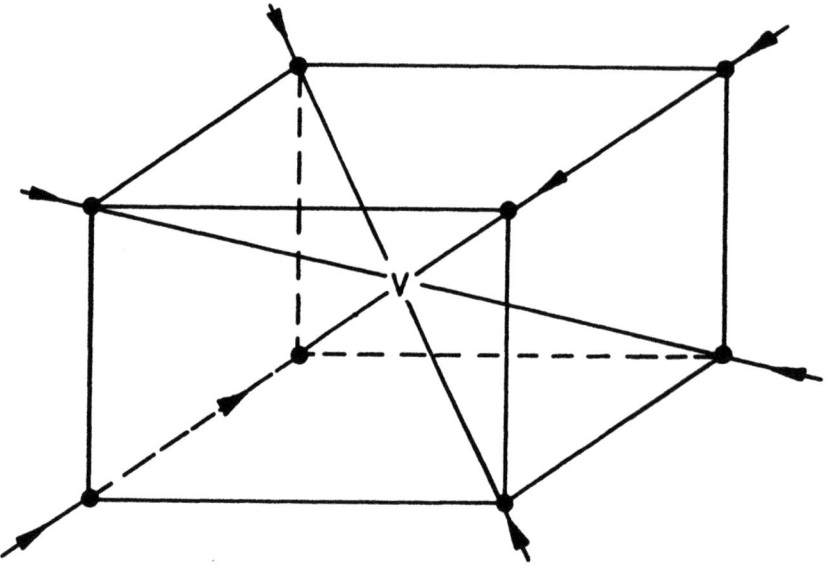

FIG. 1.

Body-Centred Cubic Vacancy with 1st Neighbour Forces.

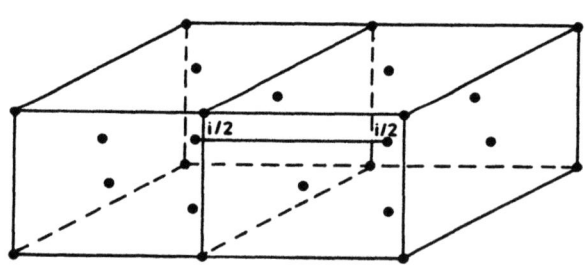

FIG. 2.

Face-Centred Cubic "Dumb-bell" Interstitial. i/2 indicate the two halves of the dumb-bell.

$$C = \exp\left[\frac{S_f}{k}\right] \exp\left[\frac{-E_f}{kT}\right] \qquad (2.1)$$

where T is the temperature (°K) and k is Boltzmann's constant. For the vacancy E_f^v is defined as the energy required to take a matrix atom out of the crystal and put it down on the surface of the crystal; it follows that the corresponding volume change during the formation of the vacancy is

$$\Delta V_f^v = \Omega - V_r^v \qquad (2.2)$$

where Ω is the atomic volume and ΔV_r^v is the relaxation volume change arising from the atomic distortion field around the vacancy. The interstitial is formally created by taking an atom from the surface of the crystal and forcing it into an interstice within the crystal matrix; the associated volume change is thus

$$\Delta V_f^i = -\Omega + \Delta V_r^i \qquad (2.3)$$

where now, in general, the interstitial relaxation volume change ΔV_r^i will be much larger than the relaxation volume change for the vacancy. This large distortion field associated with the interstitial has many important consequences; for example it ensures that such interstitials will suffer a preferential interaction with dislocations compared with the vacancies. Such a preference occurs in irradiated materials when such point defects are produced by displacement events and can lead to the technologically important phenomenon of void growth leading to reactor component swelling [8]. These point defect-dislocation interactions are discussed in section 4.

In f.c.c. metals the vacancy formation energy E_f^v is approximately [9] 1 eV (Al : 0.75 eV, Cu : 1.1 eV, Ag : 1.08 eV, Au : 0.98 eV) and the vibrational entropy factor $\frac{S_f^v}{k} \sim 1$.

It follows that the equilibrium vacancy concentration can become very large and will approach

$$C_v \simeq 10^{-3} \text{ vac/atom} \qquad (2.4)$$

at the melting point in these metals. In the b.c.c. refractory metals the vacancy formation energies are much higher [9] (Mb : $\sim$5eV, W : 3.3 eV) and in the alkali metals they are very low (Na : 0.42 eV); however equation (2.4) is reasonably correct as a maximum thermal vacancy concentration for all materials. The relaxation volume strain for the vacancy $e_v^0 = \Delta V_r^v/\Omega$ is, for the metals, usually ≤ 0.5. In contrast to these typical vacancy values the interstitial formation energy $E_f^i \simeq 5$ eV and the relaxation volume strain $e_i^0 = \Delta V_r^i/\Omega$ is greater than 1 and, in certain metals such as iron [10], can approach 2. It follows from 2.1 that the equilibrium interstitial concentration is quite negligible - even at the melting point, and therefore self interstitial properties can usually only be studied under irradiation conditions and then a huge supersaturation of interstitials may be present. In such irradiation conditions the large e_i^0 has an important effect on the ability of the interstitial to interact with the dislocations (see section 4).

Impurity atoms can either occupy an interstitial site or replace a matrix atom (substitutional site). In either case there is a distortion field associated with the point defect that leads to an interaction with nearby dislocations [11]. The magnitude of this field for all types of point defect can be estimated using various models ranging from the relatively crude isotopic continuum model with a spherical inclusion representing the point defect [12] to a full atomic model [13] with spatially distributed forces, around the position of the defect, representing its presence. The relation between these various models of point defects and a discussion of their relative accuracy has been given by Bullough, Norgett and Webb [14]; it is found that the anisotropic continuum model can only give an acceptable description of the distortion field for distances greater than 2.5 x (lattic parameter) for the vacancy in copper. There is no reason to think that the continuum representation would be superior in other materials or for other defects. In fact the much greater distortion field for the interstitial resulting from closer atomic proximities near the defect suggests the two models will only produce similar fields at even greater distances from the interstitial.

A full atomic treatment for the atomic configuration around a point defect would require a self consistent analysis of the electron redistribution around the defect with the resulting variations in the effective interatomic forces. Such fundamental calculations, including the ion relaxations, are only just being attempted [15,16] and it would not be appropriate to discuss them here. However, several relatively "simple minded" atomic calculations have been made [17] that do illustrate features to be expected from the more rigorous investigations. These involve constructing, or choosing an existing suitable interatomic potential for the perfect crystalline lattice. The defect is then introduced by imposing suitable body forces on the atoms around the position of the defect. This force array is defined at the centre of a large parallelepiped of atoms and with

suitable boundary conditions on its surfaces the atoms are all
relaxed by solving numerically the set of coupled Newton's
equations of equilibrium. If the body forces are carefully
defined as the gradient of a potential function then the procedure
can be used to discuss the distortion fields around both the
intrinsic and impurity point defects. Thus it can and has been
used to investigate the effects of local force constant changes
on the interactions with other defects [18,19] as well as the
atomic distortion fields. Such local force constant changes near
the defect have also been successfully modelled using an elastic
inclusion model where the inclusion has different harmonic (elastic) properties [20]. This will be discussed further in section
4 in connection with the inhomogeneity interaction between a
point defect and a dislocation. The usual distortion fields lead
to the well known size effect interactions which are also discussed
in section 4.

2.2 Dislocations

The line defects or dislocations can be created by suitably
aggregating the intrinsic point defects and such a synthesis will
help to maintain the defect hierarchy upon which we wish to
structure this section. Thus interstitials, when created by
irradiation for example, are supersaturated and will usually
rapidly come out of solution to form planar aggregates of
interstitials [8,9]. These planar aggregates are simply discs of
extra atomic layers and the perimeter of such discs is an interstitial dislocation loop. Similarly vacancies can aggregate to
form vacancy dislocation loops in exactly the same fashion [21];
both loops are shown schematically in figure 3. A dislocation may
be defined as the line defect in a crystal that separates regions
of a surface across which there is a discontinuity of atomic
displacement or lattice translation. The magnitude and direction
of this displacement is termed the Burgers vectors of the dislocation. The fact that a dislocation is essentially a
"displacement" defect whereas the point defect is a 'force defect'
is most important as we shall see when we discuss the general
topic of force and residual fields in section 4. This transition
from force defects to displacement defects which occurs when point
defects aggregate to form dislocation loops is a field theoretic
way [22,23] of simply saying that when the point defects form
the loops most of the previous point defects are transformed into
"good lattice" displacements and only the perimeter region of
the loop contributes to the loop distortion field.

Unfortunately the two dislocations shown in figure 3 would
be unable to slip; they have formed and grown by a climb process
involving the absorption of point defects. Dislocations that
can move by slip are also important and such glissile dislocations

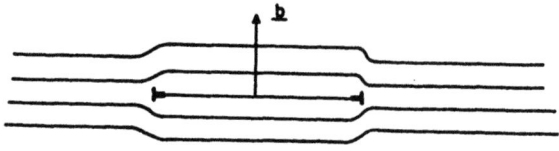

FIG. 3a.

A cross-section of a circular interstitial dislocation loop. It consists of a platelet of interstitial point defects whose perimeter is the edge dislocation as indicated by the usual dislocation symbol. The Burger's vector $\underline{b}$ is orthogonal to the plane of the loop and such a dislocation is, in general, sessile. This is the usual morphology of interstitial aggregates in crystalline materials.

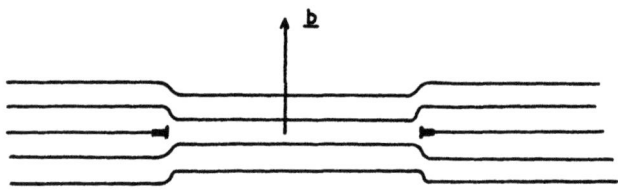

FIG. 3b.

A cross-section of a circular vacancy dislocation loop. It consists of a platelet of vacancy point defects whose perimeter is the edge dislocation as indicated by the usual dislocation symbol. The Burger's vector $\underline{b}$ is orthogonal to the plane of the loop and such a dislocation is, in general, sessile. In the absence of gases the two dimensional loop is the usual morphology of vacancy aggregates in crystalline materials; when suitable gas is present the vacancies can also aggregate as three dimensional voids.

are defined as lines on the slip plane of a crystal structure that separate a region of the slip plane that has slipped a complete slip distance from another region on the same slip plane where no slip or only partial slip has occurred.

A general dislocation is defined at any point along its length (see Fig. 4) by two vectors: $\underline{t}$ and $\underline{b}$. $\underline{t}$ is the unit vector defining the direction of the dislocation line (the tangent vector) and $\underline{b}$ is the slip vector associated with the dislocation (the so-called Burgers vector). For a glissile dislocation, $|\underline{b}|$ is usually equal to the interatomic spacing in the closepacked direction in the slip plane or a partial component of this quantity. Note that $\underline{b}$ is a global vector and $\underline{t}$ is the local tangent vector. In general, the angle between $\underline{t}$ and $\underline{b}$ is arbitrary, but two particular types of dislocation should be highlighted.

The pure edge dislocation [24] for which the vectors $\underline{t}$ and $\underline{b}$ are perpendicular. If the vector $\underline{b}$ lies in the slip plane then we have a glissile edge dislocation; if it does not, as in figure 3, the dislocation is sessile and it can only move through the crystal by climb processes. The edge dislocation is characterized as the termination of an extra plane of atoms as depicted in figure 3. If the vectors $\underline{t}$ and $\underline{b}$ are parallel we have a pure screw dislocation. This dislocation is a little more difficult to visualize and was first introduced as a crystallographic defect by Burgers [25] in 1939. It is not confined to a particular slip plane, although if $\underline{b}$ is a complete slip vector it may preferentially dissociate on a particular slip plane and be, thereby, constrained to that plane. The screw dislocation is characterized by the way in which its presence transforms the parallel lattice planes which it cuts into a single helicoidal surface.

Clearly, to make the definition of a particular dislocation unique a rule or convention has to be invented to establish the relative sense of the two vectors $\underline{t}$ and $\underline{b}$ Fig. 4; also for $\underline{b}$ to have a unique magnitude it must be defined as a vector of a perfect reference crystal.

The convention we shall use is due to Bilby, Bullough and Smith [26] and was introduced in 1955. It may be conveniently written FS/RH where FS = $\underline{b}$ is the closure failure (finish-start) of a circuit made in a perfect reference lattice which was constructed by making a one-to-one directional correspondence with each lattice step of a closed circuit surrounding the dislocation in the dislocated crystal. The sense of the circuit is deduced from the right-hand screw rule (RH). The formal procedure is depicted in figure 5.

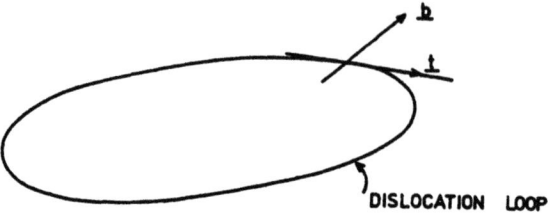

FIG. 4.

Dislocation Loop with its local tangent vector t and Burgers vector b.

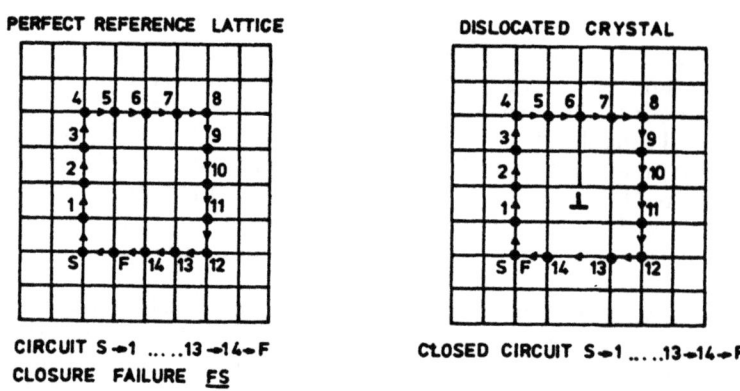

FIG .5.

The Bilby, Bullough and Smith FS/RH true Burgers vector convention The closed circuit is taken around the dislocation and then repeated in one-one directional correspondence, in the perfect crystal.

In the next section we shall calculate the energy of these dislocations and for immediate comparison with the point defect formation energies quoted above we comment here that the energy/atom plane of a dislocation is about 7 eV. This means that the dislocation is never in thermodynamic equilibrium; the crystal will always reduce its dislocation density if it can, either by annealing them out of the crystal or by geometrically arranging them to decrease the overall energy of the dislocation configuration (polygonization etc.)

2.3 Interfaces

We have just commented that, given the chance, an assembly of dislocations will try to adopt its lowest energy configuration. An example of such a low energy configuration is the simple tilt boundary or low angle tilt boundary defined as a parallel array of like, straight edge dislocation each one above the other as shown in figure 6. Similarly a general grain boundary can always be formally synthesized as a suitable array of parallel line dislocations. The hierarchy from point-line-surface defect is thus clear.

Bilby and Bullough and Smith [27,28] have shown how the concept of surface dislocation can be usefully introduced to define the structural features of general crystalline interfaces. A surface dislocation may be simply defined as a surface discontinuity separating volumes of material where the lattice strain is different; thus, in the simple tilt boundary, for example, the lattice rotation undergoes a discontinuity across the boundary. A surface dislocation is characterized by its normal ν and an associated dislocation tensor β_{ij}, just as a line dislocation is characterized by the direction of its line and its Burgers vector. β_{ij} may be defined as the i-th component of the resultant Burgers vector of dislocation lines in the surface which are cut by unit length of a line in the surface perpendicular to the j-th direction. A concept of particular significance to the crystallography of martensitic and other phase transformations is the simple glissile surface dislocation [28] which consists of a parallel array of like glissile dislocations; a tilt boundary is a simple example and the idea has been exploited to investigate the movement of martensitic habit planes and twin boundary interfaces [29].

The surface of the crystal itself is, of course, a most important defect but largely because of its interactive effect on other defects. These external surface image effects and the general interactions between dislocations and interfaces will be discussed in section 4.

In the next section we continue with more details about the intrinsic properties of dislocations.

3. ENERGY AND DISTORTION FIELDS OF DISLOCATIONS

Because a dislocation has a long ranged distortion field its associated energy is predominently elastic strain energy in contrast to a point defect whose energy is largely electronic or highly localized. It follows that elasticity theory plays an important role in determining many of the properties of dislocations. We thus begin this section with an outline of the theory of dislocations in elastic media and then proceed to describe some of the various atomic and partially discrete models that have been used to discuss the dislocation properties that are sensitive to the atomic structure of the dislocation core.

3.1 The continuum model of a dislocation: Elastic Fields

It is convenient to present first the general result for an arbitrary shaped dislocation loop in an anisotropic elastic medium and then to particularize to the straight edge and screw dislocation in both the anisotropic and isotropic medium. We then discuss the associated volume changes and the value of the non-linear continuum models. The presentation will make full use of the Green function formalism since it enables all the tedious analysis to be subsummed in one complex function and thereby does not allow the mathematics to cloud unduly the physical clarity.

3.1.1 Linear Fields in anisotropic and isotropic continua[30]

If $G_{ij}(\underline{r})$ is the infinite body Green's tensor that satifies the "unit point force" differential equation

$$C_{lmrs} G_{ir,ms}(\underline{r}) + \delta_{il} \delta(\underline{r}) = 0 \qquad (3.1)$$

where C_{lmrs} is the elastic moduli tensor, $\delta(\underline{r})$ is the Dirac Delta function and the comma notation mean partial differentiation with respect to the Cartesian coordinate system x_i then Volterra [31] has shown that the total distortion field $\beta_{ij}(\underline{r})$ associated with an infinitesimal dislocation loop at the coordinate origin $(\underline{r} = 0)$ is:

$$\beta_{ij}(\underline{r}) = A b_p c_{pkrs} G_{ir,sj}(\underline{r}) n_k \qquad (3.2)$$

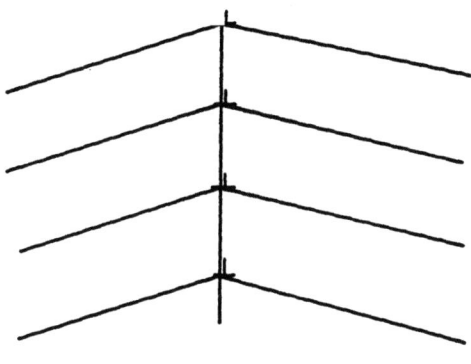

FIG. 6.

Simple Tilt boundary or Vertical Array of Edge Dislocations

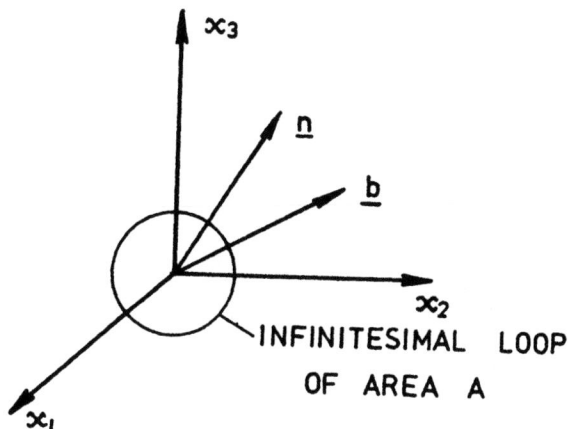

FIG. 7.

The Infinitesimal Loop with Burgers Vector b and unit normal n

The loop of area A, normal unit vector $\underline{n}$ and Burgers vector $\underline{b}$ is shown in figure 7. This result can be used to synthesize all the usual fields for finite dislocation loops. Thus if we integrate the expression (3.2) over a surface S with a boundary δS we obtain for the distortion field of the arbitrary dislocation δS the result

$$\beta_{ij}(\underline{r}) = \int_S b_p c_{pkrs} G_{ir,sj}(\underline{r} - \underline{r}') dS_k(\underline{r}') \qquad (3.3)$$

where $\underline{r}'$ is a point on S. This expression can be integrated under the integral sign and since

$$\beta_{ij} = U_{i,j} \qquad (3.4)$$

the total displacement field of the loop δS is

$$U_i(\underline{r}) = \int_S b_p c_{pkrs} G_{ir,s}(\underline{r} - \underline{r}') dS_k(\underline{r}') \qquad (3.5)$$

Finally Stokes theorem enables (3.3) to be transformed to the line integral around δS:

$$\beta_{ij}(\underline{r}) = \oint_{\delta S} \varepsilon_{jks} b_t c_{stlm} G_{il,m}(\underline{r} - \underline{r}') ds_k(\underline{r}') \qquad (3.6)$$

where now $\underline{r}'$ lies on δS and $ds_k(\underline{r}') = dx_k'$. This general result was first presented by Mura [32] and Willis [33] has shown that by ingenious manipulations of the Fourier transform representation of G in (3.6) it is possible to reduce (3.6) to a pair of repeated integrals - one around the loop δS and the other around a unit circle in reciprocal space orientated perpendicular to the loop normal. The field can thus be rapidly obtained with the aid of a suitable computer to perform the quadrature.

If the body is isotropic the Green's Tensor has the explicit form [34]

$$G_{ij}(\underline{r} - \underline{r}') = \frac{1}{16\pi\mu(1-\nu)}\left[2(1-\nu)\delta_{ij}|\underline{r}-\underline{r}'|_{,pp} - |\underline{r}-\underline{r}'|_{,ij}\right] \qquad (3.7)$$

and

$$C_{rsmn} = \frac{\mu}{1-2\nu}\left\{2\nu\delta_{rs}\delta_{mn} + (1-2\nu)[\delta_{rm}\delta_{sn} + \delta_{rn}\delta_{sm}]\right\} \qquad (3.8)$$

where μ is the shear modulus and ν is Poisson's Ratio. Substituting (3.7) and (3.8) into (3.6) yields the well known isotropic result [35]:

$$\beta_{ij}(\underline{r}) = \frac{1}{8\pi} \oint_{\delta S} \left[-\epsilon_{jkl} b_i |\underline{r}-\underline{r}'|_{,ppl} + \epsilon_{ikl} b_j |\underline{r}-\underline{r}'|_{,ppj} \right.$$

$$\left. + \frac{1}{1-\nu} \epsilon_{kmn} b_n |\underline{r}-\underline{r}'|_{,mij} \right] ds_k(\underline{r}') \qquad (3.9)$$

The stress fields associated with these loops then follow from Hookes law

$$p_{ij}(\underline{r}) = c_{ijkl} \beta_{kl}(\underline{r}) \qquad (3.10)$$

directly from (3.6) for the anisotropic medium and with c_{ijkl} replaced by (3.8) from (3.9) for the isotropic medium.

The fields for the straight dislocations can be obtained from these loop results and Willis [33] has given the explicit distortion field for the straight dislocation through the origin along the unit vector direction $\underline{l}$:

$$\beta_{ij}(\underline{r}) = -\frac{1}{\pi} \epsilon_{kjq} b_t l_q c_{lmtk} g \left\{ \sum_{N=1}^{3} \frac{(m_m + n_m W^N) N_{il}(\underline{m}+\underline{n}W^N)}{n_r \frac{\partial D}{\partial k_r} (\underline{m} + \underline{n} W^N)} \right.$$

$$\left. \times \frac{1}{y + W^N z} \right\} \qquad (3.11)$$

where $\underline{m}$ and $\underline{n}$ are orthogonal unit vectors forming a triad with $\underline{l}$ and $\overline{x}, y, \overline{z}$ are subsidiary coordinates associated with these unit vectors such that

$$\left. \begin{array}{l} x = \underline{l} \cdot \underline{x} \\ y = \underline{m} \cdot \underline{x} \\ z = \underline{n} \cdot \underline{x} \end{array} \right\} \qquad (3.12)$$

$N_{il}(\underline{k})$ is given explicitly by

$$N_{il}(\underline{k}) = \left[\text{Adj } L(\underline{k}) \right]_{il} = \left[L^{-1}(\underline{k}) \right]_{il} D(\underline{k}) \qquad (3.13)$$

"Adj" is the abbreviation for adjoint and $D(\underline{k})$ is the determinant of $L(\underline{k})$ where L is the 3 x 3 matrix of quadratic elements

$$L_{il}(\underline{k}) = c_{imls} k_m k_s \qquad (3.14)$$

and $N_{ij}(\underline{k})/D(\underline{k})$ is the Fourier transform of the Green's Tensor $G_{ij}(\underline{r})$:

$$G_{ij}(\underline{r}) = \frac{1}{8\pi^2} \iiint dk_1 \, dk_2 \, dk_3 \, \frac{N_{ij}(\underline{k})}{D(\underline{k})} e^{-i\underline{k}\cdot\underline{r}} \qquad (3.15)$$

Finally in (3.11), W^1, W^2 and W^3 are the three roots of the sextic equation

$$D(\underline{m} + \underline{n}W^N) = 0 \qquad (3.16)$$

which have positive imaginary parts. The first thing to notice is that in view of expression (3.12) the solution is entirely expressed in the laboratory reference frame x_i (which could be, for example, the primitive cube system in which the elastic constants are usually quoted in the reference books (a complex tensor transformation of c_{ijkl} is, thus, unnecessary). If the components of $\underline{u}$ relative to the $\underline{l}$, $\underline{m}$ and $\underline{n}$ directions are required, these are simply $\underline{u}\cdot\underline{l}$, $\underline{u}\cdot\underline{m}$ and $\underline{u}\cdot\underline{n}$ respectively, where $u_i(\underline{r})$ follows by directly integrating (3.11). It is easy to show that the solutions do not depend on the specific choice of $\underline{m}$ and $\underline{n}$; however, their dependence on $\underline{l}$ cannot be made explicit. Finally, the results are completely explicit; it is not, in contrast to the corresponding Eshelby, Read and Shockley solution [36], necessary to solve any sets of simultaneous equations.

The corresponding isotropic result can be deduced directly from (3.9) where now it is sensible to put the dislocation line along the x_3 axis. We find [30] for the strains around the pure edge dislocation with $\underline{b} = [b,o,o]$

$$e_{11} = \frac{b\, x_2}{4\pi(1-\nu)} \left[(3 - 2\nu)x_1^2 + (1 - 2\nu)x_2^2 \right] / r^4$$

$$e_{22} = \frac{b\, x_2}{4\pi(1-\nu)} \left[(1 + 2\nu)x_1^2 - (1 - 2\nu)x_2^2 \right] / r^4 \qquad (3.17)$$

$$e_{12} = \frac{x_1}{4\pi(1-\nu)} (x_1^2 - x_2^2)/r^4$$

and displacements

$$u_1 = \frac{b}{2\pi(1-\nu)} \left\{ (1-\nu)[\theta - \frac{\pi}{2}] + \frac{x_1 x_2}{2r^2} \right\}$$

$$u_2 = -\frac{b}{4\pi(1-\nu)} \left\{ (1-2\nu)\ln\frac{r}{r_0} + \frac{x_1^2}{r^2} \right\}$$
(3.18)

The stresses follow from (3.10) and (3.8). Similarly the strains around a straight screw dislocation lying along the x_3 axis with $\underline{b}$ = [oob] are [30]

$$e_{13} = -\frac{b}{2\pi} x_2/r^2$$

$$e_{23} = \frac{b}{2\pi} x_1/r^2$$
(3.19)

and the displacement is

$$u_3 = \frac{b}{2\pi} \left[\theta - \frac{\pi}{2} \right]$$
(3.20)

These explicit fields for the straight dislocations illustrate several important features of dislocations that are also present in the more complex anisotropic fields.

The stress and strain fields are very long ranged; they all fall off slowly as r^{-1}. In the absence of a core cut off at $r = r_0$ all the fields would diverge at $r = r_0$. This divergence is avoided by choosing an appropriate value of r_0 but its presence serves to remind one that the core properties of dislocations <u>cannot</u> be studied using linear elastic theory. An atomic or <u>semi discrete</u> model is necessary as will be discussed subsequently in this section.

The strain energy of the dislocation is given by [34]

$$E = \frac{1}{2} \int_V p_{ij} \beta_{ij} dV$$
(3.21)

where the integration is taken outside the core cut off radius r_0. For the straight dislocations it is necessary to terminate the integration at some large distance R from the dislocation to avoid a logarithmic divergence.

Thus for the edge dislocation, solution (3.17), we find

$$E = \frac{b^2\mu}{4\pi(1-\nu)}\left[\ln\frac{R}{r_o} - \frac{3-4\nu}{4(1-\nu)}\right] \qquad (3.22)$$

and for the screw dislocation, solution (3.19):

$$E = \frac{b^2\mu}{4\pi}\ln\frac{R}{r_o} \qquad (3.23)$$

Physically, this divergence of the strain energy of a straight dislocation at large R is very important though somewhat misleading. It is important because it emphasizes what a large strain energy is associated with a dislocation and that it is the dominant part of the total free energy of the defect.

The strain energy per atom plane threaded by the dislocation is approximately Eb which for $\mu \sim 10^{12}$ dyn cm^{-2}, $b \sim 2 \times 10^{-8}$ cm, $r_o \sim b$ and $R \sim 1$ cm has the numerical value

$$Eb \sim 7 \text{ eV/atom plane} \qquad (3.24)$$

for any of the various dislocations (the logarithmic term always dominates for long straight dislocations). Such a large energy value completely eliminates the possibility that such dislocations could be generated thermally. As we emphasized in section 2 dislocations are not a thermodynamic equilibrium defect; the crystal will dispose of them if it possibly can. It is misleading because all crystals are finite; there is no such thing as an infinite crystal and thus R is always limited. Furthermore the external surface is traction free and its presence reduces the dislocation strain energy somewhat (see section 4). The divergence of the strain energy as $r_o \to 0$ is non-physical; the energy expressions are only valid for a core radius r_o sufficiently large to exclude stresses and strains that violate the linear Hookes Law. As we comment above, atomic models are necessary to penetrate the high strain region of the core.

It is easy to show that the linear elastic fields associated with a dislocation will not lead to a volume change. The proof is as follows:

Suppose we have a dislocation distribution in a finite body V with surface S then if the distortion field generated by the dislocations is β_{ij} we have the volume change*.

$$\Delta V = \iiint_V \beta_{ii}\, dV \qquad (3.25)$$

Since we have linear theory

$$\beta_{ii} = S_{iikl}\, p_{kl} \qquad (3.26)$$

where p_{ij} is the internal stress and S_{ijkl} is the elastic compliance tensor.
Thus

$$\Delta V = S_{iikl} \iiint_V p_{kl}\, dV; \qquad (3.27)$$

now

$$(p_{kr}\, x_l)_{,r} = p_{kl} + p_{kr,r}\, x_l \qquad (3.28)$$

and since the stress field is divergence free (zero body forces):

$$p_{kr,r} = 0 \qquad (3.29)$$

we may express (3.27) in the 'divergence' form

$$\Delta V = S_{iikl} \iiint_V (p_{kr}\, x_l)_{,r}\, dV \qquad (3.30)$$

From Gauss's theorem and the fact that the external surface S is traction free:

$$\Delta V = 0 \qquad (3.31)$$

Thus a distribution of dislocations in a <u>linear</u> continuum will result in <u>zero</u> volume change.

---

*Note this β_{ij} is strictly speaking the residual field and is incompatible; i.e. no corresponding displacement field exists. The compatible field is obtained by adding the plastic distortion field to this residual field. For details of such defect source analysis see the work of Kroner [22] and Simmonds and Bullough [23].

3.1.2 Non-linear Fields

The only known, generally valid, non-linear dislocation solution for an anisotropic continuum is that due to Willis [37]. The solution was obtained by a perturbation procedure and uses the field theoretic result* that

$$\beta_{mn}(\underline{r}) = -\int c_{ijkl}\, G_{nk,l}\,(\underline{r} - \underline{r}')\, \varepsilon_{rmj}\, \alpha_{ri}\,(\underline{r}')\, dV'$$
$$- \int G_{nk}\,(\underline{r} - \underline{r}')\, f_{k,m}\,(\underline{r}')\, dV', \qquad (3.32)$$

where the integrals are over the whole space with respect to $\underline{r}'$ and G_{ij} is the usual Green's Tensor satisfying equation (3.1), $\underline{r}$ is the solution of the system of equations

$$-\varepsilon_{rks}\, \beta_{kj,s} = \alpha_{rj} \qquad (3.33)$$

$$c_{ijkl}\, \beta_{kl,j} = f_i \qquad (3.34)$$

For a linear distortion field β_{ij} equations (3.33) and (3.34) define the local dislocation tensor density α_{ij} and the body force f_i. Conceptually α_{ij} gives a measure of the resultant Burgers vector of the dislocations that thread an area element in the final dislocated state. If this area element has components ΔS_r then the resultant Burgers vector of lines cutting ΔS is

$$\Delta b_j = \alpha_{ij}\, \Delta S_i \qquad (3.35)$$

The tensor α_{ij} is directly related to the torsion tensor of a non-Riemannian manifold; these geometrical aspects which yield the relation (3.33) and provide a conceptual basis for the continuously dislocated state are discussed in a review by Bilby [5] and elsewhere [7].

For the linear solution $f_i = 0$ in (3.34) since there are no physical body forces present and thus the linear distortion field is, from (3.32)

$$\beta'_{mn}(\underline{r}) = -\int c_{ijkl}\, G_{nk,l}(\underline{r} - \underline{r}')\, \varepsilon_{rmj}\, \alpha_{ri}\,(\underline{r}')dV' \qquad (3.36)$$

To obtain the second order solution $\beta^2_{mn}(\underline{r})$ Willis [37] expands the stress p_{ij} and the dislocation tensor α_{ij} to second order in the distortions:

$$p_{ij} = c_{ijkl}\, \beta_{kl} + d_{ijklmn}\, \beta_{kl}\, \beta_{mn} + \qquad (3.37)$$

---
*See also Simmonds and Bullough [23] who discuss this and higher order field decomposition results.

$$\alpha_{ij} = -\varepsilon_{rks}(\beta_{kj,s} + \beta_{ij}\beta_{ki,s} + \ldots) \tag{3.38}$$

and writes
$$\beta_{kl} = \beta'_{kl} + \beta^2_{kl} + \ldots \tag{3.39}$$

When (3.39) is substituted into (3.38) and from the divergence of (3.37), $\beta^2_{mn}(\underline{r})$ is given by:

$$\left.\begin{array}{l} c_{ijkl}\beta^2_{kl,j} = -[c_{ijkl}\beta'_{kl} + d_{ijklmn}\beta'_{kl}\beta'_{mn}]_{,j} \\ -\varepsilon_{rks}\beta^2_{kj,s} = \alpha_{rj} + \varepsilon_{rks}\beta'_{kj,s} - \beta'_{kj}\alpha_{rk} \end{array}\right\} \tag{3.40}$$

These now have the general form of (3.33) and (3.34) where the first order solution is replacing the previous zero body force and the second order solution then follows directly from (3.32).

The solutions for discrete dislocations can be obtained from this general result by suitably constraining the tensor α_{ij} to a single line. For example for a single screw dislocation along the x_3 axis we may write

$$\alpha_{ij} = b\,\delta_{i3}\,\delta_{j3}\,\delta(x_1)\,\delta(x_2) \tag{3.41}$$

and the perturbation procedure can be reasonably easily followed.

Several aspects of this solution should be clarified.

a) It is the only procedure that includes full non-linear anisotropy. A stress function method has been developed by Pfliederer, Seeger and Kroner [38], but suffers from the serious objection that it is only appropriate for an isotropic continuum. Since the errors in neglecting linear anisotropy are at least as significant as any non-linear correction there seems little point in taking the trouble to obtain the non-linear solution within an overall isotropic approximation.

b) It is a tedious perturbation procedure and is very sensitive to the conditions assumed at the core of a discrete dislocation. Thus whereas the linear dislocation field only diverges at the core as the core radius tends to zero, the second order solution will diverge everywhere when such a limit is taken. Thus a detailed atomic treatment of the core will always be necessary even when reasonably long ranged non-linear properties are required. This is particularly true of the volume change which we saw was zero for the linear field and requires a non-linear calculation to estimate its value.

c) It is appropriate to comment here that it might be thought that such a non-linear analysis would enable the discrete atomic region to be reduced to a few atoms near the core and thereby facilitate the atomic simulation procedure. In our experience [39] this is not so. The numerical and analytic effort involved in getting the required non-linear solution is not justified by a commensurate reduction in the size of the core region. To achieve a real advantage it would be necessary to go beyond the second order perturbation which would require a prodigious mathematical effort. It is easier to sacrifice some machine time by retaining a fairly large atomic region. This point will be discussed further in section 3.3.

3.2 The Partially Discrete Models of Dislocations

The two partially discrete models that we shall discuss are the Peierls-Nabarro [40,41] (or horizontal-cut) model and the van der Merwe [42] (or vertical-cut) model of the straight edge dislocation. The advantage of such models is that they provide a qualitative framework for the understanding of some of the core or atomic properties of dislocations without the hidden complexity of the full atomic treatments. This analytic advantage is, of course, somewhat obviated by the necessary physical simplifications inherent in the models.

3.2.1 The Peierls-Nabarro Model [40,41]

The basic model is described here for an elastically isotropic simple cubic lattice in order to simplify the representation as much as possible and thus expose the essential principles of the model. Generalizations of the model to account for real crystal structures and elastic anisotropy are quite straight forward and will be mentioned subsequently.

The model for an edge dislocation in a simple cubic lattice is shown in figure 8. The formal glide plane is $y = 0$ and

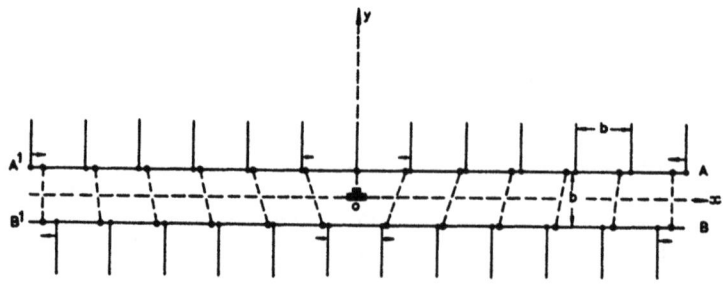

FIG.8.

The edge dislocation in a simple cubic lattice.

elastic half spaces are terminated on the planes AA' and BB'; that is at the centres of the glide plane atoms and a non-Hookean slab of width b (the slip vector and atomic spacing) joins the two half spaces. The symmetrical configuration indicated in the figure may be thought of as having arisen by cutting the perfect crystal into two halves along the $y = 0$ plane, then displacing the upper half crystal rigidly a distance b/2 in the x direction and re-welding the two half crystals. The initially perfect reference positions of the atoms in the two half crystals are indicated in the figure together with the atomic displacements that will occur on joining the two halves. The equilibrium configuration of the glide plane atoms is defined by a balance between the interatomic forces across the slip plane, which tend to align the corresponding atoms into the perfect crystal arrangement away from the centre of the dislocation and cause compression of the lattice immediately above the dislocation line and tension immediately below, and the elastic resistance of the two half spaces to such surface displacements.

If a tangential shear stress $p_{xy}(x)$ is applied to the surface A'A of a semi-infinite isotropic elastic block then it is a well known result in elasticity theory that the stress and the resulting displacement $u(x)$ in the x direction are related by the Hilbert transform:

$$p_{xy}(x) = \frac{\mu}{\pi(1-\nu)} \int_{-\infty}^{\infty} \frac{du(t)}{dt} \frac{dt}{t-x} \tag{3.42}$$

where μ is the shear modulus, ν is Poisson's ratio and the integral is understood as a Cauchy Principle value. By taking Hilbert transforms of (3.42) and integrating, the surface displacement $u(x)$ may be expressed in terms of the shear:

$$u(x) = -\frac{(1-\nu)}{\mu} \int_{-\infty}^{\infty} p_{xy}(t) \ln |t-x| \, dt \tag{3.43}$$

The stress $p_{xy}(x)$ on A'A is balanced by the shear forces due to the atoms in the opposite slip plane B'B and Peierls [40] and Nabarro [41] assumed that this force was a sinusoidal function of the relative displacement function defined by the relation

$$\begin{aligned} \phi &= b/2 + u - \bar{u} \simeq b/2 + 2u(x) \quad (x > 0) \\ \phi &= -b/2 + u - \bar{u} \simeq -b/2 + 2u(x) \quad (x < 0) \end{aligned} \tag{3.44}$$

where u is the compressive displacement of the upper surface A'A and $\bar{u}$ is the expansive displacement of the lower surface B'B. Since registry is achieved across the glide plane at large distances from the dislocation

$$u(\infty) = -\bar{u}(\infty) = b/4 \qquad (3.45)$$

and the sinusoidal function has the form

$$p_{xy} = \frac{\mu}{2\pi} \sin \frac{2\pi\phi}{b} = -\frac{\mu}{2\pi} \sin \frac{4\pi u}{b} \qquad (3.46)$$

where the amplitude of (3.46) ensures that Hooke's Law is satisfied across the glide plane for small ϕ. On equating (3.42) and (3.46) we obtain the well known Peierls-Nabarro integral equation for the unknown displacement field $u(x)$

$$-\sin \frac{4\pi u(x)}{b} = \frac{2}{1-\nu} \int_{-\infty}^{\infty} \frac{du}{dt} \frac{dt}{t-x} \qquad (3.47)$$

The integral equation (3.47) is non-linear, singular and not of standard form, nevertheless it possesses the very simple exact solution [40].

$$u(x) = -\frac{b}{2\pi} \tan^{-1}\left[\frac{x}{\sigma}\right] \qquad (3.48)$$

where

$$\sigma = b/2(1-\nu) \approx 0.75\, b \qquad (3.49)$$

is known as the half width of the dislocation. The corresponding shear stress distribution on the surface A'A follows from (3.46) or (3.42) with $u(x)$ given by (3.48)

$$p_{xy}(x) = \frac{\mu\sigma}{\pi} x/(x^2+\sigma^2) \qquad (3.50)$$

The form of the displacement is shown in figure 9 and the half width σ is defined as the distance from the centre of the dislocation in which the displacement in the glide plane has fallen to half its asymptotic (large x) value.

The shear stress given by (3.50) and the corresponding shear strain ($e_{xy}=p_{xy}/2\mu$) in the glide plane have the maximum respective values at $x=\sigma$ of

$$\begin{aligned} p_{xy}(\max) &= \mu/2\pi \\ e_{xy}(\max) &= 1/4\pi \end{aligned} \qquad (3.51)$$

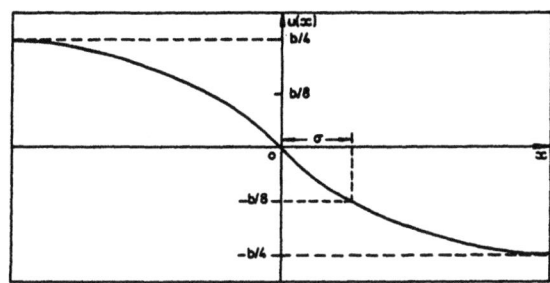

FIG.9. The half-width σ of the Peierls-Nabarro dislocation
The edge dislocation in a simple cubic lattice. A'A and B'B are adjacent slip-planes, b is the Burgers vector and the reference state is indicated. The Peierls-Nabarro model.

and though large they are certainly smaller than the corresponding elastic solution of Burgers (equation (3.17)) with its singularity at the dislocation core. In fact an important feature of the Peierls-Nabarro model is that it removes the core singularity. However the core singularity can also be avoided by simply removing a cylinder of material of radius b (say) with the dislocation line along its axis; a suitable stress field to annul the surface tractions on the core surface can then be added to the Burgers field (3.17). When this is done we find that a "hollow" edge dislocation of core radius b has lower stresses and strains around it than does the Peierls-Nabarro dislocation with the simple sinusoidal law of force in (3.46). The tensile strain $e_{xx}(x)$ in the surface of the lower block B'B has the form

$$e_{xx}(x) = \frac{b\sigma}{2\pi}/(x^2 + \sigma^2) \tag{3.52}$$

and has a maximum value of

$$e_{xx}(\max) = \frac{(1-\nu)}{\pi} \simeq 20\% \tag{3.53}$$

immediately below the dislocation ($x=0$, $y= -\frac{b}{2}$). This strain is too large to be consistent with the assumption of a linear Hooke's Law in the elastic block; in fact it is even larger than the strain at the corresponding point in the Burgers elastic solution. The tensile stress corresponding to (3.53) is

$$p_{xx}(\max) = \frac{2\mu}{\pi} \tag{3.54}$$

and exceeds the theoretical tensile strength ($\simeq \mu/30$ of many materials. These very high stresses and strains in what are assumed to be <u>elastic</u> regions is a serious criticism of the Peierls-Nabarro <u>model</u> when the sinusoidal law is used. The solution does not describe a reasonable atomic configuration around the dislocation core and should certainly not be used for this purpose. Since the non-Hookean forces were allowed only across the glide plane the strains across the glide plane are the only acceptable values as we approach close to the dislocation. However even across the glide plane the sinusoidal law has led to a relative displacement that varies much too rapidly with x. The width of the edge dislocation is only $\simeq 1.5b$ which makes the displacement function $u(x)$ in figure 9 hardly better than the simple step function

$$u(x) = -b/4 \quad x > 0$$
$$ = +b/4 \quad x < 0 \tag{3.55}$$

which is the discontinuous displacement appropriate to the corresponding elastic solution. The constraint to Hookean behaviour immediately below the dislocation should, if anything, lead to <u>excessive</u> relaxation in shear along the glide plane with a consequent exaggerated width; the fact that the actual width is so small thus emphasizes the complete inadequacy of the sinusoidal law. Furthermore the Hookean constraint below the dislocation means that relaxation can only occur by spreading the shear in the glide plane and micro-cracking normal to the glide plane is not permitted (see however the van der Merwe model discussed below which does allow such tensile relaxation).

Foreman, Jaswon and Wood [43] have devised a simple procedure for the parametric modification of the sinusoidal law which yields an alternative displacement and shear to (3.48) and (3.50) of

$$u(x) = -\frac{b}{2\pi}\left[\tan^{-1}\frac{x}{\sigma a} + (a-1)\frac{x\sigma}{x^2+a^2\sigma^2}\right] \tag{3.56}$$

and

$$p_{xy}(x) = \frac{\mu}{\pi}\left\{\frac{x\sigma}{x^2+a^2\sigma^2} + 2a(a-1)\frac{x\sigma^3}{(x^2+a^2\sigma^2)^2}\right\} \tag{3.57}$$

respectively, where a is an arbitrary parameter greater than unity. The shear (3.57) is the result of evaluating (3.42) with the displacement $u(x)$ given by (3.56). This integration is facilitated by using the operator form of (3.56)

$$u(x) = -\frac{b}{2\pi}\left[1 - (a-1)\frac{\partial}{\partial a}\right]\tan^{-1}\frac{x}{a\sigma} \tag{3.58}$$

With the relative displacement across the glide plane defined by (3.44) it may be verified that the displacement (3.56) and shear (3.57) imply the required Hookean relation, independent of the parameter a

$$p_{xy} = \mu \phi/b \tag{3.59}$$

when $\phi \to 0$ (i.e. across the glide plane at large distance from the dislocation). The implied shear law of force may be parameterized in the form [43]

$$\frac{2\pi}{\mu} p_{xy} = \frac{1}{2a^2} \left\{ 2(2a-1) \sin \theta - (a-1) \sin 2\theta \right\}$$

$$\frac{2\pi \phi}{b} = \theta - \frac{a-1}{a} \sin \theta \tag{3.60}$$

where θ is related to x by

$$\theta = 2 \cot^{-1}(x/a\sigma) \tag{3.61}$$

and the relation given by (3.59) for small θ is now clearly evident.

Foreman, Jaswon and Wood [43] showed that the force law for a=4 was close to that expected for a bubble raft (1.3 mm diameter) and did indeed predict a dislocation width in quite good agreement with the observed width of the dislocation in the bubble raft. Such a force law, with its lower amplitude than the sinusoidal law, not only increases the width of the Peierls-Nabarro dislocation, but it also leads to much lower (and thus more acceptable) stresses and strains in the elastic regions. The principle objection to the simple Peierls-Nabarro model is the choice of a sinusoidal shear law of force across the glide plane - an objection that can be removed by the above parametric modification of the force law.

Recently Kroupa and Lejcek [44] have devised a generalization of this parametric procedure which enables the relative displacement function $\phi(x)$ to be calculated when an <u>arbitrary</u> shear law is given. Following Foreman [45] a second harmonic may also be included in the law of force to describe one dimensional dissociation; the authors [44] have used this procedure to study certain planar dissociated dislocations in α-Fe. However, the Hookean constraint close to the dislocation remains a serious criticism of the model and we shall see in subsequent sections how more complex lattice models can avoid such constraints.

The energy of the Peierls-Nabarro dislocation consists of two parts: (a) the elastic strain energy of the two half spaces and (b) the energy of misfit across the glide plane.

(a) The elastic strain energy

By applying the divergence theorem to the volume integral of the energy density, we obtain

$$E_{el} = \int_{-L}^{L} p_{xy}(x)\, u(x)\, dx \qquad (3.62)$$

where L is large and represents the size of the crystal containing the dislocation and the factor of ½ has been omitted to account for the energy in each half space, which must, by symmetry, have the same elastic energy. From (3.48) and (3.50)

$$E_{el} = \frac{b^2 \mu}{2\pi^2(1-\nu)} \int_0^{L/\sigma} \frac{x \tan^{-1} x}{1+x^2}\, dx$$

$$\to \frac{b^2 \mu}{4\pi(1-\nu)} \ln \frac{L}{2\sigma} \quad \text{(for large L)} \qquad (3.63)$$

(b) The misfit energy

This is the energy stored in the anelastic slab of thickness b that represents the glide plane and is given by

$$E_m = \frac{b\mu}{4\pi^2} \int_{-\infty}^{\infty} \left[1 + \cos \frac{4\pi\, u(x)}{b}\right] dx \qquad (3.64)$$

which, from (3.48) yields

$$E_m = b^2 \mu / 4\pi(1-\nu) \qquad (3.65)$$

Foreman (see reference [46]) has given a simple proof that (3.65) is the integral form of the misfit energy for arbitrary shear law of force across the glide plane. The total energy of the Peierls-Nabarro edge dislocation is thus

$$E = E_{el} + E_m = \frac{b^2 \mu}{4\pi(1-\nu)} \left\{ \ln \frac{L}{2\sigma} + 1 \right\} \qquad (3.66)$$

and by comparing this with the corresponding expression for the elastic energy of the Burgers edge dislocation (equation 3.22), we see that we may crudely associate the core radius of the latter with 2σ the width of the Peierls-Nabarro dislocation. If the parameter a is introduced, then the reduction of energy of the dislocation is approximately given by replacing σ in (3.66) by $(a\sigma)$.

An important aspect of the Peierls-Nabarro model is that it was the first model that could be used to give an estimate of the stress to move a dislocation (the critical shear stress). Peierls [40] and Nabarro [41] argued that the discrete nature of the glide plane atoms could be utilized by replacing the integral in the misfit energy expression (3.64) by the analogous infinite sum with the centre of the dislocation displaced a fraction α of the slip distance from the symmetrical configuration in figure 1. This misfit energy can then yield the force required to cause such a dislocation movement and from the stationary value of this force the critical stress or so called Peierls stress follows.

The misfit energy when the dislocation is displaced αb is

$$E_m = \frac{b^2\mu}{8\mu^2} \sum_{n=-\infty}^{\infty} \left\{ 1 + \cos 2\left[\tan^{-1} 2(1-\nu)(\alpha + \frac{n}{2})\right] \right\} \quad (3.67)$$

This sum can be evaluated exactly to yield

$$E_m = \frac{b^2\mu}{4\pi(1-\nu)} \frac{\tanh[2\pi/(1-\nu)]}{1 - \cos 4\pi\alpha \, \text{sech}[2\pi/(1-\nu)]} \quad (3.68)$$

and the Peierls stress P is given by

$$P = F_{max}/b$$

where

$$F_{max} = -\frac{1}{b}\left[\frac{dE_m}{d\alpha}\right]_{max}$$

which from (3.68) leads to

$$P \simeq \frac{2\mu}{1-\nu} \exp[-2\pi\sigma/b] \quad (3.69)$$

More generally when the parametric force law (3.60) is used the Peierls stress becomes

$$P = \frac{2\mu}{1-\nu} \exp\left[-2\pi\frac{a\sigma}{b}\right] \left\{ 1 + 2\pi(a-1)\frac{\sigma}{b} + 4\pi^2 \frac{(a-1)^2\sigma^2}{ab^2} \right\} \quad (3.70)$$

Unfortunately the simple derivation of E_m leading to (3.69) is not entirely satisfactory* and in fact $(dE_m/d\alpha)$ has a <u>maximum</u> value at the symmetrical positions ($\alpha=0$ and $\frac{1}{2}$) and a <u>minimum at</u> the intermediate asymmetrical position ($\alpha=\frac{1}{4}$). The correct misfit energy should have a single maximum (at $\alpha=\frac{1}{2}$) and minimum values at $\alpha=0$ and 1. The Peierls stress is then given by the slope at the point of inflection at $\alpha = \frac{1}{4}$. Vitek [47] has shown that such a misfit energy follows (though not analytically) if the independent summations in (3.67) on each side of the glide plane are replaced by a summation of the interaction energy in pairs across the glide plane. The calculation yields a Peierls stress with a dependence on width that is very similar to (3.69).

Whatever point of view one takes concerning the details of the Peierls stress calculation for the edge dislocation the important qualitative result emerges, namely that the Peierls-Nabarro model with all its inherent inadequacies can be used to demonstrate that a small finite stress is required to move a dislocation (the Peierls stress) and that this stress decreases rapidly with increasing spread of misfit (width) across the glide plane.

The isotropic simple cubic model is not able to relate directly to real materials, but several attempts to extend the model to include both elastic anisotropy and real glide plane configurations have been made. Leibfried and Dietze [48] have applied the model to discuss the degree of dissociation and the Peierls widths for a number of face centred cubic metals.

Perhaps the most unfortunate feature of this application of the Peierls-Nabarro model to face centred cubic metals is the fact that the shear law of force only acts between nearest neighbours across the glide plane. In a face centred cubic lattice the stacking fault energy (on the {111} plane) depends on the interatomic interactions beyond the second neighbour distance; the model cannot therefore <u>implicitly</u> include the magnitude of the stacking fault energies within the shear law and the repulsive forces between the partials must therefore be imposed <u>explicitly</u>. The complete lattice models are not necessarily subject to short range interatomic forces and therefore a potential can be used that is itself consistent with the known stacking fault energies as discussed in section 3.3. In materials with the body centred lattice the stacking fault energies are usually very large and do depend on the short range interactions. In this sense the Peierls-Nabarro model is somewhat more appropriate for such lattices and Heinrich et al [49] have used the variational procedure of Liebfried and Dietze [48] to calculate

---

*A detailed criticism of this aspect of the Peierls-Nabarro estimate of P has been given by Hirth and Lothe [1] and more recently by Christian and Vitek [47]

the Peierls widths and energies of dislocations on {110} glide planes in a body centred cubic lattice. They have systematically increased the atomistic region separating the two elastic half spaces and find that the single atomic layer glide plane model seriously underestimates the Peierls width of the dislocations; the usual single atomic layer yeilds a width $\omega = 0.60b$ and with three atomic layers and greater the width increases to $\omega = 0.90b$.

3.2.2 The Van der Merwe Model [42]

This model, introduced by van der Merwe [42], is particularly appropriate for the study of tensile relaxation (rather than the shear relaxation in the glide plane as in the Peierls-Nabarro model) just below the extra half plane of the edge dislocation. As in the Peierls-Nabarro model it consists of two elastic half spaces separated by an anelastic slab of thickness b across which non-Hookean forces may act. However, in this case the slab passes through the dislocation in a direction orthogonal to the glide plane and a <u>tensile</u> interatomic force law is allowed to prevail between the <u>atoms in</u> the surfaces of the elastic blocks across the non-Hookean slab. The model is sketched in figure 10 where the

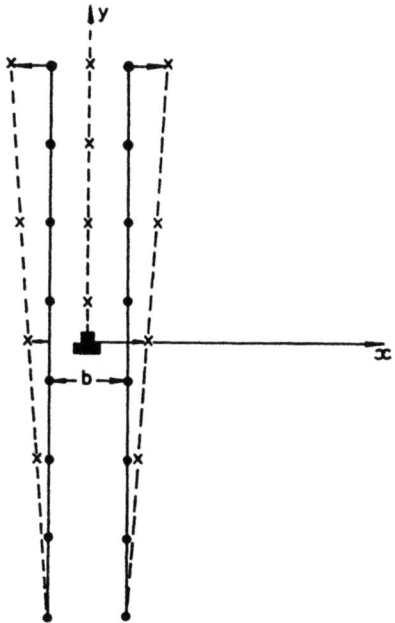

FIG. 10.

The edge dislocation in a simple cubic lattice. b is the Burgers vector and the reference state is indicated. The Van der Merwe model.

extra half plane of atoms, pushed into the x=0, y>0 region between the elastic blocks, is indicated by crosses; the other crosses indicate the final atomic positions in the dislocated state.

An integral equation similar to the previous equation (3.47) can be set up to define the equilibrium condition for the atoms in the surfaces of the elastic blocks; it is

$$\frac{\mu}{\pi(1-2\nu)} \sin \frac{4\pi u(y)}{b} = -\frac{\mu}{\pi(1-\nu)} \int_{-\infty}^{\infty} \frac{du(t)}{dt} \frac{dt}{t-y} \quad (3.71)$$

where a sinusoidal <u>tensile</u> law of force across the non-Hookean slab has been assumed and $u(y) \equiv u(\frac{b}{2},y)$ is the surface displacement of the right hand block in the x direction. The exact solution of (3.71) is

$$u(y) = \frac{b}{4} + \frac{b}{2\pi} \tan^{-1}(y/\delta) \quad (3.72)$$

where δ is the <u>vertical</u> half width of the dislocation given by

$$\delta = b(1-2\nu)/4(1-\nu) \simeq b/8 \quad (3.73)$$

This half width is only one sixth of the corresponding half width of the Peierls-Nabarro dislocation. It follows that the sine-law is even worse for the present model than it was for the Peierls-Nabarro model. It does not provide a good description of the atomic configuration - the anisotropic elastic solution is certainly preferable. Furthermore, although the singularity at the dislocation has been again removed, a simple "hollow core" elastic model could do this also. However the symmetrical tensile interatomic law of force across the vertical slab is obviously a very crude approximation; it is clearly worse than the equivalent assumption of a symmetrical shear law of force across the glide plane in the Peierls-Nabarro model. Such a symmetrical tensile law takes no account of the fact that atoms are, in general, more difficult to push together than to pull apart. The model has been modified by Bullough [50] who devised a parametric technique, analogous to the parametric modification of the shear force law of Foreman et al [43], which enabled a class of asymmetrical tensile force laws to be used. Thus the van der Merwe model can be used to study dislocations in materials where the Peierls width is extremely small and relaxation by micro-cracking could be reasonably expected. However, as in the case of the Peierls-Nabarro model, the model must exaggerate the feature it is being used to study; in this case because Hooke's Law is being enforced across the glide plane only a distance b/2 from the dislocation.

The ideal partially discrete model for the edge dislocation would appear to be the quarter plane model where both the glide plane and the plane of the extra half plane are replaced by non Hookean slabs but it appears that the mathematical complexity involved in the elasticity of quarter plane deformation precludes the existence of simple analytic solutions.

3.3 Lattice Models of Dislocations

In this section we highlight some of the additional insight into dislocation properties that can be gained by using a full atomic model. The procedure requires the simultaneous relaxation of a large atomic crystallite with boundary conditions imposed upon its surfaces that suffice to define the presence of a dislocation within the crystallite. In practice to study dislocations it is necessary to use a crystallite containing many thousands of atoms which have to be relaxed to their configuration with minimum potential energy. To study such many body relaxation processes requires the use of a large fast digital computer and sophisticated numerical relaxation techniques. Fortunately appropriate computers are now available and very efficient numerical methods have been developed for the rapid solution of the many thousands of coupled differential equations involved.

In such studies the essential physical ingredient is the interatomic potential and considerable effort is now being made into the construction of suitable potentials for different materials. Important as it obviously is it is not really appropriate to review such potential studies in this discussion of lattice models of dislocations. It will suffice to emphasize the dislocation properties that have been studied that are reasonably insensitive to the potential. These features are, of course, essentially qualitative and geometrical in nature and will be illustrated by direct reference to the studies on copper. For a rather complete discussion of this topic the reader is referred to reference [17].

To simulate a straight edge dislocation a large parallelepiped of freely interacting atoms is set up so that the atoms form a perfect lattice (face centred in the case of copper) which is in equilibrium under a suitable interatomic potential at the observed lattice parameter. The response of this crystallite must be such that all the eigenvalues of the dynamical matrix are positive - the lattice is then stable against homogeneous deformation. Furthermore because dissociation can occur, and must be allowed to occur if the system wishes, at the dislocation core with separation of the total dislocation into two partial dislocations separated by a stacking fault, it is essential that the crystal is stable against heterogeneous shear; thus the possible stacking fault energies must all be positive. The consequences of a negative fault energy on the stability of a dislocation core are usually catastrophic; the dislocation simply dissociates completely across the crystallite since the repulsion of the partials is no longer balanced by an attractive force from the stacking fault (positive stacking fault energy).

In the f.c.c. metals one has only one possible stacking fault to worry about and a suitable stable interatomic potential which is simultaneously consistent with all the above harmonic response properties of the metal and the fault energy can usually be constructed [17,51]. The situation is more complex in the b.c.c. metals when various different faulting is known to occur. Fortunately however the fault energies are usually very large (hundreds of ergs/cm^2) and various types of potentials can be constructed to yield satisfactory stability. The various copper dislocation calculations have all been done with empirical potentials; these are usually pairwise in form and either hold the system of atoms in equilibrium without additional constraint (a so-called central force potential of which the Morse potential is perhaps the most well known example or require additional constraints to maintain equilibrium. These additional constraints take the form of a volume dependent additional potential which is either a global force field or a localized force field around each atom.

The copper edge dislocation was first studied by Bullough[52] and by Cotterill and Doyama [53]. Bullough [52,54] used the purely repulse Born-Mayer potential in which overall equilibrium is maintained by simply holding the external surfaces of the crystallite in the elastically anisotropic dislocation configuration such as defined by the integral of the distortion field (3.11). However though this calculation is perhaps historically interesting it cannot be claimed to provide a good representation of the core of the copper dislocation. The Born-Mayer potential decays exponentially and yields an almost negligible second neighbour interaction; since the stacking fault energy in an f.c.c. lattice depends only on deviations of the third neighbours from perfect stacking it follows that the Born-Mayer potential implies an almost negligible stacking fault energy (less than 1 erg /cm^2). On the other hand Cotterill and Doyama [53] used the central Morse potential with a cut off distance chosen to yield a reasonable stacking fault energy ($\sim$ 30 ergs/cm^2). However this calculation also is really only of historical interest since the Morse potential (being in equilibrium) implies that the Cauchy relations should be obeyed between the elastic contacts; they are certainly not obeyed in copper and thus the harmonic response of the perfect crystal is not correct. In addition the sensitivity of the stacking fault energy to the choice of truncation distance of the potential was rather too great to instil much confidence in the use of the Morse Potential. However both these calculations demonstrated that such dislocation simulation was feasible.

The more recent studies of Norgett, Perrin and Savino [53] and Perrin, Englert and Bullough [56] used the more physically acceptable potential for copper constructed by Englert, Tompa and Bullough [57]. This is a non-equilibrium pair potential, with an implicit volume dependent component, that was specially constructed from a set of spline polynomials. This potential gives an excellent empirical description of the complete harmonic response of copper and is fitted to both the stacking fault energy and certain relevant point defect formation energies. The result of this study was a predicted dissociation of the edge dislocation of 32 Å dropping to 15 Å for the pure screw dislocation. This absolute value of the dissociation and the relative variation in going from pure edge to pure screw dislocation configuration was in rather beautiful agreement with the weak beam transmission electron microscopy observations of Stobbs and Sworn [58] of such dislocations. Furthermore by using the actual atomic configurations to compute the diffracted images Perrin and Savino [59] showed that the deduced values of the stacking fault energies using an interpretation based on elasticity theory were much too small. The essential ductility of copper was also explicable from the computed Peierls widths of the partials; these were found to have increased from $\sim 2b$ for the elastic model to just over $5b$ for the atomic model. The Peierls stress for copper should thus be extremely low (see equation (3.69) for the Peierls-Nabarro model result). The model was also used to study point defect-dislocation interactions and we shall refer to some of these results in the next section.

Finally we introduced this sub-section by emphasizing the great importance of using physically reliable interatomic potentials and we believe the point is sufficiently important to re-emphasize it again. Really quantitative studies, using atomic models, must await the construction of fundamentally reliable potentials. Such potentials, for transition metals for example, are several years off, but there is no doubt that once they are available the presently available sophisticated relaxation techniques will enable dislocation properties such as flow stress and interactions with other defects to be studied and understood from first principles. In the meantime perhaps the most important developments of the simulation procedure have been made in the treatment of the boundary conditions. Sinclair [60] has shown how to use the general linear anisotropic elastic solution to parameterize the displacement field outside the atomic region. By combining these "parameters" with the real atomic coordinates in the overall relaxation procedure he has obtained the best linear representation of each dislocation studied. Thus the anharmonic core field is represented by a linear field away from the core. In this way, for example, he could demonstrate that a cube edge dislocation in iron has an effective elliptical dilatational source just below the slip plane. Apart from the fact that such detailed core features must be peculiar to the particular dislocation and material the identification of this additional linear field should facilitate the interpretation of the behaviour of such dislocations. Finally, as we have commented previously (section 3.1.2) the non-linear continuum field does not seem to be particularly helpful [39] in reducing the size of the atomic region and thereby reducing the overall numerical effort.

4. FORCES ON DISLOCATIONS

In this section we shall largely revert back to the use of continuum results and refer to atomic model results only when the "atomic" feature is particularly relevant. The interactions of dislocations with the applied field and with other internal and induced fields is the topic of this final section.

4.1 The External Stress

When a body V is subjected to an external stress an elastic distortion field $\beta_{ij}^F(\underline{r})$ is induced within it; the superscript F indicates that this field is force induced and thus a single valued displacement field $U_i^F(\underline{r})$ exists and [23]

$$\beta_{ij}^F(\underline{r}) = U_{i,j}^F(\underline{r}) \tag{4.1}$$

If the same body contains a residual elastic distortion field $\beta_{ij}^R(\underline{r})$ arising from the presence of dislocations then this field will satisfy the zero body force equilibrium equations

$$c_{ijkl}\,\beta_{kl,j}^R(\underline{r}) = 0 \tag{4.2}$$

and a displacement potential U_i^R does not exist. The interaction energy between these two fields is

$$E_I = \int_V c_{ijkl}\,\beta_{ij}^F(\underline{r})\,\beta_{kl}^R(\underline{r})\,dV \tag{4.3}$$

which, if we integrate by parts after using (4.1) yields

$$E_I = -\int_V U_i^F(\underline{r})\,c_{ijkl}\,\beta_{kl,j}^R(\underline{r})\,dV = 0 \tag{4.4}$$

from (4.2).

This result means that the force field and the defect residual field are fundamentally orthogonal. It does not mean that a distribution of body force or surface tractions fail to interact with dislocations in the body. In fact an interaction does occur because the external agent maintaining the applied body forces or surface tractions must perform work when the body suffers the residual distortion $\beta_{ij}^R(\underline{r})$. Thus, the physical interaction energy has the form

$$E_I = \int_V u_i^R(\underline{r}) f_i^F(\underline{r}) dV \qquad (4.5)$$

where the displacement field $u_i^R(\underline{r})$ has been made single-valued by suitably cutting the body V. In general, a well-behaved, single-valued displacement field $u_i^R(\underline{r})$ does not exist over all V. This particular point is easily illustrated when $\beta_{ij}^R(\underline{r})$ is the distortion field arising from a single straight dislocation, say a screw dislocation along the x_3 axis. Then, from (3.20), we see that the displacement field is given by

$$u_i(\underline{r}) = \frac{b}{2\pi} \delta_{13} (\theta - \frac{\pi}{2}) \qquad (4.6)$$

which is multivalued, and to evaluate uniquely an integral of the type (4.5) the range of θ must be specified modulo 2π; that is, a cut must be introduced at some angle θ. As a matter of fact, when we introduce a displacement field (multivalued) for a dislocation, what we have actually done is to replace the dislocation distortion field by that arising from an equivalent topological weld [23,30].

If we consider a body V containing an arbitrary dislocation loop with Burgers vector $\underline{b}$ and a shape defined by the variation of the local tangent vector $\underline{t}(\underline{r}')$, where $\underline{r} = \underline{r}'$ define points on the loop, then from the change of the interaction energy (4.5) when the loop changes shape under the influence of the force induced stress field

$$p_{ij}^F(\underline{r}') = c_{ijkl} \beta_{kl}^F(\underline{r}') , \qquad (4.7)$$

the force on unit length of the dislocation at the point $\underline{r}'$ can easily be shown to have the general Cartesian components [61]

$$F_i(\underline{r}') = - \varepsilon_{ijk} t_j(\underline{r}') b_r p_{kr}^F(\underline{r}') \tag{4.8}$$

For an external stress p^F will usually be constant and from this general result we can see that the glide force per unit length on a loop may be written

$$F_g = \tau b \tag{4.9}$$

where τ is the resolved applied shear stress in the direction of $\underline{b}$ and the force F_g acts at each point of the loop in the glide plane and orthogonal to the loop direction $\underline{t}$.

It is this force that expands glide loops, operates Frank Read Sources and generally is responsible for the overall shape change of crystalline bodies by glissile movement of dislocations. The general result (4.8) is necessary to discuss other driven dislocation movements such as climb when thermal conditions permit diffusive motion of point defects. This will be further discussed below when we consider the dislocation-point defect interactions. Finally the result (4.8) is also the force per unit length acting on a dislocation due to the internal stress field of another defect; these defect interactions will now be discussed.

4.2 Interactions with other Defects

The results in this section are all derived from the fundamental expression for the interaction energy between two defect fields [35]

$$E_I = \int_V p_{ij}^1(\underline{r}) \, e_{ij}^2(\underline{r}) dV \tag{4.10}$$

This is simply the energy required to create the defect (2) in the field of (1) or vice-versa. If u_i^2 is the displacement field corresponding to the strain field e_{ij}^2 then (4.10) can be written

$$\begin{aligned} E_I &= \int_V p_{ij}^1(\underline{r}) \, u_{i,j}^2(\underline{r}) dV \\ &= \int_V (p_{ij}^1(\underline{r}) \, u_i^2(\underline{r}))_{,j} \, dV \end{aligned} \tag{4.11}$$

since, from (4.2), if (1) is a dislocation field

$$p_{ij,j}^1 = 0 \tag{4.12}$$

Applying Gauss's Theorem to (4.11) yields

$$E_I = \int_S p_{ij}^1(\underline{r}) u_i^2(\underline{r}) n_j dS \qquad (4.13)$$

If defect (2) is also a dislocation then since

$$p_{ij}^1 n_j = 0$$

on the external surface of the body, the integral (4.13) should be evaluated over the two surfaces of the cut required for the creation of defect (2) and necessary to permit the displacement field u^2 to exist. We can now discuss the dislocation interactions with various particular defects.

4.2.1 Dislocation-dislocation interactions

The force per unit length on dislocation (1) due to the field of dislocation (2) follows from the derivative of (4.13) and has the same general form as (4.8); that is

$$F_i(\underline{r}) = - \varepsilon_{ijk} t_j^1(\underline{r}) b_r^1 p_{kr}^2(\underline{r}) \qquad (4.14)$$

where, in general, the stress field $p_{kr}^2(\underline{r})$ of an arbitary dislocation loop in an anisotropic medium is given by (3.6) with the simple application of Hooke's Law:

$$p_{kr}^2(\underline{r}) = c_{krij} \beta_{ij}^2(\underline{r}) \qquad (4.15)$$

It thus involves a line integral around dislocation (2) and the total force on dislocation (1), if required, is the line integral of $F_i(\underline{r})$ around dislocation (1).

When the dislocations are long and straight we obtain several interesting results. Thus consider two straight skew dislocations (1) and (2) as shown in figure 11 with respective line directions $\underline{t}^1$, $\underline{t}^2$ and Burgers vectors $\underline{b}^1$, $\underline{b}^2$. Bullough and Sharp [62] have shown that if

$$\underline{t}^1 = [\sin \alpha, 0, \cos \alpha] \qquad (4.16)$$

then, from (4.14) and the field of dislocation (2) given by (3.17) and (3.19) the force per unit length at the point ($r \sin \alpha$, 0, $r \cos \alpha$) on dislocation (1) has the Cartesian components:

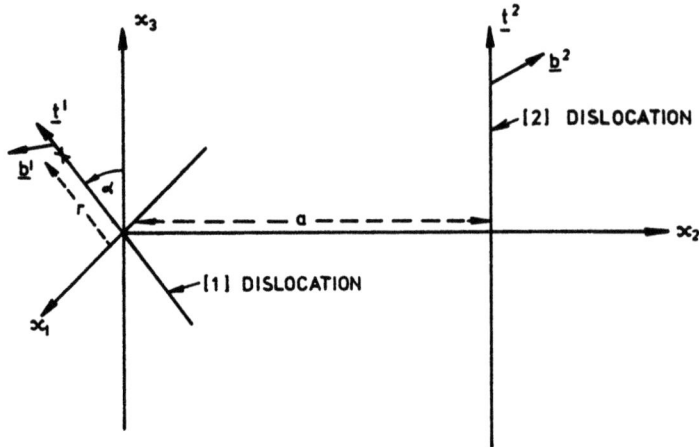

FIG. 11.

The pair of skew dislocations a distance 'a' apart with directions $\underline{t}^1$, $\underline{t}^2$ and Burgers vectors $\underline{b}^1$, b^2 respectively.

$$\begin{aligned}
F_1 &= [\mu|2\pi(1-\nu)]L^2\cos\alpha\{b_1^{(1)}b_1^{(2)}rK\sin\alpha - b_1^{(1)}b_2^{(2)}aK \\
&\quad - b_2^{(1)}b_1^{(2)}aK + b_2^{(1)}b_2^{(2)}rM\sin\alpha\} + [\mu|2\pi]b_3^{(1)}b_3^{(2)}Lr\cos\alpha\sin\alpha, \\
F_2 &= -[\mu|2\pi(1-\nu)]L^2\cos\alpha\{b_1^{(1)}b_1^{(2)}aN + b_1^{(1)}b_2^{(2)}rK\sin\alpha \\
&\quad + b_2^{(1)}b_1^{(2)}rK\sin\alpha - b_2^{(1)}b_2^{(2)}aK\} - [\mu|2\pi]\{b_3^{(1)}b_3^{(2)}La\cos\alpha \\
&\quad - b_1^{(1)}b_3^{(2)}aL\sin\alpha - b_2^{(1)}b_3^{(2)}rL\sin^2\alpha\} + [\mu\nu|\pi(1-\nu)] \\
&\quad \times\{b_3^{(1)}b_1^{(2)}aL\sin\alpha + b_3^{(1)}b_2^{(2)}rL\sin^2\alpha\}, \\
F_3 &= -[\mu|2\pi(1-\nu)]L^2\sin\alpha\{b_1^{(1)}b_1^{(2)}rK\sin\alpha - b_1^{(1)}b_2^{(2)}aK \\
&\quad - b_2^{(1)}b_1^{(2)}aK + b_2^{(1)}b_2^{(2)}rM\sin\alpha\} - [\mu|2\pi]b_3^{(1)}b_3^{(2)}rL\sin^2\alpha
\end{aligned}$$

(4.17)

where $K=K(\alpha)=r^2\sin^2\alpha-a^2$, $L=L(\alpha)=[r^2\sin^2\alpha+a^2]^{-1}$,
$M=M(\alpha)=r^2\sin^2\alpha+3a^2$ and $N=N(\alpha)=3r^2\sin^2\alpha+a^2$.

When $\alpha=0$ these components reduce to the well known results for parallel dislocations:

$$\left.\begin{aligned}F_1 &= \frac{\mu}{2\pi(1-\nu)a}(b_1^{(1)}b_2^{(2)}+b_2^{(1)}b_1^{(2)})\\ F_2 &= -\frac{\mu}{2\pi(1-\nu)a}(b_1^{(1)}b_1^{(2)}+b_2^{(1)}b_2^{(2)})-\frac{\mu}{2\pi a}b_3^{(1)}b_3^{(2)}\\ F_3 &= 0\end{aligned}\right\} \quad (4.18)$$

When $\alpha = 90°$, the two dislocations are perpendicular and expressions (4.17) reduce to

$$\left.\begin{aligned}F_1 &= 0,\\ F_2 &= [\mu|2\pi]L(\pi/2)\{b_1^{(1)}b_3^{(2)}a+b_2^{(1)}b_3^{(2)}r+[2\nu|(1-\nu)]\\ &\quad \times[b_3^{(1)}b_1^{(2)}a+b_3^{(1)}b_2^{(2)}r]\},\\ F_3 &= -[\mu|2\pi(1-\nu)]L^2(\pi/2)\{b_1^{(1)}b_1^{(2)}rK(\pi/2)-b_1^{(1)}b_2^{(2)}aK(\pi/2)\\ &\quad -b_2^{(1)}b_1^{(2)}aK(\pi/2)+b_2^{(1)}b_2^{(2)}rM(\pi/2)\}\\ &\quad -[\mu|2\pi]b_3^{(1)}b_3^{(2)}rL(\pi/2),\end{aligned}\right\} \quad (4.19)$$

where $K(\pi/2)=r^2-a^2$, $L(\pi/2)=[r^2+a^2]^{-1}$
and $M(\pi/2)=r^2+3a^2$, $N(\pi/2)=3r^2+a^2$.

These general results for two straight dislocations enable the following conclusions to be drawn for parallel dislocations

1. Two like screw dislocations always repel with a force/unit length

$$\underline{F} = \left[0, -\frac{\mu b^2}{2\pi a}, 0 \right] \quad (4.20)$$

2. Two parallel screws with opposite Burgers vector likewise attract.

3. Two like edge dislocations each with $\underline{b} = [b_1, b_2, 0]$ interact with the force

$$\underline{F} = \frac{\mu}{2\pi(1-\nu)a} \left[2b_1 b_2, -(b_1^2 + b_2^2), 0 \right] . \quad (4.21)$$

This configuration is shown in figure 12 where d defines the separation of the two glide planes, t is a measure of the displacement of the two edge dislocation from a vertical orientation and θ is the orientation of the $x_2 x_3$ plane to the glide plane.

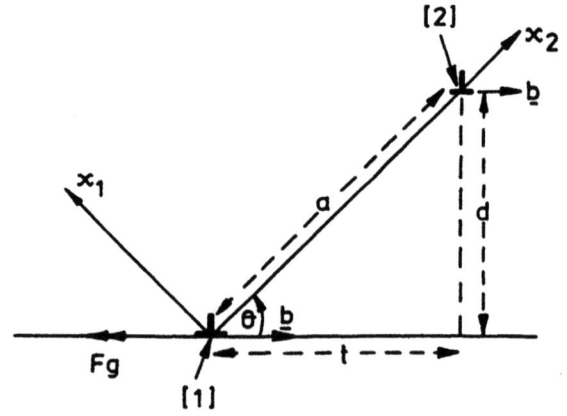

FIG. 12.

Two parallel edge dislocations a distance 'a' apart lying parallel to the x_3 axis. The parallel glide planes of the two dislocations are a distance d apart and t is their vertical displacement. F_g is the glide force on [1] due to [2].

Equation (4.21) then yields the force per unit length on (1) in the glide direction [sin θ, - cos θ, 0] to be

$$F_g = \frac{\mu b^2}{2\pi(1-\nu)} \frac{t(t^2-d^2)}{(t^2+d^2)^2} \tag{4.22}$$

Thus when t < d the glide force changes sign and will tend to pull dislocation (1) into vertical alignment with dislocation (2). This is the polygonization force and such dislocation rearrangement lowers the total energy of the crystal; when a crystal cannot actually get rid of its dislocations this kind of geometrical rearrangement is often the best it can do.

To obtain the stress required to glide one dislocation past another we must obtain the maximum value of F_g for constant d as we vary t. F_g is stationary when

$$t = \pm d \ [3-2\sqrt{2}]^{\frac{1}{2}} \simeq \pm \ 0.42d \tag{4.23}$$

Thus the force exerted on the dislocation, τb, from the external stress τ must equal F_g at this value of t; that is

$$\tau \simeq \frac{\mu}{8\pi(1-\nu)d} \ . \tag{4.24}$$

These large shear stresses required to drive a dislocation past an adjacent parallel dislocation are the fundamental reason for work hardening and since Taylor's [24] original theory of work hardening have remained of crucial importance in any explanation of hardening.

When two non-parallel dislocations are driven towards each other the actual interaction may be preceded by local cross slipping of either. These large, but relatively short ranged, forces are evident from (4.19) and have been discussed in detail by Bullough and Sharp [62] ; they may well be the explanation of profuse dipole formation during the early stages of deformation [63].

When two dislocations do pass through each other a jog is formed on each dislocation; i.e. the dislocation line has an atomic step in it equal to the Burgers vector of the other dislocation. If the Burger's vector of the interacting dislocation lies in the glide plane of the other dislocation, then the jog on the latter will not impede its glide since it will have the same glide plane as the dislocation itself; also a pure edge dislocation cannot have its glide motion obstructed by the presence of jogs on it. In general such jogs are a source or sink for point defects and provide an important source of point defects during the deformation of a crystal.

From the simple result (4.18) for parallel dislocations we see that like glissile dislocations in the same glide plane repel one another by a force that varies inversely with their separation. It is this repulsion that enables dislocations created by a Frank-Read or other dislocation source to be piled up at obstacles such as grain boundaries. Such pile ups of dislocations under a constant shear stress can be shown to be in equilibrium under the applied stress and their mutual repulsions when their distribution is defined by the roots of certain orthogonal polynomials [64]. The mathematical detail of such distributions is not very important, the essential feature of such arrays of dislocations is that they can produce much longer ranged internal stresses, falling off as $r^{-\frac{1}{2}}$ rather than the simple dislocation fall off of r^{-1}. The slip band actually behaves very much like a freely slipping crack and the "crack tip" has the expected $r^{-\frac{1}{2}}$ singularity. Such groupings of dislocations have been incorporated into theories of work hardening although it must be admitted that the direct evidence for the operation of dislocation sources of this type is not as profuse as one might have expected.

4.2.2 Point Defect Interactions

From the numerous catalogued interactions between point defects and dislocations we shall select and discuss only the three which we consider to be dominant:

a) The size effect interaction

b) The inhomogeneity interaction

c) The applied stress induced inhomogeneity interaction.

For a more complete discussion but not including (c) the review by Bullough and Newman [65] on this subject should be consulted.

Bullough and Willis [66] have recently presented a treatment that combines all these three processes into a single analysis. It will be convenient to present their results. The point defect is simulated as an embedded spherical inclusion with elastic constants μ^*, K^* differing from the matrix constants μ and K. As discussed briefly in section 2 the misfit strain e_{ij}^* is a measure of the strength of the point defect or atomically a measure of the net forces imposed on the neighbouring atoms to simulate the presence of the point defect. The different inclusion elastic constants represent the fact that the "atomic springs" (force constants) connecting the point defect to the matrix atoms will differ from the springs in the perfect matrix. By a rather ingenious cutting and rewelding thought procedure and defining an equivalent inclusion with a transformation strain, Eshelby [20] obtained various relevant expressions for the interaction energy between such an inclusion and an applied field. Bullough and Willis [66] followed a similar argument and obtained the general expression for the interaction energy between the inclusion and the applied strain field e_{ij}^A:

$$E_I = -\frac{V}{2}\left\{\frac{2\mu(\mu-\mu^*)}{\mu^*\beta-\mu(\beta-1)}\,'e_{ij}^A\,'e_{ij}^A\right.$$

$$\left. + \frac{K(K-K^*)}{K^*\alpha-K(\alpha-1)}(e^A)^2 + \frac{2KK^A}{K^*\alpha-K(\alpha-1)}\,e^A e^*\right\} \quad (4.25)$$

where $'e_{ij} = e_{ij} - \frac{1}{3}e\,\delta_{ij}$ is the deviatoric part of the strain, the quantities α, β have been defined by Eshelby [20] as

$$\alpha = (1+\nu)/3(1-\nu)$$
$$\beta = 2(4-5\nu)/15(1-\nu) \quad (4.26)$$

and V is the volume of the inclusion.

In general we can write

$$e_{ij}^A = e_{ij}^d + e_{ij}^e \quad (4.27)$$

where e_{ij}^d is the strain field due to a nearby dislocation and e_{ij}^e is the strain created by the application of an applied stress to the crystal. The complete interaction between a dislocation and the point defect when the body is subjected to an applied stress then follows from (4.25). The general expression is very complex and we will therefore only present two simple results to illustrate the form and magnitude of these interactions. Before doing so we will particularize the point defect to a lattice vacancy by setting

$$\mu^* = K^* = 0 \quad (4.28)$$

It is by no means obvious that such a local <u>void</u> is an accurate model of an atomic vacancy but it is reassurring that this approximation actually yields a total vacancy formation energy that is the order of 1 eV and thus reasonably consistent with the observed values in many metals. With this approximation (4.25) becomes

$$E_I = -3(1-\nu)V\mu\left\{\frac{5}{7-5\nu}\,'e_{ij}^A\,'e_{ij}^A + \frac{(1+\nu)}{6(1-2\nu)^2}(e^A)^2 + \frac{2\mu}{3(1-2\nu)}e_v^o\,e^A\right\} \quad (4.29)$$

for the vacancy-applied field interaction, where e_v^o is the actual relaxation volume strain in the vacancy: $e_v^o = \delta v/v$. If we consider the body to be under a uniaxial tension τ orientated relative to the dislocation line and vacancy as shown in figure 13

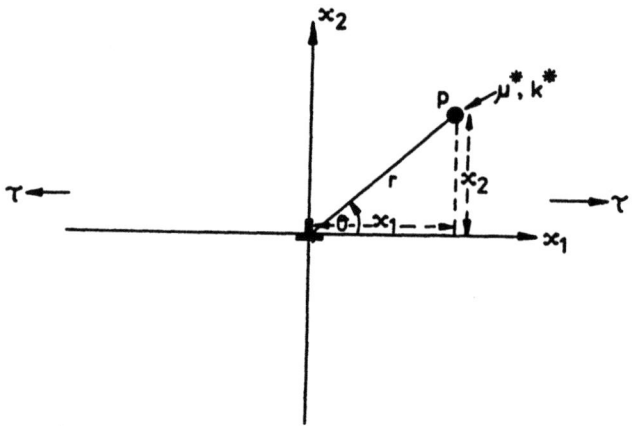

FIG. 13.
A body under uniaxial tensile stress τ containing an edge dislocation with $\underline{b}=[b,o,o]$ along the x_3 axis and a point defect at the point $P(x_1, x_2)$ shown.

then e_{ij}^e has the non-zero components

$$\left.\begin{array}{l} e_{ij}^e = \tau/2\mu(1+\nu) \\[1em] e_{22}^e = -\nu\tau/2\mu(1+\nu) = e_{33}^e \end{array}\right\} \quad (4.30)$$

The total applied strain at P due to the edge dislocation along the x_3 axis and the external strain is therefore, from (3.17) and (4.30):

$$e^A = \frac{(1-2\nu)}{2\mu(1+\nu)}\tau - \frac{b}{2\pi}\frac{(1-2\nu)}{1-\nu}\frac{x_2}{r^2}$$

$$'e_{11}^A = \frac{\tau}{3\mu} - \frac{b}{12\pi(1-\nu)}\frac{x_2}{r^4}\left[(7-2\nu)x_1^2 + (1-2\nu)x_2^2\right]$$

$$'e_{22}^A = -\frac{\tau}{6\mu} + \frac{b}{12\pi(1-\nu)}\frac{x_2}{r^4}\left[(5+2\nu)x_1^2 - (1-2\nu)x_2^2\right]$$

$$'e^A_{33} = -\frac{\tau}{6\mu} + \frac{b(1-2\nu)}{6\pi(1-\nu)} \frac{x_2}{r^2}$$

$$'e^A_{12} = \frac{b}{4\pi(1-\nu)} \frac{x_1}{r^4} (x_1^2 - x_2^2)$$

(4.31)

Substituting these applied strains into (4.29) yields the total interaction energy

$$E_I = E_I^{SE} + E_I^{SI} + E_I^{I} \qquad (4.32)$$

where E_I^{SE}, the size effect interaction, is

$$E_I^{SE} = V\mu \frac{b}{\pi} e_v^0 \frac{\sin\theta}{r} \qquad (4.33)$$

E_I^{SI}, the stress induced interaction, is

$$E_I^{SI} = \frac{33}{32} V\tau \frac{b}{\pi} \frac{\sin\theta}{r} \qquad (4.34)$$

and E_I^I, the inhomogeneity interaction, is

$$E_I^I = -\frac{135}{256} V\mu \frac{b^2}{\pi^2} \frac{1}{r^2} \left[1 - \frac{22}{45} \sin^2\theta \right] . \qquad (4.35)$$

The constant terms, independent of the dislocation-point defect separation, in (4.29), have been ignored and the three components of E_I are given explicitly for $\nu = \frac{1}{3}$.

If we replace the edge dislocation in figure 13 by a screw dislocation with Burgers vector $b = [o, o, b]$ then from (3.19) and (4.29) the interaction has only the inhomogeneity non zero component

$$E_I^I = -\frac{15}{64} V\mu \frac{b^2}{\pi^2} \frac{1}{r^2} \qquad (4.36)$$

The screw dislocation thus has a weak, short ranged, attractive interaction with a vacancy, whereas the edge dislocation has a dominant size effect interaction (4.33) which ensures that vacancies adjacent to such dislocations will preferentially migrate (or drift) to the compressed side ($0 < \theta < \pi$) of the dislocation; for a vacancy the relaxation strain e_v^0 should be negative. On the other hand an interstitial (intrinsic or impurity) should migrate to the tensile side of the edge dislocation ($\pi < \theta < 2\pi$), since the corresponding relaxation strain should be positive. In comparison with this size effect component the two other interactions (4.34) and (4.35) are, in general, negligible. However when the "differential" loss of point defects to different fractions of the dislocation network is important, the stress induced interaction (4.34) may be very important in its own right; this interaction has recently been proposed as the fundamental explanation of irradiation creep [67]. The direct attractive inhomogeneity interaction (4.35) between the edge dislocation and the vacancy is probably rarely important but it has been suggested as responsible for the kinetics of recovery in quenched-deformed metals. However such interpretations must be doubted in view of the relative magnitude of (4.35) compared with (4.33); there is no justification for putting $e_v^0 \equiv 0$ for a vacancy. The appearance of the interaction (4.35) at very short range has been demonstrated in the copper-dislocation atomic simulations. For the screw dislocation, of course, there is only the inhomogeneity interaction and so the attractive interaction (4.36) will be important. However it is dangerous to put too much credence in the relevance of such second order ($0r^{-2}$) interactions since the present discussion is based on a rather crude series of approximations which we may rather negatively challenge:

a) Elastic isotropy is assumed. Real materials are generally anisotropic (Tungsten is a notable exception) and only for conditions of precise isotropy is the first order interaction zero between the screw dislocation and a point defect.

b) The spherical inclusion model for the point defect is not really accurate. It is particularly poor for the intrinsic interstitial where the crystal point symmetry is lost and a split dumbell configuration occurs. Such departures from spherical symmetry will cause coupling with the non-principal strains in $'e_{ij}^A$ and thereby complicate the angular dependence of this interaction energy.

c) If both these criticisms are removed the basic long ranged (r^{-1}) nature of the size effect interaction remains and no doubt influences the migration of point defects in the vicinity of dislocations. It is, for example, responsible for such diverse phenomenon as void swelling in irradiated materials [8] and for the yield point drop in carbon steels etc. [65]. However even this spatial variation must be taken with some caution since it will only, in reality, be present very close to the dislocations; in the regions between the dislocations the overlapping stress fields of the dislocations must severely truncate the interaction field and midway between adjacent dislocations the gradient of E_I must be zero. In principle it is only present in the form (4.33) when the defect-dislocation separation is less than the local radius of curvature of the dislocation and so it is probable that only a few diffusive jumps of the point defect are influenced by the interaction.

d) The use of a continuum inclusion model to represent modified interatomic forces and force constants around the point defect is acceptable only for long ranged interactions[14]. The inhomogeniety interaction energies such as (4.35) only become appreciable at short ranges where precisely the approximation is weakest. However, atomic calculations for copper have shown that this interaction does manifest itself when a point defect is in the nearest neighbour positions to a dislocation core [68].

4.2.3. Second Phase Particles

The results of the previous section were all obtained on the assumption that the variation of the dislocation stress field across the inclusion could be neglected. Clearly if the inclusion represents a second phase particle embedded in the matrix it is not necessarily of atomic dimension and the spatial variation of dislocation field over its volume must be included when we discuss the interaction between it and a dislocation.

We will not discuss the mathematics required to ensure appropriate continuity of traction across such precipitate-matrix interfaces. A full discussion including both the misfit and inhomogeneity interactions for such finite particules has been given by Willis [69] based on the pioneering studies of Eshelby. It will suffice to comment that both the interactions referred to in the previous section for atomic point defects are present and indeed for a spherical particle the size effect interaction (4.33) is exact in an isotropic body because the volume integral of a harmonic function over a _spherical_ volume is exactly equal to the value of the function at the centre of the sphere multiplied by the volume of the sphere [70].

In addition to these long range spatial interactions with particles the dislocation mobility can be impeded by the direct resistance to glide offered by the particle. If the particle is completely crystallographically coherent with the matrix then the dislocation can glide right through the particle and a glide band can literally slide the particle into two parts. In this case resistance to the dislocation motion is offered by a difference in dissociation of the dislocations on those parts that enter the particle matrix compared with the dissociation in the host matrix. In general a dislocation has to cross-slip around a particle and the slip-climb processes required are usually quite complex and involve the generation of point defects by subsequent non-conservative motion of jogs.

When a dislocation gets very close to a macroscopic particle its interaction resembles that between a plane surface and a dislocation. We will discuss aspects of this interaction in the next section.

4.3 Interactions with Surfaces

The interactions of dislocations with surfaces, such as internal interfaces and the free surface of the crystal, are reasonably simple to calculate when the surface curvature is either zero or cylindrical. We will therefore focus the discussion in this section on to these two explicit situations and thereby avoid confusing the physical features of the interactions with undue mathematical complexity. In general, of course, the field of an arbitrary dislocation in a finite body of arbitrary shape can be formally given in terms of integrals of the appropriate finite body Green's tensor [23] in exactly the same way that the infinite body results were expressed in section 3.1. However, apart from its formal attractiveness, the real usefulness of such a procedure is somewhat limited since the Green's tensor for a finite body is extremely complex and only a few very special such tensors are available in the literature. Furthermore to satisfy the equilibrium conditions in a genuine finite body requires surface tractions suitably distributed to balance the internal point force together with a stabilizing couple to prevent rotation of the body; both these requirements have a degree of arbitrariness that is difficult to remove* by plausible physical argument.

4.3.1 The free surface

Since real crystals have surfaces it is important to know the form and magnitude of the forces that act on a dislocation when it approaches a surface. The source of the dislocation-free surface interaction can be seen as follows: consider a dislocation in an infinite elastic medium; if a cut is made parallel to the dislocation, thereby producing a semi-infinite region with a dislocation in it, then the dislocation stress field will lead to the presence of non-zero tractions on the surface of the region. These tractions have to be annulled by equal and opposite tractions in order

---

*Without labouring the point it is worth remarking that the physical uniqueness of the finite body Green's tensor is obscure and for defect problems the derivative of the Green's tensor is perhaps the best elementary tensor to adopt since the double force only is required to synthesize defects and then these equilibrium problems are avoided since the double force is self stable.

that the surface be stress free (and hence be, by definition a true free surface). These necessary applied tractions interact with the dislocation stress field and are the direct cause of the dislocation-free surface interaction. This formal way of evaluating the stress field round a dislocation in a finite medium by simply modifying the corresponding solution for an infinite medium was first suggested by Volterra [31] and this procedure together with the modern methods of Fourier transform theory [71] may be used as a direct method of evaluating the stress field [72]. The alternative method is simply to "guess" a stress function which automatically satisfies the vanishing traction boundary conditions and is the required dislocation solution [73].

An edge dislocation parallel to and a distance ξ from the free plane surface $x_1 = 0$ is shown in figure 14. If the dislocation has

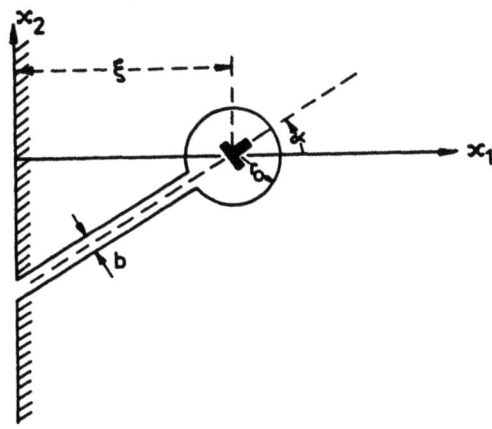

FIG. 14.

The edge dislocation adjacent and parallel to a plane free surface. In the axes shown its Burgers vector is $\underline{b}=[b \sin \alpha, b \cos \alpha, 0]$ and it is located a distance ξ from the surface $(x_1 = 0)$.

the Burgers vector

$$\underline{b} = [b \sin \alpha, - b \cos \alpha, 0] \qquad (4.37)$$

then Deitze [73] has given the general solution. The elastic energy per unit length along the dislocation stored in the half space is

$$E = \frac{b^2}{4\pi} \frac{\mu}{1-\nu} \ln \frac{2\xi}{r_0} \qquad (4.38)$$

and is independent of α. If $\alpha = 90°$, the stresses have the explicit form

$$p_{11} = A\left\{-\frac{x_2\{3(x_1-\xi)^2+x_2^2\}}{\{(x_1-\xi)^2+x_2^2\}^2} + \frac{x_2\{3(x_1+\xi)^2+x_2^2\}}{\{(x_1+\xi)^2+x_2^2\}^2} + \frac{4\xi x_1 x_2\{3(x_1+\xi)^2-x_2^2\}}{\{(x_1+\xi)^2+x_2^2\}^3}\right\}$$

$$p_{22} = A\left\{\frac{x_2\{(x_1-\xi)^2-x_2^2\}}{\{(x_1-\xi)^2+x_2^2\}^2} - \frac{x_2\{(x_1+\xi)^2-x_2^2\}}{\{(x_1+\xi)^2+x_2^2\}^2}\right.$$

$$\left. + 4\xi\, x_2\frac{(2\xi-x_1)(x_1+\xi)^2+(3x_1+2\xi)x_2^2}{\{(x_1+\xi)^2+x_2^2\}^3}\right\} \quad (4.39)$$

$$p_{12} = A\left\{\frac{(x_1-\xi)\{(x_1-\xi)^2-x_2^2\}}{\{(x_1-\xi)^2+x_2^2\}^2} - \frac{(x_1+\xi)\{(x_1+\xi)^2-x_2^2\}}{\{(x_1+\xi)^2+x_2^2\}^2}\right.$$

$$\left. + 2\xi\frac{(\xi-x_1)(x_1+\xi)^3+6x_1(x_1+\xi)x_2^2-x_2^4}{\{(x_1+\xi)^2+x_2^2\}^3}\right\}$$

where $A = b\mu/2\pi(1-\nu)$. The first term is the stress appropriate to the edge dislocation without the free surface; the second term is the stress resulting from the image dislocation at $x_1 = -\xi$ and the third term is the stress required, in addition to an image dislocation, to annul the surface tractions p_{11} and p_{12}.

The stress tending to move the dislocation towards the free surface by glide p_g is simply the p_{12} stress in (4.39) with the self stress (1st term) removed and evaluated at the dislocation, at $x_1 = \xi$, $x_2 = 0$. Thus:

$$p_g = -\frac{A}{2\xi} \quad (4.40)$$

Similarly for the edge dislocation with the orientation shown in figure 14, the corresponding shear stress is:

$$p_g = -\frac{A \sin \alpha}{2\xi} \quad (4.41)$$

The stress acting on the dislocation to move it by climb, p_c, towards the surface is the tensile stress in the slip direction: that is

$$p_c = \frac{A \cos \alpha}{2\xi} \quad (4.42)$$

Note that both the stresses (4.41) and (4.42) only depend on the image dislocation; for these stresses the third terms in (4.39) are irrelevant.

The corresponding stress field for an edge dislocation inside a cylinder but displaced from its axis, as shown in figure 15,

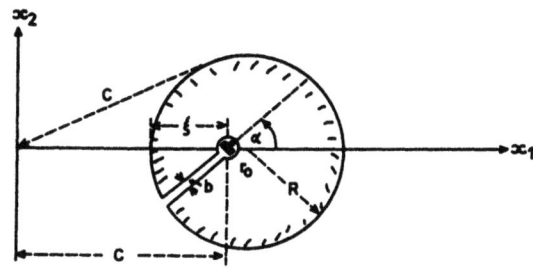

FIG.15.

The edge dislocation in a cylindrical body parallel to but displaced from the axis of the cylinder.

has also been given by Deitze [73]. If R is the radius of the cylinder and ξ is the distance between the surface of the cylinder and the dislocation then its elastic energy per unit length is

$$E = \frac{b^2 \mu}{4\pi (1-x)} \ln \frac{\xi(2R-\xi)}{r_0 R} \qquad (4.43)$$

This, not very well known solution, is appropriate to dislocations in whiskers. Again the stress pulling the dislocation by climb or glide to the surface can easily be found. The dislocation is in a position of unstable equilibrium at the centre of the cylinder.

The situation for the screw dislocation is very much easier, since, if the dislocation lies parallel to the free surface there will be only one traction to annul on the surface compared with two for the above plane strain situations. Again a direct method using Fourier transforms can be used, but in fact it is reasonably easy to guess the correct displacement field. If we replace the edge dislocation in figure 14 by a screw dislocation of identical strength we obtain the solution

$$u_3 = -\frac{\lambda}{2\pi} \left\{ \tan^{-1} \frac{x_1-\xi}{x_2} - \tan^{-1} \frac{x_1+\xi}{x_2} \right\} \qquad (4.44)$$

and thus

$$p_{13} = -\frac{\lambda\mu}{2\pi} \left\{ \frac{x_2}{(x_1-\xi)^2+x_2^2} - \frac{x_2}{(x_1+\xi)^2+x_2^2} \right\}$$

$$p_{23} = \frac{\lambda\mu}{2\pi} \left\{ \frac{x_1-\xi}{(x_1-\xi)^2+x_2^2} - \frac{x_1+\xi}{(x_1+\xi)^2+x_2^2} \right\} . \tag{4.45}$$

The elastic energy is identical to (4.38) with $\nu = 0$ and the shear stress tending to glide the screw towards the free surface is given by (4.40), again with $\nu = 0$.

The explicit solution for the screw dislocation in the cylinder, as shown in figure 15, is

$$u_3 = -\frac{b}{2\pi} \left\{ \tan^{-1}\frac{x_1-c}{x_2} - \tan^{-1}\frac{x_1+c}{x_2} \right\} \tag{4.46}$$

and

$$p_{13} = -\frac{b\mu}{2\pi} \left\{ \frac{x_2}{(x_1-c)^2+x_2^2} - \frac{x_2}{(x_1+c)^2+x_2^2} \right\}$$

$$p_{23} = +\frac{b\mu}{2\pi} \left\{ \frac{x_1-c}{(x_1-c)^2+x_2^2} - \frac{(x_1+c)}{(x_1+c)^2+x_2^2} \right\} \tag{4.47}$$

and the elastic energy

$$E = \frac{b^2\mu}{4\pi} \ln \frac{\xi(2R-\xi)}{r_0 R} \tag{4.48}$$

and

$$c = \frac{\xi(2R-\xi)}{2(R-\xi)} .$$

The shear stress p_g acting on the dislocation for glide on the $x_2=0$ plane is, from (4.47)

$$p_g = -\frac{b\mu}{2\pi} \frac{(R-\xi)}{\xi(2R-\xi)} \tag{4.49}$$

which is zero when the dislocation is at the centre of the cylinder $\xi = R$ and increases as the dislocation approaches the surface of the cylinder ($\xi=0$ or $\xi=2R$).

The physical consequences of these attractive interactions to free surfaces are obvious. If the critical shear stress for glide is approximately $10^{-3}\mu$ then results such as (4.40) or (4.49) suggest that a dislocation free zone about 100b wide should be present adjacent to the free surfaces of crystals. Such localized loss of dislocations across free surfaces undoubtedly happens [8] at the large voids in nickel when irradiated with fast neutrons to a high dose; this dislocation loss mechanism is probably responsible for the saturation of swelling in many materials. Unfortunately external surfaces of crystals (as distinct from internal surfaces such as voids) are usually contaminated with oxide etc. and such surface layers with differing moduli from the host can dominate the interaction processes. In general a surface layer with a modulus exceeding the host modulus will repel dislocations and vice-versa - the softest being the perfectly free surface just discussed.

If the dislocation actually cuts the free surface then the elasticity problem is a little more difficult in detail, but no different in principle. Thus if the edge dislocation is normal to a free surface a two dimensional distribution of tensile stress has to be annulled. This is the famous Boussinesque problem and can be solved by a particular combination of potential solutions of the three dimensional displacement equations [74] (with zero body forces) or by using the Galerkin vector method [75]. The emergent screw dislocation is, in principle, more difficult since there are two shear tractions to annul. However, one of the potential solutions referred to above has just the right symmetry to satisfy both traction conditions simultaneously. Thus, in practice, the screw dislocation is again very simple.

4.3.2 The Internal Surface

The general situation is depicted in figure 16 where we con-

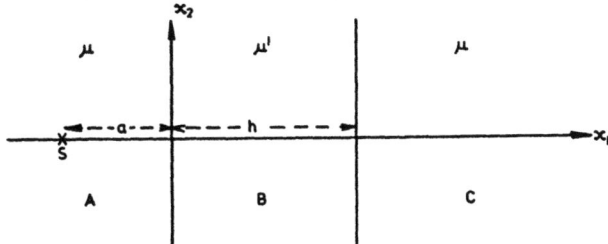

FIG. 16.

A dislocation 'S' in a matrix with shear modulus μ, a distance a from a plane lamella of width h with a shear modulus μ'.

sider a dislocation at S a distance 'a' from a plane lamella B of width h with a modulus μ' differing from the matrix A and C modulus μ. If S is a straight screw dislocation parallel to the x_3 axis then standard Fourier transform methods [71] can be used to construct a solution of the elasticity problem that yields a dislocation at S together with continuity of traction and displacement at the two interfaces. The general solution is rather complex but from it follows the glide shear stress acting on the dislocation at $x_1 = -a$ due to the presence of B:

$$p_g = \frac{\mu(\mu^2 - \mu'^2)b}{2\pi} \int_0^\infty \frac{\sinh \xi h \, e^{-2\xi a} d\xi}{(\mu^2 + \mu'^2) \sinh \xi h + 2\mu\mu' \cosh \xi h} \quad (4.50)$$

This expression requires numerical quadrature for its evaluation for arbitrary μ'/μ and h. However we note its consistency with the previous solution (4.40), with $\nu = 0$, when $\mu' = 0$.

When $h \to \infty$, (4.50) becomes

$$p_g = \frac{\mu(\mu - \mu')b}{4\pi a(\mu + \mu')} \quad (4.51)$$

and represents the interaction between a screw dislocation and a grain or twin boundary. The lattice rotation across $x_1 = 0$ creates an effective modulus change across this interface relative to the laboratory axis $(x_1 x_2 x_3)$ orientation. Thus if

1. $\mu > \mu'$

the dislocation is attracted to the interface

2. $\mu < \mu'$

the dislocation is repelled from the interface. Finally if the region B is effectively rigid ($\mu' \to \infty$) then (4.50) yields the maximum repulsion

$$p_g = -\frac{\mu b}{4\pi a} \quad (4.52)$$

which also follows from (4.51) with the same limit.

REFERENCES

1. Hirth, J.P., and Lothe, J., <u>Theory of Dislocations</u>, McGraw-Hill, New York, 1968.
2. Van Bueren, H.G., <u>Imperfections in Crystals</u>, North-Holland, Amsterdam, 1961.
3. Nabarro, F.R.N., <u>Theory of Crystal Dislocations</u>, Clarendon Press, Oxford, 1967.
4. Friedel, J., <u>Les Dislocations</u>, Gauthiers-Villars, Paris, 1956.
5. Bilby, B.A., <u>Prog. Solid Mech.</u>, 1, 329, 1960.
6. Kondo, K., <u>RAAG. Mem. 1</u>, 458, 1955.

7. Bullough, R., and Simmons, J.A., In *Physics of Strength and Plasticity*, Argon, A.S., M.I.T. Press, Mass., p.47.
8. Bullough, R., and Nelson, R.S., *Physics in Technology*, 5, 12, 1974.
9. Thompson, M.W., *Defects and Radiation Damage in Metals*, Camb. Univ. Press, Cambridge, 1969, p.37.
10. Johnson, R.A., *Phys. Rev.*, 134, 1329, 1964.
11. Cottrell, A.H., and Bilby, B.A., *Proc. Phys. Soc.*, 62, 49, 1949.
12. Bilby, B.A., *Proc. Phys. Soc.*, A63, 191, 1950.
13. Bullough, R., and Hardy, J.R., *Phil. Mag.*, 17, 833, 1968.
14. Bullough, R., Norgett, M.J., and Webb, S., *J. Phys. F.: Met. Phys.*, 1, 345, 1971.
15. Perrin, R.C., Taylor, R., and March, N.H., *A.E.R.E. Research Report*, TP.603, 1974.
16. Finnis, M.W., Unpublished Work, 1975.
17. Bullough, R., and Tewary, V.K., In *Dislocation Theory - a Collective Treatise*, Nabarro, F.R.N., Dekker, New York, 1975, Chapter 5.
18. Bullough, R., and Tewary, V.K., In *Interatomic Potentials and Simulation of Lattice Defects*, Gehlen, P.C., Beeler, J.R., and Jaffee, R.I., Plenum, New York, 1972, p.155.
19. Tewary, V.K., *A.E.R.E. Research Report*, TP.388, 1969.
20. Eshelby, J.D., *Proc. Roy. Soc.*, A241, 376, 1957.
21. Eyre, B.L., *J. Phys. F.: Met. Phys.*, 3, 422, 1973.
22. Kroner, E., *Arch. rat. Mech. Ann.*, 4, 273, 1960.
23. Simmons, J.A., and Bullough, R., In *Fundamental Dislocation Theory*, Bullough, R., Simmons, J.A., and DeWit, R., N.B.S. Spec. Pub. No. 317, 1970, p.89.
24. Taylor, G.I., *Proc. Roy. Soc.*, A145, 362 and 388, 1934.
25. Burgers, J.M., *Proc. Acad. Sci. Amst.*, 42, 293, 1939.
26. Bilby, B.A., Bullough, R., and Smith, E., *Proc. Roy. Soc.*, A231, 263, 1955.
27. Bilby, B.A., In *Defects in Crystalline Solids*, The Physical Society, 1955, p.124.
28. Bullough, R., and Bilby, B.A., *Proc. Phys. Soc.*, 69, 1276, 1956.
29. Bilby, B.A., and Christian, J.W., *Monog. Ser. Inst. Metals*, 18, 121, 1955.
30. Bullough, R., In *Theory of Imperfect Crystalline Solids*, I.A.E.A., Vienna, 1971 p.101.
31. Volterra, V., *Ann. Sci. Ecole. Norm. Supérieure*, 24, 401, 1907.
32. Mura, T., *Phil. Mag.*, 8, 843, 1963.
33. Willis, J.R., *Phil. Mag.*, 21, 931, 1970.
34. Love, A.E.H., *Theory of Elasticity*, Camb. Univ. Press, Cambridge, 1892.
35. DeWit, R., *Solid State Physics*, 10, 249, 1960.
36. Eshelby, J.D., Read, W.T., and Shockley, W., *Acta Met.*, 1, 251, 1953.
37. Willis, J.R., *Int. J. Engng. Sci.*, 5, 171, 1967.

38. Pfliederer, H.A., Seeger, A., and Kröner, E., Z. Naturforschg., 15a, 758, 1960.
39. Bullough, R., and Sinclair, J.E., Unpublished Work, 1974.
40. Peierls, R., Proc. Phys. Soc., 52, 34, 1940.
41. Nabarro, F.R.N., Proc. Phys. Soc., 54, 256, 1947.
42. Van der Merwe, J.H., Proc. Phys. Soc., A63, 616, 1950.
43. Foreman, A.J.E., Jaswon, M.A., and Wood, J.K., Proc. Phys. Soc., A64, 156, 1951.
44. Kroupa, F., and Lejček, L., Czech. J. Phys., B22, 813, 1972.
45. Foreman, A.J.E., Acta Met., 3, 322, 1955.
46. Nabarro, F.R.N., Adv. in Phys., 1, 269, 1952.
47. Christian, J.W., and Vitek, V., Rep. on Prog. in Physics, 33, 307, 1970.
48. Leibfried, G., and Dietze, H.D., Zeits. f. Physik, 131, 113, 1951.
49. Heinrich, R., Schellenberger, W., and Pegel, B., Phys. Stat. Sol., 39, 493, 1970.
50. Bullough, R., Ph.D. Thesis, Sheffield University, Sheffield, 1955.
51. Hardy, J.R., and Flocken, J.W., C.R.C. Critical Reviews in Solid State Sciences, 1, 605, 1970.
52. Bullough, R., In Nature of Defects in Crystals, Melbourne, Australia, 1965.
53. Cotterill, R.M.J., and Doyama, M., Phys. Rev., 145, 465, 1966.
54. Bullough, R., Bull. Amer. Phys. Soc., 10, 323, 1965.
55. Norgett, M.J., Perrin, R.C., and Savino, E.J., J. Phys. F.: Met. Phys., 2, L73, 1972.
56. Perrin, R.C., Englert, A., and Bullough, R., In Interatomic Potentials and Simulation of Lattice Defects, Gehlen, P.C., Beeler, J.R., and Jaffee, R.I., Plenum, New York, 1972, p.509.
57. Englert, A., Tompa, H., and Bullough, R., In Fundamental Aspects of Dislocation Theory, Bullough, R., Simmons, J.A., and DeWit, R., N.B.S. Spec. Pub. 517, 1970, p.273.
58. Stobbs, W.M., and Sworn, C.H., Phil. Mag., 24, 1365, 1971.
59. Perrin, R.C., and Savino, E.J., J. Microscopy, , , 1974.
60. Sinclair, J.E., J. Appl. Phys., 42, 5321, 1971.
61. Peach, M.O., and Koehler, J.S., Phys. Rev., 80, 436, 1950.
62. Bullough, R., and Sharp, J.V., Phil. Mag., 11, 605, 1965.
63. Tetelman, A.S., Acta Met., 10, 813, 1962.
64. Eshelby, J.D., Frank, F.C., and Nabarro, F.R.N., Phil. Mag., 42, 351, 1951.
65. Bullough, R., and Newman, R.C., Rep. on Prog. in Physics, 33, 101, 1970.
66. Bullough, R., and Willis, J.R., Phil. Mag., 31, 855, 1975.
67. Heald, P.T., and Speight, M.V., Phil. Mag., 29, 1075, 1974.
68. Bullough, R., and Perrin, R.C., In Dislocation Dynamics, Rosenfield, A.R., Hahn, G.T., Bement, A.D., and Jaffee, R.I., McGraw-Hill, New York, 1968, p.175.
69. Willis, J.R., Quart. J. Mech. and Appl. Maths., 17, 157, 1964.
70. Bullough, R., and Newman, R.C., Prog. in Semicond., 7, 101, 1964.

71. Sneddon, I.N., Fourier Transforms, McGraw-Hill, New York, 1950.
72. Bullough, R., Unpublished Work, 1960.
73. Dietze, H.D., Dissertation, Stuttgart, 1953.
74. Green, A.E., and Zerna, W., Theoretical Elasticity, Clarendon, Oxford, 1960.
75. Steketee, P., Cad. J. Phys., 36, 198, 1958.

DISCUSSION

Comment by T. E. Fischer:

Considering the influence of elastic strains on the movement of dislocations, one could imagine the following mechanism contributing to chemisorption effects on the plasticity of semiconductors: chemisorption produces a charged surface; this is compensated by an opposite space change; the space charge field results in compression of the surface layer. Extension of this layer depends on doping (50 Å to a micron). Is this reasonable?

Reply:

I think this is a very important question. Such deep compression would certainly have important effects on the mobility of the dislocations in each regions. Detailed calculations are needed to answer the question and should certainly be undertaken.

Comment by M. V. Swain:

For ionic crystals with charged jogs, what are the comparative values of the electrostatic and elastic interaction forces?

Reply:

From memory, I recall that if the jog concentration is 1/2 and all the jogs are negatively charged, then electrostatic interaction energy could be easily ten times any plausible size effect interaction energy.

Comment by J. Lothe:

May the electric field in the surface layer produce effective modulus changes by the second-order piezoelectric effect?

Reply:

Yes. I can see the field gradients will be important in the

surfaces of such materials. I suppose the magnitude of the effective modulus change would be small but an accurate estimate would be worth having.

Comment by S. Weissmann:

If a material is neutron irradiated and contains small precipitates with coherency strains, do you believe that the internal surfaces between precipitates and matrix or the coherency strains are the dominant factors for attracting vacancies and interstitials to the surfaces? By aging one can increase the surface area but simultaneously the coherency strains will decrease. Which is the more important factor in attracting the point defects to the sinks, surface area or coherency strains?

Reply:

I doubt if the actual migration of such point defects to the precipitates is much influenced by the coherency strains. The migration is probably random; however, when a defect arrives in the close vicinity of the precipitate then the kind of sink offered it by the precipitate will depend on the defect nature of the precipitate-matrix interface.

Comment by E. Kröner:

This is a comment concerning your core judgement on the use of nonlinear elasticity theory in dislocation problems. For illustration let me speak about the volume changes due to dislocations. I distinguish between the total volume change and the volume change per unit volume, i.e., the dilatation field, due to the dislocation. Because the core conditions are essential for the total volume change elasticity theory, linear or nonlinear, is an adequate tool in this case only if combined with lattice theoretical considerations of the core regions. This was shown recently by H. Suzuki for screw dislocations in bcc-lattices.

The dilatation field of a dislocation does not depend on the core conditions when measured some distance from the core. It comes out as zero in the case of screw dislocations if linear isotropic elasticity theory is applied. The linear anisotropic theory gives a contribution the average around the dislocation of which varies. Nonlinear isotropic elasticity theory leads to an axially symmetric dilatation field which decays as $1/r^2$ with distance r from the dislocation. This field plays an important role in the scattering of phonons and electrons by dislocation, i.e., for the phenomenon of thermal and electrical resistance. It provides an example in which nonlinear elasticity theory turns out to be the most appropriate tool. I am sure that other examples can be found.

Reply:

I agree. I only question the accuracy of the non-linear perturbation solution and the large effort required to get it. I think the nonlinear field can be got more easily using numerical lattice relaxation methods which provide a more explicit model of the core configuration-from whence the effect comes-than can be got from the purely continuum model with an idealized cylindrical core, etc.

Comment by D. L. Davidson:

Please comment on what discrete lattice modeling of surfaces using semiempirical potentials gives. For example, what is the effect of the surface on the position of the surface atoms? Do they locally relax or expand the interatomic spacing, and are the calculated surface energies reasonable and in agreement with any experimental values?

Reply:

Such potentials should not be used to study surface energies or surface relaxations in view of the electron redistribution that occurs at metal surfaces. I should refer you to the recent work of Finnis (Published in J. Phys. F: Metal Physics 1974) for fundamental calculations of the surface relaxations in aluminium. However, some of the semi-empirical potentials are in reasonable agreement with experimental surface energies, for example Johnson's 'IRON' potential and a 5th neighbor potential due to Bullough and Perrin (see Fundamental Aspects of Dislocation Theory: NBS, publication 1970).

Note Added in Proof:

An alternative form of the result (3.11) which has certain computational advantages has been given by Barnett [phys. stat. sol. $\underline{49}$, 741 (1972)] and discussed by Barnett and Lothe [Phys. Norvegica $\underline{7}$, 13 (1973)]. It involves expressing the Green function derivative in (3.6) as a single polar integral which in turn enables the distortion field to be expressed as a single line integral and thereby avoids solving the rather troublesome sextic equation (3.16).

INELASTIC DEFORMATION AND FRACTURE OF CRYSTALLINE SOLIDS

A. S. Argon

Department of Mechanical Engineering,
Massachusetts Institute of Technology,
Cambridge, Massachusetts 02139, U.S.A.

ABSTRACT. The mechanisms and phenomenology of inelastic deformation and fracture in crystalline solids are reviewed. Attention is drawn to possible surface and interface sensitive effects that result from the non-local nature of crystal plasticity, and from requirements of compatibility of deformation between aggregates in heterogeneous plastic media.

1. INTRODUCTION

Our task in this chapter is two-fold. First, we wish to consider the mechanisms of inelastic deformation of crystalline materials and how these mechanisms individually or in combinations give rise to the macroscopic inelastic behavior of polycrystalline material in bulk. Second, we wish to repeat this task for the fracture of polycrystalline materials. In both instances we wish to take due note of possible surface effects and indicate under which circumstances surface films can play disproportionately large "catalytic" effects, and under which circumstances their effect will remain small. Since this chapter is largely of a tutorial nature, and since the ground to be covered is large, we will necessarily have to be brief. We will, however, make up for this brevity by giving as much as possible an adequate number of additional references which the reader will be advised to consult to obtain a deeper level of understanding and appreciation.

2. INELASTIC DEFORMATION OF CRYSTALLINE SOLIDS

2.1 Mechanisms of inelastic deformation in solids

Inelastic deformation in crystalline matter can occur by a variety of mechanisms which may produce strain either independently or in combination with each other. We will consider these mechanisms below briefly in order of decreasing stress and increasing temperature.

2.1.1 Mechanisms of low temperature plasticity

a) <u>Deformation in initially perfect crystals</u>. In a perfect crystal a constant applied shear stress will find all portions of the material equally strong. Inelastic deformation in such a material can only be initiated when the shear stress overcomes the ideal lattice resistance for shear translation everywhere simultaneously by reaching a level of about $\mu/30$ for close packed metals and as high as $\mu/6$ for directionally bonded elements and compounds [1], where μ is the shear modulus. If the perfect crystal can undergo twinning or a martensitic shear transformation, inelastic strains can be produced at somewhat lower stresses by these mechanisms. In all these instances increasing the temperature does not significantly reduce the ideal strength for shear translation, twinning or martensitic shear. Stated in other words, the free energy for formation of a sheared, twinned, or martensitically transformed nucleus as a separate phase has a high stress dependence, making the process only activable by an almost temperature independent threshold stress [2]. Twinning and martensitic shears are, of course, observed also at lower stresses in imperfect crystals, where they appear to be initiated in local regions of high internal stress as is normally the case in strain hardened crystals.

b) <u>Dislocation glide at low temperatures</u>. In the low temperature range where diffusional processes are too slow, imperfect crystalline matter deforms by motion of dislocations. Such motion in structural alloys is obstructed by either an inherent lattice resistance or by more or less localized obstacles which pin down dislocations. The flexibility of dislocations permits them to overcome such obstacles locally in a substantially uncoupled manner [3-5]. The local free energy changes in unpinning are often small enough to permit significant thermally assisted overcoming of obstacles. The details of these processes have been widely discussed over the past decade and have recently been evaluated and formalized by Kocks, et al. [6] who find that the inelastic strain rate for both the lattice resistance mechanism and for overcoming localized obstacles can be given by

$$\dot{\gamma} = \dot{\gamma}_G \exp(-\Delta G/kT) \tag{2.1}$$

where

$$\dot{\gamma}_G = b\, a_o\, \rho_m\, \nu_G \tag{2.2}$$

is the pre-exponential factor. In these equations b is the magnitude of the Burgers vector, a_o the area swept out by a released dislocation segment, ρ_m the time average number of released, mobile dislocation segments per unit volume, ν_G the frequency factor of a single release event, and ΔG the activation free enthalpy for the local release process. Detailed considerations [6] show that in the pre-exponential term, a_o and ρ_m can be dependent on both the obstacle structure and the applied shear stress. Since the exponential term easily dominates the behavior, the dependence of the activation free enthalpy ΔG on obstacle structure and applied stress is of paramount importance. For the two cases of inherent lattice resistance and localized obstacles the activation free enthalpy can be given more explicitly [6] as follows:

$$\Delta G = \frac{\mu b^3}{2}\sqrt{\frac{\hat{\tau}_\ell}{\mu}} \left[1 - \left(\frac{\sigma}{\hat{\tau}_\ell}\right)^{3/4}\right]^{4/3} \quad \text{(inherent lattice resistance)} \tag{2.3}$$

$$\Delta G = \alpha\frac{\mu b^3}{2}\left(\frac{w}{b}\right)^n \left[1 - \left(\frac{\sigma}{\hat{\tau}_o}\right)^{1/2}\right]^{3/2} \quad \text{(localized obstacles)} \tag{2.4}$$

where $\hat{\tau}_\ell$ and $\hat{\tau}_o$ are the glide resistances of the slip planes at 0°K for the inherent lattice resistance and localized obstacle mechanisms respectively, μ is the shear modulus, σ the applied shear stress(*), w the average particle size acting as a localized obstacle. For certain classes of obstacles which produce hardening by internal disordering or misfit stresses $\alpha = 0.02$ and $n = 2$, while for obstacles which produce hardening by differences of stiffness $\alpha = 0.2$ and $n = 1$. The effective pre-exponential factors $\dot{\gamma}_G$ for the two cases are normally in

---

(*)In all the developments in this chapter σ, with or without subscripts, will represent applied stresses, whether they be normal stresses or shear stresses, while τ, with or without subscripts or superscripts, will represent deformation resistances, which are material properties.

the range of 10^9 sec^{-1} for normal laboratory tensile experiments [6]. Although the values of the various parameters quoted here are the best choices, some of the behavior of specific alloys and non-metallic compounds can fall outside these ranges. In this case the best approach has been to take the forms of stress dependence of the activation free enthalpy as given by eqns. (2.3) and (2.4) but use experimental data for the 0°K flow stress, and for temperature and strain rate dependence of the flow stress to determine empirically the best choices for $\hat{\tau}$, for the scale factor of the activation free enthalpy, and for the pre-exponential factor [7].

For the case of hardening by second phase particles of size w, the slip plane glide resistance $\hat{\tau}_o$ is proportional to the square root of the volume fraction c of the second phase obstacles. If the slip plane obstacles are forest dislocations or other immobile clusters of previously mobile slip dislocations, the density of obstacles in the slip plane will increase with plastic strain, increasing the effective c, and produce strain hardening. Such strain hardening produces a rise in glide resistance proportional to the square root of the dislocation density, whether it is of a purely statistical nature, resulting from accumulation of dislocations by interaction of dislocations with each other, or of a geometrical nature, resulting from accumulation of dislocations due to impenetrable second phase obstacles. The actual effectiveness of dislocation storage of each plastic strain increment can, however, be quite complicated and path dependent. It has been the subject of many imaginative, but often non-unique theories of work hardening which will not be of general concern to us except insofar as they relate to effects of surfaces, and interfaces, which we will discuss in greater detail below.

c) <u>Glide against phonon drag</u>. When dislocations move in a lattice at a finite temperature their motion is retarded by lattice thermal waves, or phonons. This retardation τ_p produces a glide resistance which is linear in velocity, v, given by

$$\tau_p = B \frac{v}{b} \qquad (2.5)$$

$$B \simeq \frac{kT}{\Omega \nu_D} \qquad (2.6)$$

The form given for the so-called drag coefficient B holds in the classical range of behavior above the Debye temperature, where Ω is the atomic volume and ν_D the Debye frequency [6]. This resistance increases with increasing temperature and acts on dislocations in free flight between obstacles. The superposition of this resistance with the obstacle controlled glide resistance has been treated by several investigators [6].

d) **Connection between glide resistance and plastic resistance**. The **glide resistance** which a single dislocation encounters in its slip plane through interaction with the lattice, with localized obstacles, or with thermal motions is rarely reflected directly into a macroscopic **plastic shear resistance** of a monocrystal. Not only do mobile dislocations bunch up to overcome effects of large scale clustering of obstacles in the slip plane but they also interact with each other. The problem is in general complex for inhomogeneously distributed slip plane obstacles. When the obstacles, however, are distributed at random, the variations in the glide resistance in the plane are minimized and so is the bunching of the mobile dislocations. In this case the mobile dislocations interact simultaneously with the characteristic slip plane obstacle and each other. The net effect is usually an increase in the stress exponent of the pre-exponential factor, to a value between 4 - 5 [8]. This point must be kept in mind when attributing experimentally determined stress dependences of the plastic strain rate to properties of the slip plane obstacles.

2.1.2 Mechanisms of elevated temperature plasticity. At temperatures where diffusion becomes rapid ($T \gtrsim 0.4\,T_m$) other mechanisms of deformation become possible.

a) **Climb controlled creep**. Easy generation and motion of vacancies imparts to edge dislocations a new degree of freedom of climb out of their slip planes. In this manner dislocations not only can overcome slip plane obstacles and continue to move, but also more readily annihilate each other to counteract the rise of dislocation density due to strain hardening. Both of these effects make steady state of plastic deformation under a constant stress possible, where the internal dislocation structure maintains itself at a steady state as well. How this steady state is approached through transient creep and how the steady state dislocation structure is kept in equilibrium have been extensively studied, but no fully satisfactory mechanistic understanding has yet emerged [8]. Phenomenologically, however, it has been possible to characterize the steady state creep strain rate in shear for all pure metals by an equation [9]

$$\dot{\gamma} = 4.3 \cdot 10^6 \frac{D \mu b}{kT} \left(\frac{\sqrt{3}\sigma}{\mu} \right)^n \qquad (2.7)$$

where D is the self diffusion constant of the metal. The exponent n is 4.4 for bcc metals, 5.25 for hcp metals, and varies from 4.5 to 5.5 in fcc metals with variation in the parameter $\mu b/\Gamma$ from 50 to 300 in which Γ is the stacking fault energy, and σ is the applied shear stress.

In solid solution alloys where a solute atmosphere can be dragged along with edge dislocations, it often turns out that this process is more difficult than climb in the pure lattice, and therefore acts as the rate controlling mechanism. In such alloys, which have been labelled as Class I, subgrains do not form and the creep law is governed by the mutual interaction between atmosphere dragging edge dislocations that can glide or climb with equal ease (or difficulty) [10]. This results in a creep law for high stacking fault alloys which has the same form as eqn. (2.7) in which the exponent is $n = 3$ and the coefficient is given by $(kT/\mu b^3)^2 / 8\, c_o\, \varepsilon_a^2$ where c_o is the concentration of the solute and ε_a its size misfit parameter.

The various aspects of dislocation creep have been reviewed by Bird, Mukherjee, and Dorn [9] to which the reader is referred for more detailed information.

b) <u>Diffusional creep</u>. At elevated homologous temperatures and at very low stresses where dislocation mobility is low, inelastic deformation in polycrystals can occur entirely by diffusional transport of matter either through the grains [11] or along the grain boundaries [12] without any accompanying distortions of the lattices of the crystal grains. For reasons of compatibility the process requires rigid body motion of grains and must therefore be accompanied by some grain boundary sliding [13]. The net shear strain rate $\dot{\gamma}$ under a shear stress σ can be readily computed [13] by dimensional considerations alone and is

$$\dot{\gamma} = 42 \frac{\sigma \Omega}{kTd^2} \left[D_v + \frac{\pi \delta}{d} D_b \right] \tag{2.8}$$

where Ω is the atomic volume, d the grain size, δ the effective thickness of a grain boundary, D_v and D_b the diffusion constants for lattice diffusion and boundary diffusion, respectively. Evidently this mode of deformation is important for fine grained material at elevated temperature, and of the two components the grain boundary component dominates over the lattice component in small grained material. One characteristic feature of this deformation is its inherent resistance to deformation localization: this is a direct result of the linear dependence of the strain rate on stress [14].

c) <u>Complex mechanisms of deformation</u>. Very fine grained polycrystalline metals are known to exhibit a behavior at elevated temperatures which is known as superplasticity [15] in which very large and stable deformations are obtained at relatively low strain rates without any appreciable change in grain shape. This suggests that the deformation involves flow of grains past each other, undergoing transient shape changes to permit the squeezing of one grain by others. A theory based on this model has been developed by Ashby and Verrall [16] which accounts for most of the known effects.

2.1.3 Deformation mechanism diagrams. Depending upon the level of applied shear stress, temperature, grain size and other internal parameters such as dislocation density and precipitate type and volume fraction, of the previously discussed mechanisms different ones can dominate the inelastic deformation process. Ashby and Frost [7] have determined the range of dominance of each deformation mechanism in the shear stress-temperature plane for constant structure preserving deformation processes, and have plotted diagrams showing contours of constant shear strain rate. Except in the low temperature deformation range where a constant structure is often unattainable due to strain hardening, the diagrams are useful in giving an overall view of the inelastic behavior of a solid. Their greatest use, however, is in showing that the inelastic behavior of nearly all crystalline solids falls within a narrow band, when the shear stress is given in units of the shear modulus, the temperature in units of the melting temperature, and the strain rate is normalized with the diffusivity of the material at its melting point. Figure 1 shows such a set of normalized curves for a variety of crystalline solids assembled by Ashby and Frost [7].

2.1.4 The role of pressure in inelastic deformation. Since inelastic deformations involve changes of shape it is almost self evident that pressure cannot by itself produce any significant inelastic deformation. Pressure can, however, affect inelastic deformation in various ways classifiable as either di-elastic effects that act through the effect of pressure on the elastic constants, and par-elastic effects where pressure produces some local inelastic deformations itself, that average to zero shape change. Di-elastic effects include the changed interactions of an obstacle such as a particle with a dislocation, either due to the changed modulus of the particle or the changed line tension and core properties of a dislocation. On the other hand, local elastic (or even plastic) strains resulting from either the different compressibilities of second phase particles compared with that of the matrix, or from the elastic anisotropies of crystal grains themselves in polycrystalline samples are examples of par-elastic effects. Such par-elastic effects can influence inelastic flow through these internal elastic strains directly or through the locally enhanced dislocation density resulting from the partial plastic dissipation of such local elastic strains produced by the applied pressure. In some instances, especially related to di-elastic interactions, pressure and negative pressure

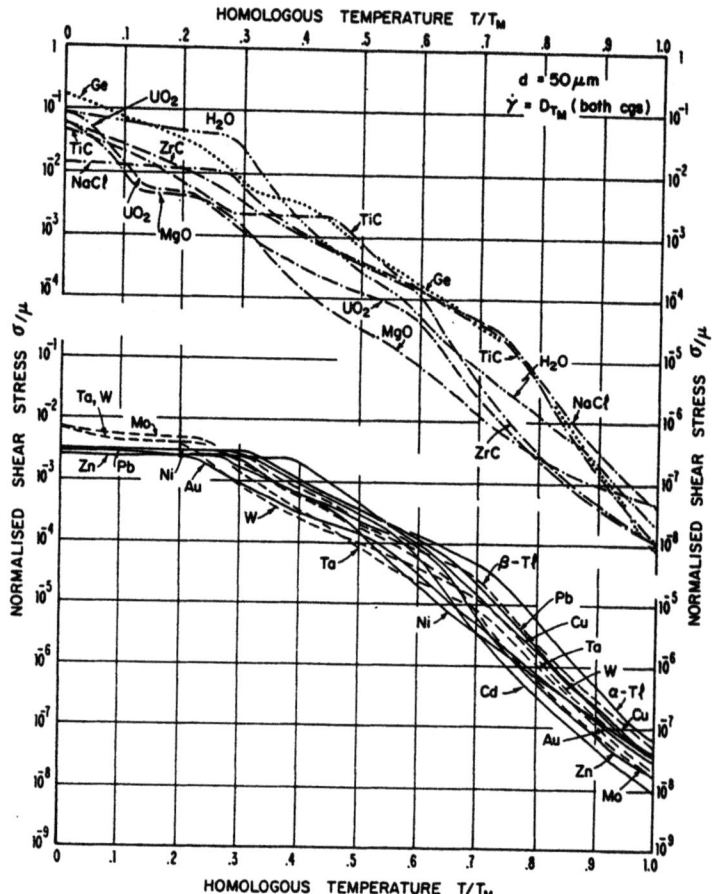

Fig. 1. "Deformation mechanism diagrams" showing the normalized inelastic deformation resistance of a large variety of crystalline solids plotted against homologous temperature, for a given normalized (with respect to the diffusivity at the melting point) shear strain rate (from Ashby and Frost [7], courtesy of M.I.T. Press).

can result in opposite net effects on the inelastic deformation resistance of a polycrystal, giving rise to so-called strength differential (S-D) effects.

2.2 Intercrystalline and interphase constraints on inelastic deformation

2.2.1 Local vs. non-local plasticity. In continuum mechanics pure crystalline solids are treated, as a first approximation, as homogeneous continua. When the solid contains heterogeneities but its behavior in bulk is of interest, the solid is still treated as homogeneous on the global scale as long as resolution of stress and strain is not of interest over volume elements of the order of the spacing of the heterogeneities. Even in cases where the local behavior of the solid is demanded on a scale of the dimensions of the heterogeneities the problem is often recast as one of "micro-mechanics" where the heterogeneities and the surrounding matrix are considered as continua of uniform properties stuck together at their common interfaces where certain boundary conditions of continuity of tractions and displacements are to be obeyed. When only elastic behavior is of interest notions of continua are acceptable down to atomic dimensions where they break down. In plasticity deformation results from dislocation motion and the notion of a continuum with smoothly varying properties breaks down at a much larger scale. Hence, the "point" in continuum plasticity often has to be a volume element of rather large proportions. Thus, crystal plasticity is inherently <u>non-local</u> on a fine scale where dislocations can be emitted from a source at a certain part of the crystal while the slip band produced by them can produce strain at a totally different part. Plasticity can be considered <u>local</u> only on a scale of volume elements which are mechanistically non-interacting. A conservative, upper limit, choice for the "point" size where continuum plasticity becomes applicable is the slip line length. Since the slip line length is physically limited by the grain size the latter often becomes a conservative choice for polycrystals. The slip line length, however, can be limited further by the presence of second phase particles. Furthermore, it becomes progressively shorter with strain hardening and increasing dislocation density [17]. A lower limit to the "point" size could be defined as the smallest volume element in which an independent elementary slip operation can be performed, i.e. having the dimensions of the mean dislocation spacing. In most cases the proper "point" size to be chosen for continuum plasticity will be somewhere in between these two limits. The recognition of non-locality of plastic deformation and the choice of the "point" size is of particular importance with regard to heterogeneities and hard surfaces which often have effects on mechanical behavior in excess of their volume fraction and what can be predicted by continuum theory.

2.2.2 Deformation of polycrystals. In single crystals of pure metals, by choosing certain orientations for straining, it is often possible to develop slip on only one slip system in the

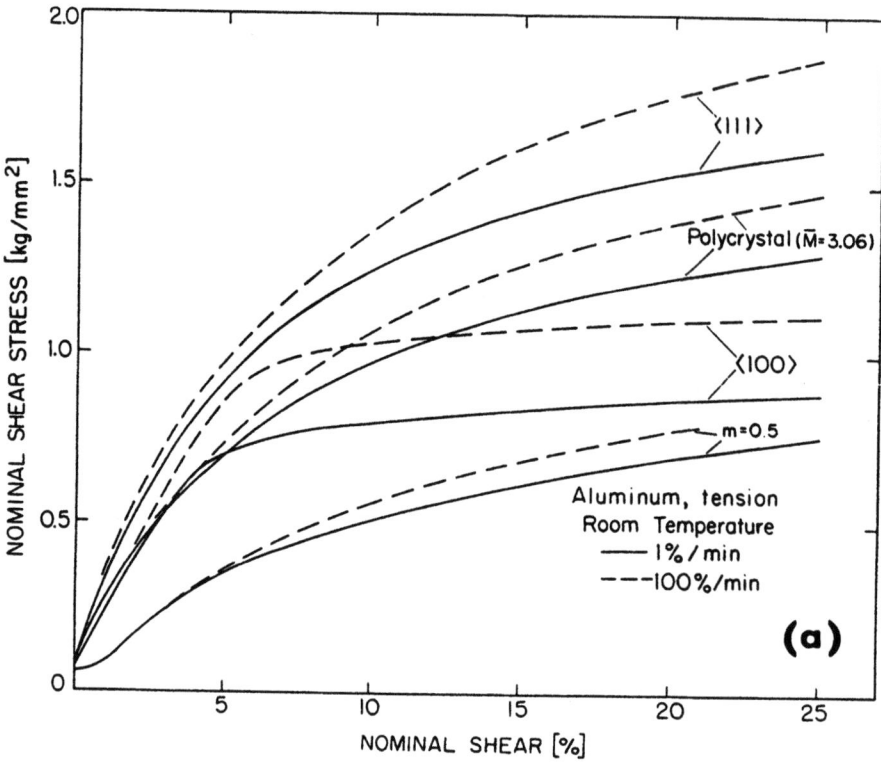

Fig. 2. Shear stress, shear strain curves for various single crystals and a random polycrystal of: a) pure aluminum and, b) Ferrovac E iron, m = 0.5 is the so-called Schmid factor giving the ratio of the resolved shear stress on a slip plane for single slip to the tensile stress on the single crystal (from Kocks [20], courtesy of ASM).

entire crystal. This gives rise to the most non-local nature of plastic flow where slip bands can in many fcc metals travel entirely across a crystal of cm. dimensions. The resolved shear stress for slip in such crystals is generally very low. In a polycrystal the required compatibility of strain between grains does not permit every grain to slip only on its most highly stressed slip system. Instead, activity on a multiplicity of slip systems in each grain is required. This has raised the question of the relationship between the resolved shear stress for slip in single crystals and the tensile yield stress of a random polycrystal. This question was addressed first by Taylor

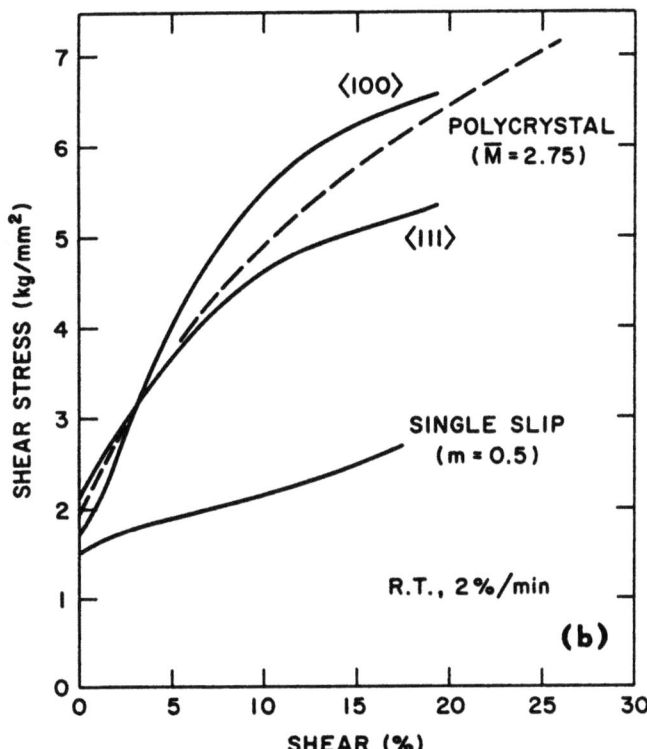

[18] and later more thoroughly by Bishop and Hill [19]. The problem is handled by considering that the tensile plastic strain in each grain is the same as the average tensile plastic strain in the polycrystal, finding for each grain orientation a set of five independent slip systems capable of giving the required average tensile plastic strain with a minimum of plastic work dissipation, summing the dissipated plastic work over all grain orientations and equating this to the plastic work done by the average tensile stress in extending the specimen to the average tensile plastic strain. The resulting analysis gives the so-called Taylor factor of the ratio of the yield stress in tension of the polycrystal to the resolved shear stress for slip in a single crystal to be $\overline{M} = 3.06$ for both fcc and bcc metals, considering only the slip elements of {111}<110> and {110}<111> for these metals respectively. Since bcc metals can undergo slip also on other planes containing the <111> direction in a mode called pencil glide, a lower Taylor factor of $\overline{M} = 2.75$ is possible (for a discussion of this and many other aspects of the

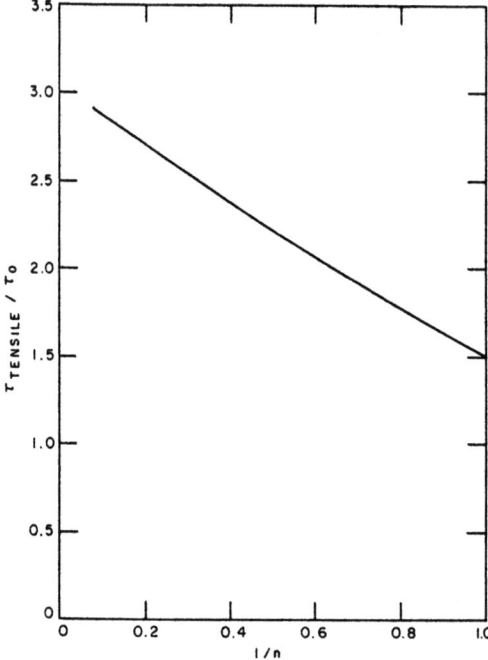

Fig. 3. Ratio of the tensile creep resistance $\tau_{TENSILE}$, of a polycrystal to the creep resistance τ_o of a single slip system (after Hutchinson [21], to be published).

deformation of polycrystals the reader is referred to Kocks [20]).
The analysis of Taylor discussed above in which slip systems are considered non-interacting pertains at best only to the flow stress of non-hardening polycrystals.(*) When polycrystals are deformed further they strain harden more rapidly than single crystals undergoing single slip because of the interaction of the multiplicity of slip systems in each grain which is necessary for compatible deformation. A similar situation holds in single crystals of high symmetry orientations where more than one slip system becomes active from the beginning, resulting in rapid rates of strain hardening. This has led to the observation by Kocks [20] that the stress strain curves of polycrystals should be obtainable by averaging over the stress strain curves of only a few single crystals of high symmetry orientations. The validity of this observation is shown in the curves of figs. 2a and 2b, de-

---

(*) The initial yield stress in shear for polycrystals is no different from that of single crystals. The Taylor factor only applies to the condition where all grains have yielded and a condition of true flow has been achieved. This requires a plastic strain of the order of about 4 times the yield strain [21].

picting the behavior of fcc aluminum and bcc iron single crystals of <110> and <111> orientations relative to the behavior of the respective polycrystals and the behavior of a single crystal undergoing only single slip.

The Taylor-Bishop-Hill analysis for time independent plastic yielding of a polycrystal has been generalized and applied to dislocation creep by Hutchinson [21], where a power-law type (Norton's law) steady state creep given by eqn. (2.7) is assumed to represent the behavior of an individual slip system. The averaging procedure using a so-called self consistent approach due to Hill [22] where each grain is considered as a heterogeneity in a matrix of average behavior results in "Taylor factors" which decrease with decreasing n as shown in fig. 3 for an fcc polycrystal.

2.2.3 Heterogeneities in plasticity. Heterogeneities have important effects on plastic flow of crystalline solids. When both the size of the heterogeneities and their mean spacing are comparable to the "point" size of the theory of continuum plasticity, their effects are completely non-local and cannot be understood by continuum theory (see also Ashby [23]). The interaction of such heterogeneities with dislocations has been elucidated first by Orowan [24], and its various subtleties have now been thoroughly worked out [6]. Such small heterogeneities, in principle, give rise to a temperature dependent flow stress obtainable from eqn. (2.4) where the maximum slip plane glide resistance $\hat{\tau}_o$ can be given by

$$\hat{\tau}_o = \left[\frac{2\mathcal{E}}{b\ell}\right]\left[\frac{\hat{K}}{2\mathcal{E}}\right]^{3/2} = \left[\frac{2\mathcal{E}}{br}\right]\sqrt{\frac{3c}{2\pi}}\left[\frac{\hat{K}}{2\mathcal{E}}\right]^{3/2} \qquad (2.9)$$

in which $\mathcal{E} \approx (\mu b^2/2)$ is the dislocation line energy, ℓ the mean spacing of the heterogeneities in the slip plane, r their average radius, c their volume fraction, and $\hat{K}$ their peak resistive force to dislocation motion as they are being cut. Consider the following typical case of hard shearable precipitates with a volume fraction of 0.001, having average radii $r/b = 10$ and shear strength of $\tau_p/\mu = 0.01$ for which $\hat{K} = 2 r b \tau_p$. If the glide resistance of the pure crystal were negligibly low, as it might be in fcc metals, then the continuum theory would predict a shear strength of the crystal with such a particle dispersion of only $\tau_p c$ based on a rule of mixtures. Equation (2.9) can be regrouped to read

$$\hat{\tau}_o = (\tau_p c)\sqrt{\frac{12}{\pi c}\frac{r}{b}\frac{\tau_p}{\mu}} \qquad (2.10)$$

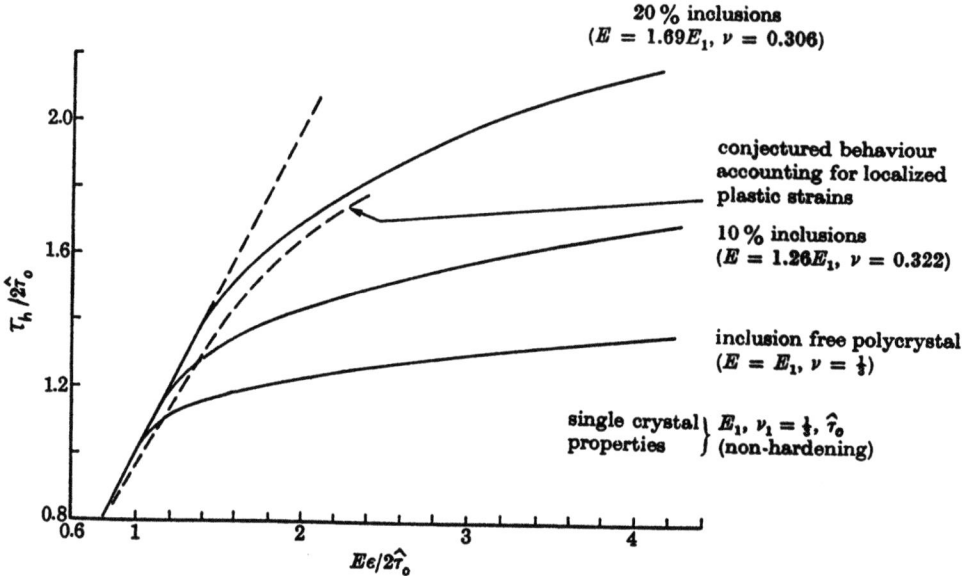

Fig. 4. Ratio of the tensile plastic resistance τ_h of the heterogeneous polycrystalline composite with an fcc matrix and containing rigid spheres to the tensile plastic resistance $2\hat{\tau}_o$ of the single crystal (from Hutchinson [25], courtesy Royal Society of London).

where the term under the square root sign can be interpreted as the amplification resulting from the non-locality of crystal plasticity. Evaluation of this term gives an amplification factor of 19.5[*] for the case considered.

When the size of the heterogeneities is very much larger than the "point" size of continuum theory their effect on the overall plastic resistance of the heterogeneous alloy becomes more readily predictable by continuum theory, provided certain non-vanishing interaction effects are recognized. Let us consider

---

[*] Caution is required in using eqn. (2.10) in which r cannot be indefinitely increased because when $\bar{K}/2\mathcal{E} \approx 1.0$ dislocations can circumvent the particles without cutting them. Thus, the maximum value that r/b can take for eqn. (2.10) to hold is 50 for the case considered, at which time the amplification factor reaches fully a value of 43.5. Beyond this point the shear strength of the alloy becomes independent of the shear strength of the particle, and decreases with increasing particle size.

two cases: a) rigid, ideally plastic, randomly distributed spherical heterogeneities built into an otherwise linearly strain hardening fcc polycrystal, b) rigid, ideally plastic, thin surface layer on a non-hardening polycrystal.

a) <u>Hard spherical heterogeneities in a strain hardening plastic continuum</u>. The problem of randomly dispersed rigid spheres in a non-hardening fcc polycrystal was considered by Hutchinson [25] using Hill's [22] self consistent analysis for heterogeneities. As shown in fig. 4 Hutchinson finds that for reasons of compatibility rigid spherical particles "harden" the polycrystal by merely making initiation and development of plastic flow more difficult. In fact, when the volume fraction of particles reaches a level of 0.4 - 0.5 where a non-yielding "skeleton" gets established, the polycrystal locks, and plastic deformation is no longer possible. The corresponding problem of greater interest where the matrix is a strain hardening plastic continuum and the particles are rigid ideally plastic has not yet been solved in a comparably rigorous manner. An approximate solution has been considered by Ashby [26] in which the required misfit dislocation loops between the plastically deforming matrix and the rigid particles are uniformly dispersed in the volume of the matrix to compute the average work hardening without considering the additional local hardening in the regions of high strain gradient around the particles. A more self consistent, but still approximate solution to this problem can be obtained by using the local plastic flow model of Argon, et al. [27], shown in fig. 5, where the developing misfit between the initially rigid spherical heterogeneity and the linearly hardening plastic continuum with a hardening rate of θ_{II} is dissipated in a set of local cylindrical plastic zones of inhomogeneous deformation. The result of such an analysis (details can be furnished upon request) is that an interface stress concentration of 1.3066 develops above the flow stress in shear of the homogeneous plastic continuum deformed to the same level of shear strain. The effect of the rigid particles on the average flow stress τ_h in shear of the heterogenous assembly having a particle volume fraction of c is

$$\frac{\tau_h}{\tau_o} = 1 + \frac{3}{4}\left(\frac{c}{1-c}\right)\sqrt{1 + \left(\frac{\beta}{2}\right)^2} \qquad (2.11)$$

where $\beta = \sqrt{2\sqrt{2}}$ is a numerical constant, and τ_o is the flow stress in shear of the homogeneous reference matrix undergoing the same shear strain as the heterogeneous assembly. It is assumed that $\tau_o = \theta_{II}\gamma$ holds, so that the traction on the particles steadily increase until they too yield when this traction reaches τ_p, the shear strength of the particles, at which point the slope of the stress strain curve abruptly decreases, roughly in proportion to the volume fraction of the particles. At the

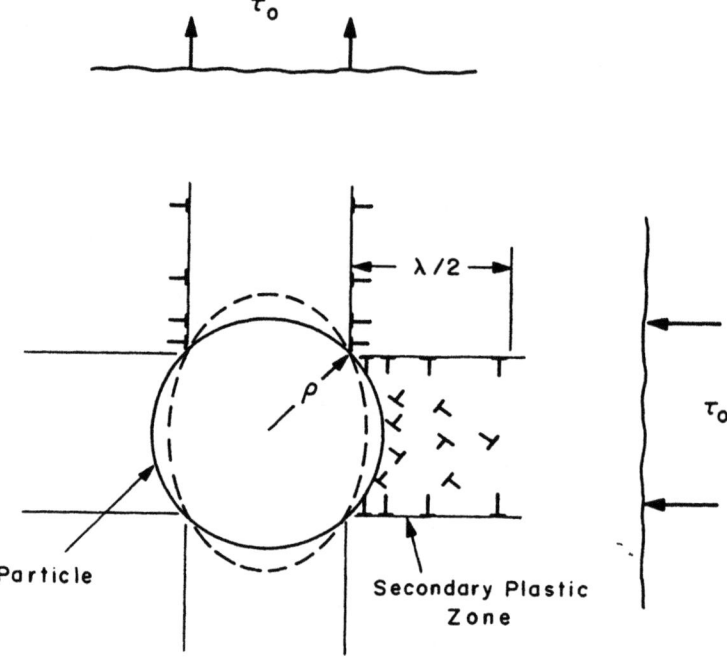

Fig. 5. Modelling inhomogeneous plastic deformation around non-yielding cylindrical particles by secondary plastic zones in which strain hardening plugs are pressed in or retracted from cavities with wall friction (after Argon, et al. [27]).

point when the particles yield

$$\tau_o = \frac{\tau_p}{\sqrt{1 + \left(\frac{\beta}{2}\right)^2}} \quad \text{(at yield of particles)} \quad (2.12)$$

Since particle induced strain hardening has been an intimate part of the model, no comparison is possible with an upper bound analysis of a non-hardening matrix. Examination of eqn. (2.11), however, reflects directly that the second term on the right side of the equation represents the additional strain hardening of the matrix due to the presence of the particles.

b) **Hard surface film on strain hardening plastic continuum.**
No important mechanical distinction can be attributed to the
smooth surface of a homogeneous plastic continuum. In real homogeneous polycrystalline solids, however, surfaces can promote
early initiation of plastic flow due to a variety of effects such
as: surface stress concentrations, lack of grain boundary constraints, presence of slight bending components in the applied
stress field, etc. [28]. In addition, it has now been well established [29] that the strain hardening efficiency of near-surface portions of a solid is lower than its interior, producing,
generally, a surface layer of lower plastic resistance in comparison to that of the interior.

It is, however, amply documented that in many instances experiments suggest the presence of surface layers of higher plastic resistance than that of the interior [30]. These appear to
be the result of hard, thin surface films of distinctly different
physical properties, or in some instances relate to an enriched
obstacle structure to slip resulting from a different thermomechanical treatment of the near-surface material [31-33]. The
effect of such thin surface films is more dramatic in single
crystals in which the non-locality effects are strong.

We begin by considering a simple case of a rectangular bar
of a non-hardening plastic continuum, as shown in fig. 6a, of
thickness $2H$, having a plastic resistance in shear of τ_o
with surface layers of thickness h having a non-hardening plastic resistance of τ_f ($\tau_f > \tau_o$). The tensile plastic resistance τ_h of this assembly can be obtained readily from an upper
bound analysis for shear at $45°$ to the tensile axis as shown
in fig. 6a.

$$\tau_h = 2\tau_o \frac{H}{H + h} + 2\tau_f \frac{h}{H + h} \tag{2.13}$$

just as would be obtainable from a rule of mixtures. Alternatively one might consider the penetration of the hard surface layer
by a band of concentrated slip in the soft interior, as shown in
fig. 6a as the dislocation pile-up. Using the exact solution for
this very problem given by Bilby, et al. [34] the very same solution of eqn. (2.13) is obtained. It is well known, however, that
the mechanical effect of hard surface films is far greater than
the volume fraction average given by eqn. (2.13). The clue lies
in the additional local strain hardening which surface films enforce in their vicinity. As shown in fig. 6b, compatibility of
deformation requires that the strains ε_{zz}, ε_{yy}, ε_{yz} be continuous across the interface between the film and the substrate.
If the film is harder than the substrate, additional slip systems
are enforced in the substrate to satisfy the conditions of compatibility. This produces intersecting slip in an interlayer of
thickness h_i in the substrate which causes rapid strain hardening

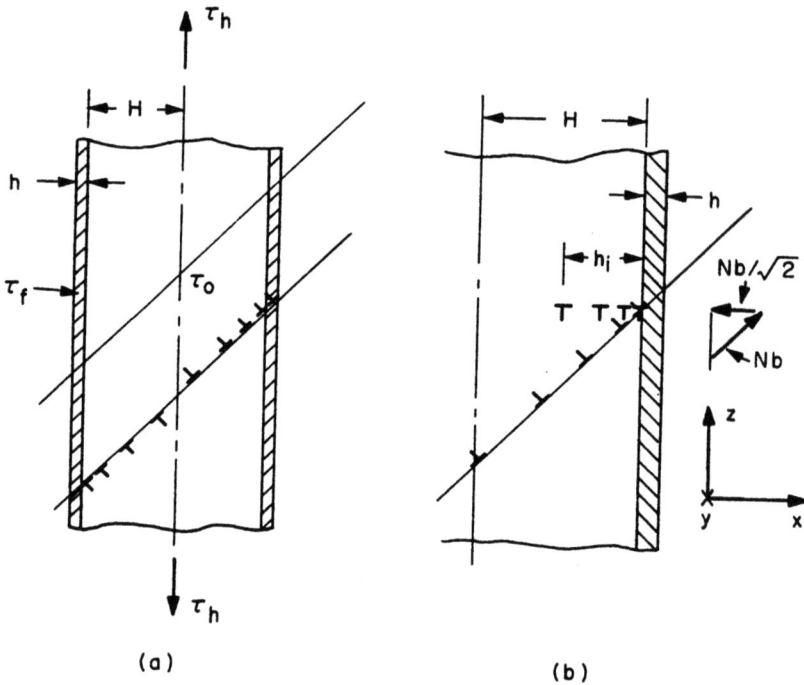

Fig. 6. Rectangular bar with hard surface film: a) penetration of a plastic slip zone into the film without considering catalytic strain hardening effects due to compatibility, b) where such effects are considered.

that adds to the load carrying capacity. A rough estimate of this effect could be obtained by considering the length of the pile-ups h_i of compatibility dislocations required to eliminate the surface steps of extent Nb which would otherwise be formed by the main slip bands in the substrate as shown in fig. 6b. Elementary considerations pertaining to dislocation pile-ups give readily that

$$h_i = (H/\sqrt{2})\left(\frac{h}{h+H}\right)\left(\frac{\tau_f}{\tau_o} - 1\right) \qquad (2.14)$$

Assuming now that the plastic resistance of the layer h_i to be the average of the interior and the film, one can equally readily find the total tensile plastic resistance τ_h of the heterogeneous assembly as (including the intermediate layer of changing plastic resistance)

$$\tau_h = \frac{2H \tau_o}{H + h} \left\{ 1 + \frac{1}{2\sqrt{2}} \left(\frac{h}{h + H} \right) \left(\frac{\tau_f}{\tau_o} - 1 \right)^2 \right\} + \frac{\tau_f h}{H + h} \quad (2.15)$$

Considering an instructive and fairly typical example for which the ratios $(h/H) \simeq 10^{-4}$, $(\tau_f/\tau_o) \simeq 10^2$ we find that the induced strain hardening in the substrate under the film raises the effective plastic resistance of soft interior core by nearly a factor of 1.35. Although the above computation is very approximate, it serves to demonstrate the catalytic effect of a hard surface film on the plastic resistance of a soft but strain hardenable matrix.

2.3 Phenomenological constitutive equations for inelastic deformation

Considering that pressure by itself does not produce plastic deformation, the multi-axial stress condition for plastic deformation of an isotropic polycrystal can be given as one in which the state of stress deviates from a pure pressure by a critical amount, or what amount to the same, where the root mean square shear stress becomes equal to a critical value. In both instances this critical value can be taken as the plastic resistance τ in a pure shear experiment. This leads to the condition:

$$s = \sqrt{\frac{1}{6} \left[(\sigma_{22} - \sigma_{33})^2 - (\sigma_{33} - \sigma_{11})^2 - (\sigma_{11} - \sigma_{22})^2 \right] + \sigma_{23}^2 + \sigma_{13}^2 + \sigma_{12}^2} = \tau$$

(2.16)

where s is known as the deviatoric shear stress or the equivalent shear stress. The yield condition can be imagined as a surface in six dimensional stress space which is symmetric about all axes and is open along a loading path describing pure pressure $\sigma_{11} = \sigma_{22} = \sigma_{33} = p$, $\sigma_{23} = \sigma_{13} = \sigma_{12} = 0$. This criterion was stated independently by a number of investigators separately but is now normally referred to as the Mises [35] yield criterion. Although the Mises yield criterion was originally stated for time independent plastic flow, it can also be generalized for time dependent plastic flow and steady state creep if the plastic shear resistance is considered to be a function of temperature and the equivalent shear strain rate, in a form given, e.g. by eqn. (2.7). The equivalent shear strain rate $\dot{\gamma}$ mentioned in this connection is defined on a basis of equivalence of plastic work dissipation rate where $s \dot{\gamma}$ is chosen in just such a way that it equals $\tau \dot{\gamma}$, the rate of plastic work dissipation in a pure shear experiment. When defined in this manner it becomes

$$\bar{\dot{\gamma}} = \sqrt{\frac{2}{3}\left[(\dot{\epsilon}_{22}-\dot{\epsilon}_{33})^2 + (\dot{\epsilon}_{33}-\dot{\epsilon}_{11})^2 + (\dot{\epsilon}_{11}-\dot{\epsilon}_{22})^2\right] + \dot{\gamma}_{23}^2 + \dot{\gamma}_{13}^2 + \dot{\gamma}_{12}^2}$$

(2.17)

Furthermore, the yield criterion of eqn. (2.16) also holds in the presence of moderate amounts of strain hardening - as long as such hardening does not result in appreciable texture development. In this case the parameter that measures the strain hardening is the equivalent shear strain obtained from the time integral of eqn. (2.17) over the entire deformation history. It is assumed that the effect of the equivalent shear strain on the equivalent plastic shear resistance is the same as the shear strain on the plastic shear resistance in a pure shear experiment, i.e.

$$\tau = \tau(\dot{\gamma}, \gamma, T) = \tau(\bar{\dot{\gamma}}, \bar{\gamma}, T) \qquad (2.18)$$

Operationally the tensile stress-strain curve is usually used to characterize both the strain hardening and the strain rate dependence of the plastic resistance. The conversions between the equivalent tensile stress $\bar{\sigma}$ and the deviatoric shear stress s, and between the two equivalent strain rates $\bar{\dot{\epsilon}}$ and $\bar{\dot{\gamma}}$, is, by eqns. (2.16) and (2.17)

$$\bar{\sigma} = \sqrt{3}\, s \qquad (2.19)$$

$$\bar{\dot{\epsilon}} = \bar{\dot{\gamma}}/\sqrt{3} \qquad (2.20)$$

The type of strain hardening behavior outlined here in which the yield surface grows isotropically, describable by one parameter only, is known as <u>isotropic hardening</u>. This is the most realistic asymptotic behavior of the material reached after transient and anelastic behavior, that can occur upon changes in deformation direction or strain rate, has disappeared. In some instances when only small strains are of interest after a change of deformation direction, a rigid translation of the yield envelope without change of size or shape is assumed. This is known as <u>kinematic hardening</u>.

When strain hardening produces texture development and plastic anisotropy, the isotropic yield criterion given by eqn. (2.16) no longer holds and can be replaced in many cases by an anisotropic but still symmetric yield criterion of Hill [36]. In strongly textured materials asymmetrical yield criteria akin to those of single crystals may have to be used [37].

Starting from the physical model of slip by block-like translation of crystal lamellae over each other [14], it is possible to demonstrate that upon yielding under a multi-axial state of stress, the various components of plastic strain rate will form a generalized six dimensional vector which is parallel to the outward normal vector to the six dimensional yield surface Q (which for the special symmetrical case of eqn. (2.16) is $Q = \bar{\sigma} - \sqrt{3}\tau = 0$) at the point representing the stress state. This fact, which can also be proved more generally [14], is known as the associated flow rule and results in the following set of equations.

$$\frac{\dot{\varepsilon}_{11}}{\bar{\dot{\varepsilon}}} = \frac{\partial Q}{\partial \sigma_{11}} = \frac{\sigma_{11} - (1/2)(\sigma_{22} + \sigma_{33})}{\bar{\sigma}} \qquad (2.21a,b,c)$$

$$\frac{\dot{\gamma}_{23}}{\bar{\dot{\varepsilon}}} = \frac{\partial Q}{\partial \sigma_{23}} = 3\frac{\sigma_{23}}{\bar{\sigma}} \qquad (2.21d,e,f)$$

In the above equations, while the first set of equalities are perfectly general, representing any yield surface, the second set of equalities have made use of the special symmetrical and isotropic yield condition of Mises given by eqn. (2.16). The first set of equalities also indicate that the yield surface acts as a flow potential.

2.4 Standard tests for measuring the inelastic deformation resistance

The inelastic response of crystalline solids is generally desired as a constitutive equation which describes the fundamental deformation resistance of the material in shear as a function of shear strain, shear strain rate, and temperature. Its multi-axial response is then generally expected to obey the isotropic conditions of yield and flow described in the preceding section. On the other hand, the physically more meaningful constitutive equation

is best considered as a shear strain rate that results when at
temperature T, a shear stress σ, is applied to a given state
of the solid, having a certain dislocation density and other sets
of slip obstacles that could be characterized as a deformation
resistance τ (a material property of state), obtainable by
means of a standard test - for example, a yield test at 0°K.
The forms of equations discussed for the various deformation
mechanisms in Section 2.1 are of this type. In this formulation
plastic strain is not considered as a proper parameter of state,
it is merely an integral of the plastic strain rate over the
duration of the experiment, which, because of additional anelastic
strains occurring at deformation reversals and transients that
are not yet fully understood, is determinable only to within an
additive constant. Although plastic shear strain (or its generalization - the equivalent plastic strain) is not the preferred
parameter for describing the plastic resistance that represents
the sum total effect of the internal slip obstacles, it remains
one of the most important means of altering it.

Based on the above considerations we write the preferred
form of the constitutive equation for inelastic deformation as

$$\dot{\gamma} = \dot{\gamma}(\sigma, \tau, T) \qquad (2.22)$$

As long as the material remains untextured, and isotropic,
we assume that the scalar form of eqn. (2.22) also gives the
generalized behavior of the material where the shear strain rate
and applied shear stress stand for the equivalent shear strain
rate and the deviatoric shear stress. When the material is textured and plastically anisotropic, the plastic resistance is not
usually describable by a scalar and more advanced considerations
are necessary [8].

2.4.1 *The tension test.* In spite of its various shortcomings the tension test remains as the most widely used test to
determine the plastic resistance of a crystalline solid. In
regions of behavior where there are no localizations of deformation the tension test at low temperatures, performed at a constant strain rate (usually constant extension rate) gives reasonably reproducible and unambiguous results. Since the state of
stress is known from overall equilibrium, no special analysis is
required to determine the fundamental plastic resistance. When
the behavior becomes inhomogeneous, either due to material or
mechanical instabilities such as formation of Lüders bands and
necks, the interpretation of the tension test becomes more difficult, requiring some analysis to determine the fundamental plastic resistance (see e.g. Kocks, et al. [6] for determination of
the strain rate sensitivity from Lüders band velocities,

McClintock and Argon [14] and Argon, et al. [38] for the determination of plastic resistance in necked bars in non-strain hardening and strain hardening solids).

Uninterrupted tension tests at different strain rates and temperatures ($T < 0.4\,T_m$) give the strain rate dependence and temperature dependence of strain hardening. The differences between the flow stress for corresponding plastic strains in such curves generally do not give the strain rate and temperature dependence of the flow stress for the reason, given earlier, that strain is not a proper parameter characterizing the internal obstacle structure. To obtain these dependences of the flow stress, the strain rate or the temperature has to be changed abruptly at a given point in the straining experiment and the change in flow stress at that current obstacle structure be recorded. In many cases it is possible to represent phenomenologically the increment in strain rate resulting from an increment in flow stress at constant structure by power functions over small ranges of flow stress, although as discussed in Section 2.1, more appropriate forms of the connection involve Arrhenius expressions with stress dependent exponents. It is also possible to analyze the stress and temperature dependence of the strain rate by means of the thermodynamical and kinetic theory of slip [6] to deduce more fundamental parameters describing the obstacles on the slip plane.

2.4.2 The stress relaxation test. When a plastic straining experiment in tension at a constant strain rate is suddenly stopped and the total extension of the specimen is held constant plastic deformation will continue in the specimen at a decreasing rate as the elastic strain is converted into plastic strain and the stress steadily relaxes. Since the total amount of strain which occurs in such a stress relaxation experiment is of the order of the elastic strain in the specimen (provided a stiff testing machine is used) it can be assumed that there will be no significant alteration of the internal obstacle structure during the test. Therefore, the particular functional relationship between the decreasing stress in the specimen and the plastic strain rate can be taken as the fundamental stress dependence of the plastic strain rate for a certain obstacle structure that could be characterized, for example, by the initial flow stress with which the stress relaxation experiment was started. Hart, et al. [39] find that all stress relaxation experiments performed on one metal at low temperature ($T < 0.4\,T_m$) give a family of curves relating $\ln \sigma$ to $\ln \dot{\varepsilon}$ that can all be translated along a slanted line to form one continuous basic curve, as shown in fig. 7, regardless of the initial level of flow stress from which the stress relaxation experiment was started. Such behavior implies that one parameter is sufficient to describe the obstacle state of the material or its collective plastic resistance [40]. Furthermore, the above mentioned shift property is consistent

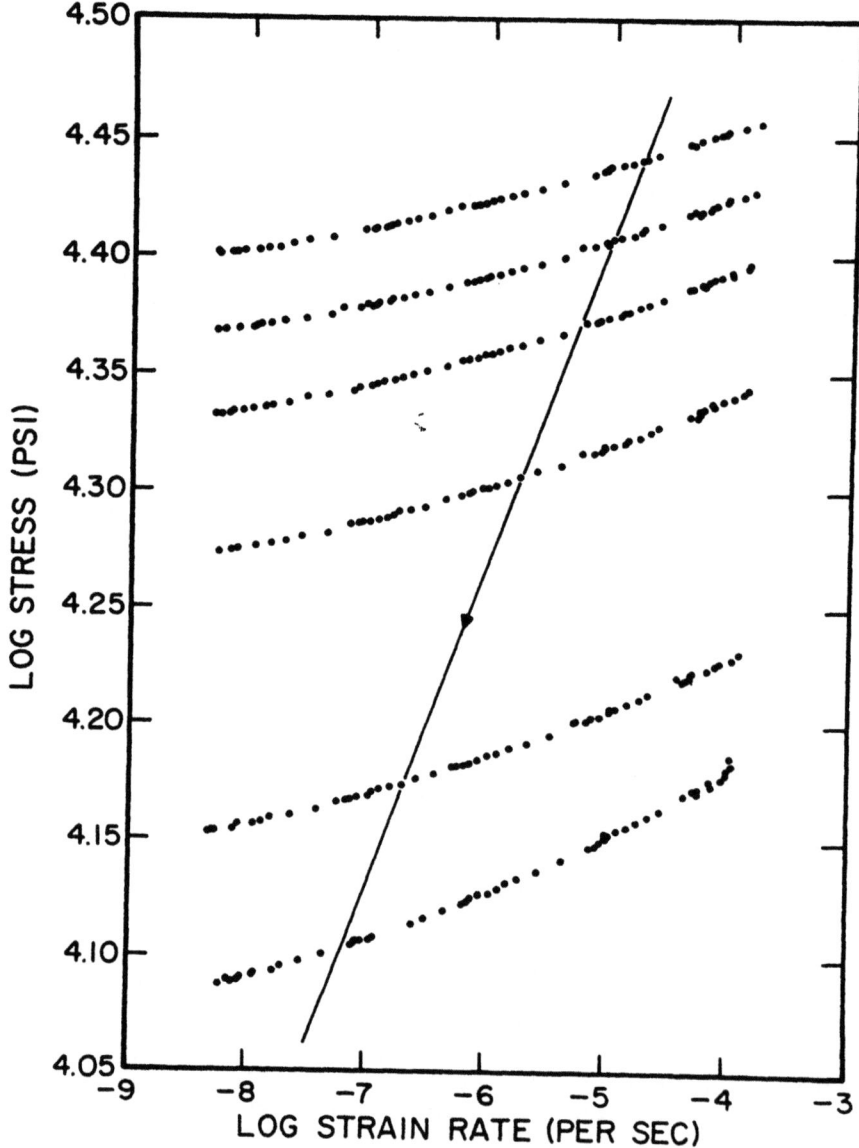

Fig. 7. Decrease in plastic resistance with decreasing strain rate in a stress relaxation experiment on pure niobium (from Hart, et al. [39], courtesy of M.I.T. Press).

with a form of constitutive equation given by

$$\dot{\gamma} = \dot{\gamma}_o \left(\frac{\sigma}{\mu}\right)^n \exp[-\Delta\Gamma(\sigma/\hat{\tau})] \qquad (2.23)$$

where $\Delta\Gamma(\sigma/\hat{\tau})$ has the required inverse temperature dependence of an Arrhenius expression [8], n is an empirical exponent usually of the order of 4, and $\hat{\tau}$ is the single parameter whose different values can characterize the different curves in fig. 7. Comparison of eqn. (2.23) with eqns. (2.1 - 2.4) indicates that $\hat{\tau}$ is the threshold slip plane resistance that represents the sum total mechanical effect of the slip plane obstacles.

2.4.3 The hardness test. A quick, convenient, and usually very informative test is the hardness test in which a very hard indenter is plastically indented into the surface of a plastic solid under a standard indentation force. The depth or the projected area of the indentation is measured. In the more meaningful tests the hardness pressure consisting of the ratio of the indentation force in Kg to the area of the indentation in mm^2 is reported as the hardness number (Brinell, Vickers, Knoop hardness tests). Since the indentation process enforces a shape change at the surface, as in polycrystal plasticity, slip on many systems is simultaneously activated, whether the indented surface is a single crystal or polycrystal. Hence the hardness pressure reflects a certain amount of strain hardening akin to that in a polycrystal and, therefore, reflects the plastic resistance of a polycrystal. Unlike in the tension test, the plastic strains in the hardness test are strongly inhomogeneous, and are surrounded by an unyielded elastic half space which imposes a plastic constraint in the form of a hydrostatic pressure that does not affect the local plasticity, but significantly adds to the hardness pressure required to make the indentation. Both theory [36] and experiments [41] show that only 1/3 of the hardness pressure reflects the plastic resistance of the indented solid. In addition, the strain hardening associated with the indentation correlates with the flow stress of the polycrystal stress strain curve at a strain of about 0.1 [14], which is near the tensile strength in most structural alloys. Hence a first order estimate of the tensile strength of a polycrystal can be obtained as 1/3 of the recorded hardness pressure.

In very high strength structural alloys and non-metals the elastic strain at yield is often very substantial, altering the nature of the plastic indentation from that of a "plowing-up" type in nearly rigid-plastic metals to that of a "sinking-in" type, having a strain field resembling a nearly spherical misfit in the high yield solids. Marsh [42] has used the elastic-plastic solutions of an expanding spherical cavity [36] as a guide

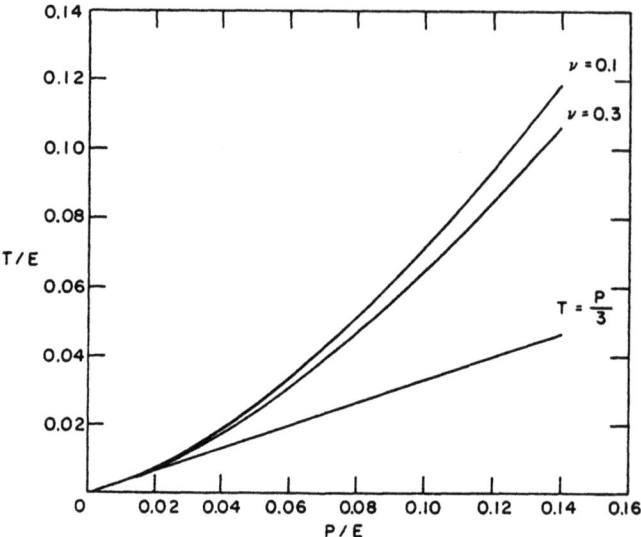

Fig. 8. Relation between tensile strength T and hardness pressure P for materials with high plastic resistances (large accompanying elastic strains) shown by the curves vs. the relation for a rigid plastic idealization, shown by the straight line (after Marsh [42]).

to determine how the hardness pressure relates to the yield strength (most high strength materials have very limited capacity to strain harden) in such solids. His solution is presented in fig. 8 in graphical form relating the ratio of the experimentally measured hardness pressure, to the Young's modulus, to the ratio of the yield strength to the modulus. Evidently the straight line giving the behavior of a rigid plastic material ceases to hold above elastic strains of 0.01 .

Needless to say, the three simple tests described above make up only the very basic core of tests useful to characterize the plastic resistance of solids. For a description of other useful tests the reader is referred to the many standard tests published, for example, by the ASTM [43].

3. FRACTURE IN CRYSTALLINE SOLIDS

It is almost self evident that for fracture to occur, a crack must propagate across a stressed solid. Therefore, the understanding of the fracture process is inextricably linked to a need to understand the mechanics of stress and strain around cracks.

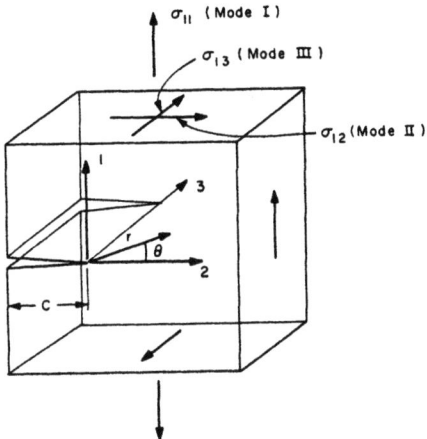

Fig. 9. The three modes of loading a cracked solid which will produce high concentrations of stress.

3.1 Inherent brittleness or ductility

Certain solids are known to be brittle and fragile while others are ductile and tough. Kelly, et al. [44] have pointed out that the basis of this division lies in the type of atomic bonding making up the solid. Consider an atomically sharp crack in an otherwise perfect elastic solid as shown in fig. 9. When a tensile stress σ_o is applied across the crack as shown in the figure, normal stresses and shear stresses are concentrated at the crack tip according to the well known Mode I solution. At the tip of the crack all stresses exhibit a singularity. In an inherently brittle solid the tensile stress at the crack tip will reach the ideal cohesive strength before the shear stress reaches the ideal shear strength, to propagate the crack in a brittle manner without any plastic dissipation. In an inherently ductile material the opposite occurs. From the Mode I loading solution one can readily show that the maximum tensile stress σ_n and maximum shear stress σ_s are given respectively by [14],

$$\sigma_n = \sigma_o \sqrt{\frac{c}{2r}} \quad \text{(at } \theta = 0\text{)} \tag{3.1}$$

$$\sigma_s = \frac{\sigma_o}{2} \sqrt{\frac{c}{2r}} \quad \text{(at } \theta = \pm \frac{\pi}{2}\text{)} \tag{3.2}$$

Equations (3.1) and (3.2) show that the ratio $\sigma_n/\sigma_s = 2$ at the crack tip. Hence, if the ratio τ_{COH}/τ_{SH} of the cohesive strength to the ideal shear strength is much greater than 2, the material is inherently ductile. Table 3.1 gives the ratio of these strengths for some clear-cut cases.

Table 3.1 Cohesive Strengths and Ideal Shear Strengths of Some Solids [1]

Material	τ_{COH} GN/m^2	τ_{SH} GN/m^2	τ_{COH}/τ_{SH}	
Diamond	205	121	1.69	
Silicon	32	13.7	2.34	Brittle Solids
Al$_2$O$_3$	46	16.9	2.72	
Copper	39	1.29	30.23	
Silver	24	0.77	31.17	Ductile Solids
Gold	27	0.74	36.49	

The results given in Table 3.1 are selected for their consistency. Deciding upon inherent brittleness or ductility for other materials is not as clear cut. This is because the initiation of plastic flow from a sharp crack is more complicated [45] than discussed above, and because the cohesive strengths and ideal shear strengths of most materials have not been calculated and are not reliably measurable. On both accounts improvements await developments in discrete atom models of crack tips, which are not yet reliable enough to be of much predictive value [46].

The simple consideration for brittleness or ductility described above has interesting applications in known cases of environment induced embrittlement. It is well known that the effective cohesive strength of solids could be significantly reduced by surface active agents: the case of silica glass and water vapor is a well documented example. In such cases the ratio of the cohesive strength to the ideal shear strength can drop from a value well above 2 to a value below 2 resulting in embrittlement. Although well documented specific examples of this phenomenon are rare, many cases of intergranular embrittlement are probably of this type.

3.2 Concentrations of stress and strain at crack tips

The singular concentration of stress at tips of very sharp cracks in elastic solids as shown in fig. 9 in modes of loading I, II, and III by boundary tractions σ_{11}, σ_{12}, σ_{13} is well documented (see e.g. [14]). Detailed solutions also exist for discontinuities that are less severe and of non-singular nature (see e.g. Neuber [47]). When the solid can undergo plastic flow, such flow is initiated from the tips of cracks and spreads out with increasing applied stress. Analytical solutions of elastic-plastic crack tip problems for elastic, non-hardening plastic continua have been obtained for the longitudinal shear mode (Mode III) [48,49]. For the more interesting Modes I and II of loading only restricted solutions exist: for yielding to be confined only to the plane of the crack [50,51]; for fully plastic situations in tension and bending (see e.g. McClintock [52] and Knott [53]), including crack tip blunting [54]; for a material obeying a Ramberg-Osgood type of power law hardening ($\bar{\varepsilon} = (\bar{\sigma}/\sigma_o)^n$) in deformation theory, for Modes I and II, by Hutchinson [55,56] and Shih [57]. The sets of solutions due to Hutchinson show that considering a material with power law strain hardening as described above, the singular concentration of stress and strain at very sharp cracks can be given by

$$\sigma_{ij}(r, \theta) = \bar{\sigma}_o \left(\frac{J}{\bar{\sigma}_o I_n} \right)^{\frac{1}{1+n}} \frac{1}{r^{\frac{1}{1+n}}} \tilde{\sigma}_{ij}(\theta) \qquad (3.3)$$

$$\varepsilon_{ij}(r, \theta) = \left(\frac{J}{\bar{\sigma}_o I_n} \right)^{\frac{n}{1+n}} \frac{1}{r^{\frac{n}{1+n}}} \tilde{\varepsilon}_{ij}(\theta) \qquad (3.4)$$

where: the coordinate system refers to that given in fig. 9, I_n are tabulated numerical constants [56,52], the θ dependent functions, normalized with the equivalent stress, are available in graphical form [55,56,52]. In eqns. (3.3) and (3.4) J is the so-called path independent energy-momentum integral first derived by Eshelby [58] as a tensor giving the generalized force on a crack (or heterogeneity) in a non-linear material due to any set of traction producing agencies external to and acting across a closed surface constructed around the crack – it is a function of the level of external stress and sets the scale of the local stresses. It has been re-derived independently by Rice [59] who has pointed out its utility as a critical fracture criterion to be discussed below. Equations (3.3) and (3.4) demonstrate an important fact that the product of the singular terms in corresponding stress and strain elements always goes as $1/r$ regardless of the exponent of the constitutive equation, and that as the exponent increases from $n = 1$ to $n = \infty$, representing a whole range of behavior, from elastic to non-hardening plastic, the singularity in the stress elements decreases in strength, while the opposite occurs for the strain elements.

The value of stress and strain distributions around cracks is of fundamental importance in the operational understanding of fracture. Our treatment here has touched only on some of the most important features. For more detailed discussions the reader is referred to some of the more informative references [47, 49, 52, 53] for further study.

3.3 Fracture by brittle cleavage

3.3.1 Brittle fracture instability. When a crack of length $2c$ is introduced in a very wide elastic plate under a tensile stress σ its free enthalpy (Gibbs free energy) will undergo a change.

$$\Delta G = 4\alpha c - \frac{\pi c^2 \sigma^2}{E} \tag{3.5}$$

where the first term represents the increase in surface free energy and the second the additional work done by the tensile stress due to the increased compliance of the plate, α is the surface free energy, and E the Young's modulus. Cracks below a certain critical size c^* will increase the free enthalpy while those above c^* will decrease it. The free enthalpy will have a maximum value at $c = c^*$ where the system of cracked plate and externally applied tractions are in a state of unstable equilibrium, i.e.

$$\left(\frac{\partial \Delta G}{\partial c}\right)_\sigma = 0 \quad \text{at} \quad c = c^* = \frac{2}{\pi} \frac{\alpha E}{\sigma^2} \tag{3.6}$$

Equation (3.6) gives the stress which will make any existing crack of length $2c$ propagate [60].

$$\sigma = \sqrt{\frac{2\alpha E}{\pi c}} \tag{3.7}$$

If brittle fracture under a stress σ were to be initiated in a perfect, crack free plate by thermal fluctuations, the latter would have to produce a crack of half length given by eqn. (3.6) for which an activation energy per unit thickness can be calculated

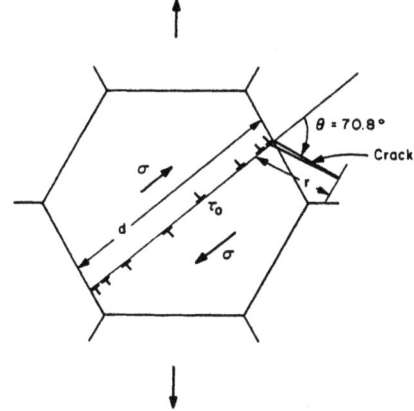

Fig. 10. Nucleation of a crack from a pile-up of dislocations.

by substituting eqn. (3.6) into (3.5) to give

$$\Delta G^* = \frac{4}{\pi} \frac{\alpha^2 E}{\sigma^2} \simeq \frac{1}{(4\pi)^2} (Ea^2)\left(\frac{E}{\sigma}\right)^2 \tag{3.8}$$

In eqn. (3.8) the second equality has resulted from a consideration that in most strong solids $\alpha \simeq Ea/8\pi$, where a is the lattice parameter. Hence for a stress even as high as $\sigma/E \simeq 0.01$ an activation energy of about 10^3 eV would be required to produce brittle fracture in a plate consisting of only one atomic layer. This indicates that in a homogeneously stressed solid brittle fracture either results from pre-existing cracks or from cracks that are produced by purely mechanical processes involving inhomogeneous strains.

3.3.2 Nucleation of cracks by inhomogeneous deformation.
Locally very large stresses can be generated by inhomogeneous slip or twinning in semi-brittle polycrystalline materials, that could nucleate cracks. Consider e.g., as shown in fig. 10, a slip band (the same considerations would hold for a twin band) in a grain of size d, of a semi-brittle material in which slip is hard to nucleate but relatively easy to propagate at a shear stress τ_o. When such a slip band is arrested at a grain boundary, stresses will be concentrated there as in the case of the crack under Mode II loading. The tensile stress at the end of the slip band is maximum across a plane making an angle of 70.8° with the extension of the slip plane, and is

$$\sigma_{\theta\theta} = \frac{1}{\sqrt{3}} \sqrt{\frac{d}{r}} (\sigma - \tau_o) \qquad (3.9)$$

Stroh [61] showed that a crack can nucleate along a length r of this plane of maximum tensile stress for which the stress in eqn. (3.9) is the crack propagation stress given in eqn. (3.7) for c = r/2 . This gives a stress condition for nucleating a crack of dimensions d as

$$\sigma = \tau_o + \sqrt{\frac{12 \alpha E}{\pi d}} \qquad (3.10)$$

provided that the shear stress concentration at the boundary does not initiate plastic flow in the neighboring grain, i.e. the material be semi-brittle. The various microstructural aspects of this problem have recently been reviewed by Averbach [62].

In semi-brittle materials with limited ductility, catastrophic fracture would require that eqn. (3.7) be satisfied where, however, α may involve considerable additional plastic work of tearing at grain boundaries. When the material is more ductile, the fracture criterion may still resemble the form given in eqn. (3.7) where, however, the terms will have drastically different meaning. We will discuss such fracture criteria below in Section 3.4.5, after we discuss the process of ductile fracture immediately below.

3.4 Ductile fracture and ductile rupture

3.4.1 Ductile rupture. Inherently ductile materials, when perfectly free of heterogeneities, will fail in tension by localization of deformation in a tensile neck soon after a load maximum has been reached [63]. The part will then literally flow apart by coming to a point. The shapes of the neck can often be calculated [14], and give the contribution of the inhomogeneous deformation to overall extension. Such localization of deformation into ductile rupture is of interest to the metal forming industry. It is of interest to us only as a process that can occur on a microscopic scale in ductile fracture, as we will discuss below.

3.4.2 Cavity formation in ductile fracture. The above ideal process of large scale rupture usually does not occur in structural metals, hardened by second phase particles, and often containing undesirable inclusions. They undergo ductile fracture which consists of formation of internal cavities from second phase particles or inclusions followed by their growth and linking as the sample is deformed further. Cavity formation occurs by either internal fracture or interface decohesion at second phase particles under the action of the rising interface tractions produced by

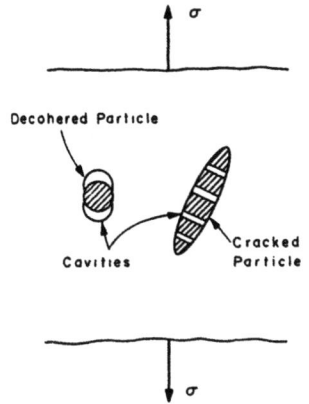

Fig. 11. Cavity formation from equiaxed and elongated non-deformable particles in a deforming ductile matrix.

the plastically deforming medium around the particles as shown in fig. 11. Recent experimental and theoretical studies [27, 64, 65] have established that in well processed structural metals cavity formation occurs only when the interface tractions reach values of the order of 0.01 E due to inhomogeneous plastic flow and strain hardening around the particles as discussed earlier in Section 2.2.3. The tractions rise more rapidly with increasing plastic strain on particles with large aspect ratio, and in plastic media with rapid strain hardening. Table 3.2 gives the computed stress concentrations on interfaces of plastic matrices and rigid particles [65] which have been experimentally verified to some degree [64].

Table 3.2 Maximum Tensile Interface Tractions t_n (at points A) on Rigid Particles in Plastic Non-Hardening and Linearly Hardening Matrices with Shear Resistance τ, Deformed in Pure Shear

Type of Particle	Non-Hardening Plastic Matrix	Linearly Hardening Matrix
A⊘A ←p→ ↑p	$t_n = \frac{3}{2} p$ $(p = \tau)$	$t_n = 2p$ $(p = \tau)$
2h ↓p ←A▬▬▬A→ ⊢ 2ℓ ⊣ ↑p	$t_n = p$ $(p = \tau)$	$t_n = \frac{3}{2} p \left[1 + \frac{2\pi(\ell/h)}{3[\ell n(\ell/h) - 0.1477]} \right]$ $(p = \tau)$

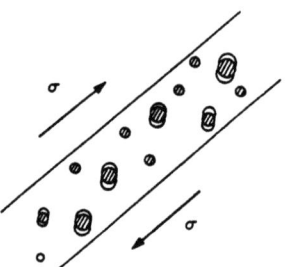

Fig. 12. Localization of deformation in a shear band due to cavity formation.

3.4.3 Cavity growth and localization at low temperature.
Except in the case of very elongated plate-like particles which undergo early internal fracturing, the decohesion of equiaxed particles occurs only after substantial levels of plastic strain of the order of unity, where the strain hardening rate of the matrix will have fallen to low levels. In such instances deformation can localize into cavitational bands almost as soon as cavities form. Consider as shown in fig. 12 a band in which a volume fraction c of cavities have formed. The total shear resistance which this band can offer is

$$\sigma = (1 - c)\tau \tag{3.11}$$

For the deformation to localize immediately upon cavity formation, the change in shear traction with increasing shear strain in the band must be zero, i.e.

$$d\sigma = (1 - c) \frac{d\tau}{d\gamma} d\gamma - \tau \frac{dc}{d\gamma} d\gamma = 0 \tag{3.12}$$

This requires that

$$\frac{dc}{d\gamma} \geq \frac{1}{n\gamma_f} \tag{3.13}$$

where n is the non-linear strain hardening exponent introduced in Section 3.2 and γ_f is the plastic strain at which the cavity formation begins. Generally, $n \approx 10$, $\gamma_f \approx 1$, requiring that $dc/d\gamma \geq 0.1$, which is readily possible.

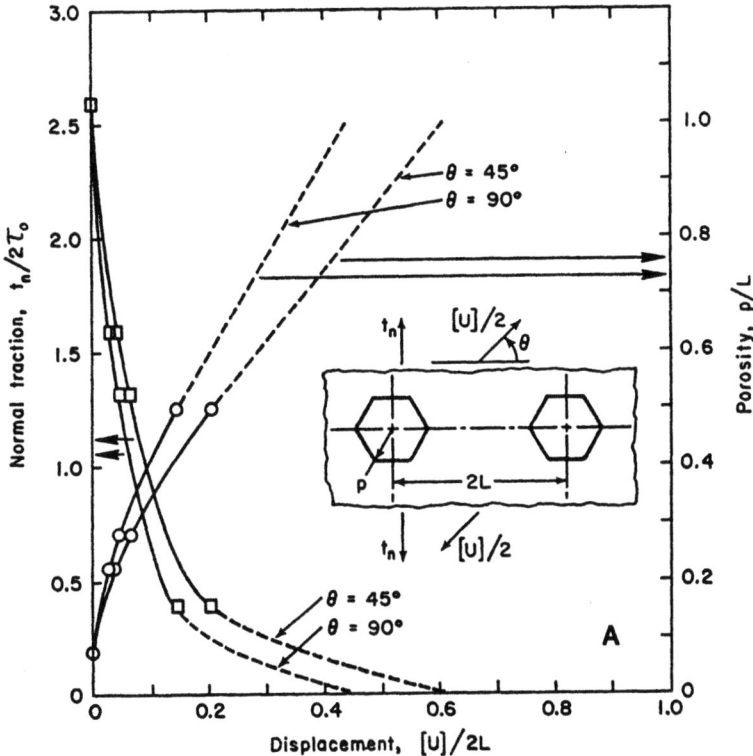

Fig. 13. a) Decrease of normal traction with plastic displacement due to plastic cavity growth and increase of porosity. b) decreased plastic work for ductile separation with increase in initial porosity (from Nagpal, et al. [66], courtesy of Noordhoff).

Once the holes have nucleated with or without accompanying strain localization, the holes can grow, primarily under the action of the negative pressure component of the stress field. Nagpal, et al. [66] have considered the interaction and growth by plastic flow of rows of cavities of various shapes situated in a plane, by obtaining the slip line field between them. Figure 13a shows the decreasing normal component of the traction across the plane and increasing porosity with increasing displacement applied at the distant boundaries, while fig. 13b shows the reduction of the total specific plastic work to fracture (by rupture of the ligaments between the holes) as a function of the initial porosity.

Ductile fracture by localized flow between voids, particularly in the form of cavitational shear bands, is found frequently in high strength structural alloys where the feature is commonly

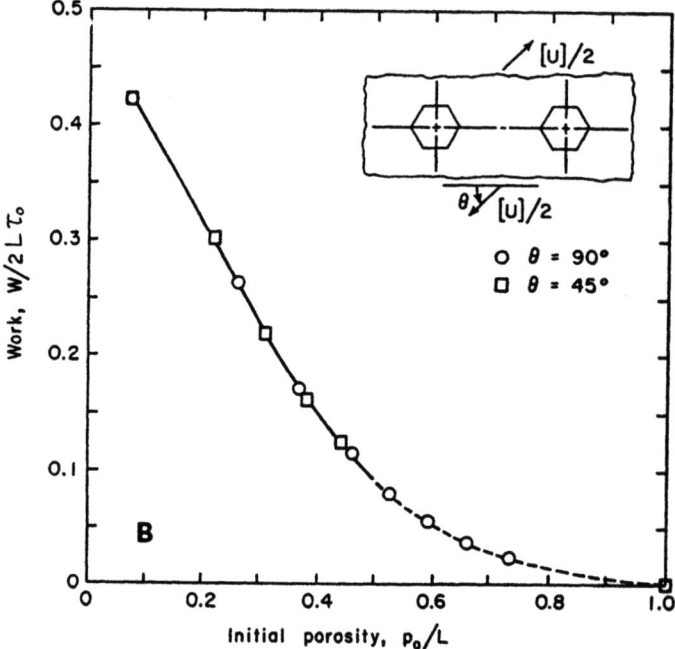

Figure 13b.

known as a void sheet [67].

3.4.4 Cavity formation and growth in creeping materials.
At elevated temperatures, as discussed earlier in Section 2.1.2, creep in polycrystals involves some grain boundary sliding. As shown in fig. 14, second phase particles along a boundary inhibit the sliding along a grain boundary and produce large local stress concentrations on the particle surface [68] that can result in cavity formation at the interface. Once such cavities form, creep deformation can localize between the holes and produce creep rupture. Thus, in a material undergoing power-law creep according to Norton's law ($d\bar{\varepsilon}/dt = (\bar{\sigma}/\sigma_o)^n$) an approximate expression for the rate of increase of porosity can be derived by considering a series of touching cylindrical regions about each hole, expanding under the negative pressure component of the normal traction t_n. This gives (details can be furnished upon request)

$$\frac{d\beta}{dt} = \frac{\beta(1-\beta^2)}{2 \cdot 3^{(n-1)/2}} \left[\frac{t_n/\bar{\sigma}_o}{n(1-\beta^{2/n})} \right]^n \qquad (3.14)$$

where $\beta = P/L$ is the porosity in the plane. Integration of eqn. (3.14) gives the total time to rupture under constant traction.

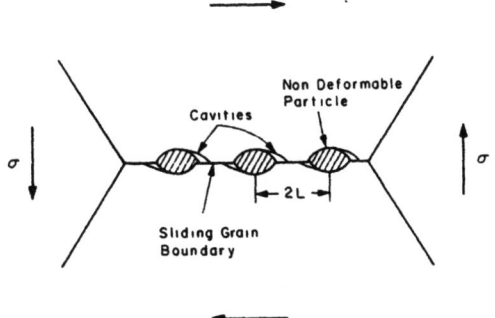

Fig. 14. Formation of cavities at interfaces of second phase particles on a sliding grain boundary at elevated temperature.

A somewhat different type of local creep rupture can occur under lower stresses by diffusional matter flow alone. In this case the pores assume equilibrium shapes as shown in fig. 15. Vacancies are generated along the grain boundary between pores. A vacancy concentration gradient is established to smooth out the otherwise high stress concentrations at the tips of the pores, and a quasi-steady vacancy flux is established between the regions between the pores and the pore surfaces that makes the pores grow as they retain their shape. It is readily shown by elementary arguments that the rate of increase of porosity in the plane is given by

$$\frac{d\beta}{dt} = \frac{\Omega \delta D t_n}{L^3 \beta (1 - \beta) kT \sin \theta} \tag{3.15}$$

which can be integrated immediately to give the total rupture time t as

$$t = \frac{L^3 kT \sin \theta}{6 \Omega \delta D t_n} \tag{3.16}$$

where the symbols were all defined in Section 2.1.2, and fig. 15.

3.4.5 Instability in ductile fracture. As discussed in Section 3.2, when plastic flow occurs with or without strain hardening at crack tips, the strength of the stress concentration is reduced while that of the strain is increased. In the non-hardening idealization a plastic zone of size R develops in front of

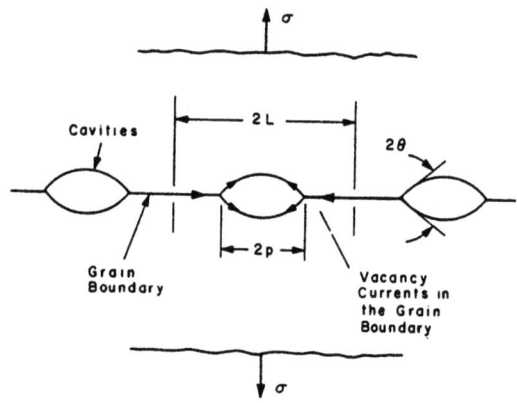

Fig. 15. Diffusional growth of cavities by grain boundary and surface currents of vacancies.

the crack which by analogy with the case of Mode III loading is (for $\sigma < 0.3\,Y$, i.e. for so-called small scale yielding [49])

$$R \simeq c\left(\frac{\sigma}{Y}\right)^2 = \frac{1}{\pi}\frac{K_I^2}{Y^2} \tag{3.17}$$

where $K_I = \sigma\sqrt{\pi c}$ is the so-called Mode I stress intensity factor which sets the scale of the stress distribution in linear elastic crack tip problems. Since singularity in stress at the crack tip is completely eliminated in the non-hardening case, the strain singularity becomes very strong and goes as $1/r$ to give an equivalent strain distribution inside the plastic zone (by analogy with the Mode III problem)

$$\frac{\bar{\varepsilon}}{\bar{\varepsilon}_y} \simeq \frac{R}{r} \tag{3.18}$$

If fracture is to occur by a mode of cavity nucleation and plastic micro-rupture of ligaments between the holes, several mechanistic fracture criteria can be stated.

An "upper-bound" criterion (on toughness) of homogeneous plastic cavitation due to McClintock [14] states that the local equivalent strain to fracture is given by

$$\bar{\varepsilon}_f = \left(\frac{n-1}{3n}\right) \frac{\ln(\beta_o)}{\sinh\left[\frac{(n-1)}{n} \frac{\sqrt{3}\,\sigma_T}{Y}\right]} \qquad (3.19)$$

where n is the strain hardening exponent as previously defined (note that $(n-1)/n = 1$ for a non-hardening material), β_o the initial volume porosity (formed by decohesion of second phase particles at a mean spacing 2L) and σ_T the negative pressure component of the local stress field. This would lead to a stress condition for fracture based on the attainment of $\bar{\varepsilon}_f$ in a region $r \approx L$ and would give by the use of eqns. (3.17-19)

$$K_{IC} = \sqrt{\pi\,\bar{\varepsilon}_f\,E\,Y\,L} \quad , \text{ for } (\sigma < 0.3\,Y) \qquad (3.20a)$$

$$\sigma = \sqrt{\frac{\bar{\varepsilon}_f\,E\,Y\,L}{c}} \qquad (3.20b)$$

where K_{IC} is the critical stress intensity factor for fracture instability(*).

Another local fracture criterion is a critical crack tip opening displacement (CTOD), as shown in fig. 16, that derives from the model of micro-rupture between a series of holes. As is clear from fig. 13a the critical plastic displacement CTOD for complete separation is very nearly equal to the mean center-to-center spacing between the holes of the initial porosity, i.e.

$$\text{CTOD} = 2L \qquad (3.21)$$

Since the plastic crack tip opening displacement (by analogy with the Mode III problem) is

$$\text{CTOD} = \varepsilon_y R \qquad (3.22)$$

(*)The conditions given in eqns. (3.20a-b) are only asymptotically correct for marginally ductile, high strength materials where they are conditions for final fracture instability. In more ductile materials these conditions are only correct for <u>initiation</u> of the fracture process at the crack tip. Final instability can often be significantly delayed [14].

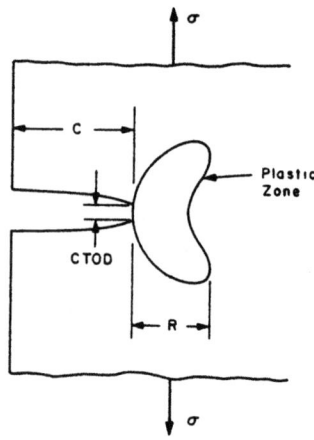

Fig. 16. Tensile crack with plastic zone in Mode I loading.

one can state another stress condition for ductile fracture as

$$K_{IC} = \sqrt{2\pi Y E L} \quad , \quad \text{for } (\sigma < 0.3 Y) \tag{3.23a}$$

$$\sigma = \sqrt{\frac{2 Y E L}{c}} \tag{3.23b}$$

When the material is very ductile and the small scale yielding restriction of $\sigma < 0.3 Y$ is no longer obeyed, criteria such as (3.20a) and (3.23a) may still be operationally useful but are, strictly speaking, no longer applicable. Another criterion that shows some promise in this range of larger scale yielding is one based on a critical value of the so called J integral discussed in Section 3.2 (as long as the deformation is monotonic). It will be recalled that, by its definition, the J integral represents the generalized force due to all mechanical agencies (i.e. external tractions, internal stresses, heterogeneities, etc.) external to the crack that tend to make it propagate. Hence, it is natural to think of a critical value of a J integral J_c which propagates the crack to fracture. The operational details of how this criterion is actually to be applied is beyond the scope of this chapter. The interested reader is referred to Knott [53] for such a discussion.

Monotonic fracture at elevated temperatures can be described along similar lines by utilizing the local fracture criteria given in eqns. (3.14-3.16) by re-casting them into local strain or displacement criteria.

3.5 Fatigue

Crystalline solids capable of deforming plastically at least on a local scale can undergo fracture by nucleation and progressive growth of cracks under cyclic stresses at levels less than general yield. Vast numbers of studies over the past century have established that under cyclic stressing some slip bands develop a high degree of reversibility of deformation which produces a certain unique dislocation structure in them and produces progressive surface roughening or grooving. Fatigue cracks eventually nucleate from such <u>persistent slip bands</u> in a manner which is still not thoroughly understood (for reviews see [69, 70, 28]). In structural alloys fatigue cracks often nucleate from interfaces of second phase particles [70]. At elevated temperatures in everything but the hardest vacuum ($\sim 10^{-10}$ torr) sliding grain boundaries take the place of persistent slip bands [71]. In the vast majority of cases fatigue cracks nucleate from free surfaces and are sensitive to a great many surface related effects (see [28]).

Once cracks have nucleated inside persistent slip bands they may continue to grow for a while in the plane of the band. Such growth along planes of maximum shear is labelled Stage I crack growth. The growth, however, soon switches to a plane normal to the principal tensile direction and continues until final fracture. This phase of growth is labelled Stage II crack growth, which appears to have two fundamentally distinct mechanisms. The basic mechanism of fatigue crack growth in ductile metals is by accumulation of new crack tip surfaces produced by the opening mode of each cycle. The length s of new surface created at the crack tip in the opening mode of a cycle is of the order of the cyclic crack tip opening displacement, which is

$$s \simeq (CTOD)_{CYCLIC} = \varepsilon_y R_{CYCLIC} \qquad (3.24)$$

$$R_{CYCLIC} = \frac{1}{\pi} \frac{(\Delta K_I)^2}{(2Y)^2} \qquad (3.25)$$

where ΔK_I is the range of the cyclic stress intensity factor and the factor 2 multiplying the yield stress results from the need to fully reverse plastic flow at a crack tip which makes the local material act as if it had a yield stress of 2Y (per cyclic reversal). Since the newly generated surface length s will either oxidize or at best, in a high vacuum, is only partially reversible due to the microscopically inhomogeneous nature of plastic flow, it is reasonable to expect that the crack growth per cycle will be a certain fraction α of the length of the new surface exposed. This gives an ideally plastic crack growth rate of

$$\frac{dc}{dn} = \alpha s = \frac{\alpha}{4\pi} \frac{(\Delta K_I)^2}{EY} \qquad (3.26)$$

that is obeyed by most ductile metals at an intermediate stress range where the concept of a stress intensity factor applies. When the crack tip opening displacement given by eqns. (3.24) and (3.25) becomes of the order of the plasticity "point size" discussed in Section 2.2.1 the crack growth rate decreases more sharply with decreasing ΔK_I as given by eqn. (3.26) and results in a threshold stress intensity factor below which the growth ceases for all practical purposes.

In semi-brittle materials which can undergo cleavage (most bcc metals near or below their so called ductile-brittle transition temperature [14]) fatigue crack growth rates can accelerate to levels above that given by eqn. (3.26) by large cleavage jumps during each cycle [72]. When the fatigue crack becomes long enough to satisfy one of the monotonic ductile fracture criteria given by eqns. (3.20a,b) or (3.23a,b) final fracture will result.

The growth of fatigue cracks by either of these two mechanisms just discussed is sensitive to environmental effects that produce oxidation, intergranular embrittlement, lowering of cleavage strength, etc.. Such effects and the many other subtleties of fatigue crack nucleation and growth are again outside the scope of this chapter. The interested reader is referred to reviews on the subject [73, 74].

3.6 Other modes of fracture

The types of fracture discussed above are relatively ideal and straightforward. There are other and more complex modes of fracture of great technological importance. Corrosion fatigue is where corrosion and fatigue interact destructively to accelerate fracture beyond what a linear additivity of the two effects may predict [75]. Stress corrosion cracking or static fatigue are time dependent fracture processes involving slow crack growth aided by environmental effects, all under constant stress [76, 77]. Liquid metal embrittlement generally produces intergranular time dependent crack growth under constant stress in the presence of some liquid alkali metals [78]. Fracture by hydrogen embrittlement [79] affects particularly high strength steels. Many of these phenomena will be discussed in detail by others in this volume, permitting us not to go any further. For more detailed additional treatment of these subjects the reader is referred to the many special chapters in the multi-volume Treatise on Fracture [74].

3.7 Variability and statistical effects in fracture

Since fracture results in association with a microstructural feature such as cracks, grain boundaries, second phase particles, all of which have certain distributions in size, shape, spacing, etc., there is an inherent variability in the fracture of solids. This variability is strongest when plasticity is limited. Hence, the statistical theories of brittle fracture are well developed [14, 80]. There are, however, other statistical considerations in the fracture of semi-brittle materials and quasi-homogeneous materials fracturing by ductile fracture, fatigue and heterogeneous materials such as composites [81]. Once again, even a superficial discussion of these effects is outside the scope of this chapter. The interested reader is referred to at least the references cited above.

3.8 Tests to characterize fracture behavior

The standard tensile test still remains as one of the most widely used tests to characterize fracture behavior of engineering solids. The fracture toughness, fatigue behavior, and other fracture modes under static loading conditions require separate and specially designed tests, many of which are becoming standards and are being described by the ASTM [43]. For a discussion of some of the more widely used special tests the reader is referred to treatises by Knott [53] and Kobayashi [82].

REFERENCES

1. Kelly, A., Strong Solids, Clarendon Press, Oxford, 1966.
2. Christian, J. W., The Nucleation of Mechanical Twins and of Martensite, in Physics of Strength and Plasticity, Argon, A. S., Ed., M.I.T. Press, Cambridge, Mass., 1969, p. 85.
3. Kocks, U. F., Phil. Mag., 13, 541, 1966.
4. Kocks, U. F., Can. J. Phys., 45, 737, 1967.
5. Argon, A. S., Phil. Mag., 25, 1053, 1972.
6. Kocks, U. F., Argon, A. S., and Ashby, M. F., Thermodynamics and Kinetics of Slip, in Progress in Materials Science, Chalmers, B., Christian, J. W., and Massalski, T. B., Eds., Pergamon Press, Oxford, vol. 19, 1975, p. 1.
7. Ashby, M. F. and Frost, H. J., The Kinetics of Inelastic Deformation Above 0°K, in Constitutive Equations in Plasticity, Argon, A. S., Ed., M.I.T. Press, Cambridge, Mass., 1975, p. 117.
8. Argon, A. S., Physical Basis of Constitutive Equations for Inelastic Deformation, in Constitutive Equations in Plasticity, Argon, A. S., Ed., M.I.T. Press, Cambridge, Mass., 1975, p. 1.

9. Bird, J. E., Mukherjee, A. K., and Dorn, J. E., Correlations Between High Temperature Creep Behavior and Structure, in *Quantitative Relation between Properties and Microstructure*, Brandon, D. G. and Rosen, A., Eds., Israel University Press, Haifa, 1969, p. 255.
10. Takeuchi, S. and Argon, A. S., *Acta Met.*, submitted.
11. Nabarro, F. R. N., in *Reports of a Conference on the Strength of Solids*, Physical Society, London, 1948, p. 75.
12. Coble, R. L., *J. Appl. Phys.*, 34, 1679, 1963.
13. Raj, R., and Ashby, M. F., *Met. Trans.*, 2, 1113, 1971.
14. McClintock, F. A. and Argon, A. S., *Mechanical Behavior of Materials*, Addison-Wesley, Reading, Mass., 1966.
15. Avery, D. H., and Backofen, W. A., "A Structural Basis for Superplasticity" *ASM Trans. Quart.*, 58, 551 (1965)
16. Ashby, M. F. and Verrall, R. A., *Acta Met.*, 21, 149, 1973.
17. Mader, S., *Z. Physik*, 149, 73, 1957.
18. Taylor, G. I., *J. Inst. Metals*, 62, 307, 1938.
19. Bishop, J. F. W., and Hill, R., *Phil. Mag.*, 42, 414, 1951; 42, 1298, 1951.
20. Kocks, U. F., *Met. Trans.*, 1, 1121, 1970.
21. Hutchinson, J. W., Bounds and Self-Consistent Estimates for Creep of Polycrystalline Materials, Harvard University Report, DEAP S-13, 1975.
22. Hill, R., *J. Mech. Phys. Solids*, 13, 89, 1965.
23. Ashby, M. F., The Deformation of Plastically Non-Homogeneous Alloys, in *Strengthening Methods in Crystals*, Kelly, A. and Nicholson, R. B., Eds., Halsted Press, New York, 1971, p. 137.
24. Orowan, E., in *Symposium on Internal Stresses in Metals and Alloys*, (Discussion), Institute of Metals, London, p. 451, 1948.
25. Hutchinson, J. W., *Proc. Roy. Soc.*, (London), A319, 247, 1970.
26. Ashby, M. F., *Phil. Mag.*, 14, 1157, 1966.
27. Argon, A. S., Im, J., and Safoglu, R., *Met. Trans.*, 6A, 825, 1975.
28. Argon, A. S., Effects of Surfaces on Fatigue Crack Initiation, in *Corrosion Fatigue*, Devereux, O. F., McEvily, A. J., and Staehle, R. W., NACE, Houston, Texas, 1972, p. 176.
29. Fourie, J. T., *Phil. Mag.*, 17, 735, 1968.
30. Kramer, I. and Demer, L. J., Effects of Environment on Mechanical Properties of Metals, in *Progress in Materials Science*, Chalmers, B., Ed., Pergamon Press, Oxford, 1961, vol. 9, p. 131.
31. Andrade, E. N. da C., and Henderson, C., *Phil. Trans. Roy. Soc.*, (London), A244, 177, 1951.
32. Rosi, F. D., *Acta Met.*, 5, 348, 1957.
33. Hauser, J. J. and Chalmers, B., *Acta Met.*, 9, 802, 1961.
34. Bilby, B. A., Cottrell, A. H., Smith, E., and Swinden, K. H., *Proc. Roy. Soc.*, (London), A279, 1, 1964.
35. von Mises, R., *Goettinger Nachr., Math-phys. Klasse*, 1913, p. 582.

36. Hill, R., *Plasticity*, Clarendon Press, Oxford, 1950.
37. Chin, G. Y., Development of Deformation Textures, in *Constitutive Equations in Plasticity*, Argon, A. S., Ed., M.I.T. Press, Cambridge, Mass., 1975, p. 431.
38. Argon, A. S., Im, J., and Needleman, A., *Met. Trans.*, 6A, 815, 1975.
39. Hart, E. W., Li, C. Y., Yamada, H., and Wire, G. L., Phenomenological Theory: A Guide to Constitutive Relations and Fundamental Deformation Properties, in *Constitutive Equations in Plasticity*, Argon, A. S., Ed., M.I.T. Press, Cambridge, Mass., 1975, p. 149.
40. Hart, E. W., *Acta Met.*, 18, 599, 1970.
41. Tabor, D., *Proc. Roy. Soc.*, (London), A192, 247, 1948.
42. Marsh, D. M., *Proc. Roy. Soc.*, (London), A279, 420, 1964.
43. ASTM Standards on Testing Engineering Materials, (1944 Standards and Various Later Supplements), ASTM, Philadelphia.
44. Kelly, A., Tyson, W. R., and Cottrell, A. H., *Phil. Mag.*, 15, 567, 1967.
45. Rice, J. R. and Thomson, R., *Phil. Mag.*, 29, 73, 1974.
46. Gehlen, P. C., Hahn, G. T., and Kanninen, M. F., Atomic Crack Simulation in Fracture Mechanics, in *Proc. 3rd Intern. Fracture Conf.*, (Munich), vol. 2, *Fracture Theory*, Verein Deutscher Eisenhüttenleute, Düsseldorf, 1973, p. I-243.
47. Neuber, H., *Theory of Notch Stresses*, J. W. Edwards, Ann Arbor, Mich., 1946.
48. McClintock, F. A., *J. Appl. Mech.*, 25, 581, 1958.
49. Rice, J. R., Mathematical Analysis in the Mechanics of Fracture, in *Fracture*, Liebowitz, H., Ed., Academic Press, New York, 1968, vol. 2, p. 191.
50. Dugdale, D. S., *J. Mech. Phys. Solids*, 8, 100, 1960.
51. Bilby, B. A., Cottrell, A. H., and Swinden, K. H., *Proc. Roy. Soc.*, (London), A272, 304, 1963.
52. McClintock, F. A., Plasticity Aspects of Fracture, in *Fracture*, Liebowitz, H., Ed., Academic Press, New York, 1971, vol. 3, p. 48.
53. Knott, J. F., *Fundamentals of Fracture Mechanics*, Halsted Press, New York, 1973.
54. McClintock, F. A., Crack Growth in Fully Plastic Grooved Tensile Specimens, in *Physics of Strength and Plasticity*, Argon, A. S., Ed., M.I.T. Press, Cambridge, Mass., 1969, p. 307.
55. Hutchinson, J. W., *J. Mech. Phys. Solids*, 16, 13, 1968.
56. Hutchinson, J. W., *J. Mech. Phys. Solids*, 16, 337, 1968.
57. Shih, C. F., Small-Scale Yielding Analysis of Mixed Mode Plane-Strain Crack Problems, in *Fracture Analysis, ASTM STP 560*, ASTM, Philadelphia, 1974, p. 187.
58. Eshelby, J. D., The Continuum Theory of Lattice Defects, in *Solid State Physics*, Seitz, F., and Turnbull, D., Eds., Academic Press, New York, 1956, vol. 3, p. 79.

59. Rice, J. R., *J. Appl. Mech.*, 35, 379, 1968.
60. Griffith, A. A., *Phil. Trans. Roy. Soc.*, (London), A221, 163, 1920.
61. Stroh, A. N., *Proc. Roy. Soc.*, (London), A223, 404, 1954.
62. Averbach, B. L., Some Physical Aspects of Fracture, in *Fracture*, Liebowitz, H., Ed., Academic Press, New York, 1968, vol. 1, p. 441.
63. Argon, A. S., Stability of Plastic Deformation, in *The Inhomogeneity of Plastic Deformation*, Reed-Hill, R. E., Ed., ASM, Metals Park, Ohio, 1973, p. 161.
64. Argon, A. S., and Im, J., *Met. Trans.*, 6A, 839, 1975.
65. Argon, A. S., *J. Eng. Mat. Technology*, in the press.
66. Nagpal, V., McClintock, F. A., Berg, C. A., and Subudhi, M., Traction Displacement Boundary Conditions for Plastic Fracture by Hole Growth in *Foundations of Plasticity*, Sawczuk, A., Ed., Noordhoff International, Groningen, 1972, p. 365.
67. Rogers, H. C., *Trans. Met. Soc. AIME*, 218, 498, 1960.
68. Lau, C. W. and Argon, A. S., work in progress, to be published.
69. Laird, C., and Duquette, D. J., Mechanisms of Fatigue Crack Nucleation, in *Corrosion Fatigue*, Devereux, O. F., McEvily, A. J., and Staehle, R. W., Eds., NACE, Houston, Texas, 1972, p. 88.
70. Grosskreutz, J. C., Fundamental Knowledge of Fatigue Fracture, in *Proc. 3rd Intern. Fracture Conf.*, (Munich), vol. 1, *Plenary Lectures*, 1973, p. PL V-212.
71. Argon, A. S., Turner, A. P. L., and Ucisik, H., work in progress, to be published.
72. Wright, R. N., and Argon, A. S., *Met. Trans.*, 1, 3065, 1970.
73. Plumbridge, W. J., *J. Mater. Sci.*, 7, 939, 1972.
74. Liebowitz, H., Ed., *Fracture*, Academic Press, New York, vols. 1-7, 1968-1973.
75. Duquette, D. J., A Review of Aqueous Corrosion Fatigue, in *Corrosion Fatigue*, Devereux, O. F., McEvily, A. J., and Staehle, R. W., Eds., NACE, Houston, Texas, 1972, p. 12.
76. Uhlig, H. H., Stress Corrosion Cracking, in *Fracture*, Liebowitz, H., Ed., Academic Press, New York, 1971, vol. 3, p. 646.
77. Wiederhorn, S. M., Environmental Stress Corrosion Cracking in Glass, in *Corrosion Fatigue*, Devereux, O. F., McEvily, A. J., and Staehle, R. W., Eds., NACE, Houston, Texas, 1972, p. 731.
78. Westwood, A. R. C., Preece, C. M., and Kamdar, M. H., Adsorption-Induced Brittle Fracture in Liquid Metal Environments, in *Fracture*, Liebowitz, H., Ed., Academic Press, New York, 1971, vol. 3, p. 590.
79. Johnson, H. H., Environmental Cracking in High-Strength Materials, in *Fracture*, Liebowitz, H., Ed., Academic Press, New York, 1971, vol. 3, p. 680.
80. McClintock, F. A., Statistics of Brittle Fracture, in

Fracture Mechanics of Ceramics I, Concepts, Flaws, and Fractography, Bradt, R. C., Hasselman, D. P. H., and Lange, F. F., Eds., Plenum Press, New York, 1974, p. 93.

81. Argon, A. S., Statistical Aspects of Fracture, in Fracture and Fatigue, Broutman, L. J., Ed., Academic Press, New York, 1974, vol. 5 (of series entitled Composite Materials) p. 153.

82. Kobayashi, A. S., Experimental Techniques in Fracture Mechanics, Iowa State University Press, Ames, Iowa, and SESA, Westport, Conn., 1973, p. 1.

DISCUSSION

Comment by H.G.F. Wilsdorf:

It might be of interest to this group to hear about another void initiation mechanism in relation to ductile fracture which was found to operate in high purity FCC metals and alloys, and which does not require the presence of second phase particles.

A comparison of inter-void spacings with inter-particle spacings made on 304 austenitic stainless steel showed that the average void density was 100 times larger than the average particle density. The inter-void spacings decrease with increasing fracture stress which indicates the operation of a dislocation-vacancy mechanism [1]. In situ fracture experiments on silver crystals inside a high voltage electron microscope [2] suggest that the initiation of voids may occur by pseudo-cleavage. Video tape recordings have shown that slits parallel to <110> and approximately 0.1 μm long form within 1/15 sec. about 1 μm in front of the propagating crack. The crystal volume where the hole originates is the most highly workhardened region in the specimen and is under a high triaxial stress. It is thought that under these conditions a supersaturation of vacancies exists which may lead to clusters or micro-void formation; these then could trigger the slit-like microcrack. This phenomenon is reproducible at intermediate elongation rates.

Our results have led us to the conclusion that void initiation can occur in high purity metals and alloys whenever the second phase particle density is below a critical value which will depend on a number of variables.

[1] R.W. Bauer and H.G.F. Wilsdorf, Scripta Met. $\underline{7}$, 1213 (1973).

[2] R.L. Lyles, Fr. and H.G.F. Wilsdorf, Acta Met. $\underline{23}$, 269 (1975).

Reply:

Professor Wilsdorf's observations lend an additional dimension to the ductile fracture process under very high flow stresses and negative pressures. In most commercial alloys ductile fracture is governed by decohesion at interfaces of second phase particles or by their fracture. These phenomena occur at relatively lower local flow stresses and negative pressures than those which must have prevailed in Professor Wilsdorf's experiment. Of course, when the material is almost totally free of second phase particles and if large negative pressures exist in the flow field to prevent plastic rupture by flowing apart, the mechanisms observed by Professor Wilsdorf should govern ductile fracture by cavity formation and linking.

Comment by I. R. Kramer:

It is well known that the crack propagation rate of metals, (Al, Cu, Al alloys, steels, etc.) is decreased when the specimens of these metals are fatigued in vacuum (10^{-1} torr and lower). How is this fact taken into account in your equations?

Reply:

The simple mechanism of fatigue crack growth based on irreversible accumulation of new surface generated during each cycle gives an asymptotic limiting behavior for very ductile metals. In this mechanism the crack growth rate is proportional to the square of the applied stress. In such metals effects of non-corrosive environments are minimal, and the effect of a vacuum, outside of crack flank re-welding effects is also minimal. In other instances when the crack growth rate is proportional to a larger power of stress than two, environmental effects play a role in local interface decohesion, stray cracking at the crack tip, etc. For such materials removal of these environments by a vacuum produces marked reductions in crack growth rate for a variety of special reasons. Reference to these effects has been made in the text of my contribution (75, 76, 77).

Workshop Session 1

SURFACE EFFECTS IN UNIAXIAL TENSION AND FATIGUE

Chairman: H. Mughrabi

Surface Effects in Uniaxial Plastic Deformation

Z.S. Basinski

Division of Physics,
National Research Council of Canada,
Ottawa, Canada, K1A 0R6

Table of Contents

1. Introduction
2. The Deformation Parameters
3. Surface Influence via Crystal Size
4. Comparison of Surface and Bulk Properties
5. Structural Evidence
6. Some Dislocation Models for the Surface
7. The Unloading Yield Point
8. Effect of Surface Removal on the Stress-Strain Curve
9. Surface Stresses
10. Conclusions

1. Introduction

In the present context we shall consider the surface only as a limiting boundary to a finite sample, and try to determine whether any perturbations due to the presence of this bounding surface significantly influence any of the parameters involved in a description of plastic deformation. The physical structure of the surface, its condition (whether for example it is a boundary between the crystal and essentially vacuum, as in Au, or an intermediate layer of different material, as in Al), etc., will be considered only when relevant to the deformation process.

The literature on the influence of the surface region on the deformation of crystalline solids is extensive but far from systematic. Existing experimental data very often cannot be intercompared because different parameters have been controlled (or remained unspecified) during the experiments, and also because of the diversity of materials used. A coherent picture of the influence of the surface is rather elusive. Some controversies have arisen from apparently irreconcilable experimental evidence, for example, although it is generally agreed that the flow stress is not uniform over the cross section of a deformed crystal, many papers present evidence that the surface region is softer than the central part, while others indicate that a harder (debris) layer is formed. In contrast, other apparent disagreements have sprung from diametrically opposing interpretations of the same phenomenon, for example, the unloading yield point has been explained in terms of both the soft surface-hard core and the surface debris layer models.

In view of the amount of literature on the subject, and the number of controversial points which have arisen from it, we shall concentrate on tensile deformation of face centred cubic metals. Primarily we shall examine the reported experimental facts and comment on some of the disagreements. In the light of this, some models and interpretations will be discussed.

2. The Deformation Parameters

Since the nature of the surface is often investigated by measuring how the deformation parameters change with, for example, size, or removal of surface material, it is advisable at the outset to outline the deformation and describe what some of these parameters represent.

The deformation of metal crystals is usually described in terms of a three stage stress-strain curve. Stages I (easy glide) and II (rapid hardening) are described as linear, and stage III as parabolic. The rate of hardening in stage I, θ_I, is of the order of 1 kg/mm² and θ_{II} is of the order of 12 kg/mm² for Cu crystals. The 3-stage curve is a very useful frame of reference, but only a simplification of the real behaviour of a crystal. Figure 1 shows the rate of hardening as a function of resolved shear strain for 3 nominally identical Cu crystals deformed at 4.2°K, 77°K and room temperature. It can be seen from this figure that regions of strictly constant hardening rate, as would be expected from the so-called linear stress-strain stages, do not in fact exist. At all temperatures "stage I" hardening appears as a minimum. After this minimum, the hardening rate increases rapidly (the transition region), and then levels off at approximately 12 kg/mm^2 to give "θ_{II}". This is reasonably well-defined at 4.2°K and even at 77°K, but at room temperature no truly constant region appears, and the curve is best described as a broad maximum.

Most of the experiments on the surface region to be described here were carried out at room temperature, mainly on Cu, Al and Au. Figure 1 shows that definition of θ_I and θ_{II} for Cu at room temperature is going to be a highly subjective process. For Al, room temperature is a much higher temperature (with respect to its melting point), separation into stages is less well-defined and stage III begins even earlier.

Another parameter which has been used widely is ε_I, the extent of stage I. This has been defined both as the strain at which the rate of hardening begins to increase (ε'_I), and as the strain at the point of intersection of extrapolations through stages I and II (ε''_I). Figure 1 shows that neither of these definitions can be interpreted unambiguously. They will in any case lead to differences in value, especially when there is a prolonged transition between stages I and II.

3. Surface Influence via Crystal Size

The mechanical parameter which varies most with size, at lease in relatively thin crystals, is the extent of easy glide, ε_I (see section 2 for discussion of parameters).

Suzuki, Ikeda and Takeuchi (1956) found that for Cu crystals <2 mm in diameter, ε''_I increased from 20% to ∼50% as the diameter decreased from ∼1.8 mm to ∼.2 mm. The data of Garstone, Honeycombe and Greetham (1956) showed that Cu crystals with diameters ranging from ∼10 mm to ∼5 mm gave easy glide of ∼8-15%, indicating that the trend for easy glide length to increase with decreasing crystal diameter continues in these thicker crystals.

For Al crystals the situation is not completely clear. Lücke (1957) reported no dependence of the length of easy glide on size for crystals from 5-.5 mm in diameter, but Suzuki (1957) indicated that a size effect becomes apparent only for diameters <.5mm. However, McKinnon (1959) using Al crystals of identical size (2×7 mm^2 cross section) and similar tensile axis orientation, but with various face orientations so that the glide path length available to primary edge dislocations, L_E, was different in different crystals, showed that the size per se was not important but that L_E determined the length of easy glide. For smaller L_E, ε_I was longer.

In more recent work crystals of particular orientations were usually used. Some of the terms used in describing the geometry are illustrated in Figure 2.

Nakada, Kocks and Chalmers (1964) showed that the amount of easy glide increases with decreasing L_E in Au crystals having L_E from $\sim$3 mm to $\sim$9 mm.

Fourie (1967) carried out a systematic study on thin Cu crystals with L_E $\sim$1.2-.05 mm. The extent of easy glide, ε'_I, increased with decreasing L_E. Fourie also established that the dimension L_S has no influence on ε'_I.

The extent of stage II, ε_{II} has been examined by Nakada, Kocks and Chalmers (1964) for Au crystals. The limited data presented could indicate that ε_{II} is constant with size, but Nakada et al. emphasise that the scatter rules out drawing definite conclusions.

Fourie (1968) reported a variation in the stress extent, $\Delta\sigma_{II}$, and strain extent, $\Delta\varepsilon_{II}$, of stage II in Cu crystals for which L_E varied from .27 mm to 28.6 mm. $\Delta\sigma_{II}$ reached a maximum for $L_E = 5$ mm while $\Delta\varepsilon_{II}$, from the data presented, appears to reach a maximum at $L_E \simeq 2$ mm. Presumably the data have been misprinted since in the absence of corresponding variation in θ_{II} this behaviour is difficult to understand. Both $\Delta\sigma_{II}$ and $\Delta\varepsilon_{II}$ decrease sharply at lower values of L_E and gradually at higher values of L_E. On the basis of these results and other arguments (see Fourie 1968) Fourie concludes that stage II is purely a consequence of the surface regions. Figure 1 shows that for Cu at room temperature the variation in hardening rate with strain in the vicinity of 12 kg/mm^2 (i.e. "Stage II") is sufficiently rapid so that a region of constant (or approximately constant) hardening is not precisely defined, even for these 3 mm thick crystals ($L_E \approx 4.5$ mm) which, according to Fourie's work, should exhibit about the maximum extent of stage II. Since the extent of the approximately linear region increases below room temperature (see Fig. 1), its dependence on crystal size might be more profitably studies at these temperatures.

Another difficulty is that for the wide crystals which become necessary for size dependence experiments, the deformation near to and far from the grips differs appreciably, so that grip effects should be critically studied.

The rates of hardening in stages I and II, θ_I and θ_{II}, have been examined as a function of crystal size in Al, Cu and Au. Slight increases in θ_I with increasing size have been reported for Al (McKinnon, 1959), Au (Nakada et al., 1964) and Cu (Garstone et al., 1956; Suzuki et al., 1958). Fourie (1967) describes θ_I as constant with size for Cu crystals having $L_E <$ 1.2 mm. but his stress-strain curves for larger crystals indicate a possible increase with increasing size (Fourie, 1968). Similarly θ_{II} is reported as increasing slightly with size in Al (McKinnon, 1959) and Cu (Garstone et al., 1958), or as approximately constant with size in Au (Nakada et al., 1964). Fourie's (1967, 1968) data (Figure 3) show that θ_{II} is approximately constant with size except for crystals with L_E smaller than about .2 mm when the hardening rates are much lower. This sharp drop is not surprising since here L_E is only of the order of a few times the mean free path of primary dislocations. The connection with mean free path could be checked by doing similar systematic work at low temperature where "stage II" extends to higher flow stress, and therefore shorter mean free paths, so that the sudden drop in θ_{II} with size would be expected at lower L_E values.

All experiments on the influence of size on the extent of stage I and the rates of hardening indicate that dislocation loss (from E faces) plays an important role. It leads to the delayed onset of stage II and, in the extreme case of crystals with glide path lengths not many times larger than the mean free path, to a decreased rate of dislocation accumulation in stage II.

4. Comparison of Surface and Bulk Properties

In the work on the effect of the surface on mechanical properties, some attention has been paid to the question of whether during plastic deformation, the surface regions are representative of the bulk material, and if not, whether they are softer or harder than the bulk.

Fourie (1968) developed a technique in which a large parent crystal of rectangular cross section ($\sim(2 \times 10)$mm^2) is predeformed and then sectioned into crystal slices whose individual flow stresses represent the flow stress profile across the wide face of the parent crystal. The work was carried out mainly on pure Cu, but the results were found to hold also in Cu 5.8% Al (Fourie and Dent, 1972). Fourie found that when the wide crystal face contained the primary Burgers vector, (i.e. is the cross glide plane, or an S face, see Fig. 2) then after stage I, slices from the surface are softer than those from the central part of the crystal. Figure 4 shows Fourie's flow stress gradients at various prestresses collected onto one figure. These are represented as a function of L_E, i.e., the trace of the primary plane on the cross glide plane. The dashed curve shows the gradient across the face of a crystal with different geometry, in which the wide face was an E face and therefore represents the flow stress profile along L_S, the path length available to primary screw dislocations. The effect here is small compared to that along L_E.

After it is fully developed in stage II, the flow stress gradient moves rigidly up with increasing prestress (Fig. 4), as concluded by Fourie (1967a) from work on much thinner crystals. The flow stress in the core, i.e. the whole cross-sectional area except for approximately 2 mm depth from the E surfaces, is uniform and slightly higher than the macroscopic flow stress.

Murphy (1969) carried out indentation hardness tests across the S faces of large flat Cu crystals such as those used by Fourie, and found results consistent with his. Also, indentation hardness around the perimeter of a cylindrical Cu crystal, indicated that the surface is softer in the region where the

Burgers vector is at right angles to the circumference. In this latter type of test however, the change in orientation around the circumference may play some part in determining the results.

Fourie's direct flow stress measurements thus indicate that during the deformation process, the E surfaces become softer than the core region of the crystal. A flow stress gradient develops, and is important only along the direction L_E.

Evidence for the conflicting view that a harder (debris) layer forms in the surface regions stems largely from experiments in which the surface is continuously removed by electropolishing during deformation (Kramer and Demer 1961, Kramer 1963).

In their experiments on Al, Kramer and Demer (1961) reported that with increasing rate of metal removal, R, the extents of stages I and II increased and the corresponding rates of hardening θ_I and θ_{II} decreased. Furthermore the rates of hardening changed reversibly with R. Also, when R was increased from a lower to a higher value they noted a sudden load drop which they interpreted as elongation due to a dislocation "pop-out" phenomenon. No drastic load changes occurred on suddenly reducing R.

In similar experiments on Au, Kramer (1963) found θ_I and θ_{II} to decrease with increasing R, but did not observe the catastrophic load drop when R was increased, as in the case of Al. He explained this on the basis of the difference in surface condition in the two metals. When R is increased, the absence of the oxide skin on Au precludes the appearance of the pop-out phenomenon. However, continuous removal of the debris layer affects the value of the hardening rate, and changes in the rate of its removal give rise to the reversible changes in θ with R.

Kramer (1965) also carried out intermittent polishing experiments on Al in which the deformation was stopped at intervals, the specimen was unloaded, and the surface envelope removed electrolytically. He reported that on reloading the polished specimen, the initial flow stress at which plastic flow began was always lower than the prepolishing value. With increasing thickness removed, the decrease in flow stress, $\Delta\sigma$, increased, until after a certain amount of removal, $\Delta\sigma$ became constant. Kramer concluded that this thickness corresponds to the depth of the debris layer (60μ for Al). The thickness of this layer remained constant with strain, but the surface layer stress (i.e. Δσ after removal of 60μ or more) increased with increasing strain.

Essentially, these two types of experiment i.e. Fourie (1967, 1968), Fourie and Dent (1972) and Murphy (1969) on the one hand, and Kramer and Demer (1961) and Kramer (1963, 1965), on the other hand, represent the main base for the controversy which has arisen as to whether the surface regions become preferentially softer or harder than the core during deformation. Other supporting work, on both sides, has been published, some of which will be discussed later.

At present we shall concentrate on trying to find reasons for experimental data being in apparent conflict.

Nakada (1965) suggested that a temperature rise during electropolishing may have important influence on the deformation characteristics of metals in such experiments as those of Kramer. However, Kramer and Demer state that the temperature, as measured by thermocouples at the centre and surface of the specimen did not change when a load drop was taking place (i.e. when R was increased), and Feng and Kramer (1965) report a temperature rise of <5°C during electropolishing, as measured by a thermocouple at the specimen surface.

Nakada (1965) measured the temperature rise, and change in specimen length, while polishing in a methanol-perchloric acid solution cooled to -5°C. He placed a thermocouple in an axial hole drilled 2.5 cm into the specimen, and detected a temperature rise of 70°C accompanied by a length change corresponding to that amount of thermal expansion. Unfortunately Nakada's electrolyte and current density were not the same as Kramer's, so that no direct comparison can be made.

The conditions of power dissipation and cooling by the electrolyte during electropolishing can be rather complicated, depending sharply on the existence and properties of the anodic viscous layer. It is impossible therefore to satisfactorily resolve the question of the temperature effect on the flow stress during deformation except by simultaneously recording in situ the flow stress, rate of metal removal (i.e. current density) and temperature for each electropolishing condition.

Figure 5 shows a simultaneous plot of flow stress and temperature for a Cu crystal which in the first part of the test, was intermittently electropolished (in a nital bath surrounded by ice-water). After electropolishing, the temperature was changed during deformation by periodically changing the temperature baths around the specimen. The temperature was measured by a copper-constantan thermocouple placed in an axial hole drilled in the specimen to a depth of about 1.5 cm. The figure shows that only a small effect is associated with the onset of metal removal (A, Fig. 5). The main effect occurs when an anodic layer forms and the temperature increases sharply. The upper curve C was obtained by correcting the flow stress by the reversible temperature dependence (Basinski, 1959). Both in the case of temperature change by electropolishing, and bath temperature change in the absence of polishing, the total effect can be accounted for purely by the variation of specimen temperature. The only difference between the two cases is in the initial load drop when the temperature was raised by electropolishing. In this case the load drop represents the thermal expansion of the specimen only, against the rest of the apparatus (tensile machine cage, etc.). When the temperature is changed by changing the bath temperature both the specimen and the rest of the apparatus respond, resulting in a smaller effect.

An even more spectacular correlation between flow stress and temperature was found by looking at simultaneous measurements in Al crystals polished both in the current-controlled and voltage-controlled polishing modes. The electrolyte was the same as that used by Kramer and Demer (1961). In an attempt to control the current density at the highest value used by Kramer and Demer, an unstable situation arose. As in the case of Cu, a relatively small change (A) in flow stress occurred when the current was first applied. The larger change coincided with the temperature rise following formation of the anodic layer. At higher temperatures however, the layer broke down, causing a very sharp drop in temperature. Due to the voltage limitation of the power supply used, the current (i.e. R) decreased before the peak in temperature was reached. The rate of polishing and temperature were thus not in phase (Fig. 6), while the flow stress oscillations very clearly followed the temperature changes.

Kramer (1964) has also reported that the strain rate increases during electropolishing at constant stress, which he interpreted as resulting from the decrease in activation energy accompanying the removal of the surface debris layer. The change in activation energy was derived using an equation of the form $\Delta U = - kT\Delta \ln \dot{\varepsilon}$. This interpretation of course assumes no temperature change. In view of the large increase in temperature observed during electropolishing, it should be instructive to examine the magnitude of temperature change which would be necessary to account for Kramer's observation. We can write, $\Delta \ln \dot{\varepsilon}/\Delta T \approx -\left(\frac{\partial \ln \sigma}{\partial T}\right)_{\dot{\varepsilon}} / \left(\frac{\partial \ln \sigma}{\partial \ln \dot{\varepsilon}}\right)_T$. Taking, from the present work, $(\partial \ln \sigma / \partial \ln \dot{\varepsilon})_T$ to be 3×10^{-3} for Cu and 2.6×10^{-3} for Al, and $(\partial \ln \sigma / \partial T)_{\dot{\varepsilon}}$ to be 5.5×10^{-4} for Cu and 10^{-3} for Al (Figs. 5 and 7), and, using Kramer's apparent changes of activation energy of 1400 cals (Cu) and 3300 cals (Al) per mole, we obtain the temperature rise ΔT required to account for Kramer's observations to be $\sim 13°C$ for Cu and $\sim 15°C$ for Al, at his highest polishing rate of .26 a/cm². These temperature increases are smaller than those observed in the present work, therefore the increase in strain rate during electropolishing would be expected simply from the rise in temperature.

In the voltage controlled mode for electropolishing in Al, and also for the bath temperature changes without polishing, the total effect of both the specimen elongation and the flow stress was accounted for simply by the temperature changes (Fig. 7). As for Cu, the small increase in flow stress in the temperature-corrected curve (C) following cessation of polishing, corresponds to extra work-hardening arising from the thermal contraction of the specimen.

5. Structural Evidence

Structural evidence on the nature of the surface regions has been obtained by TEM, etch pitting and X-ray techniques.

In Al and Cu polycrystals, Swann (1963, 1966) showed that at the surface, the dislocation density is lower and the cell formation less well-developed than in the interior. For Al single crystals, Tabata and Fujita (1972) examined the surface and interior throughout the deformation and found that from the beginning of stage I onwards, the surface density is always lower than the interior. They indicated the possibility of a slightly higher surface density in the preyield region.

In Cu crystals, Basinski and Dove (1962) noted that there is little correlation with slip lines and the dislocations in the surface regions, and that the region immediately below a slip line is usually denuded of dislocations. Fourie (1970) and Mughrabi (1970) studied the distribution and density of dislocations at successive depths below the E surfaces of Cu crystals. Fourie reported that as the surface is approached, the appearance of the dislocation arrangements is typical of earlier stages of deformation. Mughrabi found that from the transition region onwards, there is a deficit of dislocations at the surface. For primary dislocations, the ratio of interior to surface densities approaches ~ 2.5 late in stage II then does not change appreciably. The depth of the region of depleted density is about 200μ at 1.7 kg/mm^2. For secondary dislocations the density near the surface also appears to be lower but the ratio of surface and bulk densities is about the same thorughout the deformation.

Examination of etch pit density as a function of depth below the $(1\bar{1}\bar{1})$ surface can give a measure of the variation along L_S. The dislocations visible as etch pits will be primary edges and a fraction of the secondary dislocations.

Livingston (1962) and Young (1962) observed that the etch pit density and distribution at the surface are not significantly different from the bulk. Block and Johnson (1969) found no change in density with depth in Cu crystals in stage I. Kitajima, Tanaka and Kaieda (1969) offer conflicting evidence, reporting an enhanced surface density in weakly deformed Cu crystals. This apparent discrepancy could be a result of the effect of the etchant. Even after thorough washing with methyl alcohol and water, the etchant leaves a strongly adherent film clearly visible in the electron microscope (Basinski and Basinski, unpublished). Cu crystals which were etched and then lightly fatigued developed a higher surface dislocation density, while for crystals which were etched, electropolished and fatigued, the density did not change significantly with depth (Basinski and Woods, unpublished).

Variation in etch pit density along L_E can be investigated by observing any change with distance from the edge of the face (i.e. rather than with depth). In some preliminary work, Basinski and Basinski (1975) examined etch pits across the face of a (2×20)mm^2 Cu crystal deformed to $\sim.6$ kg/mm^2. Although quantitative etch pit densities were not determined, qualitative examination showed that the etch pit density was very much lower at the edges, and the distribution was much more like that normally observed at lower flow stresses. At 1.5-2 mm from the edge the general structure was the same as in the centre regions of the face.

Kramer and Balasubramian (1973) etched successive depths below the surface of polycrystalline Fe3%Si after an unspecified mechanical polishing procedure ending with abrasion with diamond paste. For depths to about .1 mm the etch pit density was higher than in the interior.

The etching studies thus probably indicate that for single crystals of Cu, no significant changes in (primary) dislocation density are detectable in the L_S direction in the earlier stages of deformation. The softer surface and gradient along L_E are detectable in etch pit studies on the cross glide face.

The electron microscope studies seem to be in agreement with the soft surface situation where the dislocation density along L_E decreases as the surface is approached. No evidence has been found of a debris layer on any surface even in Al, except possibly in the preyield region.

TEM observations on core material show that in deformed Cu crystals the density of primary screw dislocations is insignificant compared to the total density (Basinski, 1964; Essmann, 1965; Steeds, 1966; Mughrabi, 1971a). This is attributed to annihilation of screws inside the crystal during deformation (Basinski, 1964), which would account for the observation that L_S, and the S surfaces, play no important role in the context of surface effects.

Examination of deformed Cu crystals by Berg-Barrett X-ray topography has shown that a curvature develops in the primary glide planes such that near the surfaces they are bent (about [12$\bar{1}$] perpendicular to the Burgers vector [101] in the primary plane (11$\bar{1}$)) in opposite senses at opposite faces. The amount of lattice rotation changes with deformation and eventually its sign reverses (Mughrabi 1970, 1971). Although Mughrabi's lattice bending as a function of flow stress is shown schematically, unfortunately no quantitative data have been explicitly stated. However, he shows Berg-Barrett topographs of two regions of a crystal deformed to 1.2 kg/mm^2 in which the angle of incidence differs by 4° (Mughrabi 1970), in one, the central crystal region is strongly reflecting while in the other the surface region is reflecting, this may then be taken to be the lattice tilt at 1.2 kg/mm^2, (at least for this particular sample).

6. Some Dislocation Models for the Surface

Fourie (1968) has discussed his flow stress gradients along L_E in terms of a model which considers primary edge dislocations only. Dislocation sources are distributed throughout the crystal. The surfaces are not sources of dislocations, i.e. no dislocations enter the crystal from the surface, and, they present no barrier to dislocations approaching from the interior. For crystals thicker than twice the primary edge dislocation mean free path, on this model, there will be a deficit of positive dislocations at one face, and negative dislocations at the other face. The model predicts that the surface regions will be less deformed (i.e. the glide plane less rotated towards the tensile axis) than the core. For crystals which are thin compared to the mean free path, an approximately uniform dislocation density is expected from the model.

Mughrabi (1971) describes a model similar to Fourie's, but in addition shows the consequences of the operation of surface sources in thin and thick crystals. Briefly the resulting lattice tilts are in the same sense for both surface and inner sources in thin crystals (< twice primary edge mean free path) and for inner sources in thick crystals; while for surface sources in thick crystals, the rotations are opposite in sense, such that the surface is more highly deformed than the interior.

The differences in the sense of the lattice rotations given by surface source operation in thin and thick crystals on this model appear to result from the physical meaning attached to the term mean free path. The situation in thin crystals where most of the dislocations stop further from their surface of origin than from the opposite face, can arise only if the mean free path is essentially the same for all dislocations, which presupposes a correlation between a source and the place where the dislocation eventually stops. It is very likely especially in stage I, where obstacles are distributed essentially randomly, that a better definition would be in terms of the probability of a dislocation being stopped per unit length of travel. Under those conditions, the density would decay exponentially from the entry surface, and distinction in lattice curvatures between thick and thin crystals would be lost.

Mughrabi (1970) shows schematically his lattice tilts and how they change with increasing flow stress. In terms of his models, the crystals used were thick. At low flow stresses the lattice tilts are shown such that at the surface the crystal is more highly deformed than the core (rotated more nearly parallel to the tensile axis). A similar observation was earlier reported by Chalmers and Davis (1957) on Al crystals. With increasing deformation, the rotation relative to the core decreases and, at 10 kg/mm^2 appears to have reversed in sign. On Mughrabi's (1971) dislocation model for thick crystals these rotations indicate that surface source operation determined the sense of lattice rotations up to a flow of 10 kg/mm^2.

In a Cu crystal deformed to 1.2 kg/mm^2, the lattice was observed to be bent by about 4° (see previous section). From Mughrabi's stress-strain curve, the corresponding strain is ~.15 which together with his tensile axis orientation, indicates a total rotation of the primary planes of ~4°. Thus, if the surface deformed in single glide, the core would have to deform either elastically, or in polyslip with no net crystal rotation.

Mughrabi (1971) finds that the observed lattice tilts at a macroscopic flow stress of 1.2 kg/mm^2 indicate a local excess density of primary edge dislocations of one sign of ~2×10^7/cm^2 at each face. The model on which the calculation is based unfortunately is not specified in detail. If we assume that the rotation is 4° (as indicated by the rotation between his surface and core topographs), the spacing of edge dislocations in a tilt boundary which would give rise to this rotation would be $\sim 2.5 \times 10^{-7}$ cm. If these dislocations were distributed within a depth of 2 mm, we would obtain Mughrabi's figure os 2×10^7 dislocations/cm^2. Although this dislocation density sounds low (compared with average densities) closer examination of the geometry shows that their presence would cause a very high flow stress gradient. Figure 8 shows schematically the lattice bending due to a distribution of one sign of edge dislocations. These are formally equivalent to superposition of dislocations R being responsible for lattice rotation but no long range stress, and S which result in a sharp gradient in tensile stress. With the density necessary to give a rotation of 4°, the corresponding tensile stress would be ~ 920 kg/mm^2.

On a similar argument, the rotations corresponding to the flow stress gradients of about 1 kg/mm^2 indicated in Fourie's work would be of the order of 15 seconds of arc.

Thus although the rotations may represent the initialisation of the processes eventually responsible for the flow stress gradients, the models obviously cannot be discussed purely in terms of primary dislocations, but gradients in secondary slip must also be considered. The net Burgers vector of dislocations responsible for the rotation must be nearly normal to the surface if large elastic stresses are to be avoided.

On purely geometrical grounds it seems impossible to distinguish between the surface acting as a source of dislocations or a preferential sink for dislocations from the interior. In either case, we should be left with an excess of dislocations of the same sign (Figure 8). However, if the surface acts initially as a preferential source, the density of dislocations in the surface region should be higher, while if it acts as a preferential sink, it should be lower than the mean density. The latter effect might therefore be more important if the surface region is found to be softer than the interior. Another consideration arises from the likelihood that trapped dislocations or groups will influence subsequent capture of other dislocations over a finite volume of the crystal. Since the dislocations which leave the crystal can be regarded as having moved an infinite distance, they will not contribute to further work hardening and therefore a soft region will develop near the surface.

Turning now to the actual surface, it has been suggested by Latanision, Sedriks and Westwood (1973) that an ideal clean surface can provide a barrier to the egress of dislocations having a Burgers vector component normal to it, since a slip step has to be created, requiring extra surface energy. Latanision et al. concluded, using an expression for the dislocation self energy (Nabarro 1952), that perfect edge dislocations may leave the surface freely, but that it would be energetically favourable for partial dislocations to remain inside the crystal. A more recent estimate (see Nabarro, 1967) of the elastic energy is an order of magnitude higher than the early one, so that on energy considerations alone, the surface would not act as a barrier to dislocation egress. (Since surface steps are not formed on S surfaces this effect could have no bearing on the results of Kitajima et al.) Essentially the same conclusion could be obtained from Head's (1953) calculations, which show that as long as the elastic constants of the medium outside the crystal are lower than those of the crystal, the dislocation will be attracted to the boundary.

7. The Unloading Yield Point (u.y.p.)

Some of the work directed towards learning more about the surface layer has centred on investigations of the u.y.p. phenomenon. It has been discussed in terms of both the softer and harder surface envelope models.

(i) Existing Data

Briefly, when a tensile test is interrupted at a flow stress σ_a and the crystal is unloaded to $<\frac{1}{2}\sigma_a$, on reloading plastic deformation does not take place until a flow stress $\sigma_b > \sigma_a$ (Haasen and Kelly, 1957; Makin, 1958; Fourie, 1970a). The effect is absent in stage I and very early stage II, after which it gradually increases with increasing deformation. The u.y.p. is a very small effect; the maximum normally observed being $(\sigma_b - \sigma_a)/\sigma_a \simeq 0.015$.

There is not complete agreement in the literature on the effect of surface removal on the u.y.p.. Haasen and Kelly (1957) found that electropolishing off the surface of an Al crystal did not affect the size of the u.y.p.. Birnbaum (1961) made the observation that etching has no effect on the u.y.p. in Cu crystals provided that the cross-section remains uniform along the gauge length, any inhomogeneities lead to disappearance of the u.y.p. due to premature yielding in the thinner regions.

Fourie (1970a) studied changes in the u.y.p. after removing various thicknesses from the surface by chemical polishing, and found that the observed unloading yield point decreased with increasing thickness of the surface envelope removed. Removal of ∿50% of the cross-sectional area reduced the u.y.p. to about a quarter of its original value. Fourie concluded that the u.y.p. is a surface phenomenon and is due to the near-surface flow stress gradient, in accordance with the suggestion of Brydges (1968) that an inhomogeneous flow stress distribution can lead to reverse plastic flow on unloading and therefore extra hardening on reloading.

Feng and Kramer (1965), although they also found that surface removal removes the u.y.p., indicate that much thinner envelopes need to be removed in order to eliminate the u.y.p. completely, i.e., 60μ for Al and Cu, and 125μ in Au. They conclude that removal of the debris layer causes the decrease in initial flow stress upon reloading, i.e. removal of the u.y.p. ($+\Delta\sigma_y$). (The thickness of the debris layer was previously estimated to be 60μ in Al from measurement of the negative change in initial flow stress after unloading ($-\Delta\sigma$), (Kramer, 1965). However, since the u.y.p. forms only as a result of unloading, these experiments can only indicate the formation of a hard surface during the unloading event, and throw no light on the situation in a normal tensile test. In fact the experiments of both Fourie, and Feng and Kramer, suggest that during unloading the surface becomes harder than it was before, and that this hardening is transient. Feng and Kramer's curves (their Fig. 3) indicate that the surface and interior had the same strength before unloading, since there seems to be no permanent average flow stress change resulting from surface removal, which in the case of Al and Cu amounted to about 8% and in Au about 15% of the cross-sectional area. If the pre-polishing surfaces had been hard, a significant permanent decrease in flow stress should have been observed.

Similarly, in the case of Fourie's Cu crystals, if the E surfaces had been softer, then a significant displacement of the stress-strain curve upwards should have taken place after surface removal. Unfortunately no information on this point can be obtained from Fourie's published work since portions of the load-elongation curves before polishing are not presented.

(ii) Preliminary discussion of u.y.p.

Fourie (1968a) suggested that when a hard core-soft surface crystal is unloaded, the surface would be driven into compression by the larger elastic stress component in the interior. Brydges (1968) estimated, on his non-uniform flow stress model and using Fourie's (1968) flow stress gradients that an u.y.p. of 5% could be obtained if the flow stress gradient totally disappears after the surface compression. Such a mechanism would require an extremely high rate of hardening to be associated with back deformation. If, in the unloaded crystal, back plastic flow is to remove the flow stress gradient then the increased flow stress on reloading would correspond to the elastic stress gradient, i.e. the rate of hardening would equal the shear modulus, μ. For an u.y.p. of $\sim 1.5\%$ to result from such a process, the required hardening rate would be $\sim \mu|3$, about a hundred times higher than the rate observed even in the 'rapid hardening' region. Fourie's flow stress profiles in any case were obtained from unloaded crystals, and therefore do not represent conditions under load, as assumed by Brydges.

Since Fourie's flow stress profiles in stages II and III move rigidly upwards with flow stress (Fig. 4), it would be expected that $\sigma_b - \sigma_a$ would be constant with increasing deformation. The data however (Makin 1958, Fourie 1970) indicate rather that the fractional value $(\sigma_b - \sigma_a)/\sigma_a$ tends to be independent of flow stress. Obviously some additional postulates are necessary to explain this.

(iii) Some Additional Experiments

In all cases reported, the percentage of crystal cross-sectional area removed to diminish or eliminate the u.y.p. was always very much larger than the percentage value of the yield point itself $((\sigma_b - \sigma_a)/\sigma_a)$. It therefore seems advisable to examine more closely Birnbaum's observation that uneven removal is responsible for loss of the u.y.p.

We carried out (Basinski and Basinski, unpublished) some simple experiments all of which tended to support Birnbaum's conclusion. Electropolishing in situ in the tensile machine always eliminated the u.y.p. whether E or S surfaces were removed. Deformation on reloading was observed to proceed by Lüders band propagation from the top to bottom of the crystal, indicating nonuniformity. More careful polishing, after removal from the machine, enabled a larger thickness to be polished away before the u.y.p. disappeared; the best uniformity was achieved by using Mitchell's polishing method (Mitchell, Chevrier, Hockey and Monaghan, 1967). The behaviour of the yield point was not sharply different when E or S surfaces were removed (see Fig. 9).

To illustrate elimination of the u.y.p. by small nonuniformity, one S face was deliberately thinned locally by <.05 mm. (<1.5% of the cross-section locally) and the u.y.p. disappeared.

(iv) Size Dependence

Although we never succeeded in completely retaining the u.y.p. after polishing, the difficulty of achieving absolutely uniform surface removal, especially from the irregular surfaces of deformed crystals, indicates that failure to do so may be responsible for a large part of the observed effects. Size dependence of the u.y.p., without surface removal, should throw some light on the role of the surface in this effect.

Fourie (1970) studied systematically the dependence of the u.y.p. on crystal size in crystals for which only the edge dislocation path length varied, for crystal widths from ∼5-20 mm. His data are shown in Fig. 10. The magnitude of the u.y.p. appears to be not markedly dependent on size. For the thinnest crystals the u.y.p. is somewhat smaller, but does not change appreciably for crystals thicker than ∼2 mm.

We measured the u.y.p.'s in two $(2 \times 20)\text{mm}^2$ crystals oriented such that the S faces were wide in one crystal and narrow in the other. Both crystals gave u.y.p.'s of the same magnitude, in agreement with Fourie's size-dependence results.

It is difficult to reconcile these data with a model where the u.y.p. results from a flow stress gradient at the E faces. For very thin crystals the relative surface gradient is highest whereas the magnitude of the u.y.p. is lowest. For crystals having $L_E > 5$ mm, the flow stress gradient data show that only the width of the homogeneous core increases with increasing width. The u.y.p. would therefore be expected to be inversely proportional to L_E, whereas Fig. 10 shows that it is essentially independent of crystal width.

Also, such a model would predict that slices cut from a parent deformed crystal would show no u.y.p. if they came from the core region of uniform flow stress, but a disproportionately high yield point when cut from the surface regions of steep gradient. No such effect was reported by Fourie (1968).

8. Effect of Surface Removal on the Stress-Strain Curve

The effect of surface removal on the stress-strain curve should provide information on possible flow stress gradients. The objections to the surface removal technique, which were relevant when it was being used to investigate a very small effect such as the unloading yield point, would not be so important in this context.

The evidence on Al seems to be rather inconclusive. Sumino and Yamamoto (1961) and Nakada and Chalmers (1962, 1964) reported a permanent decrease in flow stress following surface removal, whereas the data of Kramer (1965) and Feng and Kramer (1965) show no permanent change in flow stress. Since Al can recover even near room temperature, Kramer (1965a) suggested that Nakada and Chalmers' effect could be explained by recovery due to the temperature rise during electropolishing. Recovery processes in Al can be sensitive to temperature changes, purity of the crystals, etc., more systematic work is necessary before firm conclusions can be reached.

In the case of Au, Nakada and Chalmers (1964) reported that the flow stress after surface removal usually increases. They attributed this to possible mishandling and concluded that the flow stress profile of a deformed Au crystal is constant throughout the specimen cross section.

In experiments on the effect of surface removal on the u.y.p. in Cu crystals, we observed that, while removal of both E and S surfaces diminished the u.y.p., in the case of E surface removal only, a permanent increase in flow stress resulted. Figure 9 shows stress strain curves for Cu crystals which were unloaded and reloaded several times; at the places marked by arrows ~.1 mm of material was removed from each of one pair of faces while the crystal was unloaded. Figure 9a represents S face removal, and 9b, E face removal. It can be seen that a significant permanent flow stress increase occurs after E face removal, but for S removal, a continuous curve could be drawn through the various sections. This effect would be expected from Fourie's flow stress profiles (Fig. 4). Removal of the soft region at the E surfaces should result in an increase in the mean flow stress, but, since the S surface regions have a flow stress not much different from the core, their removal should not appreciably affect the mean flow stress.

From the increase in flow stress after E face removal and the area of the cross section removed, the mean flow stress in the portion removed can be calculated. Table 1 lists some values obtained.

TABLE 1

Prestress σ_a kg/mm^2	Mean Flow Stress of Material Removed, σ_s kg/mm^2	$\sigma_a - \sigma_s$ kg/mm^2	Thickness Removed from each E face. mm.
1.02	.38	.64	.13
1.5	.67	.83	.08
2.54	1.39	1.15	.12
3.02	2.03	.99	.10
3.52	2.46	1.06	.10

The data indicate that with increasing deformation, the difference between the surface and core regions tends to a constant value. The data in Table 1 agree very well with Fourie's flow stress profiles (Fig. 4).

Fourie (1975) and Mughrabi (1975) have also observed a permanent increase in flow stress after surface removal, however, the whole envelope was removed, so that no distinction was made between E and S surfaces. By attributing the whole increase to the removal of the E surfaces, they obtain good agreement with the observed flow stress gradients.

One puzzling aspect of these observations is that successive surface removal events, after intermediate strain increments, show that the soft surface reforms, as previously shown by Fourie (1970b), but no corresponding decreased hardening rate is observed in the stress-strain curve (Fig. 9b).

9. Surface Stresses

The relative hardness of the surface and core regions of crystals during deformation must be reflected in the elastic deformation in those regions and the corresponding residual stresses in unloaded crystals. A softer surface will result in compressive stress while a harder surface will result in tensile stress in the surface region after unloading.

Although no work seems to have been carried out on single crystals, the results of Kolb and Macherauch (1967) on polycrystalline Al, Cu and Ni indicate that in all cases the tensile stress (surface) as measured by X-rays in the specimens under load was appreciably smaller than the macroscopic flow stress measured mechanically. Unloaded samples invariably showed compressive surface stresses. These workers also determined the flow stress profile of the internal stress by repeated polishing and measuring; they show a sharp profile extending only a fraction of a mm. into the sample.

We (Basinski and Gallace, to be published) made some preliminary determinations of the stress gradients in unloaded Cu single crystals by measuring the bending due to slow dissolution of one face. The specimen was rigidly mounted at one end and the deflection of the free end was measured by means of a sensitive LVDT. Figure 11 shows the stress as a function of distance from the E and S faces. In both cases compressive stresses were observed, although the effect is much stronger at the E faces.

These results serve as further evidence for the existence of softer E surface regions during the deformation. However quantitative comparison would be difficult since the sense of the stress has been reversed (on unloading) and the Bauschinger effect would be expected to decrease the magnitude of the observed compressive stresses. Furthermore, these measurements correspond to effectively zero strain rate, back creep after unloading should thus further reduce the flow stress gradient.

10. Conclusions

The study of the influence of crystal size on the deformation parameters shows that the extent of stage I increases with decreasing glide path length for primary edge dislocations, L_E; the glide path length for screw dislocations, L_S, plays no significant role. Variation in the rates of hardening θ_I and θ_{II} appears to be minimal, except that θ_{II} decreases appreciably for small crystals in which L_E is of the same order of magnitude as the primary dislocation mean free path.

A flow stress gradient develops during deformation such that the regions near the E surfaces are softer than the core. This conclusion is supported by direct structural observations and by the residual stresses measured in unloaded crystals.

Flow stress changes observed during continuous polishing experiments, and the elongation of the specimen, which have been interpreted as indicating the existence of a hard surface layer, can be shown to result from temperature changes which accompany the polishing action, at least after the first few percent of strain.

The weight of evidence favours preferential escape of edge dislocations from the E surfaces, while S surfaces play a minimal role in the deformation process.

The absence of a marked dependence of the unloading yield point on size, especially for thick crystals, indicates that it is unlikely to be a surface phenomenon.

REFERENCES

Basinski, Z.S., 1959, Phil. Mag. 4, 493.

Basinski, Z.S., 1964, Disc. Far. Soc., No. 38, p. 93.

Basinski, Z.S. and Basinski, S.J., 1975, to be published.

Basinski, Z.S. and Dove, B.D., 1962, Symposium on the Role of Substructure in the Mechanical Behaviour of Metals, Orlando, Florida, ASD-TDR-63-324, p. 227.

Birnbaum, H.K., 1961, Acta Metall., 9, 320.

Block, R.J. and Johnson, R.M., 1969, Acta Metall., 17, 299.

Bridges, W.T., 1968, Scripta Metall., 2, 557.

Chalmers, B. and Davis, R.S., 1957, Dislocations and Mechanical Properties of Crystals, J. Wiley, New York, p. 232.

Essmann, U., 1965, Phys. Stat. Sol., 12, 723.

Feng, C. and Kramer, I.R., 1965, Trans. Met. Soc. A.I.M.E., 233, 1467.

Fourie, J.T., 1967, Phil. Mag., 15, 187.

Fourie, J.T., 1967a, Can. J. of Phys., 45, 777.

Fourie, J.T., 1968, Phil. Mag., 17, 735.

Fourie, J.T., 1968a, Scripta Metall., 2, 63.

Fourie, J.T., 1970, Phil. Mag., 21, 977.

Fourie, J.T., 1970a, Phil. Mag., 22, 923.

Fourie, J.T., 1970b, 2nd Int. Conf. on Strength of Metals and Alloys, Asilomar, A.S.M., Vol. II, 441.

Fourie, J.T., 1975, This Conference.

Fourie, J.T. and Dent, N.C.G., 1972, Acta Metall., 20, 1291.

Garstone, J., Honeycombe, R.W.K. and Greetham, G., 1956, Acta Metall., 4, 485.

Haasen, P. and Kelly, A., 1957, Acta Metall., 5, 192.

Head, A.K., 1953, Phil. Mag. 44, 92.

Kitajima, S., Tanaka, H. and Kaieda, H., 1969, Trans. Japan Inst. Met., 10, 12.

Kolb, K. and Macherauch, E., 1967, 58, 238.

Kramer, I.R., 1963, Trans. Met. Soc. A.I.M.E., 227, 1003.

Kramer, I.R., 1964, Trans. Met. Soc. A.I.M.E., 230, 991.

Kramer, I.R., 1965, Trans. Met. Soc. A.I.M.E., 233, 1462.

Kramer, I.R., 1965a, Trans. Met. Soc. A.I.M.E., 233, 1451.

Kramer, I.R. and Balasubramian, N., 1973, Acta Metall., 21, 695.

Kramer, I.R. and Demer, L.J., 1961, Trans. Met. Soc. A.I.M.E., 221, 780.

Latanision, R.M., Sedriks, A.J. and Westwood, A.R.C., 1973, Structure and Properties of Metal Surfaces, Honda Memorial Series on Materials Science, No. 1, p. 500.

Livinston, J.D., 1962, Acta Metall., 10, 229.

Lücke, K., 1957, Dislocations and Mechanical Properties of Crystals, John Wiley, N.Y., p. 549.

Makin, M.J., 1958, Phil. Mag., 3, 287.

McKinnon, N.A., cited in, Clarebrough L.M. and Hargreaves, M.E., 1959, Progress in Metal Physics, 8, 1.

Mitchell, J.W., Chevrier, J.C., Hockey, B.J. and Monaghan, J.P., 1967, Canadian Journal of Physics, 45, 453.

Mughrabi, H., 1970, Phys. Stat. Sol., 39, 317.

Mughrabi, H., 1971, Phys. Stat. Sol., (b), 44, 391.

Mughrabi, H., 1971a, Phil. Mag., 23, 897.

Mughrabi, H., 1975, This Conference.

Nabarro, F.R.N., 1952, Advances in Physics, 1, 332.

Nabarro, F.R.N., 1967, Theory of Crystal Dislocations, Clarendon Press, Oxford.

Nakada, Y., 1965, Trans. Met. Soc. A.I.M.E., 233, 244.

Nakada, Y. and Chalmers, B., 1962, J. Appl. Phys., 33, 3307.

Nakada, Y. and Chalmers, B., 1964, Trans. Met. Soc. A.I.M.E., 230, 1339.

Nakada, Y., Kocks, U.F. and Chalmers, B., 1964, Trans. Met. Soc. A.I.M.E., 230, 1273.

Steeds, J.W., 1966, Proc. Roy. Soc., A292, 343.

Sumino, K. and Yamamoto, M., 1961, J. Phys. Soc. Japan, 16, 131.

Suzuki, H., 1957, Dislocations and Mechanical Properties of Crystals, John Wiley, N.Y., p. 549.

Suzuki, H., Ikeda, S. and Takeuchi, S., 1956, J. Phys. Soc. Japan, 11, 382.

Swann, P.R., 1963, Electron Microscopy and Strength of Crystals, Interscience, p. 131.

Swann, P.R., 1966, Acta Met., 14, 900.

Tabata, T. and Fujita, H., 1972, J. Phys. Soc. Japan, 32, 1536.

Young, F.W., 1962, J. Appl. Phys., 33, 963.

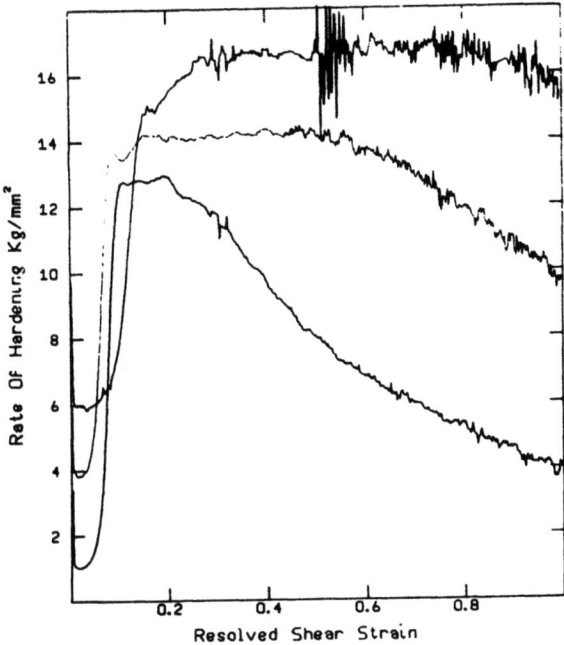

Figure 1. Rate of hardening as a function of resolved shear strain for Cu crystals deformed at 4.2°K (upper curve), 77°K (middle curve), and 295°K (lower curve). The curves for 4.2 and 77 are displaced upwards by 2.5 and 5 kg/mm² respectively. Initial orientation 10° from [110] towards [211]; cross-sectional area 3×3mm²; strain rate 2×10^{-4}/sec.. All data were normalised to the shear modulus at 0°K.

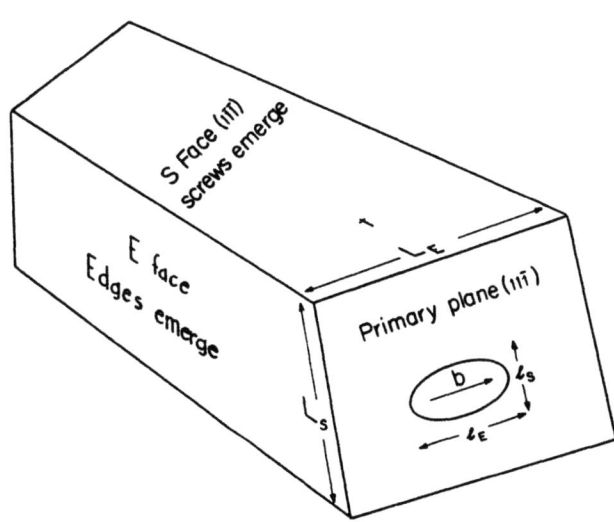

Figure 2. Schematic diagram of a crystal sectioned on the primary glide plane. L_E and L_S are the glide path lengths available to primary edge and screw dislocations. L_E and L_S illustrate the mean free paths, b is the Burgers vector.

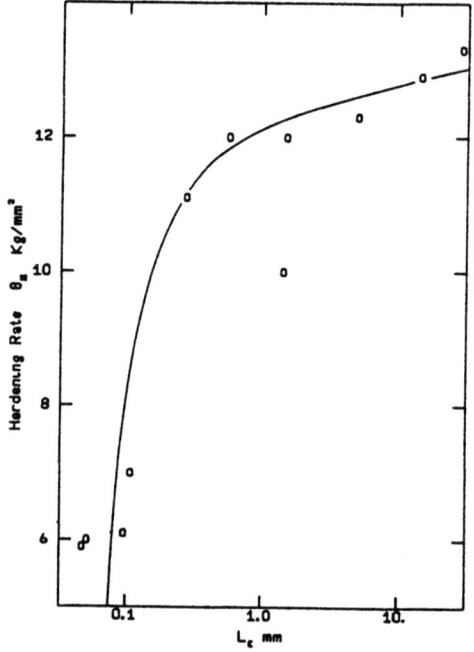

Figure 3. Rate of hardening in stage II as a function of L_E for Cu crystals deformed at room temperature. (Fourie, 1967, 1968).

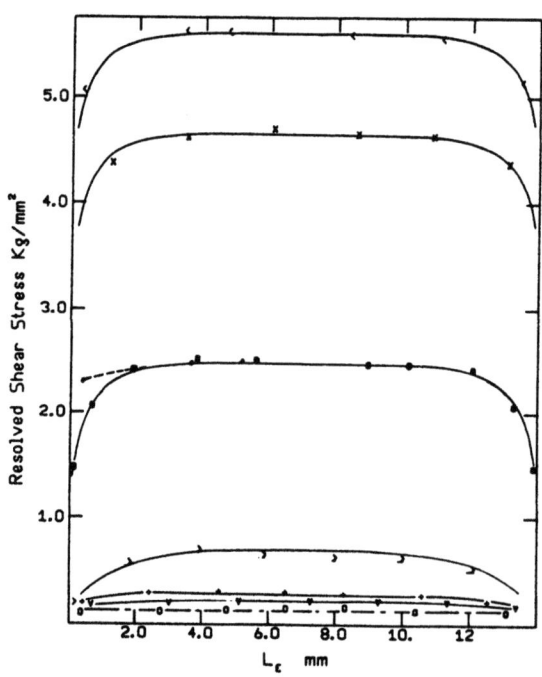

Figure 4. Flow stress distribution in Cu crystals (prestrained at room temperature) as a function of L_E. Dashed curve only: profile along L_S. The lower broken line represents the undeformed crystal. (Fourie, 1968).

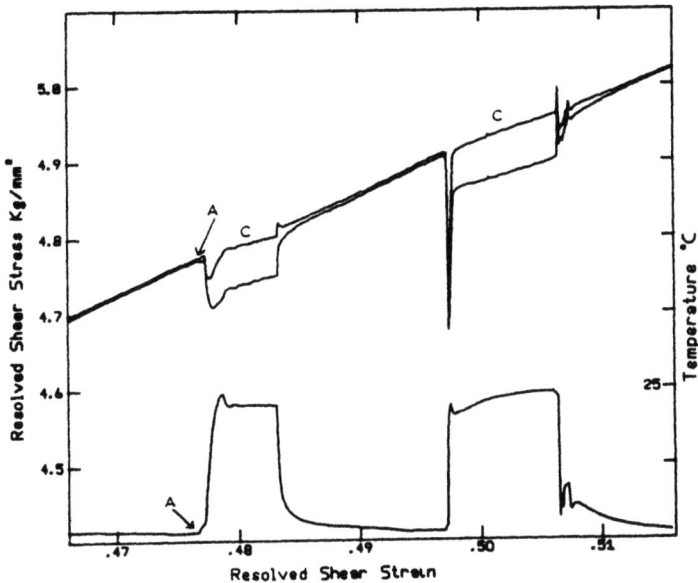

Figure 5. Portion of stress-strain curve for a Cu crystal. The temperature was changed either by electropolishing (left) or bath change (right). Lower curve, temperature; intermediate curve, observed stress-strain relation; upper curve, (C), flow stress corrected for temperature ($d\ln\sigma/dT = 5.5 \times 10^{-4}$). A indicates current switched on. Current controlled mode, current density, $.18 a/cm^2$.

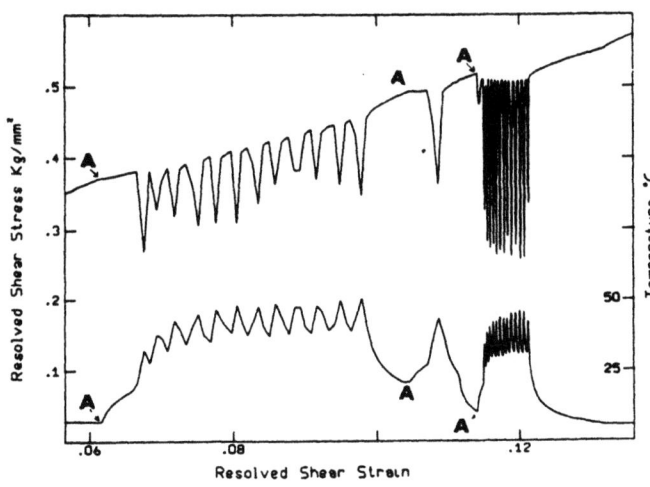

Figure 6. Correlation between temperature changes and observed flow stress in an Al crystal. For $\varepsilon < .11$ the strain rate was 2×10^{-4}/sec, above this strain the rate was 2×10^{-5}/sec. Points A, current switched on. Current controlled mode, current density $.26 a/cm^2$.

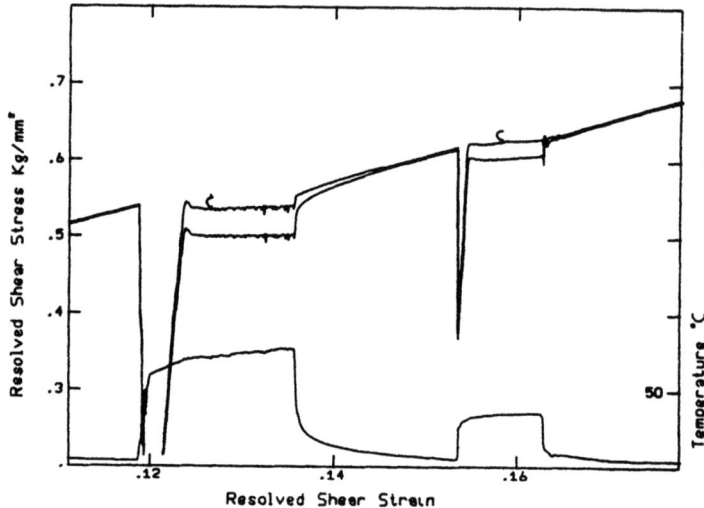

Figure 7. Al crystals, experimental details as for figure 5. $d\ln\sigma/dT = 1\times10^{-3}$. Voltage controlled mode, current density $\sim.36a/cm^2$. Measured extension corresponding to load drop when current switched on, 95µ; extension calculated from temperature rise, 86µ.

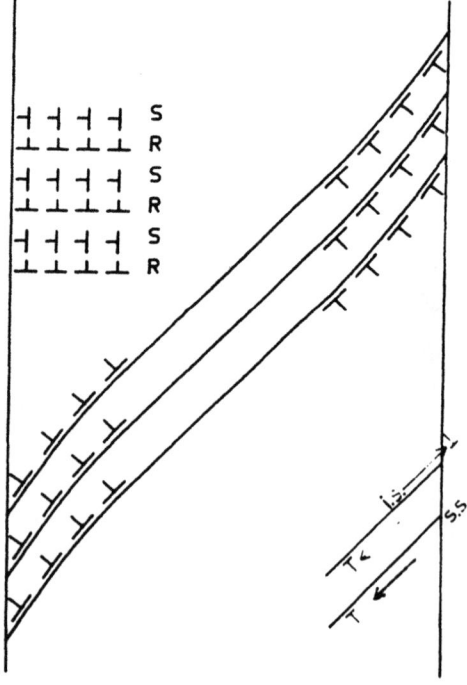

Figure 8. Lattice rotations and long range stress field associated with primary edge dislocations of one sign. Dislocations R responsible for rotations and dislocations S responsible for long range stress. Lower right: Equivalent resultant dislocation configuration due to operation of internal sources (i.s.) and surface sources (s.s.).

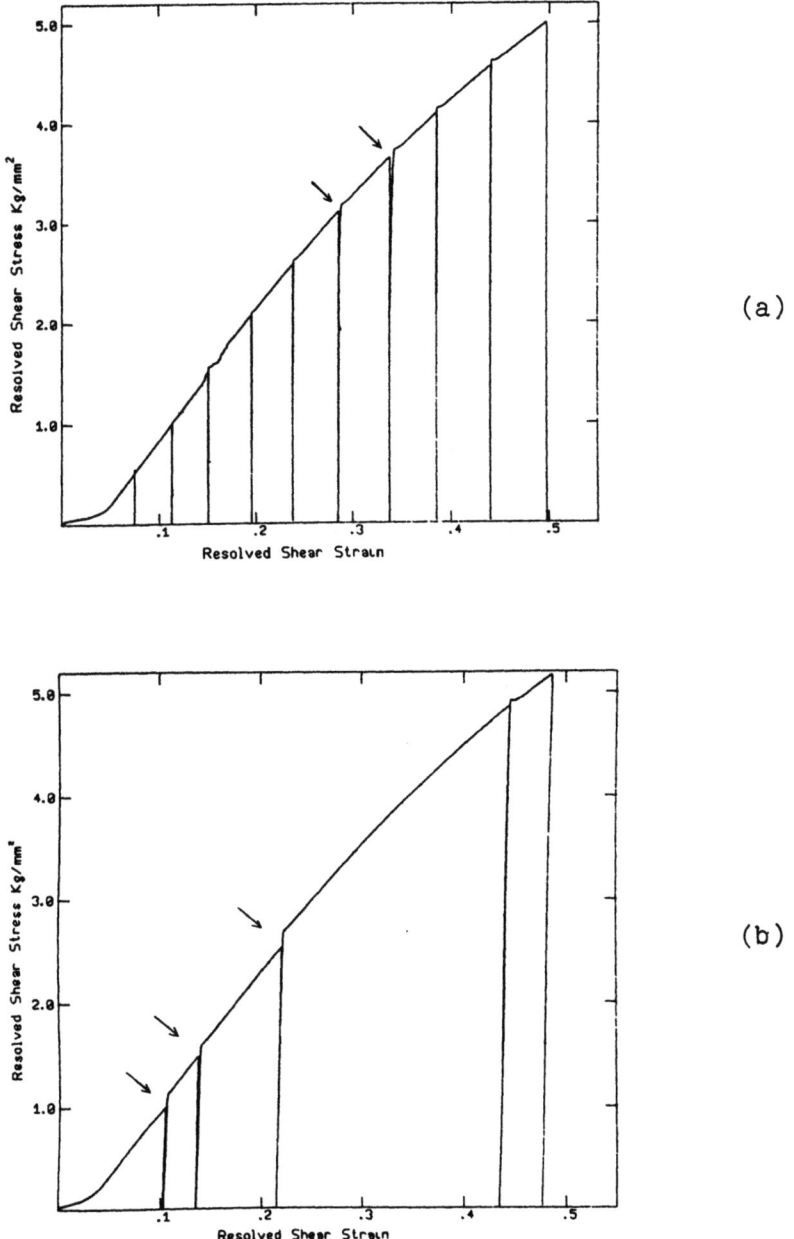

Figure 9. Stress-strain curves for Cu crystals with intermittent unloading. Arrows mark places where ∿.1 mm was polished away from two opposite faces of unloaded crystals. (a) S surfaces polished, (b) E surfaces polished.

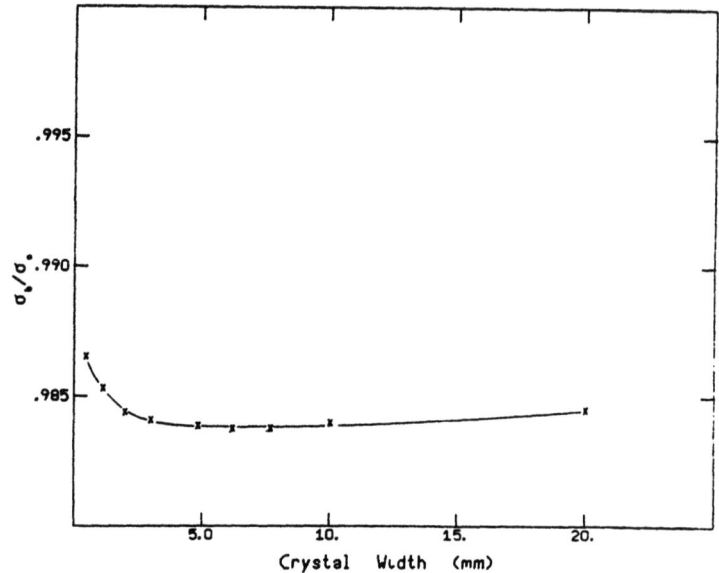

Figure 10. Fourie's (1970) data for unloading yield point as a function of crystal thickness for Cu crystals. The geometry was such that $L_E \approx 1.4w$; L_S was constant for all crystal sizes.

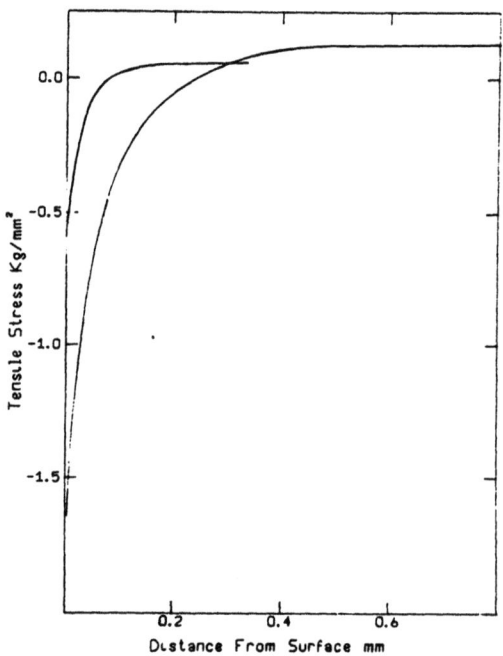

Figure 11. Internal stress as a function of distance from the surface, obtained from crystal bending due to polishing away one face. For the curve showing the more marked effect, E face polished. For the other curve, S face polished. Mean tensile stress before polishing, 11.25 kg/mm².

DISCUSSION

Comment by I. R. Kramer:

Basinski attempted to explain the changes in flow stress when the specimens are polished during deformation. It is noted that the temperature increase in the specimen was very large, in the order of 25°C. Under these conditions anodized layers are formed that will obscure and, at least complicate the analysis. In our experiments, we took care to choose solutions so that the temperature rise was 2°C. Under this condition the changes in flow stress could not be explained in terms of a temperature rise. In the same context, Basinski attempts to explain the "pop-out", that is the sudden drop in load when the current density is increased from a low to a high value. Firstly, we could not account for the drop in load by a temperature increase. Using high speed recording we noted that the drop occurred in about 10^{-4} sec. By heating the specimen we could not account for the drop in load. Further gold crystals do not show the effect. It therefore must be said that the entire effect cannot be explained by a temperature increase. In an experiment using a specimen of 7075-76 aluminum in bending, a dislocation "pop-out" occurred only when the tension side of the specimen was polished at an increased current density. The compression side was covered with a stop-off lacquer. When the tension face was covered and the compression side was polished using the same sequence of current change, the flow curve flattened but no "pop-out" was observed. Further evidence that the effects are not due to increased temperatures may be found in the observations that the extent of Stage I of the Al single crystals increased with increased current density. It is known that increased temperatures decrease the extent of Stage I.

Dr. Basinski stated in private communication that his specimens were held in insulated grips and the electrolytic cell was controlled by the voltage. In our case we provided thermal conduction paths and heat sinks and the solution was stirred. Basinski's solution was not stirred. The observations that the had delay periods after increasing the voltage indicates that thick films were formed. In our case no delay period was noted as to be expected because we used current control.

Reply:

I would like to point out that the temperature rise referred to by Dr. Kramer was not postulated to attempt explanation of the sudden elongation of the specimens and the changes in flow stress; it was an experimental observation. Temperature increases of this magnitude have been reported previously by Nakada (1965) but he used a methanol-perchloric acid solution to polish Al, rather than methanol-nitric as used by Kramer and co-workers (Kramer and Demer,

1961; Kramer, 1963; Kramer, 1965; Feng and Kramer, 1965). In the present work all available details of Kramer's polishing conditions were followed. A methanol-nitric solution (2 parts methanol to one part nitric acid as quoted by Feng and Kramer) cooled to 0°C, and a current density of .259 a/cm^2 as used by Kramer and Demer. In addition, however, the specimen temperature was measured continuously throughout the deformation, by means of a thermocouple inside the specimen, placed in a hole drilled axially into it at one end. Feng and Kramer (1965) and Kramer (1965) checked the temperature at the specimen surface, and found almost no increase (<5°C) during polishing. Kramer and Demer (1961) also checked the temperature at the surface and found no increase during the time the drop in load occurred.

In our work the specimen temperature and the flow stress were monitored simultaneously while the specimen was deformed in a polishing bath. Figures 6 and 7 show that both in the current-controlled and voltage-controlled modes in Al, the specimen behaviour correlates with temperature not with current density (i.e., rate of metal removal). Fig. 5, representing the flow stress and temperature during deformation of Cu using the current controlled mode of polishing, shows that here also the flow stress correlated with temperature. Furthermore Figs. 5 and 7 show that the temperature and flow stress plots are similar when the change is effected not by switching the current on and off, but by changing the temperature of the surrounding both.

Dr. Kramer's statement that the load drop due to dislocation pop-out occurring when the current density in increased takes place in about 10^{-4} sec. is difficult to understand. At his highest current density (.259 a/cm^2), the rate of removal of Al calculated from the electrochemical equivalent and density is 8.94×10^{-6} cm/sec (i.e. 2.11×10^{-4} inches/minute not 2.5×10^{-4} inches/minute on each side as used by Kramer). In 10^{-4} sec the thickness removed is thus only 8.94×10^{-10} cm; the region controlling pop-out is thus less than one thirtieth of an atomic layer in depth. Kramer's own figure for rate of metal removal would bring this up to 1/27 of an atomic layer.

It is rather difficult to comment on Dr. Kramer's experiments on 7075-T6 alloy in bending. However, tentatively, any heating occurring during polishing the tension side would tend to increase the curvature, whereas the opposite effect would be expected on the compression side. The effect should be more observable in a low thermal conductivity alloy such as this than in a pure metal. Experiments in which the temperature is measured are obviously needed before drawing firm conclusions, however, Dr. Kramer's interpretation appears to require that the debris layer would form only on the tension side.

Dr. Kramer obviously entirely misunderstood our private communication. I pointed out to him that his specimens were held in insulated grips (Kramer and Demer, page 781, "The specimens were electrically insulated from the tensile machine by the use of phenolic resin specimen holders...") whereas in my experiments the specimens were held in steel grips with a much higher thermal conductivity. Electrical insulation was provided by waxing the exterior of the assembly. In my experiments, therefore, heat losses through the grips and to the surrounding solution should have been greater.

As already stated, most of the work was carried out using the current controlled mode, not voltage control as implied by Dr. Kramer.

Comment by F. C. Frank:

When Basinski refers to emergence of a dislocation at a surface I understand him to mean the emergence of a dislocation line parallel to that surface: but I could describe an edge-dislocation, normal to the S-surface, as emergent at that surface. The language used in imperfect. When Basinski says "pop-out" for the former situation, his language is more perfect.

Reply:

I entirely agree.

Comment by J. P. Hirth:

I agree with Professor Frank that the usage of screw-edge surfaces is confusing: indeed, we see that both possible conventions are used. As an alternative, which literally focusses on the mechanistic role of creation of new-surface in deformation, I suggest a notation with respect to a perfect reference crystal with perfect singular surfaces. The surface where dislocation glide produces new ledges could be called the L-surface. The surface containing the Burgers vector, and where new surface can be created only by climb could be called the C-surface.

THE ROLE OF SURFACE-ENVIRONMENT INTERACTIONS ON CYCLIC DEFORMATION

D. J. Duquette, H. Hahn, and P. Andresen

Materials Engineering Department, Rensselaer
Polytechnic Institute, Troy, New York 12181

INTRODUCTION

Early work on the effects of environment on premature failure of metals and alloys suggested that corrosion environments create stress concentrations in the material surface in the form of pits,[1] or that cyclic stresses created preferential corrosion paths due to localized deformation[2-4] either in the metal proper[2,3] or through ruptured films.[4] Subsequent studies however, have indicated that these views are overly simplified and that environmental interactions may actually affect the mode and extent of surface deformation which preceeds crack initiation. For example, Duquette and Uhlig have suggested that surface connected slip in steels is accelerated by dissolution and that, if dissolution rates are sufficiently low, the deleterious effect of environment can be eliminated.[5] Additionally, there also appears to be an upper bound of dissolution rate where larger rates do not significantly cause further degradation of fatigue resistance.[6]

This study discusses the results of a series of experiments conducted on a high purity metal and an alloy with significantly different slip characteristics but similar chemical behavior - Cu and Cu-7.8% Al.

EXPERIMENTAL

Large grained (0.1 mm) specimens of 99.99% Cu and a Cu-7.8% Al alloy of equivalent purity were cyclically stressed at 30 Hz in a tension-tension mode in laboratory air and in

0.5 N NaCl solution with an applied anodic current of $100\mu A/cm^2$.

Two polycrystalline specimens were cyclically stressed in air and solution (with an applied anodic current) to 1% of total life, re-electropolished and tested to failure. These specimens were periodically replicated with cellulose acetate tape and the replicas were examined to determine the development of persistent slip bands as a function of the number of applied cycles. Additionally, single crystals of each material were grown and tested under equivalent conditions to obtain a better understanding of the slip character of the materials as affected by aggressive environments. Specimens were examined by light and electron optics to determine slip characteristics and crack initiation modes.

RESULTS

Polycrystalline pure copper is relatively unaffected by dissolution, while the CuAl alloy shows a marked degradation in fatigue life. In fact, in single crystals, fatigue lives are significantly extended for the pure material while the alloy shows a significant decrease in life (Table I). A comparison of

TABLE I

FATIGUE LIFE VERSUS ENVIRONMENT

Material	Peak Stress	Environment	Average Life
Cu (polycrystal)	140 MN/m^2	air	10^6 cycles
Cu (polycrystal)	140 MN/m^2	0.5N NaCl (anodic current)	5×10^5 cycles
Cu-Al (polycrystal)	254 MN/m^2	air	1.5×10^6 cycles
Cu-Al (polycrystal)	254 MN/m^2	0.5N NaCl (anodic current)	2.5×10^5 cycles
Cu-Al (single crystal)	180 MN/m^2	air	1.1×10^6 cycles
Cu-Al (single crystal) (same orientation)	165 MN/m^2	0.5N NaCl (anodic current)	8×10^5 cycles
Cu (single crystal)	65 MN/m^2	air	10^6 cycles
Cu (single crystal) (same orientation)	65 MN/m^2	0.5N NaCl (anodic current)	5×10^6 cycles

the surface of specimens exposed to air and to NaCl solution shows that, for both the pure metal and the alloy, fatigue generated surface slip is broadened by dissolution and that slip bands are preferentially attacked. Additionally, cyclically induced secondary slip is suppressed and preferential grain boundary grooving is observed (Figure 1). Failure in both materials is primarily intergranular.

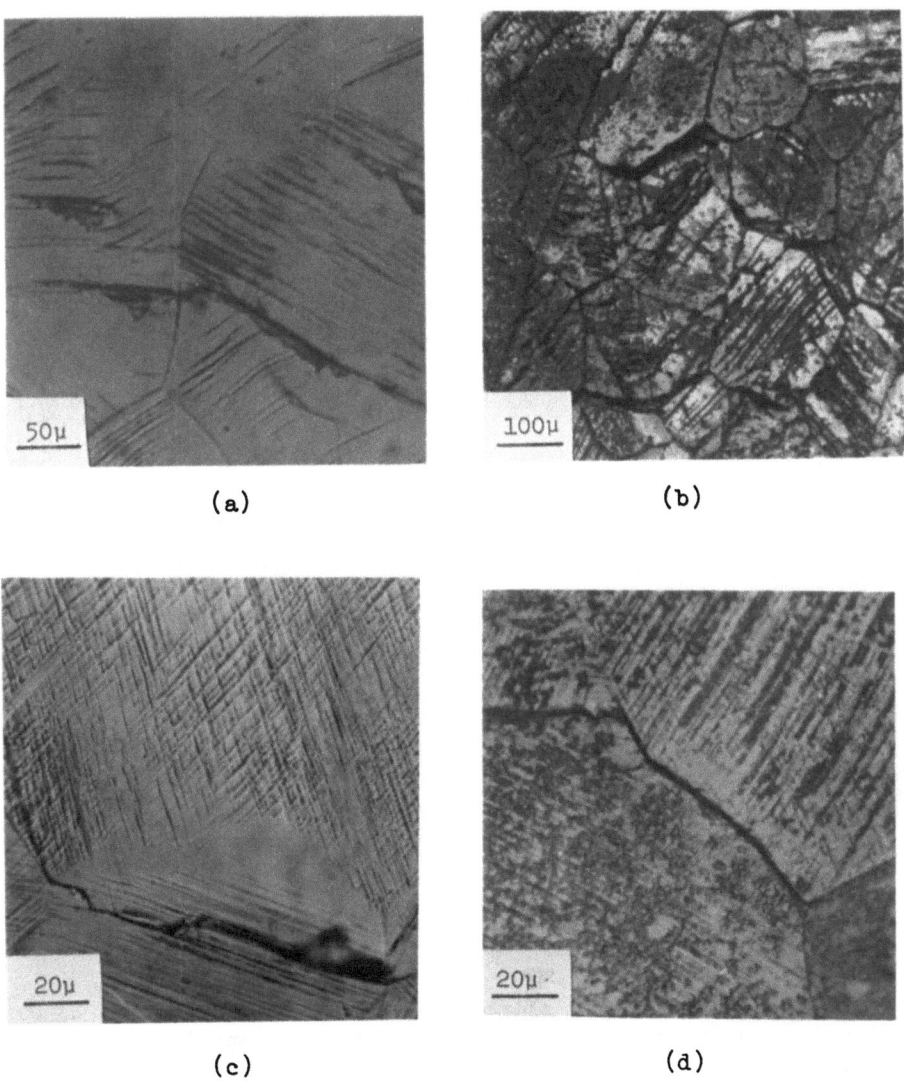

Figure 1 - Fatigue generated slip (a) Cu in air (b) Cu in NaCl solution (applied anodic current) (c) Cu-7.8Al in air (d) Cu-7.8Al in NaCl solution (applied anodic current).

The effect of dissolution on the character of the persistent slip bands is still more clearly shown for the specimens which were intermediately electropolished. In this case, continued cyclic stressing in air produces only a few intense persistent slip bands which eventually lead to transcrystalline crack initiation (Fig. 2a). Surface dissolution during cyclic stressing on the other hand results in a relatively uniform distribution of shallow slip offsets and transcrystalline crack initiation is suppressed (Fig. 2b).

In the single crystals of Cu exposed to air, widely spaced broad bands of intense slip containing large persistent slip bands are observed. Under conditions of anodic dissolution, slip is more generally distributed and the persistent slip bands are considerably less intense (Fig. 3).

In the alloy single crystals, exposed to air, only sharp relatively shallow slip bands are observed and only a single fatal crack occurs. The specimen exposed to the solution shows deep crystallographic grooving along slip traces and numerous secondary crack are observed (Fig. 4).

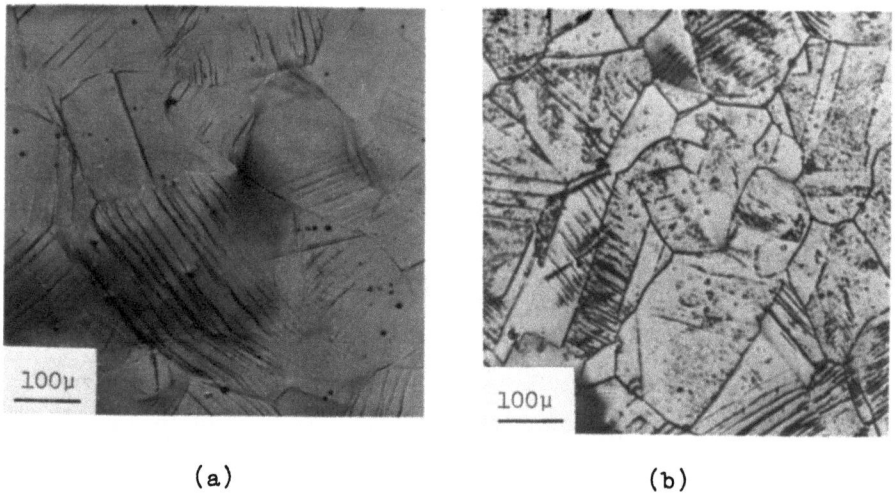

(a)   (b)

Figure 2 - Fatigue generated slip after intermediate electropolishing (a) Cu in air, (b) Cu in NaCl solution (applied anodic current).

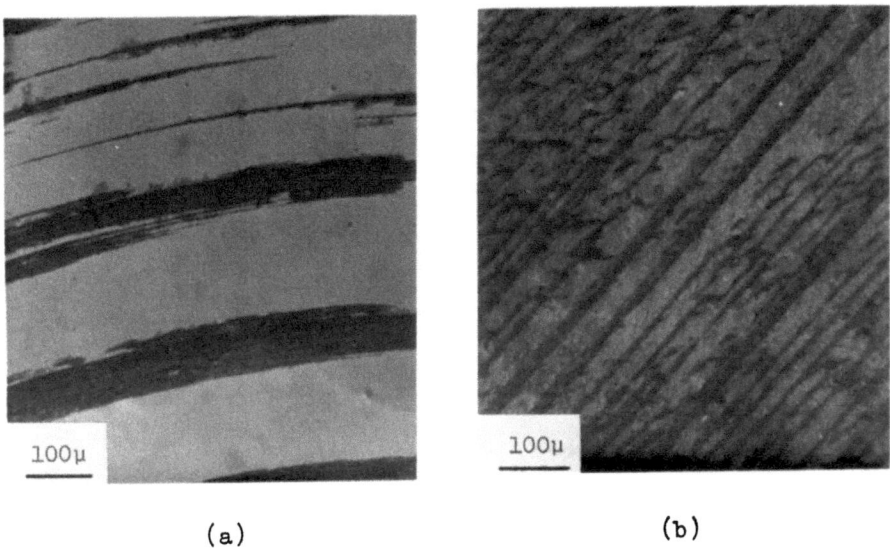

Figure 3 - Fatigue generated slip in single crystals of Cu (a) in air (b) in NaCl solution (applied anodic current).

Figure 4 - Cross sectional view of fatigue generated slip in Cu-7.8Al (a) in air (b) in NaCl solution (applied anodic current).

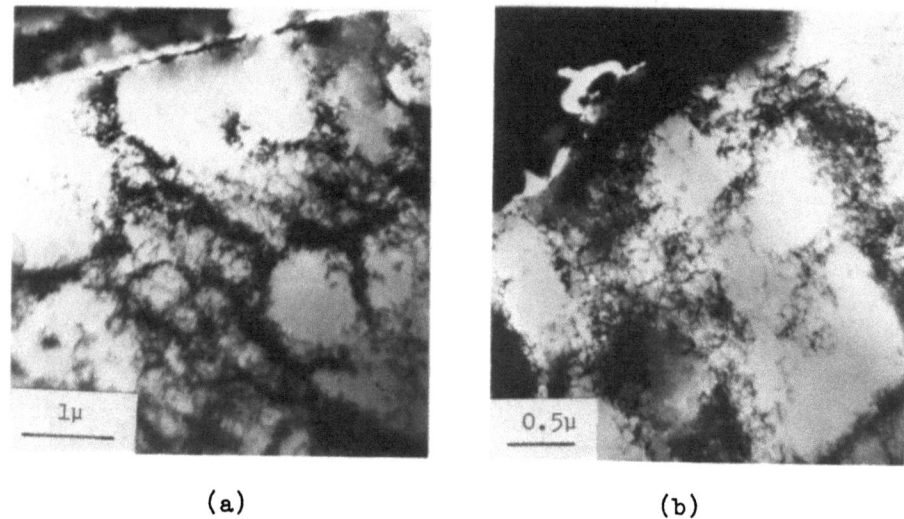

(a) (b)

Figure 5 - TEM of Cu surface after 100% of total fatigue life (a) in air (b) in NaCl solution (applied anodic current).

Figure 5a shows TEM micrographs of the free surface of the Cu after 100% of fatigue life in air and shows the cell structure normally associated with high applied stresses. The cell structure extends to the free surface of the metal with only a small decrease in dislocation density relative to the interior of the material. Exposure to the aggressive environment during cyclic deformation results in less fully defined cell structure at the metal free surface with the normally observed cell structure being retained on the interior of the metal (Fig. 5b).

DISCUSSION

The intergranular failure of both the polycrystalline Cu and Cu-Al alloy have been discussed elsewhere[7,8,9] and will not be repeated here. Rather, in accordance with the theme of the conference, the effect of the environment on the surface connected cyclic plastic deformation of the materials will be discussed.

It is apparent that the role of aggressive environments on cyclic deformation and fatigue failure is a strong function of the slip character of the metal or alloy under consideration. For example, pure copper, which exhibits a wavy slip character and a cellular type of dislocation substructure[10] shows either very little effect of surface dissolution (for polycrystalline materials) or an increase in fatigue life (for single crystals). Cu-7.8%Al, on the other hand, which exhibits a highly planar slip character due to the low stacking fault energy of the alloy,[11] shows a severe degradation in fatigue life for both polycrystalline and single crystal specimens.

Both materials show preferential attack of emerging slip bands, a reduction in the local density of primary slip band offsets, and a suppression of secondary slip offsets. These results can be understood if it is assumed that the material surrounding emerging dislocations is in a higher energy state than the surrounding material and is thus more susceptible to dissolution by the aggressive environment. While it has been shown that there is only a small change in the equilibrium electrochemical potential (and accordingly free energy) associated with deformation, the kinetic energy of atoms associated with mobile dislocations would lower the energy required to dissolve the metal (depolarization in electrochemical terms). Thus, preferential attack of emerging slip bands can be expected to occur. For both materials this dissolution results in an effective "softening" of mean surface regions and the hardening associated with cyclic saturation does not occur. Thus, for the pure copper, the well defined cell structure which is a precursor of crack initiation does not develop. Additionally, subsequent plastic strain can be accommodated in the non-saturated primary slip bands. This process has the effect of producing larger slip offsets and obviates the necessity to accommodate some of the slip in secondary slip bands. Additionally, primary surface slip offsets appear larger and broader and crack initiation is effectively suppressed. For polycrystalline materials, this results in a shift of the crack path to grain boundaries which are preferentially grooved by the dissolution process and cracking occurs because of geometric effects. For single crystals the crack initiation process is significantly delayed. The delay in crack initiation can be associated both with the surface softening effect and with the broadening and blunting of persistent slip band clusters which form intrusions in the metal surface. Thus, surface stress relief is caused by the dissolution process.

For the alloy, on the other hand, slip bands are very narrow, and while secondary slip is suppressed, preferential dissolution is confined to these narrow channels of material resulting in sharp grooves in the alloy surface. These sharp grooves act as stress concentrators thus resulting in poor fatigue resistance upon exposure to aggressive environments. Thus, these results strongly support the suggestion that the slip character of metals and alloys have a profound effect on subsequent corrosion fatigue behavior.

Revie and Uhlig have suggested that enhanced creep of polycrystalline copper wires under conditions of anodic dissolution is caused by the diffusion of dissolution generated divacancies which assist in dislocation climb, and that this process may be applicable to cyclic deformation.[12] The results of this study suggest, however, that their observations of enhanced steady state creep, which also involves the generation of a stable cellular dislocation substructure, might be better explained by the rearrangement of surface connected dislocation cells which are unstable due to the dissolution process and thus increase surface plasticity by a softening effect. This suggestion is further supported by their reported size effect which indicated that larger wires show a smaller effect than smaller wires.

ACKNOWLEDGEMENTS

The authors would like to acknowledge the support of the U.S. Army Research Office - Durham.

REFERENCES

1. McAdam, D. J., Jr. and Geil, G. W., Pitting and its effects on the fatigue limit of steels corroded under various conditions, Proc. Am. Soc. Test. Mat., 41, 696, 1928.
2. Whitwham, D. and Evans, U. R., Corrosion fatigue - the influence of disarrayed metals, J. Iron and Steel Inst., 165, 76, 1950.
3. Stubbington, C. A. and Forsyth, P. J. E., Some corrosion fatigue observations on a high purity aluminum-zinc-magnesium alloy and commercial D.T.D. 683 alloy, J. Inst. Metals, 90, 347, 1962.
4. Ryabchenkov, A. V., Electrochemical mechanisms of corrosion fatigue of steel in electrolyte solutions, Zhur. Fiz. Khim., 26, 542, 1952.
5. Duquette, D. J. and Uhlig, H. H., Effect of dissolved oxygen and NaCl in corrosion fatigue of 0-18% carbon steel, Trans. Am. Soc. Metals, 61, 449, 1968.
6. Duquette, D. J. and Uhlig, H. H., The critical reaction rate for corrosion fatigue of 0.18% carbon steel and the effect of pH, Trans. Am. Soc. Metals, 62, 839, 1969.
7. Masuda, H. and Duquette, D. J., The effect of surface dissolution on fatigue crack nucleation in polycrystalline copper, Met. Trans., 6A, 87, 1975.

8. Andresen, P. L. and Duquette, D. J., The effect of electrochemical dissolution on fatigue crack initiation and propagation in Cu-7.8%Al, submitted to Met. Trans. for publication.
9. Hahn, H. and Duquette, D. J., The effect of surface dissolution on fatigue deformation and fracture of copper, submitted to Acta Met. for publication.
10. Laufer, E. E. and Roberts, W. N., Dislocation structures in fatigued copper single crystals, Philos. Mag., 10, 883, 1964.
11. Feltner, C. E. and Laird, C., Cyclic stress strain response of f.c.c. metals and alloys; II. Dislocation structures and mechanisms, Acta Met., 15, 1633, 1967.
12. Revie, R. W. and Uhlig, H. H., Effect of applied potential and surface dissolution on the creep behavior of copper, Acta Met., 22, 619, 1974.

DISCUSSION

Comment by R. A. Oriani:

Is not vacancy injection consistent with your observations?

Reply:

Dr. Oriani is referring to a model which has been advanced by Revie and Uhlig (Ref. 12) to explain accelerated creep of Cu under conditions of anodic dissolution. I do not believe that this model can adequately explain our results since I can conceive of no driving force for divacancy diffusion to dislocations. Additionally, the fatigue process is known to produce large vacancy concentrations and the excess vacancies which might arise from surface dissolution would not be expected to significantly alter the total vacancy concentration. Finally, the dislocation substructure produced by fatigue is a very complex one consisting of many dislocation intersections and a very large number of vacancies would have to be produced at the surface to significantly alter it.

Comment by R. W. Staehle:

Our measurements show that dynamic straining of nickel single crystals in the film-free case produces no significant increase in the reaction rate. Any significant changes from a base value is probably more related to healing films.

Reply:

It is my opinion that the monotonic stress/strain behavior cannot be compared directly with cyclic deformation. The time scale of the two types of experiments is significantly different and the mode of deformation cannot be directly compared. Additionally, in the fatigue case the presence or absence of a film is not critical to the proposed model which only suggests that material associated with emergent dislocations is preferentially dissolved, either due to energetic considerations or fresh exposure to the environment.

A SURFACE EFFECT SPECIFIC TO CYCLICALLY DEFORMED
B.C.C. METALS - THE SURFACE ROUGHENING OF POLY-
CRYSTALLINE α-IRON DURING CYCLIC DEFORMATION

H. Mughrabi

Institut für Physik,Max-Planck-Institut für
Metallforschung,Stuttgart,Federal Republic
of Germany

ABSTRACT. Deformation of b.c.c. metals in alternating tension and compression causes shape changes because of the so-called asymmetry of slip.During push-pull fatigue of α-iron,initially round single crystals become elliptical;polycrystalline specimens develop a characteristic surface roughness (with the periodicity of the grains) related to the differences in the shape changes of neighbouring near-surface grains.Associated incompatibilities and notch effects at the grain boundaries give rise to intergranular fatigue crack initiation.

1. INTRODUCTION

The mechanical behaviour of b.c.c. metals differs in many respects from that of metals having other crystal structures (for recent reviews,cf.[1,2]).One specific property of b.c.c. metals is the so-called asymmetry of slip (cf.[3]) : dislocations of one and the same Burgers vector glide on different sets of slip planes in tension and compression respectively. This inequality of slip systems in tension and compression gives rise to shape changes during cyclic deformation, even under conditions of net zero strain.

The characteristic features of these shape changes are studied best on single crystal specimens [4,5,6]. The essential results of such observations are summarized in Section 2. In the case of polycrystals, the situation is more complex. Shape changes of individual grains in the interior are controlled by the stringent boundary conditions which provoke multiple slip resulting in small net shape change [5,6,7]. Besides, shape changes in the interior are not accessible to direct observation. On the other hand, grains located at the surface can develop more extensive shape changes because of the relaxed boundary conditions. The importance of these shape changes lies in the fact that the surface is a preferential site for fatigue crack initiation [8]. Observations pertaining to this "surface effect" and their consequences are presented in Section 3.

2. OBSERVATIONS ON α-IRON SINGLE CRYSTALS

The shape changes occurring during cyclic deformation in axial tension and compression were investigated in detail on high-purity α-iron single crystals having "single slip" orientations. Round specimens of initial diameter $d_o \sim 3$ mm were deformed at room temperature at prescribed plastic strain range $\Delta\varepsilon_{pl}$ and at constant cyclic (plastic) strain rate $\dot{\varepsilon}_{pl} = 2\Delta\varepsilon_{pl}\nu$ (ν: frequency). Changes in the shape of the cross section were detected by measuring the dimension d, defined in Fig. 1, as a function of azimuthal angle φ with a high-accuracy micrometer gauge after deformation to certain values of cumulative plastic strain $\varepsilon_{pl,cum} = 2 N \Delta\varepsilon_{pl}$ (N : number of cycles). The azimuthal angle φ is defined with respect to the projection of the primary Burgers vector $\underline{b}_p$ onto the cross section. The principle of the measurements is illustrated in Fig. 1.

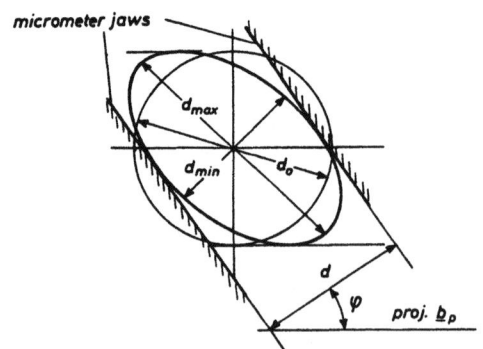

Fig. 1. Principle of measurements.

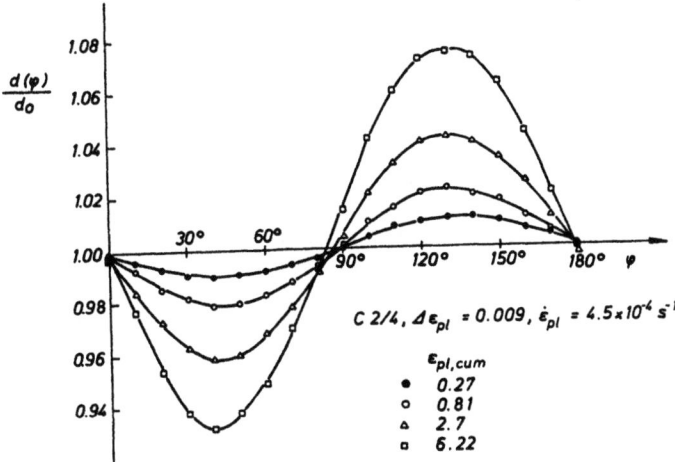

Fig. 2. Example of measurements of $d(\varphi)$ after cyclic deformation to various values of $\varepsilon_{pl,cum}$

An example of measurements of $d = d(\varphi)$ normalized with respect to d_o is shown in Fig. 2. The observed deviations of d/d_o from unity correspond to the development of an almost exactly elliptical cross section with the major and minor axes d_{max} and d_{min} under angles of $\varphi \sim 135°$ and $\sim 45°$ respectively. With increasing deformation the ratio d_{max}/d_{min} increases.

The development of an elliptical cross section has also been demonstrated on niobium single crystals and is now well understood [5,6]. The magnitude of the shape changes depends on specimen orientation [4,5] and experimental conditions [6]:
a) Shape changes are larger for single than for multiple slip orientations.
b) The rate of development of shape changes as a function of $\varepsilon_{pl,cum}$ is highest at the beginning of deformation and decreases subsequently.
c) For a given value of $\varepsilon_{pl,cum}$, shape changes are larger at a high than at a low plastic strain amplitude.
d) Shape changes are more pronounced at high than at low strain rates (or alternatively: at low than at high temperatures, still to be demonstrated).

Observations b) to d) have found an interpretation in terms of cyclic micro- and macrostrains [6], in analogy to unidirectional micro- and macrostrains [9,10], involving the motion of predominantly edge and edge as well as screw dislocations respectively. The listed dependences of shape changes on experimental details are valuable when turning to the case of polycrystals, where such detailed quantitative observations are not possible.

3. OBSERVATIONS ON α-IRON POLYCRYSTALS

The work on polycrystals was performed on cylindrical high-purity α-iron specimens of 2 mm diameter, having a mean grain size of ca. 100 μm. The surfaces of these specimens were polished smoothly before cyclic deformation and were investigated by scanning electron microscopy (S.E.M.) after deformation.

After a small cumulative plastic strain (~1) under conditions of cyclic deformation that were expected to lead to severe shape changes ("high" $\Delta\varepsilon_{pl}, \dot{\varepsilon}_{pl}$), the originally shiny surface became increasingly dull. This could be recognized with the naked eye. The corrugation of the surface that was responsible for this change in the appearance of the surface was revealed best by S.E.M. at low magnifications. Fig. 3 gives an impression of the surface topography at a value of $\varepsilon_{pl,cum}$ at which the tensile peak stress showed first indications of a

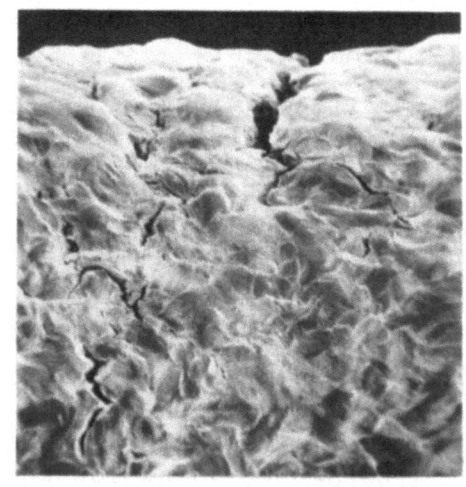

100 μm

specimen axis

Fig. 3. Low-magnification S.E.M. picture of surface topography ($\Delta\varepsilon_{pl} = 0.015, \dot{\varepsilon}_{pl} = 4.5 \times 10^{-3}$ s^{-1}, $\varepsilon_{pl,cum} = 41$)

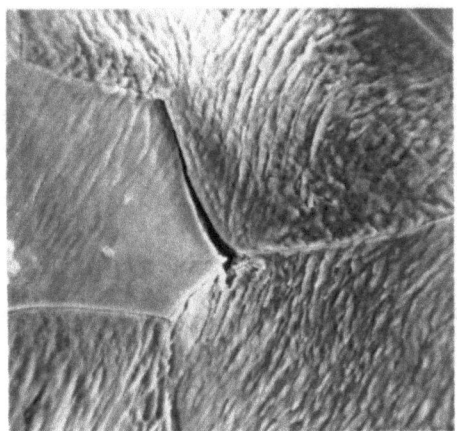

Fig. 4. Micrograph of grain boundary crack initiating transcrystalline crack. Details as in Fig. 3

decrease (N ~ 1400). In addition to the fine-scale roughness due to closely spaced slip bands a long-wave hilly topography can be recognized that is clearly correlated with the periodicity of the surface grain-structure. Microcracks can frequently be observed at grain boundaries lying in the valleys between protruding grains. This is not surprising, since at these sites the grain boundaries are weakened by notch effects. We attribute this behaviour to the differences in the shape changes of neighbouring grains [7].

The relative abundance of <u>fatigue crack initiation</u> at grain boundaries is associated with the severeness of the shape changes, as was clearly demonstrated recently [7]. Under conditions that caused less pronounced shape changes, fatigue crack initiation occured almost exclusively in the transcrystalline mode. This suggests that the incompatiblility of deformation at grain boundaries in connection with notch effects is responsible for intergranular crack initiation.

Another important observation concerns <u>fatigue crack propagation</u>. Even under conditions of predominant fatigue crack initiation at grain boundaries, propagation occurred in the mixed mode or largely transcrystalline, indicating that, once a microcrack had been formed, the stress concentrations at the crack tip were the controlling factor. In support of this view, Fig. 4 shows the initiation of a transgranular crack at the tip of a grain boundary microcrack.

4. SUMMARY

α-iron single crystals and surface grains of α-iron polycrystals have been observed to undergo characteristic shape changes during cyclic deformation.These shape changes were related to the asymmetry of slip of b.c.c. metals.In low-cycle fatigue tests on polycrystals,the differences in shape changes of neighbouring surface grains provoked fatigue crack initiation along grain boundaries.A final assessment of the significance of this effect with respect to fatigue failure of b.c.c. metals must await more detailed studies in dependence on grain size,prior deformation,plastic strain amplitude,strain rate and temperature.

ACKNOWLEDGMENTS

The encouragement of Prof. Dr. A. Seeger and Dr. M. Wilkens and fruitful discussions with Dr. Ch. Wüthrich are deeply appreciated.Sincere thanks go to Fräulein S. Kühnemann and Dr. P. Schlüter for their able support in the S.E.M. studies.

REFERENCES

1) P.B. Hirsch,Proc. of Int. Conf. on Strength of Metals and Alloys.Suppl. Trans. Jap. Inst. Metals $\underline{9}$, XXX,(1968)
2) J.W. Christian,Proc. 2nd Int. Conf. on Strength of Metals and Alloys,Asilomar:ASM,1970,p. 31
3) B. Šesták,3rd Int. Symp. Reinststoffe in Wissenschaft und Technik,Dresden 1970,Akademie-Verlag,Berlin, 1972,p. 221
4) H.D. Nine,J. Appl. Phys. $\underline{44}$,4875,(1973)
5) R. Neumann,Z. Metallk. $\underline{66}$,26,(1975)
6) H. Mughrabi and Ch. Wüthrich,to be submitted to Phil. Mag.
7) H. Mughrabi,accepted for publication in Z. Metallk.
8) C. Laird and D.J. Duquette,Corrosion Fatigue,NACE, Houston,1972,p. 88
9) H.D. Solomon and C.J. McMahon,Jr.,in:Work Hardening, edited by J.P. Hirth and J. Weertman,Gordon and Breach,New York 1968,p. 311
10) A.Seeger and B. Šesták,Scripta Met. $\underline{5}$,875,(1971)

DISCUSSION

Comment by A. S. Argon:

The effect which you have invoked in based on the anisotropy of the Peierls stress which should disappear completely in iron above 100°C for the strain rates which you have used. This should produce a sharp decrease in the rate of roughening above about 100°C. Have you tried this?

Reply:

No, since this would at the same time change tne behavior of the carbon atoms present and their interaction with dislocations (dynamic strain ageing). What we have shown is that deformation at a strain rate lower by a factor of about 100 does lead to much smaller shape changes and this is basically equivalent to what you suggest.

DISLOCATION MOBILITY IN THE SURFACE LAYER

O. Lohne

Institute of Physical Metallurgy,
University of Trondheim, Norway

ABSTRACT. By applying x-ray topography when loading ribbon shaped, pure aluminium single crystals it is shown that the orientation of the glide planes relative to the external specimen surfaces plays a significant role in the beginning of plastic deformation. Dislocation energy considerations may account for this feature. Another factor influencing the observed dislocation mobility may be the way in which the specimens are prepared. Most experiments are done using electropolished samples. When new surfaces are created the dislocation configuration close to the surface changes. This may make dislocations in the surface layer less pinned and when stressed observed to have a higher mobility than internal ones.

INTRODUCTION

Pre-yield phenomena have been of great interest for many years and a lot of experiments are reported. Using etch-pit techniques on copper crystals many investigators [1-5] have observed appreciable dislocation motion and multiplication much before yielding. By using Borrmann X-ray topography Young and Sherril [6-8] have made direct observations of dislocations in lightly deformed copper single crystals with low dislocation density. Lang X-ray topography was used by Nøst and Nes [9] to study dislocation formation due to stress in pure aluminium single crystals at elevated temperatures and by Fremist and Champier [10] to study the same at room temperature. They have all demonstrated that X-rays may be a useful tool in revealing configurations of gliding dislocations.

Most reports seem to conclude that dislocation sources appear to be at or near the crystal surfaces. Dislocation multiplication mechanisms are also reported to operate more frequent in surface regions than in bulk material. In spite of these investigations several aspects of pre-yield plastic deformation of fcc metals, such as the effect of a free surface and details of dislocation formation, motion and multiplication, are still not fully understood. The present paper, being partly a summary of some previously published experiments [11-13] is an attempt to get additional information on these aspects.

EXPERIMENTAL

Two specimens, 1 mm thick, were made by the strain-anneal method of zone-refined aluminium. After the last, critical deformation the specimens were mounted in a furnace at about 500°C before cooled to room temperature at a rate of about 5 degrees per hour. Then the specimens were electropolished, removing about 30 μm on each side, before mounting on a Lang camera. X-ray topographs showed that the specimens were single crystals with a dislocation density of approximately 500 cm^{-2}.

The desired stress was applied by elongating a calibrated spring hooked in a hole in the lower end of the specimen. The external load was increased in steps of about 1.6 g and Lang topographs were taken at each stress level. Because of long exposure time the stress was usually increased once a day.

Due to the way of mounting, the specimens were a bit skew suspended. However, the amount of bending could be accounted for and the stresses given are therefore the total maximum stress.

RESULTS

Fig. 1 is a $(00\bar{2})$ projection topograph and shows part of crystal 1 in the unstressed state. Most of the grown-in dislocations are ending in the surfaces. They are lying in random directions and are partly straight and partly curved. Some of the grown-in dislocations which later act as sources of new dislocations are indicated by arrows.

Fig. 2, which is a (022) topograph, shows that when the spring was fastened to the specimen an uncontrolled stress pulse induced new dislocations looking like straight "tails" on the grown-in dislocations. And when the controlled resolved shear stress was raised to 5.7 ± 1.4 g/mm^2, fig. 3, the new dislocations lengthened.

In fig. 2 and 3 the activated slip plane was seen nearly edge on. In fig. 4, which shows the first stages of dislocation movement in crystal 2, the glide plane is seen not so much edge on. It is here seen that the new dislocations elongate by pushing a short end segment ahead leaving a dislocation line nearly parallel to the surface. The end segments are lying in the dislocations glide plane and have typically a length of about 100 µm.

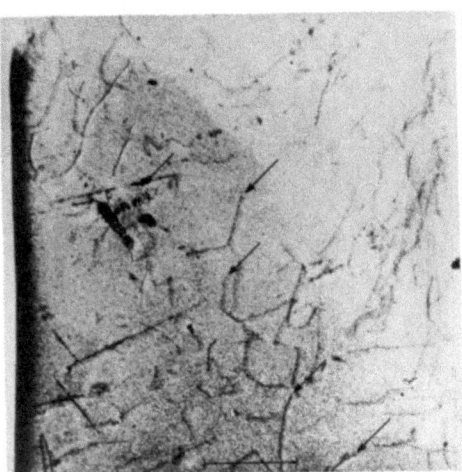 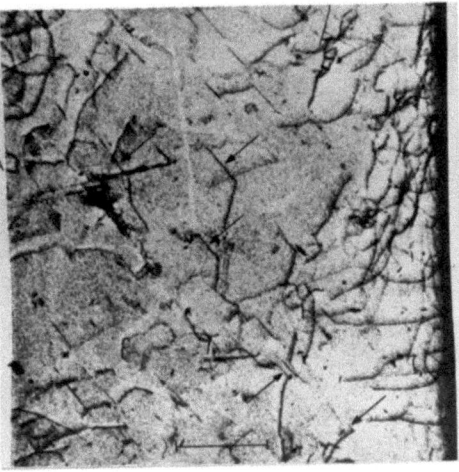

Fig. 1 shows crystal 1 in the unstressed state. (00$\bar{2}$) reflection. Scale mark 1 mm.

Fig. 2 shows crystal 1 after an uncontrolled stress pulse induced straight "tails" on some of the grown-in dislocations. (See arrows in fig. 1 and 2). (022) reflection.

In crystal 1 the new dislocations appear as straight lines crossing the specimens. Near the edge surface, however, they cross-slip making a less acute angle with the edge surface than the primary planes do, fig. 5. A typically length of a cross-slipped segment is 100 µm.

When the stress is increased, the density of dislocations is increased. This is previously reported [11-13] to take place both by more end segments of grown-in dislocations being released and by operation of single-ended Frank-Read sources at places where the new dislocations are being pinned, e.g. near the edge surface where cross-slip has taken place and at places where new dislocations have reacted with grown-in ones.

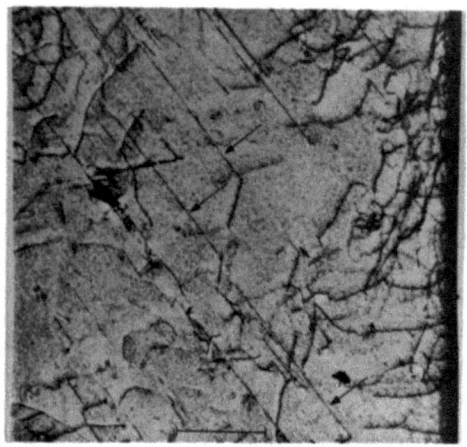

Fig. 3. When the stress is raised to 5.7 ± 1.4 g/mm^2 the new dislocations lengthen. (022) reflection.

Fig. 4 shows the dislocation movement in crystal 2. The activated glide planes are here seen under an angle of about 15 degrees. Scale mark 100 μm.

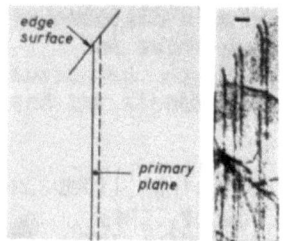

Fig. 5 shows cross-slip close to the edge surface. Scale mark 100 μm.

Fig. 6 shows part of crystal 2 in different reflections. It is here seen that three slip systems are activated. These are not those having the three greatest Schmid factors but are those having dislocation end segments which are least pinned [13].

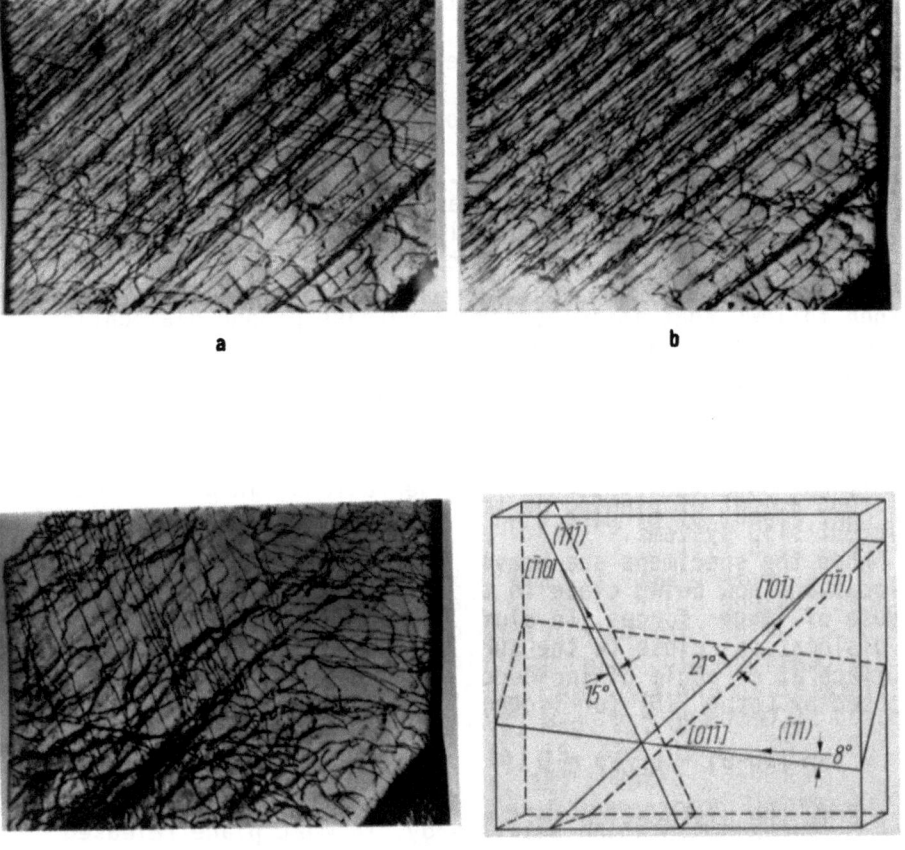

Fig. 6 shows crystal 2 in a: (11$\bar{1}$) reflection,
b: ($\bar{2}$02) reflection and c: (1$\bar{1}$1) reflection. d shows
planes and directions of activated slip systems.

DISCUSSION

Fig. 1-3 show that end segments of grown-in dislocations act as sources for new dislocations. In the first stages of dislocation formation the new dislocations can be traced back to grown-in ones. Due to dislocation multiplication this may be more difficult at higher stresses. It is only the end segments which are activated. The rest of the grown-in dislocations look rather stuck.

In this experiment this feature could be ascribed skew suspension resulting in greater stress close to the surface than in the internal material. However, making an analog experiment in copper Young [7] has also observed that new dislocations lengthen by glide of a short segment ending in the surface leaving a long dislocation line nearly parallel to the surface. He also found that grown-in dislocations are not mobile, and in his experiment skew suspension could be ruled out. The reason why the end segments have much higher mobility than the rest of the grown-in dislocations is therefore presumely not nonaxial loading.

A Frank-Read source close to the surface can be activated at half the stress of an internal one of the same length [14]. It then follows that it is easier to release pinned dislocation segments at the surface if the distribution of pinning points is the same through the specimen. And if the dislocation pinning is the same for dislocations belonging to different slip systems, the topographs would show highest dislocation activity in the surface of the slip systems having the greatest Schmid factors. This is not observed. Instead the observations can be related to different amount of pinning of the dislocations belonging to different slip systems.

When the specimens are heavily electropolished the grown-in dislocations now being close to the surface have to rearrange because of image forces. In aluminium it has been shown [13] that new dislocations close to the surface shorten themselves by hitting a surface at an angle of incidence, θ, according to the equation proposed by Lothe [15] for isotropic crystals

$$E \tan(90-\theta) - \partial E/\partial \theta = 0$$

where E is the dislocation self energy. In what plane released grown-in segments of screw dislocations are rearranging, during and after electropolishing, whether in the primary or the cross-slip plane, depends upon which give rise to the energetic most favourable position. If the energy difference between a position in the primary plane and a position in the cross-slip plane is small, the shortening may take place in both planes successively and thereby creating jogs which pin the end segments. When the energy considerations greatly favour one of the positions the shortening process is presumed to take place in only one plane and thereby creating few jogs on the end segment.

Therefore, in heavily electropolished specimens the end segments are, due to image forces, pulled from the grown-in pinning points like impurities and jogs. And the end segments which obtain the lowest jog density after the rearrangement will be those which may be activated at low stresses. This will depend upon the orientation of the atomic lattice relative to specimen surfaces. And this is also found in experiments [8, 13].

In lightly electropolished samples the dislocation positions obtained during the heat treatment will be close to the equilibrium positions after electropolishing. In this case few or only the outermost segments will be pulled from their pinning points. Loading such samples will then result in higher stresses necessary for releasing dislocations than in heavily polished samples. However, when released the end segments having few jogs will move over large distances, and dislocation multiplication should be observed without increasing the stress above releasing stress. This is what is observed by Vellaikal and Washburn [5] in copper.

REFERENCES

1. Young, F.W., Jr., J. Appl. Phys. 32, 1815 (1961); 33, 963 (1962); 33, 3553 (1962).
2. Tinder, R.F. and Washburn, J., Acta Met. 12, 129 (1964).
3. Petroff, P. and Washburn, J., J. Appl. Phys. 37, 4987 (1966).
4. Marukawa, K., J. Phys. Soc. Japan 22, 499 (1967).
5. Vellaikal, G. and Washburn, J., J. Appl. Phys. 40, 2280 (1969).
6. Young, F.W., Jr. and Sherrill, F.A., Can. J. Phys. 45, 757 (1967).
7. Young, F.W., Jr., Dislocation Dynamics (McGraw-Hill, New York, 1968) p 313.
8. Young, F.W., Jr. and Sherril, F.A., J. Appl. Phys. 42, 230 (1971), 43, 2949 (1972).
9. Nøst, B. and Nes, E, Acta Met. 17, 13 (1969).
10. Fremiot, M. and Champier, G., Int. Conf., Haifa 1969, p 91.
11. Rustad, O. and Lohne, O., Phys. Stat. Sol. (a), 6, 153 (1971).
12. Lohne, O. and Rustad, O., Phil. Mag. 25, 529 (1972).
13. Lohne, O., Phys. Stat. Sol. (a) 18, 473 (1973), 25, 709 (1974).
14. Fischer, J.C., Trans AIME 194, 531 (1952).
15. Lothe, J., Fundamental Aspects of Dislocation Theory, Nat. Bur. Standards, Special Publ. 317, Vol. I, 1970, p 11.

DISCUSSION

Comment by H. Mughrabi:

For aluminium, room temperature is not a low temperature. You usually have more than one slip system. These slip systems intersect each other. Also, you observe under load. So the question arises whether you have some creep occurring immediately after applying the increased load, causing a rearrangement of dislocations. I suspect this, since some of your dislocations are not as nicely curved as one might expect in the stress-applied state.

Reply:

As the exposure time was long, of the order of 10 hours, the technique did not distinguish between dislocation movements during loading or in the first minutes after. The observations showed that most of the grown-in dislocations did not move during the load increments. When comparing dislocation bow-outs with stress this must therefore only be done with new dislocations. Having this in mind I think there is a fair correspondence between dislocation bow-outs and applied stress.

Comment by S. Kitajima:

Are the secondary dislocations near the surface also elongated below the surface?

Reply:

Yes.

Comment by H. J. Engell:

How does the oxide surface layer formed on Al in humid air within some seconds affect your results?

Reply:

The surface oxide layer is supposed to influence the dislocations to a depth beneath the interface of about the same magnitude as the thickness of the surface layer. The answer is therefore that the major part of the new dislocations are not affected by the surface oxide.

Role of Free Surface in Yielding and Easy Glide Deformations of Highly Perfect Copper Crystals

Sadakichi Kitajima

Department of Nuclear Engineering, Faculty of Engineering, Kyushu University, Hakozaki, Fukuoka-city, Japan

ABSTRACT. First, the effects of surface removals both on stress-strain curve and on activation volume in easy glide deformation of highly perfect copper crystals are mentioned. Secondly, characteristics of the dislocation distribution near the surface and in the interior, are described. It is discussed that the surface dislocations of secondary slip system previously elongated below the surface, disturb the motion of multiplied primary dislocations. Last, it is shown that in crystals with edges dislocation multiplications occur from fresh sources secondarily produced near the crystal edges.

1. Specimens and Experimental Methods

Data on the highly perfect copper crystals [1,2] in the present report were summarized in Table 1. Crystals of two types, A and B were used. A-type crystal is parallelpiped with one pair of surfaces parallel to {111} and its axial orientation lies at <123> or near <110>. This {111} surface contains the Burgers vector of the primary slip system. Hereafter this surface is called E-surface and another surface S-surface (see Figs.4,5 and 6). B-type crystal's cross-section is a parallelogram and its four surfaces are all parallel to {111}. So its axial orientation is always <110>. Livingston's etchant [3] and modified Young's one [4,5,6] were used, according to the purpose. The former distinguishes between plus edge dislocation and minus one and the latter between edge and screw.

2. Experimental Results and Discussions

Table 1

Crystal	Shape	Orient.	Size (mm)	No. of S.B.	ρ_0 (cm^{-2})	τ_m (g·mm^{-2})	τ_y (g·mm^{-2})
No.1231	A	<123>	6x5x50	0	1.4x10^3	5.0	—
No.1232	A	<123>	5x5x67	0	2.0	9.8	9.8
No.1233	A	<123>	7x7x68	3	2.0	9.3	9.3
No.1234	A	<123>	8x8x55	2	2.0	11.1	11.1
No.1235	A	<123>	6x8x55	0	2.6	—	11.0
No.1101	A	<110>∿2°	5x4x40	1	1.7	11∿12	11∿12
No.1102	B	<110>	3x3x32	0*	0.3*	8.3	—

No. of S.B.: number of sub-boundary, ρ_0: initial dislocation density, τ_m: shear stress of dislocation multiplications. τ_y: critical resolved shear stress. *: in the observed area.

Fig. 1 [7] shows remarkable effects of surface removals on shear stress (τ)-shear strain (γ) curves at room temperature in crystals having single glide orientation <123>. Definite thickness of surface layer was intermittently removed in each crystal by electropolishing every 0.18%γ at unloaded states. Surface removals increased easy glide strain γ_I and decreased its work-hardening rate θ_I. Such conditions were accelerated with increasing the removing rate, $\dot{d}$, (Table 2 [7]). Further, it should be noted that remarkable serrations carrying sharp drops in load appeared in early part of load-elongation curve when $\dot{d}$ was high, for example, 33 μm/0.18%γ [7]. These facts indicate that surface layer controls the work-hardening in easy glide deformation.

Table 2

Crystal	θ_I/G (x10^{-4})	γ_I (%)	Rate of surface removal, $\dot{d}$
No.1233	1.7	4.2	0
No.1234	2.1	5.5	0
No.1232	1.0	9.7	33 μm/0.18%γ
No.1235	1.4	6.7	20 μm/0.18%γ

The effect of surface removal on the activation volumes of moving dislocations was examined as a function of γ by stress-relaxation test in the crystal with <123> orientation. It was found that the activation volumes in the easy glide deformation of No.1235 of which surface was removed at a rate of $\dot{d}$=20μm/0.18%γ were always larger than those in No.1234 elongated without surface removals, as shown

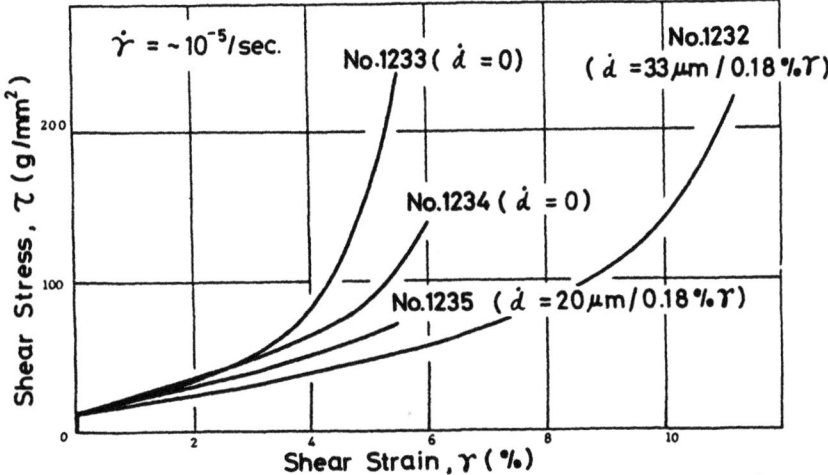

Fig.1 Effects of the surface removal and its rate ($\dot{d}$) on stress-strain curves of highly perfect copper crystals with single glide orientation <123>.

in Fig.2 [7].

The difference in morphology of dislocation arrangements near the as-deformed E-surface was examined in detail between No.1234, directly elongated to γ=1.44 % without removing surface layer, and No.1235, deformed to γ=1.66 % with repeating surface removals at a rate of $\dot{d}$=20 μm/0.18%γ. Most of the multiplied dislocations in both specimens showed a form of streamer: procession of edge dislocation dipoles of primary slip system (Fig.3). The surface removals, however, increased both the length and the interval of streamers [7]. These facts, obtained in easy glide deformations under a constant macroscopic strain rate, indicate that (1) the number of dislocations emitted in a unit from working dislocation source increases with removing the surface layer and as a result (2) the number of active slip bands decreases.

In order to know the effect of tensile orientation on the dislocation arrangements near the E-surface, Nos.1231 and 1101 were deformed to just the beginning of dislocation multiplications and then dislocations emerging on E-surfaces were examined as a function of the distance D from the original E-surface.

On the original E-surface of No.1101 a large number of short streamers [8] were observed, which were distributed rather randomly. But some of them disappeared even in the course of etching for five seconds for revealing etch pits, due to crossing of constituent unlike edge dislocations [8]. Such an instability of streamers was strikingly

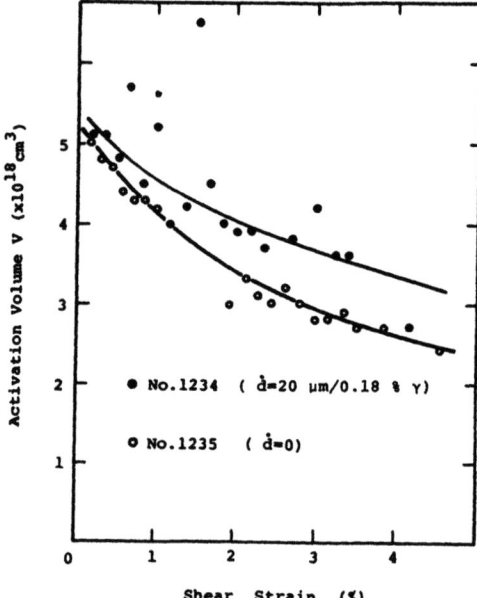

Fig.2 Effects of surface removals on activation volumes in easy glide deformation of highly copper crystals with single glide orientation.

recognized near the original E-surface when thin surface layer (for example; 20 μm) was taken away [8]. So with increasing the total thickness removed, Observable streamers decreased in number, and the forms of dislocation arrangement changed into di-and multi-poles randomly distributed [7]. On the contrary, in the inside distant more than 250 μm from the original E-surface, dislocation multiplication was recognized to occur in band-like regions parallel to the primary slip plane in which the streamers rarely existed [7]. In contrast, in No.1231 the dislocation multiplication occured in band-like regions even on the as-deformed E-surface [8] and most of them increased their width with increasing of D near the original E-surface. But in the interior with D > 100 μm the width did not change remarkably [7]. Dislocation arrangement in such interior bands was similar to that of interior bands of No.1101 [7]. Figs.4 and 5 schematically indicate the conditions of dislocation multiplications just after yielding, in Nos.1101 and 1231, respectively.

Above-mentioned orientation dependence of the dislocation arrangement near the surface should be connected with the degree of activation (lengthening) of surface dislocations of secondary slip systems [7]. Usually, among surface dislocations belonging to various kinds of slip systems, some surface dislocations with large Schmid's factor predominate in motion [8,10,11]. So, such surface forest dislocations

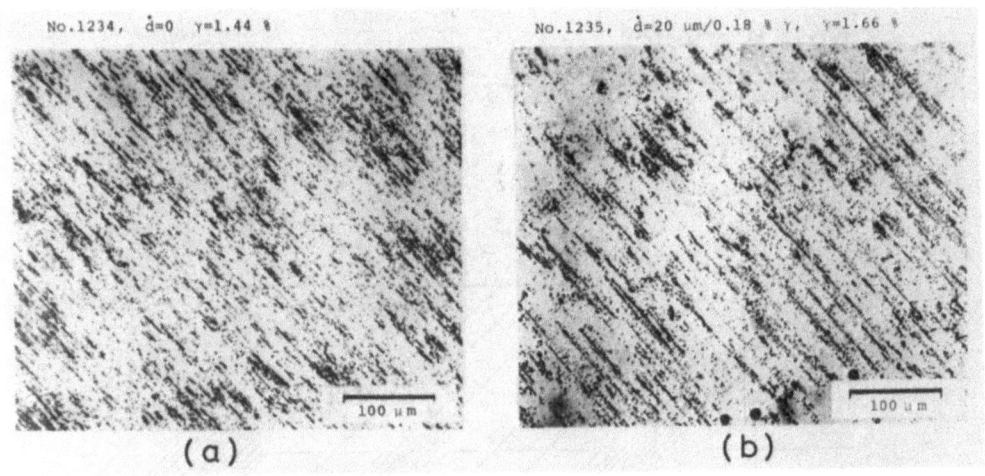

Fig.3 Difference in morphology of dislocation arrangements on the as-deformed E-surfaces between No.1234 crystal, directly elongated to γ=1.44 % without surface removal, and No.1235 crystal deformed to γ=1.66 % with repeating surface removals at $\dot{d}$=20 μm/0.18%γ.

should preferentially elongate beneath the surface and as a result they prevent multiplied primary dislocations from their motion. On the as-deformed E-surface of No.1101, tips of several short streamers were confirmed to lie along a trace of the critical slip plane on which a surface dislocations of the critical slip system was previously elongated [8].

In No.1231 some thin regions where dislocations were multiplied were frequently confirmed to grow wide and long on the E-surface [8]. But when the multiplied dislocations moving within thin region had to travel through high dislocation density area on their ways, this thin region never grew to be wide and long [8]. Therefore the narrower multiplied region near the surface of No.1231 can be understood still on the basis of preferential activation (lengthening) of surface dislocations of the secondary slip systems.

From the experimental works and discussions mentioned above, it is concluded that the motion of multiplied primary dislocations are strongly disturbed with intersecting surface forest dislocations preferentially elongated near the surface. Moving dislocations need waiting time to cut through the forest dislocations. This time would produce a chance of streamer formation with meeting with unlike dislocations coming from opposite side.

If a surface or a near-surface source distant from crystal edges is responsible for yielding and easy glide deformation, multiplied dis-

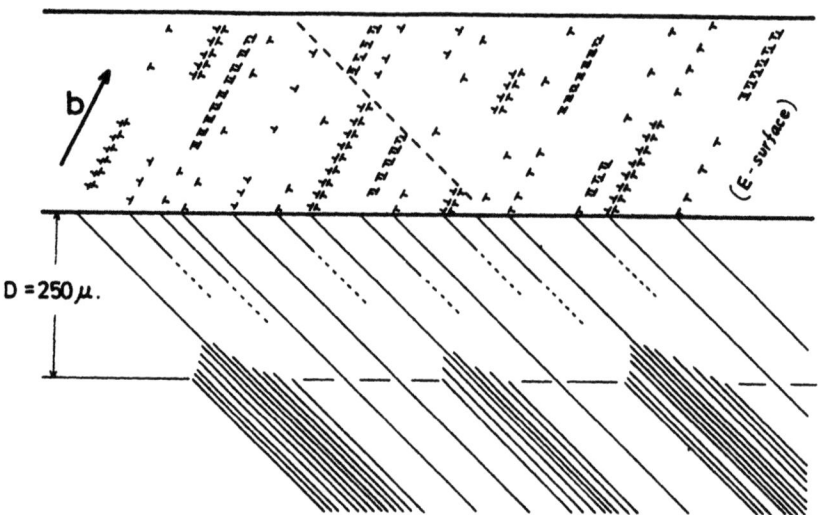

Fig.4 Illustration of dislocation in No.1101 crystal deformed to just the beginning of multiplication at τ_m=12 g/mm^2. On the original E-surface long streamers (procession of edge dislocation dipoles) are randomly distributed. They are unstable and some of them disappear in the course of etching for reavealing pits. In the interior distant more than D= 250 μm dislocation multiplications occur in band-like regions. In such regions a lot of di-and multi-poles of edge dislocations are randomly distributed.

locations of opposite signs should be observed on the same slip plane at either side of the source. In order to examine this point a B-type crystal, No.1102 with sharp edges was employed [12]. The multiplied edge dislocations lying on the single slip plane and emerging on a {111} surface were all same in sign. On the other hand, it was impossible at present to differentiate plus screw dislocations from minus ones by etch pit technique. So, in order to determine the configurations of dislocations multiplied on the single slip planes, variations in the arrangements of multiplied dislocations on the single slip planes were examined on four {111} surfaces with intermittent removals of surface layer. As a result, dislocation sources responsible for multiplication were concluded to locate near crystal edges [12].

Besides the above-mentioned experiments, the following experimental works have been carried out by the author and co-workers: (i) dislocation motion over long distance under stresses less than τ_m [8,

No 1231 crystal

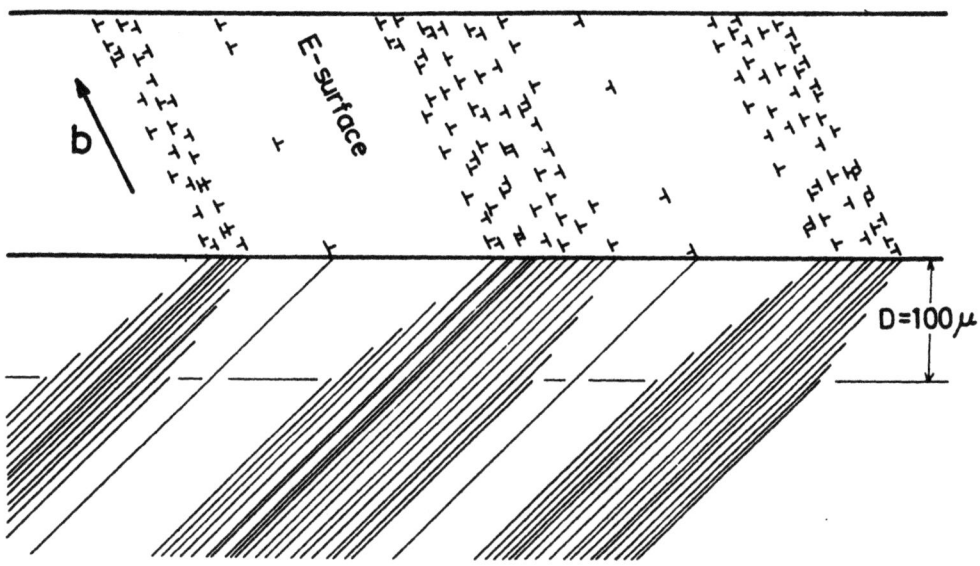

Fig.5 Illustration of band-like dislocation multiplication in No.1231 crystal deformed to $\tau = 7$ g/mm^2. "Dislocation-multiplied" regions increase their widths till D=100 μm near the surface but they are constant in width in the interior with D>100 μm. In dislocation-multiplied regions a number of dislocation-dipoles and multipoles exist randomly.

10,11], (ii) rapid increase in number of grown-in edge dislocations moved at stress levels a little less than τ_m [6], (iii) existence of super jogs [6] and (iv) motion of glissile jogs [6]. Taking into account these works and Young's [13], the dislocation source responsible for at least yielding is concluded to be introduced below the E-surface as a consequence of preferential motion (lengthening) of a grown-in primary surface dislocation with jogs, as shown in Figure 6. Namely, provided that the grown-in surface dislocation is edge dislocation and emerges on E-surface and its jogs are glissile, the grown-in surface dislocation would turn into a glissile surface screw dislocation having no jogs and emerging on S-surface. Such a fresh screw surface dislocation would work as a dislocation source working for dislocation multiplication.

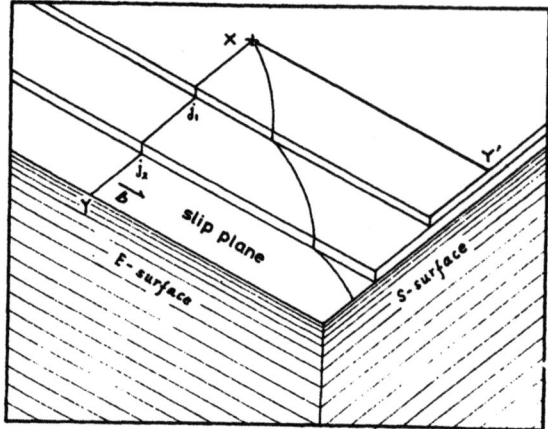

Fig.6 Transformation of a jogged edge dislocation source Xj_1j_2Y emerging on E-surface into a straight screw dislocation XY'emerging on S-surface. XY' would be a dislocation source working for multiplication.

REFERENCES

1. S. Kitajima, M. Ohta and H. Kaieda, J. Japan Inst. Metals (in Japanese), 32, 164, 1968.
2. S. Kitajima, M. Ohta and H. Tonda, J. Crystal Growth, 24/25, 521, 1974.
3. J.D. Livingston, J. Appl. Phys., 31, 1071, 1960.
4. F.W. Young, Jr., J. Appl. Phys., 32, 192, 1961.
5. K. Marukawa, J. Phys. Soci. Japan, 22, 499, 1967.
6. H. Kurishita, Y. Sakurai and S. Kitajima, to be published.
7. H. Tonda and S. Kitajima, to be published.
8. S. Kitajima, H. Tonda and H. Kaieda, Suppl. to Trans. JIM, 9, 740, 1968.
9. S. Kitajima, H. Tanaka and H. Kaieda, Trans. JIM, 10, 10, 1969.
10. M. Ohta, S. Kitajima, K. Aoyagi and H. Kaieda, J. Japan Inst. Met. (in Japanese), 35, 929, 1971.
11. M. Ohta, Y. Miura and S. Kitajima, Trans. JIM, 14, 256, 1973.
12. H. Kurishita, Y. Sakurai and S. Kitajima, to be published.
13. F.W. Young, Jr., Dislocation Dynamics, ed. by A.R. Rosenfield et al., McGraw-Hill, New York, 313, 1968.

DISCUSSION

Comment by J. T. Fourie:

The region of flow stress investigated by Prof. Kitajima is in the range 10–20 gm mm^{-2}, and the sources of dislocations can be considered to be acting relatively independently of the surrounding dislocation structure. Therefore, it is probably not reasonable to compare this system with that which prevails in the case of the soft surface effect where the flow stress is in the range of 100 gm mm^{-2} (end of stage I) to 5 kg mm^{-2}.

Reply:

Our hard surface effect would be valid at least till the end of stage I, because our results clearly show that the activation volumes in stage I are always larger in case of intermittent surface removal than in the case of no surface removal, irrespective of the shear strain. Shear stresses in stage I of our crystal are in the range 10–70 g/mm^2. These would be small, compared to your crystal. But this stress range difference is due to the different initial dislocation density, and it would not give an effect on the characteristics of stage I. So, if the as-deformed surface is maintained without being removed in the cutting processes of your as-deformed parent crystal, you would obtain the hard surface effect even at the end of stage I. But in stage II, not only secondary slip but also deformation bands appear. So the circumstances might be different from that in the stage I, as you say.

Comment by A. S. Argon:

You have presented evidence for dislocation groups of the same sign at the beginning of stage I deformation based on the black-white diffrentiation of etch pits. Are you aware of the fact that in etching copper with Livinston's etch all pits initially etch black whether the dislocations are positive or negative? Black-white differentiation occurs only when the size of the pits exceeds 5 μm. What was the size of your pits?

Reply:

When the differentiation between plus and minus edge dislocations was needed, we used large pits. I don't know the critical size. But the size of black-white pits constituting a streamer is about 5 μm, as you say.

MICROCINEMATOGRAPHIC STUDIES OF THE DEVELOPMENT OF SLIP BANDS

Ch. Schwink and H. Neuhäuser

Institut A für Physik, TU Braunschweig, Germany

ABSTRACT. The development of slip bands on the surface of metal crystals can be resolved by microcinematography enabling to measure the rate of increase of the slip bands both in depth and in length. From the data we deduce the average velocity of the dislocation groups, the mutual distance of moving dislocations and the total number of active groups on the crystal. For n-irradiated Cu, α-brass and pure Cu the connection of the surface properties with those in the interior is discussed.

1. THE MICROCINEMATOGRAPHIC (MC) METHOD

The macroscopic deformation by slip processes of dislocations on slip planes does not consist in the movement of many isolated dislocations statistically distributed in the crystal. Rather groups of dislocations move in a correlated way on one or a few neighbouring slip planes generating the well-known slip bands. Number, extension and density of the moving dislocation groups are characteristic for each material and depend on deformation rate and temperature. Thus, for an understanding of the dynamics of macroscopic deformation the processes of cooperative glide are of basic importance [1].

Microcinematography of slip bands allows the immediate observation of the movement of dislocation groups. Its simple principle is to study the growth of slip bands in depth and length by a film camera through a light microscope <u>during</u> the deformation of the crystal. Experimental details are given in [1,2,3].

As Fig.1 shows a fully developed slip band in general consists of serveral slip lamellae each of which comprises 20 - 100 dislocations distributed on a few slip planes. EM surface replicas allow to deduce the step height S and the breadth B of the total slip band

and in favourable cases height S_L and breadth B_L of the individual lamellae.

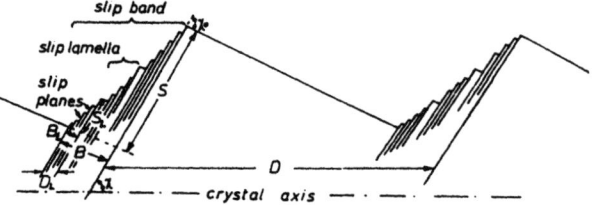

Fig. 1 Fine structure of slip bands

For the cinematography dark field illumination of the growing bands is applied. The slip lamellae representing ultramicroscopic objects scatter the incident light and produce an intensity profile on the film. With increasing step height this intensity increases and can be evaluated photometrically frame by frame. The density-time-curve of Fig.2 gives an example of band growth [cf.3]. The stepwise structure reflects the formation of successive slip lamellae.

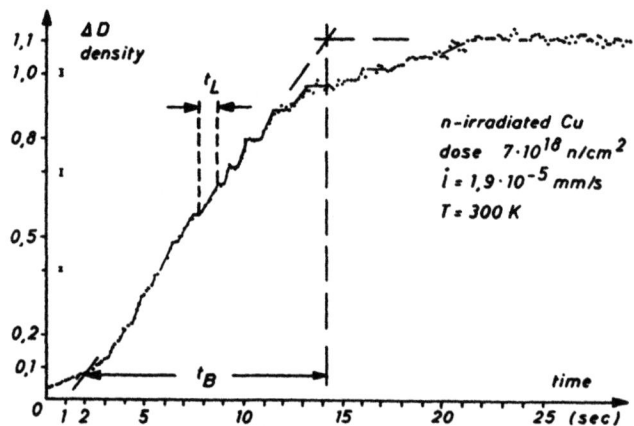

Fig. 2 Slip band growth (Lüders zone) from photometrically evaluated films, frame rate 7,5 fr/s.

From such curves representing the growth in depth of slip bands we obtain two import quantities: The time of formation of the whole band, t_B, and of a single lamellae, t_L. A third information we can get from the MC method is the growth rate in length, v_x, of bands. To this end the development of the same band is measured at two different positions Δx apart, the corresponding density-time-curves being shifted in time scale by Δt. From this we get immediately

$v_x = \Delta x/\Delta t$.

In the following B, S, t_B etc. denote average values determined from measured distributions of these quantities. Their scatter is large due to the statistical nature of the slip process itself.

While the formation times, t_B and t_L, are measured using low frame rates of about 10 - 50 fr/s, the determination of v_x requires much higher film speeds of about 200 - 500 fr/s. Frame rates up to 5000 fr/s were already applied [3]. The resolution of the MC method achieved by us may be characterised by the following data [1,3]: Least detectable step height changes (depending on film speed): 20...100 Å; least detectable growth in length (Δx): 10 μm; highest resolution in time: 0,2 ms.

2. SURFACE PROPERTIES OBTAINABLE BY THE MC METHOD ALONE

Slip band studies yield immediate informations on properties of the surface region of the investigated crystal. Among the informations which can be obtained by the MC method alone we discuss the most important ones.

2.1 The average velocity of the dislocation groups, v_x.

The growth rate in length of the slip bands immediately gives the mean velocity of the dislocation groups (of given character) in the lamellae [3,1]. If v_x is controlled by the thermally activated overcoming of obstacles by dislocations, the effective free enthalpy, ΔG, and the activation volume, V, can be deduced from it: The waiting time, t_w, of a dislocation before the obstacle gives with the usual Arrhenius-ansatz and the mean obstacle spacing l the mean velocity v:

$$v = l/t_w = 1/t_o \exp(\Delta G/kT) = b\nu_D \exp(-\Delta G/kT).$$

Here $1/t_o = b\nu_D$ is used [4], b = Burgers vector, ν_D = Debye frequency. In the simplest case $v = v_x$, yielding [3,1]:

$$\Delta G = kT \ln(\nu_D b/v_x). \tag{1}$$

With $\Delta G(\tau_s) = G_o - V\tau_s$, τ_s = effective stress, the activation volume V is (taking account of the Friedel relation) [3,1]:

$$V = 1,5 \, kT \, d(\ln v_x)/d\tau. \tag{2}$$

Thus, the MC method allows to determine ΔG and V for the surface region alone, while the known macroscopic methods - based on measurement of stress variations induced by changes of deformation rate and temperature [5] - yield average values for the crystal as a whole.

2.2 The average dislocation spacing within a group, d.

At the crystals front surface we have $\dot{S}_L = b\dot{n}$ ($\dot{S}_L$ = growth rate of a slip lamellae in depth, $\dot{n}$ = frequency of dislocations reaching the surface at this place) and $\dot{n} = v_x/d$, where d = average distance parallel to the surface between successive dislocations projected into one slip plane [3,1]. It follows $d = bv_x/\dot{S}_L$.

If the dislocations are of pure screw character which on the front surface seems to be frequently the case [3], the observed v_x equals the velocity v_s of a group of screw dislocations. Because of $\dot{S}_L = S_L/t_L$, where lamellae step height S_L and formation time t_L are measurable quantities, d can be completely determined by experiment. In all cases investigated so far d≫b resulted, as to be expected [3,1]. Therefore, we define the <u>local</u> (surface) areal density ρ of mobile dislocations as $\rho = 1/hd$, h = crystallographic distance of neighbouring slip planes.

2.3 The number of active dislocation groups, N_{aL}.

Within a time interval Δt of strain the total number of slip bands on the crystal surface increases by ΔN_B simply to be counted under a light microscope (LM). With the mean formation time t_B of a single band we get the number n_{aB} of slip bands which are active at any moment during the deformation, $n_{aB} = t_B \Delta N_B/\Delta t$ [2].

Quite analogous, the number of active slip lamellae (dislocation groups) within each single, active band is $n_{aL} = t_L \Delta n_L/\Delta t = t_L S_B/S_L t_B$ ($n_L = S_B/S_L$ = mean number of lamellae per band) and can also be measured, if the growing lamellae are resolvable in time [1]. Now, the total number of active groups is $N_{aL} = n_{aB} n_{aL}$.

The mean width of a lamellae on the surface being D_L (cf. Fig.1) the active lamellae taken all together cover a length $l_a = N_{aL} D_L$ measured in axis direction of the crystal. Within this length slip in the surface region may be considered homogeneous. Therefore, we call l_a "active crystal length" [6]. The normally used average dislocation density (in the surface region) then is $\bar{\rho} = \rho l_a/l_o$, l_o = crystal length.

Till now only the cinematographic method allows to determine the local movable dislocation density and the number of active dislocation groups resp. the active crystal length separately.

Example. Representative data for Cu crystals of medium orientation irradiated with 2×10^{18} neutrons/cm^2 and deformed into the Lüders zone region (τ = 2,3 kp/mm^2) with a rate $\dot{l}$ = 1,8×10^{-3} mm/s at T = 300 K are:
EM: S_B = 4500 Å, S_L = 225 Å, D_L = 80 Å;   LM: N_B = 5 s^{-1};
CM: t_B = 1,8 s; t_L = 5,5 10^{-2} s; v_x = 1,6 mm/s.

Calculation: $d = bv_x t_L/S_L = 1$ (μm); $\rho = 1/hd = 4,7 \cdot 10^{11}$ (cm^{-2});
$N_{aL} = N_B t_B t_L S_B / S_L t_B = 5,5$; $l_a = N_{aL} D_L = 440$ (Å).

3. CONNECTION BETWEEN SURFACE AND BULK PROPERTIES

All quantities discussed so far relate to the surface region of the crystals. The answer to the further reaching question whether and which conclusions can be drawn from surface properties to corresponding ones in the interior depends on the material considered. Relevant examples are: n-irradiated Cu, 70/30 α-brass and pure Cu.

n-irradiated Cu. Various EM and TEM investigations give evidence that the slip processes in the interior don't differ from those near the surface. Perhaps most convincing are the observations showing that the cleared channels which appear in deformed n-Cu are correlated with surface slip steps indicating that the "slip bands are a true representation of the bulk deformation behaviour" [7]. Another argument corroborating this fact is that the effective free enthalpy, ΔG, determined by the surface method (v_x) agrees rather well with that deduced from the bulk method [3].

70/30 α-brass. The comparison of TEM and slip line results [8] as well as of ΔG-values from bulk and surface studies [9] favour the view that also in α-brass interior and surface region behave nearly equally as to the plasticity. Values for v_x, d, N_{aL} were determined and found similar to those for n-Cu [1,9].

But even in the case of a radial variation of plastic properties of cylindrical crystals further reaching conclusions can be drawn [1], if just the structure and distribution of obstacles controlling the shear stress is the same throughout the sample volume. Then, the activation volume V is independent of the sample radius r and given by

$$V = 1,5kT \ln(v_2/v_1)/(\tau_2 - \tau_1), \tag{2'}$$

where the differentials of equ.(2) are replaced by measurable differences. The difference of the stresses τ_2 and τ_1 belonging to deformation rates $\dot{l}_2$ and $\dot{l}_1$ is - at least to a good approximation - independent of r, even if there is a radial stress gradient. Therefore, v_1/v_2 will be constant over the crystal cross section, too. That means that variations of the velocity of dislocation groups induced by strain rate changes are near the surface the same as in the interior. Consequently, variations of the mean dislocation spacing (cf.2.2) and of the number of active groups (and of active crystal length) (cf.2.3) will in the interior be the same as near the surface and can be determined from experiments at the surface alone.

Pure Cu. In crystals of medium orientation a radial gradient of the stress as well as of the dislocation density develops during deformation. In stages II and III the latter according to TEM-studies increases from surface to the interior [10], slip line studies [11] of ρ_{sl}

and L_{sl} (the mean slip line density and length developed in a given strain interval) yield an increase of ρ_{sl} to the interior, too, and further $\rho_{sl}L_{sl}$ = const. over the sample cross section. In stage II the activation volume connected with the intersection of forest dislocations by primary ones, is inversely proportional to τ. Thus, we have according to equ.(2'): $\Delta\tau/\tau \sim \ln(v_2/v_1)$, $\Delta\tau = \tau_2 - \tau_1$. V and with it $\Delta\tau/\tau \sim \Delta\rho_s/\rho_s$ (ρ_s = density of secondary dislocations)[1] are now expected to vary with r. $\Delta\tau/\tau$ and $\Delta\rho_s/\rho_s$ can be determined by static measurements like those in [10]. Suppose one could determine $\Delta\tau/\tau = f(r,\tau)$ in this way, then the combination with the dynamically at the surface measured ratio v_2/v_1 would yield $v_2/v_1 = g(r,\tau)$ (and additionally taking acount of $\dot{l} \sim \rho l_a v$ = const $\rightarrow (\rho l_a)_1/(\rho l_a)_2 = g(r,\tau)\dot{l}_1/\dot{l}_2$). That means, static measurements together with dynamic ones at the surface alone can in principle afford informations about velocities and activation parameters in the interior. Especially, if the mean athermal stress τ_G within the surface region were known, $\Delta G_o = \Delta G - V(\tau - \tau_G)$ otherwise inaccessible could be evaluated from surface v_x-measurements.

Till now, systematic quantitive results about the increase of l_a and τ (resp. ρ_s) into the interior are lacking, v_x-studies being rather difficult for pure Cu were not yet undertaken. To perform such investigations seems to be of high interest.

REFERENCES

1. H. Neuhäuser, Habilitationsschrift, Braunschweig 1975.
2. Ch. Schwink and H. Neuhäuser, phys.stat.sol. 17, 35 (1964).
3. H. Neuhäuser and R. Rodloff, Acta Met. 22, 375 (1974); phys.stat.sol. (a)24, 549 (1974).
4. W. Frank, M. Rühle and M. Saxlova, phys.stat.sol. 26, 671 (1968).
5. e.g. G. Schoeck, phys.stat.sol. 8, 490 (1965).
6. Ch. Schwink, phys.stat.sol. 18, 557 (1966).
7. J.V. Sharp, rad. effects 14, 71 (1972).
8. H. Neuhäuser, J. Koropp and R. Heege, Acta Met. 23, 441 (1975).
9. H. Neuhäuser, H. Flor, in preparation.
10. J.T. Fourie, Phil.Mag. 17, 735 (1968); 21, 977 (1970). H. Mughrabi, phys.stat.sol. 39, 317 (1970); (b)44, 391 (1971).
11. N. Himstedt and H. Neuhäuser, Scripta Met. 6, 1151 (1972).

THE PLASTIC FLOW STRESS IN THE CORE AND SURFACE
REGIONS OF DEFORMED CRYSTALS

J.T. Fourie

National Physical Research Laboratory, C.S.I.R.,
Pretoria

ABSTRACT. The pre-macro-yield region of load-strain curves of
previously deformed crystals is analysed. Two distinct linear
regions in the pre-yield region can be predicted for crystals
unloaded but not polished, while three such linear regions are
expected for crystals unloaded and then polished electrolytically
or chemically. The slopes of linear regions and the load values
at which abrupt changes in slope occur can be used for calculating the flow stress distribution in the crystal. The results
obtained by this indirect method agree with those from the
direct method of crystal sectioning and thus confirm the
presence of a soft surface effect.

Notation

τ — resolved shear stress
τ_p — resolved shear stress in parent crystal
τ_c — resolved shear stress in core
τ_s — resolved shear stress in surface
σ — engineering stress
σ_p — engineering stress in parent crystal
σ_c — engineering stress in core
σ_s — engineering stress in surface
F — total load on crystal
F_c — load on crystal core and yielded surface at core yield
F_s — load on surface region and core at surface yield
m — slope of curve
Δr — radial thickness of material polished away.
$\emptyset_p$ — cross-sectional area of parent crystal
$\emptyset_c$ — cross-sectional area of core region
$\emptyset_s$ — cross-sectional area of surface region

1. INTRODUCTION

The question of non-uniform strain hardening in f.c.c. metal single crystals has been under discussion for a number of years. A phenomenon known as the soft-surface effect was discovered in 1966 by Fourie [1] and subsequently investigated by this author [2,3] by direct measurement of the flow stress in crystals sectioned from a larger pre-deformed crystal. These results were subsequently confirmed by Mughrabi [4] who also expanded on theoretical aspects of the concept in later papers. However, Kramer [5] deduced from the results of experiments which involved electrolytic polishing of crystals after deformation with subsequent further deformation, that the surface region was harder than the interior. The crucial assumption which he appears to have made can be explained with reference to Fig. 1. It is clear

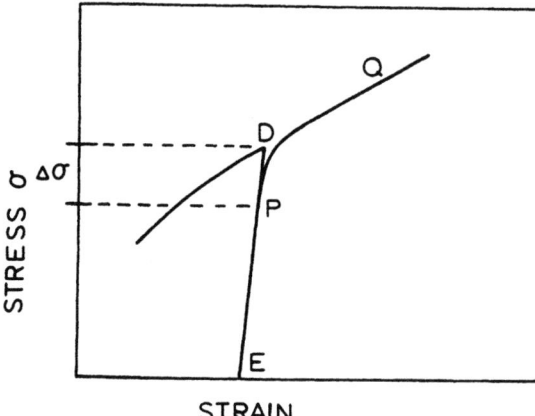

Fig. 1. The plastic yielding of a pre-strained crystal after electrolytic polishing.

that the stress-strain curve of the crystal deviates from the linear elastic region EP at the point P, which is well below the point of unloading (D). Kramer [5] assumed that the whole crystal yielded at P and therefore that electrolytic polishing of the surface had reduced the initial flow stress by an amount $\Delta\sigma$. From this assumption he then concluded that the pre-strained crystal (before polishing) must have had a thin, hardened layer of dislocation "debris" very close to the surface and that this layer controlled the flow stress of the bulk crystal.

It will be shown in this paper that Kramer's basic assumption is incorrect. Furthermore it will be shown that a careful observation of the pre-macro-yield phenomena not only confirms the soft-surface concept but also provides a new independent method for calculating the flow stress distribution in a pre-strained crystal.

2. EXPERIMENTAL

Single crystals of copper were grown from the melt of a 99.999% pure material from the American Smelting and Refining Company. Two geometrical crystal shapes were grown - a rectangular shape with cross-section 2 mm x 10 mm and 80 mm long, and a dumb-bell

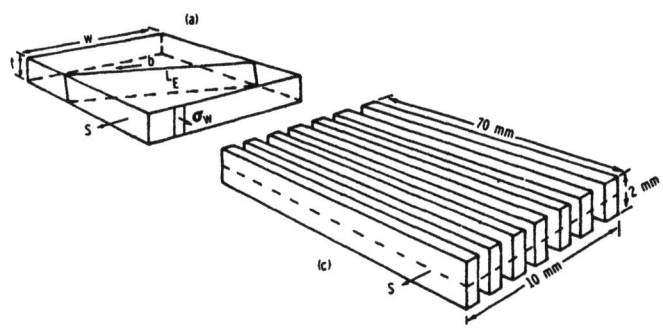

Fig. 2. The orientation of crystals used in this investigation and the procedure of sectioning.

shape with a gauge length of 50 mm and a diameter of 5mm.

Rectangular crystals were sectioned into component crystals (Fig. 2) after straining by means of an electrolytic jet technique which is described in detail in reference [6]. The rectangular cross-sectional areas of the sectioned crystals were obtained by accurate weight and length measurements. Each sectioned crystal was pulled individually to determine its flow stress. The general accuracy of the technique is demonstrated by the consistency of the results.

In the case of the dumb-bell-shaped crystals, various thicknesses of surface layer were removed after tensile deformation had been applied. Chemical polishing was employed for some crystals while electrolytic polishing was used for others. The thickness of the layer removed from the cylindrical surface was calculated by measuring the gauge length and weight loss. The thickness of the layer could be determined to an accuracy of 10^{-4} mm. After removal of the layer, the dumb-bell-shaped crystals were pulled again to determine the overall flow stress of the remaining crystal.

The tensile machine employed was an Instron TT-M-L machine equipped with Fleischer-type grip attachments which minimised grip effects during deformation. The rectangular crystals were clamped at their ends whereas the dumb-bell-shaped crystals were held by split grips which matched the dumb-bell ends exactly.

3. RESULTS

In Figs. 3 and 4 high resolution load versus strain curves of rectangular pre-strained crystals which had been pulled to a resolved shear stress of 10.83 N mm^{-2} are shown. Both crystals were immediately unloaded after reaching this point. The curve in Fig. 3 was obtained by pulling one of these crystals (without

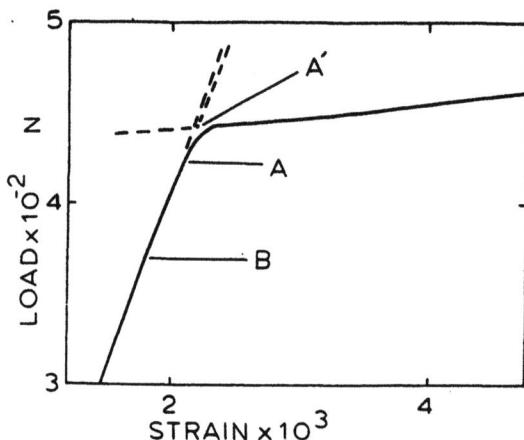

Fig. 3. A high-resolution load-strain curve of a rectangular crystal in the pre-macro-yield region.

any further treatment) to macro-plastic yield. It should be noted that the curve deviates from the elastic curve at point B, but that the new segment of curve BA is also linear. At point A there is a further sharp deviation and the curve then becomes one of a continuously changing slope. The curve in Fig. 4 was obtained after the unloaded crystal was electropolished to remove a surface layer of 0.014 mm, and then repulled to the point of macro-plastic yielding. Here it should be noted that there are now two distinctly linear segments which deviate sharply at points C and B. The non-linear deviation occurs again at point A. The curve in Fig. 5 was obtained in the same way from a crystal which was chemically polished after unloading. The thickness of surface layer removed was 0.021 mm and the linear deviations are more pronounced.

In Fig. 6 the flow-stress distribution in rectangular crystals which were pulled to different values of the resolved shear stress, is represented in the form of histograms. The ratio τ/τ_p is given as a function of position along the width of the crystal, where τ is the resolved shear stress in the sectioned

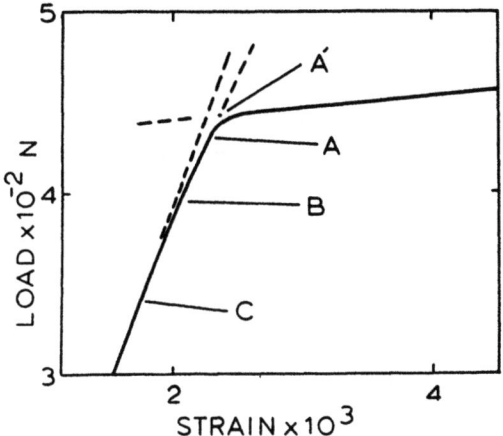

Fig. 4. A high-resolution load-strain curve of a rectangular crystal in the pre-macro-yield region after electrolytic polishing.

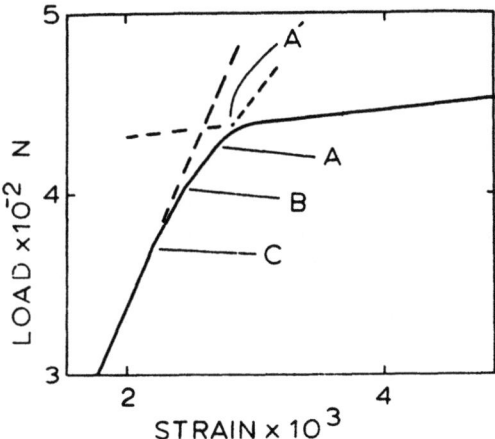

Fig. 5. A high-resolution load-strain curve of a rectangular crystal in the pre-macro-yield region after chemical polishing.

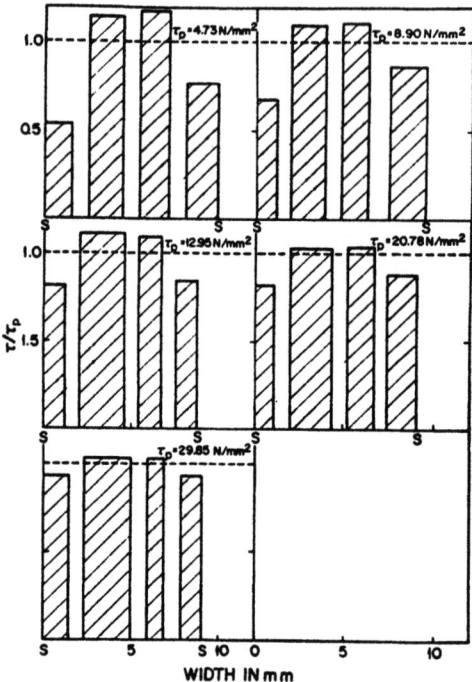

Fig. 6. Histograms of the flow stress distribution in deformed rectangular crystals.

component (see Fig. 2) and τ_p that in the parent crystal before it was unloaded for sectioning. It should be noted that τ/τ_p for the core region decreases with stress whereas that for the surface increases, both approaching the value $\tau/\tau_p = 1$.

In Fig. 7 the results of experiments on cylindrical dumbbell-shaped crystals are represented. Here τ is the resolved shear stress for plastic flow in a pre-strained crystal after a surface layer of thickness Δr had been polished away (τ_p is the same as before). It is clear that τ/τ_p increases with Δr and further that there is no difference between the effects of chemical and electrolytic polishing.

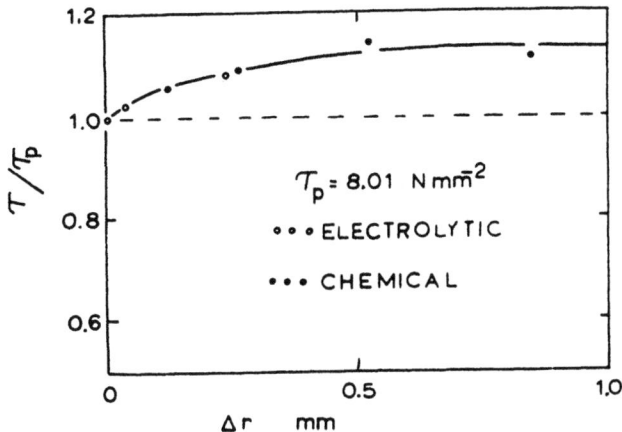

Fig. 7. The effect of removing a surface layer of thickness Δr by polishing of a pre-strained crystal, on the ratio τ/τ_p

4. DISCUSSION

The existence of a soft-surface region has been proved by direct means, in numerous investigations, by electrolytic slicing of crystals [1,2] and by transmission electron microscopy [3,4]. An indirect method for measuring the flow stress distribution quantitatively will now be considered.

In Fig. 8 the cross-section of a cylindrical crystal is represented. It is assumed that there are two regions in the crystal – the core region with cross-section $\emptyset_c$ and the surface region with cross-section $\emptyset_s$. The plastic flow stress of the two regions are σ_c and σ_s with $\sigma_s < \sigma_c$. The cross-section of the whole crystal is

$$\emptyset_p = \emptyset_c + \emptyset_s . \tag{4.1}$$

When such a crystal is subjected to tensile strain, only the overall <u>load</u> is recorded and the curve obtained cannot be interpreted in a simple manner. In the linear elastic region OB, the stress will rise uniformly in the crystal and the stress-strain curve will have the same form as the load-strain curve. However, when the load reaches the value F_s, where

$$F_s = \sigma_s \emptyset_p , \tag{4.2}$$

that portion of the cross-section indicated by s will begin to

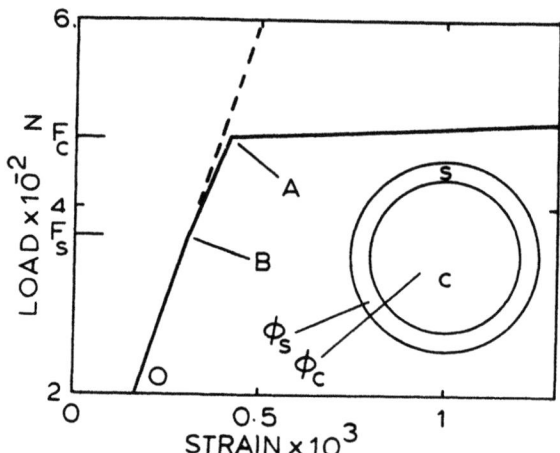

Fig. 8. Load versus strain curve in pre-macro-yield region for crystal with two distinct regions of different flow stresses.

yield plastically for any increased load. Thus a sharp decrease in slope will occur at this point. The measured load-strain curve will continue to be linear, however, because the core region will continue to deform elastically (with the same rate of increase in stress) and because the slope of the plastic load-strain curve of the s-region is negligible and will not interfere with the elastic portion. It should be emphasised that the slope of the elastic load-strain curve BA changes abruptly at B because the effective <u>elastic</u> cross-section then becomes

$$\phi_c = \phi_p - \phi_s .$$

The elastic <u>stress</u> in the core, however, continues to rise at the same rate as in the portion OB of the curve. When the load reaches F_c where

$$F_c = \sigma_c \phi_c + \sigma_s \phi_s , \qquad (4.3)$$

the core will yield plastically.

The curve in Fig. 9 will now be considered. Firstly, the physical significance of polishing away part of the surface should be appreciated. Mughrabi [4] has shown (his Fig. 5) that the dislocation density increases from about 0.8×10^7 mm^{-2} to

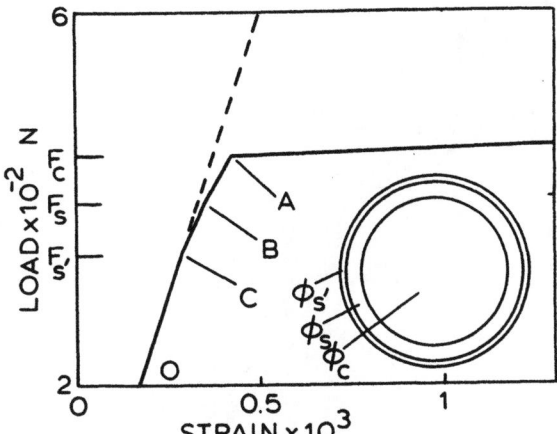

Fig. 9. Load versus strain curve in the pre-macro-yield region for a crystal with three distinct regions of different flow stresses.

1.9×10^7 mm^{-2} over the first 0,1 mm below the surface. Hence, when the surface layer is polished away, a section is effectively made through material of higher dislocation density. There will thus be an increase in density of surface sources which will be capable of yielding at low stresses. The evidence for this can be observed directly in Fig. 10. In this instance a crystal was pre-strained into stage II, sectioned through the core by electrolytic slicing, and then pulled again. It is apparent that at the point where dislocation cell walls intersect the newly exposed surface, a great deal of slip occurs in a very localised region. Such localisation is possible only if the slip occurs at a much lower stress level than the flow stress of the surrounding material.

In view of these considerations it is reasonable to postulate an additional soft region at the surface of a crystal which has been polished chemically or electrolytically after being plastically strained. This is shown diagramatically in Fig. 9. Using similar arguments to those used in relation to Fig. 8, it can be shown that two distinct linear deviations from the initial curve OC can be expected, and that the following relations are valid:

$$F_{s'} = \sigma_{s'} \phi_p \tag{4.4}$$

$$F_s = \sigma_s (\phi_s + \phi_c) + \sigma_{s'} \phi_{s'} \tag{4.5}$$

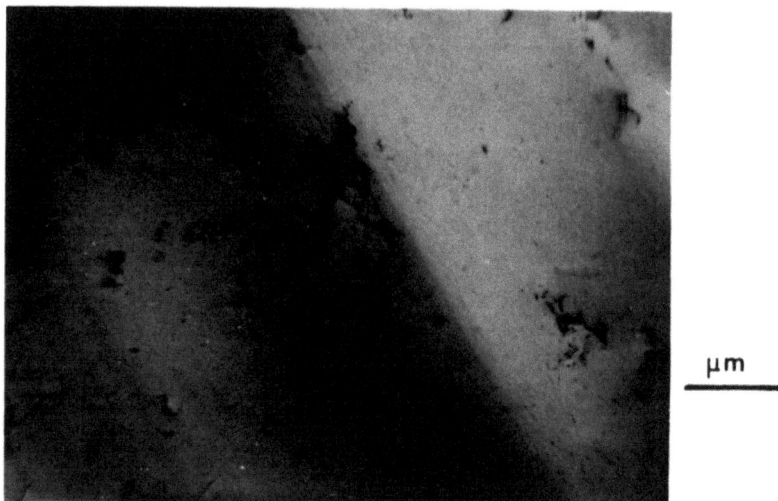

Fig. 10. Transmission electron micrograph of a crystal showing slip instability at a newly exposed surface.

$$F_c = \sigma_c \emptyset_c + \sigma_s \emptyset_s + \sigma_{s'} \emptyset_{s'} \qquad (4.6)$$

where $F_{s'}$ represent the load at which the newly exposed surface sources will begin to operate in the surface region with cross-section $\emptyset_{s'}$. The other symbols are defined as before.

In addition, if $\emptyset_p$ is known (it can be measured directly), the other $\emptyset_i$ can be obtained from the ratios of the slopes of the elastic curves. Hence, if $m_{s'}$, m_s and m_c represent the slopes of curves OC, CB and BA, respectively (in Fig. 9), it follows from the general relation

$$m = \left(\frac{\Delta F}{\Delta \epsilon}\right) = \frac{(\Delta \sigma)\emptyset}{\Delta \epsilon} \qquad (4.7)$$

and the fact that the <u>stress</u>-strain slope is the same for all three elastic regions, that

$$\frac{m_{s'}}{m_s} = \frac{\emptyset_p}{\emptyset_p - \emptyset_{s'}} \qquad (4.8)$$

and

$$\frac{m_s}{m_c} = \frac{\emptyset_p - \emptyset_{s'}}{\emptyset_c} \qquad (4.9)$$

The quantities $F_s{'}$, F_s, F_c, $m_s{'}$, m_s and m_c can be measured from the recorder chart of the test and the equations thus evaluated. Table I contains the values calculated from the tests shown in Figs. 3 - 5.

TABLE I

Numerical values of flow-stress distribution quantities

Relates to Figure:	mm Δx	mm^2 $\emptyset_p$	mm^2 $\emptyset_s{'}$	mm^2 $\emptyset_s$	mm^2 $\emptyset_c$	Nmm^{-2} σ_p	$\frac{\sigma_s{'}}{\sigma_p}$	$\frac{\sigma_s}{\sigma_p}$	$\frac{\sigma_c}{\sigma_p}$	$\frac{\sigma_c{'}}{\sigma_p}$
3	0	19.7	-	1.5	18.2	21.7	-	0.85	0.98	1.03
4	0.014	19.7	1.1	2.8	15.8	21.7	0.78	0.91	1.02	1.06
5	0.021	18.5	3.4	4.0	11.1	23.1	0.85	0.95	1.04	1.09

Δx is the thickness of surface layer polished away.

The values $\sigma_c{'}/\sigma_p$ in the last column of Table I are obtained if the value for F_c is taken at A' instead of A (see Figs. 3, 4.) This is the logical position since there will always be some pre-yielding when the load for total plastic flow in the crystal is approached. Hence if the value is taken at A where the deviation from linearity occurs, one can expect a value which is slightly lower than the actual value.

The important points to note about the curves in Figs. 3 - 5 are:

(a) That they do show the distinct linear regions predicted by the arguments presented in this discussion, and which were based on the concept of a soft-surface region.

(b) That the calculations provide values which are in reasonable agreement with the direct measurements given in Figs. 6 and 7.

(c) That if a hard surface, controlling the plastic-flow stress were present, an electrolytically or chemically polished crystal would show a pre-macro-yield behaviour devoid of distinctly linear regions in the load-strain curve.

It can further be shown that the histograms in Fig. 6 are accurately consistent. Using the equation derived by Fourie [2], namely

$$\tau_p \emptyset_p = \tau_{c1} \emptyset_{c1} + \tau_{c2} \emptyset_{c2} + \tau_{s1} \emptyset_{s1} + \tau_{s2} \emptyset_{s2} \qquad (4.10)$$

where τ_p, τ_{c1}, τ_{c2}, τ_{s1} and τ_{s2} are the resolved shear stresses measured in the parent crystal, the core sections and the surface regions respectively, while $\emptyset_{s1}$ and $\emptyset_{s2}$ are the cross-sections of the two outer surface sections and $\emptyset_{c1}$ and $\emptyset_{c2}$ are taken collectively as all of the material between the surface sections such that

$$\frac{\phi_p - (\phi_{s1} + \phi_{s2})}{2} = \phi_{c1} = \phi_{c2},$$

the value of τ_p can be calculated from the values for individual components in the histograms. The results are given in Table II.

TABLE II

The consistency of direct measurements of flow-stress distribution

τ_p measured Nmm^{-2}	τ_p calculated Nmm^{-2}	$\dfrac{\tau_{calc} - \tau_{meas}}{\tau_{meas}}$ %
4.64	4.58	-1.27
8.73	8.75	+0.22
12.70	13.04	+2.70
20.38	20.54	+0.77
29.27	29.87	+2.04

5. CONCLUSIONS

It has been shown that the pre-macro-yield behaviour of pre-strained crystals, can be predicted accurately on the basis of a physically reasonable model. Calculations based on the model are in reasonable agreement with direct measurements of flow-stress distributions in crystals. On this basis it can be concluded that the basic assumption of Kramer [5] illustrated in Fig. 1, namely, that the whole crystal yields at point P, is incorrect. The concept of a hard dislocation "debris" layer thus becomes untenable. Instead, it has been shown that the concept of a soft surface region is in agreement with measurement of flow stress distribution by both direct and indirect means.

REFERENCES

1. Fourie, J.T., Can. J. Phys., 45, 777, 1967.
2. Fourie, J.T., Phil. Mag., 17, 735, 1968.
3. Fourie, J.T., Phil. Mag., 21, 977, 1970.
4. Mughrabi, H., Phys. Stat. Sol., 39, 317, 1970.
5. Kramer, I.R., Trans. AIME, 233, 1462, 1964.
6. Fourie, J.T., Corrosion Fatigue, National Association of Corrosion Engineers, Houston, 1972, p. 1462.

DISCUSSION

Comment by I. R. Kramer:

Dr. Fourie has repeated some of the experiments we described some 10 years ago. In brief we showed that when high purity metals, Al and Cu, were strained and unloaded and then allowed to relax, upon reloading the plastic flow began at a stress $\Delta\sigma$ lower than the stress from which the specimen was first unloaded. $\Delta\sigma$ increased with time until it was equal to the same value as that obtained when the surface layer was removed chemically. It was also shown that after relaxation the activation volume increased. The data indicated that the dislocation density or, better, obstacles to the motion of dislocations were decreased. For copper it was found that the apparent activation energy was about 3400 cal/mole and the surface layer was depleted of excess dislocations in about 48 hours. On aluminium the surface layer was eliminated in 4 hours. This relaxation rate of course is very important in observations including TEM because all of the excess dislocations in the surface layer will be eliminated. It may be important to note that in our experiments we did not solely rely on the stress-strain curve to determine $\Delta\sigma$. This is a very important point since Fourie reported that after removing the surface layer, plastic flow started at about the stress from which the specimen was unloaded. This is of course to be expected on a "soft" surface layer concept. In some of our work we used the appearance of the slip band to denote the initiation of plastic flow. For Al monocrystals, Fe-3% Si alloys and 2014 T-6 aluminium alloy, in all cases the plastic flow initiated at a stress lower than the unloading stress. In the case of the aluminium monocrystals the surface layer stress, σ_s, was the same as that obtained through the use of the stress-strain curve. By measuring the hardness of strained Cu monocrystals it was noted that the hardness decreased as the surface layer was removed.

Reply:

A crystal which is deformed into stage II, unloaded and reloaded within 1 minute shows a $\Delta\sigma$ (this paper Fig. 1). I should thus suggest that the interpretation given in the present paper, namely that it is due to the surface region yielding before the core, is the more logical.

Comment by T. E. Fischer:

We have observed inhomogeneities on cleaned GaSb which disappeared at room temperature over a period of a few hours. We interpreted these in terms of cleavage-induced strains. This shows a) that cleavage does produce strains and b) that annealing occurs at room temperature. Also tungsten tips in the Field-Ion

Microscope show appreciable distortions of the crystalline structure if, after having applied the large electric field (and stresses), one removes the field and lets the tip warm up to room temperature for about one hour. Since many experiments are done on copper which is softer, I wonder if stress-relief and dislocation motion at room temperature would not cause changes in mechanical properties during the experiments.

Reply:

Appreciable annealing by dislocation climb in copper does not occur before a temperature of about 300°C is reached and since the specimen was kept at a temperature close to room temperature, annealing should not have occured. However, it is realized that any interference with the condition of the specimen by unloading, polishing or slicing must inevitably cause some rearrangement of dislocations. For this reason all the experiments are designed so that inter-comparison of specimens and/or sliced components is possible. It is then usually possible to judge whether the re-arrangement had had a significant effect on subsequent measurements. For example in the present paper, the flow stress in the parent crystal is calculated from the flow stresses measured in the sliced components. The results of this calculation in Table II indicate that the rearrangement did not have a significant effect from the point of view of this particular measurement.

Comment by Z. S. Basinski:

In order to explain the effect of the deviation of the load-elongation curve during reloading from the linear elastic line you must assume that some plastic deformation must have taken place during the unloading-reloading cycle. If the soft regions flowed back, the reloading line is in fact displaced backwards and after the region of forward plastic flow you rejoin the load-elongation curve at the point before unloading. Your interpretation of the additional effect observed after polishing the crystal is of softening of the newly exposed free region. How do you explain rapid rehardening of the surface region?

Reply:

In the case of polishing away some of the surface, we have a twofold effect. Firstly, we polish away some of the soft surface of the crystal, so that the overall mean stress of the regions ϕ_S and ϕ_C in Fig. 9 is now higher than had been the case for the regions ϕ_S and ϕ_C in Fig. 8. Thus, if we considered the regions ϕ_S and ϕ_C on their own in the case of the polished crystal, the point A in Fig. 9 would actually have been higher than it would have been with no polishing. However, when we add the region $\phi_{S'}$ to the outside of the crystal, the final mean stress at point

A could either be lower or higher than before polishing depending on the amount Δr polished away. For very small Δr the stress at point A will probably be slightly lower if there had been no unloading yield point. However, the presence of the unloading yield point masks this effect and by the time enough material has been removed to reduce the unloading yield point effectively, the relative magnitudes of $\phi_{s'}$, ϕ_s and ϕ_c are such that point A continues to lie above the value expected for the unpolished specimen. In view of this complicated situation it is unnecessary to suggest a very high rate of workhardening in the $\phi_{s'}$ region.

Comment by I. G. Greenfield:

Electrolytic polishing of surfaces of crystals tends to change the topography of the surface. Etch pits, surface roughening, rippled surfaces, etc., are frequent occurrances. Some of these surface irregularities can lead to stress intensification factors. Has there been a study of surface topography during or after the experiments in your and other experiments associated with the dissolution of surfaces?

Reply:

There have not been any systematic studies in this regard. However, as long as the surface is not contaminated I should not expect surface irregularities to play a significant role.

AN ANALYSIS OF THE UNLOADING YIELD POINT WITH REFERENCE TO THE PRE-MACRO-YIELD REGION

J.T. Fourie

National Physical Research Laboratory, CSIR,
South Africa

ABSTRACT. Experimental results are presented to show that additional strain hardening occurs in the surface region during unloading. An analysis is made of the stress-strain behaviour of a crystal in the pre-macro-yield region. It is shown that the unloading yield point follows logically from a small amount of extra strain hardening in the surface.

Notation

σ_c — engineering stress in core
σ_s — engineering stress in surface
σ'_s — engineering stress in surface after unloading
F_c — load on crystal core and yielded surface at core yield
F_s — load on surface region and core at surface yield
F'_s — load on surface region and core at surface yield after unloading
$\emptyset_p$ — cross-sectional area of parent crystal
$\emptyset_c$ — cross-sectional area of core region
$\emptyset_s$ — cross-sectional area of surface region
Δd — thickness of layer polished from surface

1. INTRODUCTION

Following the discovery of the soft-surface effect in plastically deformed crystals [1], Brydges [2] pointed out that the compressive stresses set up during the unloading sequence could cause plastic flow and thus harden the surface region of the crystal. However, the hardening of the surface during unloading should be considered in a relative sense. The fact that the surface undergoes some hardening does not mean that it becomes harder than the core of the crystal, but rather that the flow-stress distribution changes from what it was in the dynamic state before unloading (A) to that after unloading (B) in Fig. 1.

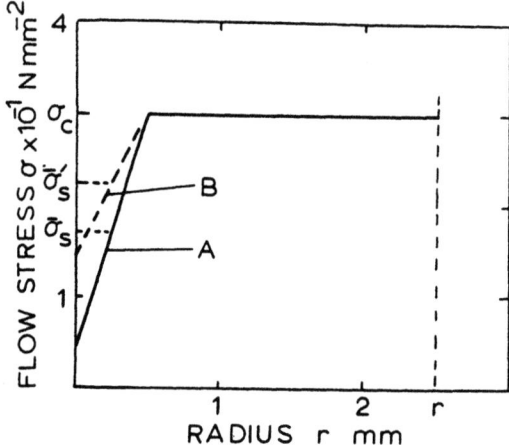

Fig. 1. The change in flow-stress distribution resulting from unloading effects. (A) represents the flow-stress distribution in the dynamic state and (B) that in the unloaded state.

Brydges [2] envisaged that the unloading yield-point effect could be attributed to the increased hardness in the surface region, and Fourie [3] confirmed some of his deductions experimentally. It was shown that the unloading yield point, which is only about 1.2% of the flow stress of the crystal, continued to decrease with the thickness of layer (Δd) polished from the surface. This decrease was continuous and was observed for Δd values in excess of 0.5 mm. In the following, some experimental results will be presented which show directly that unloading does cause an alteration in the flow-stress distribution. Further, an analysis will be made on the same lines as in the previous paper [4] of the pre-macro-yield region. It will be shown how a small rise in the flow stress of the surface region can explain the unloading yield-point effect quantitatively.

2. EXPERIMENTAL RESULTS

Two identical single crystals of copper were pulled to a final resolved shear stress of 12.70 Nmm^{-2}. In the one case the crystal was unloaded only at 12.70 Nmm^{-2} for slicing into sections as described in [4]. In the other case the crystal was given an extra unloading sequence at 6.3 Nmm^{-2}.

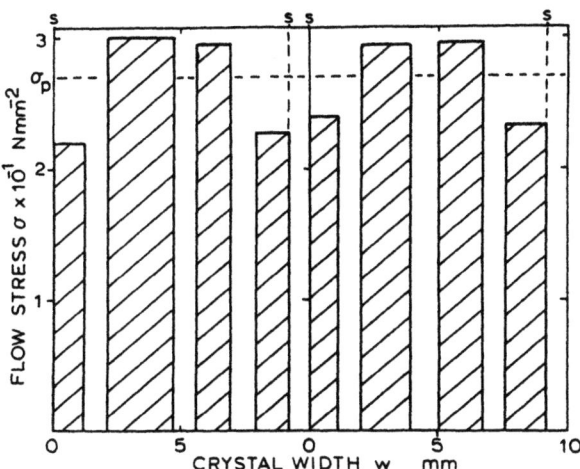

Fig. 2. Direct evidence of surface hardening during unloading of a crystal.

The flow-stress measurements on the components of the sectioned crystals are represented in the form of histograms in Fig. 2. The crystal which was subjected to two unloading sequences shows an average increase of about 6% in the flow stress of the surface region as against that in the crystal which was unloaded once only.

3. DISCUSSION

In analysing the pre-macro-yield behaviour of a crystal which has been unloaded from the dynamic state of deformation, the following assumptions are made:

(a) That in the dynamical state, the following parameters apply

$$\phi_p = 20 \text{ mm}^2$$
$$\phi_c = 16 \text{ mm}^2$$
$$\phi_s = 4 \text{ mm}^2$$

Where ϕ_p, ϕ_c and ϕ_s are the cross-sections of the whole crystal, the core region and the surface region, respectively. Also,

$$\sigma_c = 30.0 \text{ Nmm}^{-2}$$
$$\text{and } \sigma_s = 15.0 \text{ Nmm}^{-2}$$

where σ_c and σ_s are the flow stresses in the core and surface regions, respectively.

(b) That in the unloaded state the surface region is 10% harder than in the dynamic state due to reverse plastic flow. The flow stress in the core region remains unchanged. Thus

$$\sigma_c = 30.0 \text{ Nmm}^{-2}$$
$$\text{and } \sigma'_s = 16.5 \text{ Nmm}^{-2}$$

If an idealised case is considered, where a crystal is unloaded without altering the dislocation structure or flow stress distribution at all, then from equation (4.3) in reference [3] the total load when the crystal is returned to the dynamic plastic-flow state is given by

$$F_c = \sigma_c \phi_c + \sigma_s \phi_s$$

and the load on the whole crystal when the surface region yields, by

$$F_s = \sigma_s \phi_p.$$

For the real case, where the dislocation structure and flow stress distribution is altered by unloading the macro-plastic yield point on reloading occurs at a total load

$$F'_c = \sigma_c \phi_c + \sigma'_s \phi_s$$

and as before

$$F'_s = \sigma'_s \phi_p.$$

From the values assumed above for the various parameters,

$$F_c = 540 \text{ N}$$
$$\text{and } F'_c = 546 \text{ N}.$$

This represents a 1.1% increase in yield point.

In Fig. 3 a graphical representation is given of the stress-strain curves of the crystal in the pre-yield region. Here A represents the idealised stress-strain curve in which there has been no hardening during unloading, whereas B corresponds to the calculations above for the real crystal which did harden upon unloading. The changes in slopes at P and P' are calculated as explained in reference [4].

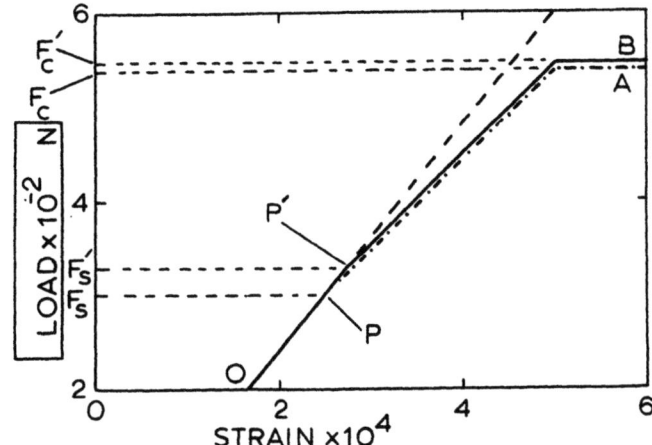

Fig. 3. The unloading yield-point effect as it develops in the pre-macro-yield region.

4. CONCLUSIONS

It has been shown experimentally that the surface region does harden during unloading sequences. However, it remains considerably softer than the core. An analysis of the pre-macro-yield region of the stress-strain curve shows that an unloading yield point follows logically from a small amount of extra strain hardening in the surface.

REFERENCES

1. Fourie, J.T., Phil. Mag. 17, 735, 1968.
2. Brydges, W.T., Scripta Met. 2, 557, 1968.
3. Fourie, J.T., Phil. Mag. 22, 923, 1970.
4. Fourie, J.T., this conference proceedings.

NEW STUDIES OF SURFACE EFFECTS IN DEFORMED COPPER
SINGLE CRYSTALS

H. Mughrabi

Institut für Physik, Max-Planck-Institut für
Metallforschung, Stuttgart, Federal Republic
of Germany

ABSTRACT. Simple theoretical arguments show that single slip deformation of f.c.c. metal single crystals leads to a higher dislocation density in the interior than in the near-surface regions. In order to establish the mechanical and microstructural states at different depths from the surface, a variety of experiments have been performed on copper single crystals deformed into stage II. The results imply preferential hardening of the core.

1. INTRODUCTION

Previous experiments on deformed copper single crystals have provided evidence that the core hardens preferentially [1,2,3,4]. These results and their interpretation have been criticized [5,6]. For this reason the problem has been reinvestigated. Some general theoretical arguments have been derived. In order to characterize the mechanical and microstructural states of copper single crystals deformed into stage II as a function of distance from the surface, a number of independent conventional experimental techniques and also some less frequently used criteria have been applied.

2. THEORETICAL CONSIDERATIONS

The origin of the development of a gradient of the flow stress over the cross section, mainly in the direction of the primary Burgers vector $\underline{b}_p$, has been asso-

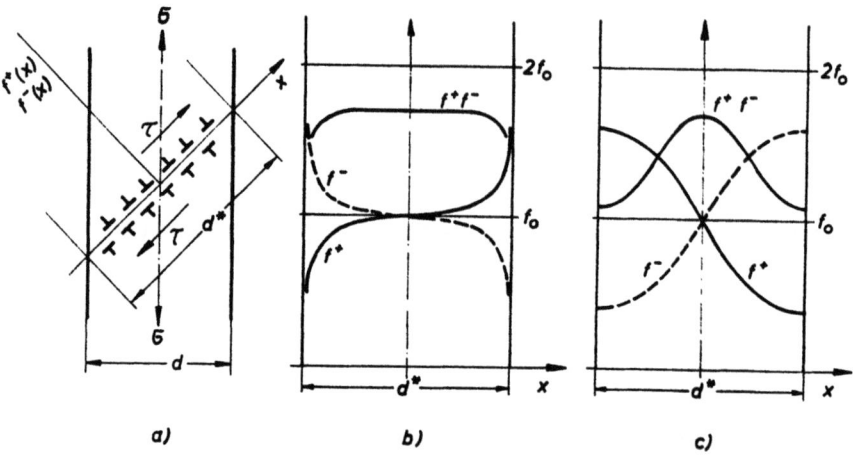

Fig. 1. Illustration of dislocation model. a) Introduction of coordinate system. σ: tensile stress, τ: resolved shear stress. b) f^+, f^- and f^+f^- for interior dislocation sources. c) as b) for surface dislocation sources. Arbitrary units.

ciated with stage I deformation [2,4]. In stage I mainly edge dislocation dipoles form and accumulate, screw dislocations annihilating by cross slip [7]. We can hence confine ourselves to the one-dimensional problem of primary edge dislocation flow. The coordinate system is chosen so that the x-axis runs parallel to $\underline{b}_p$, cf. Fig. 1a). Dislocations of positive sign are assumed to travel in the positive x-direction under the action of σ (resp. τ). Denoting the densities of positive and negative mobile edge dislocations by ρ^+ and ρ^- and the corresponding dislocation current densities by f^+ and f^- respectively, the resolved shear strain rate $\dot\gamma$ can be written as:

$$\dot\gamma = (\rho^+ + \rho^-) v b_p \qquad (1a)$$
$$= (f^+ + f^-) b_p , \qquad (1b)$$

where v is the mean dislocation velocity. In a crystal of infinite thickness $f^+ = f^- = f_o$, i.e. $\dot\gamma = 2f_o b_p$. In a crystal of finite thickness steady-state dislocation flow is described by

$$\frac{df^+(x)}{dx} = q - c_1 f^+(x) f^-(x), \qquad (2)$$

q denoting the differential increase of f^+ resulting

from the activity of uniformly distributed dislocation sources, while the second term describes the differential decrease of f^+ by the immobilization of dislocations which are trapped as dipoles. The trapping probability is contained in the constant c_1. It is sufficient to consider f^+, since, for reasons of symmetry, $f^+(x)=f^-(-x)$. With the subsidiary condition

$$f^+(x) + f^-(x) = 2 f_0 \qquad (3)$$

eq.(2) can be integrated with appropriate boundary conditions. In cases of practical interest f^+ (and f^-) are found to be either monotonically increasing or decreasing functions of x. For this reason the product $f^+ f^-$ which is a measure for the (local) rate of dipole accumulation has its maximum value at x=0. This is shown qualitatively for the case of uniformly distributed dislocation sources in Fig.1b) and for surface sources (q=0) in Fig.1c).

The increase of the local density ρ of dislocations stored in the form of edge dipoles with increasing resolved shear strain γ can be expressed as

$$\frac{d\rho(x)}{d\gamma} = c_2 f^+(x) f^-(x), \qquad (4)$$

where c_2 is a constant. Eq.(4) implies that ρ increases linearly with increasing γ.

The important conclusion that a higher dislocation density is expected to accumulate in the centre of the specimen than in the near-surface region during stage I deformation does not appear to be sensitive to the particular deformation model chosen. It suggests that stage II deformation will be initiated locally in the centre, while the near-surface regions still deform in stage I [1,4].

3. EXPERIMENTAL STUDIES

3.1 Determination of the flow stress distribution

Fourie [2] determined the flow stress gradient over the cross section from the flow stresses of component crystals sliced out of predeformed crystals. The danger of bending stresses arising in the component crystals because of the flow stress gradient [6] is avoided, if a defined surface layer is removed all around the circumference of the specimen. The mean local flow stress

$\overline{\tau}_{f,loc}$ of the removed surface layer is then given by

$$\overline{\tau}_{f,loc} = \frac{\overline{\tau_{f,1}} A_1 - \overline{\tau_{f,2}} A_2}{A_1 - A_2}, \qquad (5)$$

$\overline{\tau_{f,1}}$ and $\overline{\tau_{f,2}}$ being the mean flow stresses and A_1 and A_2 the cross sections before and after removal of the surface layer respectively.

This procedure was applied repeatedly to a copper single crystal of 3.89 mm diameter, deformed to a mean flow stress of 1.17 kg/mm² at room temperature. Surface layers were removed by chemical dissolution. The new cross sections were determined by accurate measurements of the major and minor axes of the elliptical cross section and by precise weighing. Corrections were made for the small work-hardening that occurred when the new flow stresses were determined (altogether 0.13 kg/mm²).

Fig.2 shows the obtained values of $\overline{\tau}_{f,loc}$ plotted in the direction of the minor axis, i.e. very close to the direction of the projection of b_p onto the cross section. These measurements confirm Fourie's results with regard to the existence of a flow stress gradient, its sign, the magnitude of the effect and its penetration depth.

The local work-hardening coefficient was found to be 7.7 kg/mm² in the outermost layer and 13.5 kg/mm² elsewhere, indicating that part of the outermost layer was not yet deforming in stage II.

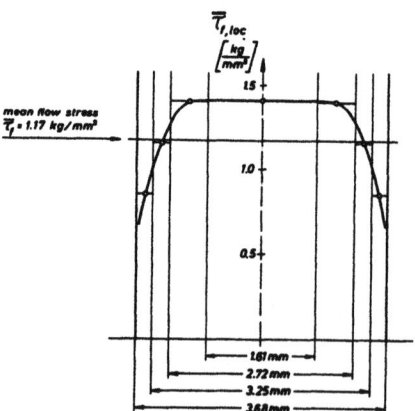

Fig. 2. Distribution of mean local flow stresses.

3.2 Electron microscopic observation of slip lines

A number of copper single crystals were deformed to various shear strains γ into stage II at 78 K. The existing slip lines were removed by electropolishing away a surface layer of thickness t ≲ 2 µm. After an incremental deformation (Δγ~0.04) palladium-shadowed carbon replicas were prepared. Subsequently, a surface layer of t~1 mm was removed by chemical dissolution and electropolishing. After another deformation step carbon replicas from the new surfaces were prepared. The carbon replicas were studied by transmission electron microscopy (T.E.M.).

In a preliminary evaluation, the mean lengths $\overline{L'}$ (t≲2 µm) and $\overline{L}$ (t~1 mm) of randomly selected prominent slip lines were determined from independent counts of three persons. In analogy to Mader's [8] presentation, Fig.3 shows $1/\overline{L'}$ and $1/\overline{L}$ plotted versus γ. The results can be summarized by

$$\overline{L'} \sim \Lambda'/\gamma \tag{6}$$

and $\overline{L} \sim \Lambda/\gamma$, (7)

with $\Lambda' \sim 10 \times 10^{-4}$ cm and $\Lambda \sim 3.6 \times 10^{-4}$ cm.

In agreement with Mader [8], it is found that slip line lengths decrease with increasing deformation. As estimated before [4] and as confirmed in less detailed investigations [9,10], $\overline{L'}$ is observed to be systematically lower than $\overline{L}$. The ratio Λ'/Λ agrees with the earlier predictions [4]. Thus, if slip line lengths reflect dislocation slip paths in the vicinity of the investigated surface, then the interior corresponds to a more advanced state of work hardening than the original surface layer.

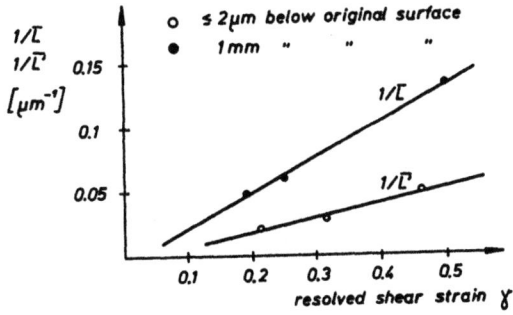

Fig. 3. Reciprocal slip line lengths as a function of γ.

3.3 X-ray Berg-Barrett study of the gross microstructural features

In order to obtain a survey of the microstructural features, the Berg-Barret X-ray technique was chosen because of its large field of view and its (small) penetration depth of some μm [11]. A copper single crystal having the orientation and the cross section indicated in Fig.4 was deformed at room temperature to $\overline{\tau_p}$=1.135 kg/mm². Slices of about 3 mm thickness were cut parallel to the primary glide plane (111) and irradiated at ambient temperature with fast neutrons (>0.1 MeV) to an integrated flux of 2×10^{18} n/cm² in order to stabilize the dislocation arrangement [12]. The faces A and B of these parallelepipeds and the (111) plane were investigated using the $(\overline{2}00),(\overline{2}02),(0\overline{2}2),(\overline{1}3\overline{1})$ and (222) reflections. Series of topographs were taken from the original surfaces and after stepwise removal of surface layers.

Figs.5 and 6 give examples of topographs of the surface B before (Figs.5a,6a) and after removal of a

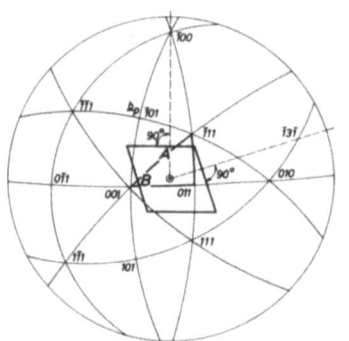

specimen axis in centre

lateral dimensions ca. 3.5 mm

Fig. 4. Specimen orientation and geometry

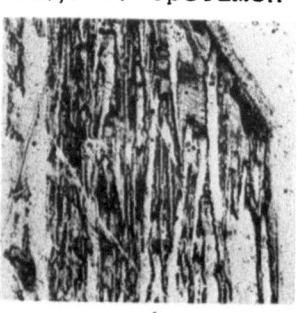 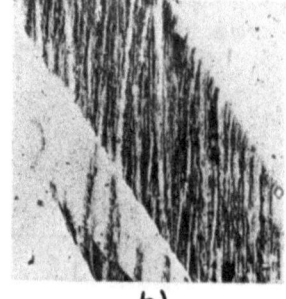

trace of primary slip plane vertical

0.5 mm

a)   b)

Fig. 5. Side B,$(\overline{1}3\overline{1})$-reflection. a) t=0; b) t=0.65 mm.

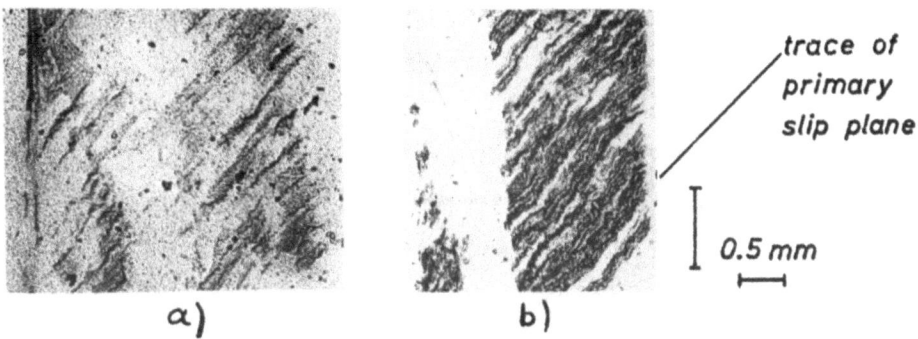

Fig. 6. Side B', (0$\bar{2}$2)-reflection. a) t=0; b) t=0.65 mm.

surface layer (t=0.65 mm). Typical spacings of the lines of contrast are found to be about three times smaller in the interior than at the original surface. Typical features of stage II microstructure (cf.[11,13]) are only found in the interior. This applies to the dislocation layers parallel to (111), revealed by extinction contrast in Figs.5 and 6, and also to the excess dislocations of the same sign in front of kink walls and multipole bundles, giving rise to cusps along the lines of extinction contrast in Fig.6 through displacement contrast (cf. [11,13]). At the original surface the topographs are similar to those found in the bulk of crystals deformed into the transition from stage I to stage II [13].

Again, we conclude that the surface regions have experienced less work hardening than the crystal core.

3.4 Evaluation of local stresses acting on dislocations

In our earlier studies [3], dislocation arrangements were stabilized in the stress-applied state and studied by T.E.M.. Since then, the method of evaluation of dislocation curvatures as a measure for the locally acting stress τ_{loc} [12] has been refined and extended [14,15]. In stage II, τ_{loc} consists of the (local) sum of the external and the spatially fluctuating long-range internal stress field [12,16] whose amplitude is almost equal to the external stress [15]. Accordingly, τ_{loc} fluctuates spatially between zero and roughly twice the external stress.

T.E.M. micrographs were taken from (111)-sections of specimen MÜ 1 (cf.[3]) that had been deformed at 78 K to $\overline{\tau_f}$=1.2 kg/mm^2 and subsequently irradiated at 4.2 K in the stress-applied state. The spatial distribution of

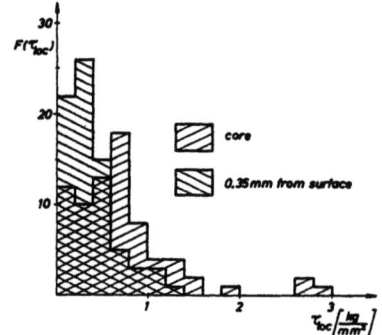

Fig. 7. Frequency distributions of local stresses.

radii of curvature $r_\perp$ of primary edge dislocations was determined carefully. Using $\tau_{loc}=0.42/r_\perp$ ($r_\perp$ in µm, τ_{loc} in kg/mm^2) [14], the corresponding frequency distribution $F(\tau_{loc})$ was obtained.

Such distributions of τ_{loc} are shown in Fig.7 for a specimen from the interior and for another which was 0.35 mm away from the surface, measured in the direction of $\underline{b}_p$. In such distributions, high values of τ_{loc} are underestimated systematically [15]. Nevertheless, it is easily verified that values of $\tau_{loc}<0.6$ kg/mm^2 occur more and higher values less frequently in the near-surface region than in the interior. From other details of Fig.7 we can conclude that the local values of both external and internal stresses are almost twice as high in the interior as in the near-surface region. This implies that the core of a crystal carries a higher stress than the surface region and that internal stresses are higher in the core. Both conclusions indicate that the core work-hardens preferentially.

It should be pointed out that the described procedure of low temperature deformation followed by neutron irradiation at an even lower temperature excludes the possibility that an existing "surface debris layer" could have relaxed [5,6].

4. SUMMARY

The presented studies indicate, both from a theoretical point of view as well as on the basis of a variety of specifically designed experiments, that deformation of f.c.c. single crystals in single slip leads to a preferential work hardening of the core. The mechanical and microstructural observations have been shown to be compatible with this view and support the earlier findings.

Admittedly, the emphasis of these investigations has been placed on the more detailed evaluation of surface effects having a penetration depth of typically some 100 µm. In cases, where the immediate vicinity of the surface was accessible, no evidence of a higher local dislocation density was obtained.

ACKNOWLEDGEMENTS

The kind support of Prof. Dr. A. Seeger and Dr. M. Wilkens is deeply appreciated. The expert help of Fräulein M. Rapp in the replica studies and Frau I. Klicpera's co-operation in the evaluation are gratefully acknowledged. Some of this work was supported by the Deutsche Forschungsgemeinschaft.

REFERENCES

1) J.T. Fourie, Can. J. Phys. 45, 777, (1967).
2) J.T. Fourie, Phil. Mag. 17, 735, (1968).
3) H. Mughrabi, phys. stat. sol. 39, 317, (1970).
4) H. Mughrabi, phys. stat. sol.(b) 44, 391, (1971).
5) I.R. Kramer and A. Kumar, Scripta Met. 3, 205, (1969).
6) I.R. Kramer and N. Balasubramanian, Acta Met. 21, 695, (1972).
7) U. Essmann, Acta Met. 12, 1468, (1964).
8) S. Mader, Z. Phys. 149, 73, (1957).
9) J.T. Fourie and N.C.G. Dent, Acta Met. 20, 1291, (1972).
10) N. Himstedt and H. Neuhäuser, Scripta Met. 6, 1151, (1972).
11) M. Wilkens, Can. J. Phys. 45, 567, (1967).
12) U. Essmann, phys. stat. sol. 3, 932, (1963).
13) B. Obst, H. Auer and M. Wilkens, Mat. Sci. Eng. 3, 41, (1968/69).
14) H. Mughrabi, Proc. of 3rd Int. Conf. on Strength of Metals and Alloys, Cambridge, 1973, p. 410.
15) H. Mughrabi, Constitutive Equations in Plasticity, edited by A.S. Argon, M.I.T. Press, in press.
16) A. Seeger, J. Diehl, S. Mader and H. Rebstock, Phil. Mag. 2, 323, (1957).

DISCUSSION

Comment by F.R.N. Nabarro:

How far does your dislocation flux model account for the lattice rotations you reported in an earlier paper?

Reply:

The model predicts, in addition to the stored distribution of dislocation dipoles, an excess density of dislocations of one sign near the surface (Fig. 1) and hence a lattice bending of the primary glide plane. The sense of the bending depends on whether surface sources (Fig. 1c) or interior sources (Fig. 1b) dominate. Thus the model could explain lattice bendings in both senses qualitatively. Quantitatively, the predicted lattice bendings is much smaller than the observed one which, as Dr. Basinski points out in his survey lecture, cannot be predominantly due to elastic bending resulting from nothing but primary dislocations.

Comment by J. F. Prins:

This investigation, and the paper by Dr. Fourie, show that in Cu single crystals the surface layer is depleted of dislocations relative to the interior of the crystal. Obviously dislocations which arrived at the surface, escaped through it. The question remains whether dislocations escape at the same rate at which they arrive at the surface during deformation. If for some reason (e.g., energy needed to create surface steps) the escape rate is lower than the arrival rate, it should have an effect on the workhardening, especially in stage I where low workhardening prevails. It is then conceivable that electrolytic polishing during deformation will affect the escape rate and accordingly the workhardening in stage I.

Reply:

It is difficult to tell whether the effect you envisage plays a detectable role in practice. In any case, this effect would be superimposed on another effect, namely that the workhardeing rate is _not_ uniform over the cross section, since the near-surface region continues deforming in stage I after the core has begun deforming in stage II. So one could just expect an additional complication.

INFLUENCE OF SURFACE PLASTICITY ON THE DEFORMATION BEHAVIOR AND
FRACTURE MODE OF SILICON

S. Weissmann

College of Engineering, Rutgers University, New
Brunswick, New Jersey 08903, U.S.A.

ABSTRACT. Silicon was selected as a model material to study the
deformation behavior and fracture mode as a function of applied
stress and deformation temperature. A sensitive combination
method of structural analysis was applied which is capable of revealing the distribution of microplastic regions and residual,
elastic strains not only in the bulk material but also in relation to the crystal surface. A localized microplastic zone about
30 μ in size initiated the cleavage type of fracture at deformation
temperatures up to about 600°C, and no other microplastic zones
were formed which could constrain the release of the elastic strain
energy at the crack tip. With increasing deformation temperatures
microplastic zones were formed, predominantly at surface sites,
which constrained the residual elastic strains. Deformation of
smooth crystals into which a precipitate structure was introduced
generated above the ductile-brittle transition temperature microplastic zones, which were formed by dislocation sources situated
at surface sites and not by dislocation sources associated with
precipitates.

1. INTRODUCTION--STUDY OF MODEL MATERIAL BY COMBINATION METHOD OF
   STRUCTURAL ANALYSIS

The importance of surface layer effects on the mechanical properties of crystalline materials has received particular attention
in recent years and has aroused considerable controversy [1-9].
Various investigators studying surface dissolution related surface
effects not only to tensile properties [5,9-11] but also to creep
[1,12,13] and to fatigue properties [14,15].

It became apparent to the author that in order to investigate the special contribution which surface layers may make to mechanical properties, it would be desirable to select a model material which possesses properties that would meet many stringent requirements. Such a model material could then be subjected to a recently developed combination method of structural analysis which is capable of disclosing sensitively the interrelationship between microstructure and mechanical properties [16-18]. Silicon was chosen as such a model material for the following reasons.

1) Crystals can be obtained dislocation-free so that the initial stages of deformation and fracture initiation can be ideally studied. Moreover, the initial perfection of the crystals allows for a topographic visualization of the microplastic zones and residual strains introduced by deformation. This was made possible by means of one of the component elements of the combination method, namely, Pendellösung Fringe (PF) X-ray topography.

2) At elevated temperatures (above 60 percent of the absolute melting temperature), silicon as well as other semiconductor crystals become increasingly ductile and behave like metal crystals. At lower temperatures they are extremely brittle, behaving like typical ceramic materials. Consequently, the phenomenon of ductile-brittle transition can be ideally elucidated.

3) In these crystals the dislocations are practically immobile at lower temperatures, so that the dislocation configuration of the plastic zones developed at higher temperatures becomes frozen-in when the specimen is rapidly cooled. Consequently, the characterization of the plastic zone by transmission electron microscopy finds ideal conditions in semiconductor crystals.

The element of the combination method which initiated the structural investigation was PF topography. The great sensitivity of the method is based on the dynamical interaction of the X-ray wave fields inside perfect crystals and the disturbance of this interaction in the presence of lattice defects. Applying this method to wedge-shaped crystals, the microplastic zones can be disclosed by the destruction of PF, while the elastic strains can be revealed by the distortions of PF [16,17].

Using the combination method in a synergistic manner, the lattice distortions are quantitatively assessed by double-crystal X-ray diffractometry. The dislocation structure in the plastic zones is disclosed by transmission electron microscopy and the defect structure of the fracture surface is characterized by scanning electron microscopy. The synergistic exploitation of the component elements in structural combination methods was described recently by the author [19].

2. EXPERIMENTAL RESULTS

2.1 Effect of microplasticity on elastic strain distribution and fracture mode

When a transmission X-ray topograph of a wedge-shaped, dislocation-free silicon crystal is taken in such a way that only the Bragg-diffracted beam of the stationary crystal is recorded, a Lang section topograph is obtained such as that shown in Fig. 1a. When a traverse specimen motion, coupled to a film motion, is carried out, the projection topograph of Fig. 1b with its characteristic striped PF pattern is obtained. The PF periodicity is controlled by the extinction distance ξ_g, which is defined as the depth periodicity of the Pendellösung oscillation measured along the normal to the X-ray entrance surface of the crystal. Since the disturbance of the PF pattern is used as a sensitive, diagnostic indicator of lattice defects, the projection topograph is very useful for the mapping of microplastic zones and the mapping of the elastic strain distribution. Large specimen volumes can be easily scanned and, if necessary, the entire specimen volume can be surveyed quite readily.

When a V-notched specimen was subjected to tensile deformation in the temperature range at which brittle cleavage type of fracture occurred along {111} planes, fracture was always preceded by the formation of a small microplastic zone at the crack tip, commensurate with the slip activity as shown in Fig. 2b. The important

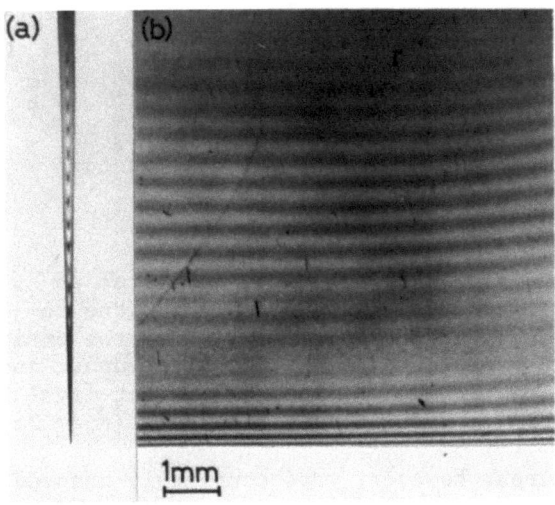

Fig. 1. X-Ray topographs showing PF in silicon crystal.
(a) Section topograph; (b) projection topograph.

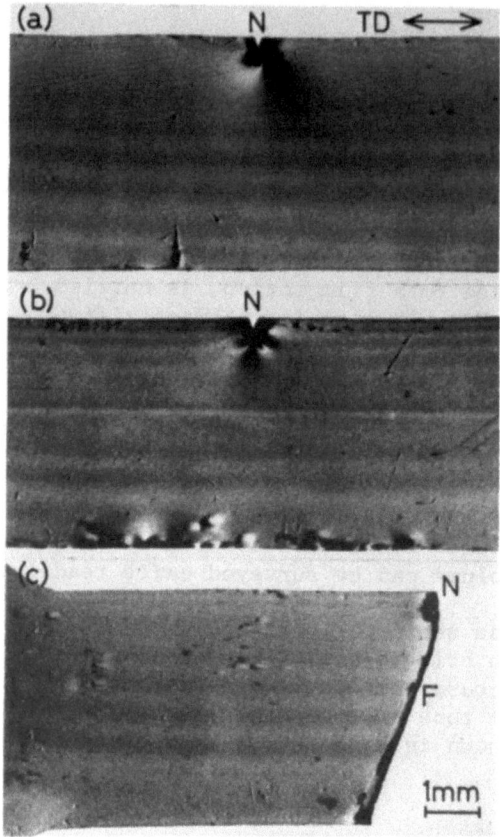

Fig. 2. X-Ray topographs showing disturbed PF due to microplastic zone induced in notched crystal at 600°C in tension. (a) $\sigma_n = 1.9$ kg mm^{-2}; (b) $\sigma_n = 2.7$ kg mm^{-2}; (c) fractured at $\sigma_{nf} = 3.0$ kg mm^{-2}. N = notch; TD = tensile direction [110]; F = portion of fracture surface.

aspect concerning this cleavage mode of fracture is the total absence of any other microplastic zone which could constrain the release of the elastic strain energy at the crack tip. Once the strain energy was released, the crack propagated in a shock-like manner and the two fractured portions of the crystal behaved as perfect crystals, as shown by the undisturbed course of the PF pattern (Fig. 2c).

The microstructural features, however, were completely changed when the specimen was deformed at more elevated temperatures. In addition to the enlarged plastic zone at the crack tip, plastic zones were formed predominantly at the surfaces. These zones are manifested in the topograph of Fig. 3 by black areas, signifying

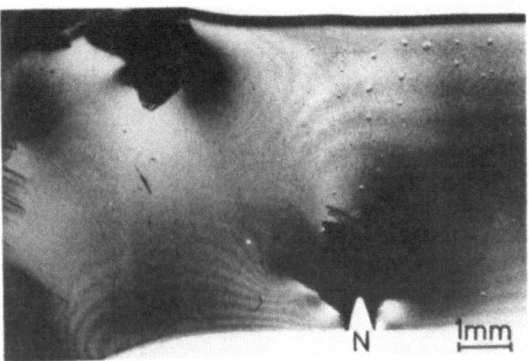

Fig. 3. X-Ray topograph showing distribution of residual strains constrained among microplastic zones in notched silicon bent at 800°C. N = notch.

total destruction of PF. These plastic zones constrained the residual, elastic strains which are characterized by the distortions and bending of the PF pattern. It was shown by transmission electron microscopy that the microplastic zones, formed at the surface, represent zones which were work-hardened by multiple slip activity [17]. In fact, their constraining effect on the residual, elastic strains in the interior of the material depended on their work-hardening characteristic. It was also shown that when the material was annealed at high temperatures, viz. 1000°C, the elastic strains disappeared accompanied by a softening of the work-hardened plastic zones and by a sweeping of dislocation loops emanating from the boundaries of the zones into the regions of elastic strains. Thus the residual, elastic strains could be relieved only by invasion of microplasticity emerging from the softened plastic zones [17].

It was also shown that the problem of notch-sensitivity was intimately connected with the formation of work-hardened plastic zones at the specimen surface and with the constraining influence of these zones on the elastic strain distribution in the interior. Thus, while the fracture stress in smooth crystals decreased with increasing temperature, as expected, in notched crystals it had a very low value in the temperature range where brittle cleavage fracture occurred (<600°C). Increasing the deformation, the fracture stress of notched crystals rose to a maximum value at about 650°C (notch-brittle transition temperature). This ascent in the temperature dependence of fracture stress of notched crystals was shown to be intimately associated with the generation of residual, elastic strains constrained by the plastic zones at the surface. By contrast, in the temperature range of total, brittle cleavage fracture (<600°C) the constraining influence was absent, since microplastic

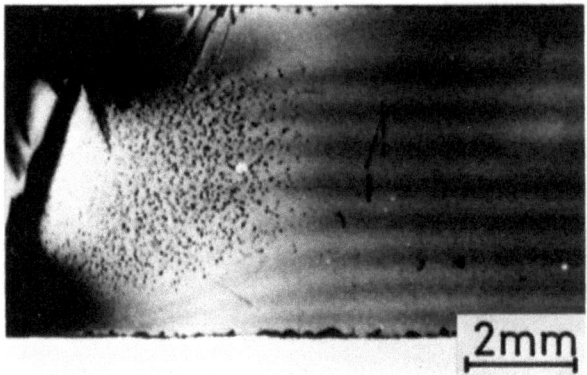

Fig. 4. X-Ray topograph of wedge-shaped, unnotched crystal containing gold-induced defects bent at 800°C.

zones were not yet formed and the fracture stress exhibited a low value [17].

2.2 Deformation response of surface layers in silicon crystals containing a precipitate defect structure in the bulk

A precipitate defect structure was introduced into the bulk of silicon crystals by diffusion of gold at elevated temperatures. It was shown by transmission electron microscopy that the fine precipitated particles of gold gave rise to interstitially faulted dislocation loops which were stable even at prolonged annealing at 1000°C [18]. The purpose of introducing the precipitate defect structure was to test whether the surface would still retain its propensity of generating dislocations if the bulk of the material would be made more susceptible to plastic deformation by stress-raisers in the form of interstitially faulted loops. Tensile and bending deformation performed on smooth and notched crystals at temperatures above the brittle-ductile transition temperature, e.g. 800°C, gave rise to profuse generation of dislocations emanating from surface sites rather than from the precipitates in the interior, as shown in Fig. 4. The slip activity at surface sites destroyed totally the PF pattern in these regions, while in regions further removed from the plastic zones the PF pattern emerged again and the fringe spacings regained their regularity. Moreover, it may be seen that in areas occupied by the precipitates the PF pattern was virtually undisturbed. The results of this competitive deformation experiment, in which the relative deformation response between surface and bulk was tested, appear to imply that the surface sites still retain their special propensity of generating dislocation sources, even though the susceptibility of the bulk material may have been raised by the introduction of stress-raisers. Indeed, the crystal surface, when viewed on an atomic scale, contains steps which may be regarded as stress-raisers in the form of micronotches. This concept of surface roughening finds support in the results of recent studies dealing with the equilibrium properties of crystal surface steps by Monte Carlo simulation methods [20]. These studies showed that above a critical temperature T_R surface roughening sets in and surface steps are being formed without increasing the surface free energy.

REFERENCES

1. Revie, R. W. and Uhlig, H. H., Acta Met., 22, 619, 1974.
2. Kramer, I. R., Scripta Met., 8, 1231, 1974.
3. Goel, V. S., Scripta Met., 3, 465, 1969.
4. Kramer, I. R. and Kumar, A., Scripta Met., 3, 209, 1969.
5. Duquette, D. J., Scripta Met., 3, 513, 1969.
6. Fourie, J. T., Scripta Met., 2, 63, 1968.
7. Billelo, J. C. and Metzger, M., Scripta Met., 2, 581, 1968.
8. Fourie, J. T., Scripta Met., 2, 629, 1968.
9. Block, R. J. and Johnson, R. M., Scripta Met., 3, 511, 1964.
10. Kramer, I. R., Trans. AIME, 233, 1467, 1965.
11. Fourie, J. T., Phil. Mag., 22, 923, 1970.
12. Kramer, I. R., Trans. ASM, 60, 310, 1967.
13. Sing, V., Roag, V. V. P. K. and Rama, P., Scripta Met., 5, 525, 1971.
14. Kramer, I. R. and Kumar, A., Met. Trans., 3, 1223, 1972.
15. Masuda, H. and Duquette, D. J., Met. Trans., 6, 87, 1975.
16. Weissmann, S., Tsunekawa, Y. and Kannan, V. C., Met. Trans., 4, 376, 1973.
17. Tsunekawa, Y. and Weissmann, S., Met. Trans., 5, 1585, 1974.
18. Tsunekawa, Y. and Weissmann, S., Mat. Science Eng., 17, 51 1975..
19. Weissmann, S., Metallography: A Practical Tool for Correlating the Structure and Properties of Materials, ASTM Special Technical Publication, Philadelphia, 1974.
20. Leamy, H. J. and Gilmer, G. H., J. Crystal Growth, 24/25, 499, 1974.

DISCUSSION

Comment by A. S. Argon:

In relation to Professor Weissmann's presentation I should like to point out that one must distinguish the yielding of two types of materials: a) mobile dislocation starved materials such as silicon at room temperature, aged low carbon steel, impure magnesium oxide, etc., and b) pure ductile metlas such as copper and the like. In the first class of materials plastic flow initiates always at the surface because of surface stress concentrations, etc. Initiation of plastic flow in these materials can be aided by sprinkling SiC and scratching. The initiation and morphology of plastic flow in the second group of materials is much less surface sensitive. One must, therefore, be cautious in drawing conclusions from experiments of one of these types of materials to answer questions of the other type.

Reply:

It is admittedly always difficult to make predictions of applicability of observed phenomena from a single case to a general case. It may be pointed, however, that at elevated deformation temperatures silicon and germanium exhibit the deformation characteristics similar to that of copper. (H. Alexander and P. Haasen, Solid State Phys. Vol. 22, pp. 27-158, 1968). It was shown by R. Pangborn and A. Gysler, two coworkers in my laboratory, who deformed silicon crystals 10% plastically at 650, 700, and 800°C, that the X-ray rocking curves decreased significantly with distance from the surface in an anologous manner. Evidence of this recent work was presented at this conference. Deformation carried out at 800°C falls well in the region in which silicon deforms in a ductile manner.

ON THE RESIDUAL STRESSES IN PLASTICALLY DEFORMED
POLYCRYSTALS AND COMPOSITES

E. Kröner

Institut für Theoretische und Angewandte
Physik der Universität und Max-Planck-Institut für Metallforschung, Stuttgart

ABSTRACT. In this note an observation is described which is thought to be relevant for the interpretation of those X-ray experiments, mentioned by Dr. Basinski in his lecture at this Institute, that aim at the determination of residual stresses near the surface of plastically deformed polycrystals and composites.

1. Statement of problem

The particular conditions dislocations encounter near the crystal surface suggest that plastic deformation near the surface proceeds somewhat differently from that in the interior of the material. Valuable information about this phenomenon can be obtained by X-ray measurements of lattice parameters because their changes indicate elastic strain [1]. In disordered composites and polycrystals in which the elastic moduli c_{ijkl} are distributed over the plastically deformed specimen in a nonhomogeneous manner there exists a peculiar effect that must be taken into account when conclusions are to be drawn from the mentioned measurements: the ensemble-averaged elastic residual strain $\langle \varepsilon_{ij} \rangle$ of a plastically deformed and then unloaded specimen may be nonvanishing although the average residual stress $\langle \sigma_{ij} \rangle$ is always zero due to a theorem of Albenga*. In the following we shall derive a quantita-

---
* For reference see E. Kröner [2].

tive description of this statement.

2. The macroscopic residual strain effect

As has been discussed in ref.[3] and elsewhere the residual stress field $\sigma_{ij}(1)$ which remains at point 1 after unloading a specimen that had undergone a plastic strain $\varepsilon_{kl}^P(2)$ in its points 2 can be written as an integral over the volume V of the specimen:

$$\sigma_{ij}(1) = - \int dV_2 \, \Delta_{ijkl}(1,2) \, \varepsilon_{kl}^P(2) \, . \tag{1'}$$

Here, $\Delta_{ijkl}(1,2)$ are the components of the modified (or stress-)Green tensor which, in principle, can be determined from the elastic moduli $c_{ijkl}(1)$, given the shape of the surface of the (elastically inhomogeneous) body. In a self-evident abridged notation eq.(1') becomes

$$\underline{\sigma} = - \underline{\Delta} \, \underline{\varepsilon}^P . \tag{1}$$

For simplicity we restrict ourselves to situations in which the tensor fields $\underline{\sigma}$, $\underline{\varepsilon}$ and $\underline{\varepsilon}^P$ are statistically (i.e. macroscopically) homogeneous random functions so that their ensemble averages $\langle \underline{\sigma} \rangle$, $\langle \underline{\varepsilon} \rangle$ and $\langle \underline{\varepsilon}^P \rangle$ are constant in the specimen. Let us now perform a Gedanken-experiment in which we assume that the specimen undergoes a plastic strain $\underline{\varepsilon}^P$ that is correlated with the distribution of the elastic moduli. It will shortly be proved that the elastic strain $\underline{\varepsilon}$, too, is correlated with the elastic moduli. If this is true then $\langle \underline{c} \, \underline{\varepsilon} \rangle = \langle \underline{c} \rangle \langle \underline{\varepsilon} \rangle + \langle \underline{c}' \underline{\varepsilon}' \rangle$ with $\langle \underline{c}' \underline{\varepsilon}' \rangle \neq 0$. Here, as throughout this paper, the prime denotes the deviation from the mean, i.e. $\underline{c}' = \underline{c} - \langle \underline{c} \rangle$, $\langle \underline{c}' \rangle = 0$.

Multiplying eq.(1) by $\underline{c}' \underline{s}$, where $\underline{s} = \underline{c}^{-1}$ is the elastic compliance tensor, we obtain after averaging

$$\langle \underline{c}' \underline{\varepsilon} \rangle = - \underline{\underline{\Lambda}} \underline{\varepsilon}^P = - \langle \underline{\underline{\Lambda}} \rangle \langle \underline{\varepsilon}^P \rangle - \langle \underline{\underline{\Lambda}} \, \underline{\varepsilon}'^P \rangle , \quad \underline{\underline{\Lambda}} \equiv \underline{c}' \underline{s} \, \underline{\Delta} . \tag{2}$$

Since $\underline{\Delta}$ applied to a constant results in zero we have $\langle \underline{c}' \underline{\varepsilon} \rangle = - \langle \underline{\underline{\Lambda}} \, \underline{\varepsilon}'^P \rangle$ which is not zero if $\underline{\varepsilon}^P$ is correlated with $\underline{c}$ and zero otherwise. This observation proves our former statement.

Applying now Albenga's theorem $\langle \underline{\sigma} \rangle \equiv \langle \underline{c} \underline{\varepsilon} \rangle = 0$ we find that

$$\langle \underline{c} \rangle \langle \underline{\varepsilon} \rangle = - \langle \underline{c}' \underline{\varepsilon} \rangle \neq 0 \quad (\underline{\varepsilon}^P \text{correlated with } \underline{c}). \tag{3}$$

This is the mathematical manifestation of the residual strain effect. It also implies that the mean values of the residual stress and strain are not connected by an effective Hooke's law. In eq.(3) we can replace $\underline{\varepsilon}$ by $\underline{\varepsilon}'$. Written in cartesian coordinates and solved for $\langle\underline{\varepsilon}\rangle$ we obtain

$$\langle \varepsilon_{ij} \rangle = -\langle A_{ijkl}\, \varepsilon'_{kl} \rangle, \quad A_{ijkl} \equiv \langle c \rangle^{-1}_{ijmn} c'_{mnkl} \qquad (4)$$

where A_{ijkl} is a convenient dimensionless measure of the fluctuations of the elastic moduli.

3. Estimate of the effect under the assumption of a perfect sign correlation

For the rigorous calculation of the right hand side of eq.(4) one needs information about the details of the plastic deformation and the distribution of the elastic moduli. Since such a calculation would require a research program of its own we content ourselves with an estimate that is based upon the assumption of a perfect sign correlation. This assumption is rather special but not unrealistic at all. We explain it as follows: Consider a residual strain state in which all components of $\langle \varepsilon_{ij} \rangle$ except one, $\langle \varepsilon_{11} \rangle$ say, vanish. Then eq.(4) yields the following expression for $\langle \varepsilon_{11} \rangle$:

$$\langle \varepsilon_{11} \rangle = \langle A_{1111}\varepsilon'_{11} \rangle + \langle A_{1122}\varepsilon'_{22} \rangle + \langle A_{1133}\varepsilon'_{33} \rangle$$
$$+ 2(\langle A_{1123}\varepsilon'_{23} \rangle + \langle A_{1131}\varepsilon'_{31} \rangle + \langle A_{1112}\varepsilon'_{12} \rangle). \quad (5)$$

There are five more equations that have a zero on the left hand side. The assumption of perfect sign correlation means that in eq.(5) either the positive ε'-values coincide with the positive A-values and correspondingly the negative quantities or that the opposite is true. Because this assumption allows enough freedom for the quantitative distribution of the ε'_{kl} there are many choices to satisfy the mentioned additional five equations.

We believe that the perfect sign correlation is not unrealistic because the A_{ijkl} specify increased or decreased resistance against plastic flow. For instance, the critical shear stress increases with the shear modulus in a great variety of materials. We are therefore inclined to speak of normal behaviour (in contrast to anomalous) in the case when the ε'_{kl} and A_{ijkl} have the same sign.

Under the assumption of the perfect sign correlation eq.(4) can be transformed into

$$|\langle \varepsilon_{ij} \rangle| = \langle |A_{ijkl}| \, |\varepsilon'_{kl}| \rangle \tag{6}$$

where, for instance, $|\varepsilon'_{kl}|$ is the modulus of ε'_{kl}. The lowest possible value for the average on the right hand side is obtained if correlations other than the sign correlations are absent. Then we can factorize the average. If we do not make this assumption then the result of the factorization will be a lower bound:

$$|\langle \varepsilon_{ij} \rangle| \geq \langle |A_{ijkl}| \rangle \langle |\varepsilon'_{kl}| \rangle . \tag{7}$$

In most of the real situations the given information about the elastic moduli suffices to obtain $\langle |A_{ijkl}| \rangle$. However, the actual calcultation, though elementary, may be quite involved.

4. Application to 2-phase composites and to polycrystals with constituents of cubic symmetry

In order to give simple examples we restrict ourselves to materials which are macroscopically isotropic in their elastic properties and have the same compression modulus throughout the specimen. Then $\langle |A_{ijkl}| \rangle$ become the components of an isotropic tensor, and therefore have the form

$$\langle |A_{ijkl}| \rangle = a \delta_{ij} \delta_{kl} + b I_{ijkl} \tag{8}$$

where $I_{ijkl} = (\delta_{ik}\delta_{jl} + \delta_{il}\delta_{jk})/2$ and a,b are positive constants. Due to the assumption about the compression modulus only the deviator part of eq.(7) remains. It has the form

$$|\langle e_{ij} \rangle| \geq b \langle |e'_{ij}| \rangle \tag{9}$$

if e is the deviator of ε. We now give two applications:
(i) The first concerns 2-phase 50/50 composite materials with compression moduli $\kappa_1 = \kappa_2$ and shear moduli $\mu_1 = \mu_2$. The assumption of perfect sign correlation can be physically real in this case. One easily derives that $b = |\mu_1 - \mu_2|/(\mu_1 + \mu_2)$. If μ_1 and μ_2 are rather different then b comes close to 1.
(ii) The second application concerns polycrystalline aggregates of cubic crystallites. A rough estimate leads to $b \approx |\mu - \nu|/(\mu + \nu)$ where, in Voigt's notation,

$\mu \equiv c_{44}$; $\nu \equiv (c_{11}-c_{12})/2$. For copper, $b \approx 1/2$, for α-iron $b \approx 1/3$. This estimate has been derived by using the relationship

$$(A_{ij}) = \frac{2A}{5} \begin{pmatrix} -2 & 1 & 1 & & & \\ 1 & -2 & 1 & & & \\ 1 & 1 & -2 & & & \\ & & & 1 & & \\ & & & & 1 & \\ & & & & & 1 \end{pmatrix}, A \equiv \frac{5(\mu-\nu)}{6\mu + 4\nu} \quad (10)$$

where A_{ij} are the Voigt matrix elements of the tensor A_{ijkl}.

We note in passing that in eq.(9) $\langle |e'_{ij}| \rangle = \langle \sqrt{e'_{ij}{}^2} \rangle$ is close to $\sqrt{\langle e'_{ij}{}^2 \rangle}$ (no summation in these expressions) which is the standard deviation of the residual (deviator) strain.

5. Internal stresses of the 1st, 2nd and 3rd kind

Before coming to an end with this consideration let us recall U.Wolfstieg and E.Macherauch's reassessment of the internal stresses of a polycrystal in terms of the so-called stresses of the 1st, 2nd and 3rd kind [1]. The stresses of the 1st kind are large scale stresses which have the same sign over domains containing several or many grains. Those of the 2nd kind change the sign from grain to grain and those of the 3rd kind fluctuate around the mean stress in a grain. The stresses treated in the above investigation are clearly those of the 2nd kind. The stresses deciding whether the region near the surface is elastically elongated or compressed with respect to the interior are just as clearly of the 1st kind. We shall come back to this difference immediately.

6. Residual strain of the 1st kind

We now return to our original problem. The standard X-ray experiment described in ref.[1] examines the region near the surface of a plastically deformed body and gives us numerical values $\langle e_{ij} \rangle_{exp}$ and $\langle |e'_{ij}| \rangle_{exp}$ which might not coincide with the corresponding values in eq.(9) because this equation takes into account the stresses of the 2nd kind only. Restricting ourselves to the above specified normal situation, i.e. when $\langle e_{ij} \rangle \leftarrow b \langle |e'_{ij}| \rangle$, we conclude that the part of the average elastic strain which corresponds to the stresses of the 1st kind is

$$\langle e_{ij}\rangle_{1st} \geq \langle e_{ij}\rangle_{exp} + b\langle |e'_{ij}|\rangle_{exp}. \qquad (11)$$

It follows that $\langle e_{ij}\rangle_{1st}$ can be positive even if $\langle e_{ij}\rangle_{exp}$ is negative. In other words, the internal stresses of the 1st kind near the surface could be tensional even when the experimental lattice strain is compressional along the axis of a specimen which had undergon plastic elongation.

7. Conclusion

The residual strain effect discussed in this note means that a plastically deformed polycrystal or composite possesses, after unloading, an overall residual strain the volume average of which does not vanish except if the elastic moduli are constant throughout the specimen. This residual strain has nothing to do with the surface conditions of the material but it must be taken into account in the interpretation of X-ray measurements (and also of strain gauge measurements) of the lattice strains near the surface. The fact that our main quantitative statements were inequalitites rather than equalities implies that, on the grounds of such measurements, the separation of the residual strains into parts which correspond to internal stresses of the 1st and 2nd kind is not possible. The problem remains then to find the equations which replace the inequalities, for instance no.(11), in a more complete description of the plastic deformation of polycrystals and composites. Unfortunately, this problem appears to be quite involved.

REFERENCES

1. U.Wolfstieg and E.Macherauch, Chemie-Ingenieur-Technik 45, 760, 1973.
2. E.Kröner, Kontinuumstheorie der Versetzungen und Eigenspannungen, in: Erg.Ang.Math. 5, Springer Verlag, Berlin-Heidelberg-NewYork, 1958, Appendix.
3. E.Kröner, On the Physics and Mathematics of Self-Stresses, in: Topics in Applied Continuum Mechanics, ed.by J.L.Zeman and F.Ziegler, Springer Verlag, Wien-NewYork, 1974.

Workshop Summary: Surface Effects in Uniaxial Tension and
                   Fatigue

Prepared by H. Mughrabi

Discussion Leaders: H. Mughrabi, T. E. Fischer, A. S. Argon,
                     A.R.C. Westwood

Recorders: W. R. Tyson, I. G. Greenfield, G. Champier,
                     N. H. Macmillan

1) Origin of Controversy:

In two of the workshop groups the opinion was voiced that the hard vs. soft surface layer controversy was more the result of a lack of common language and interpretations than of disagreement between experimental observations. For example, the generalization of conclusions based on one specific type of experiment was criticized. It was suggested that, depending on material, degree of deformation, etc., and also on whether studies were confined to the immediate vicinity of the surface or extended well into the interior, the surface region could be found to be softer or harder than the interior. The need for well-defined statements, also with regard to the limitations of the applied techniques, was recognized.

2) Definition of Terms:

Micro- and macroyielding should be distinguished clearly. For example, the macroyield stress after a certain amount of unrestrained plastic flow should not be compared with the stress where the first deviation from linearity in the stress-strain curve is observed after unloading and subsequent reloading, especially if the stress-strain curve of the continuation experiment coincides with the extrapolation of the initial stress-strain curve. This holds also for experiments involving unloading, surface removal and reloading. Proper attention should be paid to transient and local flow, as opposed to general yielding.

3) Necessary Experimental Precautions:

In mechanical testing extreme care should be taken to eliminate bending due to misalignment and elastic stress gradients. In surface removal experiments the much discussed problem of temperature rises should be met by reliable temperature control and measurement. The possible influence of surface films enforcing other slip systems of the substrate should be kept in mind, especially in the case of metals like aluminum.

4) Discussion of Earlier Experimental Work:

One group made an attempt to discuss critically the conclusiveness of some of the earlier work.

A consensus was achieved that the investigations on weakly deformed copper single crystals by Kitajima, who concluded that the surface layer was harder than the core, should not be compared directly with the work on more strongly deformed crystals by Fourie or Mughrabi, who came to the opposite conclusion. A shift of the picture with increasing deformation was considered possible.

The existing evidence regarding Kramer's debris layer was summarized. Kramer's avaluation of flow stresses after reloading was criticized (cf.2). It was pointed out that at this conference three studies on copper single crystals (Basinski, Fourie, Mughrabi) had shown that reloading after removal of a surface layer led to a higher flow stress. In lack of convincing direct evidence, the proposed structure and properties of the debris layer as concluded from mechanical measurement involving surface removal or relaxation were considered. An explanation was sought for Kramer's finding that the surface layer stress relaxed appreciably even at room temperature (for Al and somewhat less for Cu). The question was raised whether point defects rather than dislocations could be responsible for the debris layer and its relaxation at room temperature, since a recovery mechanism involving dislocation rearrangement or annihilation appeared improbable.

The survey lecturer (Basinski) criticized the interpretation of Mughrabi's X-ray studies which indicated a bending of the primary glide plane of deformed copper single crystals (for details, see survey lecture). Mughrabi agreed that the observed bending was too large to be predominantly elastic and conceded that part of the effect could be due to grip effects. Basinski suggested the effect was caused by secondary dislocations. It was agreed that, because of the smallness of the expected effects, highly accurate and systematic studies were desirable.

5) Discussion of Dislocation Behaviour in Vicinity of Free Surface:

One group discussed and agreed that in nearly all known instances dislocation motion will occur first near the surface for reasons of image forces, larger dimensions of the pre-existing network near the surface, surface stress concentrations and bending effects. Nevertheless, since large-scale plastic flow requires that the entire cross section become plastic, the initiation of dislocation motion near surfaces has only minor effects on the yield stress for general plastic flow.

In another group the effect of image forces and possible variations of the lattice parameter in the immediate vicinity of the surface were discussed. Concerning the latter effect, it had been reported by the surface scientists at this meeting that the extent of lattice distortions near close-packed surfaces is very small. In the light of these observations it appeared unlikely that the phenomenon suggested by Fleischer some time ago, namely that a change in lattice parameter at a surface could hinder the egress of dislocations through that surface, would occur in clean metal crystals.

It was pointed out that the range over which the image force would influence the behaviour of a dislocation approaching a surface varies as $\mu b/4\pi\tau_c$ (μ: shear modulus, b: Burgers vector), where τ_c is the stress required for dislocation motion. Thus, in annealed pure metal crystals, the range of influence of the surface could be quite extensive, whereas in a hard covalently bound crystal the influence would extend to much smaller depths. It was suggested that this fact might account in part for the difference between observations on aluminum (Lohne) and silicon crystals (Weissman).

The tendency of dislocations to cross slip in the vicinity of a free surface was considered in the sense that the jogs created by such cross slip would render the near-surface dislocations less mobile and lead to surface hardening. The loss of dislocations at free surfaces, especially if freshly exposed, was also considered. The analogy with the loss of dislocations during the thinning of transmission electron microscopy specimens, resulting from the disturbance of the equilibrium configurations, was also pointed out. This effect would make the newly exposed surface layer softer than it had been in the bulk.

6) Suggested Methods of Future Approach:

Two groups came to the conclusion that some of the key experiments pertaining to the hard vs. soft surface layer controversy should be repeated under well-defined conditions, paying special attention to the critical points outlined above (cf. also 7). Where possible, these experiments should be performed by investigators not directly involved in the present controversy.

In another group it was suggested that the experiments performed so far had not provided evidence suitable to resolve the controversial issues. Hence, future studies should resort to new, more specific testing methods.

The exploitation of the now available strong X-ray sources (synchrotron, DESY, Hamburg) and high-flux reactors (Grenoble) was proposed. The X-ray sources would permit critical experiments on surface layers, whereas neutron diffraction studies would be applicable to bulk ferromagnetic materials, e.g., nickel.

Investigations on b.c.c. metals were expected to be able to shed some light on the dependence of surface effects on the dislocation mode of deformation since, at low temperatures, screw dislocations in b.c.c. metals are much less mobile than edge dislocations. In view of the fact that the influence of point defects (e.g., vacancy injection) on surface phenomena is still very unclear, experiments should be performed at different temperatures, e.g., at low temperatures, where vacancies are immobile, or at higher temperatures, where vacancies can anneal.

A number of hopefully critical experiments were suggested by the various groups (cf. also 7). Studies on single crystals which can be strained extensively in single slip (e.g. cadmium) and subsequent experiments on crystals of copper (or aluminum, with some reservations) oriented for duplex slip were proposed. It was pointed out that, since residual stresses are unambiguously related to the elastic strains, the measurement of elastic strains in the near-surface region could establish clearly whether the surface region of deformed, unloaded crystals is in a state of tension or compression. Depending on whether the core or the near-surface region work harden preferentially, the surface layer would be expected to be in compression or in tension respectively. Attention was drawn to the fact that, as a matter of principle, in plastically deformed polycrystals and composites an overall residual stress exists the volume average of which does not vanish. Hence, special care is required to take this effect into account in the interpretation of X-ray measurements (cf. contribution to this conference by Kröner).

7) Application of Surface Science Tools:

The desire to achieve a better characterization of surfaces in critical experiments was brought out clearly by all the groups. One group in fact concentrated its attention to those issues of uniaxial plastic deformation and fatigue of metals to which the concepts and methods of surface physics are applicable.

In assessing the scope and limits of relevance of surface physics to plasticity it was noted that the effects of the surface on electronic properties, crystal structure and gross composition due to chemisorption or segregation (in contrast to impurity concentrations) extend over a few atomic layers only. By contrast, the plastic behaviour can be influenced by the presence of a surface to a depth up to one millimetre. Thus it appeared evident that surface physics and chemistry are relevant to crack initiation and propagation and to the initiation of slip steps, but not to the hard vs. soft surface layer controversy.

It was pointed out that effects of <u>environment</u>, and therefore presumably of <u>chemisorption</u>, on metal fatigue are well known. For example, the fatigue lives of metals tested in vacuum are several times longer than in air. It would hence be very desirable to observe the chemical composition at the very tip of the crack. Such direct experiments were not considered feasible at the present time.

Some indirect experiments were suggested with the hope of obtaining some guidelines for the elaboration of models. Studies of <u>crack propagation</u> under very pure gas atmospheres and parallel measurements of chemisorption on clean flat surfaces of the same metal were proposed. The aim would be to search for correlations between crack propagation and chemisorption phenomena. On the other hand, it was suggested that <u>crack initiation</u> should be studied in ultrahigh vacuum on surfaces that are cleaned and characterized with the analytical methods of surface physics to establish the effect of microstructure and chemical composition of the surface. A general shortcoming of the tools of surface physics was pointed out, namely that they do not permit investigation of defects in the surface structure.

In uniaxial deformation, the study of the initiation of <u>slip steps</u> and its dependence on the chemistry of the surface were named as a potential field of study. Tensile testing in an ultrahigh vacuum apparatus, where the sample is accessible to treatment and surface characterization, were considered promising. Preliminary studies utilizing such an equipment were reported which indicated that gases escape from the sample as it is deformed. Finally, in a discussion of possibilities for examining slip steps, it was noted the LEED has some limited capabilities for the observation and measurements of steps.

8) Some General Conclusions:

The complexity of the problems of interest was generally reflected in the almost frightening number of details that were raised in the discussions. Nevertheless, some valid criticism of earlier work did come out of these discussions. The lack of a common consensus on what has been done and what should be done represents a challenge to those working in the field. Future progress will demand a careful assessment of what is feasible at the present time. Critical experiments will presumably have to be simple in principle in order to avoid additional unforeseen complications.

In the way of an <u>interdisciplinary approach</u>, some promising areas for future work have been indicated. For example, it can be hoped that the important field of fatigue which was only touched at this meeting might benefit in the near future from the present degree of understanding in plasticity and surface science.

Workshop Session 2

THE INFLUENCE OF SURFACE FILMS ON MECHANICAL PROPERTIES

Chairman: P. Neumann

THE INFLUENCE OF SURFACE FILMS ON MECHANICAL PROPERTIES

H. G. F. Wilsdorf and G. E. Ruddle*

Department of Materials Science, University of Virginia,
Charlottesville, Virginia 22901 U.S.A.
*CANMET, Department of Energy, Mines and Resources,
Ottowa, Canada KIA OGI

ABSTRACT. The variables that influence the mechanical behavior of metals covered with thin films have been enumerated. In particular, factors affecting dislocation glide in the surface region have been characterized in terms of image stresses taking into account both the free surface and the accommodation dislocation network at the substrate-film interface.

It is generally found that surface films on metals increase the yield stress. This can be expected since misfit dislocations at the interface represent obstacles to the motion of glide dislocations. Occasionally, however, it has been reported that thin films did not affect mechanical properties or even reduce the yield stress. The latter phenomenon is being explored further and new experimental evidence for the effect is being provided. A reduction of yield stress of 50% or more was measured on copper crystals electro-plated with 600 Å films of nickel, gold or α-brass, all of which are epitaxial overgrowths. The complexity of the microstructure in the substrate-thin film composite has been described and the consequences on dislocation glide explored.

Finally, an example of a system which exhibits an increase of the critical resolved yield stress, τ_y, is given. Here, diffusion layers less than 1 μ thick caused τ_y to increase by a factor of two, a consequence of the hindering of glide dislocations by accommodation dislocation networks.

INTRODUCTION

Research pertaining to the influence of surface films on the mechanical properties of solids has continued to command a great deal of attention during the past decade. However, despite the progress made in understanding surface phenomena as well as in conducting more refined experiments, no unifying concept to explain the effect of surface layers on mechanical properties has come to the fore. Instead, more questions have been raised than answered.

The initial idea that one has to consider a material with bulk properties on which the free surface had been replaced by a surface film turns out to be unrealistic. Even the introduction of interfacial concepts has been of small help in that a complete characterization of an interface on the basis of experimental data is very difficult to obtain. It has become clear that it is virtually impossible to prepare crystalline bulk substrates for film coating without leaving contaminants on their surfaces or without introducing gases during the preparation process, both of which can be expected to affect the behavior of dislocations. Consequently, it must be concluded that available experimental data cannot be evaluated simply by assuming a two-phase system, namely substrate and film, provided they consist of two chemically different materials. Instead, the surface contamination of the bulk substrate must be taken into account together with the actual surface film which also will be affected at its solid-gas interface by the environment during a mechanical test.

However, in the absence of experimental data taking these physical conditions into account, our treatment of the subject will follow conventional lines which in a number of cases offer a satisfactory insight into the mechanisms operating at the microstructural level. That the plastic behavior of the crystalline bulk substrate can be significantly altered by the presence of a thin solid surface film has generally gained interpretation in terms of dislocation movement between the coating and substrate. This paper will outline the effects on dislocation mobility that are imposed on a crystalline material by the addition of a thin solid coating.

LITERATURE REVIEW

The first notable review article on the subject of this lecture, written by Kramer and Demer in 1960 [1], makes still good reading on account of its well conceived organization, critical discussions, and extensive illustrations. Two reviews by Machlin [2] and Westwood [3] complete the information gained during the exploratory phase of the subject matter.

In the seventies two excellent reviews by Latanision and Westwood [4] and by Latanision, Sedriks, and Westwood [5] have appeared which provide updated and exhaustive information; up to 250 literature citations, including environmental influences have been listed [4]. A wide variety of materials was covered, as well as various modes of deformation, including creep and fatigue.

It was thought to be more in the spirit of this Conference to concentrate on a few selected topics which, hopefully, may prove helpful to further our understanding of how thin surface coatings may produce wide ranging mechanical effects. The reader will find the use of the review articles cited above more than satisfactory.

FILM TYPES AND FACTORS INFLUENCING MECHANICAL BEHAVIOR

Most solids are naturally covered with adsorbed layers or solid films. Our concern, however, is with experiments which were designed to evaluate the effect of solid films on certain mechanical properties. Most of the work has been done with films that can be classified as follows:

Oxide films, thickness	$10^{-6} - 10^{-4}$ cm
Hydroxide films	$10^{-6} - 10^{-4}$ cm
Vacuum evaporated films	$10^{-6} - 10^{-4}$ cm
Electro-deposited films	$10^{-6} - 10^{-4}$ cm
Diffusion layers	$10^{-5} - 10^{-3}$ cm
Inherent polycrystalline layers	$10^{-4} - 10^{-3}$ cm

Within each film type certain characteristics should be known, including the following: single or polycrystalline nature of film; whether epitaxial or not; thickness; ductile or brittle; nature and magnitude of inherent stresses; continuous film or ruptured; nature of inherent lattice defects. Clearly, microstructural details are of utmost importance when considering the film-substrate interface, such as misfit, dislocation arrangements, and impurity distribution.

With few exceptions, the effects of thin films on the composite behavior have resulted in an increase of the yield stress and a change in the workhardening rate for tensile elongations, a decrease in the creep rate, and an increase in fatigue life under suitable conditions. Explanations of these phenomena have invoked the barrier effect of surface films to dislocation egress or their effect on the operation of surface or near surface sources. A number of interrelationships between substrate and film obtain which have considerable influence on dislocation movement: chemical composition, crystal structure, orientation relationships, shear modulus, and stacking fault energy. Additional factors are, amongst others, the geometrical shape of the bulk specimen, the method of substrate surface preparation, and the contamination at the substrate-film interface as well as the film-atmosphere interface.

FACTORS AFFECTING DISLOCATION GLIDE IN THE SURFACE REGION

Fundamental interrelated factors which affect dislocation glide in the surface region of the composite are (1) the critical yield stresses, (2) the shear moduli, and (3) the stacking fault energies of the component metals, and (4) the misfit of the composite interface. Values of the important parameters are given for the various metal components in Table I as they apply to unpublished experimental results which will be reported in a later section.

Critical yield stresses given in Table I are for bulk single crystals and form a range of values according to variation in crystal purity, size, orientation, and perfection (i.e., dislocation density and distribution). The shear stress τ is related to shear strain γ by the shear modulus G as

$$\tau = G\gamma \qquad (1)$$

in linear elasticity theory. Given that elastic strains in the surface film are sufficiently small, the decrease in G from the bulk value is small [6], and hence the corresponding decrease in the inherent critical yield stress τ_c for the surface film is very small. However, when the surface film is thin relative to the substrate thickness, the effects of the other factors enumerated above on the yield stress for the total composite can be comparatively much larger.

If, for example, the stacking fault energy of one of the components is low, then the corresponding restriction on the climb and cross-slip movements of dislocations will influence dislocation mobility in the surface region of the composite.

Table I Values of Important Parameters for the Components of the Film-Substrate Systems

Parameter		Substrate		Surface film			
		Copper		Nickel		Gold		Alpha-brass
Lattice Parameter	(Å)	3.608 [1]		3.517 [1]		4.070 [1]		3.677 [2]
Misfit	(f × 100%)			2.59		-11.35 [ ]		-1.88
Shear modulus	(Kg/mm^2)	4.9 × 10^3 [3]		7.6 × 10^3 [3]		2.8 × 10^3 [3]		3.6 × 10^3 [4]
Poisson ratio		0.35 [3]		0.32 [3]		0.42 [3]		0.36 [5]
Critical yield stress	(Kg/mm^2)	0.02 - 0.25 [6]		0.33 - 0.75 [7]		0.09 [8]		1 - 1.5 [6]
Stacking fault energy	(ergs/cm^2)	40 - 160 [7] 73 [9]		60 - 450 [7] 400 [9]		10 - 50 [7] 55 [9]		7 - 30 [3, 7, 8]
Surface energy	(ergs/cm^2)	1725 [9]		1725 [9]		1485 [9]		1725 [5]

[1] Handbook of Chemistry and Physics.
[2] ASM Metals Handbook.
[3] J. Friedel, "Dislocations."
[4] P. R. Strutt, B. H. Kear, and H. G. F. Wilsdorf, Acta Met. 14, 611 (1966).
[5] Assumed.
[6] From the literature.
[7] W. J. McGregor Tegart, "Elements of Mechanical Metallurgy."
[8] H. G. van Bueren, "Imperfections in Crystals."
[9] J. P. Hirth and J. Lothe, "Theory of Dislocations."

Dislocations lying in the surface region of the composite have imposed on them image stresses which are related to the changes in shear modulus at the free surface and at the substrate-film interface. Head [7] has applied the method of images to an infinite isotropic elastic medium, in which a plane boundary defines two regions of different shear moduli, to obtain an approximate linear elastic solution for the image stress of a screw dislocation lying parallel to the boundary. The dislocation situated in the medium of modulus G at distance r from the elastically harder medium of modulus G' experiences a repulsive force per unit length,

$$\left(\frac{G' - G}{G' + G}\right) \frac{Gb^2}{4\pi r} , \qquad (2)$$

where b is the magnitude of the Burgers vector.

The case of the screw dislocation in a semi-infinite medium bounded by a plane free surface is obtained by letting $G' = 0$, and thus the dislocation is attracted to the free surface by the force per unit length,

$$\frac{Gb^2}{4\pi r} . \qquad (3)$$

Head then extended the analysis to a semi-infinite medium with the free surface parallel to the boundary defining regions of different shear moduli, effectively the case of a thin film on a semi-infinite substrate. The solution in this case involved an infinite set of images describing the image stress of a screw dislocation lying in the substrate parallel to the boundary and the free surface. The more complicated interaction of an edge dislocation with the boundary and free surface was not readily solvable.

In order to obtain an approximate estimate of the image stress on a dislocation lying in the film, or in the substrate, for the present discussion, the solution for the image stress at an internal boundary in an infinite medium (Equation (2)) is superimposed with the solution for the image stress at the free surface (Equation (3)). This form of solution is found to be approximately equivalent to the first two and most significant terms in the infinite series solution derived by Head for a screw dislocation. Thus the adopted solution provides a satisfactory estimate for the discussion, accurate to within about 10 to 15% for a screw dislocation. The accuracy of the solution is less for an edge dislocation [7]. More extensive mathematical treatments have been given to this problem by others [8], [9], [10].

The case of an infinitely long edge dislocation lying parallel to the free surface on a slip plane inclined at angle α to the surface normal is considered (Figure 1). Prins [11] has shown that the surface image force of attraction on the dislocation is given by the stress

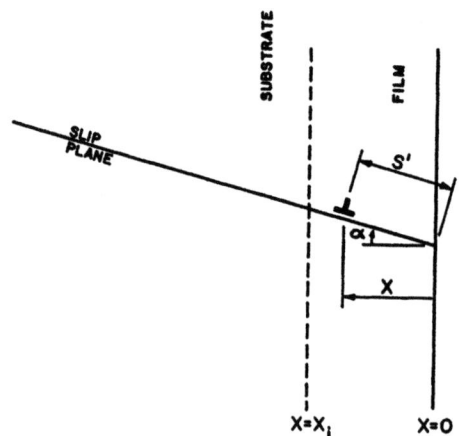

Fig. 1. Geometry of edge dislocation lying parallel to the surface on a slip plane inclined to the surface.

$$\tau_s = \frac{Gb}{4\pi(1-\nu)S'} = \frac{Gb \cos \alpha}{4\pi(1-\nu)X} , \qquad (4)$$

where S' is the distance from the surface to the dislocation along the slip plane, and X is the normal distance from the surface to the dislocation. Letting the normal distance from the surface to the interface be X_i, and assuming that the geometrical factor cos α similarly applies at the interface, the solution for the interface image stress in an infinite medium is combined with Equation (4).

Thus the net stress required to move the dislocation inward from the free surface is

$$_j\tau = {}_j\tau_s + {}_j\tau_i$$

$$= \frac{G_j b_j \cos \alpha}{4\pi(1-\nu_j)X} + (-1)^j \left(\frac{G_2 - G_1}{G_2 + G_1}\right) \frac{G_j b_j \cos \alpha}{4\pi(1-\nu_j)(X - X_i)} , (5)$$

where the subscripts i, s, 1, and 2 denote interface, surface, film, and substrate, respectively, and assuming that the net image stress imposed on the dislocation is the critical stress necessary for its inward movement. For $X < X_i$, i.e., when the dislocation is lying in the surface film, let $j = 1$ in Equation (5); for $X > X_i$, the dislocation is lying in the substrate, and j = 2. Equation (5) without the $(1 - \nu_j)$ term in the denominators may be similarly used for the case of a screw dislocation parallel to the interface and surface.

The net image stresses may be graphically represented in terms of the experimental results to be presented by letting $X_1 = 600$ Å, letting $\alpha = 19.5°$ for the screw-surface geometry (Figure 5), using appropriate values of a, G, and ν from Table I; and using Burgers vector magnitudes of $0.866 \frac{a}{2}[110]$ and $0.5 \frac{a}{2}[110]$, respectively, for the edge and screw components of a dislocation parallel to the screw surface. Both the surface image stress $_i\tau_s$ and the combined surface and interface stress $_i\tau$ (Equation 5) have been plotted for a nickel film (Figure 2) and for a gold film (Figure 3) on copper substrates. The stress necessary to move the dislocation toward the free surface may be obtained simply by a mirror reflection of the plot about the zero stress axis. The image stresses for the edge-surface geometry ($\alpha = 45°$) would be approximately 0.84 of the magnitude of those shown for the screw surface in Figures 2 and 3. (Screw and edge surfaces are specific substrate geometries shown in Figures 5 and 6.) The image stresses for the case of an alpha-brass film on a copper substrate are similar to those shown in Figure 3.

From the graphical representation of the effect of the combined image stress, it is clear that the egress of dislocations through the surface region would be most likely to occur in a system such as gold-plated copper. In a system such as nickel-plated copper, it appears unlikely that egressing dislocations could be forced through the repulsive effect of the interface, or that dislocations could be injectd inward from the free surface force of attraction, without the aid of an unusually high stress.

The real substrate-film composite crystal is considerably more complex than the hypothetical continuum and boundary conditions assumed for the linear elastic analysis [7], [12]. The

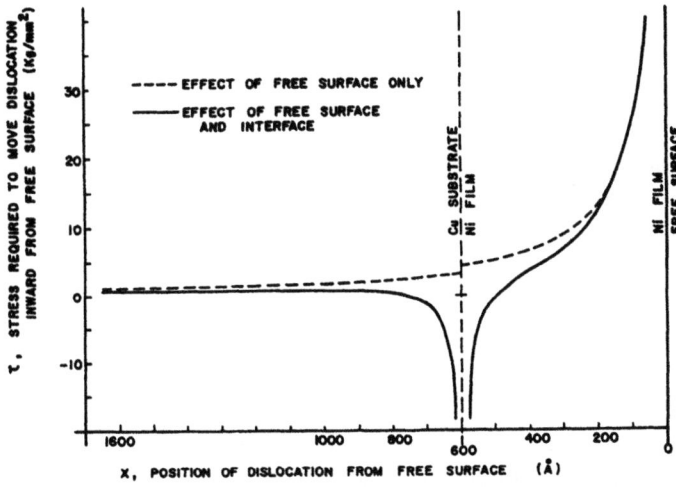

Fig. 2. Effect of surface and interface image stresses on a dislocation lying parallel to the screw (111) surface in a nickel-copper composite.

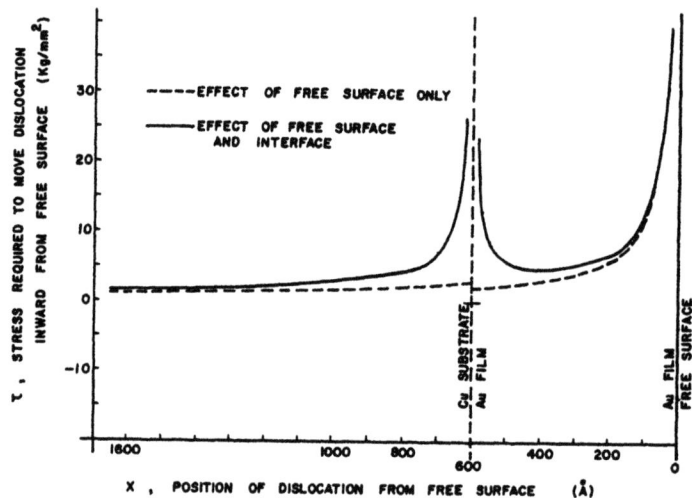

Fig. 3. Effect of surface and interface image stresses on a dislocation lying parallel to the screw (111) surface in a gold-copper composite.

analysis breaks down for the interaction of the non-linearly elastic core region of the dislocation with the interface and the free surface [12]. An epitaxial surface film is fully or partially coherent with the substrate, and therefore the elastic properties may not be discontinuous at the interface, consequently the image stress function may not be discontinuous at the interface. Additionally, the presence of interfacial dislocations in a partially coherent interface would alter the stress conditions, and hence would alter the interaction with a glide dislocation. Further, the thin film coherently formed on a bulk substrate would not be expected to have bulk elastic properties. The shear modulus G_1 of the surface film has in fact been predicted to be reduced to G_1' according to the relation

$$G_1' = \frac{G_1}{(1 + e_1)^2} \qquad (6)$$

as a result of the elastic strain e_1 of the film; however, as stated earlier, the reduction is small for small strains [6]. Although their results indicated that the change in shear modulus has some effect on the passage of dislocation through the interface, Brame and Evans [13] concluded that the misfit resulting from the change in lattice parameter at the interface is the predominant factor. The nature of the substrate-film interface is determined by factors in addition to the shear moduli and lattice parameters, some of which will be discussed in the next sections.

MISFIT AT THE COMPOSITE INTERFACE

The greatest influence on dislocation glide in the surface region is due to the structural misfit at the substrate-film interface. The general nature of the misfit at an interface formed between a substrate and a film of the same crystalline structure and orientation is determined by the lattice parameters, shear moduli, and thicknesses of the components, the strength of bonding across the interface, the orientation of the interface, and the mode of film growth [14], [15], [16], [17], [18]. The influence of these factors in the accommodation of the lattice misfit by elastic strain and by misfit dislocations has been reviewed recently by Matthews [19] and by Stowell [20].

For a thin surface film epitaxially formed on a bulk substrate, all of the elastic strain of misfit is accommodated in the surface film. According to the analysis of Jesser and Kuhlmann-Wilsdorf [21], the surface film is elastically strained to match the substrate in the initial stages of growth. In systems with less than 4% misfit, the film grows continuous while in this pseudomorphic state. When the film reaches a certain critical thickness, the strain energy in the film is so large that the pseudomorphic state becomes unstable. Thereafter the equilibrium elastic strain e_1 in the film is decreased as the inverse of the film thickness h_1 by the introduction of dislocations to accommodate misfit at the film-substrate interface.

A dislocation, generated along a glide system to the interface, from hereon referred to as a "glide-in" dislocation, can accommodate misfit at the interface in proportion to the edge component of its Burgers vector which is parallel to the interface. Thus, the orientation of the interface becomes important with respect to the angle it forms with the glide systems of the crystal. Experimental results have demonstrated that screw segments of "glide-in" dislocations cross-slip to accommodate misfit in two directions in the (100) interface and in three directions in the (111) interface (refer to Matthews and Jesser [22], [15], [23], Thompson and Lawless [24], [25]).

The possible positions and signs of the edge components of "glide-in" dislocations on a given slip plane are considered with respect to the interface in Figure 4. For a dilated surface film, dislocations of both signs in the film and in the substrate would glide in the directions shown (Figure 4(a)) to accommodate misfit at the interface. Application of a tensile stress σ (Figure 4(b)) to the crystal would enhance the indicated glide of the dislocations lying in the surface film, and would oppose the indicated glide of the dislocations lying in the substrate. The opposite situation exists for a surface film in elastic compression (Figure 4(c)). Application of a tensile stress σ (Figure 4(d)) would enhance the glide of the dislocations lying in the substrate, and would oppose the glide of the dislocations lying in the surface film.

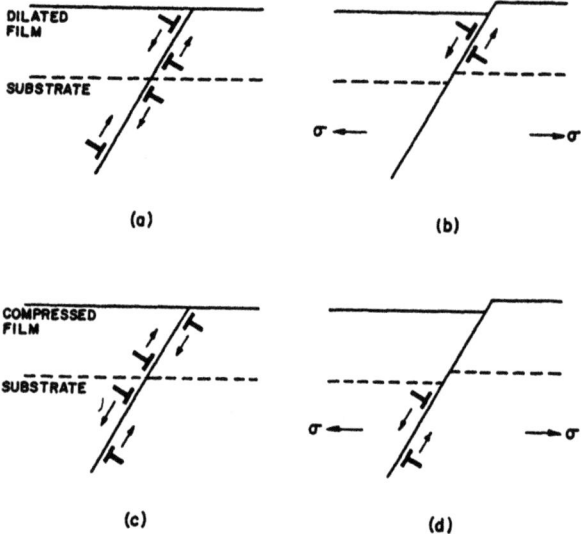

Fig. 4. Possible interactions of edge components of "glide-in" dislocations with interface for accommodation of misfit in (a) a dilated film and (c) a compressed film. Compatibility of glide-in dislocations with applied tensile stress σ is shown for (b) the dilated film and (d) the compressed film.

Recently, a number of experiments have been carried out on single crystals with special axis and surface orientations. In order to aid in the discussion of the plastic behavior of some of these systems, two typical cases will be given in the following section.

GLIDE SYSTEMS RELATIVE TO THE SCREW AND EDGE SURFACES

The relations of the glide systems to the screw and edge surfaces are to be established for the consideration of mechanisms of dislocation glide into the interface in these two specific orientations. In the context of the experiments to be discussed, a face-centered-cubic substrate crystal is considered.

The four (111) glide planes in the crystal are represented by sides 1, 2, 3, and 4 of the regular tetrahedron, shown relative to the screw surface in Figure 5 and are appropriately indexed. The Burgers vectors of the three glide systems on a given glide plane j are then represented by the edges of the tetrahedron side as b_{j1}, b_{j2}, and b_{j3}. For example, the glide systems on the

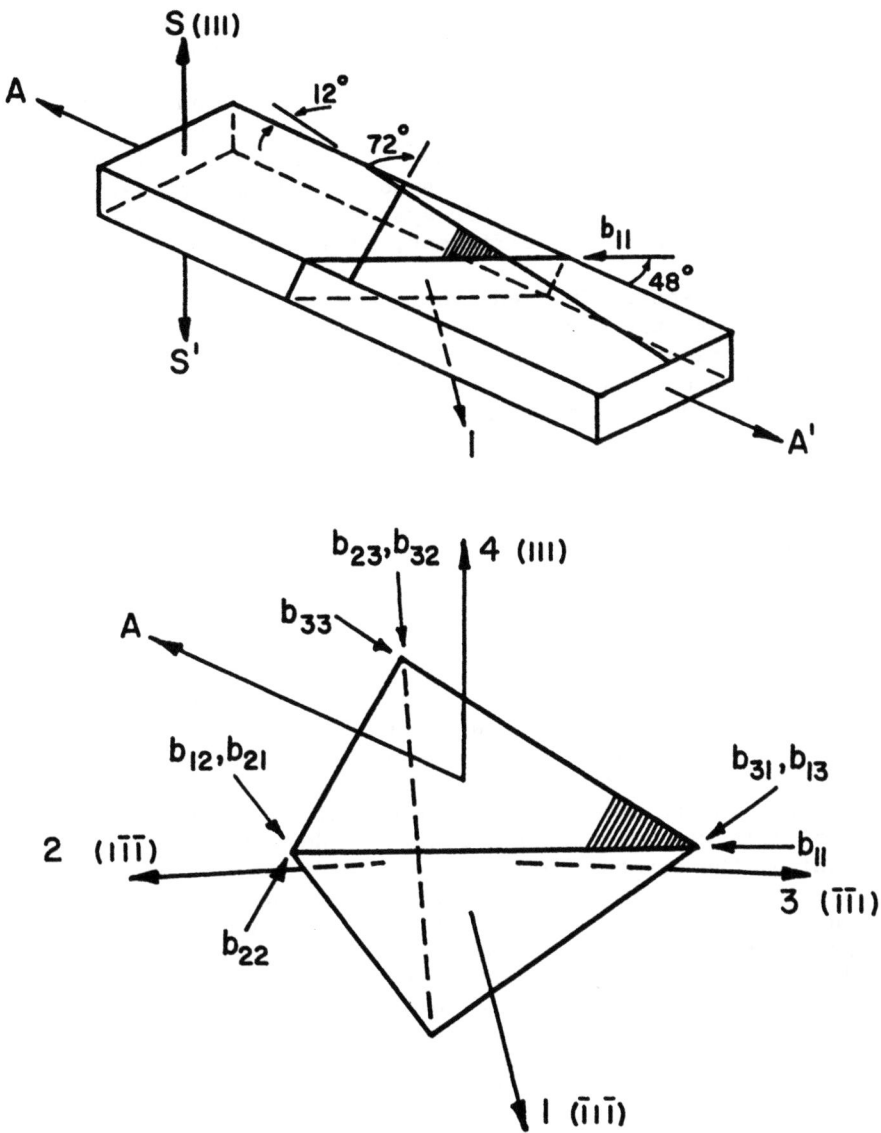

Fig. 5. Illustration of geometry of glide systems relative to the screw surface (S).

primary plane 1 are b_{11}, b_{12}, b_{13}, where b_{11} is the primary glide system. Note that screw dislocations on the primary system are parallel with the screw surface S. A more complete analysis relating the glide systems to the tensile axis A-A' would include determination of the Schmid factors [26].

In the regular symmetry of the screw-surface geometry, it is readily seen that the glide systems b_{ij}, $i = j$, are not effective in the accommodation of misfit at the (111) interface since dislocations on these glide systems have no edge component parallel to the interface. All of the glide systems b_{ij}, $i \neq j$, are equally effective, about 29% efficient, in accommodating (111) interfacial misfit.

Similarly presented in Figure 6 is the geometry of the glide systems relative to the edge surface. In this case, edge dislocations on the primary system b_{11} are parallel with the edge surface E. With the edge-surface geometry, all glide systems b_{ij} are effective, by different amounts, in the accommodation of misfit.

EFFECT OF NICKEL, GOLD AND α-BRASS FILMS ON THE YIELD STRESS OF COPPER CRYSTALS

A. Experimental Data

Taking into account the various factors influencing the mechanical properties of metals through surface films, a series of experiments was designed and carefully carried out on 99.999 copper crystals which had a rectangular cross-section of 2 mm x 6 mm with their large surfaces in the screw or edge orientation (Figures 5 and 6). The crystal axis was oriented for single glide. After electrolytical polishing, the crystals were plated immediately without drying either with Ni, Au or α-brass; the details of the experimental procedures have been given elsewhere [26]. In view of the objective of the study, thin films were chosen, namely 600 Å thick, which provided films epitaxed on the substrate with a low equilibrium elastic strain, e_1, in the order of 0.5% or less [24],[27].

Since the yield stress is most sensitive to the perfection of the crystals, all specimens were carefully examined by the Berg-Barrett technique. Low angle sub-boundaries as well as twist boundaries have been observed and their orientations and dimensions measured on the screw and edge surfaces.

The relation of the twist boundary to the screw and edge surfaces is shown by the orientation of the (111) plane in Figures 5 and 6. For both surface geometries the relations and interaction of primary glide dislocations (Burgers vector b_{11}) with the twist boundaries would be the same. The twist boundaries, forming an angle of 36° with the edge surface, would be acting relatively more as barriers in the glide path of edge dislocations to the

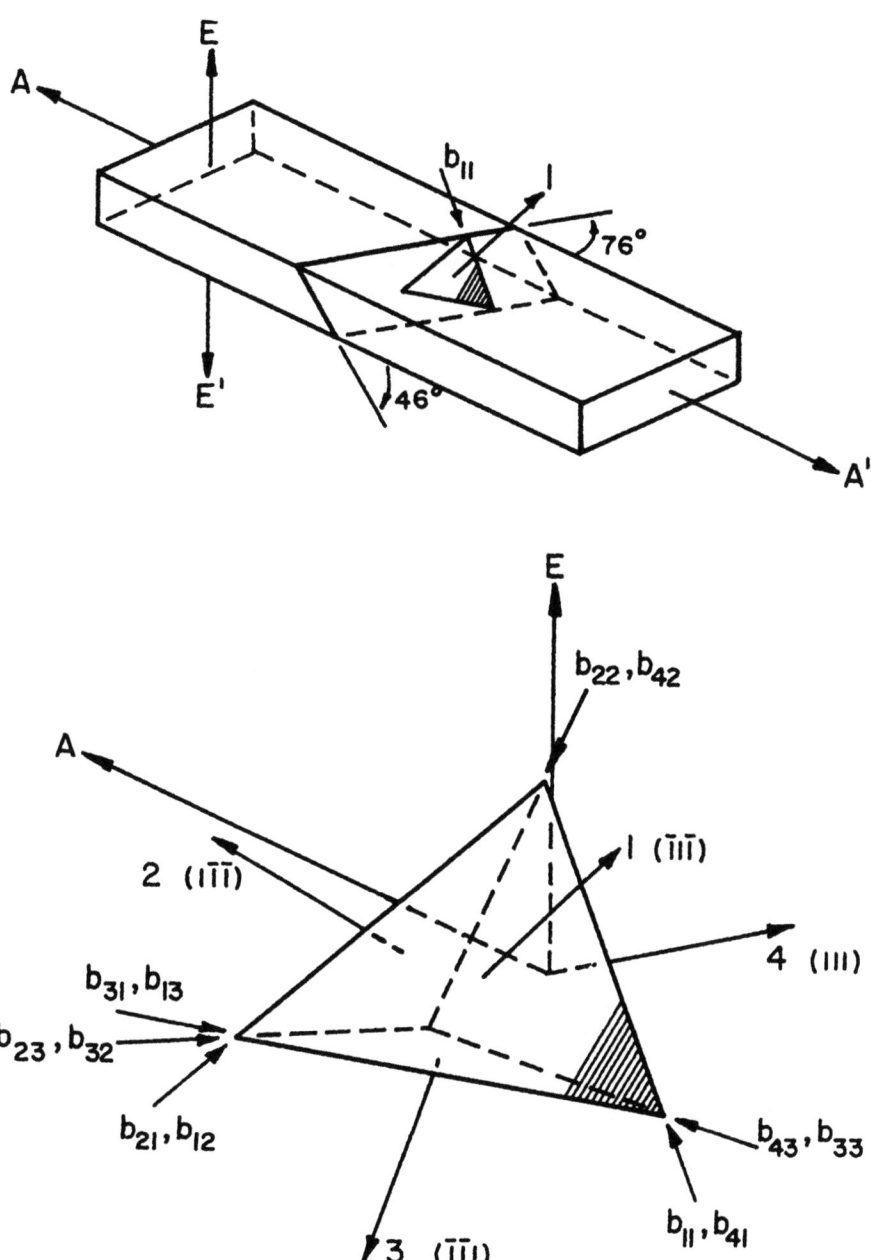

Fig. 6. Illustration of geometry of glide systems relative to the edge (E) surface.

surface. In the screw surface geometry, the boundaries are more steeply inclined, at an angle of 71°, to the screw surface.

B. Results

Most work on the surface sensitivity of the mechanical behavior of metals shows that the presence of solid surface films generally increases the yield strength of single crystals, although occasionally no influence of such films has been found [28]. The present study, which has not been published previously, provides evidence that under certain well controlled experimental conditions the yield stress may be lowered by amounts of 50 or 60% in comparison to film free crystals of identical orientation and shape.

Throughout the study great care was taken to compare crystals which had comparable sub-structures. As will be seen from Figures 7 - 10, the yield stress of unplated crystals varied depending on the substructure. Consequently, two groups of crystals were defined, those with twist boundaries and those without, the latter containing only low angle sub-boundaries. For each group, the measurements are given separately for crystals with edge or screw surface respectively.

First, crystals which do not contain twist boundaries will be discussed. In Figure 7 we see that the plating of Ni and α-brass on the screw surface reduces the yield stress substantially and this result is in agreement with data published earlier for Ni [29]. However, as shown in Figure 8, this effect is not observed for edge surface crystals.

In Figures 9 and 10 are presented the results for crystals which contain twist boundaries. As expected, the stress level for all crystals is higher. The most significant observation, however, is the consistent behavior of the yielding of Ni-plated crystals for both surface orientations. In contrast to Au and α-brass plated crystals the yield stress is unchanged from the value for an unplated crystal within the error limits, and for the screw surface crystals the average resolved shear stress for stage one is above the unplated crystal value with an increasing slope. The yield behavior for Au and α-brass plated crystals is consistently lower for both orientations. The most significant drop in yield stress is found for edge surface crystals; a gold plated crystal, E 13, yielded at 25 g/mm^2 which amounts to nearly 1/4 of the yield stress of the unplated comparison crystal.

C. Discussion

In trying to understand the results presented above, the differences between the three composites rather than their common characteristics should be enumerated. Ni produces a dilated film on

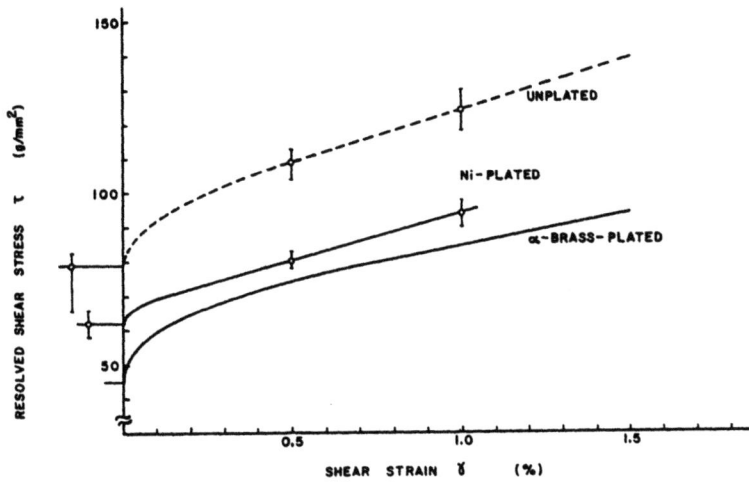

Fig. 7. Average stress-strain data of groups of screw-surface crystals including crystals which do not contain twist boundaries.

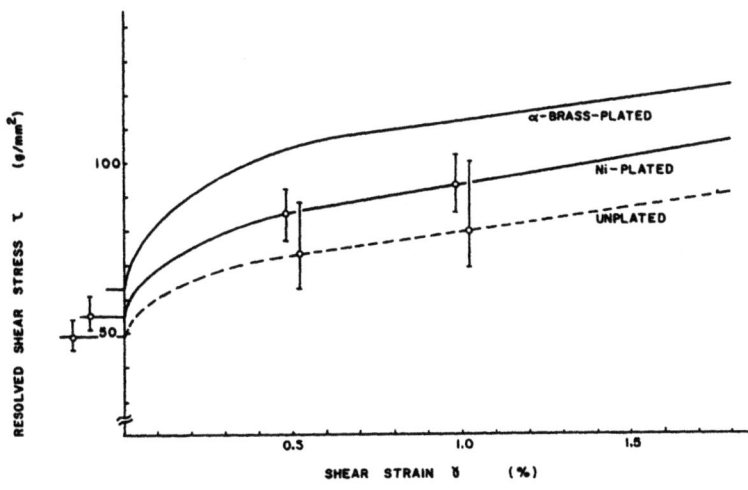

Fig. 8. Average stress-strain data of groups of edge-surface crystals including crystals which do not contain twist boundaries.

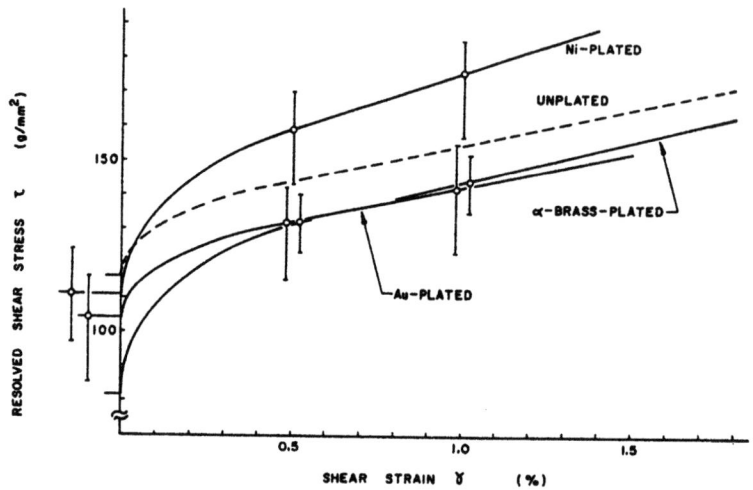

Fig. 9. Average stress-strain data of groups of screw-surface crystals including crystals which contain twist boundaries.

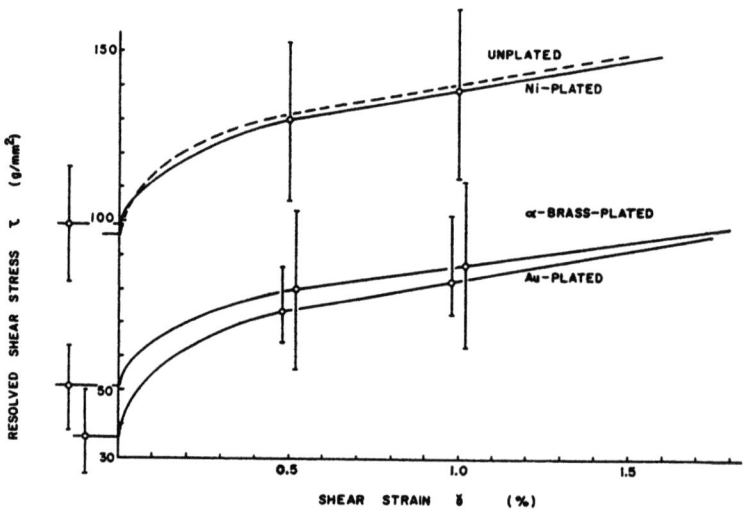

Fig. 10. Average stress-strain data of groups of edge-surface crystals including crystals which contain twist boundaries.

Cu but has a higher shear modulus G. With Au and α-brass a compressed film is formed and G is smaller than that of Cu. Clearly, these relationships change the forces which act on dislocations in the film and on those in the substrate crystal near the interface. The magnitude of the misfit between film and substrate is 2.59% for Ni, -11.35% for Au, and -1.88% for α-brass (Table I).

Further, differences in the interface and plating will have been most likely introduced by the experimental conditions. The following factors will influence the film and physical interface considering the plating process only: current density, electrode geometry, crystallographic orientation of the substrate surface, the presence of addition agents or impurities in the electrolyte solution, the nature of the anions in the solution, the nature of the cations being deposited or in solution, solution concentration, pH of solution, temperature, and agitation of solution.

Even within one set of crystals plated with the same metal, significant differences must exist on account of the atomic smoothness of the screw surface in comparison to the edge surface. While it will be impossible to polish a crystal surface 6 x 50 mm into a single plane of atomic smoothness, it is obvious that one can expect the (111) screw surface to be smoother than the edge surface which is approximately parallel to (210). Thus the surface steps must cause the interface to be a mixed one in regard to crystallographic orientation and no models are available to describe a stepped interface in terms of misfit dislocations. In this connection it should be noted that the continuation of substrate dislocations into the film through a stepped interface presents a most intriguing problem. Finally it should be realized that the free film surface is not necessarily the copy of the stepped interface. As a result of "bunching" due to uneven growth of the deposit layers on the substrate surface facets [30], the free surface of the composite crystal is likely less smooth than the interface, the surface steps being fewer and deeper (Figure 11).

How the mobility of glide dislocations is affected near the interface and in the surface film has been discussed earlier on the basis of elasticity theory. It is equally important, however, to note that there are significant differences between dislocations in the substrate and those present in the plated film. The largest number of "new" dislocations will be located at the interface. Vermaak and van der Merwe [31], [32] and others, and more recently, Shinohara and Hirth [33] have shown that accommodation networks are stable under applied stresses and provide a strong obstacle to glide dislocations at room temperature. Details of interrelationships between accommodation dislocations and glide systems also have been given in an earlier chapter. The reason for their resistance to glide motion is their geometrical arrangement in contrast to substrate dislocations which resist glide on account of impurity pinning. Obviously, unpinned dislocations can move at lower stresses and the lowering of the yield stress observed could be explained if a sufficient density of "new" dislocations would

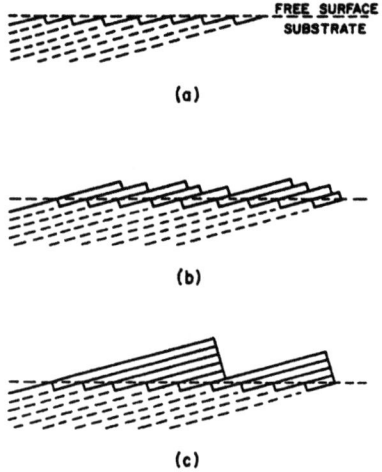

Fig. 11. Illustration of the bunching process involving (a) the original faceted substrate surface, (b) the growth of deposit layers on the substrate surface, and (c) the bunching of deposit layers to form larger surface facets.

be available, and if it is assumed that near-surface dislocations are responsible for the onset of plastic flow. There seems to be sound evidence for the latter contention as provided by a study of surface effects in aluminum by Fabiniak and Kuhlmann-Wilsdorf [34].

Two types of dislocations free of impurity pinning suitable for glide motion are being introduced by an electro-plated epitaxial film: (i) dislocations that form irregular links in the accommodation network; (ii) grown-in dislocations as continuation of substrate dislocations. As has been mentioned above, the interface between substrate and plating is not perfectly flat even for screw surfaces while for edge surfaces the interface must consist of a multitude of steps. Thus, it can be concluded that the accommodation network to a large extent is three-dimensional. Towards the outer sides of the network, in particular, it will have many irregular links, the length of which must be greater than the network; since they are not impurity pinned, they can be expected to glide at lower stresses than substrate dislocations. This reasoning holds also for grown-in dislocations in the film which are the only ones to penetrate the network. Through their motion, they are capable of "unzipping" their pinned portions in the substrate and therefore can glide at low stresses. The applied stress acting on these two types of dislocations is increased locally by stress raisers in the free surface in the form of surface steps due to

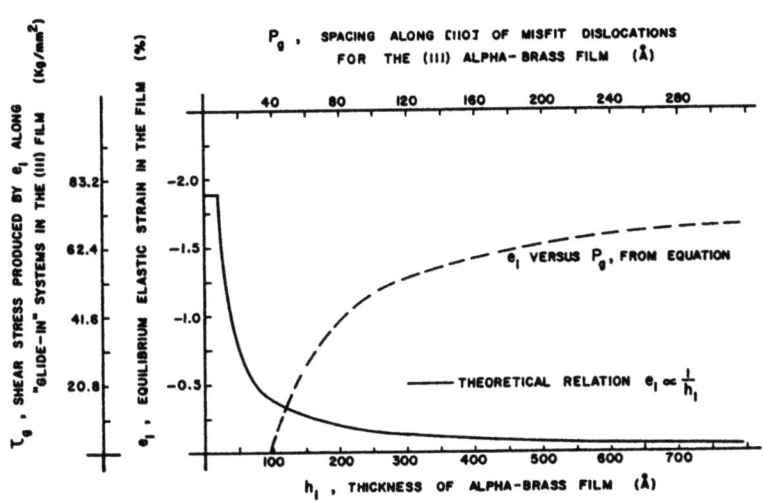

Fig. 12. Graphical analysis for discussion of the alpha-brass-plated copper composite.

"bunching" [30] and by the inherent stresses present in the deposit [27]. Figure 12 shows the results of converting the residual strain e_1 in the platings into a shear stress τ_g acting on "glide-in" systems [24], [26], [35].

In conclusion, it can be said that an epitaxial overlayer of suitable thickness can provide dislocation arrangements at the substrate-film interface and in the film itself which contain sufficient numbers of dislocations that are capable of moving at stresses substantially lower than "pinned" dislocations in the unplated crystal. The availability of such dislocations are expected to be rather uniform over the surface and cooperative effects in regard to their glide motion are likely to occur.

THE EFFECT OF DIFFUSION LAYERS ON THE TENSILE DEFORMATION OF METAL CRYSTALS

The results given in the previous chapter shall be complemented by reporting a study which shows that a specific surface treatment can hinder glide dislocations from leaving the crystal resulting in an increase of the critical resolved shear stress.

Greenfield and his associates followed up on earlier studies by Rosi [28] and Adams [36] who had doped copper crystals with silver and zinc respectively, thereby changing the chemical composition of a thin surface layer. In a series of papers Greenfield et al. [37], [38], [39] described systematic experiments on copper crystals with surface alloy layers containing either platinum, gold, nickel or silver. These elements were electro-plated or vacuum evaporated onto the copper ranging in thickness between 120 Å and 1,200 Å. Following was a heat treatment designed to produce a concentration gradient. The depth of the diffusion layers was in the order of a few microns.

Figures 13 and 14 show the increase of the critical resolved shear stress, τ_c, and the changes of slope in the easy glide range for copper crystals doped with silver. The thickness of the electro-plated layer was 900 Å; for the temperature treatments see Nakagawa and Greenfield [40]. For crystals oriented with the tensile axis parallel to [001], τ_c increased from $84 g/mm^2$ to $201 g/mm^2$ and the slope for the crystal with the highest τ_c was approximately three times steeper in comparison with the unplated crystal. Note that τ_c has a maximum which happens to occur when the plated film has been dissolved almost completely into the crystal.

The mechanisms that may lead to an increase of τ_c are primarily related to accommodation dislocation networks and solid solution hardening. Steep compositional gradients cause accommodation networks, an example of which is shown in Figure 15. The mesh size changes with the diffusion treatment; for the study reported here the increase of the mesh size, h, as a function of diffusion distance is plotted in Figure 16. One sees that h may change by a factor of about ten. Equally important is the distribution of accommodation dislocations. Figure 17 illustrates the effect of heat treatments on the concentration gradient, C_{Ag}, the dislocation density, ρ, and the location of dislocations in relation to the interface. After determining their Burgers vectors Nakagawa and Greenfield determined that they were not glissile on the primary and the conjugate slip planes. Thus the authors concluded that the dislocation network was largely responsible for the increase of τ_c in crystals with a (001) orientation by blocking glide

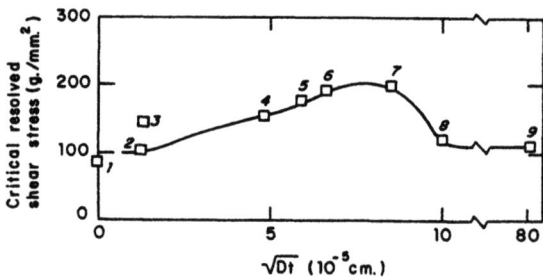

Fig. 13. Critical resolved shear stress as a function of diffusion parameter $\sqrt{Dt}$ for Ag-doped Cu crystals [40].

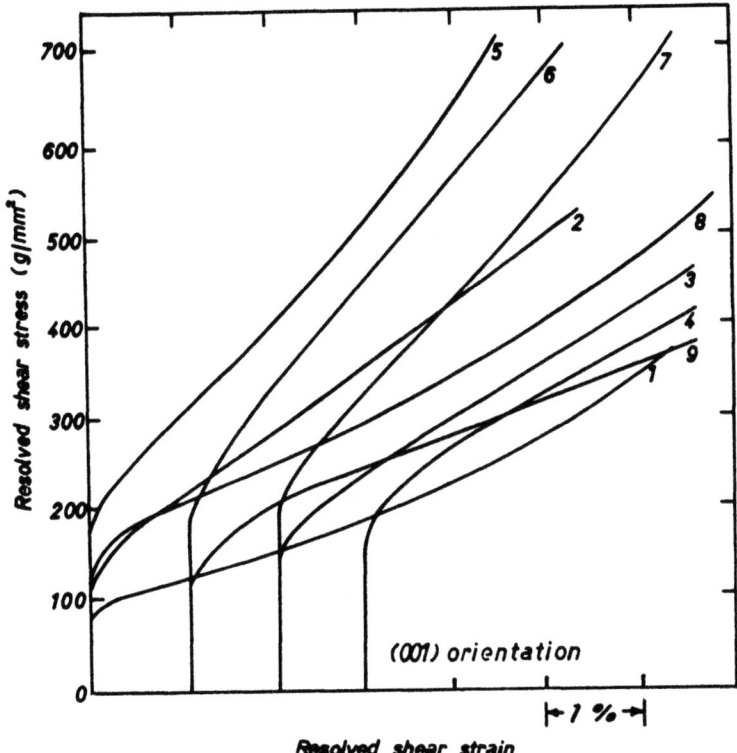

Fig. 14. Resolved shear stress-strain curves for (001) oriented Ag-doped Cu crystals [40].

dislocations. An investigation of (110) crystals, however, yielded results that required a more complex mechanism on account of glissile components in the dislocation network.

In conclusion: it has been demonstrated that accommodation dislocations caused by diffusion layers can be effective barriers to the egression of glide dislocations, although other hardening mechanisms may contribute to the increase of the yield stress.

CONCLUSIONS

In general, thin surface films have a significant influence on the yield stress, τ_y, and other mechanical properties. Their presence causes in most cases an increase in τ_y, often doubling its value,

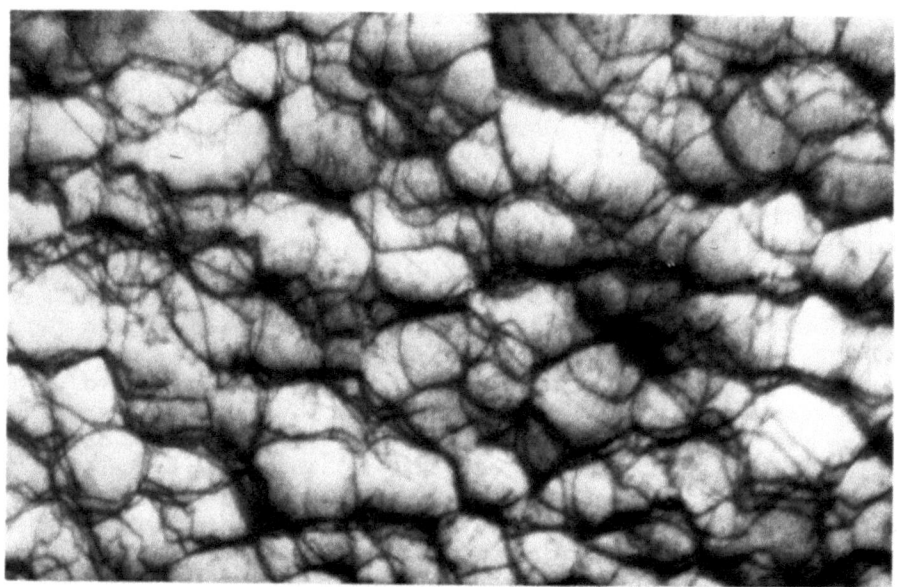

Fig. 15. Accommodation dislocation network resulting from the diffusion of Pt into Cu crystal. 34,000 x. Stereomicrographs show that network is three-dimensional [37].

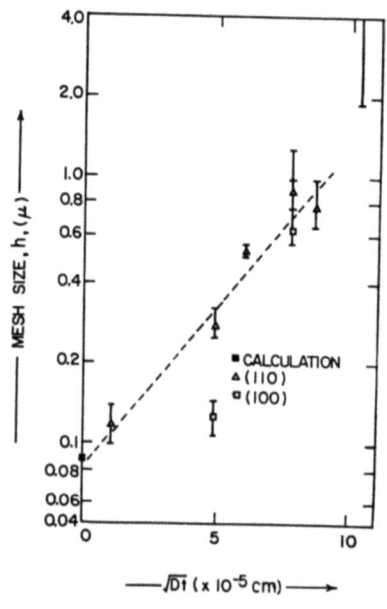

Fig. 16. Increase in dislocation network mesh size as a function of $\sqrt{Dt}$ [40].

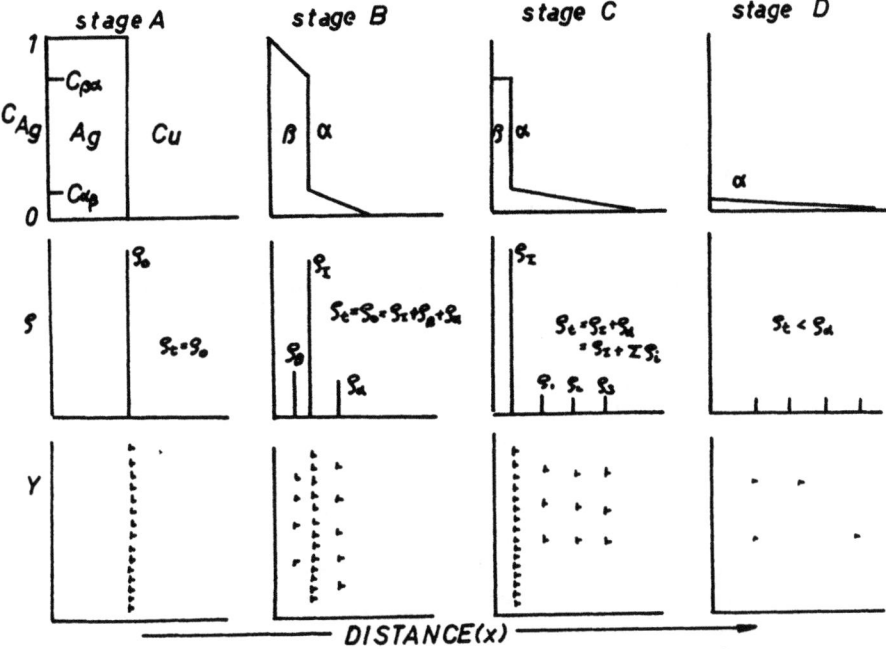

Fig. 17. Accommodation dislocation distribution in Ag-doped Cu crystals [40]. Stage A: as deposited; B and C: before and after β-phase is homogenized, respectively; D: after depletion of the β-phase.

for the reason that they provide obstacles to glide dislocations attempting to leave the crystal or trying to move into the crystal. However, a recent study shows that electro-plated epitaxial films on single crystals may also lower τ_y by 50% or more. Although no unifying concept is available to account for all the phenomena observed so far, confidence is being expressed that dislocation theory can provide adequate explanations.

The awareness that thin films not only can increase the yield stress but are also capable of decreasing τ_y will provide a new perspective to these complex phenomena. Through refinements of experiments and theory a better insight should be achievable during the coming decade.

REFERENCES

1. I. R. Kramer and J. Demer, Progr. Mat. Science 9, 133, 1961.
2. E. S. Machlin, Strengthening Mechanisms in Solids, ASM Metals Park, Ohio, 1962, p. 375.
3. A. R. C. Westwood, Environment-Sensitive Mechanical Behavior, Gordon and Breach, New York, 1966, p. 1.
4. R. M. Latanision and A. R. C. Westwood, Adv. in Corrosion Sci. and Techn., Plenum Press, 1970, Vol. I, p. 51.
5. R. M. Latanision, A. J. Sedriks and A. R. C. Westwood, Struct. and Prop. of Metal Surfaces, Maruzen Co., Tokyo, 1973, p. 500.
6. W. A. Jesser and D. Kuhlmann-Wilsdorf, Phys. Stat. Sol. 19, 95, 1967.
7. A. K. Head, Phil. Mag. 44, 92, 1953.
8. Y. T. Chou, Phys. Stat. Sol. 17, 509, 1966.
9. G. H. Conners, Int.J. Engng. Sci. 5, 25, 1967.
10. I. G. Greenfield, J. Scient. Ind. Res. 32, 521, 1973.
11. J. F. Prins, D. Sc. Dissertation, U. Virginia, 1967.
12. J. P. Hirth and J. Lothe, Theory of Dislocations, McGraw-Hill, New York, 1968, pp. 68, 85, 89, 131.
13. D. R. Brame and T. Evans, Phil. Mag. 3, 971, 1958.
14. W. A. Jesser, Ph. D. Dissertation, U. Virginia, 1966.
15. J. W. Matthews and W. A. Jesser, Acta Met. 15, 595, 1967.
16. J. H. van der Merwe, Phil. Mag. 7, 1433, 1962.
17. J. H. van der Merwe, J. Appl. Phys. 34, 117, 123, 1963.
18. H. van der Merwe, Single Crystal Films, Pergamon Press, Oxford, 1964, p. 139.
19. J. W. Matthews, Epitaxial Growth, Acad. Press, New York, 1975, Vol. B, p. 559.
20. M. J. Stowell, Epitaxial Growth, Acad. Press, New York, 1975, Vol. B, p. 437.
21. W. A. Jesser and D. Kuhlmann-Wilsdorf, Phys. Stat. Sol. 19, 95, 1967.
22. J. W. Matthews, Phil. Mag. 13, 1207, 1966.
23. W. A. Jesser and J. W. Matthews, Phil. Mag. 15, 1097, 1967.
24. E. R. Thompson, D. Sc. Dissertation, U. Virginia, 1966.
25. E. R. Thompson and K. R. Lawless, Appl. Phys. Letters 9, 138, 1966.
26. G. E. Ruddle, D. Sc. Dissertation, U. Virginia, 1969.
27. E. R. Thompson and K. R. Lawless, Electrochim. Acta 14, 269, 1969.
28. F. D. Rosi, Acta Met. 5, 348, 1957.
29. G. E. Ruddle and H. G. F. Wilsdorf, Appl. Phys. Letters 12, 271, 1968.
30. K. R. Lawless, J. Vac. Sci. and Techn. 2, 24, 1965.
31. J. S. Vermaak and J. H. van der Merwe, Phil. Mag. 10, 785, 1964.
32. J. S. Vermaak and J. H. van der Merwe, Phil. Mag. 12, 453, 1965.
33. K. Shinohara and J. P. Hirth, Phil. Mag. 27, 883, 1973.

34. R. C. Fabiniak and D. Kuhlmann-Wilsdorf, *Environment-Sensitive Mechanical Behavior*, Gordon and Breach, New York, 1966, p. 147.
35. R. W. Vook and F. Witt, *J. Appl. Phys.* 36, 2169, 1965.
36. M. A. Adams, *Acta Met.* 6, 327, 1958.
37. L. C. deJonghe and I. G. Greenfield, *Acta Met.* 17, 1411, 1969.
38. W. R. Patterson and I. G. Greenfield, *Acta Met.* 19, 123, 1971.
39. Y. G. Nakagawa and I. G. Greenfield, *Acta Met.* 21, 335, 1973.
40. Y. G. Nakagawa and I. G. Greenfield, *Acta Met.* 21, 367, 1973.

ACKNOWLEDGEMENTS

We wish to thank Professor Kuhlmann-Wilsdorf for suggesting the mechanism of "unzipping" dislocations as a possible explanation for the reduction of yield stress through thin surface films.

It is gratefully acknowledged that the research reported in this paper was sponsored by the Metallurgy Branch, Office of Naval Research, Arlington, Virginia.

DISCUSSION

Comment by R. W. Staehle:

With respect to the reproducibility of experiments involving especially electroplating, there are three important caveats:
1. Plated layers can sustain a wide range of compressive and tensile stresses depending upon the parameters of plating.
2. Plated layers may contain networks of microcracks with varying depths and spacing.
3. During electroplating there may be a simultaneous reaction which interferes—e.g., water is often reduced to form hydrogen, some of which enters the substrate material on plating.

Reply:

You made three interesting points.
1. The stresses in electroplated Ni films on {111} and {100} surfaces of Cu were investigated and determined by E. R. Thompson (D.Sc. Dissertation, Univ. Virginia, 1966). We have been using his plating technique and used his data in our evaluation.
2. Our films were examined by shadowed negative replicas in the electron microscope after straining the composite 1-3%. Although the lateral resolution of the replicas is about 30 Å, no evidence for the presence of microcracks in the plating was found.
3. I cannot think of any purely chemical reaction that could possibly affect the behavior of dislocations so as to reduce the yield stress. Hydrogen, in particular, would be expected

to reduce the dislocation mobility.

Comment by H.-J. Engell:

In electrochemical production of monocrystalline metal layers, the dislocation structure of the substrate is not reproduced by the layer. Dislocations with a Burgess vector without a component normal to the surface do not enter into the layer, other dislocations give rise to spiral growth and thus are reproduced. However, new dislocations can be generated where the growth pyramids meet. Thus, the dislocation structure of the overlayer can very much depend on the type of metal plated and the plating conditions. Further, plating of Ni and brass is always accompanied by some hydrogen uptake due to equilibrium reasons, and this hydrogen may well have considerable influence on the mechanical properties.

Reply:

Your remarks are touching a very important point. As a matter of fact, our proposal that a thin epitaxial film provide more or less uniformly distributed dislocation sources near the surface requires the presence of dislocations other than misfit dislocations. Clearly, accomodation networks with link lengths of 25 Å, as in the case for the Au plated composite, would never be able by themselves to satisfy the above requirement. Therefore, irregularities in the misfit dislocation network on the one hand, and dislocations introduced possibly by the mechanisms described by you on the other hand, are the prerequisite to the formation of a three-dimensional dislocation arrangement that could provide the sources needed to explain the lowered yield stress. The introduction of hydrogen will have an effect on the mechanical properties of the composite, in that the mobility of dislocations will be reduced. Consequently, it may lead to a competing process regarding the proposed near-surface dislocation source operation and as such would explain why in some cases electroplated films show an increase of the yield stress, or at least no reduction.

Comment by J. H. van der Merwe:

1. Has the difference on the influence of misfit dislocations with Burger's vectors in and inclined to the interface been investigated?

2. Has the case where the initial layer is pseudomorphic and misfit dislocations are introduced subsequently been investigated?

Reply:

To 1: To my knowledge, this has not been investigated.

To 2: This case was found to be true for the electroplating of Ni on Cu (E. R. Thompson and K. R. Lawless, Appl. Phys. Letters $\underline{9}$, 138 (1966)). Here the Ni forms a continuous film already at a thickness of 10 Å.

Comment by Ch. Schwink:

If one is interested in the internal stresses of Ni-films one can make profitable use of magnetic measurements, the hysteresis of nickel being highly stress sensitive.

Reply:

Thank you. This is a very fine suggestion.

SURFACE DAMAGE RESULTING FROM FILM CRACKING

R. M. Johnson

Department of Mechanical Engineering, The University
of Texas at Arlington, Arlington, Texas

ABSTRACT. Film cracking induced surface damage of metal crystals
is described. Quantitative relations between flow stresses and
dislocation substructures in high purity Cu crystals are given.
Observations of weakening as well as strengthening of crystals by
crack induced damage are presented. Finally, the combined effects
of plastically constraining surface films and crack induced damage
are briefly discussed.

1. INTRODUCTION

The effects on the dislocation arrangements in the surface regions
of copper single crystals resulting from cracking of electro and
vacuum deposited metal films was first reported in 1968 [1]. Etch
pitting studies on (111) faces of 99.999 Cu previously plated with
Cr and then subsequently removed by dissolution in dilute HCl revealed patterns of high densities of dislocations beneath the positions of transverse cracks produced in the films during straining.

A conclusion of that work [1] elaborated upon by Block and
Metzger [2] stated that the release of the residual and applied
stresses during film rupture with the attendant plastic deformation (surface damage) in the substrate could account for the large
strengthening effects. This conclusion has been challenged by a
work by Pridans and Billelo [3] in which the plastic constraint
imposed on the surface regions of the substrate by the film is held
responsible for the observed strengthening. Their observations of
increased activity of secondary slip systems progressively from the
plated surfaces inward is convincing support for the importance of
plastic constraint in uncracked films. In the cracked films,

however, the relative importance of plastic constraint and crack induced damage on the deformation characteristics of the substrate is still not clear. Some effects of crack induced damage in crystals stripped of the film before straining have been reported recently [4] in which certain configurations of dislocations apparently lower flow stresses in the microyield region. Further effects of this type of damage will be described following some information on the characterization of crystals by their subgrain size.

2. EFFECTS OF SUBGRAIN SIZE

It has been long recognized that initial dislocation substructure is important in determining flow stresses in single crystals. Usually, however, quantative data is lacking and the standard procedure is to eliminate this effect by comparing results from samples from the same crystal or to compare results from samples of similar substructure.

The information in Table 1 was obtained for oriented crystals grown from Cu of initial purity of 99.999+, annealed and tested as described previously [1]. In each relation obtained below for the unplated crystals, the flow stress τ is in g/mm^2 and the subgrain size x found by a random lineal intercept method on etch pitted surfaces is in mm. Linear regression was used in each case.

At a resolved shear strain of 10^{-4}:

$$\tau_o = 16.38 + 5.35(x^{-1})$$

The sample size was 9 and the standard deviation was 2.22g/mm^2.

For the "resolved easy glide stress"

$$\tau_1 = 19.91 + 6.05(x^{-1})$$

The sample size was 7 and the standard deviation was 3.37g/mm^2.

For the flow stress at the end of Stage I found from the intersection of the extrapolations of the linear portions of Stages I and II:

$$\tau_2 = 89.07 + 5.90(x^{-1})$$

The sample size was 6 and the standard deviation was 7.25g/mm^2.

The subgrain sizes ranged from 0.12 mm to 0.93 mm for the samples utilized to determine the above relations. These values represent an average subgrain size over the approximately 2.5 cm gage length of the tensile sample. These samples were cut from

TABLE I

Crystal	Surface Condition	λ (Deg.)	ϕ (Deg.)	Subgrain Size (mm)	τ_0 Observed (g/mm^2)	τ_1 Observed (g/mm^2)	τ_2 Observed (g/mm^2)	a_2 (cm/cm)	θ_I (g/mm^2)	θ_{II} (g/mm^2)
3F3	unplated	43	50	0.34	31.3	36.7	110	0.0529	1290	10970
4B3	unplated	49	50	0.13	55.8	72.0	138	0.0626	1070	10650
4C1	unplated	49	50	0.61	22.8	27.0	109	0.0937	940	9000
4D5	unplated	49	50	0.93	22.2	----	----	----	----	----
4F4	unplated	49	45	0.85	25.9	27.8	----	----	----	----
5B1	unplated	46	47	0.79	20.0	----	----	----	----	----
5F4	unplated	46	47	0.12	61.7	65.9	137	0.0432	1580	12680
8B3	unplated	55	40	0.25	40.6	43.4	105	0.0631	940	10000
10E3	unplated	52	41	0.63	26.0	32.3	90	0.0520	1110	11640
3F4	0.4μ Cr removed	43	50	0.29	30.8	37.5	101	0.0468	1320	13680
4C4	0.5μ Cr removed	49	50	0.25	34.5	39.1	98	0.0538	1050	10770
3F5	0.4μ Cr	43	50	0.40	49.0	66.8	283	0.0535	3980	11270
4C2	0.8μ Cr	49	50	0.51	80.2	112.0	408	0.0680	4492	8370

595

crystals originally 25 cm in length which exhibited at most a range in subgrain size over their whole length of 0.37 to 1.83 mm in any 2.5 cm segment. The average variation in subgrain size over a long crystal ranged from 0.36 to 1.25 mm.

Other information obtained from the crystals utilized above includes the following:

The mean value for the slope of stage I, θ_I = 1155 g/mm^2.

The sample size was 6 and the standard deviation was 245g/mm^2.

The mean value for the slope of Stage II, θ_{II} = 10,824 g/mm^2.

The sample size was 6 and the standard deviation was 1279g/mm^2.

The mean value for the ratio of the Stage II to Stage I slopes (θ_{II}/θ_I) = 9.53 with a standard deviation of 1.06.

The slope of Stage I was found to be dependent on initial orientation (or inversely related to the length of Stage I, a_2). The slope of Stage II was less affected by orientation. The initial dislocation densities which ranged from 0.20 x 10^5/cm^2 to 2.35 x 10^5/cm^2 did not appear to affect any of the properties listed.

3. EFFECTS OF FILM CRACKING

Preliminary studies indicate that films which cracked before or during removal of the film from the substrate crystal prior to straining influenced the slope of Stage I little. However, flow stresses were lowered throughout Stage I for crystals which had 0.4μ and 0.5μ Cr coatings deposited and then removed prior to straining. τ_2 was lowered approximately 8g/mm^2 and 14g/mm^2 respectively for the 0.4μ and 0.5μ coated crystals compared to the values calculated using equation 3.

Of the greater significance was the unexpected observation that the slope of stage II was affected in these crystals. θ_{II} was <u>increased</u> in each case from that observed for the unplated control sample cut from the same large crystal. For the 0.4μ Cr coated and removed sample, θ_{II} was found to be 13,860 g/mm^2 compared to 10,970 g/mm^2 for the control. For the 0.5μ Cr coated and removed sample θ_{II} was 10,770 g/mm^2 compared to 9000 g/mm^2 for its control sample.

By contrast the slopes of Stage I were greatly increased by films which crack during straining as indicated by the observations on crystals tested with coatings of 0.4 Cr and 0.8 Cr. The initial flow stresses were increased as well as has often been reported. These films seem to affect the slope of Stage II little as would be expected from most theoretical analyses.

4. DISCUSSION

It is obvious that the effects of crack induced damage and of plastic constraint cannot be easily separated in a test in which a film progressively cracks during deformation of the substrate. Also the patterns of cracks developed under the different conditions of film removal prior to testing and under uniaxial extension produce differing patterns of stress induced dislocation networks. In the few tests performed to examine the effect of crack induced damage alone the results indicate that the effects of the damage and of the constraining film may in some cases even be in opposition. That is, the crack induced damage at least at certain levels of intensity and orientation tends to weaken the crystal in Stage I and produce increased hardening in Stage II.

5. CONCLUSIONS.

The information on crack induced damage presented here needs to be examined further by careful experimental studies utilizing sufficient samples to lend more confidence to the observations. The crystals to be used should be characterized in terms of initial dislocation substructure as well as by the usual methods.

Investigation of plastically constraining films and their ability to constrain the effects of crack induced damage as well as to produce it should be pursued. If indeed it is established that initial surface conditions producing crack induced damage can affect deformation and hardening well into Stage II this factor will have to be considered in explaining a variety of phenomena.

ACKNOWLEDGEMENT

This work was supported in part by Organized Research, University of Texas at Arlington and by the U.S. A.E.C.

REFERENCES

1. R.M. Johnson and R.J. Block, Acta Met $\underline{16}$ 831 (1968).
2. R.J. Block and M. Metzger, Phil Mag $\underline{19}$ 599 (1969).
3. J. Pridans and J.C. Billelo, Acta Met $\underline{20}$ 1339 (1972).
4. R.M. Johnson, Scripta Met $\underline{8}$ 965 (1974).

EFFECT OF ANODIC OXIDE FILMS ON LOW TEMPERATURE MECHANICAL BEHAVIOR OF NIOBIUM SINGLE CRYSTALS

V.K. Sethi and R. Gibala

Case Western Reserve University
Cleveland, Ohio 44106
U.S.A.

ABSTRACT. The effect of thin ($\lesssim 1500$Å) anodic oxide films on the mechanical behavior of single crystals of niobium at low temperatures (T $\lesssim 0.15 T_M$) was investigated. Oxide films affect mechanical behavior in two ways: the yield stress is reduced and the stress-strain curves are serrated over an appreciable range of strains. When oxide-coated specimens are also pre-strained into stage I at 300°K, the serrations observed at low temperatures disappear, the flow stress is further reduced, the ductility is increased, and a three-stage work hardening behavior occurs. A model involving generation and motion of non-screw dislocations from the oxide-metal interface is used to explain the results.

1. BACKGROUND

The mechanical behavior of bcc metals at temperatures T below ~ 0.15 of the melting temperature T_M is characterized by several important features: (a) a large (often ≥ 30-fold) increase in flow stress as T→0°K; (b) a sharp decrease in ductility over the same temperature range; (c) a pronounced asymmetry of slip [1]. These and other related features can be explained in terms of two important strengthening mechanisms: dislocation-interstitial solute interaction and intrinsic lattice (Peierls-Nabarro) strengthening.

The tetragonal distortion of an interstitial solute in bcc crystals interacts strongly with both screw and edge dislocations. The screw dislocation-interstitial interaction causes a very rapid change in yield stress at low temperatures with just small ppm-level additions of interstitial in solid solution [2]. This interaction affects all of those features mentioned above. However, of greater importance is the fact that bcc metals have large temperature-dependent <u>intrinsic</u> flow stresses at low temperatures. There is at least a 20-fold difference in yield stress of unalloyed niobium tested at 300°K and at 77°K [2-4].

The intrinsic contribution to the strength of bcc metals has been successfully rationalized in terms of a high Peierls-Nabarro stress of screw dislocations [1]. The core of the screw dislocation has a characteristic three-fold symmetry arising from its localized dissociation into three non-coplanar partial dislocations on intersecting {112} or {110} planes. There are no such complicated core structures of edge and mixed dislocations, and their high mobility permits microstrain deformation at stresses much lower than the macroscopic yield stress [5]. Thus the plastic flow of bcc metals is understood in terms of a two-dislocation model: non-screw dislocations move easily at relatively low stresses and give mainly microstrain deformation until they are exhausted into the screw orientation. Thereupon macroflow commences and is controlled by the movement of comparatively immobile screw dislocations at much higher stresses.

The preceding discussion suggests that the macroflow stress of bcc metals at low temperatures could be very low if it were possible to avoid deformation effected by the movement of screw dislocations. We have investigated this possibility and have found such a method of altering the basic dislocation dynamics of bcc metals. The method involves injection of a high density of mobile non-screw dislocations into a bcc metal by prestrain at temperatures $T \lesssim 0.15 T_M$ coupled with the introduction of an efficient multiplication source of non-screw dislocations, viz. a thin, highly self-stressed surface oxide film [6].

The remainder of this paper reports details of the experiment and several results on niobium single crystals. We have also obtained identical results on tantalum [3,4].

2. EXPERIMENTAL METHODS

The experimental procedures to prepare high purity single crystals of niobium have been reported in detail elsewhere, cf. [2-4]. The crystals are $\sim$ 2.6mm diam. and $\sim$ 13mm long with the tensile axis oriented parallel to [321] for single slip. The 300°K/4.2°K resistivity ratio of the crystals is $\sim$ 2000. All mechanical testing was done in tension at a resolved shear strain rate of 5.5×10^{-4} sec^{-1}.

It is well-known that prestrain of bcc metals at temperatures $T \gtrsim 0.15 T_M$ introduces mainly non-screw dislocations and can enhance the extent of microstrain by non-screw dislocation movement at $T < 0.15 T_M$ [1,5,7]. For niobium, 300°K is a convenient "high" prestrain temperature. If the prestrain is limited to stage I deformation in crystals oriented for single slip, one obtains predominantly edge dislocations in various dipole and multipole configurations [8].

It is also well-known that deposition of surface films, particularly oxides on metallic substrates, introduces large stresses in the films and in the substrate by epitaxy and by stresses associated with misfit dislocations [6,9]. That large stresses are generated by surface oxides on niobium has been demonstrated by Pawel and Campbell [10]. They observed extensive bending, corresponding to stresses well above the low temperature flow stress, of thin ($\sim$0.4mm thick) foils of niobium oxidized on one surface. In thick crystals, which are elastically and plastically constrained by their size and by uniform deposition of the oxide over the entire surface, such a surface film can serve as a nucleation source for dislocations during deformation. We have employed anodically deposited oxide films in these experiments because of the simplicity and control offered by the technique [11]. Oxide films of thickness $\sim$100-1500Å were easily and reproducibly deposited over the course of the investigation.

3. EXPERIMENTAL RESULTS

We have made extensive measurements of the yield stress of niobium single crystals at low temperatures as a function of test temperature, strain rate, prestrain and oxide thickness [4]. The stress-strain curves given in Fig. 1 represent a simple summary of our findings.

The stress-strain curve for uncoated, unprestrained niobium (curve A) is typical of results obtained by several investigators on high purity niobium. Prestraining this material ~10% at 300°K decreases the yield stress and reduces the rate of work hardening at large plastic strains (curve B), but the differences between A and B are not much greater than those encountered among various investigations on pure niobium.

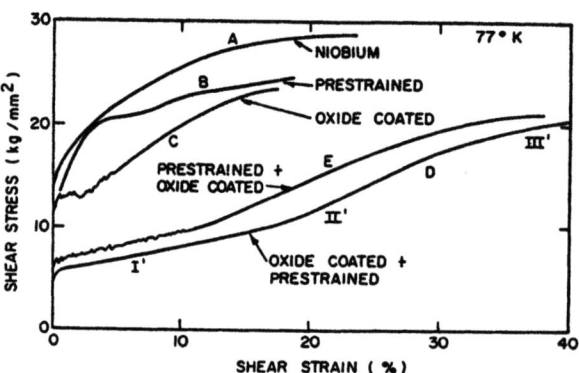

Fig. 1. Effects of oxide coating and prestraining on mechanical behavior of Nb single crystals.

When unprestrained niobium is coated with a 650Å oxide film, the yield stress is reduced (from 18kg/mm^2 to 12kg/mm^2 at 2% shear strain) and the stress-strain curves are initially serrated over several percent strain, as in curve C. If the oxide-coated niobium is also prestrained into stage I at 300°K (curve D), there is an additional substantial decrease in flow stress, a three-stage work hardening behavior developes, and the ductility, measured as the extent of uniform elongation, is greatly increased. Curve E is included to show that the order in which the oxide coating and prestraining are applied to the material is not significant. Crystals that are first prestrained and then oxide-coated exhibit slightly higher flow stresses and the serrated flow characteristic of unprestrained, oxide-coated materials.

The experiments depicted in Fig. 1 have been performed at several test temperatures below 300°K. Fig. 2 shows the dramatic effect of oxide coating plus prestraining by comparing the temperature dependence of the resolved shear stress (RSS) at 2% shear strain for pure niobium and the same material with a 650Å oxide film and prestrained 10% at 300°K. At all temperatures below ∿ 200°K there is a two- to three-fold difference in the RSS of the two materials. Note that at high temperatures (T ≳ 300°K), the oxide film hardens rather than softens niobium, as often observed in fcc crystals, cf. [12].

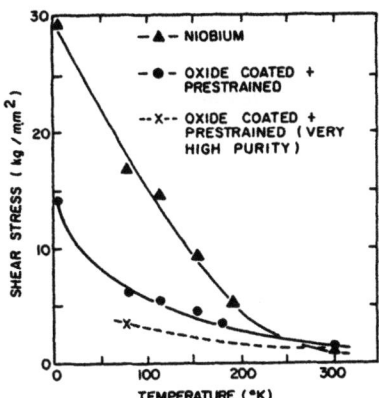

Fig. 2 Temperature dependence of the yield stress of uncoated Nb and oxide-coated+prestrained Nb of different purities.

Fig. 2 also includes some results obtained on the effect of substrate purity. Niobium crystals were outgassed for several days in ultra-high vacuum at temperatures as close to the melting point as possible to reduce residual interstitial impurities to a minimum. These crystals were oxide-coated and prestrained in the same manner as other materials, and were tested at 77°K. The mechanical behavior was much the same as indicated in Fig. 1, except that the flow stress was reduced by an additional 40-50%. The RSS at 2% strain at 77°K was in the range 3-4 kg/mm^2 for these crystals.

Several experiments were performed to establish that the behaviors observed in Figs. 1 and 2 were caused by the coating/prestrain treatment and not a result of contamination or other artifactual effects. Fig. 3 gives results from experiments involving removal of oxide films from deformed specimens which exhibited the pronounced softening. If the low temperature deformation is interrupted and the oxide film is removed by etching, the subsequent flow stress and stress-strain behavior at the same temperature are quite similar to that for prestrained, <u>uncoated</u> specimens.

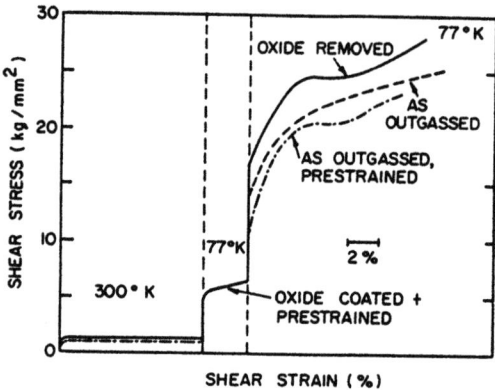

Fig. 3. Effect of film removal on the softening behavior caused by oxide coating + prestraining.

4. DISCUSSION

It is most reasonable to interpret the above results in terms of the model experiment hypothesized earlier: a high purity bcc metal can be deformed macroplastically at very low stresses at low temperatures if movement of non-screw dislocations is responsible for the deformation. Thus prestrained niobium (B in Fig. 1) deforms at only moderately reduced stresses (by 0-4 kg/mm^2) because the pre-injected non-screw dislocations are quickly exhausted into the screw orientation. The exhaustion process can better be avoided in oxide-coated niobium for which the self-stressed oxide generates large tensile residual stresses in the substrate, allowing nucleation of non-screw dislocations during deformation. The serrated flow in curve C of Fig. 1 is probably a manifestation of heterogeneous activation of surface sources. However, again the reduction in flow stress is not extremely large because the surface sources alone can not adequately supply non-screw dislocations across the entire glide plane of the specimen.

The combination of prestraining and oxide coating is so effective in reducing the flow stress and increasing ductility because it eliminates the difficulties associated with each of the operations performed separately. Prestraining introduces a high initial density of mobile dislocations across the entire glide plane that are not produced during deformation by oxide coating alone. Oxide coating supplies fresh non-screw dislocations during deformation that prestrain alone can not.

The dominant role of non-screw dislocation motion in the proposed model requires that the three-stage work hardening observed in curves D and E in Fig. 1 be interpreted differently than in normal work hardening theories. These stages are labeled I', II' and III' in Fig. 1 to distinguish them from the usual designations I, II and III. The specific interpretations of stages I', II' and III' are the basis of another investigation in which we have examined the dislocation substructures of these materials.

The additional softening observed in the very high purity niobium in Fig. 2 shows that interstitials interact strongly with edge dislocations and stop their movement. This behavior is no different than observations made on the importance of edge dislocation-interstitial interaction in microstrain deformation of uncoated bcc metals [5], except that in the present work the observations and interpretations apply to the macrostrain regime.

5. SUMMARY

We have proposed that macroflow of bcc metals at low temperatures can occur at greatly reduced applied stresses and with enhanced ductility if plastic flow can be carried primarily by edge dislocations rather than screw dislocations. To test this proposal experiments have been performed on niobium single crystals which were oxide-coated to give an important source of non-screw dislocations during deformation and also prestrained at temperatures high enough that a large density of non-screw dislocations was present prior to deformation. Such materials exhibited very small temperature dependences of the flow stress and large ductilities at low temperatures. The basic concept of using surface films to enhance ductility should be applicable to many other bcc materials.

ACKNOWLEDGMENTS

This research was sponsored by the United States Energy Research and Development Administration.

REFERENCES

1) J.W. Christian, Proc. Second Int. Conf. on Strength of Metals and Alloys, ASM, Cleveland, Ohio, Vol. I, p. 31, 1970.

2) M.G. Ulitchny and R. Gibala, J. Less-Common Metals 33, 105 (1973).

3) V.K. Sethi and R. Gibala, Scripta Met. 9, 527 (1975)

4) V.K. Sethi, Ph.D. Thesis, Case Western Reserve University, Cleveland, Ohio 1975.

5) Microplasticity, Advances in Materials Research, Vol. 2, C.J. McMahon, Jr., ed., Interscience: Wiley, New York, 1968.

6) Stress Effects and the Oxidation of Metals, J.V. Cathcart, ed., AIME, New York, 1975.

7) I.M. Bernstein, Acta Met. 17, 249 (1969).

8) R.A. Foxall, M.S. Duesbery and P.B. Hirsch, Canad. J. Phys. 45, 607 (1967).

9) Epitaxial Growth, Parts A and B, J.W. Matthews, ed., Academic Press, New York, 1975.

10) R.E. Pawel and J.J. Campbell, Acta Met. 14, 1827 (1966).

11) L. Young, Anodic Oxide Films, Academic Press, New York, 1961.

12) J. Pridans and J.C. Bilello, Acta Met. 20, 1339 (1972).

DISCUSSION

Comment by A. Seeger:

The prestrained and oxide coated materials tested at 77°K show a well-defined three-stage workhardening behavior that is difficult to explain by workhardening theories applied to FCC metals and to BCC metals deformed at higher temperatures unless motion of screw dislocations is taken into account. Consequently it seems difficult to accept your proposal that movement of edge dislocations accounts for the reduced flow stresses. Perhaps the origin is in other effects such as alloy softening due to contamination by oxygen or hydrogen.

Reply:

Contamination is at most a minor background problem in these experiments. Oxygen if present in other than the surface oxide does not have adequate diffusivity to move into the substrate and have any important effects. Some hydrogen may have been introduced during anodization, but if so the concentrations are very small. Electrical resistivity measurements failed to detect its presence. We have observed dislocation substructures produced by deformation at 77°K through the three stages of workhardening and there are some important differences from what is observed by deformation of uncoated materials at high temperatures. At the beginning of stage I at 77°K the substructure is essentially that inherited from the high temperature prestrain. This preponderantly non-screw substructure is gradually exhausted into long straight screws during stage I and in early stage II. Edges continue to move and tangle extensively around the immobile screws in stage II. Stage III occurs at stresses that are above the yield stress of the uncoated material. Correspondingly, we find stage III substructures of the coated materials differ very little from the usual substructures observed for uncoated niobium at 77°K. On both materials, screw dislocations are mobile, although their movement is controlled by the Peierls-Nabarro stress.

Comment by H. Mughrabi:

The sequence in which you presented your work gave me the impression that you _knew_ before the experiment that an oxide coating would allow you to retain mobile edge dislocations so that deformation could proceed without having to make the less mobile screws move significantly. The difficulty I have in visualizing your process is that, even if your edge dislocations cannot escape but pile up at the oxide layer, you a) still lose those dislocations that become trapped in the bulk and b) have no edge dislocations running into the interior of your specimens which are necessary to ensure deformation in the interior.

Reply:

In these experiments we have tried to minimize trapping of edge dislocations in the bulk by using single crystals oriented for single slip and of very high purity ($\leq$ 10 at ppm of residual interstitials). We believe that either our surface edge sources are sufficiently homogeneously distributed over the specimen surface to allow large paths lengths for many edge dislocations or the prestrain at high temperatures is critical in "injecting" edges well into the interior to permit it to deform easily along with the surface regions being fed by the surface sources.

Comment by R. Bullough:

Do you consider the new edge dislocations are coming in some sense from the misfit dislocation arrays (similar to the work of Professor Wilsdorf on Ag deposits on Cu) or is the imposed coherency strain of the deposit stimulating dislocation sources in the matrix to emit dislocations.

Reply:

We feel that coherency-like residual stresses are probably the more important effect, primarily because it's likely that the anodic films are amorphous or at least micro-crystalline. We have looked for the proposed sources with high voltage (650 kV) electron microscopy, but thus far have not identified their nature.

THE EFFECT OF NEAR-SURFACE DIFFUSION LAYERS ON FATIGUE OF COPPER*

Irwin G. Greenfield and Ankur Purohit+

College of Engineering, University of Delaware
Newark, Delaware 19711

ABSTRACT. Thin layers of diffused solute extending from the surface have a significant influence on the fatigue properties of copper. Experiments were conducted in which concentration profiles of solute were developed in single crystals and polycrystalline fatigue specimens by diffusing elements such as nickel, gold, zinc, silver, or aluminum into copper test specimens. The fatigue life and the characteristic surface structure developed during cycling was found to be dependent upon the solute element, the diffusion treatment and the strain amplitude. The changes in fatigue lifetimes are explained by the blocking of the egress of dislocation to the surface.

INTRODUCTION

Demonstrated during this conference is the fact that minor perturbations to the surface and the near surface layers can significantly affect the mechanical properties of materials. In this paper, the major consideration is given to the region below the surface extending from a few atom distance to nearly one micron. For the experimental results described, the near surface region was altered by diffusing a second element into the bulk of the specimen in order to create a compositional gradient. Such a near-surface compositional profile can affect the glide dislocations

---

* A portion of this work has been supported by the National Science Foundation
+ Materials Science, MSD, Argonne National Laboratory, 9700 South Cass Avenue, Argonne, Illinois 60439

in several ways since (a) there will be a gradual change in elastic modulus along with an abrupt change at the surface with consequent image forces on near surface dislocations (1,2), (b) in some cases the resulting mismatch of lattice parameters will lead to the development of accommodation dislocations (3,4), and (c) the introduction of solute atoms will lead to solute hardening (5-8). In addition, a second phase could be introduced if the constitutional diagram does not represent complete solubility (3,4). Since this phase is likely to possess different properties, the movement of glide dislocations to the surface would be affected.

The forces acting on a dislocation as a result of a surface layer of different elastic constants were determined by several investigators (9,10). A simple and satisfactory solution for the screw dislocation is based on an image force model. Illustrations describing the extent of these effects for different thicknesses of second phase material in the surface layers is found in reference 2. An elastically hard coating will repel a dislocation because the strain energy of the dislocation is increased as the distance between the hard coating and the dislocation is decreased; whereas, the energy of the system is reduced when the dislocation reaches the surface. As a result of these two factors a stable position below the hard coating is achieved. This position depends upon the relative hardnesses of the coating to that of the bulk and the thickness of the coating.

With an epitaxial or topotaxial layer, a certain mismatch of lattice spacing can occur. The high strain energy in the region of mismatch can be reduced by the insertion of an accommodation dislocation network (11). Thus, compositional gradients in surface layers which expand or contract the lattice can lead to a limited depth of a three dimensional dislocation network. Spacing between the dislocation in the network (mesh size) is dependent on the steepness of the strain gradient. Moreover, the effectiveness of the accommodation dislocation network as a barrier to glide dislocations is related to the Burgers vectors of the dislocations in the network and the dislocation mesh size (3,4). The Burgers vector tends to be parallel to the surface, but frequently, minimum dislocation energy requires dislocations with components of the Burgers vectors perpendicular to the surface. These dislocations are called insufficient dislocations (12) and can lie on slip planes oriented favorably for dislocation glide. Thus, the extent of hardening from the surface will be dependent upon the Burgers vectors of the accommodation dislocation network (4). The relative importance of the hardening from solute atoms to that by accommodation dislocation networks can be evaluated by studying the orientation dependence of the surface hardening. In the silver-on-copper system investigated by Nakagawa (4), solid solution hardening was found to have a minor influence on the critical-resolved shear stress when compared to the effect of the accommodation dislocations. Since precise experiments combining the determinations of the changes in critical-resolved shear stress and the identification of the Burgers vectors of the dislocations in the accommodation network are necessary, most data available in the literature do not separate these two surface hardening mechanisms.

The mechanisms leading to fatigue failure depend upon the nucleation and growth of cracks. The development of the failure processes in cyclic deformation is usually initiated in the near surface region, thus structural alterations such as compositional gradients of these regions should affect fatigue life. The results presented in this paper are based on experiments in which surface layers were altered by developing compositional gradients from the surface in single and polycrystal specimens of copper. These data will be reported in more detail in the near future, and consequently, only the highlights of the experimental results are presented to indicate the important feature of the effects.

EXPERIMENT

The diffusion gradients were established in the surface of copper (purity of 99.999%) by first creating a thin layer of the solute element on a surface by electrolytic deposition, evaporation, etc., and then diffusion-anneal this couple in a high vacuum. The boundary conditions (13) for the diffusion couple were chosen so that the surface concentration and maximum slope of the compositional profile decreased and the depth of solute penetration increased as the time of the diffusion treatment increased. If other phases were present in equilibrium conditions, the diffusion was more complex. For comparison purposes, however, the treatments were carried out at a specific temperature and only the time of the diffusion-anneal was altered. The characteristic distance, $\sqrt{2Dt}$, was used as a measure of the diffusion annealing treatment where D was estimated (3,4,8).

Three methods of cyclic testing were used. Fixed amplitude flexure, free-end tapered cantilever beam cycled by an alternating magnetic field (13) and cyclic tension-tension were found appropriate. Whenever possible, scanning electron microscopy or transmission electron microscopy were used to identify the changes in surface and near surface regions.

RESULTS

The effects of cyclic deformation on three cantilever specimens with different surface treatments (14) are in shown in Figure 1. Compared in this figure are a copper single crystal with no surface coating, a copper single crystal with a diffused layer of zinc,

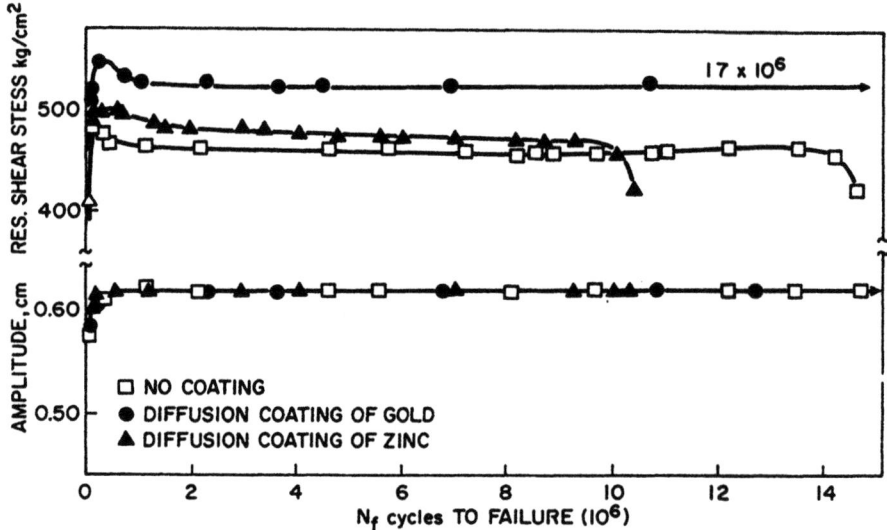

Figure 1 Cyclic deformation of single crystals of copper

and a copper single crystal with a diffused layer of gold. The constant amplitude of the deflection of 2.5 inch-tapered-cantilever beam was established after about 10^4 cycles. A higher amount of cyclic hardening was found with the diffusion treated specimen. Also, the gold treated crystal in this series of experiments has a greater fatigue life; whereas, the zinc treated specimen has the shortest. Because of the development of cracks in the last stages of fatigue, the stress necessary to retain the constant amplitude of deflection decreased, and consequently, the curves show a negative slope near failure. At the conclusion of the fatigue test the surfaces of the specimens show marked differences. For example, in the case of the uncoated copper single crystal, the surface is uniformly covered with extrusions and intrusions; in contrast, the surface of the specimen coated with a thin diffused layer of gold exhibits few deep intrusions and extrusions. Typical surface structures are shown in Figure 2 and 3.

A series of flexure-fatigue tests have been carried out to investigate the effectiveness of various solutes on fatigue lifetime. Aluminum, gold, nickel, silver and zinc were employed. It should be noted that fatigue lifetimes are dependent upon the extent of the diffusion treatment ($\sqrt{2Dt}$) and the amplitude of strain deformation. For polycrystalline specimens of copper with

Figure 2 Extrusions and intrusion after cycle deformation on the surface of a copper single crystal

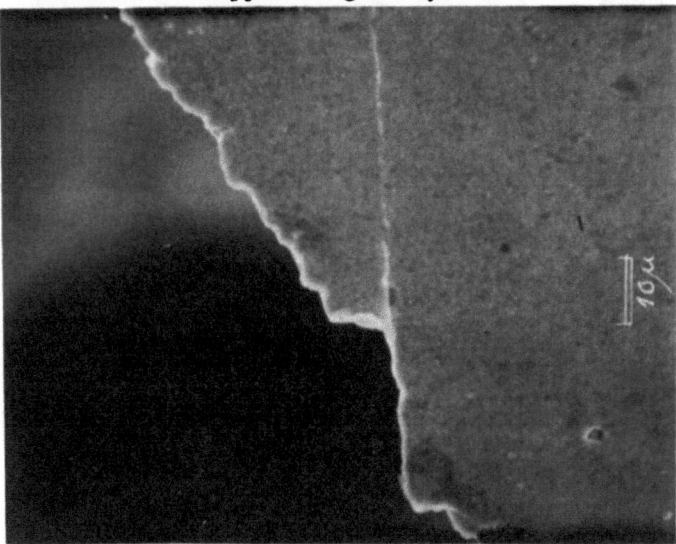

Figure 3 Relatively clear surface near the fracture of a copper single crystal with a gold diffused surface layer

a diffused coating of zinc cycled at 0.10%, 0.092% and 0.072% strain, a maximum fatigue lifetime was observed for a $\sqrt{2Dt}$ in the range of 0.2 to 0.3 microns. These data are plotted in Figure 4. These maxima are found to be higher than the lifetimes of pure copper or copper plated with the same thickness of zinc used to develop the diffusion couples. On the other hand, with nickel diffusion layers contrary results are obtained. In Figure 5, the fatigue lifetimes decreased as a result of the

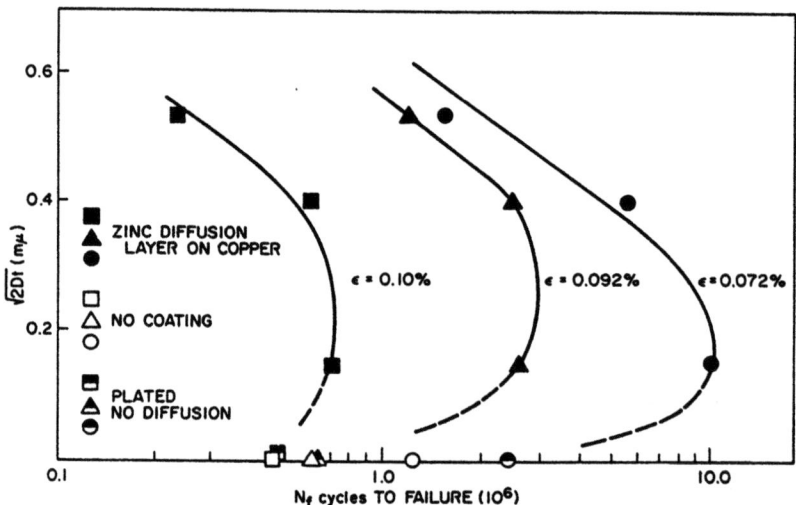

Figure 4 Effect of strain amplitude and diffusion of zinc on fatigue lifetime of copper polycrystalline specimens

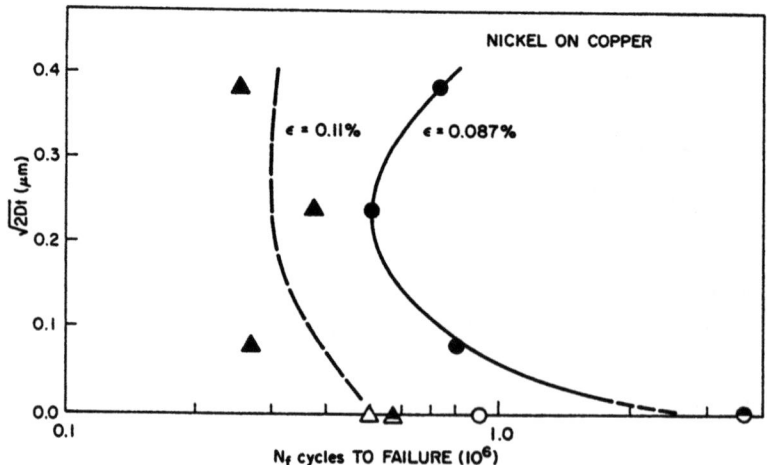

Figure 5 Effect of diffused nickel on fatigue of copper polycrystalline specimens

treatment; only the non-diffused nickel plated specimen shows an improved lifetime over pure copper. For the strain amplitudes studied, aluminum, copper, gold and silver diffusion layers tend to influence the lifetimes in a manner similar to that of the copper with a zinc diffusion layer.

A vivid example of the change in fatigue produced surface structure because of diffusion layers is shown in Figure 6. A

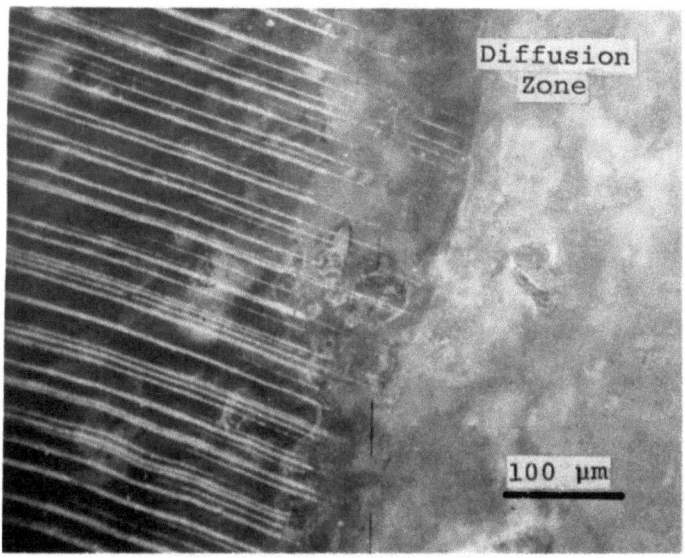

Figure 6 Single crystal of copper partly diffusion coated

single crystal of copper was partially coated with zinc and diffusion annealed. The strain amplitude and the number of cycles in both portions of this crystal are identical but the extrusions and intrusions developed during cycling were reduced drastically by the diffusion coating.

The final experimental evidence to be shown is taken from a series of copper specimens coated with different thicknesses of zinc. No annealing was applied. The fatigue lifetime as a function of thickness is shown in Figure 7. A maximum lifetime somewhat similar to that of the zinc diffusion layer specimens is noted; however, depending upon the thickness of the coating, the lifetimes may be greater or less for the uncoated copper. A partly coated single crystal specimen is shown in Figure 7. It is noted that in the untreated portion, the extrusion-intrusion

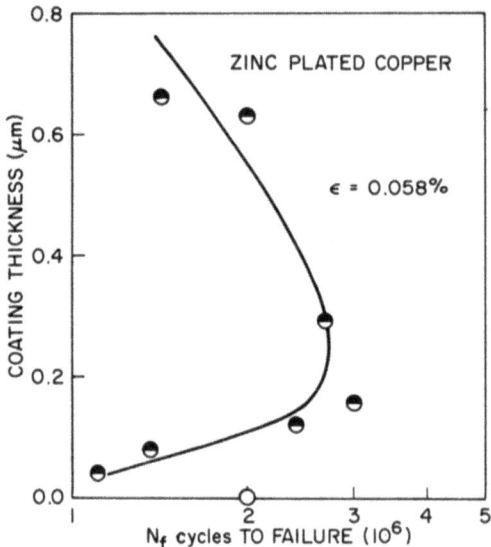

Figure 7 Fatigue lifetime as function of zinc plated thickness

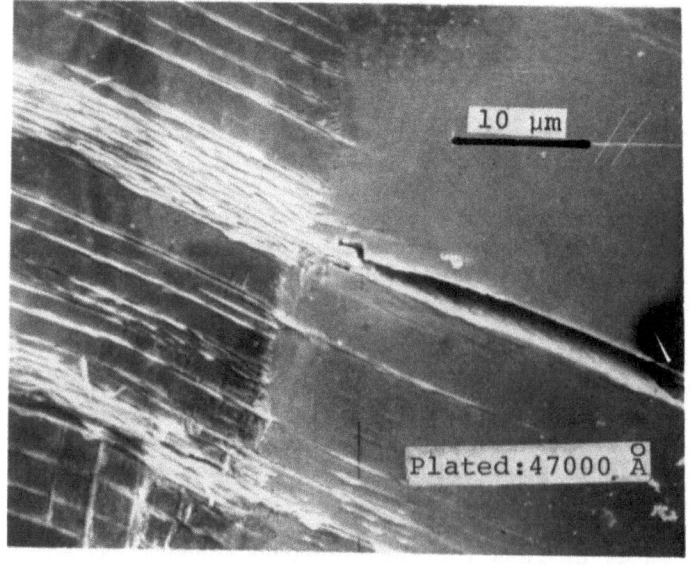

Figure 8 Partially zinc plated copper crystal

bands are clearly developed but in the zinc coated region, fewer extrusions and intrusions have formed. Moreover, at a region of these activities a fatigue crack has nucleated and grown. Conditions such as those shown in the plated region can lead to premature failure on shorter fatigue lifetimes than in pure copper.

Other surface changes also take place during cycling which can be correlated with fatigue lifetimes. For example, in Figure 9 the coating has separated from the base of material during the fatigue testing. This situation is frequently associated with a decrease in the fatigue lifetime depending upon the relationship between the stress concentration from cracks to the effect of the decohesion of the film on dislocations near the surface. Thin film transmission electron microscopy of near surface regions show an increase in dislocation density (15) directly below the blocking layer. Deeper into the bulk, however, the dislocation density is similar to that expected in a normal fatigue specimen.

Figure 9 Decohesion of a zinc film during cyclic deformation

DISCUSSION

The results of this series of experiments show that diffused coatings affect the fatigue life of copper in a signicant manner. They indicate that an increase in elastic moduli of the surface layers is not necessarily associated with a longer fatigue lifetime although the plated nickel probably corresponds to this model. Moreover, there is no one element that will uniformly improve the fatigue properties of copper. The data show, however, that compositional profiles of certain solutes lead to an increased fatigue lifetime. These results are similar to those found in tensile test experiments of diffusion coated single crystals of copper (3,4).

The increase in fatigue lifetime is usually associated with the retardation or near elimination of the formation of extrusions and intrusions on the surface (15). Nevertheless, when the specimens finally fracture there is evidence of local regions in which the fatigue surface structure characteristics appear. Thus, an effective coating tends to block the egress of dislocations (16), but under the surface layer the dislocation density and dislocations will break through to the surface then the cumulative force reaches a critical value. The extrusions-intrusions will then act as a nuclei of cracks. As in the case of the single crystal tensile studies, the barrier to glide dislocations is an accommodation dislocation network or the effect of solute atoms on the mobility of the glide dislocations.

CONCLUSIONS

1. The fatigue life was found to increase significantly if extrusion and intrusion formation is impeded by a non brittle diffused layer of solute.

2. For certain solutes, a maximum in fatigue lifetime is achieved by diffusion treatments which develop a compositional profile whose characteristic thickness is about 0.2–0.3 µm.

3. It is expected that accommodation-dislocation-network blocking and solute hardening are the controlling surface hardening mechanisms.

REFERENCES

1. Head, A. K., Proc. R. Soc., B66 793, (1953).
2. Greenfield, I. G., Jour. Sc. and Indus. Res., 32, 10, 521 (1973).
3. DeJonghe, L.C. and Greenfield, I. G., Acta Met., 19, 123 (1971).
4. Nakagawa, Y. G. and Greenfield, I. G., Acta Met., 21, 335 (1973) and 367 (1973).
5. Roscoe, R., Phil. Mag., 21, 399 (1936).
6. Rosi, F. D., Acta Met., 5, 348 (1957).
7. Adams, M. A., Acta Met., 6, 327 (1958).
8. Patterson, W. R. and Greenfield, I. G., Acta Met., 19, 123 (1969).

9. Chou, Y. T., Phys. Stat. Solids, 17, 509 (1966).
10. Conners, G. H., Int. J. Eng. Sci., 5, 25 (1967).
11. Vermaak, J. S. and van der Merwe, J. H., Phil. Mag., 10, 785 (1964).
12. Matthews, J. W., Phil. Mag., 13, 1207 (1966).
13. Jost, W., "Diffusion in Solids, Liquids and Gases", Acad. Press (1952).
14. Dadras, P., "Effect of Composition on the Fatigue Deformation Behavior", Ph.D. Dissertation, University of Delaware, 1972.
15. Greenfield, I. G., Corrosion Fatigue, NACE-2, 133 (1972).
16. Greenfield, I. G., Acta Met., 19, 857 (1971).

DISCUSSION

Comment by H. Mughrabi:

My question concerns the fatigue of your diffusion layer specimens. As you approach the surface from the interior you come from the pure copper into an alloy layer of increasing substitutional alloying content. So the question arises whether you have a systematic change in stacking fault energy. Copper is considered to be a wavy slip material, the alloys could be more of the planar slip type. Since you also mention transmission electron microscopy observations, did you notice indications of the effect or local variations in stacking fault energy, e.g., did the formation and structure of persistent slip bands depend in a distinct manner on distance from the surface?

Reply:

The intrusions and extrusions on the surface were greatly influenced by the element used to produce the diffusion layer. Of the elements used, Zn, Au, Ag and Al in solid solution reduce the stacking fault energy, whereas, nickel increases the stacking fault energy. The most extensive SEM and TEM studies which we have carried out were with gold and zinc. In both cases the diffused layer changed the extrusion and intrusion characteristics so that they resembled that of the lower stacking fault energy alloys. Because of the compositional gradient, the preparation of thin films for TEM was difficult and only regions below the diffusion layer were observable. These regions contained dislocations arrays similar to pure copper except there seemed to be higher density near the layer. As for persistent slip bands, we have not investigated their continuation into the base material. This is obviously an experiment worth doing.

Comment by H.-J. Engell:

In the case of Zn, Al and Ag on Cu, you will pass during

diffusion through a two phase region.

Reply:

The experiments which are described have several boundary conditions. The platinum, gold or nickel diffused surface layers form a continuous solid solution, and if proper annealing temperatures are used followed by rapid cooling, a continuous profile is produced. If low temperature annealing is employed or slow cooling is encountered from the high temperature annealing, ordering will occur throughout the appropriate compositional range. In fact, this phenomenon was used to evaluate the effect of thin ordered layers near the surface.

When two or more phases exist in equilibrium at the diffusion annealing temperature (such as in the cases of Ag/Cu, Al/Cu and Zn/Cu) then it is expected that a layered surface structure containing the phases would be present. The thickness of each phase will depend upon the relative diffusion rates of the solute in each phase. This boundary condition was carefully worked out for the Ag/Cu system by both calculations and electron diffraction studies (see text for reference). The diffusion annealing treatment can be adjusted so that two or only one phase is present.

With respect to the Al/Cu and Zn/Cu, the experiments have been carried out to see the effect of the coatings. The thickness of the phase layers in the composite structure near the surface cannot be theoretically predicted. At present Auger-Sputtering work is being carried out to experimentally determine the profiles.

Workshop Summary: The Influence of Surface Films on Mechanical
                   Properties

Prepared by P. Neumann

Discussion Leaders: P. Neumann, Z. S. Basinski, C. B. Duke,
                    A. Seeger

Recorders: R. Gibala, S. Weissmann, N. H. Macmillan,
                    J. J. Mills

It was pointed out that two large areas of study of surface films in connection with crystal plasticity should be distinguished. On one hand, films had been used as a research tool in order to probe the effect of the crystal surface on plasticity, for instance in connection with the question of surface sources of dislocations. The other area is that surface films are present for a variety of reasons and that we have to cope with their effects on the plastic properties. It was the general feeling, strengthened by the results of the preceeding sessions, that our knowledge of thin films on the surface of thin crystals are simply not sufficient to use them as a reliable research tool.

A thorough characterization of the surface has to precede any serious investigation concerning the influence of surface films on mechanical properties. It was suggested that methods, well established by surface scientists, should be used in the preparation of clean surfaces. These methods include electropolishing, ion bombardment (sputtering) and annealing cycles aided by Auger spectroscopy. Only when clean, reproducible surfaces are obtained is it hoped that an elucidation of the role of surface films could be established. It was agreed, however, that due to experimental difficulties such painstaking procedures are unlikely to be followed.

Dr. Duke commented on the need to fully identify the structure, orientation and composition of the substrate and in particular to identify any mobile contaminants present in particular temperature regimes. Useful techniques should include LEED, AES, and CLS. Likewise he stressed the need to characterize the deposition process with respect to deposition rate, temperature, annealing procedure, and the overlayer with respect to structure, composition, and thickness. For the latter purpose he suggested that ellipsometry might be used. Dr. Staehle recommended that one needs to characterize the degree of coverage of the surface, the film thickness, the stresses exerted on the underlying material, the degree of epitaxy, and finally the continuity of slip systems of the bulk into the surface film. It was agreed that a nondestructive method for distinguishing the mechanical state of the surface layer from that of the bulk crystal was highly needed. A number of proposals were made including such techniques as obtaining a measure of the dislocation density by means of nuclear magnetic resonance. However, it was felt that only from diffraction type experiments could we hope to get sufficiently detailed information on not only dislocation density but also on glide planes, directions of dislocations, sizes of piled-up groups etc.

Consideration was given as to the type of surface films that should be applied once the surface has been unequivocably characterized (or standardized). The question arose whether hard films or soft films would be more suitable for the study or dislocation interactions with surfaces and films. Several delegates suggested using hard outer layers such as can be obtained by ion implantation and subsequent aging or by introducing ordered precipitates used in superalloy development. It was decided, however, that such hard layers would be unsuitable if the behavior of Stage II or III of the deformation process is to be studied since cracking may soon arise because of mechanical incompatibility between the bulk and surface film.

Since the classical linear elastic theory of image forces is very often used in the consideration of mechanical properties of surface films, Prof. Kröner mentioned that there are certain formal difficulties regarding the application of image force theory. Prof. Frank pointed out that the interaction of dislocations with the surface is most easily visualized if one thinks of the surface as the boundary beyond which the strains due to a dislocation have been removed; the nearer the surface is to the dislocation the lower is the dislocation energy. The dislocation is thus always attracted to the surface.

There was more general concern, however, that the simple elasticity picture of repulsion by harder films and attraction by softer ones did not explain how misfit dislocations formed at the film/substrate interface could become glissile dislocations in the substrate. In this context it was pointed out that even amorphous films appeared to be able to inject glissile dislocations into the substrate. The question was then raised as to whether interdiffusion of two species always produced dislocations - perhaps by coalescence of vacancies generated via the Kirkendall effect. Dr. Bullough pointed out in reply that diffusion of P into Si produced many dislocations, but that addition of 10% As to the P suppressed this dislocation generation. This suggested that impurities could play an important role in the nucleation of misfit dislocations and perhaps also in their postulated evolution into glissile dislocations. It was pointed out, that surface sources can be important only under circumstances in which plasticity in the bulk is source limited. This agrees with the evidence from experiments at low critical shear stresses such as those of Prof. Kitajima and Dr. Young which suggest the importance of the surface as a source of dislocations. However, the specific configuration which acts as sources has not yet been identified. Nabarro pointed out that if the lattice parameters of film and substrate differed by 1%, the separation of the dislocations would be 100 b, and therefore, the network could not act as a source of dislocations at low stresses. He suggested that film softening would be easier to understand if the accommodation was taken up elastically. It was noted, however, that once plastic deformation started the accommodation strain would be relieved by a network of dislocations leading to a hardened surface layer. It was agreed that the problem is still unsolved. Furthermore it was pointed out that chemical stresses, that is free energy changes arising from chemical reactions at the surface, can be very much larger than the mechanical stresses that are likely to come into the play. So one should keep in mind chemical effects when discussing the origin of the surface sources.

Prof. Lothe pointed out that we are using the word surface for two different concepts. One is the surface per se which simply is the boundary of our specimen. The other concept is that of a surface layer whose mechanical and other properties may be different from the bulk. Confusion may result if these two different concepts are not clearly separated. Let us suppose for the sake of the argument that we have a thin soft surface layer and a wide hard core, and let us assume that we are polishing off from the surface soft layer. Upon subsequent plastic deformation, a new soft layer will form at the expense of the hard core. If we measure the flow stress, we will find that it has been lowered, and one school will say that we have removed a hard layer. This hard layer has in fact been removed, but it was not the surface layer--it was effectively taken away from the hard core. The situation is similar to city planning when a business road is to be widened. The land that has to be taken is most valuable if one considers the place where the old shops have been. The land owners will argue that the most valuable land has been taken away and they should be reimbursed accordingly. The city planners, however, will argue that nothing has been done than shifting back the strip of land. And nothing has been done than to take away some square meters in the back yard where the land is not very expensive. The land owners should therefore be reimbursed for the land which they lost in the backyard! It is clear from these examples that the controversy can only be solved by looking at the situation in detail.

The macroscopic stresses introduced by the film formation were agreed to be considerable. Prof. Engell pointed out that large macroscopic stresses can be produced via the Kirkendall effect as was demonstrated in the case of the formation of oxide layers on iron. In these experiments (Acta. Met. 5, 695, 1957) iron sheets which were bent before oxidation became more and more bent as a consequence of the Kirkendall effect as the oxide layer grew thicker. Dr. Schröter pointed out that differences in thermal expansion of film and substrate will give rise to thermal stresses due to heating in the electron microscope. Dr. Hirth pointed out that such stresses will usually be highly triaxial. Dr. Fisher suggested that the surface strains could be measured by applying piezoelectric film to the surface. It was agreed that in practice such experiments would be very difficult. The almost total absence of information on the dislocation substructure after deformation of film coated specimens was discussed. It appeared that the possibility already mentioned in Workshop 1, namely to study the bulk arrangement in ferro-magnetic materials by means of neutron diffraction and the surface arrangement by X-ray methods, is particularly promising.

Coating the top or side surfaces separately was discussed. It was concluded that such experiments would be valuable in assessing the role of surface films as dislocation sources, especially in experiments on bcc materials such as those reported by Gibala. In this case the coating of side surfaces would result in creation of screw dislocations which due to low mobility would not affect the stresses appreciably, whereas sources at top faces would produce edge dislocations and therefore would be expected to play an important role in the deformation process of low temperatures. If there is, however, no well defined primary slip system because of the overwhelming surface effects in thin specimens, the distinction of top and side surfaces becomes less clear.

The discussion of the technological application of surface films revealed that it was not understood, for example, why additions of rare earths improve aluminide coatings for turbine blades, why TiC coatings improve WC and Al_2O_3 cutting tools, or why the nitriding of steel should be more effective when carried out by ion implantation rather than interdiffusion. Prof. Argon did point out, however, that Al_2O_3 films appeared to improve the fatigue life of Al by suppressing surface slip step formation, hence, making crack nucleation more difficult.

On the question of what experiments should be performed, Dr. Duke emphasized that Cu, Ni, and Al were particularly suitable substrate materials in that methods for producing clean surfaces on these metals in UHV have already been established. And he added that the surface science community also had experience with the deposition of overlayers of Cu and of the alkali metals. High quality metal layers on metals can be produced also, as Prof. Engell pointed out, by electrodeposition with large exchange currents and low polarization currents. The question was also raised as to whether mechanical experiments could distinguish between overlayers of Cu on Ni, Ni on Ni, Ni on Cu, and Cu on Cu, and it was suggested that this should be tried, even though it might take a very long time to build up a layer of adequate thickness under suitable near-equilibrium conditions. Dr. Hirth and Dr. van der Merwe proposed to perform yield tests of coated samples in tension as well as in compression to check whether the effects of layers with positive or negative misfit would interchange.

Workshop Session 3

CHEMISORPTION—INDUCED VARIATIONS IN THE PLASTICITY AND FRACTURE OF NONMETALS

Chairman: W. M. Mularie

CHEMISORPTION-INDUCED VARIATIONS IN THE PLASTICITY AND FRACTURE
OF NON-METALS

N. H. Macmillan

Materials Research Laboratory, The Pennsylvania State
University, University Park, PA 16802, USA

ABSTRACT

A brief account is first given of some of the ways in which
essentially non-corrosive environments can affect the surficial
flow and fracture behavior of crystalline inorganic non-metals
and silicate glasses, and it is demonstrated that many features
of the phenomena observed are common to both crystal and glass.
The several theories proposed to explain these phenomena are
then critically reviewed, and it is shown how the sensitivity
of both the flow and the fracture behavior of crystalline inorganic non-metals to such environments derives from changes in
near-surface dislocation mobility that are themselves the result
of a chemisorption-induced, surface charge controlled, redistribution of the mobile charge carriers in the surficial
region of the solid. Finally, possible reasons for the similar
mechanical response of glass to these same environments are also
explored. In this case, however, lack of knowledge about the
chemisorptive behavior, the electronic structure and the possible
plastic response of the material involved, together with
apparently conflicting experimental evidence as to the relative
importances of chemisorption and corrosion (dissolution) in
many environments, has so far prevented the development of anything more than the most tentative and incomplete understanding
of the mechanism(s) responsible.

1. INTRODUCTION

When sufficient force is applied to a crystalline solid one of two things may happen. Either the atoms pull apart, or they slide past one another. Invariably it is the easiest of these alternatives - i.e., the one requiring the least force - that occurs first and determines whether the subsequent deformation will be brittle or ductile. In neither case, however, does deformation occur simultaneously over the whole load-bearing cross-section, even in a uniformly stressed crystal. Rather, when atoms separate they do so few by few in turn at the tip of a growing crack, and when they slide past one another they again do so sequentially, the unit process this time being one of readjustment of the cohesive bonds between a few atoms situated at the core of a glissile dislocation. The former process necessarily creates new surface, and is therefore very often directly accessible to the surrounding environment, while the latter may create new surface (in the form of surface slip steps), and sometimes occurs close enough to some surface to be influenced by the environment beyond. The result is a plethora of surface- and environment-sensitive mechanical phenomena, most of which have been or will be discussed in this series of survey lectures.

The present lecture, which is the third of the series, discusses the current ideas, incomplete and contradictory though they are in certain respects, about the influence of chemisorption on the motion of cracks and dislocations in the surficial regions of inorganic non-metals. And, because chemisorption produces certain phenomenologically similar changes in the mechanical behavior of both crystals and glasses, the term "inorganic non-metal" will be interpreted more widely than the title of this conference implies, and assumed to include amorphous silicates. This raises questions about the possible nature, extent and environment-sensitivity of plastic flow in such materials, and so a brief discussion of these topics is also included.

It should also be recognized at the outset that chemisorption is only one among many possible environmental influences on the mechanical behavior of solids. In particular, other chemical phenomena such as dissolution or diffusion of some species from the environment into the solid, and also temperature, light and electric fields, can all affect surficial (and sometimes bulk) flow and fracture behavior. Nor is it always easy to separate the influences of these several variables in any one experiment. Finally, it should be noted that for any solid in any environment changes in these same variables will in general produce corresponding changes in both surface energy and surface stress.

2. THE EFFECT OF ENVIRONMENT ON THE HARDNESS OF NON-METALS

The microhardness test occupies a central position in any discussion of environment-sensitive plastic flow in non-metals for two reasons. First, because it is easy to control the depth to which the solid is deformed in such a test, and thereby to achieve a mechanical measurement of great environmental sensitivity. And second, because this test generates large compressive stresses that tend to suppress cracking, thereby permitting significant amounts of plastic flow to occur in highly brittle materials. Indeed, the subject of this lecture really begins with Rebinder's 1928 observation that the hardnesses of various minerals depend on the environment in which they are measured [1]. And, in recognition of his discovery, and of his lifetime of pioneering work in this field, chemisorption-induced variations in mechanical behavior are often generically termed Rebinder effects.

During the mid-1960's, Westbrook et al. [2,3] at G.E. in the U.S. studied the Vickers microhardness of many crystals, both metallic and non-metallic, in "moist" air and "dry" (i.e., "water-free") toluene environments. They typically used loads of 1-25 gm so as to obtain indentations $\sim 1\mu m$ deep, and varied the loading time from one to a few hundred seconds. Their results revealed that at absolute temperatures below about half the melting temperature metallic crystals have time-independent hardnesses that are the same in both environments, but that ionic, covalent and van der Waals bonded crystals, and also silicate glasses, show time-independent hardnesses in dry toluene and lower, time-dependent hardnesses in moist air (an effect known as "anomalous indentation creep"). A recent study by Gunasekera and Holloway [4], using loading times from 10^{-3} to 10^{+5} sec, confirms Westbrook's finding that the Vickers microhardness of glass decreases with increasing loading time in moist air, and shows also that this decrease occurs more rapidly in water and less rapidly in dry toluene or in liquid paraffin dried over Na.

These results reveal clearly that the specific influence of a given chemical environment on the microhardness of any material depends very much on the nature of the interatomic bonding in that material. In particular, there is no effect of environment for crystals having metallic bonding, but there are significant effects in crystals with covalent, ionic or van der Waals bonding, and in silicate glasses. The dependence of anomalous indentation creep on bonding type is further evidenced by its anisotropy. Thus, Westbrook and Jorgensen [5], in a study of the influence of the same two environments on the microhardnesses of different faces of a variety of oxide, silicate, sulfide, fluoride, carbide and carbonate minerals, found that in almost every case softening occurred in moist air, and that for some minerals the extent of this softening differed sufficiently between two faces to reverse their relative hardnesses in moist air and dry toluene. Likewise, Hanneman and Westbrook [3] showed that antipodal surfaces normal to a polar axis in various non-centric crystals – notably III-V and II-VI compounds, SiC, boracite ($Mg_3B_7O_{13}Cl$) and quartz – had similar time-independent hardnesses in dry toluene and quite different time-dependent hardnesses in moist air.

Anomalous indentation creep takes its name from the fact that the creep rate – defined by Westbrook and Jorgensen [2] as the reciprocal of the loading time required to give a hardness in moist air one third less than that in dry toluene – <u>decreases</u> with increasing temperature. In the case of LiF, for example, the creep rate drops effectively to zero at $\sim 70°C$ [2]. And likewise, if LiF specimens are desorbed of moisture in dry argon at $\sim 70°C$ prior to testing in dry toluene, the creep rate is again zero [2].

By varying the load applied to the indenter while holding the loading time constant, Westbrook and Jorgensen [2] showed also that the ratio of the hardness of Al_2O_3 in moist air to that in dry toluene increases asymptotically to one as the depth of the indentation increases. Both this observation, and the observation that the creep rate tends to zero at large loads and long loading times, confirm that anomalous indentation creep is a near-surface effect occurring in a region no more than microns in depth.

In 1957, Kuczynski and Hochman [6] discovered what has since become known as the photomechanical effect – namely that the Knoop microhardnesses of Ge and Si, and to a lesser extent of InSb and InAs, were sensitive to light. It should, however, be noted that all of their test specimens were coated with mineral oil, acetone or alcohol for cooling purposes. Later, Westbrook and Gilman, working in air, reported [7] an analogous electro-mechanical effect – namely that the microhardnesses of Ge, Si, InSb and SiC decrease if a small (~ 1 v) potential is applied between the indenter and the specimen or if a small current ($\sim mA/mm^2$) is made to flow through the specimen.

These discoveries were followed up in several laboratories, notably in the U.S.S.R., but also at G.E. in the U.S. where Westbrook [8,9] obtained preliminary data suggesting that neither effect occurs in Ge in a dry toluene environment, but that both effects occur, and result in a similar maximum reduction in hardness, in moist air. Westbrook therefore argued that both effects were primarily chemisorptive in origin, and that both illumination and an applied potential affected the flow stress (hardness) indirectly, via modification of the chemisorption process, rather than directly. The G.E. group did not however explore this suggestion any further, for they apparently felt [10,11] that the difficulties of eliminating operator bias from the measurement of the very small indentations involved precluded either confirming the preliminary results or establishing the existence of the effects themselves to their satisfaction. It should be added, though, that (i) the critical experiments reported in References 10 and 11 did confirm in all respects the other manifestations of anomalous indentation creep reported above, and that (ii) subsequent work has confirmed the existence - though not the environmental dependence - of both the photo- and the electromechanical effect beyond any reasonable doubt (see Section 3). It should also be noted that Kimura and Hyodo [12] have reported the Knoop microhardness of soda lime glass to vary slightly with potential when measured in air, but not when measured under either water or "dry" toluene.

3. THE MECHANISM OF ANOMALOUS INDENTATION CREEP AND RELATED PHENOMENA IN CRYSTALS

A variety of mechanisms have been postulated to explain both the environment-induced changes in microhardness just discussed and related effects observed in other tests in different environments. Rebinder and co-workers, for example, have suggested that softening occurs in surface-active environments as a direct consequence of chemisorption-induced reductions in the surface free energy of the solid [1,13]. They argue that indentation of either crystals or glasses involves in part the creation of new surface area at the tips of microcracks beneath the indenter, and that the indenter can therefore penetrate further for a given load when the surface energy is reduced. A simple calculation shows, however, that the work done by an indenter carrying a 10 gm load in creating a 1 μm deep indentation is 1 erg. Now, the surface energy of an inorganic non-metal is typically ~ 1000 erg/cm^2, so this work is sufficient to create $\sim 10^{-3}$ cm^2 - i.e., $\sim 10^5$ μm^2 - of new surface. Yet the <u>total</u> areas of 1 μm deep Vickers and Knoop impressions are only ~ 10 and ~ 100 μm^2, respectively, and only a fraction of this is new surface created by the process of indentation. To absorb all or most of the work of indentation by crack formation, as would be necessary to account for the 30-50% changes in hardness observed by Westbrook et al. [2,3,5], would thus require production of a substantial network of cracks extending perhaps 10 or more indentation diameters beyond the actual indentation itself. And these cracks would have to be accessible to the environment (i.e., open at the surface) in order to give rise to any Rebinder effect. Yet little or no cracking is actually seen around microhardness impressions made at these low loads (see, for example, Figure 2(a)). Nor does it seem likely that this hypothetical cracking could all occur unseen beneath the indenter during loading (where the compressive stresses are largest and environmental access is restricted), and then close

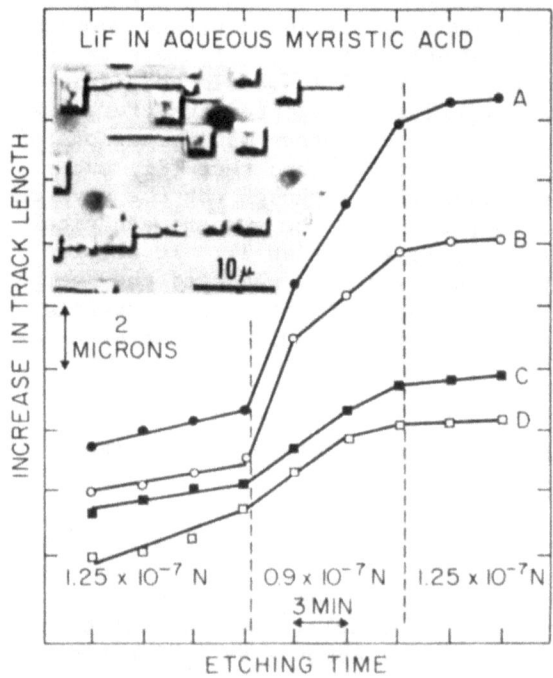

Fig. 1. Variation of length of etched dislocation tracks (inset) with time for LiF in aqueous myristic acid solutions. After Westwood et al. [15,16].

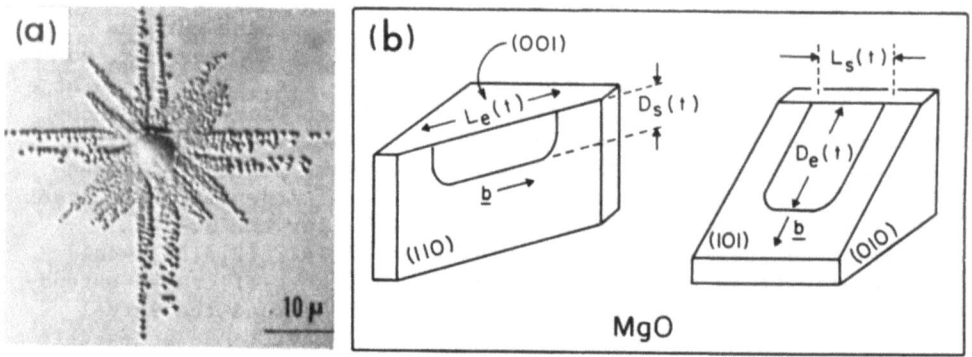

Fig. 2. Dislocation glide around a hardness impression on a {100} MgO monocrystal surface. After Westwood et al. [18,19] and Macmillan et al. [21].

up again on unloading. Indeed, recent studies show that when cracking occurs beneath diamond indenters it does so during both loading and unloading [14]. It must, therefore, be concluded that the changes of surface energy that undoubtedly result from any change of environment are secondary effects, incidental to rather than responsible for the accompanying changes in microhardness.

The corollary to the preceding argument is that 99% or more of the work done by an indenter during formation of a hardness impression is absorbed by plastic flow processes. In the case of crystalline materials, this flow occurs by movement of dislocations; in the case of glasses, by some mechanism as yet unknown. Hence, in discussing the remaining theories of the origin of Rebinder effects, all of which invoke some influence of environment on plastic flow, it will be convenient to consider initially only crystalline inorganic non-metals. Then, in the following section, it will be shown how the most successful of the theories proposed for crystalline materials may perhaps be extended to encompass non-crystalline materials also. Before beginning this discussion, however, it will be useful to describe a simple experiment which clearly demonstrates that environment can affect near-surface dislocation mobility, and to introduce an experimental technique for quantifying such mobility, albeit in rather arbitrary units.

The influence of environment on near-surface dislocation glide was first demonstrated by Westwood et al. some 15 years ago [15,16]. These authors studied the effect of dilute aqueous solutions of organic fatty acids on the rate of cross-slip of near-surface {110} <1$\bar{1}$0> screw dislocations in freshly cleaved LiF monocrystals [16]. The driving force for this glide was the residual elastic stress remaining after cleavage. The fatty acid solutions employed are dislocation etchants for LiF [15], Figure 1 inset, so dislocation mobility could be ascertained directly from measurements of the length of etched dislocation tracks as a function of time. Results for several dislocations (denoted A, B, C and D) in myristic acid ($C_{13}H_{27}COOH$) solutions are shown in Figure 1, and reveal that each dislocation moves faster in the more dilute solution. (Their different velocities merely reflect random variations of the residual stress "seen" by each).

The method of quantifying dislocation mobility was pioneered by Vaughan and Davisson [17], and subsequently used widely by Westwood and co-workers [18-21]. It is illustrated in Figure 2 for the particular case of a Vickers diamond pyramid indentation of t sec duration on a {100} MgO monocrystal surface, but extension to other cases is obvious. Figure 2(a) shows the characteristic dislocation rosette resulting from {110} <1$\bar{1}$0> glide around and beneath the actual hardness impression as it is revealed by etching; and Figure 2(b) defines the extent of such edge and screw dislocation motion both parallel to the surface, $L_e(t)$ and $L_s(t)$, and into the crystal, $D_e(t)$ and $D_s(t)$. These latter two quantities are readily determined by successively chemically polishing and etching. Dislocation mobility is then conveniently characterized by the parameters

$$\Delta X(t) = X(t) - X(2),$$

where $X = L_e$, L_s, D_e or D_s. This procedure is not as arbitrary as it might at first sight seem, for X is related to the size of the impression, and therefore to the hardness and the flow stress [22-26]. And experimentally the procedure has the great advantage that X is at once substantially larger and better defined than the impression, and can therefore be measured much more accurately.

With these remarks as background, consider now the suggestion by Hanneman and Westbrook [3] that environment-induced changes in microhardness come about because adsorbed species somehow alter the coefficient of friction between the indenter and the specimen - i.e., act as a lubricant. This might be expected to alter the stress distribution around the indenter and, because dislocation velocity varies rapidly with applied stress, could lead to changes in dislocation mobility and hardness.

Two experimental results appear to disqualify this theory, however. The first is Westbrook's finding [8] that neither changing the coefficient of friction by varying the material of the indenter nor by lubricating a diamond indenter with graphite has any effect on microhardness. The second is the demonstration by Westwood and Goldheim [20] that a particular environment influences near-surface dislocation mobility in the same manner regardless of the presence or absence of indenter lubrication effects, Figure 3. Thus, when CaF_2 monocrystals are first indented on {111} in dimethyl formamide (DMF) and then unloaded and placed in DMF-DMSO (dimethyl sulfoxide) for 20 hr, the distance ΔR_e (20 hr) that the edge segments of the half-loops created during indentation relax back towards their sources, Figure 3(b), varies with environmental composition in precisely the same manner as ΔL_e (1000), Figure 3(a).

Another possible, though rather indirect, mechanism - or, more accurately, family of mechanisms - is the "force-field effect" suggested by Machlin [27] and Westbrook [8]. This is based on the idea that there are differences between the bulk and the surface values of such properties as elastic modulus, lattice parameter, bond length and/or angle (and presumably, therefore, bond strength also, though Hanneman and Westbrook [3] choose to classify this as a separate possibility), and that these differences are at once both influenced by adsorption and themselves capable of influencing near-surface dislocation mobility in some manner un-

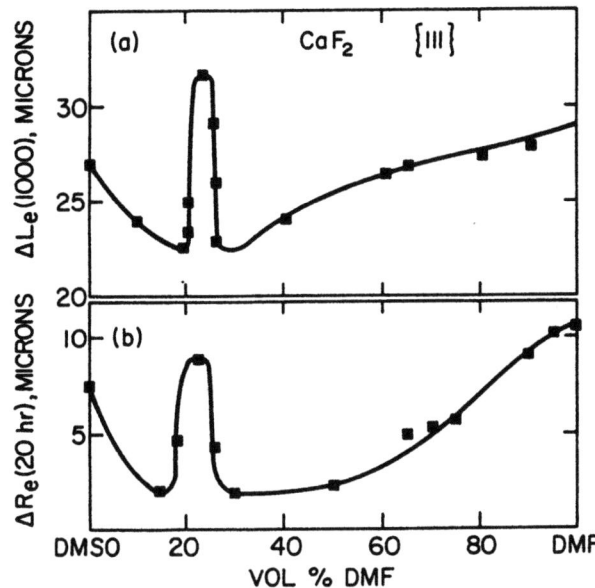

Fig. 3. Influence of DMSO-DMF environments on dislocation mobility in CaF_2 in the presence (a) and the absence (b) of possible indenter lubrication effects. After Westwood and Goldheim [20].

specified. A related suggestion is that environment-induced changes in surface stress (as distinct from surface energy [28]) are the important factor. The significant point to note, however, is that these mechanisms all involve properties of the solid that under the relevant experimental conditions (i.e., adjacent to newly-created surfaces at room temperature) are probably relatively insensitive to the impurity content of the bulk solid - simply because the diffusion coefficients of the impurity species are likely to be too low to allow them to diffuse to and preferentially accumulate at or near the surface to any significant extent. Yet Westwood, Goldheim and Lye [19] have shown that Rebinder effects in ionic crystals, at least, are very sensitive to variations in the type, concentration and state of ionization of the impurities present in the solid. Accordingly, it seems probable that force-field effects and adsorption-induced changes in surface stress, like adsorption-induced changes in surface energy, are incidental to rather than responsible for Rebinder effects.

Hence, since none of the more indirect influences of environment on dislocation motion offered as explanations of Rebinder effects can satisfactorily explain all of the experimental observations, consideration must now be given to possible mechanisms involving more direct influences. These latter may conveniently be summarized as follows:

i) dislocation egress effects [3,27] - i.e., the existence of some environment-sensitive barrier to the emergence of dislocations at the surface of the crystal;

ii) surface drag effects [3,27] - i.e., variation in dislocation mobility parallel to a crystal surface due to an environment-induced change in the ease of slip step formation when the Burgers vector is perpendicular to the surface, and in the ease of surface shear when this vector is parallel to the surface;

iii) surface anchoring effects [3,27] - i.e., an environment-induced variation in dislocation mobility arising from selective dissolution or adsorption at the points of emergence of dislocations at crystal surfaces;

iv) electronic effects [18,19] - i.e., changes in the Peierls stress and in dislocation-point defect and dislocation-dislocation interactions arising from chemisorption-induced changes in the electronic structure of near-surface dislocations and/or point defects.

Note that the Westwood, Goldheim and Lye (WGL) mechanism (iv) differs from (i) - (iii) in that it involves a "near-surface" effect, that conceivably can occur to a depth of the order of the Debye length, whereas the latter three effects all necessarily occur actually at the solid surface. It follows that mechanisms (i) - (iii) potentially can operate in any material. Mechanism (iv), however, is restricted to non-metals, because the mobile electrons in metals screen dislocations and point defects more than a few Å below the surface from any influence of adsorption. Now, recall that anomalous indentation creep occurs only in non-metals [2], and note that edge and screw dislocation mobilities in MgO, as indicated by ΔD_e (1000) and ΔD_s (1000), are environment-sensitive to depths >20 μm and >10 μm, respectively [21]. These results both suggest strongly that (iv), rather than any of (i) - (iii), is the mechanism primarily responsible for Rebinder effects. This does not imply that mechanisms (i) - (iii) do not operate in particular instances - there is obviously dissolution of LiF during etch pit formation in dilute aqueous myristic acid and formation of a hydroxide film on MgO in aqueous environments, for example - but rather suggests that their influence on dislocation mobility is minor.

The critical test of the WGL model (iv), however, is whether it can provide an explanation of _all_ of the various experimental findings discussed above. The next task, therefore, is to investigate this question. But note first that the model is for the most part based on well-established results. Thus, it is known that dislocation mobility in bulk ionic and covalent crystals is primarily determined by electrostatic interactions between dislocations and (extrinsic) point defects [29-31] and by the Peierls stress (i.e., the dislocation - lattice interaction) [32-34], respectively. Likewise, it has been shown that in semiconductor crystals, where some electronic conduction remains, chemisorption involves in part the transfer of electrons between the adsorbate and near-surface point and line defects [35,36]. Further, because of the occurrence of chemisorption and the presence of near-surface point and line defects, it is not impossible for a similar redistribution of electrons to occur in the near-surface regions of insulators such as LiF (band gap 13.6 eV [37]) and MgO (band gap 7.8 eV [38]), even though the ionic diffusion responsible for the limited high temperature conductivity of these materials is absent at room temperature. In particular, chemisorption will cause band-bending, and will give rise to localized surface states in the band gap in addition to the intrinsic surface states occurring there [39]. Likewise, point and line defects may also produce characteristic localized states in the band gap. And some of these extra states may be close enough to either the conduction or the valence band to allow some thermally activated motion of electrons or holes at room temperature. Further, dislocations emerging at the surface of the crystal may exchange charge between the surface and the interior, and dislocation glide may transport charge through the interior [40]. It is therefore at least plausible that any change in chemisorption behavior should result in redistribution of the electrons in both semiconductors and insulators to depths of the order of their Debye lengths (typically a few microns, though markedly dependent on impurity and point and line defect content). It then follows that the various electrostatic interactions between dislocations, point defects and the lattice - and hence the hardness - should also be environment sensitive to a similar depth in such materials, as found by Westbrook and Jorgensen [2].

Because any effect of chemisorption on a metal is screened out within a few Å, it follows that the fracture behavior of a metal - which involves the successive rupture of (new) surface bonds - can be very adsorption sensitive, as in liquid metal embrittlement [41], but that its microhardness - which involves plastic flow in a layer a few microns deep - cannot, except under special conditions where some other effect, such as the surface drag effect, is significant in determining the hardness [42]. Thus, anomalous indentation creep is not ordinarily observed at a clean metal surface [2]. If, however, an oxide film which can exhibit Rebinder effects is present on the metal surface, and if this oxide film plays a significant role in determining the yield and flow behavior of the metal substrate, perhaps via the dislocation-egress effect, then "pseudo-Rebinder effects" may be observed [18,43]. This would appear to be the probable resolution of the controversy as to whether or not metals actually exhibit Rebinder effects.

The WGL model likewise is consistent with the observation that anomalous indentation creep is sensitive to surface orientation, for the specific electron transfer processes occurring during chemisorption will certainly be affected by this parameter. Further, because an increase in temperature both decreases the extent of charge transfer between the solid and the absorbate (i.e., favors desorption) and more effectively screens out the effect of such charge transfer (i.e., increases the number of mobile charge carriers), the WGL model predicts that the anomalous indentation creep rate should decrease with temperature, as observed by Westbrook and Jorgensen [2].

In respect of the photo- and electromechanical effects, the WGL model implies that each can occur either in the absence of any environmental influence, as a result of direct ionization of near-surface point and line defects by incident light or an applied potential, or in the presence of a surface active environment, via light- or potential-induced changes in the near-surface electronic transitions associated with chemisorption. In the former case, both phenomena could occur throughout the volume of a sufficiently transparent insulator or semiconductor; in the latter, both are restricted to a surficial layer of the order of the Debye length in depth. And, in experimental confirmation of this view, the observation of substantial light-induced changes in the bulk bend or compressive strengths of ZnO[44], CdTe[45], CdS[46] NaCl, KCl, KBr and KI [47,48], to cite but a few recent examples, provides convincing evidence of the occurrence of the environment-insensitive photomechanical effect. Likewise, the 1000-fold increase in dislocation velocity that results when a field of 10 kV/cm is applied along the compression axis during <001> compression of an NaCl monocrystal [49] is equally convincing evidence of the existence of the corresponding electromechanical effect. In contrast, the evidence for the existence of the parallel environment-sensitive effects also predicted by the model is rather less convincing, consisting as it does of: (i) Westbrook's unconfirmed observations [8,9] that the microhardness of Ge is light- and potential-sensitive when measured in air, but insensitive when measured under toluene; (ii) Kimura and Hyodo's report [12] that the Knoop microhardness of glass, which is unaffected by potential in water or toluene, is very slightly changed by this variable in air; and (iii) the observation by Westwood, Goldheim and Lye [18] that for MgO in DMF ΔL_e (4000) is greater in daylight than darkness. Note, however, that as far as they go, these data are consistent with the WGL model, and merely show that the environment-sensitive photo- and electromechanical effects are generally substantially smaller than the corresponding environment-insensitive effects.

Note too that the WGL model is consistent with the observation [19] that the influence of environment on near-surface dislocation mobility in ionic crystals is markedly dependent on impurity content, because dislocation-extrinsic point defect interactions control the flow stress in these materials. (This realization may explain much of the "irreproducibility" of Rebinder effects reported by earlier workers). Conversely, the WGL model predicts that the corresponding environmental influence should be relatively insensitive to impurity content in materials where the Peierls stress (i.e., the dislocation-lattice interaction) controls dislocation mobility. And studies by Westwood, Macmillan and Kalyoncu [50] of the environment-sensitive hardness of Al_2O_3 confirm that this is indeed the case.

A further implication of the WGL model is that quite different adsorbates should have much the same effect on dislocation mobility provided they transfer similar numbers of electrons to or from the near-surface region of the solid. This in turn suggests that there should be some correlation between dislocation mobility and hardness on the one hand and electrical charge density in the surface double layer - which may conveniently be characterized by the ζ-potential* [51] - on the other.

That such a correlation does indeed exist was first established by Heins and Street [52], who showed that the pendulum hardnesses of AgBr in aqueous bromide environments and of AgI in aqueous iodide solutions are greatest when their respective ζ-potentials are zero. Subsequently, Westwood and co-workers, using the same experimental technique, have established that the same correlation holds also for: Al_2O_3 in aqueous NaOH and in the n-alcohols [50]; calcite in aqueous NaOH and HNO_3 [53]; quartz in buffered and unbuffered aqueous $Al(NO_3)_3$ [54]; and soda-lime glass in the n-alcohols [55], in aqueous KCl and $Th(NO_3)_4$ solutions [56], and in solutions of KI in iso-propyl alcohol [55]. The data for quartz in aqueous

---

* Recall that part of the excess charge on the liquid side of the double layer is held fixed on the surface of the solid, but that the remainder forms a diffuse layer that is relatively more mobile. Hence, when there is relative motion of the solid and the liquid, this mobile fraction is sheared off. The number and sign of the charges thus displaced can be measured with a suitable system of electrodes, and hence the potential difference between the plane of shear and the bulk liquid can be calculated. And this potential, variously called the electrokinetic or ζ-potential, in many cases provides a semi-quantitative indication of the charge density in the double layer.

Al(NO$_3$)$_3$ [54] and for soda-lime glass in the n-alcohols [55] are shown in Figures 4 and 5, respectively. It thus appears that not

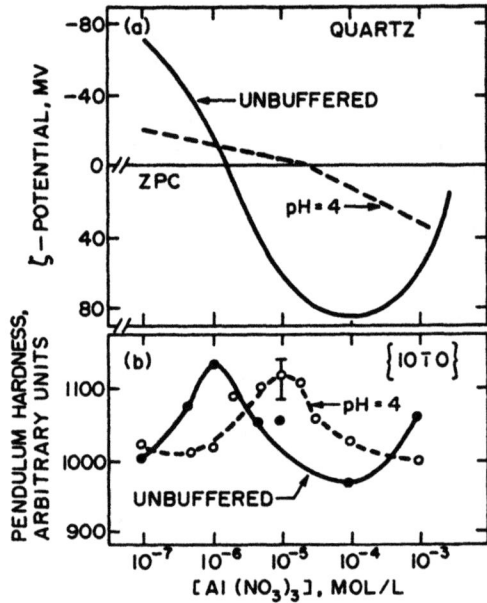

Fig. 4. Variation with concentration of ζ-potential (a) and pendulum hardness (b) for quartz in aqueous Al(NO$_3$)$_3$ solutions. After Westwood et al. [54].

only is this so-called "ζ-correlation" a generic property of all crystalline inorganic non-metals, as the WGL model suggests, but that it extends to non-crystalline materials also. The mechanistic implications of this latter finding are discussed in Section 4.

The Kuznetsov pendulum sclerometer [57] employed by Heins and Street [52] and by Westwood et al. [50,53-56] consists basically of a compound pendulum with a diamond fulcrum. In use, this is set oscillating on a test surface immersed in the environment of interest, and the pendulum hardness H is calculated from the logarithmic decrement of its amplitude of motion by the formula

$$H = t/2.303 \, (\log A(0) - \log A(t)),$$

where $A(0)$ is the initial amplitude and $A(t)$ the amplitude t sec later [58]. Qualitatively, therefore, a low value of H implies a high rate of absorption of energy by flow and fracture around and beneath the fulcrum and vice-versa. Quantitatively, however, H cannot readily be related to either dislocation mobility or the

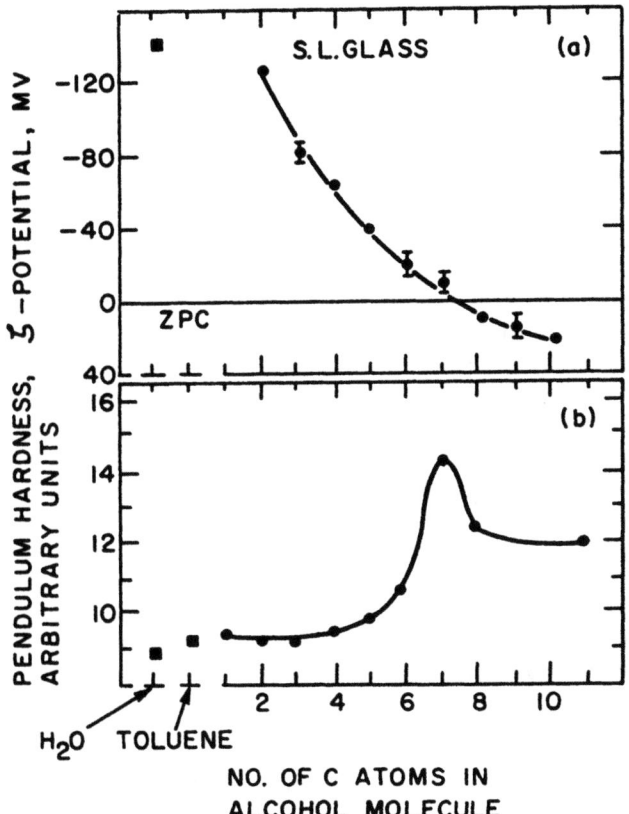

Fig. 5. Variation with molecular chain length of ζ-potential (a) and pendulum hardness (b) for soda-lime glass in n-alcohol environments. After Westwood & Macmillan [55].

more usual Knoop and Vickers measures of hardness. Accordingly, Macmillan, Huntington and Westwood [21] undertook a study of dislocation mobility as a function of ζ-potential for MgO in 10^{-2}N aqueous NaCl solutions buffered to the desired pH (in the range 10.5 - 13.5) by additions of NaOH or HCl. Their results are shown in Figure 6, and reveal that both edge and screw dislocation mobilities - as indicated by ΔL_e (1000) and ΔL_s (1000), respectively - are least when ζ = 0 and increase with increasing |ζ|. In addition, Swain, Latanision and Westwood [59] have recently shown that the Knoop microhardness of the {0001} face of an Al_2O_3 monocrystal in aqueous NaOH environments is also greatest at the zero point of charge (zpc) (i.e., when ζ = 0). It is thus apparent that the correlation established with the pendulum sclerometer is valid also for more conventional measurements of hardness, and in the case of crystalline inorganic non-metals, at least, derives from a corresponding (inverse) variation of dis-

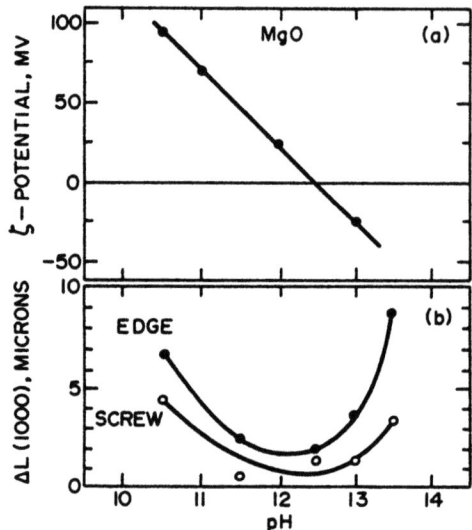

Fig. 6. Variation with pH of ζ-potential (a) and edge and screw dislocation mobilities (b) for MgO in aqueous NaCl solutions. After Macmillan et al. [21].

location mobility with surface charge.

Finally, Swain, Latanision and Westwood [59] have also confirmed a related implication of the WGL model - namely that chemisorption rather than physisorption is necessary for the occurrence of Rebinder effects. This distinction arises because charge transfer between the solid and the adsorbate is an integral feature of the WGL model, and such transfer results only from chemisorption. Now, Han, Healy and Fuerstenau [60] have shown that the zpc of Al_2O_3 in buffered aqueous NaCl solutions is shifted to a different pH by additions of oleic acid, but is unaffected by additions of lauric acid - suggesting that the former chemisorbs on Al_2O_3, but the latter merely physisorbs. Swain, Latanision and Westwood have confirmed this result. They have also measured the rate at which Al_2O_3 can be drilled and ground in these environments under appropriate standard conditions, and have demonstrated that these rates are maximized and minimized, respectively, at the zpc in all of the environments studied, regardless of whether these contain lauric or oleic acid. Hence, because these extremum values of drilling and grinding rate correspond to hardness maxima [50, 53-55,61,62], it follows that the chemisorbed oleic acid, which does alter the zpc of Al_2O_3, also alters the environmental dependence of its microhardness to match. Likewise, it follows that the physisorbed lauric acid, which does not effect the zpc, also does not affect the environmental dependence of the microhardness.

It is clear, therefore, that of all the models proposed to explain Rebinder effects in crystalline inorganic non-metals, the WGL model alone is entirely successful in explaining all of the phenomena observed. In quantitative terms, however, the origin of the ζ-correlation still remains obscure, partly because no detailed understanding yet exists of the electron transitions resulting from chemisorption, and partly because the full interrelationship of ζ-potential, surface charge and near-surface electronic structure has so far not been clarified.

4. THE MECHANISM OF ANOMALOUS INDENTATION CREEP AND RELATED PHENOMENA IN GLASSES

It was shown in Figure 5 that the hardness of soda-lime glass is not only environment-sensitive, but also obeys the ζ-correlation. It therefore seems reasonable to suggest that anomalous indentation creep and related phenomena in silicate glasses also arise from chemisorption-induced changes in near-surface flow behavior, and to ask if the WGL model can accommodate this hypothesis. Now, recall the basic tenet of this model - namely that chemisorption alters the surficial flow properties of materials because it changes the number and distribution of any mobile charge carriers they contain to depths of the order of their Debye lengths. When the WGL model is stated in this more general form, it is at once apparent that it can be extended - conceptually at least - to encompass materials showing quite different chemisorption behavior and having quite different mechanisms of plastic flow and/or electrical conduction. Accordingly, it is instructive to examine the mechanisms of chemisorption, conduction and plastic flow (if such occurs) in silicate glasses, in so far as these are yet understood, to see if a more detailed explanation for the similarity of Rebinder effects in these materials and in crystals can be developed within the general conceptual framework of the WGL model.

Consider first, then, possible similarities in the chemisorption behavior of crystals and glasses. It will be recalled that stable glasses form in materials where most of the cohesive energy is tied up in first nearest neighbor bonds, and that these bonds persist essentially unchanged in both crystal and glass. Glasses therefore characteristically retain a substantial degree of short range order and a continuous network of first nearest neighbor bonds of very similar geometry and strength to those occurring in the corresponding crystalline material [63-65]. Hence, because chemisorption is a localized process determined largely by the properties of particular active sites, similarities in chemisorption behavior between crystals and glasses of like composition are not improbable.

Next, recall that plastic flow in crystals occurs by kinks moving along dislocation lines, systematically advancing them across their slip planes without leaving any trace of their passage behind them. Consequently, there is a high degree of correlation between the unit (atomic-scale) flow processes occurring at adjacent points in a crystalline solid. And, correspondingly, some current thinking about flow in glasses appears to be moving away from the earlier idea [66,67] that this involves completely uncorrelated movements of small groups of atoms, towards a picture involving some degree of correlation between events at adjacent points in the structure, at least at low temperatures. Thus, Gilman, for example, has suggested [68,69] that flow in glasses may occur via "dislocation-like" defects which leave broken bonds, density fluctuations etc. in their wakes. And Bowden et al. [70, 71] have made the more specific suggestion that such defects, although not present initially, form at the yield stress in polymeric glasses. Likewise, Ashby and Logan [72] have postulated that flow in a glass may proceed through the agency of a defect they call a disjunction, which is somewhat akin to a dislocation with its long-range elastic stress field relaxed.

In view of the lack of long range order in glasses, however, it is presumably at the level of the unit atomic-scale flow process that the similarities of mechanism will be most evident. Now, the advance of a kink one atomic repeat distance along a dislocation in (say) crystalline quartz involves the breaking and remaking of a tetrahedral Si-O bond [73], perhaps by an electron tunneling quantum mechanically from one orbital to another as suggested by Gilman [74,75]. And, because this network of Si-O bonds persists essentially unchanged in both crystalline and glassy silicates, it is tempting to suggest that the unit flow process is also virtually the same in both cases.

In fairness, it should be added here that other authors [76,77] do not accept that plastic flow (shear) can occur in silicate glasses at room temperature, either by the motion of "dislocation-like" defects or any other mechanism. Ernsberger [77] has stated this opposing view with particular clarity. He suggests that a clear distinction can be drawn between yield by shear, which he argues conserves volume, and yield by densification, which (obviously) does not. He further suggests that the former is a "reconstructive" process, involving breaking and remaking primary Si-O network bonds, and therefore not recoverable by annealing, whereas the latter is merely "displacive", and consequently can be reversed by heating because no bonds are broken.

Experimentally, it is found that yield phenomena in glass are never entirely recoverable by annealing [77] - an observation that proponents of the densification hypothesis rationalize as due to viscous flow during annealing, and others cite as evidence of plastic shear. It seems more probable to this author, however, that the distinction between densification and plastic shear is not nearly so clear cut, and that densification would cause localized shear instabilities and shear result in densification. And in this context it is interesting to note that Gilman's suggested "dislocation-like" defects are not constrained to move at constant volume, for they can leave density fluctuations in their wakes [68]. Likewise, if these postulated defects did indeed move by a quantum mechanical tunneling process [74,75], so that their motion was stress activated rather than thermally activated, the objection that reconstructive shear cannot occur at room temperature because the Si-O bond energy >>kT might also be dismissed.

Third, note that silicate glasses, like LiF and MgO, are good insulators at room temperature. And note too that such limited bulk conductivity as they exhibit at higher temperatures is ionic in nature, and results from the diffusion of non-network monovalent cations (usually Na^+ ions [78]). These ions move very slowly at room temperature, and would therefore take a considerable time to redistribute throughout the surficial region the excess electrons (or holes) produced at the surface by chemisorption. Hence, a time lag might be expected between exposure of a glass to a surface-active environment and the occurrence of any change in its flow behavior. Yet the fact is that no such time lag has ever been observed, and substantial environment-induced changes in hardness persist both when drilling glass at bit penetration rates of tens of μm/sec [55,61,62] and when indenting it using loading times of 0.1 - 1 sec [4]. This behavior may perhaps be rationalized if it is assumed that the requisite surficial charge redistribution takes place in glasses via the same sort of "defect-controlled local electronic conduction" postulated to occur in crystalline insulators. At the present time, however, there is not enough known about the electronic structure of silicate glasses, near-surface or otherwise, to gauge the plausibility of this suggestion.

On the one hand, then, these speculations suggest that the similarities in the Rebinder effects observed in crystals and glasses of like composition could perhaps originate from similarities in chemisorption and deformation behavior at the atomic level. On the other hand, however, they fail to explain why no great difference in behavior results from the very different mobilities of the usual charge carriers in these materials.

Such considerations of mobility aside, however, it is interesting to note that the ionic conduction hypothesis alone provides any explanation for the ζ-correlation, and that only for glasses [56]. Thus, this model predicts that any excess of cations will diffuse towards the surface if the adsorbate donates electrons to the glass, and that excess anions will likewise accumulate in the surficial region if chemisorption transfers electrons in the opposite direction. Hence, because an excess of non-network ions of either sign weakens glass [79], a hardness maximum will occur at the zpc - i.e., when there is no excess charge at the surface to induce a near by build-up of non-network ions.

It is thus concluded that the WGL model provides not only a qualitative understanding of Rebinder effects in crystalline inorganic non-metals, but also a conceptual framework for an eventual understanding of the analogous effects in glasses. At the present time, however, lack of knowledge about the chemisorption, electrical conduction and possible plastic flow behavior of these latter materials prevents developing any more detailed model of their environment-sensitive mechanical behavior. Consequently, only a tentative and incomplete understanding yet exists of the reasons for the commonality of the ζ-correlation to both crystals and glasses.

5. THE EFFECT OF CHEMISORPTION ON THE FRACTURE OF INORGANIC NON-METALS.

The preceding sections of this paper have described the effects of essentially non-corrosive environments on the near-surface flow behavior of inorganic crystals and glasses, and have shown how these effects derive from the influence of chemisorption-induced variations in surface charge on the flow stress in the surficial region of the solid. It follows that both the coefficient of friction and the drilling and grinding behavior of such materials, each of which depends to a significant extent on the ease of near-surface flow, are likewise chemisorption dependent. Each of these properties is therefore amenable to some degree of environmental control. Such applications of chemisorption-induced changes in mechanical behavior form the subject of a later lecture in this series, and will not be discussed here. Note, however, that drilling, grinding and friction measurements - like the pendulum hardness measurements discussed earlier - differ from anomalous indentation creep and dislocation mobility studies in that they involve a complex interplay of both near-surface flow and fracture processes, whereas the latter involve near-surface flow processes alone. Nevertheless, as has already been shown in the case of Al_2O_3, for example, both Knoop and pendulum sclerometer measures of hardness obey the ζ-correlation [50,59]. Hence, it may legitimately be asked whether fracture also is affected by chemisorption.

For semi-brittle ceramic crystals, however, this is not a straight-forward question. These materials typically display some ductility, but lack the five independent glide systems necessary to accommodate a general change of shape. Consequently, though they often start to deform plastically, they usually fail in a brittle manner after accumulating some characteristic amount of strain determined by the deformation geometry. In the case of MgO, for example, the cleavage cracks responsible for such failure are usually nucleated at blocked slip bands [80-83]. Curve 1 in Figure 7 indicates schematically the sort of stress-strain curve that typically results. Note that the flow stress σ_{y1} is not very different from the fracture stress σ_{f1}.

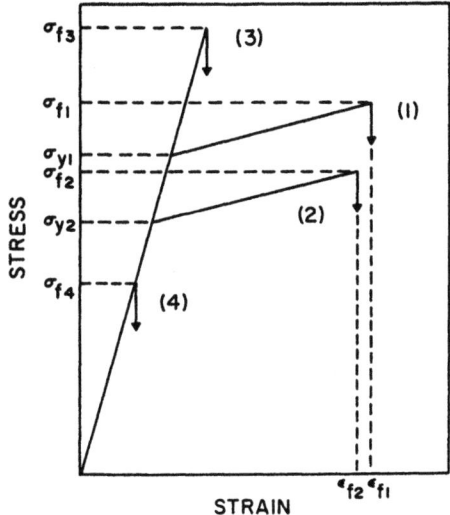

Fig. 7. Schematic of the possible influences of environment of the stress-strain behavior of a typical semi-brittle ceramic crystal.

Curve 2 in the same figure indicates the form Curve 1 might adopt if the deformation involved took place in the near-surface region of a solid immersed in a surface-active environment which reduced the flow stress from σ_{y1} to σ_{y2} - as might happen for example when using a cutting fluid during machining. Note that not only is there a corresponding reduction of the fracture stress from σ_{f1} to σ_{f2} but, because the fracture strain is relatively invariant for any particular deformation geometry (i.e., $\varepsilon_{f1} \simeq \varepsilon_{f2}$), the work of fracture is also reduced. Such flow-facilitated changes in fracture behavior are important from a practical point of view because they can be parlayed via the ζ-correlation into dramatic improvements in machining behavior [84], and because they may affect static fatigue life [84]. From the present mechanistic viewpoint, however, they merely represent further manifestations of the influence of chemisorption on near-surface flow (i.e., dislocation mobility), and as such may be understood in terms of the WGL model and Figure 7 [84]. Indeed, it is interesting to note that Ahlquist [85] has produced precisely parallel flow-facilitated changes in the fracture behavior of bulk LiF by using additions of divalent impurities to control dislocation mobility in place of chemisorption.

Curve 3 of Figure 7 shows the dramatic sort of ductile to brittle transition that can occur in the surficial region if chemisorption raises the flow stress of Curve 1 above some essentially environment-independent fracture stress σ_{f3}. The best known example of this sort of behavior is the complex-ion embrittlement of AgCl [86,87], which again can be understood in terms of the WGL model [43,86,87].

Polycrystalline AgCl exhibits a ductile, transgranular fracture in air, but becomes brittle and adopts an intercrystalline mode of failure when exposed to aqueous environments containing highly charged complex ions of either sign. Monocrystals also can be embrittled if they contain a pre-existing crack. Now, the mobile charge carriers in AgCl are not electrons, but intrinsic Frenkel defects on the cation sub-lattice - i.e., positively charged Ag interstitials (Ag_i^+) and negatively charged Ag vacancies (Ag_v^-) [88]. The former migrate preferentially towards surfaces where negative ions are being chemisorbed, and the latter towards surfaces where the adsorbing ions are positively charged. Hence, because both these species of point defect restrict dislocation mobility, the flow stress is raised both adjacent to surfaces exposed to highly charged adsorbates of either sign and also immediately ahead of the tips of any cracks accessible to such species. The resultant hardening raises the flow stress in such regions above the fracture stress, and ensures that brittle fracture becomes the initially preferred mode of deformation when chemisorption occurs. In the polycrystalline material these brittle cracks nucleate at points where slip bands are blocked at wide angle grain boundaries, Figure 8(a) [43], and then propagate along the boundary until they penetrate right through the hardened layer and are arrested by blunting in the softer material beyond. Further chemisorption immediately occurs on the newly exposed surface, however, and causes further Frenkel defects of appropriate sign to diffuse towards the crack tip and reform the embrittled layer around it. The whole cycle thus starts over again. In the monocrystal case, a precisely similar sequence initiates at the tip of the pre-existing crack.

Such a discontinuous fracture process appears to account satisfactorily for the "striae" observed on embrittled fracture surfaces of both mono- [87] and polycrystalline [43] samples, Figure 8(b) [43]. Further support for the same mechanism is provided by microhardness studies [87] which clearly reveal the near-surface hardening caused by the adsorbed species. The

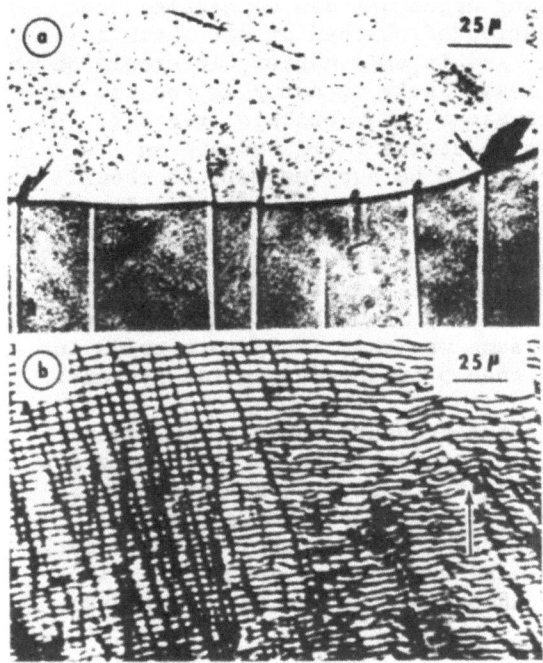

Fig. 8. Points of crack initiation (arrowed, (a)) and intercrystalline fracture surface (b) for polycrystalline AgCl strained in 6N aqueous NaCl containing $AgCl_4^{3-}$ complex ions. Arrow in (b) shows direction of crack propagation. After Westwood et al. [86,87].

strongest substantiation comes, however, from the observation [87] that the activation energy for the failure process in the temperature interval 30-100°C is in good agreement with that for Ag_i^+ diffusion when the adsorbing species are negatively charged, and with that for Ag_v^- diffusion when they are positively charged.

The final possible influence of chemisorption on the fracture behavior of a non-metallic crystal that needs to be considered is illustrated by Curve 4 of Figure 7. This reveals that a surficial ductile to brittle transition would also occur if chemisorption were to greatly reduce the near-surface brittle fracture stress σ_{f1} of Curve 1 to σ_{f4}, without much affecting the near-surface flow stress. Regardless of its apparent similarity to the previous case, however, such a change would necessarily come about via a quite different mechanism, for it represents an influence of chemisorption on crack propagation rather than dislocation motion. Such ductile to brittle transitions are well known in metals as the phenomenon of liquid metal embrittlement [41], but have not been observed in non-metals.

The extent to which the preceding arguments can be carried over to glass center around the question of whether or not plastic shear occurs in this material. And in this context it should be recognized that such shear is less likely to occur at the tip of a crack, where large tensile stresses favor brittle fracture, than in the dominantly compressive stress field beneath an indenter, where cracking is likely to be suppressed. It should also be recognized that, if glass does shear plastically, it will do so on any and all planes on which the resolved shear stress reaches the flow stress. Hence, plastic flow in glass should readily be able to accommodate a general shape change. Consequently, it is not clear whether any flow that does occur can initiate fracture in quite the same way that it appears to do in the case of crystals with less than 5 independent glide systems.

The fact remains, however, that the fracture behavior of glass is not that of an ideally brittle material. In particular, glass exhibits static fatigue (i.e., its fracture strength decreases with increasing time under load), and its fracture energy is typically several times its expected surface energy. Also, crack propagation in glass is stress- and environment-dependent. It needs hardly be added that any theory of the fracture of glass must explain these facts.

Wiedmann and Holloway [89], like Marsh [90,91] before them, suggest that the assumption of plastic flow - by some mechanism unexplained - provides the most plausible explanation, and have recently presented new evidence supporting their view that such flow is common to both the tensile (fracture) and the compressive (indentation) deformation of glass. They start by using the data of Gunasekera and Holloway [4] to obtain empirical algebraic expressions for the variation of the yield stress of float glass with load duration in air, water and liquid paraffin environments. They then show that, by combining these expressions with a limited plasticity fracture criterion, they can for each environment predict not only the variation of crack velocity with stress intensity factor, but also the static fatigue lives of both pristine and pre-cracked specimens as a function of load.

Westwood et al. [56,62] have also presented evidence suggesting that plastic flow occurs during the fracture of glass. These authors measured the fracture energy of soda-lime glass as a function of molecular chain length in both the n-alcohols and the n-alkanes, using first the double cantilever beam test [62] developed by Gilman [92] and Wiederhorn [93,94], and later the center-loaded crack technique [56] developed by Panasyuk and Kovchik [95] from an analysis by Barenblatt [96]. The double cantilever beam studies revealed no variation of fracture energy with environment, but the center-loading technique showed this energy to vary with molecular chain length in both alcohols and alkanes in precisely the same manner as the (diamond-rotary) drilling rate and the pendulum hardness [62,97]. In particular, these studies showed that in the n-alcohols both the pendulum hardness, Figure 5(b), and the fracture energy passed through a maximum at the zpc - i.e., in heptyl alcohol, Figure 5(a). Accordingly, Westwood et al. [56,84] have interpreted these results as evidence that chemisorption-controlled plastic flow can influence fracture in glasses just as in crystals.

Recently, both Frieman [98] and Wiederhorn and Evans [99] have also investigated the influence of the n-alcohols on crack propagation in soda-lime glass. Both used the double cantilever beam method, and both measured crack velcoity v in these environments as a function of the stress intensity factor K. Unfortunately, however, detailed numerical comparisons between their results and those of Westwood et al. are not possible, because the latter worked at constant Instron cross-head speed rather than constant crack velocity, and so obtained fracture energies that were effectively "averaged" over an (unknown) range of crack velocity. Both Frieman and Wiederhorn and Evans found that the plots of log v versus K obtained in the n-alcohols had the usual trimodal pattern, and looked very similar to the corresponding plots obtained earlier by Wiederhorn [100] in N_2 gas containing varying amounts of water vapor. They therefore suggest that in the two lower velocity regions of the log v versus K plot, at least, crack growth takes place by the stress-enhanced corrosion mechanism first described by Charles and Hillig [101,102] - i.e., that crack growth results from the corrosive influence of traces of water in the alcohols rather than from any influence of the alcohols themselves. (For completeness, it should perhaps be added at this point that Kimura and Hyodo [103] have put forward the precisely opposite view - namely that non-aqueous environments affect the fracture behavior of glass because they remove water from its surface). Only in the highest velocity region of the log v versus K plot do Frieman and Wiederhorn and Evans suggest that the alcohols themselves have any effect on crack propagation. And this effect, which is attributed to the different hydrodynamic pressures developed by the flow of alcohols of different viscosities to the crack tip [99], appears to vary monotonically with molecular chain length rather than to be maximized or minimized in heptyl alcohol [98].

Nor is this the end of the apparent inconsistencies in the environment-sensitive fracture behavior of glass. Thus, Westwood and Huntington [56] have shown that the fracture energy maximum obtained in heptyl alcohol can be reproduced in a suitable mixture of octyl and pentyl alcohols. This result is readily understandable in terms of the ζ-correlation, for these alcohols individually impart ζ-potentials of opposite sign to glass, Figure 5(a). Hence, when mixed together in suitable proportions they presumably produce a zero ζ-potential. And at the same time this result appears inconsistent with the Charles-Hillig theory, for the water content of an octyl alcohol-pentyl alcohol mixture would vary monotonically with composition rather than displaying an extremum value. The obvious conclusion is not to be, however, for Wiederhorn [104] finds that ζ-potential has no effect on crack propagation in glass in aqueous solutions of either $CaCl_2$, $LaCl_3$ or $Th(NO_3)_4$, even though the latter has been independently reported to significantly affect the pendulum hardness by two different authors [56,105].

A possible resolution of these apparent experimental inconsistencies is suggested by the results of McCammond et al. [106], who have recently demonstrated that the fracture strengths of glass rods in organic environments are extremely sensitive to the presence of very small amounts of impurities in those environments. These authors also report an inverse correlation between the fracture strength of glass and the polarity of the environment in which it is measured, and follow Rebinder [1,13] in suggesting that this correlation comes about because the more polar environments produce greater reductions in the surface free energy of the glass.

At the present time, therefore, it can only be concluded that the relative importances of possible chemisorption-controlled plasticity and stress-assisted corrosion (dissolution) in determining the fracture behavior of silicate glasses in different environments are not yet unequivocally established.

REFERENCES

1. P. A. Rebinder, in Proc. 6th Physics Conf., p. 29, State Press, Moscow (1928).
2. J. H. Westbrook and P. J. Jorgensen, Trans. AIME, 233, 425 (1965).
3. R. E. Hanneman and J. H. Westbrook, Phil. Mag., 18, 73 (1968).
4. S. P. Gunasekera and D. G. Holloway, Phys. and Chem. Glasses, 14, 45 (1973).
5. J. H. Westbrook and P. J. Jorgensen, Am. Mineral., 53, 1899 (1968).
6. G. C. Kuczynski and R. F. Hochman, Phys. Rev., 108, 946 (1957).
7. J. H. Westbrook and J. J. Gilman, J. Appl. Phys., 33, 2360 (1962).
8. J. H. Westbrook, in Environment-Sensitive Mechanical Behavior, Eds., A. R. C. Westwood and N. S. Stoloff, p. 247, Gordon & Breach, New York (1966).

9. J. H. Westbrook, Some Effects of Adsorbed Water on the Plastic Deformation of Non-metallic Solids, Report No. 65-RL-3975M, G.E. Res. Lab., Schenectady, New York, June (1965).
10. R. E. Hanneman and P. J. Jorgensen, J. Appl. Phys., 38, 4099 (1967).
11. R. N. Hall, in Proc. 9th Intl. Conf. on Physics of Semiconductors, Vol. I, p. 481, Acad. Sci. USSR, Moscow (1968).
12. M. Kimura and S. Hyodo, Jap. J. Appl. Phys., 11, 15 (1972).
13. P. A. Rebinder and V. Likhtman, in Proc. 2nd Intl. Conf. on Surface Activity, Ed., J. H. Schulman, Vol. III, p. 563, Academic Press, New York (1957).
14. B. R. Lawn and T. R. Wilshaw, J. Mater. Sci., to be published.
15. A. R. C. Westwood, H. Opperhauser, Jr., and D. L. Goldheim, Phil. Mag., 6, 1475 (1961).
16. A. R. C. Westwood, Phil. Mag., 7, 633 (1962).
17. W. H. Vaughan and J. W. Davisson, Acta Met., 6, 554 (1958).
18. A. R. C. Westwood, D. L. Goldheim and R. G. Lye, Phil. Mag., 16, 505 (1967).
19. A. R. C. Westwood, D. L. Goldheim and R. G. Lye, Phil. Mag., 17, 951 (1968).
20. A. R. C. Westwood and D. L. Goldheim, J. Appl. Phys., 39, 3401 (1968).
21. N. H. Macmillan, R. D. Huntington and A. R. C. Westwood, Phil. Mag., 28, 923 (1973).
22. K. Inabe, K. Emoto, K. Sakamaki and N. Takeuchi, Jap. J. Appl. Phys., 11, 1743 (1972).
23. S. A. Varchenya, F. O. Muktepavel and G. P. Upit, Sov. Phys.-Sol. State, 11, 2300 (1970).
24. G. W. Groves and M. E. Fine, J. Appl. Phys., 35, 3587 (1964).
25. R. W. Davidge, J. Mater. Sci., 2, 339 (1967).
26. J. R. Hopkins, J. A. Miller and J. J. Martin, phys. stat. sol., 19, 591 (1973).
27. E. S. Machlin, in Strengthening Mechanisms in Solids, Ch. 12, p. 375, ASM, Metals Park, Ohio (1962).
28. R. M. Latanision, N. H. Macmillan and R. G. Lye, Corrosion Sci., 13, 387 (1973).
29. W. G. Johnston, J. Appl. Phys., 33, 2716 (1962).
30. P. L. Pratt, R. Chang and C. H. Newey, Appl. Phys. Letts., 3, 83 (1963).
31. P. L. Pratt, R. L. Harrison and C. H. Newey, Disc. Faraday Soc., 38, 211 (1964).
32. T. L. Johnson, R. J. Stokes and C. H. Li, Acta Met., 6, 713 (1958).
33. O. W. Johnson and P. Gibbs, J. Appl. Phys., 34, 2852 (1963).
34. J. R. Patel and A. R. Chaudhuri, J. Appl. Phys., 34, 2788 (1963).
35. F. F. Volkenstein, Electronic Theory of Catalysis on Semiconductors, Pergamon Press, Oxford, England (1963).
36. F. F. Volkenstein, Sov. Phys.-Uspekhi, 9, 743 (1967).

37. D. M. Roessler and W. C. Walker, J. Phys. Chem. Solids, 28, 1507 (1967).
38. D. M. Roessler and W. C. Walker, Phys. Rev., 166, 599 (1968).
39. J. D. Levine and P. Mark, Phys. Rev., 144, 751 (1966).
40. J. P. Hirth and J. Lothè, Theory of Dislocations, Ch. 12, p. 376, McGraw-Hill, New York (1968).
41. A. R. C. Westwood, C. M. Preece and M. H. Kamdar, in Fracture, Ed., H. Liebowitz, Vol. III, p. 589, Academic Press, New York (1971).
42. R. M. Latanision, H. Opperhauser, Jr., and A. R. C. Westwood, in The Science of Hardness Testing and Its Research Applications, Eds., J. H. Westbrook and H. Conrad, p. 432, ASM, Metals Park, Ohio (1973).
43. R. M. Latanision and A. R. C. Westwood, in Adv. in Corrosion Sci. and Technol., Vol. 1, p. 51, Plenum Press, New York (1970).
44. L. Carlsson, J. Appl. Phys., 42, 676 (1971).
45. L. Carlsson and C. N. Ahlquist, J. Appl. Phys., 43, 2529 (1972).
46. Yu. A. Osipyan, V. F. Petrenko and G. K. Strukova, Sov. Phys.-Sol. State, 15, 1172 (1973).
47. J. M. Cabrera and F. Agulló-López, J. Appl. Phys., 45, 1013 (1974).
48. J. M. Cabrera and F. Agulló-López, J. de Physique, 34 (C-9), 253 (1973).
49. L. B. Zuev, V. E. Gromov and V. P. Sergeev, Sov. Phys.-Sol. State, 16, 1099 (1974).
50. A. R. C. Westwood, N. H. Macmillan and R. S. Kalyoncu, J. Am. Ceram. Soc., 56, 258 (1973).
51. D. J. Shaw, Introduction to Colloid and Surface Chemistry, 3nd Edn., Ch. 7, p. 133, Butterworths, England (1970).
52. R. W. Heins and N. Street, Soc. Pet. Eng. J., 5, 177 (1965).
53. N. H. Macmillan, R. E. Jackson and A. R. C. Westwood, in Proc. 15th ISRM Symposium on Rock Mechanics, Custer State Park, S. Dakota, Sept. 17-19 (1973), to be published.
54. A. R. C. Westwood, N. H. Macmillan and R. S. Kalyoncu, Trans. AIME (Mining), 256, 106 (1974).
55. A. R. C. Westwood and N. H. Macmillan, Ref. 42, p. 377.
56. A. R. C. Westwood and R. D. Huntington, in Mechanical Behavior of Materials, Vol. IV, p. 383, Soc. Mater. Sci., Japan (1972).
57. V. D. Kuznetsov, Surface Energy of Solids, pp. 45 and 74-110, HMSO, London (1957).
58. P. A. Rebinder, Ref. 57, p. 80.
59. M. V. Swain, R. M. Latanision and A. R. C. Westwood, in NSF Hard Materials Research, Vol. III, p. 32, Lehigh University, Bethlehem, Pennsylvania, July (1974).
60. K. N. Han, T. W. Healy and D. W. Fuerstenau, J. Coll. and Interface Sci., 44, 407 (1973).
61. N. H. Macmillan and A. R. C. Westwood, in Surfaces and Interfaces of Glass and Ceramics, Eds., V. D. Fréchette, W. C. La

Course and·V. L. Burdick, p. 493, Plenum Press, New York (1974).
62. A. R. C. Westwood, G. H. Parr, Jr., and R. M. Latanision, in Amorphous Materials, p. 533, Wiley, London (1972).
63. R. A. Huggins, Report of 1971 ARPA Mater. Conf., Woods Hole, Mass., Vol. 1, p. 136, Univ. of Michigan, Ann Arbor, Michigan (1972).
64. R. H. Doremus, Ann. Rev. Mater. Sci., 2, 93 (1972).
65. J. H. Konnert, J. Karle and G. A. Ferguson, Science, 179, 177 (1973).
66. S. Glasstone, K. J. Laidler and H. Eyring, The Theory of Rate Processes, McGraw-Hill, New York (1941).
67. M. H. Cohen and D. Turnbull, J. Chem. Phys., 31, 1164 (1959).
68. J. J. Gilman, J. Appl. Phys., 44, 675 (1973).
69. J. J. Gilman, J. Appl. Phys., 46, 1625 (1975).
70. P. B. Bowden and S. Raha, Phil. Mag., 29, 149 (1974).
71. A. Thierry, R. J. Oxborough and P. B. Bowden, Phil. Mag., 30, 527 (1974).
72. M. F. Ashby and J. Logan, Scripta Met., 7, 513 (1973).
73. R. D. Baëta and K. H. G. Ashbee, Am. Mineral., 54, 1574 (1969).
74. J. J. Gilman, J. Metals, 18, 1171 (1966).
75. J. J. Gilman, J. Appl. Phys., 39, 6086 (1968).
76. R. H. Doremus, Glass Science, Ch. 15, p. 281, Wiley, New York (1973).
77. F. M. Ernsberger, Ann. Rev. Mater. Sci., 2, 529 (1972).
78. R. H. Doremus, Ref. 76, Ch. 9, p. 146.
79. S. M. Cox, Phys. and Chem. Glasses, 10, 226 (1969).
80. R. J. Stokes, T. L. Johnston and C. H. Li, Phil. Mag., 3, 718 (1958).
81. J. Washburn, A. E. Gorum and E. R. Parker, Trans. AIME, 215, 230 (1959).
82. F. J. P. Clarke, R. A. J. Sambell and H. G. Tattersall, Phil. Mag., 7, 393 (1962).
83. F. J. P. Clarke, R. A. J. Sambell and H. G. Tattersall, Brit. Ceram. Soc. Trans., 61, 61 (1962).
84. A. R. C. Westwood, J. Mater. Sci., 9, 1871 (1974).
85. C. N. Ahlquist, Acta. Met., 22, 1133 (1974).
86. A. R. C. Westwood, D. L. Goldheim and E. N. Pugh, in Grain Boundaries and Surfaces in Ceramics, p. 553, Plenum Press, New York (1966).
87. A. R. C. Westwood, D. L. Goldheim and E. N. Pugh, Phil. Mag., 15, 105 (1967).
88. R. J. Friauf, J. Appl. Phys. (Suppl.), 33, 494 (1962).
89. G. W. Weidmann and D. G. Holloway, Phys. and Chem. Glasses, 15, 68 (1973).
90. D. M. Marsh, Proc. Roy. Soc., A279, 420 (1964).
91. D. M. Marsh, Proc. Roy. Soc., A282, 33 (1964).
92. J. J. Gilman, J. Appl. Phys., 31, 2208 (1960).

93. S. M. Wiederhorn, in Mater. Sci. Res., Vol. 3, p. 503, Plenum Press, New York (1966).
94. S. M. Wiederhorn, J. Am. Ceram. Soc., $\underline{52}$, 99 (1969).
95. V. V. Panasyuk and S. E. Kovchik, Sov. Phys.-Doklady, $\underline{7}$, 835 (1963).
96. G. I. Barenblatt, Adv. in Appl. Mech., $\underline{7}$, 55 (1962).
97. A. R. C. Westwood and R. M. Latanision, in Ceramic Machining and Surface Finishing, Eds., S. J. Schneider, Jr., and R. W. Rice, p. 141, NBS Special Publ. No. 348, May (1972).
98. S. W. Freiman, J. Am. Ceram. Soc., $\underline{57}$, 350 (1974).
99. S. M. Wiederhorn and A. G. Evans, Paper No. 92-B-75, American Ceramic Society Conf., Washington, D. C., May (1975).
100. S. M. Wiederhorn, J. Am. Ceram. Soc., $\underline{50}$, 407 (1967).
101. R. J. Charles and W. B. Hillig, in Symposium on the Mechanical Strength of Glass and Ways of Improving It, p. 511, Union Scientifique Continentale du Verre, Charleroi, Belgium (1962).
102. W. B. Hillig and R. J. Charles, in High Strength Materials, Ed., V. F. Zackay, p. 682, Wiley, New York (1965).
103. M. Kimura and S. Hyodo, Int. J. Fracture Mech., $\underline{8}$, 475 (1972).
104. S. M. Wiederhorn, in Annual Report of Inorganic Materials Division, p. 91, Institute for Materials Research, NBS, October (1974).
105. P. G. Fox, Soc. Glass Tech. Conf. on Strength of Glass, Sussex, England, 25-27 March (1974).
106. D. McCammond, A. W. Neumann and N. Natarajan, J. Am. Ceram. Soc., $\underline{58}$, 15 (1975).

DISCUSSION

Comment by E. D. Shchukin:

I'll say just a few words, leaving the main part for my lecture in the afternoon.

The effects under consideration are many and complicated. They have several aspects which influence each other. Hence we are to choose those methods of investigation where the role of the surface itself is important to a sufficient extent.

I'm sorry to contradict Dr. Macmillan, but Rebinder himself disliked the usual methods of indentation (usual means here under heavy loads), because great bulk deformations develop in this case, and the part of energy spent on formation of new surface is too small, in the same manner as in the case of uniaxial tensile experiments with large samples.

Of course, such experiments, when thoroughly carried out, are necessary, interesting and fruitful. However, they are not sufficient by themselves. If we want to understand the mechanism of the phenomenon, we must carry out (along with such essentially macroscopic experiments) the other ones, closer to the subject we want to study.

In this case, when the indentation load was 10g, it was too heavy to let us feel the direct influence of the surface. The latter is possible only under loads of the order of parts of a gram.

Comment by R. F. Firestone:

Aren't hardness indentations always accompanied by cracks, particularly in truely brittle ceramics like sapphire? The TEM work of Hockey and others has shown dislocations and cracks under indentations. I have observed microscopic cracking at the corners of Knoop indentations in sapphire with 50 gm loads. I would suspect that even in semi-brittle ceramics, like MgO, TEM examinations would show cracks. On the other hand, I would not suspect cracking in a truly ductile material like copper. And, it is interesting to observe that copper, in distinction to MgO and Al_2O_3, shows no surface effects or creep in Westbrook's experiments.

Reply:

My impression is that this is mainly a matter of load--and that at small enough loads one does not get cracks. In the present case, however, you would have to have cracks open at the surface (i.e., accessible to the environment) in order to get any environmental effect, and I find it hard to believe that such cracks would not be revealed by etching. I grant that I might fail to see subsurface cracks, but I fail to see how these--if they occur--could lead to any environmental sensitivity of the hardness.

Comment by F.R.N. Nabarro:

If one were to believe that indentation is accompanied by cracking, then, even if the surface energy of the crack was much less than the total work done, which is determined by plastic deformation, the argument of Westwood, Preece, and Kamdar allows us to believe that the total work is directly proportional to the energy of the newly-exposed surface. Does this not destroy the logic of your argument?

Reply:

I presume you refer to the Westwood, Preece and Kamdar explanation (in Fracture, Ed., H. Liebowitz, Vol. III, p. 589, Academic Press, New York (1971)) for the phenomenon of liquid metal embrittlement. Reduced to essentials, these authors make two suggestions. First, they follow Kelly, Tyson and Cottrell in arguing that fracture will occur when (i) Griffith's criterion is obeyed with respect to the appropriate work of fracture, and (ii) either the maximum tensile stress σ_{max} at any point in the solid reaches the ideal tensile strength σ_{th} or the maximum shear stress τ_{max} reaches the ideal shear strength τ_{th}. And second, they suggest that chemisorption of a liquid metal leads to embrittlement of a normally ductile metal because it so far reduces σ_{th} that the condition $\sigma_{max} = \sigma_{th}$ is reached at some crack or notch in the specimen not only before $\tau_{max} = \tau_{th}$, but also before τ_{max} reaches the stress required for continued dislocation motion. And, this seems reasonable because σ_{th} is determined by the bonds exposed at the tip of the crack or notch concerned, whereas the dislocations in the crystal are screened from any influence of the adsorbate by the mobile conduction electrons. It is only after the change in the mode of fracture from ductile to brittle, therefore, that the work of fracture approaches the surface free energy of the (adsorbate covered) new surface produced. Before the transition, the work of fracture is much larger. Thus, only a small part of the total work of fracture is accounted for by the energy of the newly exposed surface, and I see no contradiction with the hypothesis that chemomechanical effects in inorganic nonmetals cannot be explained in terms of an adsorption-induced reduction in surface free energy.

Comment by H.-J. Engell:

How can a charge transfer process control dislocation motion?

Reply:

We simply do not know at the present time. Rather, we came to this conference hoping that some suggestions for a detailed mechanism would be forthcoming from our colleagues in other disciplines. I suppose that a "best guess" is simply that the dislocation somehow becomes charged by virtue of acquiring electrons from or donating electrons to the adsorbate. There is evidence that this happens in semiconductors (F.F. Volkenstein, Sov. Phys.-Uspekhi , 9, 743 (1967) and Electronic Theory of Catalysis on Semiconductors, Pergamon Press, Oxford, England (1963)), and we can only assume from the similarity of the effects seen in materials of differing band gap that a similar process occurs in insulators also. But how, we do not know.

Comment by J. P. Hirth:

In insulators, at any rate, it would be easy to develop models where dislocation mobility was a maximum at the zero of ζ potential. In this connection I wonder whether the apparent minimum in mobility as measured by hardness tests is in fact a consequence of a variation in multiple slip caused by suppression of bulk dislocation nucleation in the presence of space-charge layers. The model is as follows: with no space-charge ($\zeta = 0$), bulk (near surface) nucleation of dislocations occurs in the region of maximum Hertzian stress beneath the indenter, together with surface nucleation. The spreading bulk loops interfere with the surface loops and give high hardness by intersection hardening. With space charges, bulk nucleation is suppressed relative to surface nucleation, and, despite greater point defect interaction, the absence of intersections gives greater spread of the surface loops. Detailed study of the distribution of dislocations as a function of character and position would test the above suggestion. Also, the implication is that the mobility of an isolated dislocation would be a maximum in presence of zero space charge, consistent with the well known maximum mobility of bulk dislocations at the isoelectric temperature where bulk charge effects vanish.

Reply:

This is an ingenious suggestion, and one that I shall go away and devise an experiment to investigate.

SURFACE EFFECTS ON THE YIELD POINT OF SAPPHIRE*

Ross F. Firestone and Arthur H. Heuer

University of Pittsburgh Case Western Reserve University
Pittsburgh, Pennsylvania Cleveland, Ohio

In an earlier paper [1] the authors showed that a yield drop does not always occur during the tensile deformation of sapphire, single crystal α-Al_2O_3; it can be eliminated or enhanced by previous mechanical or thermal treatments. In this paper, the results of the earlier investigation are summarized and collated with characterization of the surfaces produced by the treatments.

Cylindrical specimens of sapphire, grown by the Verneuil and Czochralski processes, were deformed by basal slip, (0001) {1120}, in air at 1500°C. The purity of the two types of sapphire was the same. The dislocation etch pit density on the basal plane was about 10^7 per cm^2 for the Verneuil and 10^3 for the Czochralski.

Some of the specimens were deformed "as received" from the manufacturer with a coarsely ground surface. Before deformation, other specimens were "reground" to a satin finish; or "vacuum annealed" at 10^{-5} Torr for 18 hours at 1870°C; or "air annealed" for 65 hours at 1800°C; or "commercial flame polished" by a single pass through an oxy-hydrogen flame; or "laboratory flame polished" by multiple passes until all visible evidence of grinding damage had been eliminated.

The deformation behavior of all specimens is shown in Fig. 1. As received Czochralski never displayed a yield drop while all other specimens invariably did. It was concluded that the yield drop in sapphire was caused by dislocation multiplication and not by unpinning from an impurity atmosphere. The variation in yielding behavior was attributed to variations in the mobile dislocation density due to surface sources introduced during grinding

---

* This work was partly supported by the Office of Naval Research.

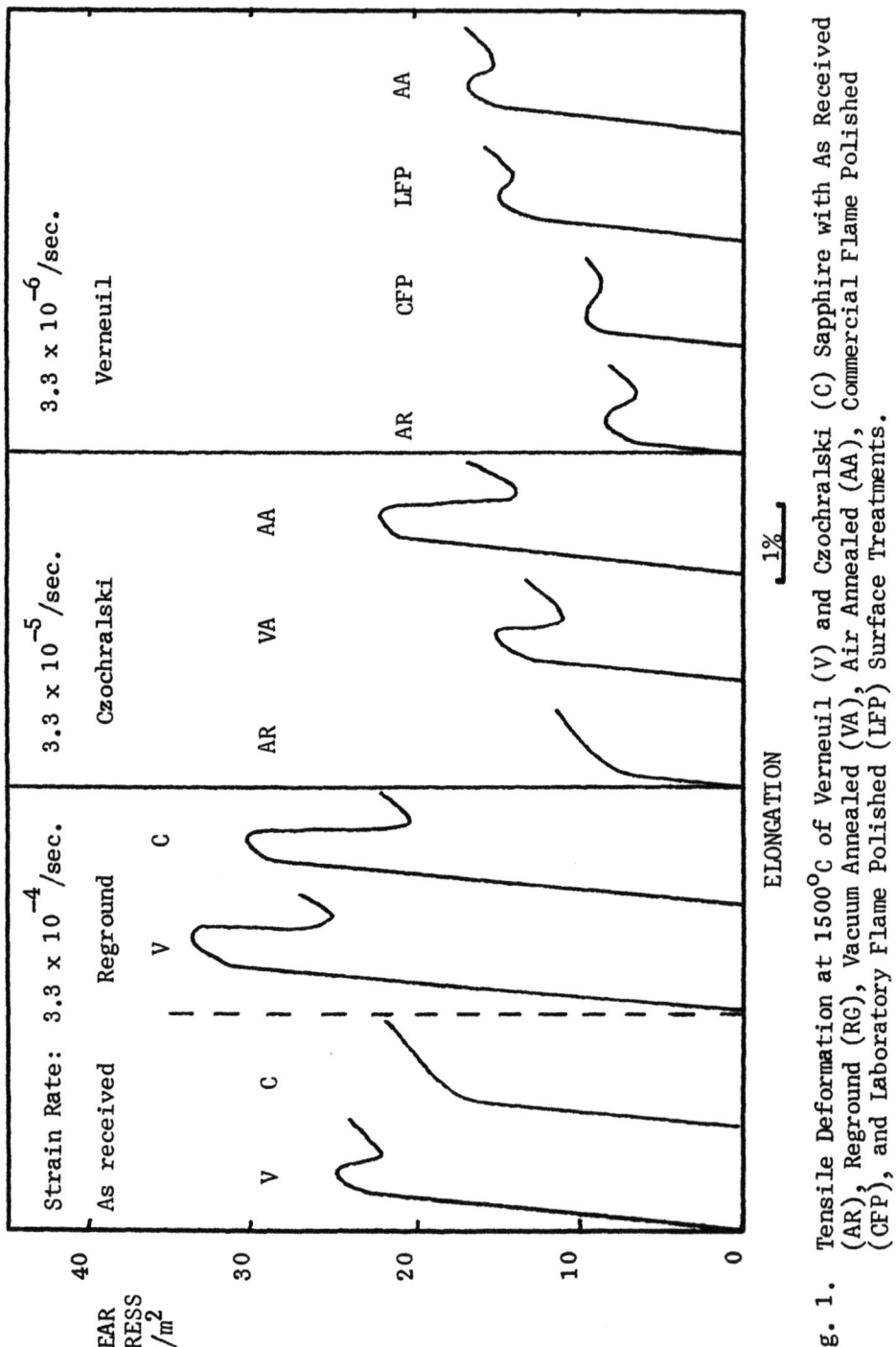

Fig. 1. Tensile Deformation at 1500°C of Verneuil (V) and Czochralski (C) Sapphire with As Received (AR), Reground (RG), Vacuum Annealed (VA), Air Annealed (AA), Commercial Flame Polished (CFP), and Laboratory Flame Polished (LFP) Surface Treatments.

and reduced to a greater or lesser extent by treatment. The absence of a yield drop in as received Czochralski was attributed to a very badly damaged surface.

Specimen surfaces were examined by optical microscopy and scanning electron microscopy. The grinding direction is vertical and the tensile axis is horizontal for all micrographs in this paper; their magnification is 800X.

The entire circumference of the specimen was examined for crystallographic control of surface features but none was found. Although sapphire is anisotropic and the hardness [2] and wear resistance [3] vary with orientation, it is uniaxial so the maximum anisotropy is between the basal and other planes [4]. Since the basal plane is always at an angle to the surface of the specimen, the variation around the circumference is small and apparently has no effect on surface features.

Examination of the specimen surfaces shows that grinding damage is indeed greater for as received Czochralski (Fig. 2) than for any other specimens. The Czochralski surfaces are covered with deep pits filled with grinding debris. The debris is tenaciously adherent and can not be removed by high pressure air blasts or prolonged ultrasonic cleaning. As received Verneuil

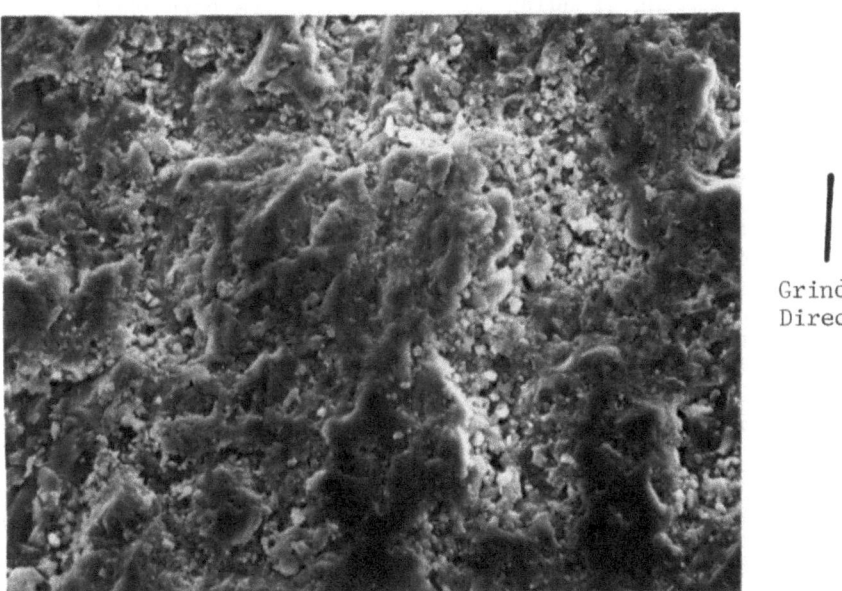

Grinding Direction

Fig. 2. SEM of As Received Czochralski Specimen Surface (800X).

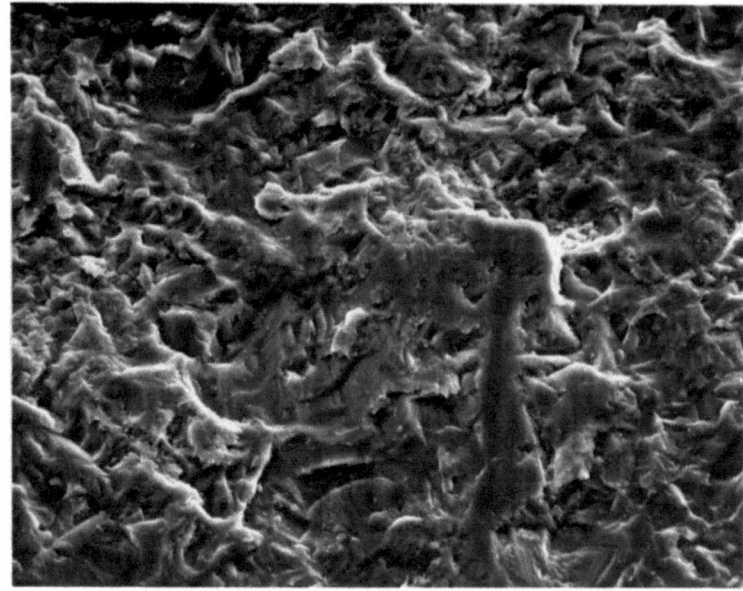

Grinding Direction

Fig. 3. SEM of As Received Verneuil Specimen Surface (800X).

surfaces (Fig. 3), on the other hand, appear to have been plastically deformed during grinding and many areas are burnished rather than fractured. There are many more cracks and steps on the Czochralski surface even though the maximum surface roughness amplitude as measured with a profilometer is 3 μm for both as received surfaces.

The badly damaged as received Czochralski surface may have been produced by insufficient coolant during diamond grinding. The adherence of debris suggests that grinding pressure was heavy and produced high temperatures which may have been sufficient to sinter the debris and also cause high thermal stresses. The burnished surfaces of the as received Verneuil suggest heavy pressure but also the presence of adequate coolant so that the debris was washed away and plastic flow rather than fracture occurred. Westwood [5] has shown that the liquid environment can greatly affect the behavior of ceramics during abrasion and make them react either brittlely or plastically.

As received specimens were reground with light pressure, a fine diamond wheel, and a copious flow of coolant. The surface roughness is less than 1 μm but, as shown by the micrograph of a reground Czochralski surface (Fig. 4), some of the pits from the previous coarse grinding were deeper than the 76 μm removed by regrinding and were not completely removed. The open star shaped crack at the upper right may be associated with a pit

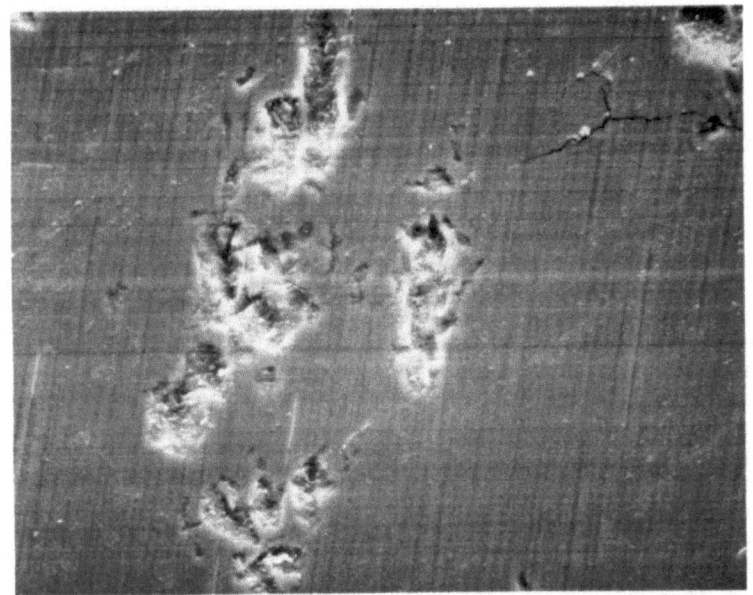

Fig. 4. SEM of Reground Czochralski Specimen Surface (800X).

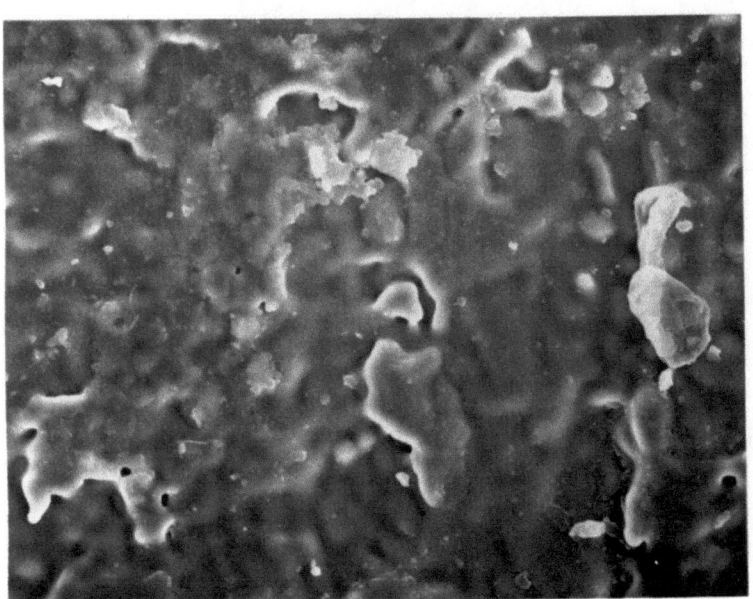

Fig. 5. SEM of Air Annealed Czochralski Specimen Surface (800X).

which has been ground away or may be due to thermal cracking. The surfaces of reground Verneuil have similar residual pits but these have smooth bottoms. There are no open cracks.

Annealing in air has a profound effect on the surface topography as the micrograph of an air annealed Czochralski shows (Fig. 5). The jagged surface relief has been rounded and the bottoms of the pits smoothed and filled in by what may be debris which has recrystallized during the prolonged high temperature annealing. The surfaces of air annealed Verneuil are similar. Unfortunately, the surfaces of vacuum annealed specimens could not be examined due to a translucent film which produced strong elemental contrast in the SEM. This film was determined to be tantalum by x-ray analysis and was probably deposited from the furnace heating elements.

The surfaces of flame polished specimens all appeared flawless when examined. When they were heated in air and in vacuum to thermally etch their surfaces, only the vacuum etched commercial flame polished specimens displayed any surface features. Ultramicroscopic examination revealed arrays of bright lines aligned with the grinding direction and with a distribution roughly corresponding to the pits on as received surfaces (Fig. 6). The lines may be cracks introduced during grinding which had not been

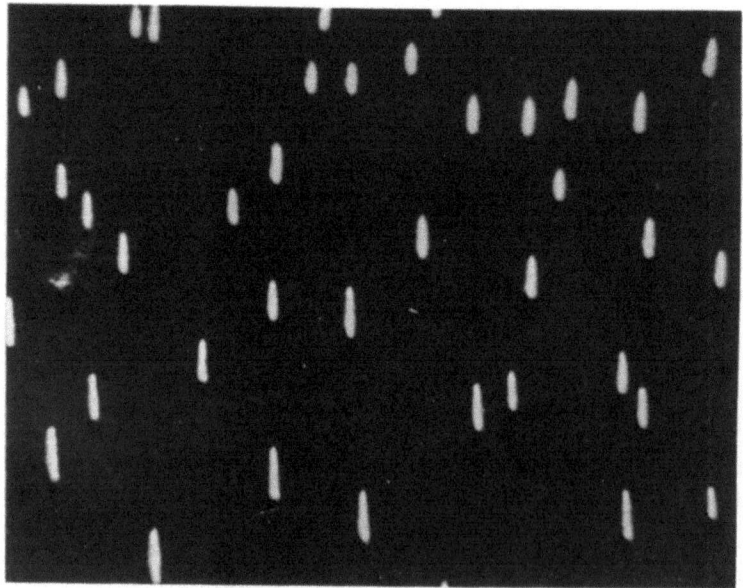

Grinding Direction

Fig. 6. Ultramicrograph of Surface of Commercial Flame Polished Verneuil Specimen after Vacuum Etching (800X).

healed by commercial flame polishing and which were enlarged by vacuum etching.

The above characterizations of the specimen surfaces support the previous conclusion that the effect of mechanical and thermal treatments on the yield point of sapphire is to change the density of surface flaws which act as mobile dislocation sources. If we rank the specimens on the basis of their yield strength, there is an inverse relation with surface damage. A badly damaged surface, such as as received Czochralski, has many cracks and steps which produce high stress concentrations [6] and are excellent dislocation sources [7]. These specimens have the lowest yield strengths and no yield drops. Various treatments increase the yield strength in proportion to the reduction in surface damage and introduce a yield drop when the rate of dislocation multiplication is insufficient to sustain the applied strain rate.

The etching of flame polished specimens shows that wide pits are not as good dislocation sources as narrow cracks since the yield strength of commercial flame polished specimens from which the pits but not the cracks have been eliminated is only little higher than of as received specimen. A similar effect is produced by vacuum annealing which may round sharp edges but enlarges cracks. Air annealing and laboratory flame polishing not only smooth the surface but also heal cracks, an effect previously reported in air annealed ceramics [8]. Hence, these two treatments produce the highest yield strength. However, surface damage mainly affects the upper yield stress; the lower yield stress is determined by lattice resistance which is changed by thermal but not mechanical treatments [1].

REFERENCES

1. R. F. Firestone and A. H. Heuer, J. Amer. Ceram. Soc. (56), 137.
2. H. Insley et alii, Amer. Mineral. (32), 1.
3. H. Winchell, ibid. (29), 399.
4. R. W. Rice, N.B.S. Spec. Pub. (348), 365.
5. A. R. C. Westwood, ibid. (348), 141.
6. D. W. Marsh, Phil. Mag. (5), 1197.
7. B. J. Hockey, N.B.S. Spec. Pub. (348), 333.
8. F. F. Lange, ibid. (348), 236.

MICRODEFORMATION OF HARD NON-METALLIC CRYSTALS BY SOFTER INDENTERS AND SLIDERS.

C.A. Brookes.

Department of Engineering Science, University of Exeter
Exeter EX4 4QF, England.

ABSTRACT. Dislocation etching techniques have been used to study the region of microdeformation produced in crystals deformed by indenters and sliders. It is observed that the volume of this dislocated zone is related to the magnitude of the applied load but is basically independent of the shape of the indenter or the size of the indentation or groove. This behaviour is maintained even when conventional diamond indenters are replaced by cones made from materials which are much softer than the deformed crystals. The initiation of visible macroscopic wear has been observed when the number of traversals of a well lubricated metal slider exceeds a critical level on a crystal which is at least five times harder than the slider. Dislocations multiply, slip steps are formed and work-hardening of the hard brittle crystal is established well before the first crack is formed leading to visible fragmentation and wear. Finally, it is observed that the critical number of traversals required to produce visible wear is dependent on the crystallographic direction of sliding.

It is well known that the measured hardness of crystalline solids is dependent on such experimental variables as the applied load (1,2,3), crystallographic orientation (4,5), shape of indenter (6), etc. Such measurements basically reflect the macroscopic deformation processes which result in the formation of the indentation. Similarly, gross plastic flow and/or fracture directly associated with a scratch have been of predominant importance in studies on friction and wear. In this paper, particular attention is paid to the extent and nature of micro-deformation, i.e. the so called dislocation zone as revealed by dislocation etch pit techniques, which extends well beyond the indentation or scratch.

1. EXPERIMENTAL.

Either a Leitz Miniload hardness apparatus or a device designed to measure the scratch hardness and friction of single crystals, and described elsewhere (7), were used for experiments carried out at room temperature. Indentations were made following the conventional procedure for hardness measurements. However, a variety of the normal hard rigid indenters were employed - i.e. Vickers, Knoop, Al_2O_3 spheres, diamond cones etc. - and sometimes a conical indenter having an included angle of 136° and made from a metallic material significantly softer than the crystal specimen. Single and reciprocating traversals were made in specific crystallographic directions, on a given crystal surface, using a sliding speed of approximately 20 mm/min. with the same cones as those used in the indentation experiments. The sliding surfaces were continuously lubricated with soluble oils.

The experiments were carried out on magnesium oxide crystals, unless otherwise stated, because the resultant distribution of dislocations could readily be observed by etching (8) on a (001) plane prepared by cleavage and then chemically polished. Whenever the etching techniques revealed that the deformation process had produced dislocations on the surface of the crystal, the depth of penetration of the dislocated zone beneath that region in the bulk of the crystal was then determined. Two methods were used for these observations. One involved removing layers from the original surface of the crystal, using chemical polishing techniques, and etching at each new level until the dislocations introduced by the deformation process were finally removed (9,10,11). Experience with this technique enabled accurate estimates to be made on the probable depth of a given zone after the first few layers had been removed - thus reducing the number of chemical polishing steps needed to determine that depth. Similar techniques have been used to study the extent of deformation in LiF crystals (10). The second method was based on cleaving through the indentation or scratch, generally on a (100) but occasionally on a (110) plane, and etching the dislocations intersecting the exposed plane.

2. DEFORMATION OF THE CRYSTALS.

The observations made in this paper were generally based on experiments using a normal load of 500 g but it is considered likely from earlier work (10,11,12) that similar effects would be observed at other loads. Fig.1 illustrates the first method used to identify the dislocated zone and Fig.2 that where the dislocations are exposed on fracture surfaces. It can be seen from Fig.1 that, whilst the anisotropy in Knoop hardness is such that <110> directions are almost twice as hard as <100>, the overall size of

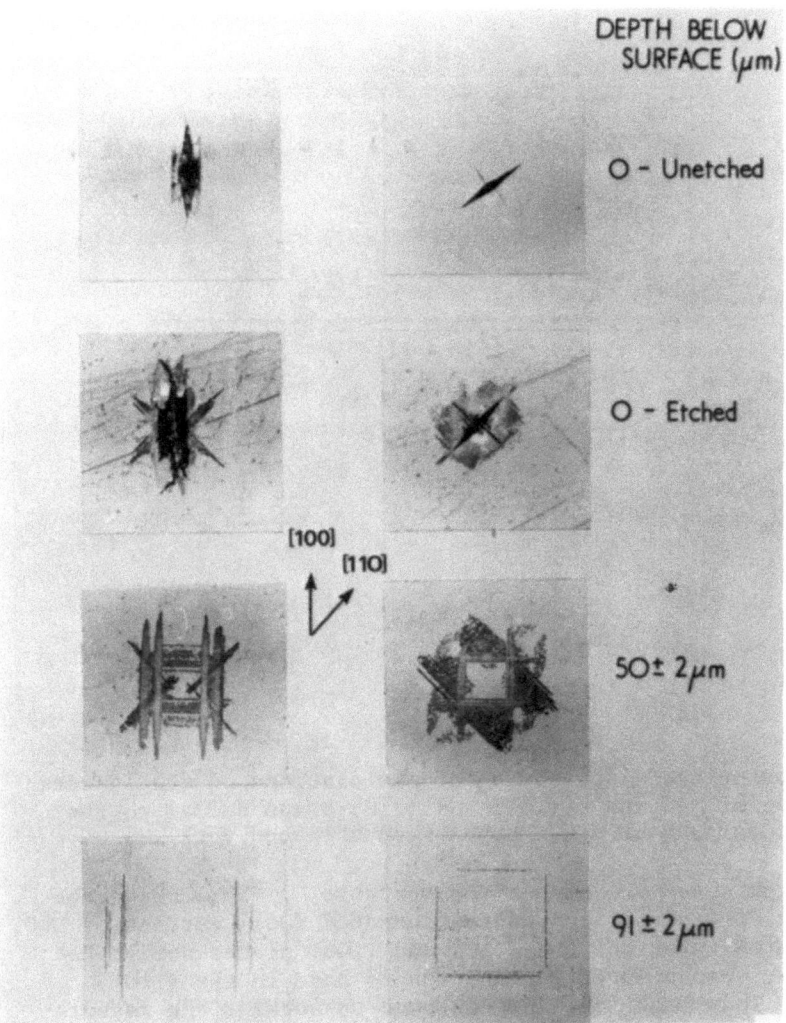

Fig. 1. Illustrating the repeated chemical polish and dislocation etching technique for Knoop indentations in [100] and [110] directions on a (001) MgO surface.

Fig.2. Showing the distribution of dislocations on the indented (001) surface of MgO and a (100) and (110) plane normal to that surface. (Courtesy of J.E. Morgan - unpublished work).

the dislocated zones are directly comparable. Fig.2 shows the distribution of dislocations on the indented (001) surface of MgO, and those which intersect the (110) and (100) planes normal to that surface, when a Knoop indentation is made in the <110> direction. This technique was employed to augment the measurements and observations based on that using repeated polishing and etching. In further experiments, using the range of hard rigid indenters listed previously, it has been observed that the volume of the resultant dislocated zones were essentially independent of the indenter shape. On the other hand, the cube of the dislocated depth, i.e. the vertical depth beneath the indented surface containing all the dislocations produced by the indentation, was directly related to the applied load.

We can now contrast the behaviour of dislocations when the normal rigid indenter or slider is replaced by one made from copper which, with a hardness of approximately 50 kg/mm^2, is significantly softer than the MgO. These cones were invariably blunted as a result of the 'indentation' process but there was no

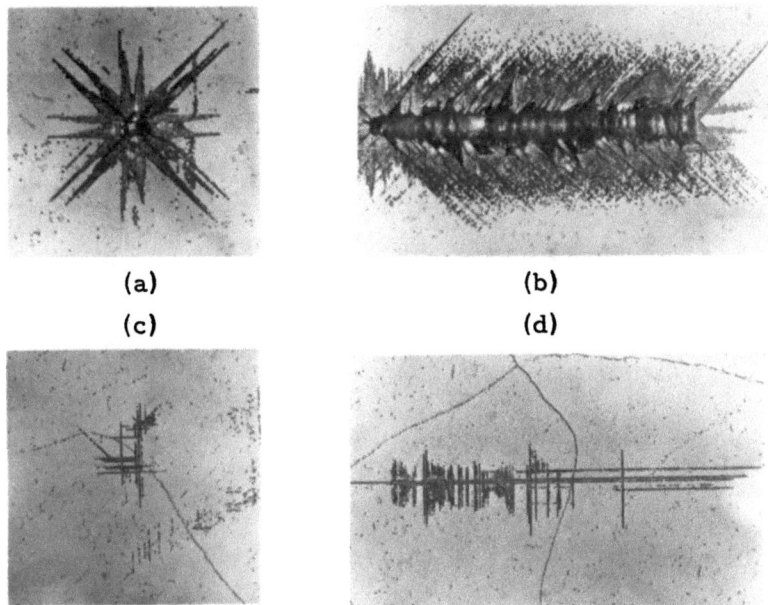

Fig. 3. Etch pits showing dislocations on a (001) MgO surface produced (a) a static 136° diamond cone; (b) a sliding diamond cone; (c) a static copper indenter; and (d) a sliding copper cone. (x75)

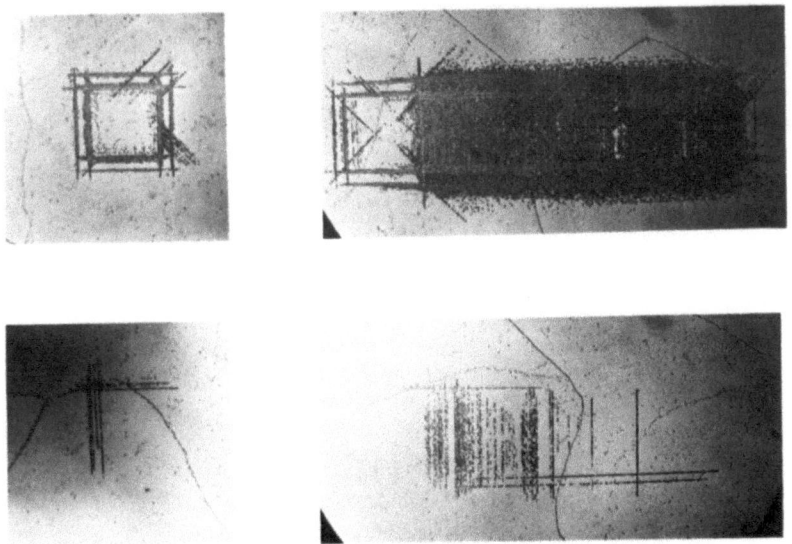

Fig. 4. The same specimens as shown in Fig. 3 but etched after chemical polishing at a level 60 microns below the original surface. (x75)

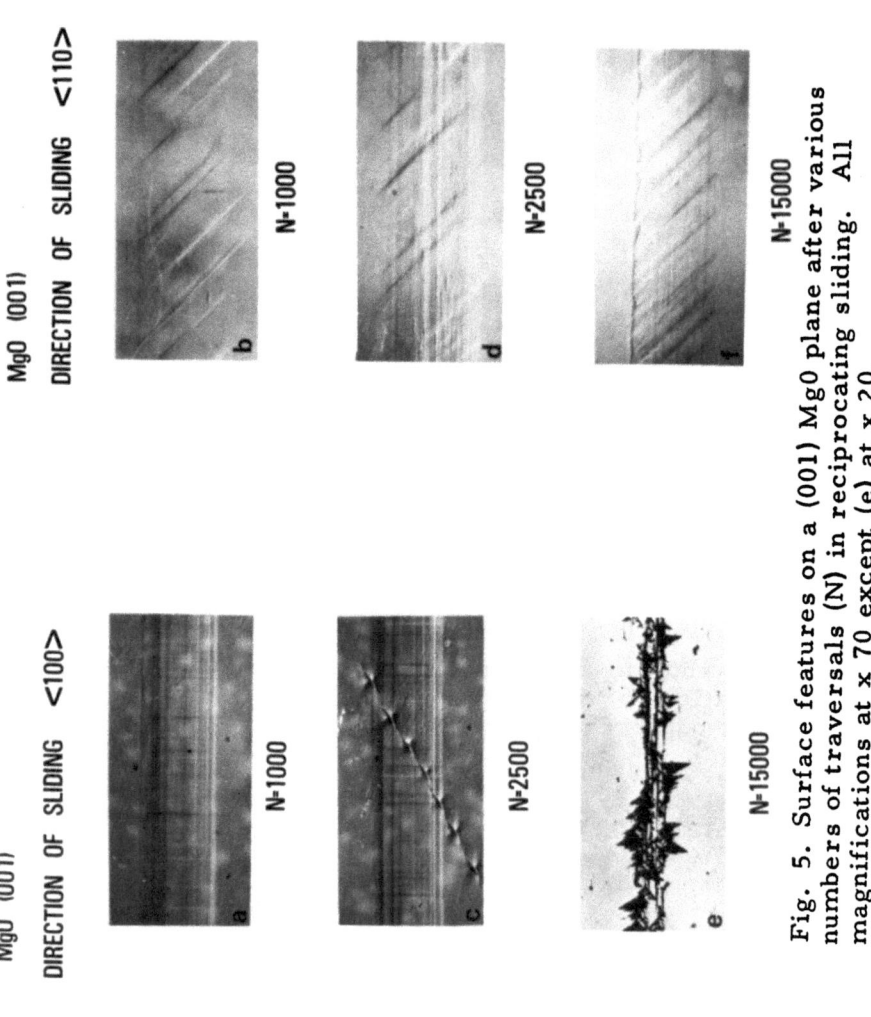

Fig. 5. Surface features on a (001) MgO plane after various numbers of traversals (N) in reciprocating sliding. All magnifications at x 70 except (e) at x 20.

apparent damage on the MgO surface in the regions of contact. However, etch pits confirmed that microdeformation had been produced by the load applied with quasi-static copper cones. The dislocations did not normally form a rosette pattern. Nevertheless, it was possible to measure the depth of the microdeformation zone and this did confirm that the dislocations penetrated almost as deep as those produced under similar conditions by the diamond cone. A few other experiments were carried out at room temperature using a copper indenter which was blunted by previous experiments at the same load. Again, there was no visible damage on the MgO surface nor any further macroscopic blunting of the copper indenter. Whilst it would appear that the load was supported elastically, the depth and distribution of dislocations in the MgO was comparable with those produced when the original undeformed copper indenter was used. Similar observations were made when the cones were used as sliders. The copper cone did not normally cause the formation of a visible wear track although metallic particles were often observed adhering to the MgO surface. However, dislocations were produced at the surface and within the bulk of the crystal, to a depth of approximately 160 microns, beneath the copper slider. This was directly comparable to the depth of the dislocated zone formed beneath a diamond cone even though the coefficient of friction for diamond was nearly ten times greater for diamond than for copper sliding on MgO. Typical dislocation patterns produced under these conditions, both on the original surface and in the bulk of the crystal, are shown in Figs. 3 and 4.

From this and earlier work (11), it is clear that microdeformation may readily occur when contact pressures are developed which are 10 - 100 times lower than the nominal hardness values. Thus, most contacting surfaces in engineering applications will be subjected to such deformation - at least in the early stages of loading. It is clear that the development of the dislocated zone, under conditions of cumulative deformation, could lead to wear of a hard solid by one which is significantly harder. This possibility has been explored by using a reciprocating brass slider on a (001) MgO surface under the same conditions of lubricated sliding as those described earlier.

The hardness of the brass slider was increased from the original value of 100 to a maximum of 250 kg/mm^2 as a result of the sliding process. Knoop hardness measurements, at 100 g in <110> directions, show that the hardness of the MgO in the contact region increases from approximately 980 to 1150 kg/mm^2 after the first hundred traversals. Continued abrasion, in the experiments for both directions of sliding, produce slip lines but no further increase in hardness. Ultimately, fragmentation and wear is observed after about 15,000 traversals when the direction of sliding corresponds to <100>. Such fragmentation has not been observed after sliding in the <110> directions. Typical stages in the deformation process are shown in Fig.5 in which a series of Knoop micro-hardness measurements across the contact region is also apparent - i.e. Fig.5 (c). Finally, similar micro-hardness measurements in a section beneath the contact region, on a (100) cleavage plane across the wear track, confirm that the depth of hardening is of the same magnitude as that of the dislocated zone.

ACKNOWLEDGEMENTS.

The author thanks his colleagues for discussion - in particular he is grateful to Mr. M.Shaw for permission to use unpublished results - and De Beers Industrial Diamond Distributors for a grant to the laboratory.

REFERENCES.

1. D. Tabor, The Hardness of Solids, Oxford - Clarendon Press, 1951.
2. H. O'Neill, Hardness Measurements of Metals and Alloys, London - Chapman and Hall, 1934.
3. C.A. Brookes, Industrial Diamond Review, 338, 1973.
4. F.W. Daniels and C.G. Dunn, Trans. Am. Ceram. Soc., 41, 419, 1949.
5. C.A. Brookes, J.B. O'Neill and B.A.W. Redfern, Proc. Roy. Soc., A 322, 73, 1973.
6. A.G. Atkins and D. Tabor, J. Mech. Phys. Solids, 13, 1965.
7. P. Green, PhD Dissertation, University of Exeter, 1974.
8. J.J. Gilman and W.G. Johnston, J. Appl. Phys., 2, 30, 129, 1959.
9. P. Harrison, PhD Dissertation, University of Exeter, 1973.
10. R.P. Burnand, PhD Dissertation, University of Exeter, 1974.
11. C.A. Brookes and P. Green, Nature, 246, 155, 1973.

DISCUSSION

Comment by A.R.C. Westwood:

You commented that you found greater "Anomalous" Indentation Creep (AIC) in Mo than Cu, etc. But Westbrook's definition of AIC is that this time-dependent, environment-sensitive hardness effect decreases with increasing temperature. Did you see such an effect? (If not, the creep behavior observed was probably conventional creep.)

Incidentally, both Westbrook and myself are aware of the orientation dependence of hardness, and so we both always oriented our indenters in some particular crystallographic direction throughout any particular series of tests. Westbrook and Jorgensen also investigated the crystal plane dependence of AIC in a number of minerals and published this work in "Anistropy in Single Crystal Refractory Compounds", Vol. 2, Plenum Press, 1968, pp. 353-360; and also in the "American Mineralogist" in 1967 or 1968. They found that this effect disappeared when the surfaces were "dry", i.e., not exposed to an active environment.

Reply:

Yes, our measurements were made at loads considerably greater than those used by Westbrook and were not intended to study AIC. Initially, we had hoped to minimize surface effects by working in the range, e.g., above 500g in MgO, where the hardness is virtually independent of the normal laod and where the dislocated zone extends more than 100 μm below the surface. However, I should point to two of our observations which might be borne in mind in future applications of hardness measurements to studies of surfaces:

(i) Indentation creep, under our experimental conditions, has been observed in many common metals and monometallic crystals. While the role of creep does increase with experimental temperatures, in most cases, there does not appear to be a simple relationship with homologous temperatures. For example, we do not measure creep in copper, at lower temperature, but the role of creep in sapphire is particularly pronounced. That observation in copper may be particularly important in the context of Westbrook's results where, again, creep was not observed in either toluene or in air. I simply point out that had he used a different metal for his work, it is possible that he would have observed AIC.

(ii) Indentation creep in MgO, on a freshly cleaved surface under toluene, occurs with the Knoop indenter oriented in the <110> direction on the (001) surface. We have not looked at this type of crystallograph effect for the Vicker indenter.

THE CHEMOMECHANICAL EFFECT IN SEMICONDUCTORS

W. Schröter and P. Haasen

IV. Physikalisches Institut und Institut für Metallphysik der Universität Göttingen und SFB 126, Göttingen-Clausthal

INTRODUCTION   It is well established that dislocations in semiconductors and ionic crystals can carry a line charge q. In the case of semiconductors, this charge results from the occupation of localized electronic states by electrons or holes.[1] In the case of ionic crystals, it is either due to a surplus of ions of one sign in the neighbourhood of the dislocation core or due to defects (jogs) on the dislocation line.[2] In the space charge region at a surface, this line charge will be modified due to its electrostatic interaction with the space and surface charges. As will be shown here for the elemental semiconductors Si and Ge, this modification may lead to a change of the dislocation mobility and thereby to a change in the plastic properties within a certain region ranging from the surface down into the bulk.

We at first review briefly the main features on the electronic states at dislocations in semiconductors, then interpret the well studied influence of doping on the dislocation mobility as an effect of the line charge of the dislocation. Finally, we derive from this connection a possible mechanism for a change of the plastic properties in the surface region.

1. ELECTRON STATES AT DISLOCATIONS IN Si AND Ge

Fig. 1 shows a 60°- dislocation in the diamond lattice ($\mathbf{4}(\underline{b},\underline{s})$ = 60°). The qualitative properties of the energy-spectrum connected with this dislocation may be derived from a simple LCAO-treatment proposed by Shockley in 1953[3] which arrives at the same conclusions as the more elaborate theoretical treatments which are available at the moment. It is also in agreement

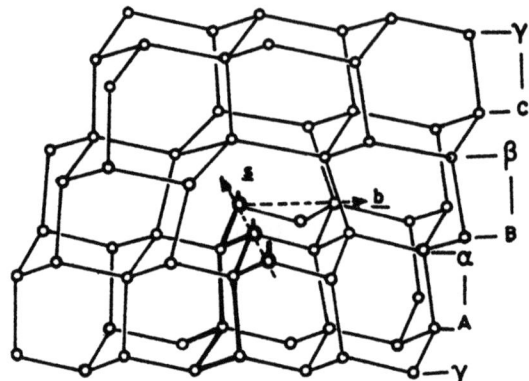

Fig. 1: 60°-dislocation in the diamond structure (**b** Burgers vector, **s** line element)

with the experimental results.

In this treatment we start with four sp^3-orbitals for each single atom. Taking the atoms to their sites in a diamond lattice and switching on the overlaps between different sp^3-orbitals (1 and 2 of nearest neighbours) in the undisturbed lattice, the degenerate atomic levels split into binding and antibinding levels (see fig.2). A broadening of the binding levels into valence band states and of the antibinding states into conduction band states occurs, when the overlap between orbitals 2 and 3 at the same atom and the overlap between orbitals at second nearest neighbours is switched on.[4,5]

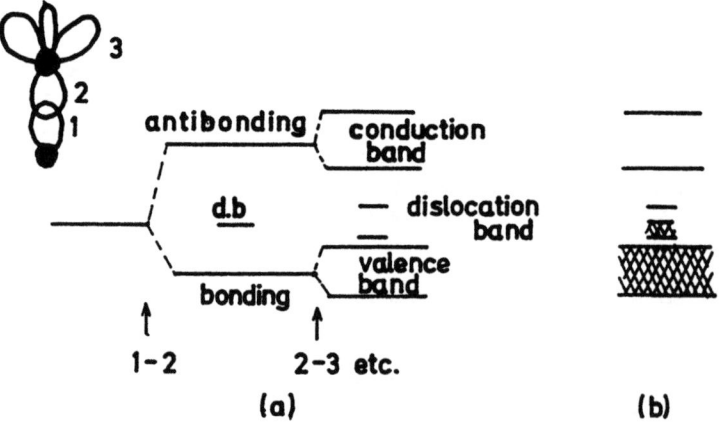

Fig. 2: (a) Different overlaps between sp^3-type orbitals leading to the bandstructure of a semiconductor with a 60°-dislocation, (b) occupation of the bands by electrons

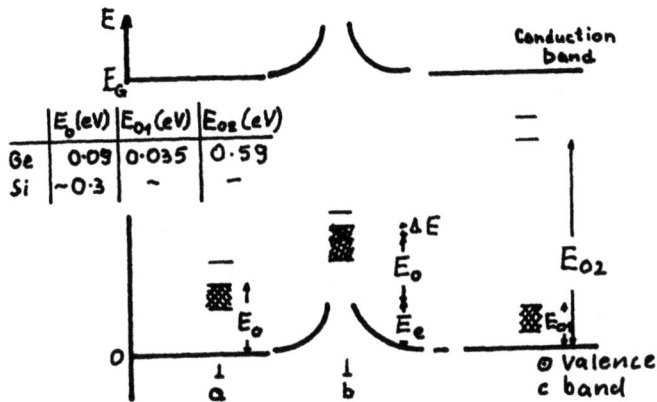

Fig. 3: Energy position of the one dimensional bands at
(a) neutral 60°-, (b) a negatively charged 60°- and
(c) a neutral screw dislocation

The atoms in the core of a 60°-dislocation miss one nearest neighbour. The levels of those orbitals (dangling bonds) remain unaffected when the overlap between orbitals 1 and 2 is switched on, but split into bands through the higher-order overlaps. As the lattice is periodic only in line direction, these states at the dislocations group into one dimensional bands. With one state and one electron per dangling bond, the one dimensional band at a 60°-dislocation is half filled by electrons when the dislocation is neutral (see fig. 3). The band structure at a 60°-dislocation is rather similar to that of a one dimensional metal.

By the same line of arguments one concludes that at a screw dislocation the states split into filled and empty bands which may be separated by a band gap and then are similar to the band structure of a one dimensional semiconductor.[6]

Electrical measurements on 60°- and screw dislocations in Ge and for 60°-dislocations in Si fully support this picture.[6,9] The parameters of the dislocation bands which have been derived from electrical and optical measurements are summarized in table 1. Lifetime and photoconductivity studies have demonstrated that in generation and recombination processes further bands at dislocations are involved.[7] But for the evaluation of the line charge q of the dislocation under conditions of thermal equilibrium, the contribution of those bands is neglegible.

Starting from the dislocation in the neutral state, any additional electron or hole brought into states at the dislocation

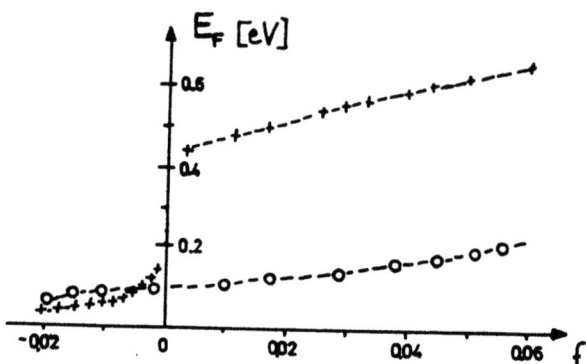

Fig. 4: Fermi level as a function of the occupation ratio
$f = q \cdot a/e$ for $60°$-dislocations ($\circ$:
$N_{disl.} = 3.9 \ 10^7 \ cm^{-2}$, $N_c = 2.6 \ 10^{13} \ cm^{-2}$, p-type)[7]
and for screw dislocations in Ge ($+$: $N_{disl.} = 3.2 \ 10^8$
cm^{-2}, $N_c = 4.6 \ 10^{14} cm^{-3}$ for p-type and $N_c = 1.9 \ 10^{15} \ cm^{-3}$
for n-type material),[9] theoretical curves are marked by
dashed lines[1]

leads to a line charge q of the dislocation. The electrostatic potential V(r) connected with this line charge is screened by charged impurities and by a rearrangement of free carriers. Under usual experimental conditions, V(r) varies slowly on a scale which measures the halfwidth of the bound core states. This fact allows for a great simplification in establishing the occupation statistics for the dislocation states. The treatment of this system, whose energy spectrum depends on the occupation, can be divided into three separate steps: calculation of (1) the electronic states at a neutral dislocation, (2) of the electrostatic potential V(r) for a given line charge q, and (3) of the fraction f of occupied states for given dislocation states which are shifted rigidly by eV(o) from their energetical position in the neutral state. The self-consistency of the treatment is controlled by the condition $q = e \cdot f/a$ (a distance between dangling bonds).[6]

The solution of Poisson's equation and the statistical treatment are available in different approximations.[6] Fig. 4 shows theoretical curves and experimental data for screw and $60°$-dislocations in Ge in the range of small positive and negative line charges.

2. DYNAMICAL PROPERTIES OF DISLOCATIONS IN SEMICONDUCTORS

If one heats Si or Ge to temperatures above one half of the

absolute melting temperature, it deforms plastically unter a stress τ. To get some insight into the microscopic processes involved, the mobility of single dislocations has been thoroughly studied as a function of stress and temperature by various authors. This is usually done by creating isolated dislocation loops in a dislocation-free crystal, marking their positions by etching techniques or X-ray topography and by measuring their displacements by a stress pulse.

The results have been analysed in terms of the dislocation velocity

$$v(\tau,T) = f(\tau) \exp\left(-\frac{Q}{K_B T}\right)$$

and have been interpreted by a model which assumes that the creation of double kinks across the Peierls potential is the rate-limiting process. Theoretical calculations of the Peierls potential and of the double kink formation energy E_{dk} in different approximations support this assumption,[10] although a consistent interpretation of the stress factor $f(\tau)$ on the basis of this model has not been given so far.[11,12]

In 1966, Patel and Chaudhuri[13] studied the dislocation mobility in heavily doped Ge-crystals and detected a new phenomenon: in a concentration range starting at the inset of extrinsic conduction and ending at the inset of solution hardening, they found that the dislocation mobility increases for As-doping and decreases for Ga-doping (fig. 5).

Further investigation of this effect in Si and Ge by several groups has shown two important features:

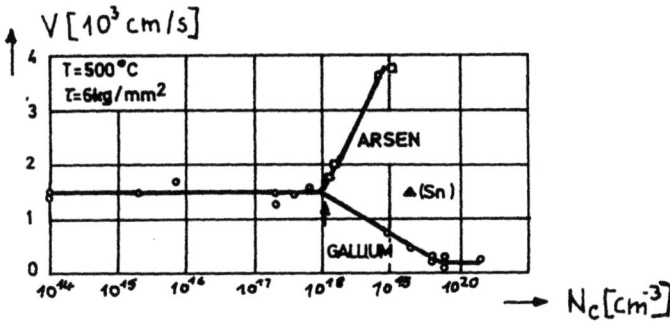

Fig. 5: Dislocation velocity v as a function of the doping concentration N_c.[13] The arrow indicates the value of the intrinsic carrier concentration.

(1) in the range marked above, the dependence of the dislocation velocity $v(\ ,T,N_c)$ on doping concentration N_c may be transformed into a dependence on the position of the Fermi levels E_F: $v(T,E_F)$, (2) the dislocation mobility increases monotonously when E_F moves towards the conductions band edge, but runs through a minimum when E_F moves towards the valence band edge.[14]

If the variation of the dislocation velocity were due to a change of the force constants between atoms in the lattice, one would expect a symmetric behaviour with respect of any deviation of E_F from its intrinsic position E_{Fi}. This has not been found. On the other hand, one may state, with all the uncertainties due to the temperature dependence of the width of the band gap, of the energetical position of any bound state with respect to the band edges in mind, that the minimum in v occurs for a value of E_F which is at the position E_o of the one dimensional band for the neutral 60°-dislocation. This is in support of a correlation between the dislocation charge q and the velocity v. v is a minimum for $q = 0$ and increases with the absolute value of the line charge.

Recently, Haasen[15] has proposed a mechanism which directly connects the dislocation velocity with the line charge. He points out that the electrostatic self energy of a system of charges on a straight line is lowered by any deviation from the straight line. The difference E in the electrostatic part of the self energy between a straight dislocation and a dislocation with a double kink in the saddlepoint configuration is proportional to q^2 (see fig. 6) and of the order of the changes in the activation energy Q of the dislocation velocity with doping. This concept together with the model for the energy spectrum at the dislocation, as outlined in section 1, can account for all presently known features of the effect, found by Patel and Chaudhuri.

3. ELECTROSTATIC INTERACTION BETWEEN SURFACE AND DISLOCATION

If a semiconductor is brought into contact with a liquid electrolyte, differences in the electrochemical potentials for electrons, holes, and the different ions between semiconductor and electrolyte will induce a charge transfer across the interface. In the interface, ions may become bound to surface atoms and electrons or holes may occupy localized levels. The final charge configuration which consists of localized charges and dipole layers at the interface, and of space charge regions in the electrolyte and in the semidonductor, must be neutral as a whole.[17,18]

The line charge of a dislocation which approaches the interface will be modified in the space charge region of the semiconductor compared to its value in the bulk. A quantitative evaluation of $q(z)$ is not available at present. According to the model out-

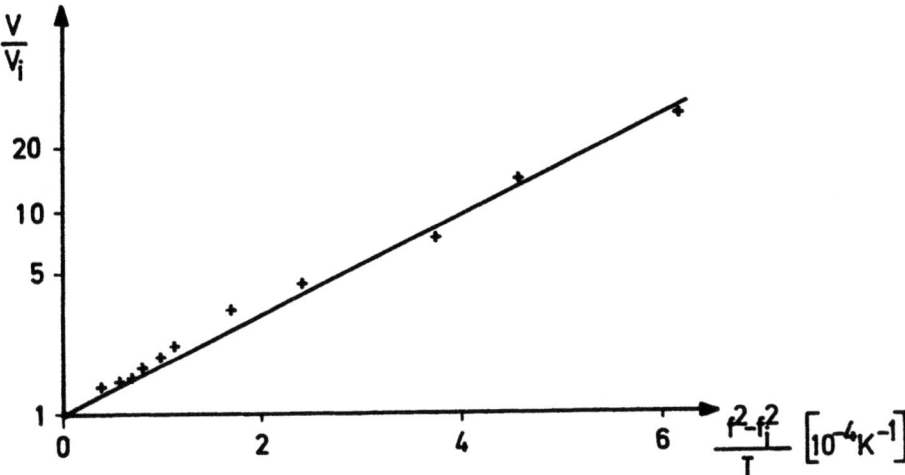

Fig. 6: Dislocation velocity v normalized to its value v_i in the intrinsic material as a function of $(f^2 - f_i^2)/T$. Data are taken from [14]. (f is the occupation ratio of the dislocation in the doped material, f_i in the intrinsic material; temperature dependence of the band gap according to [16]).

lined above, the variation of q near the interface is coupled with a change in the double kink formation rate and thereby with a change of the dislocation mobility in the surface region. The width of the region in which the dislocation velocity is affected depends on the width of the space charge region at the interface and on the mean free path of kink motion along the dislocation line.

Quantitative estimates of the expected surface influence on the dislocation mobility are possible at present only for Si and Ge, for which the electronic levels at dislocations have been thoroughly investigated. But these materials are brittle at room temperature. Possible candidates for materials which allow electrical measurements of the dislocation velocity at room temperature are some II-VI-compounds.

ACKNOWLEDGEMENTS

The authors are indebted to Christa Meier and Joachim Böttger for their help in the preparation of the manuscript.

REFERENCES

1. Labusch, R. and Schröter, W., Inst. Phys. Conf. Ser. 23, 56 (1975)
2. Whitworth, R. W., Adv. in Physics 24, 237 (1975)
3. Shockley, W., Phys. Rev. 91, 228 (1953)
4. Chadi, D.J. and Cohen, M.L., phys. stat. sol. (b) 68, 405 (1975)
5. Pantelides, S.T. and Harrison, W.A., Phys. Rev. B 11, 3006 (1975)
6. Labusch, R. and Schröter, W., Dislocations, Collective Treatise, ed. Nabarro. F.N.R. 1975
7. Schröter, W., phys. stat. sol. 21, 211 (1967)
8. Labusch, R. and Schettler, R., phys. stat. sol. (a) 9, 455 (1972)
9. Wagner, R. and Haasen, P., Inst. Phys. Conf. Ser. 23, 387 (1975)
10. Alexander, H. and Haasen, P., Solid State Physics 22, 27 (1968)
11. Schaumburg, H., Phil. Mag. 25, 1429 (1972)
12. George, A., Escaravage, C., Champier, G. and Schröter, W., phys. stat. sol. (b) 53 483 (1972)
13. Patel, J.R. and Chaudhuri, A.R., Phys. Rev. 143, 601 (1966)
14. Erofev, V.N., nikitenko, V.I. and Osvenskii, V.B., phys. stat. sol. 35, 79 (1969)
15. Haasen, P., phys. stat. sol. (a) 28 145 (1975)
16. Varshni, Y.P., Physica 34, 149 (1967)
17. Harten, H.U., Festkörperprobleme III, 81 (1964)
18. Schulte, H.-D., to be published

DISCUSSION

Comment by F.R.N. Nabarro:

Can you comment on the close correlation between the uncharged state of the dislocation and the zero of the ζ-potential?

Reply:

I do not see a direct correlation. As far as I have understood, the condition $\zeta = 0$ implies a flat-band situation. But even in this case one would need a direct correlation between the ζ-potential and the Fermi level to derive the line charge of the dislocation.

Comment by J. J. Mills:

I would like to caution against equating the flat-band condition with the $\zeta = 0$ since the ζ-potential measures a potential at some distance into the liquid and does not necessarily imply zero surface change.

Workshop Summary: Chemisorption-Induced Variations in the
                   Plasticity and Fracture of Non-metals

Prepared by W. M. Mularie

Discussion Leaders: M. W. Mularie, F.R.N. Nabarro, F. C. Frank,
                     H. C. Gatos

Recorders: M. V. Swain, J. Moskovitz, J. H. VanderMerwe,
                     H. Viefhaus

PERSPECTIVE

Professor Macmillan, in his survey lecture, has reviewed the plethora of experimental evidence for the existance of environmentally-induced variations in the mechanical behavior of nonmetals. Accordingly, a basic tenet of this workshop is that the reality of these phenomena, at least qualitatively, should be accepted. Further, Macmillan has reviewed various mechanisms postulated to explain these phenomena. However, in his view, evidence favors the existence of an electrostatic interaction which is generic to these phenomena, specifically, chemisorption - induced changes in the electronic structure of the near-surface dislocations and/or point defects. This electrostatic interaction is responsible for variations in dislocation mobility; hence, surficial flow and fracture processes. To aid in examination of this view Professor Schröter has reviewed the present model for the energy states of dislocations within the framework of the band structure of silicon. Unanimity on the electrostatic interpretation was not expected or presumed. Rather, a critical analysis of some aspect of this hypothesis was suggested as a discussion topic for the four concurrent sessions.

The following workshop summaries are presented principally in the session chairman's narrative. In these narratives, the greatest concern was to accurately portray the major issues which, unfortunately, sometimes masked individual efforts within the group process.

I. SUMMARIES

   Group A: Leader: W. M. Mularie
             Recorder: M. V. Swain

The Models:

The observed phenomena place constraints upon any model we propose. Some of the behavior the model must yield are:

1) Short Time Constant: Drilling experiments of Westwood and co-workers suggests the phenomena in particular cases occurs in $<10^{-5}$ seconds.

2) Operable in a range of solids whose properties vary from plastic (LiF) to brittle (Al_2O_3), from semiconducting to insulating (including amorphous silicates).

3) Exhibit "symmetry" in mechanical properties about a zeta potential of zero.

4) Be consistent with related phenomena such as photoplastic and electromechanical effects.

5) Operable at room temperature.

T. E. Fischer began the session by proposing a conceptual model for rationalizing charge-exchange models in insulators, where the free electron density is small and corresponding electron charge equilibration times many of orders of magnitude larger than those observed above. He stated: "Chemisorption of fluids which yield a non-zero zeta potential modifies the movement of dislocation under stress over considerable distances (microns). It is concluded that this is not a chemical effect on the surface but a result of the electric field induced by the adsorbed ions. In semiconductors, these fields can bring about a modification of the state of charge of these dislocations. In insulators, the mobile electrons and holes necessary for the space charge as well as for charge exchange with dislocations do not exist. An alternate way for expressing the same fact is to say that the energy splitting between the localized states in the dislocations and the valence band is so large that charge transfers would occur very slowly. The following model describes forces exerted on a dislocation by an electric field in an insulator in the case where no charge transfer between the perfect crystal and the dislocation occurs but electrical conductivity along the dislocation is possible. (See Figure). Depending on the kind of solid, one can imagine several mechanisms of conductivity along a dislocation. Electronic conduction can occur through the electronic states in the band gap that are associated with the defects (e.g., dangling bonds) in the dislocation, as described by Dr. Schröter. Such conductivity has, in fact, been observed in CdS. In many insulators, however, one would expect that the tight binding of electrons would not give the electrons sufficient mobility. In this case, ionic conduction can be envisaged. Pipe diffusion along dislocation is a well known fact. Ionic conduction is then no more than the drift (or asymmetric diffusion) of ions along the dislocations. Consider then an insulator with an electric field (that can result from ionic adsorption, but not uniquely) and a conducting dislocation. If no charge exchange with the matrix

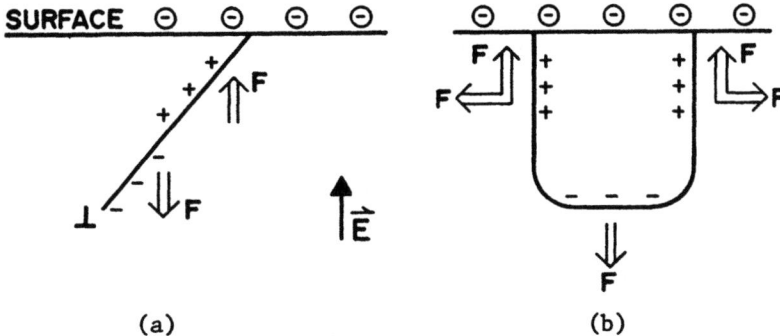

Figure 1 Schematic showing (a) Coulomb force along a dislocation and (b) possible Coulomb repulsion of two charged branches.

is possible, conductivity along the dislocation will cause local charging. The electric field then exerts a force on the core of the dislocation. This force can help it to overcome obstacles, thereby reducing the effectiveness of pinning and increases the dislocation mobility. Illumination of the sample can induce photoconductivity and increase the softening of the material. Notice that in the space charge (i.e., semiconductor) model, illumination causes a decrease of band bending, (i.e., a situation more similar to low zeta potential) and, therefore, a hardening of the sample. It should be noticed also that the conductivity model provides for increased dislocation mobility by an applied field, independent of its sign."

J. J. Mills proposed an alternative model to account for the effect of chemisorption upon the dislocation mobilities in non-metals. The electronic structure of all materials have, to varying degrees, bonding and anti-bonding energy states due to hybridization of atomic orbitals of the same symmetry but from different lattice sites. Transfer of electrons or holes from donor or acceptor adsorbate molecules to these antibonding or bonding states thus causes a local decrease in the binding energy of the lattice sites: a local decohesion of the surface atoms regardless of the sign of the charge transferred. If one presumes that this local surface decohesion facilitates the throwing of the first kink formed by the dislocation under stress, the mobility of these dislocations is increased in a manner symmetrical about the point of zero surface charge. The model has the great advantages of simplicity and generality since it can be applied to all materials. In wide band gap ionic materials, it is necessary to invoke the

Levine and Mark model for intrinsic surface states a) to increase
the covalent component of the binding energy, and b) to bring
the bonding and anti-bonding states closer to the Fermi level.

Discussion:

J. P. Hirth summarized his and other contributions as follows:
"Regarding Fischer's suggestion of charged tubes of dislocations
for ionic insulators, it is most likely, based on both theory and
experiment, that the charges on the dislocations are in localized
states corresponding to kinks, jogs, pipe vacancies or pipe
interstitials. Even in the absence of a surface field, a local
surface charge at the point of emergence of the dislocation
should develop with a compensating exponential distribution of
charge along the dislocation. Imposed surface charge effects
would superpose on this intrinsic effect (incidentally suggesting
that the zero dislocation charge potential and hence the potential
of minimum dislocation mobility could be slightly displaced from
the zero zeta potential). Importantly, however, the charged
defects on the dislocation should all <u>decrease</u> its mobility. The
vacancies, interstitials and jogs would all lead to strong local
pinning as well as longer range electrostatic interaction. The
charged kinks would in many instances be of opposite sign and
hence increase the activation energy for the Peierls - double
kink process, in direct contrast to the intrinsic semi-conductor
case discussed by Schröter.

Hence, a crucial issue in resolving the chemo-mechanical
effect for large band-gap, ionic materials is deciding whether the
response time for formation of charged defects is sufficiently
short for the above effects to occur. It is well-known that
activation energies for pipe self-diffusion are roughly 0.3 to
0.5 of those for bulk diffusion. Hence, for many halides, the
dislocation diffusion activation energies are about Q = 0.5 to
1eV. With Q = 0.7eV the diffusion mean free path at room temperature must be about $\bar{x} \simeq 10^{-5}$ cm. With the Einstein relation
this gives the relaxation time as:
$$t = \bar{x}^2/D = 10^{-10} \exp(Q/kT) \simeq 10^2 \text{ seconds.}$$
Hence, the preexponential factor for diffusion is taken as 1 cm^2/
sec. This value would be appropriate for formation of the intrinsic space charge layer. More importantly, one can consider
the situation under an imposed potential gradient of the order
of $\partial V/\partial x = 10^5$ V/cm, again for a requisite penetration distance
of $\bar{x} = 10^{-5}$ cm. Here the velocity is given by the drift equation:
$$v = \frac{D}{kT} \frac{\partial V}{\partial x} = 4 \times 10^{-4} \text{ cm/sec.}$$
or t = 0.2 sec. For t = 10^{-3} sec., one would require Q = 0.5 eV;
for t = 10^{-6} sec. one would require Q = 0.3 eV. Thus, depending
on the pertinent relaxation time of the experiment, these would
be the lower limiting values of Q for the ionic defects to charge

the dislocation.

It was noted that an important feature of Fischer's model was that the adsorption of the active specie need not occur on a dislocation site, rather any charge exchange with, e.g., point defects in the surface region will influence the potential gradient in the space charge layer hence the electrostatic interaction with the dislocation. Unfortunately time did not allow careful consideration of the Mill's model although it was questioned whether the magnitude of the splitting of the intrinsic surface states from the Bloch states would be great enough to allow any interaction with point defect or dislocation states.

Group B: Leader: F.R.N. Nabarro
        Recorder: J. Moskovitz

It was suggested that the observed phenomena might arise from one of three distinct classes of interactions: electrostatic interactions between dislocation and point defects, interactions affecting the energy of surface steps (what might be called the true Rebinder effect) and, interactions involving the elastic moduli and lattice parameter near the surface. Duke has given us evidence that changes in lattice parameter of the surface of a clean metal were extremely small but it was by no means clear that these changes would also be small in non-metals or in metals with a highly-charged adsorbed layer.

It was questioned how fully we have separated effects which were due to adsorption from those which might be due to dissolution. Among adsorption effects were the possible injection of holes or vacancies. It was pointed out that Westwood had done one very nice experiment on silver chloride in solutions which were or were not pre-saturated with silver chloride charged complexes and which were or were not contained in large vessels and vigorously stirred.

Questions were raised as to our ability to distinguish between strong chemisorption, that is to say chemisorption with high energies of adsorption and chemisorption which produced large electrical effects.

There was considerable discussion whether there was any real difference between the Rebinder type of the surface energy effects and other effects. Nabarro suggested that experiments of the kind in which each dislocation moves along the surface without exposing a new ledge would distinguish clearly between them. But Staehle and Neumann pointed out that this was an idealization of the true situation. The surface was normally undergoing dissolution and would therefore have ledges on it and indeed in many cases mobile ledges. Even if it were a perfect surface it could very well be reconstructed so that the passage of an edge dislocation would introduce a stacking fault into the reconstruction order.

Group C: Leader: F. C. Frank

Recorder: J. H. VanderMerwe

The group concentrated its discussion on space-charge effects on dislocation mobility, dealing separately with the cases of electronic and ionic semiconductors.

There was general acceptance of the electronic theory interpretation of the effect of doping on dislocation mobility in germanium, as presented by Schröter, using Haasen's idea that electric charge (of either sign) on the dislocation line tends to destabilize its straight configuration and thus lower the activation energy for double kink formation. This appeared to explain the observations in germanium in a satisfactory way, at least for the 60° dislocations. George reported to the group that his results on silicon, for n-type doping, were quite similar to those in germanium, but the results with p-type doping were rather different: this did reduce mobility but at most by a factor of two. The unexpected result had been that the enhancement of mobility by n-type doping was virtually the same for screw as for the 60° dislocation.

It was noted that when doping is used to control the position of the Fermi level relative to electronic energy bands, complicating effects of the "solution-hardening" type are inevitable: the pure effects of changing the charge state on the dislocation lines would be better seen in the near-surface space-charge layers controlled by electrolytic environments or applied electric potentials.

It was thought that just the same principles should have application to ionic semiconductors. In this case the dislocation charge is carried as an unequal number of positively and negatively charged jogs, in equilibrium with the local concentrations of charged point defects (e.g., cation and anion vacancies) which are disturbed from their distant values to form a cylindrical space-charge, screening the charge on the dislocation line. But the same principle that the charge density on the dislocation line, whether positive or negative, should promote the formation of glide double kinks ought to apply, leading to the same possibilities of enhanced dislocation mobility in a near-surface space-charge layer controlled by the environment or by applied electric fields.

There was some discussion of the effect of charge on the dislocation on dislocation-dislocation interactions, it being noted that since these electrostatic interactions have the same 1/r dependence as the elastic interactions (with less complication from angle-dependent factors) there was no great change in principle except that whereas the dislocation cannot change its Burgers vector, it can, and will, change its charge density.

Group D: Leader: H. C. Gatos

Recorder: H. Viefhaus

The results of several experimental groups discussed in this conference indicate that ambient species (e.g., water and polar organic molecules) increase the plasticity (decrease the microhardness) of non-metals (e.g., MgO and LiF)--the velocity of dislocations, introduced by the diamond indentor, increases in the presence of these species. The microhardness of non-metallic materials, such as those mentioned, reaches a maximum when the zeta-potential in the solutions, with which they are in contact, becomes zero (zero-point of charge, zpc, $\zeta = 0$).

Since these results have been obtained in numerous experiments and under broadly varying conditions, their validity, at least in a qualitative frame-work, must be accepted. An understanding of the observed effects, however, on the basis of quantitative or semiquantitative models, has not been achieved. In this workshop an attempt was made to examine these effects on the basis of reasoning stemming from certain assumptions which are not necessarily dictated by, but are consistent with the results.

Assumptions

1) The effects originate at the surface, although plasticity changes extend to significant distances beneath the surface.

2) The zpc, as determined experimentally, corresponds to the flat band condition at the surface of the non-metals.

3) At the zpc the dislocation velocity reaches a minimum (the micro-hardness reaches a maximum). This assumption is dictated by the experimental results.

Reasoning

If the change in the velocity of dislocations is due to charge interactions, between the dislocations and the space charge layer, then:

1. Change in ζ leads to change in q (charge) on the dislocation and thus to changes in the velocity of dislocations.

    (a) Let us suggest symmetry about q = 0.

    (b) If (a) is valid then it becomes necessary to understand why at ζ = 0 we must have q = 0 (at the dislocations and or in the space charge layer).

    (c) If dislocation velocity effects are due to charge interactions, then minimum velocity is expected at q = 0. On the other hand, if dislocation pinning is involved, then the dislocation velocity should be maximum at q = 0, as there will be no attraction of defects to bring about pinning.

    Symmetry about q = 0 becomes difficult to deal with particularly in extrinsic materials such as those under consideration.

2. Let us consider symmetry about

    (a) $\frac{\partial \gamma}{\partial q} = 0$ at ζ = 0, where γ is the surface tension.

    (b) $\frac{\partial (n + p)}{\partial q} = 0$ at ζ = 0, where n and p are the densities of negative and positive charge, respectively.

Consequences:

1. Change in the field must lead to change in the density of charge defects and in turn to either increased pinning (decrease in dislocation velocity) or increased charge interaction (increase in dislocation velocity).

2. Increase in temperature must lead to increased defect density and in turn to increased pinning or increased charge interaction.

3. Changes in surface charges should produce defects throughout the space charge which, in the present materials, can extend through the entire sample.

4. Illumination must decrease charge defect density and decrease dislocation velocity (if charge interactions are involved).

Essential Information

If the assumption of dislocation--defect interactions is correct, it is imperative that concentrations (and nature) of defects be known about $\zeta = 0$.

Reference Data

It is essential that we know that at $\zeta = 0$, flat band conditions exist. The energy of defects involved in the above phenomena can be determined by photovoltage spectroscopy.

Alternate Reasoning

On the basis of energetics it is difficult to explain pronounced changes in plasticity on the basis of the small energy changes associated with chemisorption (and the resulting space charge layer changes). It might be worth considering an amplification or transistor-type effects, i.e., chemisorption modulates the space charge regions which in turn modulates the high-energy processes involved in plasticity. No specific mechanisms for this type of amplification are apparent at present.

II. SUGGESTED EXPERIMENTS:

The need for proper choice of materials was prominent in the discussions. It was suggested that single crystals of the II-VI groups, such as CdTe, constitute a group with generally well-characterized electrical and mechanical properties. "Doping" of ionic solids and observing the concomitant change in dislocation mobility was suggested as a method to distinguish between elastic and electrical solution hardening effects. For example, an easy "doping" experiment would be changing the ion valency of ferrous-doped MgO by oxidation to the ferric state with compensating cation vacancies.

It was questioned whether the indentation techniques were not a rather primitive method of testing. It was emphasized that even a rather shallow indentation may transmit dislocations to depths of a hundred times as deep thus probing the bulk of the crystals. An interesting suggestion to supplement these indentation measurements was to study single crystal II-VI compounds, such as ZnO prisms, in torsion (under various environments) with the basal plane normal to the torsion axis. The resultant shear stresses would be maximum at the surface which would favor dislocation nucleation and motion. It was thought that one should make X-ray topographs of the region of indentation during the course of the time-dependent studies described by Macmillan.

The need to distinguish clearly and devise experiments which separate surface and near surface effects was stressed. Also one must distinguish between dislocation "drag", a velocity dependent effect and "pinning", a static effect. It was suggested that the time constant for the chemomechanical phenomena could be obtained from an internal damping type of experiment. This could perhaps determine whether the phenomenon is electronic or ionic in character.

Prominent in all discussions was the need to investigate the correlation between the electrokinetic zeta potential and the true surface potential and accompanying space image layer in the solid. More importantly, what more direct techniques would allow monitoring of the electrostatic profile of the surface region during mechanical measurements? It was suggested that a metal-insulator--semiconductor sandwich could be constructed in which the semiconductor space charge could be easily controlled independent of the mechanical stresses imposed by, for example, bending. It was suggested that photovoltage spectroscopy (see Gatos) would allow a convenient tool for monitoring the "band bend" profile at the surface under changing environments, as well as allow probing of the defect energies of the materials.

Workshop Session 4

ENVIRONMENTALLY INDUCED LOWERING OF SURFACE ENERGY AND THE MECHANICAL BEHAVIOR OF SOLIDS

Chairman: D. J. Duquette

ENVIRONMENTALLY-INDUCED LOWERING OF SURFACE
ENERGY AND THE MECHANICAL BEHAVIOR OF SOLIDS

E. D. Shchukin

Moscow Lomonosov State University, Chemistry Department,
Institute of Physical Chemistry of Academy of Sciences of the
U.S.S.R.

The phenomena connected with the influence of environment on mechanical properties of solids are extremely varied in their final macroscopic manifestations [1-14], often leading to contradictions (mostly only apparent ones) in their interpretation. One of the main problems in this field should be considered to be the separation of the few primary interactions, the combination of which determines the effects observed. This classification may be approached from different points of view. Thus, in terms of physicochemical peculiarities of atomic (or molecular) interaction, it is expedient to emphasize on the one hand, surface interactions(I) localized, according to the thermodynamic conditions at the interfacial boundary, in the molecular layer right near this boundary; outside of areas of critical phenomena, the thickness of such layers does not usually considerably exceed molecular sizes. On the other hand, numerous effects take place (II) which are connected with the layer of finite thickness adjacent to the boundary, this layer noticeably exceeding the monomolecular one and being actually composed of phases. One may include in this second group: (IIa) formation of solid solutions, as well as segregation of particles of a new fine-dispersed phase (i.e., appearance of "defects" in general) in the near-surface layer due to diffusion; (IIb) corrosion processes and dissolving of the near-surface layer, including the Yoffe effect, (IIc) formation of a solid film, continuous or mosaic, on the surface as a result of precipitation from the medium or chemical reaction. As a rule, one may speak in these cases about the phenomena being irreversible or not in equilibrium; removing the environment does not lead to rehabilitation of the initial conditions at the surface of the solid, and irreversible changes embrace a certain finite volume. These and other analogous cases include a very wide scale of phenomena of both increase and decrease in the resistance of the solid to deformation and fracture. In cases (IIa, c), for instance, the restraint of dislocations movement (and, generally, reconstruction and fracture of bonds) in the near-surface layer may be, in most cases, considered to be the primary consequence, but as the process develops and other factors interfere, the ultimate results may be quite different. For example, the surface crack -concentrators may either be healed or developed. The defects in the near-surface layer, including foreign atoms and dispersed particles may serve not only as obstacles to the movement of dislocations, but also as dislocation sources. In case (IIb), in particular, the kind of material removed is important more or less perfect in relation to the bulk of the sample, etc.

We ascribe to group (I) various cases of surface interactions under conditions close to the equilibrium, i.e. under conditions of thermodynamically stable coexistence of the given solid and liquid (or gaseous) phases at the original interfacial boundary [13, 14]. Removing the environment in this case restores the conditions at the surface of the solid to the initial ones. Here we find both the contact with the liquid phase, especially akin in chemical composition and structure, and adsorption from a solution or vapor, in case of both physical adsorption, which will be mainly considered next, and chemosorption. The primary consequence of the reversible adsorption, if the components interact moderately, is, generally speaking, facilitation of deformation and destruction processes connected with the development of new surface -- according to general thermodynamic regularities which predetermine the adsorption of components reducing the work of formation of the new surface. The term Rebinder effect is just the one to cover all the various phenomena of deformation and fracture facilitation, which differs from the other forms of influence of environment with their general (thermodynamic) nature, i.e., decrease in the work of new surface formation as a result of reversible interaction with the medium.

Of course, the groups of phenomena under consideration are not isolated; they may continuously pass from one to another and partly coincide. Thus, the adsorbed monolayer can manifest properties of a solid phase film; on the other hand, it is adsorption which constitutes usually the very first act in other forms of interaction with the environment, too. A characteristic "intermediate" form of interaction is stress corrosion. We may speak here then about a peculiar continuous spectrum of phenomena: from a "purely mechanical" process of deformation and fracture to "purely chemical" (of corrosion, not requiring mechanical work), with a transition through such forms as facilitation of the mechanical process of interatomic bonds, reconstruction and fracture under conditions of reversible interaction with the environment (the Rebinder effect), and acceleration of irreversible chemical processes under conditions of mechanical activation (stress corrosion).

Analysis of surface interactions of the solid under deformation with the environment, these interactions being included in the above mentioned group (I) which is the subject of our consideration and united further on with the term "adsorption interactions", according to Rebinder, allows us to emphasize the following two questions: (1) to what extent the mechanical property we consider is "superficial", that is, connected with the interactions in the surface layer itself, and (2) how one can quantitatively characterize the influence of atoms (molecules) of the environment (to be more exact, of the adsorption layer directly adjacent to the surface of the given solid) on these interactions. (This adsorption layer may be identical in its composition to the bulk of the medium -- e. g., for a one-component liquid phase -- or contain a surplus of an adsorbed component; despite the considerable differences, no absolute border exists here and one case may, generally speaking, adjoin another.)

To begin with, let us consider the strength P_C, i.e., the mechanical property which is determined in the end with interatomic bonds resistant to fracture. We may assert that, at least in absence of appreciable macroscopic residual deformations, the strength P_C is just the surface property directly connected with overcoming the cohesion between atoms which used to belong to the volume and then appear at the surface, i.e., with the work A of new surface formation. The fracture being brittle (or in numerous cases close to brittleness to a greater or smaller extent), the symbatic connection between P_C and A is approximated by the universal Griffith correlation: $P_C \sim A^{1/2}$. Since the conditions under our consideration are more or less close to equilibrium (towards the processes in the apex of the crack), the isothermic work A is determined by the free surface energy σ, and we have $P_C \sim \sigma^{1/2}$. (This correlation is true both for originally considered cavities of elliptic section and for cracks with sharp inner edge, the description of which, introduced by P. A. Rebinder, was later quantitatively developed in detail [15-17], and presumes that σ falls from the maximum constant value in the part of the crack wide enough to zero at the tip. It should be pointed out at once that, assuming some definite thermodynamic conditions of the process -- in this case, the isothermic ones -- we should keep in mind the participation of thermal fluctuations in this process and, consequently, take into account the kinetics of the process. (If the destruction is substantially plastic, or other forms of considerable dissipation of energy take place during the crack propagating, A and σ can no longer be identified but still $A = A(\sigma)$, this dependence remaining symbatic so that a sharp decrease in σ causes an even steeper fall of A.)

In their turn, the adsorption interactions close in character to the reversible ones allow one to describe changes of the ΔA value under the influence of environment as changes of $\Delta \sigma$. Let us emphasize that the latter value in cases of substantial changes of σ (i.e., of positive adsorption), which are interesting to us, is always negative. In essence, it is this result to which one brings the different schemes of macroscopic analysis of strength reduction, namely reduction, due to reversible physicochemical interaction with the environment, independent of details of the micromechanism of the phenomenon considered by one or another investigator. At the same time, we should consider that participation of the medium can essentially change the conditions of the destruction process and its microaspects: foreign atoms take part in overcoming the cohesion forces only inasmuch as the crack configuration, the state of the material in the apex of the crack, and thermal fluctuations in the inner edge part allow the foreign atoms to penetrate there in time. This is probably the most important kinetic condition of manifestation of the effect. (As is known, the effect disappears at high rates of destruction [7, 28].) In other words, the manifestation of adsorption-induced facilitation of deformation and fracture may be connected with thermally-activated processes which would be unnecessary under other conditions.

Let us consider some data characteristic of solids of different structure and different types of atomic interactions, mainly from those obtained in our laboratories at the Institute of Physical Chemistry of the Academy of Sciences of the USSR, and at the Chair of Colloid Chemistry of the Moscow University, where these investigations were being carried out for many years under the general leadership of P. A. Rebinder by V. I. Likhtman, L. A. Kochanova, L. S. Bryukhanova, E. A. Andreeva, Yu. V. Goryunov, N. V. Pertsov, A. V. Pertsov, B. D. Summ, V. Yu. Traskin and their collaborators as active participants.

(a) High-disperse fine-porous structure of magnesium hydroxide obtained by hydration-hardening of magnesium oxide. The strength of the samples is connected with cohesion in bridge contacts between particles, easily accessible to the environment. In desiccators over sulphuric acid solutions of given concentration so that the pressure, p, of water vapor is known, the equilibrium adsorption of water molecules is reached; together with the adsorption values Γ (obtained by weighing) it allows calculation of the value of

$$-\Delta \sigma = \sigma_0 - \sigma = RT \int_0^p \Gamma \, d\ln p$$

Comparison of these $\Delta \sigma$ values with the results of strength measurements displays the linear dependence of the $(P_0^2 - P_c^2)/P_0^2$ value on $\Delta \sigma$ (Fig. 1), that is, the fulfillment of the Griffith formula, the inclination giving a reasonable value of $\sigma_0 \approx 300$ erg/cm^2 in the absence of adsorption. As adsorption reaches the saturated monolayer, the strength falls twofold and the surface energy fourfold. Special NMR experiments have established that phaseous menisci -- water collars -- were absent in these experiments, so the dissolving of the bridge contacts a priori could not take place.

(b) Polycrystalline, optically transparent samples of potassium chloride prepared as a wire about 1 mm in diameter with perfect cohesion at grain boundaries. The tests were carried out in the continuous series of heptane-dioxane and dioxane-water solutions saturated with potassium chloride; the initial strength P_0 in the non-polar medium -- heptane -- is the greatest and it regularly falls with the growth of polarity of the environment (Fig. 2). The transition (with the help of the Griffith correlation $P_c^2 \sim \sigma$) from the "strength isotherm" $P_c(C)$ to the surface energy isotherm $\sigma(C)$, C being concentration, allows us to calculate the limiting adsorption Γ_m and gives the correct value of the area occupied by a molecule ($S_1 \approx 30$ Å^2).

Water, solutions of electrolytes and salt melts turn out to be highly surface-active also for a wide range of crystalline minerals and inorganic glasses.

(c) Cleavage of naphthalene crystals according to Obreimov (Gilman's version), allowed us to estimate the values of σ in air ($\sigma_0 = 60$ erg/cm^2) and in a number of liquid media: from the most polar one-water (where $\sigma = \sigma_0$) to non-polar media most akin to naphthalene, particularly benzene (in which the σ values fall 4-6 times) (Fig. 3). The comparison of these data with the results of strength measurements shows that here the Griffith formula remains approximately true.

(d) The same molecular crystals were used for measurements of strength reduction in aqueous solutions of lower- and medium-normal aliphatic acids and alcohols. It appeared that in this case the Traube-Duclos rule for the homologues of the given row is fulfilled and the transition from $P_c(C)$ to the $\sigma(C)$ isotherm, as described above, gives true values for a square per molecule in the saturated adsorption layer (Fig. 4).

(e) Analogous regularities are found qualitatively for samples of polymer materials (Fig. 5): in the series of normal aliphatic alcohols, the polyethylene (a hydrocarbon) strength falls, while the strength of polymethylmethacrylate decreases the most in the most polar media.

(f) Germanium is an example of a covalent crystal. At elevated temperatures, when the samples in air are plastic enough, their strength and deformation limit decrease sharply in contact with certain liquid metals (gold, copper, gallium; Fig. 6). Calculations in a simple local-coordination approximation, for instance, allows us to estimate σ at the boundary of the crystal with melts in equilibrium to it at various temperatures; comparison of results of mechanical tests with such calculations shows that the Griffith correlation holds true.

Similar phenomena of strength reduction as a result of contact with certain metal melts take place in the cases of graphite and diamond, aluminum oxide ceramics, etc.

(g) For solid metals, the strong surface-active media which determine great decrease in σ and P_c (under conditions of reciprocal saturation of the solid and the liquid phase and presence of a thermodynamically stable interfacial boundary) are various metal melts. There is much data about the possibility of the strength P_c being reduced by many times [3-5, 7, 13, 14, 27-34]. Only a few experimental attempts to estimate σ are known, obviously due to methodical difficulties. (Gilman, Westwood et al. used the method of cleavage of a crystal [35, 36]; Likhtman, Bryukhanova, et al. did the one of zero creep [37].) A number of theoretical calculations of σ at the boundary of a solid metal with the liquid one have been carried out [38, 39], particularly by Yushchenko; although approximate, they still prove quite definitely that σ at such a boundary may fall very considerably, within an order of magnitude or more.

So, the more or less marked reduction of σ of the solid at a boundary with environment may be considered to be a universal thermodynamic condition of a noticeable reduction of strength (within the limits of adsorption interactions). However, we had noted several times that this condition is necessary but not sufficient [13, 14, 40]. Manifestation of the effect requires several other conditions to be fulfilled; it is expedient to consider them from the point of view of questions (1) and (2), formulated above. Thus, the level of mechanical stresses and the character of the stressed state (presence of "rigid" components) plays an essential role in these reversible interactions [1, 2, 41, 42]. The effect does not occur below the temperature of solidification of the environment, but often disappears also with increase in temperature, when plasticity of the given solid sharply grows (i.e. changing of the form of the body due to reconstruction of bonds in the bulk of the sample takes place more easily due to their rupture and formation of new surface] [3-7, 30, 43]. The decisive kinetic factor is the supply of the liquid phase, conditions of its flow, and the possibility of a quick (thermo-fluctuational) process of penetration of the adsorption-active atoms into the zone, overcoming the cohesion forces, and reconstruction of bonds in the apex of the crack [44-46]. Along with this, it is usually necessary that the defects of the structure be present and be able to be the nuclei of fracture. This was confirmed also with direct observations, the surface of the body being pricked (Westwood and Kamdar [33, 36]; Denshchikova, Polukarova et al.[47]). Such experiments confirm the concept developed earlier by Likhtman and the author about two stages of fracture (gradual growth of the crack and loss of its stability), and about the role of decrease in σ in both stages.

As regards the micro-aspect of the phenomena under consideration, it should be emphasized that a definite correlation of the cohesion energy of atoms of the medium U_{AA}, of the solid U_{BB}, and of heterogeneous atoms U_{AB} corresponds to this group of interactions, which are close to the reversible ones and occur at the thermodynamically stable interfacial boundary. A large decrease in A -- the work of new surface formation -- should be observed when atoms of the medium most perfectly co-build the solid, providing the compensation for the bonds ruptured at the surface. It is reached in particular at the boundary with the related liquid phase, when the absolute values of the mixing energy are small ($U_0 = U_{AB} - 1/2 (U_{AA} + U_{BB})$); binary diagrams in this case often belong to the simple eutectic type (the Pertsov-Rebinder rule)[39, 40, 49, 50]. Large positive values of U_0 correspond to the absence of interaction, while the large negative values may mean the absence of the equilibrium interfacial boundary (or, for instance, loss of necessary migrational mobility by the atoms of the medium). At the same time, "moderate" chemical interactions (first of all the chemosorptional ones) which approximately provide compensation for the work of formation of new surface (i.e., of the order σ_0 erg γ - related to a unit of the surface)[51], even being irreversible, may also be quite reasonably included in this scheme of consideration, since the reaction is limited to the surface layer.

We have certain grounds here for the following approach. Though under conditions of irreversible chemical interactions at the boundary, we can no longer speak about the thermodynamic value of $\Delta\sigma$, the concept of reduction of ΔA -- the work of new surface formation -- as a result of interaction with the environment is still valid, since this work is present and has a certain value, and the surface being formed is stable enough; so, finally, we may speak in the same terms as above about the contribution of interaction with the medium to the reduction of strength ΔP. Thus, a continuous transition appears from strength reduction under the conditions of physical adsorption and chemosorption to phenomena of stress corrosion type [52, 53] -- as soon as there is a basic act of reconstruction or rupture of bonds as a result of mutual contribution of mechanical stresses and interaction with the atoms of the environment (assuming the participation of thermal fluctuations).

We cannot be concerned here with peculiarities of adsorption at interfacial boundaries in the solid, particularly at grain boundaries. However, it is essential to note that, under conditions of adsorption facilitation of destruction, such specific factors as: superfluous surface energy of the boundaries, i.e., the thermodynamically predetermined "weakening"; possibility of very quick migration of active atoms; preliminary (prior to the application of stress) enriching of the boundaries with them; interaction of these active atoms with admixtures adsorbed at the boundary earlier, are manifested along grain boundaries. Along with this, the possibility should be considered of adsorption activity at the grain boundaries in absence of it at the free surface; for instance, when atoms of a refractory admixture are bound strongly to atoms of the given solid, the grain boundaries appear to be strengthened [54, 55].

Among present attempts at quantitative studies of the micromechanism of strength reduction, there are not a large number adequately applicable to the atomic-molecular level.

Favorable opportunities are given here, though, by numerical dynamic-type computer experiments (Yushchenko, Grivtsov, Shchukin [56, 57]). A two-dimensional crystal (with a cavity) was chosen as the first model, with the atomic interaction "6-12" potential with the parameters of argon. The system was enclosed in a box and was characterized by a definite initial temperature (the statistically distributed impulses of atoms were given), the atoms forming a regular hexagonal lattice (Fig. 7a). The two walls could be moved apart (the stronger bonds of atoms with them were given), thus carrying out deformation at the required rate. The program provided producing coordinates and impulses of all the atoms in definite time intervals, and also values of the total potential and kinetic energies of the system and of stress from the side of the wall.

At low temperatures brittle fracture was observed (Fig. 7b): a crack emerged from the cavity-concentrator. As the temperature increased, the picture changed, and various defects were found to appear, up to severe local destruction of the lattice (Fig. 7c); the appearance of "incomplete atomic rows" (dislocations) and their movement -- i.e., transition to plastic deformation could be observed.

Suppose we introduce into the cavity adsorption-active atoms also interacting according to "6-12" law, but with smaller atomic radius belonging to some imaginary component with a lower melting point, with small positive value of the mixing energy U_0. Now, even at elevated temperatures (Fig. 8), the development of cracks may be observed, and in a number of cases quick migration of foreign atoms into the pre-destruction zone in the apex of the crack is distinctly seen. We should note the extreme variety of pictures observed, predetermined by the out-of-control details of the initial conditions. In some cases the active atoms fill the crack, helping it to grow; in others they just facilitate its nucleation (in this place the atoms may "mix" as if forming a drop of eutectic), and afterwards the crack develops with no active atoms penetrating into it, and so on. These various local acts of rupture and reconstruction of bonds in such combinations with a small number of particles are likely to be the ones to produce the macroscopic picture averaging them -- from extreme brittleness to high plasticity.

As a result, one can vividly imagine the picture of elementary acts of destruction in the surface-active environment in the following way. If the related <u>liquid phase</u> A "manages in time" to penetrate the pre-destruction zone of the material B in the apex of the crack, the facilitation of destruction is then, in essence, a transition to the other mechanism which is easier and more possible under these conditions, namely, it is a substitution of the act of rupture of bonds B-B for the act of their reconstruction, thermal fluctuations participating and assisting the atoms of the medium to get placed between the atoms of the solid phase (schematically B-B → B-A-B → B-A-A-B and so on); the reconstruction of bonds in the liquid phase takes place by means of a shear mechanism and does not require large stresses (A-A'-A" → A'-A-A", and so on); this is the essence of what we see macroscopically as the substitution of the boundary of the solid with vacuum (or vapor) for the boundary with the related liquid phase in the destruction process. Compensation of bonds at such a boundary is incomplete so that some work is needed to provide the process.

If under the given conditions there is no related liquid phase in the crack, and only an adsorption layer is formed (saturated or not), the situation may be noticeably different. In this case, the bonds must rupture, and it can be essentially facilitated only by being substituted for the rupture of bonds between atoms of the medium. This is possible provided that, first, the bonds B-B are relatively easily substituted for the A-B bonds (i.e., the A-B interaction is of the same order as B-B), and, second, the A-A interaction is considerably weaker than B-B. The above considered example with water adsorption on magnesium hydroxide corresponds to this case. The A-B bond should not be too strong; otherwise the A atoms lose the necessary migrational mobility and the effect of destruction facilitation would not take place.

It is hoped that these quantitative experiments will allow us to study the migration process of active atoms in the stressed and distorted lattice (using different types of interatomic potentials), and also to model the interactions of dislocations with the free surface and influence of adsorption on these interactions.

As regards the analysis of effect of environment (in terms of reversible physico-chemical interactions under consideration) on <u>plastic flow</u>, we should, following the chosen scheme, specify the circumstances under which the given mechanical property -- plasticity becomes "superficial". (We are speaking here mostly about such characteristics of the plastic flow as the yield point, work hardening, the creep rate, etc., though the limit deformation before fracture also should not be neglected.) We shall be interested in cases where the medium <u>helps</u> the reconstruction of bonds with its primary effect and, therefore, where the initial role of the surface amounts to deformation of various <u>obstacles</u> to the plastic deformation.

Unlike the destruction process where the long-range forces in the lattice are not, generally speaking, a necessary factor, they should be involved in the consideration of the dislocation plasticity: the surface is interesting to us as soon as it interferes with the deformation process by hindering the movement, nucleation and multiplication of dislocations, since this influence, which is due to the long-range action of elastic fields of dislocations, is peculiar to the whole of the near-surface layer up to a depth of at least the order of average dislocation segment size or even appreciably greater. It is this circumstance that determines to what extent plasticity appears to be the "superficial" property: namely, to the extent the near-surface layer of the indicated thickness (depth) determines the essential part of the total resistance of the sample to the applied stress. This takes place in many cases, not only with thin samples, but also with massive bodies under conditions of contact interactions, i.e., in friction joints, in cutting and pressing treatment processes.

(Generally speaking, strength-reducing influence may also be the initial influence of the surface. For instance, various defects peculiar to the near-surface layer of a real crystal may serve as dislocation sources, thus imparting an unusually high plasticity to this layer [58,59]. The edge of the elastic field of the dislocation near the surface determines "the mirror image force" with which the surface attracts the dislocation, helping it to leave the crystal [60,61]; this in turn, allows us to consider the near-surface dislocation segments as if they were of double length, and as if their starting stress as dislocation sources were decreased twofold, respectively, etc. However, right now, since we are analyzing the role of adsorption, we are interested in the first place in the initial strength-increasing effect of the surface.)

In many cases obstacles to the movement of near-surface dislocations from the side of the surface are caused by the presence of an outer solid film (e.g., oxide film) which can, according to the correlation of elastic constants, push dislocations into the bulk of the solid (the long-range action inverse in sign to the "mirror image force" pushing dislocations out) and, at the same time, hinder the displacement of points of emergence of dislocations along the surface, especially if a component of the Burgers vector is present perpendicular to the surface when a step of new surface should appear (short-range action, determining the fixing of points of emergence of dislocations). Initial influence of environment may consist here just of adsorption facilitation of destruction of the film, and then everything said above about the reversible influence of environment on strength is applicable. Of course, the consequences may be very different depending on whether the cracks in the film would appear to be concentrators or not, whether involvement of the near-surface layer in plastic flow would assist general plastic deformation or, vice versa, nucleation of cracks, whether the medium which is adsorption-active in respect to the film is the same in respect to the solid itself, etc. In the end, both resistance to deformation and its maximum value may decrease as well as increase, the initial act of influence of the medium being one and the same. Of course, everything said is true in the cases when there is no continuous film, but islets-obstacles formed with some impurities are present at the surface.

As regards preliminary enrichment of the near-surface layer with various impurities so that solid solution or a new disperse phase is formed, the plasticizing influence following may be connected with dissolving of this layer which contains obstacles (a case we do not consider here); of course, diffusion penetration of the medium, leading to facilitation of destruction of such under-the-surface obstacles may also be assumed, but just as a special phenomenon.

Moreover, even an increase in resistance to plastic flow under conditions of equilibrium reversible adsorption may be assumed as a complicated secondary effect in the case, for instance, when defects of the near-surface layer, determining certain fields of elastic stresses, serve as dislocation sources, while adsorption causes relaxation of these stresses. (Relaxation of this kind under conditions of chemosorption had been observed for micro-stresses in the disperse structure of magnesium oxide catalyst [62].)

The next question concerns the direct influence of clear free surface on behavior of dislocations in the near-surface layer and, primarily, the resistance it can offer to movement and nucleation of dislocations. In cases where the long-range elastic field of the dislocation is essential, the quantitative solution becomes complicated; however, in the limit case where the long-range action does not play a direct role -- for a screw dislocation normal to the surface -- the obvious potential barrier is determined by the necessity to form new cells of the surface as it is drawn ("ripped up") with the end of the dislocation. In this case, the reduction of work of formation of new surface A, i.e., reduction of free surface energy σ at the boundary with the medium, should assist the movement of dislocations at least as a primary act.

It was this very approach we had used once to explain the experiments of Rebinder, Venstrem, Taubman, Likhtman et al., in which plastic flow facilitation had been observed for metal monocrystals due to adsorption of higher alcohols and acids from a non-polar environment: decrease in the yield point and work-hardening coefficient, increase in creep rate [1, 7, 63, 64]. An analogous effect is produced also by electric polarization of samples in solutions of electrolytes (according to the Lippman equation of the electrocapillary curve $\partial \sigma / \partial \phi = -e_s$).

Later, Kochanova and the author [65] reproduced these experiments on lead single crystals 0.5 mm in diameter, trying to exclude possible extraneous influences, and compared the acceleration of creep as the potential deviated from the point of zero charge (in the cathode part so that any films were certainly absent) with the reduction of creep rate in experiments in air of samples with a surface oxide or metal polycrystalline film. In both cases the maximum effect -- approximately two times (under the conditions of maximum sensitivity of samples -- near the yield point, Fig. 9). It was explained on the basis of the concept mentioned above, taking into account the specificity of thin annealed samples with their gross dislocation structure, when the whole volume starts to "feel" the presence of the surface.

To make it definite, let a metal single crystal 2r in diameter and "l" long flow with the rate $\dot{\varepsilon}$ under the action of a tensile stress "P". Here, the work spent on development of the new surface is determined by the rate of growth of the surface $\dot{s}$ and the value of the free surface energy σ; the work per second would be

$$W_1 = \dot{s}\sigma = 1/2 \, s\dot{\varepsilon}\sigma = 1/2 \cdot 2\pi r l \dot{\varepsilon} \sigma$$

(the factor 1/2 appears since the volume of the sample should remain constant). At the same time, the stress accomplishes work $P\pi r^2 l \dot{\varepsilon}$ at the crystal. If we assume this energy is evenly distributed in the whole volume of the crystal, we might evaluate the part of this energy in the near-surface layer "b" thick ("b" is the translation constant of the lattice): this part is determined by the ratio $2\pi r b / \pi r^2$ and is equal to

$$W_2 = P\pi r^2 l \dot{\varepsilon} \, (2\pi r b / \pi r^2) = 2\pi r l \dot{\varepsilon} P b.$$

We find from these two correlations that

$$W_1 / W_2 = \sigma / 2 P b.$$

To within a factor it is the ratio of two characteristic forces: the force per unit length dislocation and the force per unit length surface. If σ = 500 erg/cm^2, b = 3.5 x 10^{-8}cm and P = 350 G/mm^2 (the data refer to the experiments with lead single crystals mentioned), we obtain W_1/W_2 = 200.

A question arising from this result is where should the energy necessary for the new surface formation be taken from. The only possible answer is as follows: at the given level of stress the energy imparted to the deformed crystal by the external force should be totally "pumped" into the superficial layer from a near-surface layer at least 200 b thick, i.e., about 0.1μ thick. Moreover, if we consider that under actual circumstances determined by the structure of the near-surface layer, the latter gives to the surface just a certain, maybe a small part of the reserved energy, it can be presumed that the depth of such a layer which provides the surface with energy reaches several μ or even tens of μ. In other words, under certain conditions, the surface potential barrier is able to influence considerably the dynamics of dislocations in a wide near-surface layer. Concepts on the nature of the given barrier were first formulated by the author in connection with the analysis of the mechanism of plasticization by surface-active media in 1958 [3,7,66]. In the simplest case it is related to the formation of new cells of the surface by $\approx$ b^2 as the end of the dislocation draws ("rips up") the surface of the crystal; the corresponding elementary work is then $\approx \sigma$b^2.

In the indicated experiments with lead single crystals (2r = 0.05 cm, l = 1.5 cm), facilitation of plastic flow under the conditions of electric polarization (and, likewise, adsorption of surface-active substances) and hindering the flow in presence of a solid film corresponds with a shift of the yield point 10-20% to smaller or greater values, respectively. This change of the yield point of the samples (namely the samples, not the material) may be interpreted as a 10-20% change of the stress acting on the whole inner part of the sample (where the surface is not directly "felt") due to a considerable change of the stress field where it is determined by the surface barrier, i.e., in a ring-like near-surface layer making up the same 10-20% of the section of the sample, or 5-10% of its radius, i.e. about 10 μ. This value corresponds quite well with the ideas on the energy balance at the surface and in the near-surface layer, which were given above.

As a mechanism by means of which the near-surface layer carries out the work connected with formation of new surface, we had offered, for instance, the following scheme. The applied stress presses at a dislocation segment of length L (in our case, on the order of μ), which is fastened with one end somewhere inside the crystal and coming out to the surface with the other. Displacing this segment (or increasing its length), the stress carries out work in the near-surface layer. Then the end of the dislocation stressed with the force $\approx Gb^2$ (this corresponds to $\approx Gb^2$ surplus free energy per unit length dislocation) "rips up" the surface, carrying out work of the order of $\approx Gb^3$ at one elementary region. In other words, liquidation of a monoatomic piece of the dislocation leads to emission of energy Gb^3, which can be used for formation of an elementary cell of the surface with its characteristic energy $b^2\sigma$. (However, this is just an evaluation using the greatest values possible; a strictor calculation requires taking into account dissipation of energy by the whole segment.) In the absence of great reduction of free surface energy, these values Gb^3 and $b^2\sigma$ characterize one and the same parameter -- i.e., the energy necessary for the rupture of interatomic bonds -- solely according to their physical sense, and hence they have values equal to about 10^{-12} erg. But under conditions of more or less considerable reduction of σ this correlation may essentially change, which manifests in facilitation of overcoming the potential barrier. Vice versa, strong surface film offers an inverse effect.

A question whether a dislocation segment provides a necessary component of the force along the surface may be answered positively if this segment is able to "curve" greatly enough under the given conditions -- i.e., if the correlation $L \approx Gb/\tau$, where $\tau = P/2$ is the shear stress, is fulfilled. For the mentioned experiments with lead single crystals, such evaluation leads, naturally, to values of the order of a μ.

The whole of such experiments well convinces in correctness of thermodynamical macroscopic concepts about the potential barrier connected with the development of the free surface, which should be overcome in the course of plastic flow. Along with this, specification of micromechanism of the phenomenon requires direct observations of dislocation behavior [10-12, 67, 68, 69, 73]. A number of interesting attempts were made in this field. Unfortunately, not in all the cases physicochemical conditions had been definite enough to draw any final conclusions.

Savenko and Kochanova in the works carried out in our laboratory [68, 70-72] applied special efforts to exclude all possible by-influences from the method. Behavior of dislocations was compared after indentation of cube facets of LiF and NaCl crystals. In the first case the surface is relatively hydrophobic -- i.e., water vapor adsorption reduces σ no more than few tens per cent; in the second case, hydrophility is strongly pronounced, so that even in usual moist air adsorption of water vapor causes σ to decrease several times. Distances covered by the leading dislocations (lengths of rays of dislocation rosettes) were measured under conditions of water vapor adsorption and when the surface of the crystals had been protected, cleaving out the samples and indentation being carried out in heptane dried with zeolyte.

Comparison of the results (Table 1) shows that in NaCl crystals distances covered by screw dislocations l_s increase about 20% as a result of water adsorption, while in LiF crystals they almost do not change. Thanks to a great number of measurements, the error here did not exceed 1-2%. The effect may seem small, but such is not the case. The depth of bedding of dislocations (i.e., the depth of the near-surface layer which appeared to be included into interaction with the surface) is large, amounting to μ and tens of μ, and there are "obstacles" in this volume which are not influenced by the surface. In other words, the conditions of the experiment are far from separation of elementary acts of interaction of dislocations directly with the surface in the points of their emergence. At the same time in "standard" experiments in dried heptane, full absence of water molecules cannot be guaranteed. Still, even in such "semi-macroscopic" experiments, facilitation of dislocations movement near the surface, the free surface energy being reduced, is found reliably and reproducibly.

In experiments with the alkali halide crystals described, this effect was distinctly observed only for screw dislocations. It is apparently determined with a well-known circumstance which is due to the slip systems geometry: screw dislocations drawing along a facet of the cube form a step of new surface while the edge dislocations do not. In the first case one may speak about rupture of bonds, and in the second, about their reconstruction. In both cases a certain barrier is to be overcome. However, for the NaCl crystals which are appreciably plastic, the first one should be considerably higher. Quite convincing in this respect is the ratio $\theta = l_e/l_s$ of the maximum distances covered by screw dislocations l_s: under small loads while indenting, when the relative role of the surface (as compared with the influence of the bulk) grows, θ appreciably increases as the load decreases; in water vapor when the surface barrier is reduced due to adsorption (and its role decreases) such growth of θ is found under loads considerably smaller than in tests in heptane.

Observation of the facilitation of edge dislocations movement is possible under some other conditions, particularly when the geometry of slip elements is different and especially when the character of interatomic bonds is of some other kind in the crystal. Thus of interest are the results obtained by Westwood, Huntington, and Macmillan [74] who described increase in mobility of edge dislocations in MgO crystals as a double electric layer formed on their surface in media with various pH. Since the effect is symmetrical in respect to the deviation of the electrokinetic potential in both directions from the zero point of charge, it may be presumed that the electrocapillary effect directly manifests here.

It is necessary to emphasize here that the concept of the Rebinder effect as facilitation of rupture and reconstruction of interatomic bonds under the influence of environmental atoms (molecules) during both deformation and fracture processes has universal character so that opposing of one or another detail of a presumably mechanism to this general approach cannot be considered to be sufficiently well-grounded.

Finally, we shall speak about practical significance of the share of participation in plastic deformation which belongs to the surface and to the near-surface layer. We shall be concerned only with some aspects of this problem -- those connected with the reversible effect of environment.

It is absolutely obvious that such participation is especially important (no need to mention films and thin samples) under the conditions of high, sharply heterogeneous stressed states localized in small volumes, taking place as a result of most various contact interactions in solids [75]. Among experimental methods these are indentation of the surface in various ways [67, 73, 76], scratching (e.g., very demonstrative electron microscopic observations of "ripping up" the surface with a needle [77]), among technological objects, for instance, all kinds of powders, not only being pressed, but also at other stages of their preparation and forming, since cohesion of particles is determined with local plastic deformations. (In connection with this, let us point out that with transition to higher dispersity of powders there manifests a "scale effect" analogous to the behavior of wiskers: the samples endure greater elastic deformation before the residual deformations, as shown in x-ray photographs of powders after pressing [78].) And, of course, the role of local surface deformations is exceptionally high in all processes of treating materials with pressure and of boundary friction [79-81, 85]. In all these cases it is established that adsorption layers may radically interfere with the process. Facilitation of plastic deformation in the near-surface layer under the influence of surface-active components of lubricants displays directly, for instance, in metal drawing and is found metallographically and with x-ray analysis as work-hardening in a thin near-surface layer (parts of a μ thick) sharply decreases. As a result, after diminishing of work-hardening and preventing the appearance of surface cracks in the sample in the zone of contact with the instrument, the metal not only requires considerably less force for drawing but also endures much greater deformation per circle, as shown by Veiler and Likhtman [79]. (This does not make impossible, as a secondary effect, the increase in dislocation density near the surface and degree of work-hardening at the next stages, e.g. under conditions of cyclic fatigue [1, 82].) Similar effects are widely spread in pressing of powders of various nature; it should be mentioned here, under conditions of compressive components' predominance and in absence of considerable tensile stresses, highly surface-active media (some of which were considered above in connection with the adsorption-induced strength lowering) may lose the role of the factor facilitating destruction and turn out to be a factor of plastification.

Under conditions of surface interactions together with the effect of plastification of the near-surface layer mentioned above, adsorption layers very often render a completely different effect: strongly adsorbed molecules of the monolayer or even incomplete monolayer, deprived of migrational mobility, prevent the contacting solids from cohesion. This effect is especially significant when the contact stresses are so high that the hydrodynamics of the viscous flow of the lubricant cannot alone provide its non-extrudability. The named factors are closely connected; identification of their role and mechanism requires that experimental methods approach the level of singular acts of interaction.

Among these methods we had used direct precise measurements of the cohesion force -- the strength of the contact between two particles brought in touch for a certain time, under the definite force of compression, in various media (in comparison with pressing of powders of the same materials). These experiments, carried out by Amelina, Yusupov et al. [83, 84], established critical conditions for the leap transition of a contact from coagulational type to the phaseous one (i.e., the one emerging at a site noticeably exceeding the size of an elementary cell, when there begins plastic flow in the zone of contact). Migrationally mobile surface-active molecules of the environment may facilitate the appearance of such deformations. On the contrary, the chemosorbed monolayer (e.g., octadecylamine on silver chloride crystals) turns out to be a strong screen: to reach cohesion one should increase the force of their compression by a factor of one or even two orders of magnitude (Fig. 10).

Another method we used was "ultramicroscratching" -- e.g., the surface of a LiF crystal with a corundum needle, the loads being so small (parts of a dyna) that no visible damages were left and etching allowed only after a certain force had been reached to find out the emerging of first dislocations along the track (Fig. 11) [68, 71]. As the force applied at the moving indentor grows, the density of dislocations along the track regularly grows as well. This method also allowed Savenko to evaluate the tangential force -- i.e., to measure the friction coefficient when its values were still very small (on the order of 10^{-3}). Such calculations make it possible to compare the deformation and adhesion components of the friction force with the values of the work spent and to approximate evaluations of the "work of plastic deformation" (of appearance of first dislocations). These two components appear to be connected with each other: uncovering of the juvenile surface as a result of ripping up the surface with dislocations enables cohesion bridges to appear; in turn this cohesion increases the tangential stresses and promotes the development of plastic deformations in the zone of contact.

An example of observations of the influence of the environment in this method may be the data on linear density of dislocations nucleating at the LiF surface after such "ultramicroscratching" in air and in an inactive lubricant (i.e., paraffine oil). Forces being small, the weak (physical) adsorption of molecules of the environment still appreciably hinders cohesion of contacting surfaces and nucleation of dislocations (Fig. 12), as compression grows the layer of such lubricant is extruded.

One may hope that in this direction -- in approaching the analysis of mechanism of influence of environment on plastic deformation at the level of elementary acts -- considerable help is provided by the above mentioned numerical experiments of dynamic type involving computers. On the other hand, in experimental investigations, physico-chemical conditions at the surface are widely varied, there are reasonable experimental methods, which allow us to identify basic primary factors.

REFERENCES

1. Likhtman, V.I., Rebinder, P.A., Karpenko, G.V., "Effect of Surface-Active Media on the Deformation of Metals", Moscow, AN SSSR, 1954; English translation N.Y., Chem. Publ. Co., 1960.

2. Karpenko, G.V., "Influence of Active Liquid Media on Endurance of Steel", Kiev, AN USSR, 1955.

3. Shchukin, E.D., Rebinder, P.A., Kolloidn.zh., 20, 645, 1958.

4. Likhtman, V.I., Shchukin, E.D., UFN 66, 213, 1958.

5. Rostocker, U., McCogie, G., Marcus, H., "Embrittlement by Liquid Metals", N.Y., 1960; Russian transl. ed. by E.D. Shchukin, Moscow, IL, 1962.

6. Kramer, I., Demer, L., "Influence of Environment on Mechanical Properties of Metals", Moscow, "Metallurgiya", 1964 (translated from English).

7. Likhtman, V.I., Shchukin, E.D., Rebinder, P.A., "Physico-Chemical Mechanics of Metals", Moscow, AN SSSR, 1962.

8. Westwood, A.R.C., in collection, "Fracture of Solids", N.Y., 1963, Russian transl. Moscow, "Metallurgiya, 1964.

9. "Environment-Sensitive Mechanical Behavior", ed. by A.R.C. Westwood and N. Stoloff, N.Y., 1966, Russian transl. Moscow, "Mir", 1969 (9a).

10. Westwood, A.R.C., in coll. "Treatise on Brittle Fracture", N.Y., 1967; or ref. 9a, p. 27.

11. Adv. in Corrosion Science and Technology, Plenum Press, 1970, p. 51.

12. Westwood, A.R.C., Latanision, R.M., "The Science of Ceramic Machining and Surface Finishing", National Bureau of Standards special publ., 348, 1972, p. 141.

13. Rebinder, P.A., Shchukin, E.D., UFN, 108, 3, 1972.

14. Rebinder, P.A., Shchukin, E.D., Progress in Surface Science, Pergamon Press, 3, N2, 97, 1972.

15. Shil'krut, D.I., DAN SSSR, 122, 69, 1958.

16. Barenblatt, G.I., PMM, 24, 667, 1960; 25, 46, 1961; 26, 329, 1962.

17. Panasyuk, V.V., "Limit Equilibrium of Brittle Solids with Cracks", Kiev, Naukova Dumka, 1968.

18. Pertsov, N.V., Shchukin, E.D., Fiz. i chim. obrabotki materialov, 2, 60, 1970.

19. Shchukin, E.D., in coll. Advance in Colloid Chemistry, Moscow, Nauka, 1973, p. 159.

20. Shchukin, E.D., Vestnik AN SSSR, 11, 30, 1973.

21. Yushchenko, V.S., Dukarevich, M.V., Chuvayev, V.F., Shchukin, E.D., Zh. F. Kh., 43, 1556, 1969.

22. Traskin, V.Yu., Pertsov, N.V., Skvortsova, Z.N., Shchukin, E.D., Rebinder, P.A., DAN SSSR, 191, 876, 1970.

23. Skvortsov, A.G., Sinevich, E.A., Pertsov, N.V., Shchukin, E.D., Rebinder, P.A., DAN SSSR, 193, 76, 1970.

24. Sinevich, E.A., Pertsov, N.V., Shchukin, E.D., DAN SSSR, 197, 1376, 1971.

25. Shchukin, E. D., Soshko, A. I., Mikityuk, O. A., Tynniy, A. N., Fiz.-khim. mekh. materialov, 7, N2, 33, 1971.

26. Pertsov, N. V., Kruchinin, V. T., Yushchenko, V. S., Shchukin, E. D., DAN SSSR 199, 391, 1971.

27. Kishkin, S. T., Nikolayenko, V. V., Ratner, S. I., ZhETF, 24, 1455, 1954.

28. Potak, Ya. M., Shcheglakov, I. M., ZhETF, 25, 897, 1955.

29. Rebinder, P. A., Likhtman, V. I., Kochanova, L. A., DAN SSSR, 111, 1278, 1956.

30. Rozhanskiy, V. N., Pertsov, N. V., Shchukin, E. D., Rebinder, P. A., DAN SSSR, 116, 769, 1957.

31. Shchukin, E. D., Pertsov, N. V., Goryunov, Yu. V., Kristallografiya, 4, 887, 1959.

32. Liquid Metals Handbook, ed. by Lyon, R. N., Washington, v. 1, 1950, v. 2, 1951.

33. Kamdar, M. H., Westwood, A. R. C., RIAS Technical Report 66-4; Phil. Mag., 15, 641 (1967); Westwood, A. R. C., Preece, C., Kamdar, M., ref. 9, 1966, p. 581; ref. 9a, 1969, p. 118 or coll. Treatise on Brittle Fracture, N. Y., 1967.

34. Nikitin, V. I., "Physico-Chemical Phenomena Accompanying Interaction of Solid and Liquid Metals", Moscow, Atomizdat, 1968.

35. Gilman, J. J., J. Appl. Phys., 31, 2208, 1960.

36. Westwood, A. R. C., Kamdar, M. H., Phil. Mag., 8, 787, 1963.

37. Likhtman, V. I., Bryukhanova, L. S., Andreyeva, I. A., Shchukin, E. D., Fiz.-khim. mekh. materialov, 6, 66, 1970.

38. Breger, A. H., Zhukhovitskiy, A. A., Zh. F. Kh. 20, 355, 1946; Wassink, R. I., J. Inst. Metals, 95, 38, 1967; Tiller, W. A., Takanashi, T., Acta Met., 17 483, 1969; Zadumkin, S. N., DAN SSSR, 101, 507, 1955; 112, 453, 1957.

39. Frumkin, A. N., Zh. F. Kh., 12, 337, 1938.

40. Shchukin, E. D., Summ, B. D., Goryunov, Yu. V., DAN SSSR, 167, 631, 1966.

41. Chayevskiy, M. I., Shatinskiy, V. F., Increase in Endurance of Steels in Agressive Environments under Cyclic Stresses, Kiev, Naukova Dumka, 1970.

42. Maximovich, G. G., Micromechanical Investigations of Properties of Metals and Alloys, Kiev, Naukova Dumka, 1970.

43. Shchukin, E. D., Kochanova, L. A., Pertsov, A. V., Kristallografiya, 8, 69, 1963.

44. Shchukin, E. D., Goryunov, Yu. V., Denshchikova, G. I., Pertsov, N. V., Summ, B. D., Kolloidn. zh., 25, 108, 1963; Shchukin, E. D., Summ, B. D., Goryunov, Yu. V., Pertsov, N. V., Pertsov, A. V., Kolloidn. zh., 25, 253, 1963; Summ, B. D., Goryunov, Yu. V., Pertsov, N. V., Traskin, V., Yu., Shchukin, E. D., FMM, 14, 757, 1962.

45. Goryunov, Yu. V., Uspekhi Khimii, 33, 1062, 1964.

46. Shchukin, E. D., Summ, B. D., in coll. Surface Diffusion and Spreading, Moscow, AN SSSR, 1969.

47. Soldatchenkova, L. S., Goryunov, Yu. V., Denshchikova, G. I., Polukarova, Z. M., Summ, B. D., Shchukin, E. D., DAN SSSR, 203, 83, 1972.

48. Shchukin, E. D., Likhtman, V. I., DAN SSSR, 124, 307, 1959.

49. Chayevskiy, M. I., Popovich, V. V., Fiz.-khim.mekh. materialov, 7, 85, 1971.

50. Pertsov, N. V., Rebinder, P. A., DAN SSSR, 123, 1068, 1958.

51. Pertsov, N. V., Volodin, Yu. A., Pertsov, A. V., Fiz.-khim. mekh. materialov, 10, N1, 41, 1974.

52. Carpenko, G. V., Strength of Steel in Corrosion Environment, Moscow, Mashgiz, 1963; Carpenko, G. V., Vasilenko, I. I., Corrosion Cracking of Steel, Kiev, Teknika, 1971.

53. Stress Corrosion Cracking and Embrittlement, N. Y., Wiley, 1956; Physical Metallurgy of Stress Corrosion Fracture, N. Y., Interscience, 1959.

54. Arkharov, V. I., Tr. In-ta fiz. metallov AN SSSR (Sverdlovsk), 8, 54, 1946; 12, 94, 1949; 14, 15, 1954; 16, 7, 1955.

55. Glikman, Ye. E., FMM, 26, 233, 1968.

56. Yushchenko, V.S., Grivtsov, A.G., Shchukin, E.D., DAN SSSR, 215, 148, 1974.

57. Yushchenko, V.S., Grivtsov, A.G., Shchukin, E.D., DAN SSSR, 219, 162, 1974.

58. Alyokhin, V.P., Gusev, O.V., Trefilov, V.I., Shorshorov, M.H., DAN SSSR, 188, 326; 188, 548, 1969; Rykalin, N.N., Shorshorov, M.H., Kudinov, V.V., Fizika i khimiya obrabotki materialov, 4, 1968.

59. Milevsky, L.S., Smol'skiy, I.L., FTT, 16, 1028, 1974.

60. Yoffe, E.H., Phil. Mag., 6, 1147, 1961.

61. Indenbom, V.L., Dubnova, G.N., FTT, 9, 1171, 1967.

62. Shishlyannikova, L.M., Suzdaltzeva, S.F., Contorovitch, S.I., Shchukin, E.D., Kolloidn. zh., 36, 612, 1974.

63. Rebinder, P.A., Venstrem, E.K., Zh. F. Kh., 19, 1, 1945.

64. Taubman, A.B., DAN SSSR, 74, 521, 1950; Zh. F. Kh., 26, 389, 1952.

65. Shchukin, E.D., Kochanova, L.A. in coll. Dynamics of Dislocations, Kharkov, FTINT AN USSSR, 1968; Likhtman, V.I., Kochanova, L.A., Leykis, D.I., Shchukin, E.D., Electrokhimiya, 5, 729, 1969.

66. Shchukin, E.D., DAN SSSR, 118, 1105, 1958.

67. Westbrook, J.H., in ref. 9a, p. 257, or in coll. Treatise on Brittle Fracture, N.Y., 1967.

68. Shchukin, E., Rebinder, P., Les phenomenes de surface dans la deformation et la fracture des solides, Seminaire de Mechanique des surfaces (V - 1971), CNRS, Paris, 1973.

69. Andarelly, G., Maugis, D., Courtel, R., Wear, 23, 21, 1973.

70. Shchukin, E.D., Savenko, V.I., Kochanova, L.A., Rebinder, P.A., DAN SSSR, 200, 406, 1971.

71. Ibid., 200, 406, 1971.

72. Savenko, V.I., Kochanova, L.A., Shchukin, E.D., Kristallografiya, 17, 995, 1972; coll. News in the Field of Microhardness Tests, Moscow, Nauka, 1974, p. 67.

73. Westwood, A.R.C., Goldheim, D., Lye, R., Treatise on Brittle Fracture, N.Y., 1967; Russian transl. s. 9a, p. 274.

74. Macmillan, N.H., Huntington, R.D., Westwood, A.R.C., Phil. Mag., 28, N4, 1973; Westwood, A.R.C., Huntington, R.D., Macmillan, N.H., J. Appl. Phys., 44, N11, 1973.

75. Fuks, G.I., Gantsevitch, I.B., Kolloidn. zh., 29, 304, 1967; Kanayev, A.A., L'vov, V.N., Veyler, S.Ya., Rebinder, P.A., DAN SSSR, 187, 314, 1969.

76. Upit, G.P., Varchenya, S.A., DAN SSSR, 178, 834, 1968; Varchenya, S.A., Spalvin', I.P., Upit, G.P., Flyorov, V.I., FTT, 10, 2195, 1968; Varchenya, S.A., Muktepavel, F.O., Upit, G.P., FTT, 11, 2841, 1969.

77. Gane, A., Bowden, F.P., J. Appl. Phys., 39, 1432, 1968; Gane, N., Cox, J., Phil. Mag., 22, 881, 1970.

78. Kontorovich, S.I., Malikova, Zh.G., Shabanova, E.A., Shchukin, E.D., Kolloidn. zh., 30, 691, 1968; Fizika i Khimiya Obrabotki Materialov, 6, 105, 1968.

79. Veyler, S.Ya., Likhtman, V.I., Effect of Lubricants on Metals Pressing, Moscow, AN SSSR, 1960.

80. Bowden, F.P., Tabor, D., The Friction of Solids, Oxford, 1950.

81. Kragel'skiy, I.V., Friction and Wear, Moscow, Mashinostroyenie, 1968, ch. V and VIII.

82. Kostetskiy, B.I., Kolesnichenko, L.F., DAN SSSR, 157, 574, 1964; Shevelya, V.V., Kostetskiy, B.I., DAN SSSR, 175, 1270, 1968.

83. Shchukin, E.D., Yusupov, R.K., Amelina, E.A., Rebinder, P.A., Kolloidn. zh., 31, 913, 1969; Yusupov, R.K., Amelina, E.A., Shchukin, E.D., Rebinder, P.A., DAN SSSR, 200 1077, 1971.

84. Amelina, E.A., Yusupov, R.K., Shchukin, E.D., Kolloidn. zh., 36, 931, 1974.

85. Rybakova, L.M., Kuksyonava, L.I., FMM, 39, 363, 1975.

TABLE 1.

Characteristics of rosette rays appearing after microindentation of NaCl single crystals; W is the variation coefficient 72.

Load P, G	In Heptane				In Air (moisture 60%)			
	l_c, mc	l_s, mc	*	**	l_c, mc	l_s, mc	*	**
0.32	48.2	17.0	2.82	39	46.2	19.7	2.36	13.3
0.54	63.8	24.2	2.64	30	64.5	29.4	2.19	7.9
0.98	90.6	37.6	2.41	18.7	87.4	40.3	2.17	6.9
1.54	117	49	2.38	17.2	111.5	54.5	2.05	~0
3.27	171.5	73.6	2.33	14.8	175.5	88.3	1.99	~0
10.4	297	146	2.03	~0	300	146	2.05	~0
19	496	249	1.99	~0	503	250	2.01	~0
40	607	298	2.04	~0	603	298	2.02	~0
Wt. %	2	2	2		2	2	2	

* $\theta = l_c / l_s$ ** $\dfrac{\theta - \theta_o}{\theta_o} \cdot 100\%$

TABLE 2.

True width of the β x-ray line (10^{-3} rad) of magnesium hydroxide powders before pressing, after pressing, pressure being $4T/cm^2$, and after demolishing the structure (thorough grinding) of the pressed samples. The value of $\beta = 16 \cdot 10^{-3}$ of a fine-disperse sample, disappearing after it is ground, corresponds to elastic microdistortions; the value of $\beta = 6.5 \cdot 10^{-3}$, preserved after grinding of a coarse-disperse sample, is determined with plastic deformations.

Sample	Dispersity (size of particles)	
	0.05 mc	3 mc
Original Powder	0,	0
Pressed Sample	16	6.8
The Powder Obtained by Grinding the Sample	0	6.5

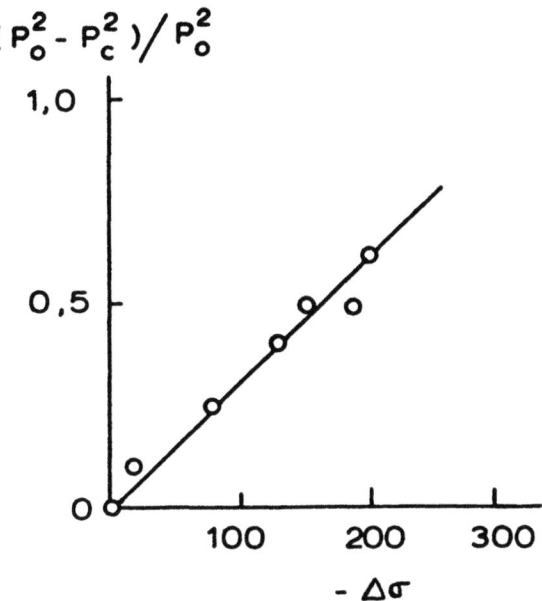

Fig. 1. Comparison of independently obtained values of decrease in strength P_c and lowering of free surface energy σ (erg/cm^2) of high-disperse fine-porous magnesium hydroxide samples in presence of water vapor [21].

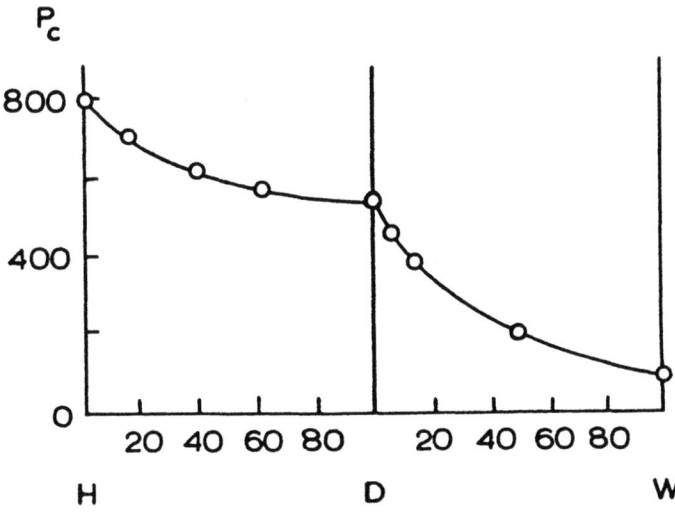

Fig. 2. Dependence of strength P_c (G/mm^2 of polycrystalline potassium chloride samples on composition of environment (wt. %) in heptane-dioxane (H-D) and dioxane-water (D-W) solutions [22].

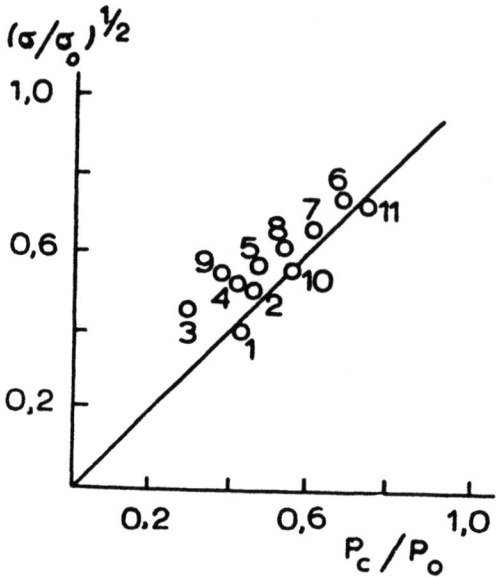

Fig. 3. Comparison of values of relative decrease in surface energy σ of naphtalene single crystals and relative decrease in strength P_c of polycrystalline naphtalene samples in various media (1 - benzene, 2 - heptane, 3 - methylene chloride, 4 - chloroform, 5 - carbon tetrachloride, 6 - methyl alcohol, 7 - ethyl alcohol, 8 - propyl alcohol, 9.- butyl alcohol, 10 - tertiary butyl alcohol, 11 - 0.2 M butyl alcohol solution in water; the straight line corresponds to the Griffith equation [23].

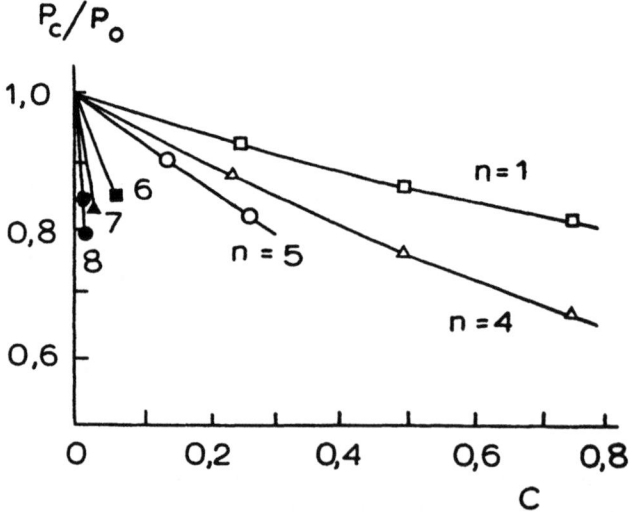

Fig. 4. Relative decrease in strength P_c of polycrystalline naphthalene samples in aqueous solutions of aliphatic alcohols $C_nH_{2n+1}OH$ depending on concentration C (mol/l) [24].

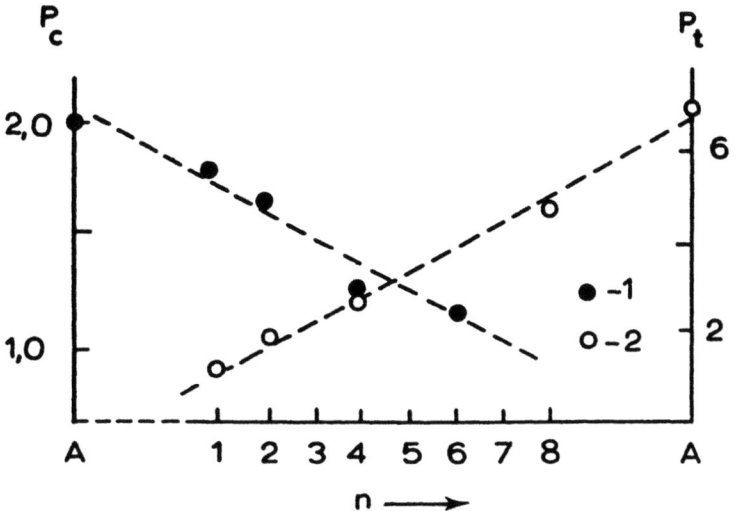

Fig. 5. Influence of aliphatic alcohols $C_nH_{2n+1}OH$ on strength P_c (kG/mm^2) of polyethylene (1) and on limit stress P_t (kG/mm^2) corresponding to the time before fracture t_f = 100 sec., for polymethylmetacrylate (2); A is the initial value in air [25].

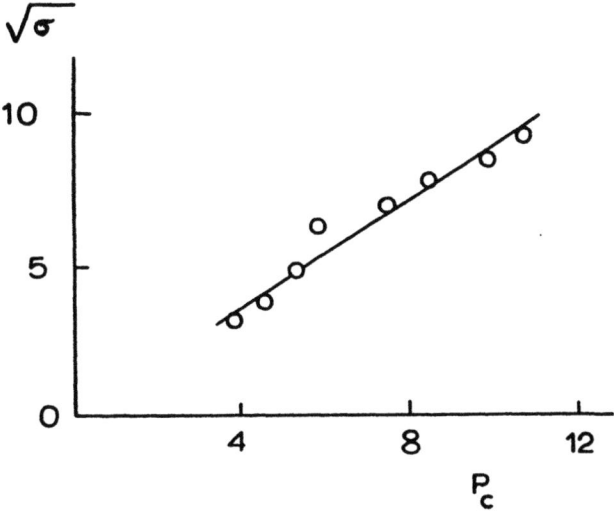

Fig. 6. Comparison of experimental values of strength (P_c, kG/mm^2) and calculated values of free interfacial energy ($\sigma^{1/2}$, erg$^{1/2}$/cm) of germanium single crystals tested at various temperatures in presence of a drop of gold (saturated with germanium) [26].

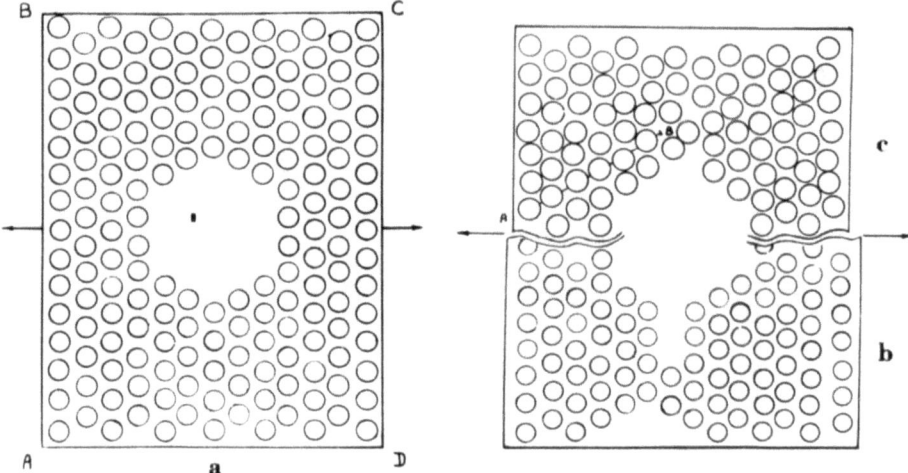

Fig. 7. Initial state of the system under consideration (a) in the center of the cavity there is a rectangle corresponding to the accuracy of calculation of atomic coordinates. Development of a brittle crack (b) and formation of a dislocation (c) according to results of numeric experiments of dynamic type, deformation carried out at low and high temperatures respectively [56].

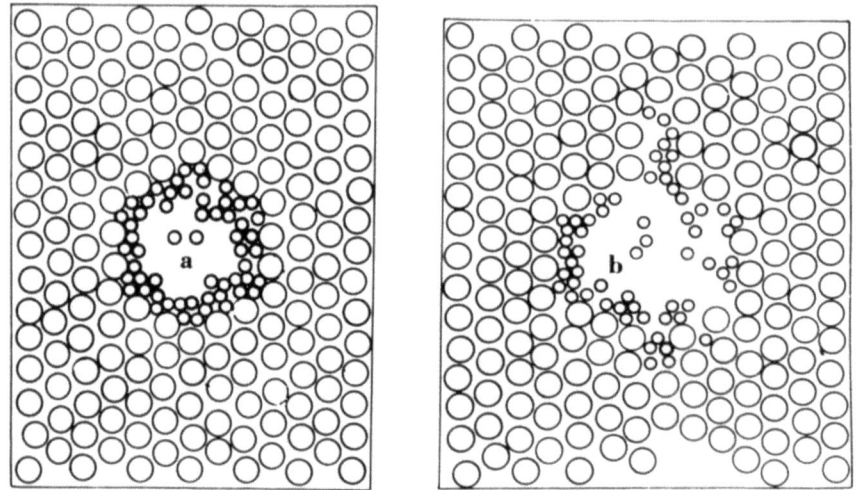

Fig. 8. Initial state -- adsorption of foreign atoms (of the environment) at the walls of the cavity (a) and formation of a crack as atoms of the medium penetrate the stressed lattice (b) [57].

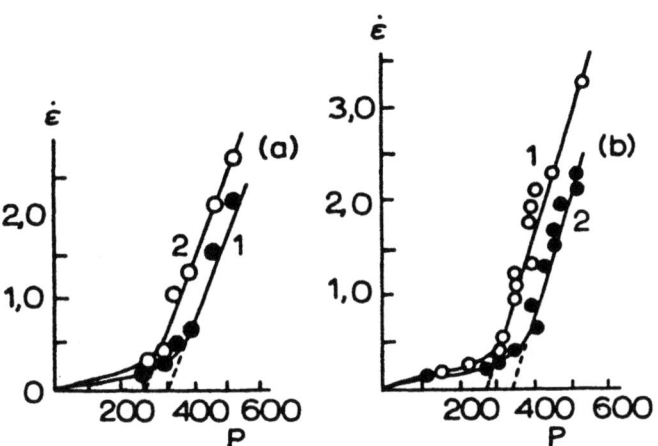

Fig. 9. Dependence of plastic deformation rate (10^{-4} sec.$^{-1}$) of lead single crystals at the initial stage of creep on the stress applied P (G/mm^2): (a) in 0.01N H$_2$SO$_4$ solutions, the potential of zero charge being ϕ = -0.70 v (1) and ϕ = -0.87 v (2); (b) in air for clear samples (1) and for the ones covered with oxide film 0.25 mc thick (2) [65].

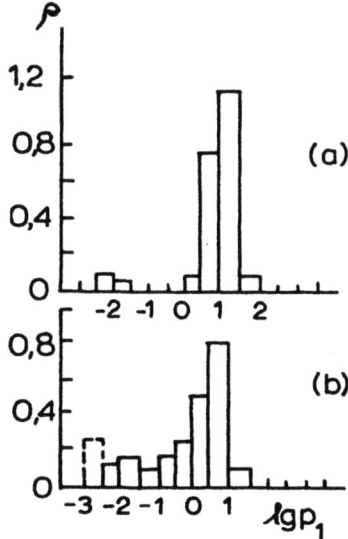

Fig. 10. Differential distribution ρ of strength of contacts (p_1, dyn) appearing between silver chloride particles being brought together in air with a force of 50 dyn (for 10 sec.) in absence (a) and in presence of octadecylamine at the surface in quantities corresponding to a monolayer (b) [84].

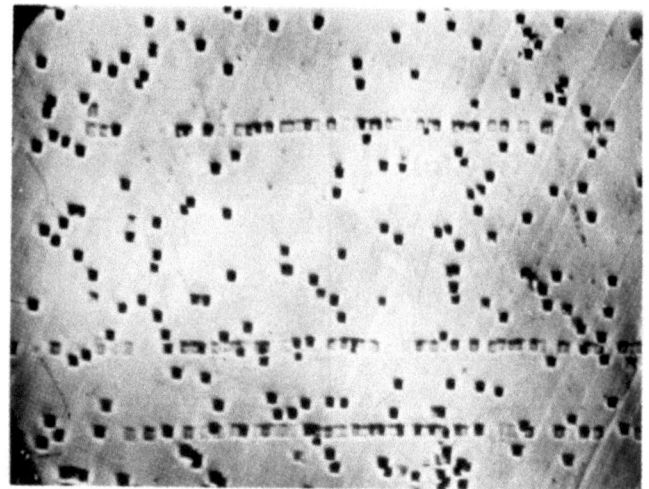

Fig. 11. Dislocation structure appearing in course of sliding of an indentor along the surface of a lithium fluoride crystal in air, the load being 15 dyn., 125 x [71].

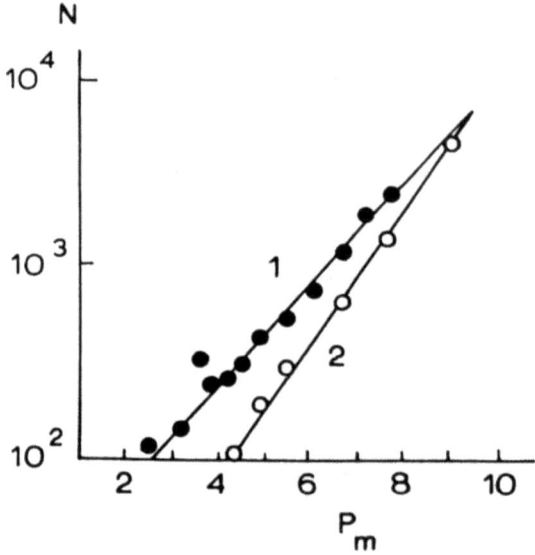

Fig. 12. Dependence of linear density of dislocations N (cm^{-1}) on the value of maximum normal stress P_m (10^9 dyn/cm^2) after sliding of an indentor along lithium fluoride surface in air (1) and in paraffine oil (2) [71].

DISCUSSION

Comment by R. A. Oriani:

Your demonstration of the fragility of zinc in contact with a liquid metal compared to its fragility in air was described by you in terms of the lowering of the free surface energy of zinc by the liquid metal. My question deals with the reference state from which the lowering, $\Delta\gamma$, is to be computed. Since the comparison was made with zinc in air, I presume that the $\Delta\gamma$ is appropriately calculated from the γ of the oxygen-covered zinc. Now, oxygen must adsorb strongly on Zn. Do you know that $\gamma(Zn/O_2) - \gamma(Zn/metal)$ is a positive quantity? This is an example of a general problem of reference state for $\Delta\gamma$, by which many people try to understand environmentally-assisted cracking. Another example is that of a crack which does not propagate in steel under a given stress in air, but will propagate under hydrogen at the same stress, even though $\gamma(Fe/O_2) - \gamma(Fe/H_2)$ is less than zero.

You have said that lowering of the surface free energy is necessary but not sufficient for environment-assisted cracking. I completely agree. You and I say that penetration to the tip of the crack by the agent from the environment is necessary. It follows then that the action of the active agent is at the strained atomic bonds before they have parted to allow the generation of a surface of lowered γ. We see that the adsorption of the active species with the lowering of γ is <u>subsequent</u> to the essential action of the active species because the $\Delta\gamma$ manifests only after bonds have been parted to create new surface atoms. One should therefore not expect that the adsorption-modified γ, placed into the Griffith equation, is the proper parametre to describe environment-assisted cracking. Instead, one should use some measure of how much the <u>force</u> necessary to part bonds is lessened by the active species <u>absorbed</u> into the highly elastically strained regions. I believe that the appropriate parametre is F_m, as affected by the active solute atoms--where by F_m, I mean the maximum in the curve of cohesive force with interatomic separation. Using ΔF_m instead of $\Delta\gamma$ is much more than a change of a language since one cannot in general expect that F_m and $\gamma^{1/2}$ will have the same functional dependence with chemical potential of the active species in the environment.

Reply:

It is hard to answer Dr. Oriani in brief. In fact, the answer is the full text of the lecture and our other articles. The essence of his objection is a misunderstanding. It is the separation of deformation and fracture processes in the adsorption-active environment into two stages: (i) formation of new surface and (ii) lowering of free surface energy (γ). In some cases this separation is possible (e.g., when the environment for kinetic or steric

reasons can not reach the tip of the crack), but in such systems
no strength reduction is observed (see, for instance, N.V. Pertsov's
and E.A. Sinevich's works on influence of organic molecules on
strength of naphtalene). On the contrary, when the strength is
reduced, the strength reduction is a single indivisible process of
facilitated rupture and reconstruction of interatomic bonds, i.e.,
of creation of new surface which already has a smaller value of
γ than that of the solid/vacuum interface.

Dr. Oriani suggests the force approach, opposing it to the
energetic one which we use. If both approaches are rigorous and
refer to the same conditions, this opposition has no basis, being
merely a question of different terms. However, the energetic
approach has its own advantages. It allows, leaving alone the
details of the molecular mechanism, comparison quantitatively of
the macrocharacteristic (strength) with measurable microparameters
(energies of interaction between atoms or molecules of the solid,
of the environment, and between them). It can be done with the
help of the Griffith equation or any other connecting P and γ,
e.g., the Petch-Stroh equation, and by calculation of γ from
energies of molecular interactions. The energetic approach allows
direct experimental control, as far as the γ and $\Delta\gamma$ values may be
measured independently of strength (for instance, from data on
adsorption).

It goes without saying that the $P^2 \sim \gamma$ correlation is not rigorously fulfilled under all conditions. However, I have emphasized
the proportionality of P^2 to the work of formation of new surface.
This is a conclusion of a rigorous thermodynamical approach analogous
to the Gibbs-Volmer theory and it is absolutely true to the extent
that the conditions of manifestation of this correlation are fulfilled. As the conditions of the process deviate from (quasi)
equilibrium, the correlation cannot be quantitatively accurate,
but P would always be symbiotic to γ. This is confirmed by all the
experimental data referring to most solids in various environments.
Along with this, wide variation of objects allowed selection of
systems where the uncertainty of conditions due to films at the
surface was excluded. The films were surely absent on naphtalene,
AgCl and in other cases when the comparison with strength in air
is completely equivalent to the one with strength in its own vapour.
We should emphasize that in many cases which were the subject of
our studies, proportionality was observed between P^2 and γ, this
showing that the Griffith equation can be widely applied.

Let us now consider the force approach. All the variants
described earlier do not take into account the thermal movement
of atoms, i.e., refer to the absolute zero of temperature, and do
not actually describe the thermally activated real process. The
approach can be permitted when analyzing usual brittle fracture,
but cannot be used when we deal with the influence of environment--

not just because thermal activation is important for the whole mechanism of transport of the active component, but also because the elementary acts of reconstruction of interatomic bonds, which involve foreign atoms, are essentially a thermally activated process. Thus the consideration of the simple 3-4 atomic models in terms of the force approach is unacceptable. Besides that, two more difficulties inevitably appear here: (i) the force approach is closely connected with the model of fracture which we are to invent, and (ii) a question necessarily arises regarding the influence of environment atoms on the interaction potential between particles of the solid, while this question cannot be solved yet.

An outstanding example of the rigorous solution uniting the force and energetic approaches is the numeric experiment carried out by Dr. V.S. Yushchenko (V.S. Yushchenko, A.G. Grivtsov, E.D. Shchukin, DAN SSSR, 219, 1, 162, 1974). The conditions of the experiments include simultaneously the non-zero temperature and the given kinetics of deformation: the ensemble of atoms is large enough (thermal fluctuations are possible), the atoms move "freely" (i.e., we do not consider the model of fracture, thus not interfering with the process). In these experiments we obtain all the force and energetic characteristics, i.e., coordinates, impulses, energies and forces of interaction of every atom. It would give nothing to put this solution down to the primitive level of F_m and ΔF_m, since the model is chosen so that the foreign atoms do not directly influence the binary potential of B-B interaction. Moreover, one of the main conclusions of the experiment is that this influence, though it may take place in some systems, is not necessary for the manifestation of the Rebinder effect. On the contrary, the adsorption activity of the environment (low values of γ, i.e., thermodynamically easy formation of new surface) together with thermal fluctuations is necessary for the manifestation of adsorption-induced strength reduction.

Comment by D. L. Davidson:

Please describe the functions used for the B-B bonds (the ones between bulk atoms), the A-A bonds (between adsorbate atoms) and the A-B bonds used in the atomic model you have described.

Reply:

The walls of the box interacted only with the nearest layer of B atoms. We used the 9-3 potential in the form

$$\phi = \varepsilon \left[(r_o/r)^9 - 3 (r_o/r)^3 \right] /2$$

where $r_o = 3.82$ Å, $\varepsilon = 6.9 \times 10^{-14}$ erg for the movable side walls, and $\varepsilon = 1.38 \times 10^{-15}$ erg for the fixed upper and lower walls. The potential of interaction with the side walls corresponds approximately to the potential field of semi-endless two-dimensional crystal.

All the atomic interactions were described by the Lennard-Jones potential

$$\phi = \varepsilon [(r_o/r)^{12} - 2 (r_o/r)^6]$$

where $\varepsilon_{BB} = 1.656 \times 10^{-14}$ erg, $r_{oBB} = 3.82$ Å, this corresponding to argon. The ε_{AA}, r_{oAA}, ε_{AB}, r_{oAB} values varied in each experiment. In our article (V.S. Yushchenko, A.G. Grivtsov, E.D. Shchukin, DAN SSSR, 219, N1, 162, 1974), Fig. 2 refers to the system where $\varepsilon_{AA} = 2.76 \times 10^{-15}$ erg, $r_{oAA} = 1.27$ Å, $\varepsilon_{AB} = 8.28 \times 10^{-15}$ erg, $r_{oAB} = 2.54$ Å; Fig. 3 shows the system with $\varepsilon_{AA} = 8.28 \times 10^{-15}$ erg, $\varepsilon_{AB} = 1.242 \times 10^{-14}$ erg, $r_{oAA} = r_{oAB} = 3.82$ Å. That is, in the first case the atoms of the second component are small and weakly connected with each other (easily movable), the mixing energy is 1.38×10^{-15} erg/bond. In the second case the atoms of the basic component and of the impurity are of the same size, the mixing energy is zero (the solutions are ideal) and the adsorption-active atoms interact with each other and with the crystal strongly enough. To compare let us point out that the kinetic energy was about 3×10^{-15} erg/atom (at 20-30°K).

Comment by R. Bullough:

Your two-dimensional atomic simulations of stressed cavities in the presence of impurity atoms were most interesting, I would like to know:

1) Does your 6-12 potential define a sensible surface energy?

2) In your simulation of a finite temperature did you find equipartition or was there a strong departure from equilibrium due to the unharmonicity of the 6-12 potential?

3) How did you fix the potential between the impurity atoms? Did the adsorption of those on the native atom surface substantially change the surface energy?

Reply:

In our experiments we considered two-dimensional systems containing about 200 atoms. The parameters of interatomic potentials are given in the answer to Dr. Davidson's question. It was in about 10^{-11} sec. after the beginning of the experiment that the distribution of particles according to their velocities was well approximated by the Maxwell equation. By this time the full energy of the system (its surplus over the minimum possible magnitude) was equally distributed between potential and kinetic energies and the heat capacity of the system was close to the doubled Bolzman's constant, i.e., the Dulong-Petit law for two-dimensional systems was fulfilled. The kinetic energy in the experiments did not exceed 10% of the binding energy of atoms and no reasons existed for noticeable display of unharmonicity.

The investigation of surface energy was not the special aim of the work. The production of results by the computer was adjusted to the micro-scale study of stress and fracture, so in fact the answers to questions 1 and 3 cannot be complete.

Special experiments reveal that surface tension in our system does exist and its value may be approximately calculated by the number of atomic bonds ruptured. We did not study the movement of surface atoms, but this is quite possible.

The parameters of the system provided that γ at the solid/liquid interface was small. Values of γ at the solid/gas interface in the two-component systems were not measured, since this parameter is not directly connected with the possibility of adsorption strength reduction under the influence of the melt. However, adsorption occurs in our case and thus the decrease of γ does take place.

Comment by C. A. Brookes:

Would you explain how you estimate the frictional force from studies on dislocation etch-pits in order to determine the coefficient of friction. How can you tell what part of the force to operate the source is due to normal (vertical) components and what part is due to transverse (horizontal) components?

Reply:

One of the methods used to estimate the friction force in the ultramicroscratching experiments was evaluation of plastic deformation of the sample per unit length of the indentor track. The deformation was estimated from the density of etch-pits along the track.

As for the question regarding which component of the force appearing in the contact during the microscratching process causes generation of dislocations, experiments with lubricants allow us in this case to give quite a definite answer. In fact, introduction of lubricant into the contact under small load results in complete disappearance of dislocations along the track. The force of pressure in these experiments did not alter, but friction in the contact noticeably decreased. The friction force in the case under consideration is connected with the rupture of adhesion bonds in the contact during the tangential movement of the indentor. So this force inevitably plays a role in the creation of dislocations along the track as well as the normal force of pressure. For details see: 1) E.D. Shchukin, V.I. Savenko, L.A. Kochanova, P.A. Rebinder, DAN SSSR, 200, N6, 1971; 2) Seminaire de Mechanique des surfaces dirige par R. Courtel, "Les phenomenes de surface dans la deformation et la fracture des solides", Par le professeur

Schukin et le professeur Rebinder. Editions metaux: 32 rue du Marechal, Joffre 78100, Saint-Gemain-en-Lasse.

Comment by C. A. Brookes:

Have you considered the possibility of secondary slip on {100} <011> systems, in addition to primary {110} <1$\bar{1}$0> slip? Could the operation of these systems in NaCl at room temperature explain the observed difference in behaviour between this material and LiF?

Reply:

It is true that at room temperatures the {100} slip in LiF crystals is less intense than in NaCl. However, this circumstance by itself does not give additional opportunities to explain why water does not influence the mobility of screw dislocations.

CHEMICAL-INDUCED VARIATIONS IN THE PLASTICITY AND FRACTURE OF METALS AND MINERALS—CHEMOMECHANICAL EFFECT

E. Gutman*

Chair of Metal Technology, Petroleum Institute, Ufa, U.S.S.R.

ABSTRACT. The chemomechanical effect as we called it [1] is a phenomenon representing a change of physico-mechanical properties and dislocation substructure (plasticity) of solids under the influence of chemical (electrochemical) reactions on its surface, causing additional dislocations flow.
This phenomenon was established and proved by means of methods of irreversible thermodynamics because irreversible processes of mechanochemical corrosion are connected with entropy production in the system. When studying the chemomechanical effect (CM-effect) we dealt with single crystal and polycrystal materials (metals and minerals). The lecture contains the following parts: 1. Chemical potential of dislocations and dislocations flow; 2. Mechanochemical effect; 3. Entropy production in mechanochemical processes; 4. Chemomechanical effect; 5. Autocatalytic process of chemical and mechanical failure of solids (stress corrosion).

NOTATION LIST

A affinity of plastic deformation process.
$\tilde{A}$ affinity of electrochemical reaction.
$\vec{b}$ Burgers vector.
I_0 flux of new surface production.
ΔG^* addition of partial Gibbs free energy of dislocation within configurational entropy.
$J = J_1 + J_2$ dissolution current (flow) density.

* Editor's Note: Professor Gutman was unable to attend this meeting. His manuscript was mailed in advance.

j_o exchange current (flow) density.
L_{ik} phenomenological coefficient irreversible thermodynamic equations.
n number of piled-up dislocations (in one group of the coplanar arrays).
N dislocation density.
m constant in equation $N=\bar{d}(\varepsilon-\varepsilon_o)^m$, ($m \approx 1$ for Fe, Ni)
$\dot{N}$ dislocations flow.
N_o primary dislocation density.
N_{max} maxima possible dislocation density.
R' gas constant per mole, $R'=kN_{max}$ (k-Boltzmann constant)
T absolute temperature.
α, β transfer coefficients.
$\bar{d}$ constant in equation $N=\bar{d}(\varepsilon-\varepsilon_o)^m$
ε plastic strain.
ε_o primary plastic strain.
$\dot{\varepsilon}$ velocity of creep.
$\dot{\varepsilon}_o$ primary velocity of steady-state creep.
Λ the coefficient is in reciprocal dependence of length of slip distance of dislocations (collision interval) from strain.
μ_D chemical potential of dislocations.
σ tension of surface layer.
τ fast in shear.

I. CHEMICAL POTENTIAL OF DISLOCATIONS AND DISLOCATIONS FLOW

Let us introduce the very notion of chemical potential of dislocations and define it as additional Gibbs free energy of crystal due to the formation of unit dislocation (having the length equal to unit) in the volume equal to unit. Considering the formation and motion of dislocations as united process of plastic deformation of crystal (cold flow) and calculating the equivalent mechanical work, we get quantitative expression of this potential

$$\mu_D = \frac{\tau}{\bar{d}^{1/m}} \qquad (I)$$

Really, according to numerous data the maxima possible dislocation density in crystals reaches $N_{max} = (0,5 \div 1) \cdot 10^{12}$ $dist/cm^3$ (then one dislocation has in average partial volume of crystal equal to $1/N_{max}$). By analogy to Avogadro constant this value may be considered as one mole of dislocations (in the same way when speaking of the mole of vacancies). The dimension of this value is $dist/mole$ and the number of moles is equal to N/N_{max}. Calculating partial work to produce one dislocation (inherent in its partial volume, i.e. multiplying space density of work by $1/N_{max}$) and then multiplying it by the number of dislocations in one mole, we get the expression (I) having the dimension *joule/mole* inherent in chemical potential as partial mole thermodynamic potential. In this connection we should note the main supposition of Zener's theory "the strain-energy model" of point defects is also equality of isothermo-isobaric straining work resulting in a certain point defect with its own thermodynamic potential. In order to establish the relationship of chemical potential of dislocations and dislocation density let us imagine homogeneous solid with dislocations as a two-component solution of dislocations in N_{max} number of possible localities (the model of "dislocation lattice"). We mind the fact of formation of some number n of piled-up dislocations in one flat group (the coplanar arrays) which in its turn increase the energy of every dislocation. Calculating the corresponding configurational entropy, we get

$$\mu_D = \frac{n\tau}{\bar{a}} = n\Delta G^* + R'T \ln \frac{N/n}{N_{max}/n} \qquad (2)$$

Chemical affinity of the process of formation and motion of dislocations is the difference of the meanings of the state function $\Delta \mu_D$ corresponding to the difference of stresses for two states of solid

$$A = \Delta \mu_D = \frac{n\Delta\tau}{\bar{a}} = R'T \ln \frac{N}{N_o} \qquad (3)$$

When dealing with "the reaction" of formation and motion of dislocations in crystal volume connected with the length of slip distance of dislocations as steady-state process having chemical affinity (3) from continuum flow condition and taking into account well-known formulaes made for velocity of creep and length of slip distance of dislocations, we get a number of dislocations formed in the unit of volume during the unit of time. We call them "dislocations flow"

$$\dot{N} = \frac{\dot{\varepsilon} N_o}{\delta \Lambda \bar{a}} \left(\exp \frac{n\Delta\tau}{\bar{a} R'T} - 1 \right) = L_{11} R'T \left(\exp \frac{A}{R'T} - 1 \right) \qquad (4)$$

2. MECHANOCHEMICAL EFFECT

Stress system of crystal influences upon the kinetics of chemical or electrochemical reactions in which a solid is participated (mechanochemical effect).

We have introduced the definition of "mechanochemical (mechanoelectrochemical) activity" of matter, which makes it possible on the base of mass action law to build kinetic equations for mechanochemical effect (M.C.-effect) earlier known from experiments. So we established that local mechanochemical activity of crystal depends upon that part of stress-tensor, which gives absolute value of spherical stress-tensor (hydrostatic stress) in the given point of crystal (irrespective of minus-plus sign of stress-uniform compression or tension). Plastic deformation of crystal changes its mechanochemical activity due to the appearance of stress fields around defects of the structure.

For calculating Gibbs free energy of crystal containing dislocations, it is necessary to express chemical potential of a solid through the value of chemical potential of dislocations. It results in a kinetic equation of mechanoelectrochemical effect caused by solid plastic deformation

$$j = j_0 \left[\exp\frac{n\Delta\tau}{\alpha R'T} \exp\frac{d\tilde{A}}{RT} - \exp\left(-\frac{\beta\tilde{A}}{RT}\right) \right] \tag{5}$$

In the case of chemical reaction it is necessary to substitute chemical affinity for $\alpha\tilde{A}$ and consider $\beta=0$ in this equation.

It follows from (5) that M.C.-effect showed the increasing on the stages of strain hardening; M.C.-effect will be considerably less on the stage of easy glide and on the final stage of dynamic recovery, when decreasing of strain hardening takes place because of the development of cross slip of dislocations and disappearance of groups of piled up dislocations, in spite of multiplication of dislocations going out to the surface. The behaviour of M.C.-effect predicted by the theory has been tested experimentally by using metals Fe, Mo, Cu, Al, steel T8-8, etc., as well as $CaCO_3$ in acid solutions and was compared with data of x-ray crystal analyses, electron micrograph technique, exo-electronic emission, differential capacity of double layer [3].

3. ENTROPY PRODUCTION IN MECHANOCHEMICAL PROCESSES

Combined irreversible processes: plastic deformation and chemical (electrochemical) reactions induce the entropy production, which is expressed as the whole sum of products of generalized flows and strengths

$$T\frac{dS}{dt} = \dot{N}A + j_1 \alpha\tilde{A} + j_2(-\beta\tilde{A}) \tag{6}$$

If solid fracture takes place and new surfaces having some surface energy are formed, the term $I_o(\Delta\bar{\tilde{\sigma}})$ representing dissipative function of this process is added to the equation (6) and is causing the Rebinder effect by the changing $\bar{\tilde{\sigma}}$ in the course of adsorbtion.

4. CHEMOMECHANICAL EFFECT

From equation (6) it follows

$$\dot{N} = L_{11} R'T (\exp \frac{A}{R'T} - 1) + L_{12} RT (\exp \frac{d\tilde{A}}{RT} - 1) \qquad (7)$$

$$\dot{J} = L_{21} R'T (\exp \frac{A}{R'T} - 1) + L_{22} RT \left[\exp \frac{d\tilde{A}}{RT} - \exp(-\frac{\beta\tilde{A}}{RT}) \right] \qquad (8)$$

Second term of the right part of the equation (7) shows the appearance of additional flow of dislocations when electrochemical (chemical) reaction is proceeding. This phenomenon was called by us "chemomechanical effect" (C.M.-effect)

$$\Delta\dot{N} = \frac{R}{R'} \dot{j}_o (\exp \frac{2d\tilde{A}}{RT} - \exp \frac{d\tilde{A}}{RT}) \qquad (9)$$

We may express C.M.-effect through the corresponding increasing of the velocity of plastical deformation under steady-state creep conditions

$$\Delta\dot{\varepsilon} = \frac{\dot{j}_o \bar{b} \Lambda R}{R'(\dot{\varepsilon}_o t + \varepsilon_o)} (\exp \frac{2d\tilde{A}}{RT} - \exp \frac{d\tilde{A}}{RT}) \qquad (10)$$

Comparison of calculations on the base of this equation according to experimental data on Cd single crystal creep in sulphuric acid solutions (state of solid active dissolving) showed good conformity. C.M.-effect was determined in our work for the first time by microexamination on $CaCO_3$ and Fe single crystals undergone microhardness indentation at first, and then crystal dissolution by acid.

The C.M.-effect was displayed in crystal dissolution in the first case by the appearance of mechanical twins and their growth from microhardness indentation, in the second one-by the appearance of jogs of slip-band on faces of microhardness indentation and reducing of microhardness under the action of anodic dissolution. Electron micrograph investigation showed form and sizes of elements of slip plastic shear appeared anew, and the kinetics of the residual microstresses relaxation. As for polycrystal materials C.M.-effect was under study using marble subjected to the previous compression in belt-apparatus and was observed in the appearance of new elements of shear. Addition of surface active inhibitors of chemical dissolution into acid electrolyte led to the disappearance of C.M.-effect. Mechanism of C.M.-effect consists in the appearance of additional flow of dislocations as the result of dissolution of surface atoms by the current of corrosion which helps the process of relaxation of dislocations both earlier accumulated before surface energy barrier and originated on the account of the start of internal (primary) sources and of new surface (induced) sources formed by dissolution of surface with the formation of monoatomic etches (two-dimensional seeds).

5. AUTOCATALYTIC PROCESS OF CHEMICAL AND MECHANICAL FAILURE OF SOLIDS (STRESS CORROSION)

There are conditions for active appearance of C.M.-effect at the top of corrosion-mechanical crack, as far as propagation of crack is defined by the properties of one crystal (transcrystalline corrosion) or of two boundary crystals (intergranular corrosion). Then C.M.-effect promotes the increasing of chemical potential of surface atoms due to the dislocations going out to the surface and stimulates M.C.-effect which in its turn increases dissolution of surface atoms and rises C.M.-effect, i.e. autocatalytic process of continuum chemical and mechanical failure on the top of the crack is developing. It leads, for example, to the increasing of the velocity of the crack in the case of stress corrosion.

This process shows ways of considerable facilitating the fracture (grinding) and mechanical dispersion of materials in different technological processes. We have investigated [3] optimal conditions of crystal for mechanical dispersion in the presence of organic and inorganic acids with inhibitors additions (which serve metallic instrument protection) and the opportunity of rigid reduction of solid mineral hardness was established.

Testing optimal dissolution on the drilling stand showed the decreasing of strength value twice as little in the case of marble drilling.

REFERENCES

I. E.Gutman, Fiziko-Chimicheskaya Mechanika materialov, 3, 264, 548, 1967 (see Soviet materials science, published by "Faraday Press").

2. C. Zener, Acta Cryst, 2, 163, 1949.
3. F. Gutman, Mechanochimiya Metallov i Zacshita ot Korrozii, Metallurgiya, Moscva, 1974.

Workshop Summary: Environmentally-Induced Lowering of Surface
Energy and The Mechanical Behavior of Solids

Prepared by D. J. Duquette

Discussion Leaders: D. J. Duquette, R. A. Oriani, C. M. Preece,
I. R. Kramer

Recorders: H. P. Bonzel, N. H. Weinberg, W. Schröter,
R. E. Cuthrell

The discussions in this workshop were primarily concerned with liquid metal embrittlement (LME), hydrogen embrittlement (HE) and the relationships between the two phenomena when interpreted in the context of the Rebinder effect. The "chemomechanical" results presented by Westwood were also discussed in one of the workshop groups. It was generally concluded that, for the time being, LME studies are practically at an impasse due to a lack of information on quantum mechanical states in a metal or alloy free surface both for clean surfaces and for surfaces which are covered with adsorbates. A strong emphasis on obtaining this type of theoretical data was urged and several experiments were suggested for further phenomenological work in the LME areas. For example, slow crack growth rate experiments to determine the effects of surface diffusion and area coverage by an adsorbate might yield some information in the relative roles of these two variables. Auger electron spectroscopy was suggested as a possible tool for these studies, particularly is vapor phase adsorbates were utilized.

Much discussion centered on the electronic state of a material which is embrittled by Rebinder effects, including the role of stress or strain at a crack tip on the electron distribution and the role of the adsorbate/metal interaction on electron bonding. For example, it was suggested that true chemisorption involving electronic bonding may not exist but, instead, strong physisorption due to electrostatic changes may in fact be the case. It was also suggested that LME may be compared to current work on transition metal surfaces where a high density of states of antibonding character is thought to be a consequence of adsorption. The population of antibonding orbitals would accordingly cause a metal/metal bond weakening.

Considerable discussion centered on the justification of thermodynamic analyses to explain Rebinder effects, LME and HE. For example, it was pointed out that the large plastic zones often associated with crack growth in metals are not amenable to analyses by the Griffith (Gibbs-Volmer) approximations. It was generally agreed that kinetic studies of surface diffusion for LME and bulk diffusion for HEM were badly needed. These studies should include the role of elastic and plastic stresses and strains on these phenomena. It was also suggested that these types of experiments would lead to a better understanding of the phenomenological aspects of these effects, but that they would yield little information on the mechanisms of bond weakening. It was re-emphasized that this type of physical understanding could best be analyzed by quantum mechanical modeling studies.

In summary it was generally concluded that, in fact, little quantitative information is available for basic understanding, and that there are a number of promising areas for interaction between surface scientists and materials scientists.

Workshop Session 5

SURFACE EFFECTS IN DISSOLUTION – RELATED EMBRITTLEMENT PHENOMENA

Chairman: R. W. Staehle

SURFACE EFFECTS IN DISSOLUTION-RELATED EMBRITTLEMENT PHENOMENA

H.J. Engell

Max-Planck-Institut für Eisenforschung, Düsseldorf, F.R.G.

ABSTRACT. In metal-electrolyte systems undergoing corrosion, stress-assisted anodic dissolution and corrosion-induced embrittlement can be observed. The first of these two cases is known as active path stress corrosion cracking (APSCC), the second is represented by hydrogen embrittlement (HE). For the understanding of active path stress corrosion cracking, mechanisms involving formation of protective surface films, their local breakdown by yielding and their repair by processes called repassivation are of major interest. A critical review of the treatment by Vermilyea, by Scully and by Staehle (involving Swann's topological aspects) is given and the implications and the further development of the model are considered.
Embrittlement of a material generally occurs if the local fracture stress σ_f is lower than the local yield stress σ_y. HE thus is to be expected if a primary or secondary effect of corrosion is a reduction of the fracture stress or an increase of the yield stress at or near the crack tip. A well developed theory proposed by Oriani is based upon the reduction of the cohesive strength of the metal by hydrogen accumulated in high stressed regions near the tip of a growing crack, thus locally giving $\sigma_f < \sigma_y$. In contrast an increase in σ_y by hydride formation at the crack tip is discussed by Gilman. At high hydrogen concentrations in the metal, hydrogen gas at very high pressures may be precipitated in microvoids. The conditions for the

action of these microvoids as critical cracks are discussed. A problem involved in active path crack propagation and in hydrogen embrittlement is the composition of the electrolyte at the tip of a growing crack and the influence of the chemical and electrochemical variables of the system upon the electrochemical situation at the crack tip. Some basic considerations referring to this problem are presented.

1. INTRODUCTION

Two limiting cases can be defined for the different phenomena of interdependence of dissolution and embrittlement of metals in contact with electrolytes: Stress-assisted anodic dissolution and corrosion-induced embrittlement. In the first case the extent and topography of surface reactions is affected by the applied stress or by the resulting strain, the properties of the bulk material some atomic diameters beyond the surface not being changed by the corrosion process. In the case of corrosion-induced embrittlement, on the other hand, the corrosion process involving dissolution changes the bulk properties and causes embrittlement of the material within a certain depth under the surface. An example of the first group of phenomena is active path stress corrosion cracking, e.g. of a gold-copper alloy in a ferric chloride solution. The best known example representing the other limiting case is hydrogen embrittlement of high strength steel in a solution containing hydrogen sulfide.
A problem involved in active path crack propagation and in hydrogen embrittlement is the composition of the electrolyte at the tip of a growing crack and the influence of the chemical and electrochemical variables of the system upon the electrochemical situation at the crack tip. In the literature there is major disagreement concerning the chemical and electrical coupling between the bulk electrolyte in contact with the outer surface of the specimen and the electrolyte in contact with the crack tip. In this context and with relation to embrittlement phenomena the formation and action of metal sponge layer by decomposition of an alloy during corrosion is of interest.

2. ACTIVE PATH STRESS CORROSION CRACKING (APSCC)

APSCC by definition is the anodic dissolution of a metal into an electrolyte along a very narrow, crack-like path, caused by silmultaneous action of a corroding process and an external or internal stress. The corrosion path can be trans- or intergranular, in the first instance following crystallographic defined directions or going simply normal to the local stress vector. In most APSCC-systems (being composed of a susceptible metal and a very special electrolyte) some plastic deformation can be observed along the crack walls. In some systems, even without internal or external stress, selective corrosion along grain boundaries or slip lines is observed. If this corrosion is increased by an applied stress, then depending on the extent of acceleration the phenomenon is called stress assisted intergranular corrosion or intergranular stress corrosion cracking. The susceuptibility of a metal to APSCC normally is increased by factors causing course slip:

- low stacking fault energy
- precipitates that can be sheared by moving dislocations
- long range and short range order
- Lüders slip
- Partevin - Le Chatelier Effect

It can be stated that almost all metal-electrolyte systems which show active pass stress corrosion cracking (APSCC) are known to form surface films. Possibly, the statement could be: There is no metal-electrolyte system exhibiting APSCC known in which the formation of surface films can be excluded. On the other hand many systems with surface films are completely free of stress corrosion susceptibility (SCS). Thus, surface films are a necessary, but not a sufficient condition for APSCC. In addition to this restriction it must be mentioned that most of the metals and alloys used for engineering constructions under corroding conditions owe their chemical stability to the formation of surface films, mainly layers of reaction products of small solubility or low rate of dissolution.
The films discussed here may be of the following types:

(a) Layers of adsorbed species of the electrolyte if they have a marked influence upon the corrosion process, particularly upon the anodic dissolution of the metal;

(b) Passive layers, ranging in thickness from one atomic diameter to several hundreds or even a thousand Angström-units;

(c) tarnishing layers built up by the products of the corrosion reaction, dense or porous, but exhibiting a marked protection upon the metal and reducing the rate of corrosion by a factor of 3 or more;

(d) undissolved or re-deposited components of the alloy, e.g. the noble component of a noble metal alloy, if they influence the corrosion reaction in the same manner as the tarnishing layers mentioned above.

The contribution of very porous reaction products by their possible action as the cathode of the corrosion process which was assumed to be of importance for SCS [1] can remain out of discussion for in most systems the metal itself will be a more active cathode than the reaction products.[2] The surface layers have to concentrate the anodic electrochemical process to the crack tip and to allow for the cathodic processes to produce a current sufficient to balance the anodic currents at the crack tip. Primarily, the model gives no indication of the cathodic current densities or the relative localization and size of anodes and cathodes. The anodic current density j_a (A/cm^2) at any of the growing cracks, referred to the area of the anodic zone at the crack tip, is given by the rate of crack propagation v_c (cm/sec) and the Faraday law,

$$j_a = v_c \cdot \frac{zF\rho}{M} \qquad (1)$$

where zF is the charge necessary to oxidize one mole of metal, ρ the density in g/cm^3 and M the atomic weight of the metal. The maximum current densities observed in the process of anodic dissolution of metals range from 10 A/cm^2 for aluminium [3] to 100 A/cm^2 for iron [4] . The maximum rate of crack propagation by such an electrochemical process, therefore, must be smaller than $5 \cdot 10^{-3}$ cm/sec. For SCC systems with higher rates of crack propagation other mechanisms have to be considered.

2.1 Formation, breakdown and reformation of surface films

Figs. 1-3 show current density-potential curves for iron in 4N KOH at 90°C and in 55% $Ca(NO_3)_2$ solution at 110°C and for 18Cr-8Ni steel in boiling $MgCl_2$ solution. In all these systems APSCC is observed. The transition from a maximum in the anodic current density to very small residual values is caused by the process of film formation. In Fig. 3 an additional strong cathodic current shifts the curve partially into the cathodic region. Thus, in this case, film formation is indicated by a transition from anodic to cathodic current densities and by the 'passivation peak'.

Fig. 3 shows that by potentiostatic measurement, that is for $dE/dt \rightarrow 0$, an anodic peak of the current density-potential curve cannot be observed. The reason is that the formation of the surface film is a time-consuming reaction, the rate of which increases strongly with the potential. In the range of the potential, where the film formation happens, the potentiodynamic curves of Fig. 3 are a superposition of the current density-potential curves of active and passive (or film-free and covered) parts of the surface of metal. Fig. 4 shows schematically this additive behaviour. If θ denotes the amount of coverage of the surface by the film, $d\theta/dt$ is the rate of film formation. Below the maximum current density, the rate of polarization dE/dt exceeds the rate of film formation $d\theta/dt$; above, the film formation is faster than the polarization. The increase in current at high potentials is due to pitting corrosion. At the film free surface, the anodic current is partially due to anodic dissolution of the metal, partially to formation of the surface compound. The relative amounts of both parts will depend on the potential as well as on the rate of polarization: The higher the rate of polarization the larger will be the amount of metal dissolved in the electrolyte.

Valuable information about the rate of film formation give current-time curves measured after a sudden change of the electrode potential from the active to the passive region.

In APSCC the role of stress is assumed to be the generation of strain, causing slip lines emerging mainly at the crack tip where the stress has a maximum. These slip lines are active at the moment of formation if the slip steps are large enough to cause a break

of the surface film which normally will be brittle
relative to the underlying metal. Thus rapid anodic
dissolution at the slip steps occurs which makes the
crack growing. Concurrent re-formations of the films
reduces and finally stops the dissolution process.
Continuous crack growth thus is only possible if the
formation of unfilmed surface by slip is fast enough
as compared with the re-passivation kinetics to keep
the dissolution at the crack tip at a sufficiently
high level. Some authors [5] assume that the film needs
some thickness or other protectiveness which is developed only after a certain time interval to reduce
the current substantially. Therefore, it is not
necessary that the crack tip stays actually bare, but
the rate of dissolution after film rupture must be
sufficient.

2.2 Further developments of the APSCC model

Quantifications of this model of APSCC are given in a
number of recent papers [5,6,7,8,9]. Scully [8]
defines a constant charge criterion as a prerequisite
for APSCC: A certain amount of charge (equivalent to
a certain amount of dissolved metal) must pass before
repassivation occurs. By including measurements of
the influence of strain rate on the quasi-stationary
dissolution of stress-corroding metal/ electrolyte
system and of the decay of the current after straining
some semi-quantitative rules correlating APSCC with
strain rate and film repair are derived. Scully's
film rupture - crack propagation - re-passivation
process is a discontinuous process, as shown by
Fig. 5, but this seems to be not an essential of his
model.
Vermilyea [7] introduces into the discussions the
fact that the crack grows into a strained matrix. The
growth process changes the distribution of stress and
strain near the crack tip and thus causes new strain
to occur at the tip. (Fig. 6). He presents a quantitative basis for his model by measuring the current
transients for iron in 35% NaOH at 85°C after high
speed deformation. ($\dot{\varepsilon} \geq 50$ sec^{-1}) by use of a drop
weight apparatus. Initial currents of up to 4 Amp/cm^2
decay according to $i = Kt^{-c}$, $2 < K < 14$, $c \approx 0,5$, and
reach one tenth of this value within about $5 \cdot 10^{-3}$ sec.
This shows that the growth of the film is very rapid,
but the film reaches its complete protective
properties only after ~ 1 sec. Within this time interval,

20 to several hundred Ångstrom of metal are dissolved. Vermilyea assumes this to be sufficient for APSCC. It seems obvious in this respect that propagation of cracks by APSCC with only small plastic deformation along the crack walls requires an amount of dissolution between successive slip events which is at least 10 x the amount of slip in any of these events.
Staehle discusses the film rupture-dissolution model[9] of APSCC using substantially the same terms as Scully and Vermilyea, but introduces the aspect of tunnel corrosion observed by Nielsen [10] and experimentally verified and interpreted by Swann [11] and other investigators.
The merits of this contribution is, that the tip of the corrosion tunnels formed after film break at the crack tip during the transient dissolution, penetrate farther into the material than an uniform dissolution would allow for the total crack front (Fig. 7). Thus at the tips of the corrosion tunnels a larger amount of plastic relaxation can occur according to Fig. 6, if brittle fracture of the material in between the tunnels is assumed to happen as part of the process of crack propagation. The formation of corrosion tunnels seems to be an assured experimental fact, and Swann [11] has given calculations making tunnel formation plausible for alloys of a noble and a less noble metal. It is open how tunnel formation in alloys forming a passive layer consisting of oxides can be understood. Possible, here the change in electrolyte composition within the crack or the tunnel must be taken into account.
This discussion may have shown that the main problem involved in a quantitative treatment of APSCC in terms of the film rupture-dissolution-repassivation model are the rates of these processes relative to another. More independent measurements of these reactions would be necessary and should be fitted into a mathematical model in which also better information upon the plastic processes at the tip of a growing crack must be introduced.

2.3 Crack geometry and intergranular APSCC

In the models mentioned above the production of unfilmed surface or surface with less protective films is due to slip processes and is, therefore, connected with the slip geometry of the metal crystals. The following dissolution may be independent of crystallographic directions, extending the crack tip hemispherically into the crystal. By the action of the different slip systems of the metal and due to the undirected dissolution, crack geometry independent on the crystallography and more or less normal to the main stress may develop. (Fig. 8)

A different situation arises if the anodic dissolution follows preferentially certain crystallographic directions, the slip bands or the grain boundaries. This may be due to the kinetics of the anodic dissolution as well as to the properties of the protective film. It is well known that the grain boundaries of different metals and alloys developing protective surface films are corroded in special electrolytes which in many cases too are producing stress corrosion cracking. E.g., unalloyed steel shows intergranular attack in nitrates if polarised anodically [12], aluminium alloys are attacked at the grain boundaries by dilute HCl [13] and many noble metal alloys suffer from intergranular attack e.g. in dilute $FeCl_3$ solutions [14]. Also attack of slip bands can be observed with noble metal alloys [15,16]

It can be assumed that the reason for this type of selective corrosion is the formation of a less-protective layer on top of regions of the metal with high concentrations of lattice defects. The film consists of an oxide in the case of steel and aluminium alloys, of a noble metal in case of noble metal alloys. If can be assumed that the noble metal forms a coherent film on top of the mixed crystal, being formed by nucleation and growth processes. At regions of high density of lattice defects, formation and growth of incoherent nuclei may be energetically favourable, and, therefore, a less protective film (metal sponge) is built up.

In metal-electrolyte systems showing preferential attack at grain boundaries or at grain boundaries and slip band already without applied mechanical stress it seems quite reasonable that APSCC follows the grain boundaries. A fissure developing along a grain boundary will at its tip interfere with slip bands if sufficient stress is applied. This tip, therefore, will be a place of lowest probability of formation of a protective layer and of highest rate of dissolution. The tricky interaction of film rupture, anodic dissolution and film repair is not a necessity in this case. Nevertheless, the model of "stress assisted intergranular corrosion", where the stress only opens up the crack to allow for the diffusion process, seems to be too simple for many SCC systems.

3. HYDROGEN EMBRITTLEMENT

Much confusion has been created in approaches for fundamental models of SCC by the fact that in many systems a clear decision was not possible or not made whether the development of brittle fractures under corroding conditions was due to APSCC or to hydrogen embrittlement HE. It now seems clear that for many systems which formerly have been classified into APSCC the real cause for the observed failures is hydrogen embrittlement. The reason for this change is the fact that it has been understood that also anodic polarisation can increase the uptake of hydrogen by locally destroying protective layers. This may happen preferentially in chloride solutions if there exists a certain pitting potential. If the potential is higher than that value pits develop by local anodic dissolution of the bare metal. Within the pits the p_H of the solution may change and the local potential can be substantially lower than the overall potential [17]. In many investigations the formation of gas (hydrogen) bubbles in the pits has been observed at overall potential above the hydrogen equilibrium potential. This holds for different types of steel as well as for aluminium alloys. It is interesting to note that this argument has been reversed by Ateya and Pickering [18]. Their argument that at cathodic polarisation the crack tip is more anodic than the metal surface due to the IR drop within the crack is correct, but it needs to be mentioned that it would be even more anodic during anodic polarisation. Therefore, it is unlikely that cathodic polarisation could favour APSCC. A situation reverse to that described for the interaction of pitting with hydrogen uptake could here only arise by cathodic polarisation of a passive metal into the region of active-passive transition which could help to stabilise active conditions at the tip.
The effect of dissolved hydrogen on the fracture strain of steel at different temperatures and strain rates is shown by Fig. 9.

3.1 The pressure fracture mechanism

Up to about ten years most investigators interpreted

HE by the assumption that by the corrosion process high concentrations of hydrogen are built up in the metal equivalent to pressures of the order of magnitude of 10^3-10^5. This hydrogen can be precipitated in microscopic voids and develops a hydrostatic pressure of the above mentioned values which is sufficient to produce brittle fracture if an undirectional stress is applied. This hypothesis is shown to be not of general validity by the observations of Hancock and Johnson [19] which was verified later by other investigators. These investigations have shown that the ductility of metals, mainly of steel, at low strain rates can be drastically lowered if the tensile test is carried out in hydrogen gas of a partial pressure as low as on atmosphere. Nevertheless, the pressure - fracture - mechanism of HE cannot be excluded generally, just if conditions prevail which make probable that high pressures of hydrogen can be developed in microvoids in the metal[20]. This is confirmed by the fact that even without an external stress cracks can be produced by cathodic charging of a metal in electrolytes or by quenching it from sufficiently high temperatures after equilibration with hydrogen gas. Both measures produce hydrogen concentrations in the lattice in equilibrium with high pressure hydrogen gas in microvoids.
Tetelman [20] gives a quantitative treatment of the pressure - fracture - mechanism which is based upon a reduction of the external stress necessary to propagate a microcrack σ_G by the internal pressure $P(H_2)$:

$$\sigma_G(H_2) = \sigma_G(o) - P(H_2)$$
$$\sigma_G(o) = 4\mu\varrho \cdot \gamma(H)/k_y d^{1/2}$$

with μ = shear modules, $\gamma(H)$ = surface energy in the presence of dissolved or gaseous hydrogen, d = grain diameter, k_y = Petsch constant and $\varrho \sim 10$ accounting for local plastic flow at the crack tip. According to this formula, $P(H_2)$ at about 10^4 Atm approaches $\sigma_G(o)$ and thus substantially contributes to crack propagation.

For any propagation of the crack increases the crack volume and, therefore, reduces $P(H_2)$, further growth needs restoration of the pressure by transport from the volume or an external source. The transport normally is slow as compared with crack propagation and, therefore, discontinuous crack propagation can be expected.

Embrittlement of a material generally occurs if the local fracture stress σ_f is lower than the local yield stress σ_y. Any explanation of hydrogen embrittlement, therefore, has to involve a mechanism reducing the fracture stress or increasing the yield stress. Tetelman's model gives an explanation for a reduction of σ_f by the internal hydrogen pressure in microcracks. The present author would like to focus the attention to the fact that high pressure in microvoids can as well increase the local yield stress: Any microvoid filled with hydrogen gas is surrounded by a stress field of the order of magnitude $P(H_2)$. This stress field interacts with the elastic stresses of the dislocations, attracting or repelling the dislocations and giving rise to an increase of the stress necessary to move the dislocations σ_y. In this model, as in precipitation hardening, the average distance between the microvoids would be determining for the increase in σ_y.

A HE model which combines the pressure-fracture concept with the fact, that even 1 Atm hydrogen gas can create loss of ductibility, is presented by Louthan [21]. This author states that the fracture processes typical of hydrogen embrittlement require extensive localized plastic deformation and are thus consistent with a model based on hydrogen transport by dislocations to create high localized hydrogen concentrations and subsequently high hydrogen pressures at microvoids and other volume defects in the lattice. These high pressures may aid in the extension of the defect by either continued plastic deformation or by brittle fracture until the internal pressure is sufficiently reduced. This dislocation-transport model is identical in most respects to the pressure - fracture - mechanism of embrittlement except that plastic deformation must precede void growth unless the specimen has been overcharged prior to testing. The observed relationship between susceptibility to embrittlement and coplanar dislocation motion in austenitic steels is consistent with the dislocation-transport model, as is the higher susceptibility to embrittlement in alloy systems with high binding energies for hydrogen-dislocation interactions.

A quantitative treatment shows the possibility that hydrogen can be pumped into the microvoid up to a pressure of 10^6 Atm at an external pressure of 1 Atm.

For dislocations can pass through the microvoids, it is open why they do not carry the hydrogen out of the voids as well as into them. It, therefore, seems to the present author that this model may better be applicable to accumulation of hydrogen in dislocations pile-ups. Then, however, it must be combined with an explanation of the influence of the enriched hydrogen upon the local fracture stress, as was done by Oriani [22] on the basis of a treatment by Troiano [23].

3.2 The lattice decohesion model

Oriani's model also involves the fact that dissolved hydrogen is accumulated in regions of high elastic stresses, e.g. within a few atomic distance from the crack tip, in response of the lowering of its chemical potential by the elastic strain of the crystal. The cohesive forces of the metal atoms $F(x)$ can be correlated with the surface energy γ by

$$\gamma = \int_{x-x_o}^{\infty} F(x)dx$$

x_o being the equilibrium atomic distance. For there is experimental evidence that adsorption of hydrogen lowers γ to $\gamma(H)$, $F(x)$ too must be decreased by hydrogen present in the lattice:

$$\gamma(H) = \int_{x-x_o}^{\infty} F(x)(H)dx.$$

By this, also the local maximum fracture stress σ_f must be lowered. The accumulation of dissolved hydrogen concentrates this effect at the crack tip. The time dependence (or strain rate dependence) of HE in this model is caused by the rate of hydrogen accumulation at the crack tip.

A very similar model has been treated quantitatively by van Leuwen [24] to express the crack velocity v as a function of the stress intensity factor K. This treatment leads to an equation describing the characteristic shape of the crack velocity versus stress intensity curve. The occurrence of a plateau velocity can be explained tentatively as the combined effect of four causes: (1) development of a plastic enclave at the crack tip, (2) blunting of the crack tip due to plasticity and corrosive attack, (3)

limited diffusion rate of hydrogen, and (4) a specific relationship between critical hydrogen concentration for failure and stress intensity. An effective crack tip radius is introduced. It determines both treshold stress intensity (K_{ISCC}) and plateau velocity (v_p).
Mainly for low hydrogen concentrations or equivalent partial pressures, where the pressure-fracture concept is not applicable, this decohesion theory of HE is most satisfactory at the moment.
Whereas Oriani's concept expects the largest effect of hydrogen in the dissolved state and a few Ångstrom units below the surface of the crack tip, Gilman [25] focusses attention to the influence of the adsorption of hydrogen at the crack tip surface. He argues, that changes in the surface environment can change the energy needed to create a dislocation and the energy needed for cleavage. Such changes can shift the balance between glide and cleavage initiation at a crack tip. Inhibition of plastic deformation increases the tendency for cleavage. Numerical estimates of the affect show that relatively small adsorption energies can have marked effects. Since hydrogen interacts strongly with most elements it can readily cause embrittlement and its interaction with iron is specifically discussed in terms of the formation of a surface hydride on iron. A very different idea is that of Beachem [26] who asserts that the essential action of hydrogen is to aid the plastic deformation processes at the lattice. Beachem argues that hydrogen causes cracking by increasing the plasticity or decreasing σ_y at constant σ_f, what is not reducing ductility. In addition one must question the atomistic mechanisms by which the alleged enhencement of plasticity could be effected by hydrogen. There is on the contrary evidence such as from strain-ageing and internal friction that hydrogen impedes dislocation motion.

4. THE COMPOSITION OF THE ELECTROLYTE AT THE TIP OF A GROWING CRACK

It is assumed that in APSCC as well as in HE the main surface reaction responsible for the growth of the crack-anodic metal dissolution or cathodic reduction of hydrogen ions - happens at the crack tip and that the partial reactions necessary for electroneutrality compensating this main reaction are spread out over the crack walls and the outer metal surface. It is obvious, therefore, that the composition of the electrolytes must change within the crack. A quantitative treatment of a simplified model in which only the reaction $H^+ + e^- \rightarrow H$ happens at the crack tip, all other reactions only at the outer surface or at a counter electrode and the crack has a flat bottom and parallel walls, was given by Ateya and Pickering [18] In this model the current density in the crack is constant. The authors present their results by plotting the local concentrations of the cations H^+ and M^+ and the anion Y^- as a function of the product of the current density i_H and the distance from the outer surface x. In their graphs they also show the potential difference between the outer surface and x. (Fig. 10).

The actual situation within a crack growing in a steel due to either APSCC or HE is somewhat different from Pickering's model, for - as pointed out by Pickering - even at cathodic polarisation of the crack tip by an external current or a differential aeration element iron is anodically dissolved at this place. The diffusion of this reaction product out of the crack needs some increase in concentration within the crack of both iron ions and anions, normally chloride ions. The highest possible enrichment is reached when the solution becomes saturated with $FeCl_2 \cdot 4 H_2O$. The p_H of the solution is regulated by the reactions

$$FeCl_2 \rightleftharpoons Fe^{2+} + Cl^-$$

$$3 Fe^{2+} + 4H_2O \rightleftharpoons Fe_3O_4 + 6H^+ + H_2$$

Using the free enthalpies of the components Pourbaix [27] calculated an equilibrium p_H value of 4,8 for this reaction at the Fe/Fe_3O_4 equilibrium potential and about 3,0 at the potential of the $Fe_3O_4 - Fe_2O_3$ - equilibrium.

It can be assumed that the potential at the tip of a growing crack is some hundred millivolts anodic with respect to the Fe/Fe^{2+} equilibrium potential. Under these conditions a minimum p_H of about 4 could be expected.

Observations of different authors using very elaborated technics in accordance with these considerations gave p_H values at the tip of growing cracks of about 4 under freely corroding conditions for different steels in 3.5% NaCl solutions [28], but between 2,7 and 4,7 in 10^{-3} M NaOH plus 10^{-3} NaCl solution [29]. Anodic polarisation shifts the p_H even to lower values [29,30]

These low p_H values can only be understood if a shift in potential towards much more noble values or an oxidation of the divalent iron by oxygen entering into the crack [31] can happen:

$$2\ Fe^{2+} + 3\ H_2O + 1/2O_2 \rightleftarrows 2\ Fe(OOH) + 4H^+$$

This reaction can establish a p_H lower than 1.

Another possibility of understanding the very low p_H values at crack tips reported in the literature is the formation of instable conditions within the crack. E.g. in corrosion pits in iron, the formation of supersaturated solutions with very low conductivity has been observed [17].

REFERENCES

1. L. Graf W. Wittlich, Werkst. u. Korr. 17 (1966) 471
2. A.C. Macrides, J. Electrochem. Soc. 108 (1961) 412
3. H. Kaesche, Die Korrosion der Metalle, p. 263, Springer Verlag, Berlin-Heidelberg-New York 1966
4. U.F. Frank, Z. Naturforschg., 4a (1949) 378
5. H.L. Logan, J. Res. Bur. Std. 48 (1952) 99
6. R.B. Diegle and D.A. Vermilyea, J. Electrochem. Soc. 122 (1975) 180
7. D.A. Vermilyea, J. Electrochem. Soc. 119 (1972) 405, Proc. Intern. Conf. Stress Corr. Crackg. and Hydr. Embrittlement, Firminy 1973
8. J.C. Scully, Corr. Sci. 15 (1975) 207
9. R.W. Staehle, Proc. Intern. Conf. Stress Corr. Cracking and Hydr. Embrittlement, Firminy 1973
10. N.A. Nielsen, Corrosion 20 (1964) 105
11. P.R., Swann, Corrosion 19 (1963) 3 and 102 Corrosion 25 (1969) 147. The Theory of Stress Corr. Crackg. in Alloy, Editor: J.C. Scully, p. 113, Nato Brussels 1971
12. H.J. Engell and A. Bäumel, Physical Metallurgy of Stress Corr. Fracture, Editor: T.N. Rhodin, Interscience Publishers, New York (1959)
13. L. Graf and Neth
14. L. Graf and W. Richter, Z. Metallkde. 52 (1961) 833
15. W.D. Robertson in: Stress Corr. Crackg. and Embrittlement, Editor: W.D. Robertson, p. 32, John Wiley and Sons, New York 1956
16. H.J. Engell, Jahrbuch der Max-Planck-Gesellschaft 1970, p. 97
17. G. Herbsleb and H.J. Engell, Z. Elektrochem. 65 (1961) 881
18. B.G. Ateya and H.W. Pickering, Internat. Conf. on Stress Corr. Crackg. and Hydrogen Embrittlement, Firminy 1973

19. G.G. Hancock and H.H. Johnson, Trans. Met. Soc. AIME 236 (1966) 513
20. A.S. Tetelman in: Hydrogen in Metals, p. 17 Editors: I.M. Bernstein and A.W. Thompson, ASM 1974
21. M.R. Louthan Jr., in: Hydrogen in Metals Editors: I.M. Bernstein and W. Thompson, ASM 1974
22. R.A. Oriani, Ber. Bunsenges. Physik. Chem. 76 (1972) 848
23. A.R. Troiano, Trans. AMS 52 (1960) 54
24. H.P. van Leuwen, Corrosion 31 (1975) 42
25. J.J. Gilman, Int. Conf. on Stress Corr. Crackg. and Hydrogen Embrittlement, Firminy 1973
26. C.D. Beachem, Met. Trans. 3 (1972) 437
27. M. Pourbaix, The Theory of Stress Corr. Crackg. Alloys, Editor: J.S. Scully, Nato Brussels 1971, p. 16
28. B.F. Brown, Cebelcor Bulletin E76 8 (1969) 1
29. M. Pourbaix, Corrosion 26 (1970) 431
30. B.F. Brown, The Theory of Stress Corr. Crackg. of Alloys, Editor: J.C. Scully, Nato Brussels 1971, p. 186
31. M. Pourbaix, The Theory of Stress Corr. Crackg. of Alloys, Editor: J.C. Scully, Nato Brussels 1971, p. 38/39

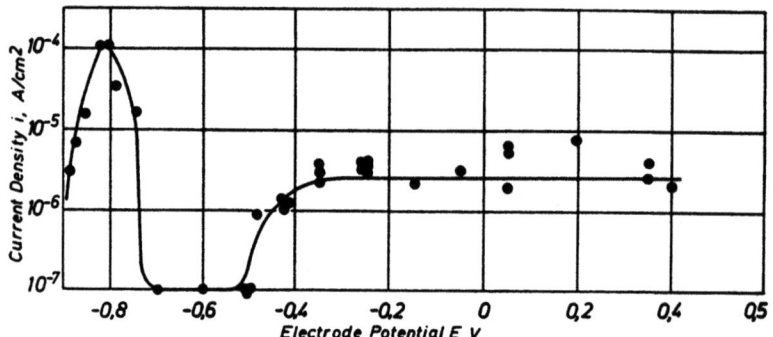

Partial Current Density - Potential Curve of the Anodic Dissolution of Iron in 4n KOH at 90°C, According Simon and Schwarz

Fig. 1

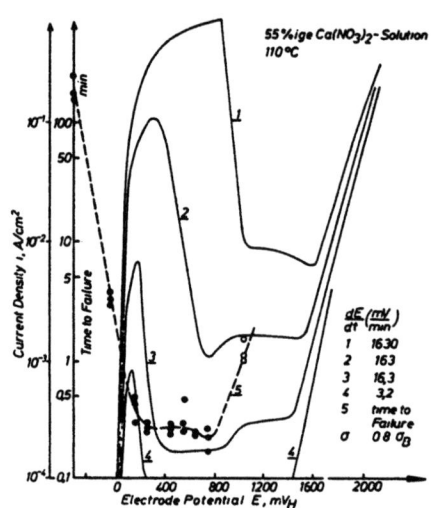

Potentiodynamic Current Density-Potential Curves (1, 2, 3 and 4), effect of the Polarization on the Time to Failure (5)

Fig. 2

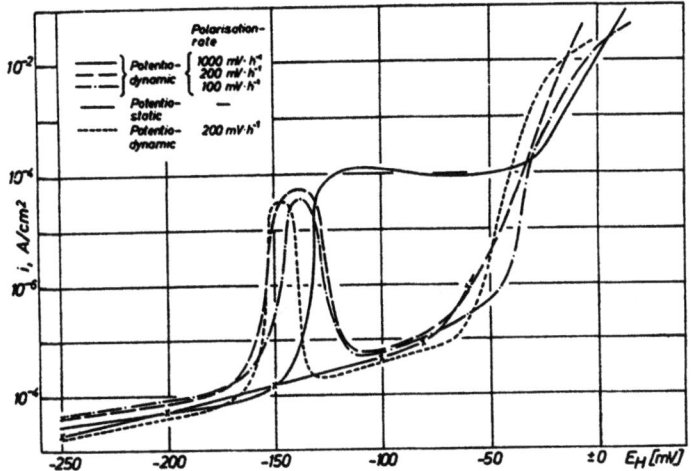

Potentiodynamic and Potentiostatic Current Density-Potential Curves of an 28Cr-9Ni-Steel in boiling 42% MgCl₂ solution, according to Ternes

Fig. 3

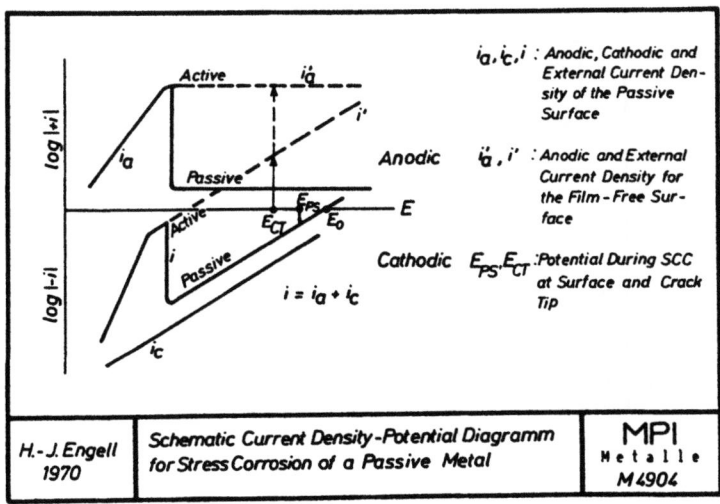

Fig. 4

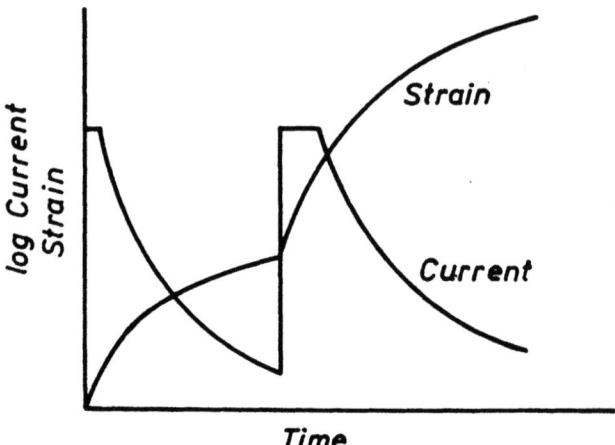

Current transients and strain at the tip of a growing crack

Fig. 5

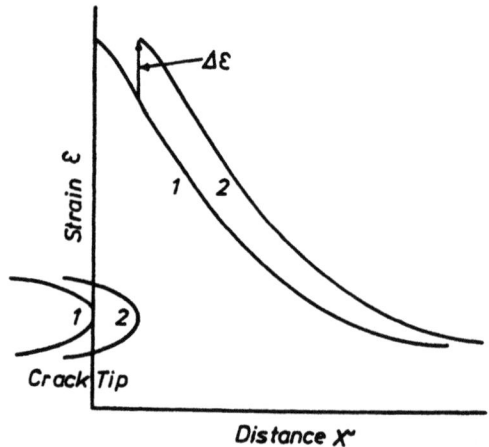

Straining in the metal caused by growth of the crack from 1 to 2

Fig. 6

Corrosion tunnels in 18-8 stainless steel
(Nielsen)

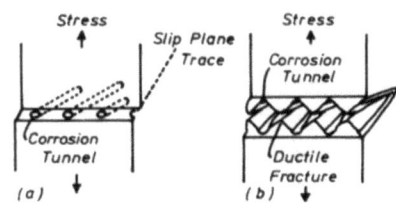

Schematic representation of tunnel formation in the stress corrosion cracking. (Swann)

Fig. 7

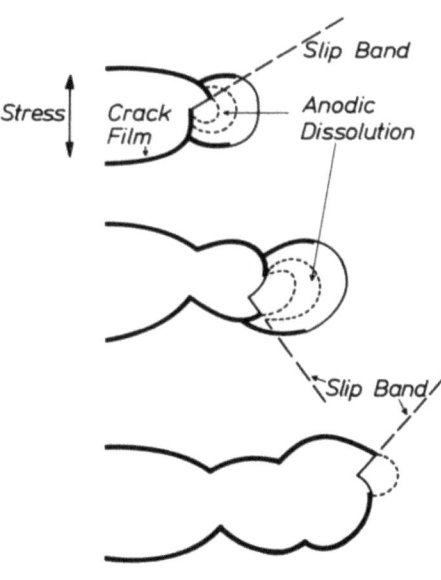

Crack Geometry

Fig. 8

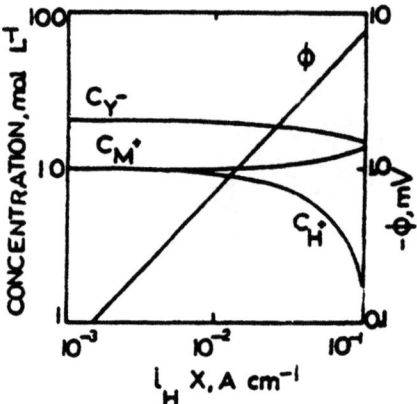

Calculated concentration and potential profiles as a function of the product of the depth x and the current density i_H during hydrogen evolution at the bottom of a crack.

Fig. 9

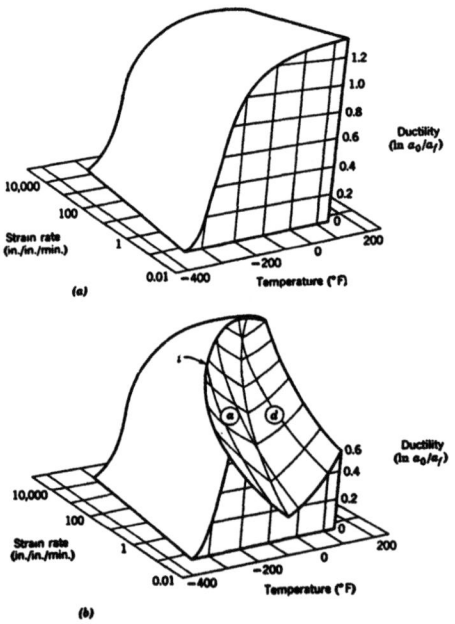

The ductility of an SAE 1020 steel as a function of strain rate and temperature: (a) as annealed, (b) as charged cathodically for 1 hr in 4 percent sulfuric acid. Curve i, in Figure 1b, bounds the range of strain rates and temperatures where embrittlement is found.

Fig. 10

DISCUSSION

Comment by C. M. Preece:

The data I have seen for hydrogen embrittlement of iron and steel could be interpreted an an interaction between hydrogen and impurities. Is there any unequivocable evidence for a reduction in the Fe-Fe bond strength?

Reply:

The reduction of the cohesive forces by hydrogen seems to be the most promising approach to quantitative understanding of hydrogen embrittlement. The correlation between hydrogen embrittlement susceptibility and "impurities" in the alloy is doubtful in my opinion.

Comment by A. Seeger:

With regard to the divancy-diffusion mechanism of Pickering and Wagner discussed in your presentation I should like to ask the following questions:

1) The estimate of the diffusion coefficient of divacancies in metals are such that a moderate lowering of the temperature should suppress the divacancy contribution decisevly. Why has the divacancy mechanism not been tested on a system where it is thought to be important by lowering the temperature?

2) Why should divacancies formed at the surface want to diffuse through the bulk of the material at all? The bulk would become strongly supersaturated, and the divacancies would tend immediately to go to the nearest and most efficient sinks. However, these should be the surface, where the divacancies were to begin with.

Reply:

If the dissolution process proceeds by transfer of atoms from a perfect crystal surface of the metal into the electrolyte, anihilation of the vacancies created in this plane needs the formation of a two dimensional pit. This nucleation process can only occur if the vacancy concentration in the surface reaches a certain supersaturation with respect to the thermodynamic equilibrium. Therefore, vacancies can diffuse to sinks in the interior of the crystal.

Comment by J. J. Mills addressed to R. A. Oriani:

I am interested in your bond decohesion model mentioned by Professor Engell. Is there an atomistic model of what happens?

If there is not, I would like to note that my theoretical model for the chemomechanical effects can also be used in metals if charge transfer is involved. That is electrons or holes from the media transferred into antibonding on bonding states in the metals causes decohesion of the surface metal atoms. According to Fischer, the electrons or holes being transferred in the solid are not smeared out over the whole solid as might be expected from simple wave mechanics, but are localized in the vicinity of the adsorbate donor or acceptor ions. Thus the shift in the Fermi level is much more substantial than one would think and substantial local decohesion can result.

Reply by R. A. Oriani:

There is not at present any atomistic model for the lowering of the cohesive force between two metal atoms by hydrogen. Your ideas on this subject are welcome and should be encouraged. I would only say that it is not clear to me to what extent one can extend what one learns about what happens on the surface to what happens at strained atomic bonds in the interior.

MECHANICAL BEHAVIOR OF OXIDE FREE STAINLESS STEEL SURFACES IN A
LOW PRESSURE HYDROGEN ENVIRONMENT*

R. E. Cuthrell

Sandia Laboratories, Albuquerque, New Mexico 87115
USA

ABSTRACT. Results are presented of an investigation of the environment-sensitive mechanical behavior of 440C stainless steel, a bearing material which has potential for applications in the "hydrogen economy." The removal of surface oxides by sputter etching in an ultrahigh vacuum system was found to soften the material, and reversible hardening/softening was observed on absorbing/desorbing hydrogen in the metal at room temperature. The presence of small amounts of carbon monoxide in the environment was observed to halt both the absorption and the desorption of hydrogen in the steel. Both contact resistance and laser interference microscopic measurements gave consistent results for the surface mechanical behavior. The latter technique showed that surface asperities were persistent under surface-surface loading. These experiments were performed in a bakeable ultrahigh vacuum enclosure in which surfaces were contacted with precise control of environmental, surface, and mechanical variables.

INTRODUCTION

The mechanical properties of metallic surfaces and of the bulk metals beneath are sensitive to the environment [1]. Dislocation pile-up is thought to occur at oxidized surfaces [2-5] and the presence of oxides substantially influences the mechanical behavior of many materials [6]. In addition, the chemical reactivity of metallic surfaces depends on the constitution of the surface. Stress corrosion cracking has been shown to occur at breaks in the

---

* This work supported by the U.S. Energy Research and Development Administration.

protective surface films on some metals [7-10]. In a high temperature hydrogen environment the protective surface oxides may be reduced and removed from metals. Hydrogen permeation into the metal may be increased as a result of the loss of the oxide barrier [11], and hydrogen embrittlement may result. Surface effects may be of significance in the implementation of new energy programs which range from coal gasification to the so-called "hydrogen economy." New materials and new surface coatings may be required.

Sensitive techniques have been developed and special ultrahigh vacuum equipment has been constructed for the study of the mechanical behavior of metallic surfaces in contact. These techniques and equipment have been described in detail elsewhere and the results of electrical contact studies and of metal-metal bonding studies using this equipment have been reported previously [12-16]. In this paper, the results are reported for the surface mechanical behavior and electrical conductance of martensitic 440C stainless steel bearing balls under load in various surface conditions and environments. Similar results are reported for titanium for comparison with the steel.

EXPERIMENTAL

A bakeable ultrahigh vacuum work chamber is shown in Fig. 1 [15]. Two metallic samples were mounted within the chamber, the system was baked in vacuo at 200°C, and base pressures of about 2×10^{-10} Torr were produced after outgassing. Oxides and organic contaminants were removed from the metallic surfaces by argon ion sputtering for 1.5 hours at pressures of about 5×10^{-5} Torr (Auger surface analyses showed barely detectable nickel contamination from the 304 stainless steel fixtures only when sputtered at significiantly higher pressures). The metallic samples were brought together using micrometers which were specially designed to minimize transverse motions, impact, and chatter. The entire vacuum system was isolated from building vibrations. A load cell and four-wire connections to the samples were used for contact resistance-load measurements. Alternatively, the upper sample was swung out of the way and the lower sample was pressed against a sapphire optical flat for measurements of surface topography and load bearing area as functions of contacting load. These measurements were made using a laser interference fringe microscope [13] which is similar to that designed by Tolansky [17]. The interference fringe pattern is essentially an optical contour map of the surface. Local fringe deviations are produced by surface roughness. The contact resistance-load and the interference fringe-load measurements were performed periodically during exposure to ultrapure hydrogen which was produced by thermal desorption from a titanium wire.

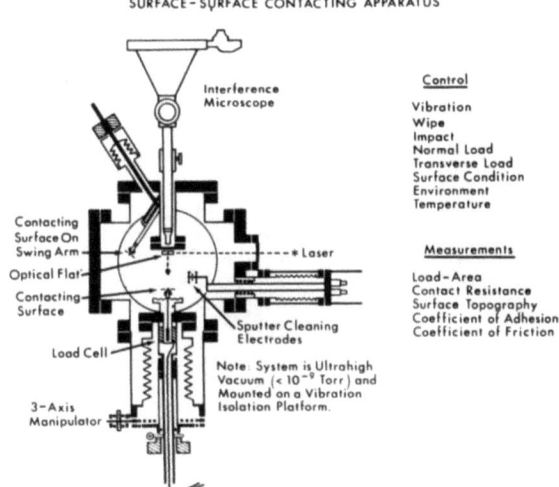

Fig. 1. Bakeable ultrahigh vacuum work chamber for surface-surface contacting.

RESULTS AND DISCUSSION

The overall area of contact between metallic samples which are pressed together can be defined approximately in terms of the elastic properties of the metals and is known as the Hertz area [18-20]. This area of contact may be microscopically discontinuous as a result of the persistence of asperities in the junction region [21]. Interference fringe topographs (upper, Fig. 2) for the 440C stainless steel bearing balls show that the Hertz area (the central dark region) increases as the load is increased against a sapphire optical flat. The dependence of the radius of this area on the applied load is shown in Fig. 2 (lower) for the oxidized surface (in air), for the surface after sputtering away the oxide, and for the sputtered surface after exposure to hydrogen. The surface is apparently softened when the oxide is removed since the Hertz area is greater under a given load than for the oxidized surface. When exposed to hydrogen, the surface region is hardened and possibly embrittled. Since no evidence of cracking was detected in SEM photographs of the surface, if embrittlement occured, the stresses did not exceed the fracture strength. On pumping the hydrogen out of the work chamber, the hardness decreased and a Hertz radius-load curve (not shown) very similar to that for the original surface was obtained. These results indicate that the hydrogen is very weakly bound in the metal and that a solution of hydrogen, rather than a stable hydride, is probably formed under the low pressure and room temperature conditions.

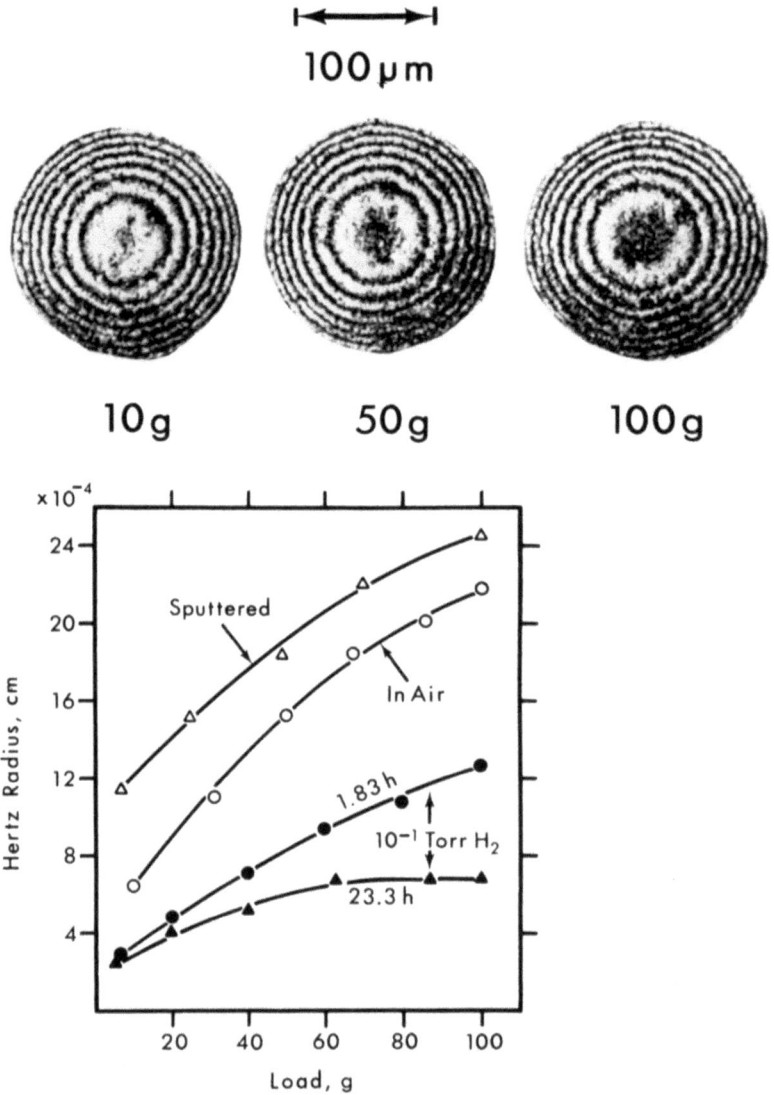

Fig. 2. Laser interference fringe patterns for 440C stainless steel (upper) show the increase in the Hertz area (central dark region) with load. The surfaces are softened on removing the oxide and hardened on exposure to hydrogen (lower).

Since the darkness of the contact area (which shows the presence of asperities) was not decreased on loading, the asperities are more rigid than the subsurface bulk, which was found to deform elastically (the uploading curves in Fig. 2 were reproduced on downloading).

The deformation of titanium (99.97%) is also elastic (upper curve, Fig. 3) provided the oxides are not removed from the surfaces. However, after sputter etching, the behavior is predominantly plastic (the Hertz area produced on uploading remains essentially constant on downloading, lower curves, Fig. 3). These data are presented by way of contract to the observations for the stainless steel and to show the capabilities of the instrumentation for studying surface mechanical effects under carefully controlled conditions.

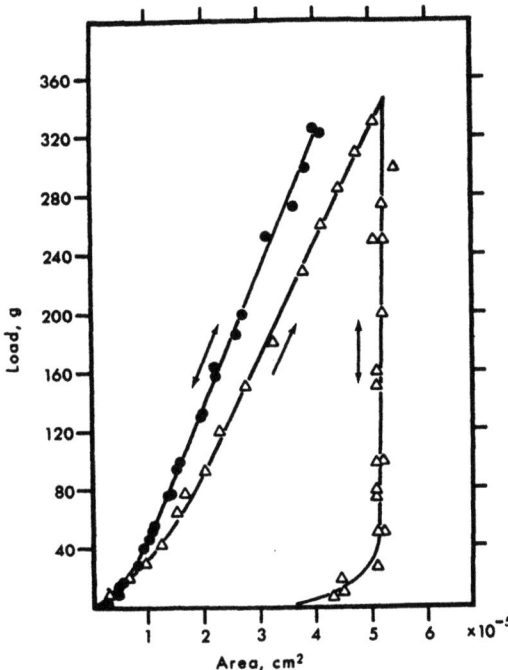

Fig. 3. The Hertz area-load curves for titanium show the elastic behavior before (upper curve) and the plastic behavior after (lower curves) the removal of the surface oxides.

Electrical contact resistance-load measurements (which are more precise than the direct measurement of contacting areas using laser interference fringe microscopy) show the same hardening/softening trends for 440C stainless steel on absorbing/ desorbing hydrogen (Fig. 4, upper). Under fixed load the Hertz area is less and the contact resistance is, therefore, greater for a harder surface. A detailed interpretation of these data in terms of the elastic modulus and of the persistent surface asperities in the contacting area will be presented elsewhere [22]. Carbon monoxide, which can be adsorbed on the surfaces, was found

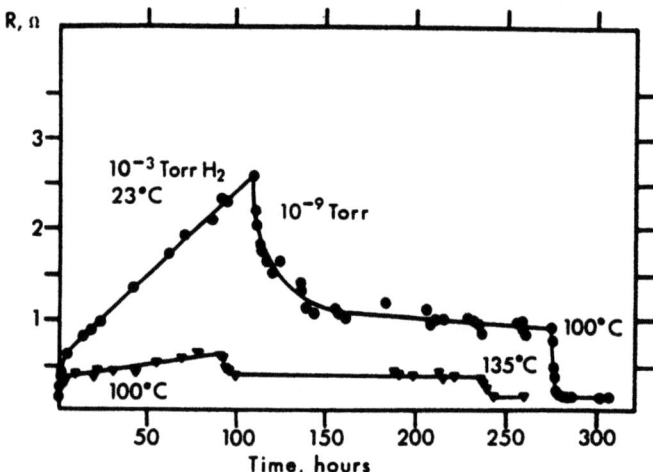

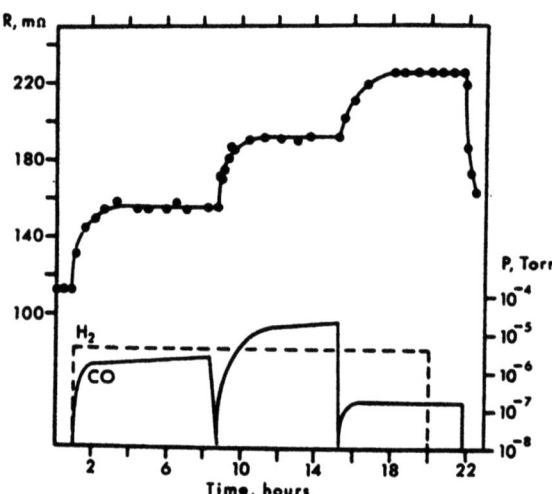

Fig. 4. Contact resistance of 440C stainless steel as a function of exposure time to a hydrogen environment (upper), and to a hydrogen environment containing carbon monoxide (lower). Repetitive closures were made under 20g closure loads for these data points.

to halt both the absorption and desorption of hydrogen in the stainless steel (Fig. 4, lower) when present in the environment at partial pressures of 10^{-7} Torr or greater. Since there are several steps which may occur at the surface before hydrogen atoms or protons can diffuse into metals, it is uncertain whether the diffusion of hydrogen is stopped by physical blockage by CO molecules at adsorption sites (a stearic effect) or the dissociation/ionization steps are hindered by the CO-metal sharing of electrons (an electronic effect).

CONCLUSIONS

The surface mechanical behavior of 440C stainless steel is significantly affected by the presence of brittle surface layers such as oxides and by environmental constituents. When the oxides are removed by sputtering the surfaces, the material is softened. Reversible hardening/softening occurs on the absorption/desorption of hydrogen in the metal at room temperature. Both contact resistance and laser interference fringe measurements as functions of the contacting load give consistent indications of this softening/hardening. Since the metal behaved elastically, it is concluded that the absorption of hydrogen increased the elastic modulus. Surface asperities were found to persist under loading, which indicates that the asperities are more rigid than the underlying bulk. Since the surfaces were free of oxides, the geometry of the asperities may be responsible for their greater rigidity. Experimental results also indicate that both hydrogen absorption and desorption are blocked by carbon monoxide when present in the environment of the surfaces at partial pressures of 10^{-7} Torr or greater.

These results may be significant in the understanding of hydrogen embrittlement phenomena, in the choice of materials and surface treatments for hydride containment in nuclear energy applications, in the choice of materials for use in the so-called "hydrogen economy," and in the understanding of the behavior of metallic surfaces in contact (especially when surface asperities are persistent on loading). The results are also significant in that they show that substantial changes in mechanical properties may accompany the absorption of hydrogen in materials which do not form stable hydrides (Fe, Mn, Cr, and Cu, the metallic constituents of 440C stainless steel, were neither predicted nor found to form hydrides [23,24]).

REFERENCES

1. A. R. C. Westwood and N. S. Stoloff, Eds., Environment-Sensitive Mechanical Behavior (Gordon and Breach, New York, 1966).
2. A. H. Cottrell, Dislocations and Plastic Flow in Crystals (Oxford, 1953), p. 54.
3. A. V. McRae and L. H. Germer, Ann. N. Y. Acad. Sci., 101, 627 (1963).
4. R. L. Fleischer, Acta Met. 8, 598 (1960).
5. I. R. Kramer and L. J. Demer, Trans. Met. Soc. AIME, 221, 780 (1961).

6. E. N. da C. Andrade and C. Henderson, Phil. Trans. Roy. Soc. A244, 177 (1951).
7. A. J. Forty and P. Humble, Phil. Mag. 8, 247 (1963).
8. A. J. McEvily and A. P. Bond, J. Electrochem. Soc. 112, 131 (1965).
9. A. J. Forty and P. Humble, Environment-Sensitive Mechanical Behavior (Gordon and Breach, New York, 1966), p. 403.
10. A. J. McEvily and A. P. Bond, Environment-Sensitive Mechanical Behavior (Gordon and Breach, New York, 1966), p. 421.
11. C. L. Huffine and J. M. Williams, Corrosion 16, 430t (1960).
12. R. E. Cuthrell and D. W. Tipping, "Electric Contacts: Equipment and Mechanics of Closure for Gold Contacts," Sandia Laboratories Research Report No. SC-RR-72-0783, December 1972.
13. R. E. Cuthrell and D. W. Tipping, J. Appl. Phys. 44, 3277 (1973).
14. R. E. Cuthrell and D. W. Tipping, J. Appl. Phys. 44, 4360 (1973).
15. R. E. Cuthrell, J. Vac. Sci. Technol. 11, 1166 (1974).
16. R. E. Cuthrell and D. W. Tipping, IEEE Trans. PHP-10, 4 (1974).
17. S. Tolansky, Multiple-Beam Interference Microscopy of Metals (Academic, New York, 1970).
18. H. Hertz, Gesammelte Werke 1. Leipzig: Barth 1895.
19. R. J. Roark, Formulas for Stress and Strain (McGraw-Hill, New York, 1954), p. 287ff.
20. S. P. Timoshenko and J. N. Goodier, Theory of Elasticity (McGraw-Hill, New York, 1934), pp. 409-420.
21. R. Holm and E. Holm. Electric Contacts (Springer-Verlag, New York, 1967), pp. 35-37.
22. R. E. Cuthrell, "Mechanical Behavior of Oxide Free Stainless Steel Surfaces in a Low Pressure Hydrogen Environment," submitted to Corrosion.
23. T. R. P. Gibb, Jr., Progress in Inorganic Chemistry (Interscience, 1962), 3, p. 315.
24. K. M. Mackay, Hydrogen Compounds of the Metallic Elements (E. & F. N. Spon, London, 1966), pp. 52-53.

A UNIFYING MODEL OF INTERGRANULAR CORROSION, INTER-
GRANULAR PENETRATION OF LIQUID METALS, STRESS
CORROSION CRACKING, AND LIQUID METAL EMBRITTLEMENT
OF SUBSTITUTIONAL SOLID SOLUTIONS

W. Frank and L. Graf

Max-Planck-Institut für Metallforschung,
Stuttgart, and Institut für theoretische
und angewandte Physik der Universität
Stuttgart, Stuttgart, Germany

ABSTRACT. Inhomogeneities in solid solutions, like
those arising from atomic misfits, can be accommodated
in part during crystal growth by the penetration of
dislocations from grain boundaries. By these dislocations and by those leaving the grains during
plastic deformation, slip steps are generated on the
grain boundaries. Both the reactivity of the grain
boundaries and the diffusivity of aqueous electrolytes
or liquid metals along them are strongly enhanced by
these slip steps. This concept permits a unifying
interpretation of intergranular corrosion, stress
corrosion cracking, and of related phenomena of solid
solutions.

1. INTRODUCTION

In the presence of aqueous electrolytes, e.g.
$FeCl_3$, substitutional solid solutions, such as CuAu
and CuZn, show intergranular corrosion [1]. It starts
at the crystal surface and penetrates along grain
boundaries into the bulk according to a $\sqrt{t}$-law (t=time
of exposure). The same is true for the intergranular
penetration of liquid metals into solid solutions [1].
These effects, which do not take place in pure metals,
indicate an extraordinarily high reactivity of the
grain boundaries of solid solutions ("static solid
solution effect").

During plastic deformation the presence of aqueous

electrolytes or liquid metals strongly reduces the
mechanical strengths of solid solutions, thus leading
to stress corrosion cracking or liquid metal embrittlement, respectively [1]. The observation that the
fracture occurs along the grain boundaries or, as in
the case of transgranular stress corrosion cracking,
starts from perturbed regions of the crystal surface
(e.g., from slip bands) implies that during plastic
deformation the reactivity of these areas is even
higher ("dynamic solid solution effect"). - Both solid
solution effects are most pronounced at a concentration
of 50 at.-%.

The similarity of these effects for various solid
solutions in either aqueous electrolytes or liquid
metals suggests that a _physical_ property of the grain
boundaries of solid solutions must be the primary
cause of these phenomena, though for their occurrence
(electro-)chemical prerequisites have to be fulfilled,
too: Whereas the structure of the grain boundaries and
its changes during plastic deformation govern the
general features of the solid solution effects, the
peculiarities of a given alloy in a given electrolyte
(liquid metal) originate from the specific (electro-)
chemical reactions taking place in this system. The
present paper does not cover details due to (electro-)
chemical aspects. It rather aims at presenting
theoretical estimates for a model which can account
for the general features of the solid solution effects,
thus laying the foundation of a unifying interpretation
of the effects mentioned in the title.

2. BASIC IDEAS [2]

It is well known that in boron- or phosphorus-
doped regions of silicon crystals, dislocations are
incorporated [3]. In this way the increase of the free
energy of the system which arises from doping-induced
inhomogeneities, such as atomic misfits, can kept low.
One may expect that in a similar way the surface energy
of the grains of metallic solid solutions can be decreased by the penetration of dislocations from the
grain boundaries. Thus already during crystal growth
slip steps are produced on the grain boundaries. It is
evident that atoms located on these steps particularly
easily react with electrolytes or liquid metals. This
explains the high reactivity of the grain boundaries
of solid solutions and hence the occurrence of intergranular corrosion or intergranular penetration of

liquid metals in these materials. These processes are limited by the diffusion of the electrolyte (liquid metal) through the grid of slip steps on the grain boundaries which connects the corrosion front (penetration front) with the crystal surface (cf. Sect. 4).

During plastic deformation additional slip steps are generated on the grain boundaries. This leads to an enhancement of both the reactivity of the grain boundaries and the diffusion rate along the grid of slip steps and gives rise to an accelerated motion of the corrosion front (penetration front) along the grain boundaries (cf. Sect. 4). The result is stress corrosion cracking or liquid metal embrittlement, respectively.

3. MATHEMATICAL TREATMENT

3.1 General remarks

For the sake of simplicity we use in the following the nomenclature of intergranular corrosion and stress corrosion cracking, although all results hold for the intergranular penetration of liquid metals and liquid metal embrittlement, too. As to the dynamic solid solution effect, the calculation is specified to intergranular stress corrosion cracking during a tensile test, in order to have well-defined conditions. It should be emphasized, however, that the characteristic features of the results remain unchanged for other deformation modes or for transgranular stress corrosion cracking.

3.2 Intergranular corrosion (static solid solution effect)

The critical shear stress τ_o of metallic solid solutions is controlled by the overcoming of solute atoms ("solid solution inhomogeneities") by dislocations. Hence, the depth Δ_o to which dislocations emitted from grain boundaries may penetrate into the grains <u>without</u> the assistance of an external stress, is a fraction $f_o(<1)$ of the average distance Λ between the solute atoms in the glide planes, i.e.,

$$\Delta_o = f_o \Lambda = f_o \frac{g_\alpha a_o}{\sqrt{c}} \tag{1}$$

(g_α = geometrical factor of the order of unity, which depends on the operating slip system; a_o = lattice constant). Here c is the atomic fraction of solid solution inhomogeneities and related to the atomic fraction c_A (c_B) of the atoms of type A(B) of a binary alloy through

$$c = c_A c_B = c_A(1-c_A) = (1-c_B)c_B . \qquad (2)$$

Since $\Delta_o \frac{c}{\Omega}$ is the number of inhomogeneities under the grain boundary unit area up to the depth Δ_o, the total length of dislocation lines in the unit area of the surface layer of thickness Δ_o is $\Delta_o \frac{c}{\Omega} \delta$ if δ is the average length of dislocation line emitted from the grain boundaries per inhomogeneity. The density n_1^o of slip steps per unit length of the corrosion front may thus be written as

$$n_1^o = g_\beta \frac{\Delta_o c \delta}{\Omega} = \frac{g_\alpha g_\beta f_o a_o \delta}{\Omega} \sqrt{c} . \qquad (3)$$

(Ω = atomic volume; g_β = geometrical factor which accounts for the inclination of the slip steps against the corrosion front and for the fact that in general the dislocations in the surface layers of the grains are not parallel to the grain boundaries).

The diffusion rate of the reactive ions of an electrolyte along a slip step is approximately given by $\sqrt{\frac{D_1}{2t}}$ (D_1 = diffusivity along a slip step, which may be considered to be a pipe with the cross-section q_1). The ion flux through one pipe is then

$$I_1(t) = q_1 \sqrt{\frac{D_1}{2t}} C \qquad (4)$$

(C = number of reactive ions per unit volume of the electrolyte). Assuming that the grid of slip steps is equivalent to n_1^o uninterrupted pipes connecting the unit length of the corrosion front with the crystal surface, the rate at which ions arrive at the corrosion front unit length at time t, t+dt is

$$R_D = n_1^o I_1(t) = \frac{g_\alpha g_\beta f_o a_o \delta q_1 C}{\Omega} \sqrt{\frac{D_1 c_A(1-c_A)}{2t}} . \qquad (5)$$

If the intergranular corrosion were exclusively reaction-controlled, the rate of reaction, R_R, would be proportional to the density n_1^o of slip steps at the corrosion front:

$$R_R = \mu_1 n_1^o = \frac{\mu_1 g_\alpha g_\beta f_o a_o^\delta}{\Omega} \sqrt{c_A(1-c_A)} \qquad (6)$$

(μ_1 = constant of reaction). The actual reaction rate, R, is obtained from (5) and (6) according to

$$R = \frac{R_D R_R}{R_D + R_R} \qquad (7)$$

and related to the velocity v of the corrosion front via

$$v = \frac{g_\gamma}{g_\delta} a_o^2 R \qquad (8)$$

($g_\delta a_o$ = interatomic distance parallel to the direction of motion of the corrosion front, $\frac{g_\delta}{a_o}$ = number of atoms per unit length of the corrosion front).

3.3 Intergranular stress corrosion cracking (dynamic solid solution effect)

The production of slip steps during plastic deformation has various causes:

(i) The application of a stress τ increases the depth $\Delta(\tau)$ to which dislocations penetrate from the grain boundaries in order to release the energy stored around inhomogeneities, i.e., f_o is increased by an amount $f_\tau^{(1)}\tau + f_\tau^{(2)}\tau^2 + \ldots$ and exceeds unity for $\tau > \tau_o (f_\tau^{(i)} = \text{const})$.

(ii) The number of inhomogeneities lying within a layer of thickness $\Delta(\tau)$ under the grain boundaries increases linearly with the shear strain a on account of a corresponding increase of the grain boundary surface. This leads to an additional increase of the slip step density by a factor of $(1 + f_a a)$; f_a = const.

Due to (i) and (ii) the slip line density at the stress τ and the strain a is

$$n_1 = \frac{g_\alpha g_\beta (f_o + f_\tau^{(1)} \tau + \ldots) a_o^\delta}{\Omega} \sqrt{c}\,(1 + f_a a)$$

$$\approx n_1^o \left(1 + \frac{f_\tau^{(1)}}{f_o} \tau + f_a a\right). \qquad (9)$$

Within a linear work-hardening stage, the work-hardening coefficient Θ in

$$\tau = \tau_o + \Theta a \qquad (10)$$

is independent of a, so that (9) takes the form

$$n_1 = n_1^o (k_\alpha + k_\beta a) \tag{11}$$

with the constants

$$k_\alpha = 1 + \frac{f^{(1)}_\tau}{f_o} \tau_o , \tag{11a}$$

$$k_\beta = \frac{f^{(1)}_\tau}{f_o}\theta + f_a . \tag{11b}$$

The subscript of n_1 indicates that the height of these slip steps is of the order of 1 b (b = dislocation strength).

(iii) Dislocations emitted from dislocation sources near the grain boundaries generate additional slip steps on the grain boundaries. The height of these steps is mb (m = number of dislocations emitted by a source before it stops to operate). Since the separation x (measured parallel to the corrosion front) of these m-fold slip steps is related to a via [4]

$$x = \frac{g_\epsilon \, mb}{a} , \tag{12}$$

their density per unit length of the corrosion front is

$$n_m = \frac{1}{x} = \frac{a}{g_\epsilon \, mb} \tag{13}$$

(g_ϵ = geometrical factor).* In tensile tests the slip rate $\dot{a}$ is kept approximately constant, so that (11) and (13) may be re-written as

$$n_1 = n_1^o(k_\alpha + k_\beta \dot{a} \, t) \tag{14a}$$

and

$$n_m = \frac{\dot{a}}{g_\epsilon \, mb} t . \tag{14b}$$

Now the calculation of R_R and R_D is straightforward. For R_R we obtain

---

*Symbols with the subscript m, e.g., n_m, q_m, D_m, refer to m-fold slip lines. They have the corresponding meanings as the symbols refering to 1-fold slip steps, e.g., n_1, q_1, D_1.

$$R_R = \mu_1 n_1 + \mu_m n_m = \mu_1 n_1^o k_\alpha + (\mu_1 n_1^o k_\beta + \frac{\mu_m}{g_\varepsilon \, mb}) \dot{a} t \quad (15a)$$

and for R_D, with

$$I_{1,m}(t,t') = q_{1,m} \sqrt{\frac{D_{1,m}}{2(t-t')}} \; C \; , \quad (16)$$

$$\begin{aligned} R_D &= n_1^o k_\alpha I_1(t) + \int_{t'=o}^{t} I_1(t,t') \frac{dn_1}{dt'} dt' + \int_{t'=o}^{t} I_m(t,t') \frac{dn_m}{dt'} dt' \\ &= n_1^o k_\alpha q_1 C \sqrt{\frac{D_1}{2t}} + n_1^o k_\beta q_1 C \, \dot{a} \sqrt{D_1 t} \\ &\quad + \frac{q_m C}{g_\varepsilon \, mb} \dot{a} \sqrt{D_m t} \; . \end{aligned} \quad (15b)$$

It is important to realize that the diffusivities entering (15b) increase exponentially with time because of the following reason: During plastic deformation n_1 and n_m and thus the probability for the overlapping of slip steps increase proportional to t. Hence, in the course of time the activation energies Q_1 and Q_m, which govern the diffusivities along 1-fold and m-fold slip steps, decrease according to

$$\begin{aligned} Q_{1,m} &= Q_{1,m}^o (1 - \sqrt{q_1} n_1 - \sqrt{q_m} n_m + \ldots) \\ &\approx Q_{1,m}^o [1 - (K_\alpha + K_\beta t)] \end{aligned} \quad (17)$$

with

$$Q_{1,m}^o = Q_{1,m}(n_1 = n_m = o) \; , \quad (17a)$$

$$K_\alpha = \sqrt{q_1} \, n_1^o \, k_\alpha \; , \quad (17b)$$

$$K_\beta = \sqrt{q_1} \, n_1^o \, k_\beta \, \dot{a} + \frac{\sqrt{q_m} \, \dot{a}}{g_\varepsilon \, mb} \; . \quad (17c)$$

As a consequence, the diffusivities take the form

$$D_{1,m} = D_{1,m}^o \, e^{-Q_{1,m}/kT} = D_{1,m}^{o*} \, e^{X_{1,m} t} \quad (18)$$

($D_{1,m}^o$ = pre-exponential factor of $D_{1,m}$, k = Boltzmann's constant, T = absolute temperature) with

$$D_{1,m}^{o*} = D_{1,m}^o \, e^{-\frac{Q_{1,m}^o (1-K_\alpha)}{kT}} \quad (18a)$$

and

$$X_{1,m} = \frac{Q_{1,m}^o K_\beta}{kT} \; . \quad (18b)$$

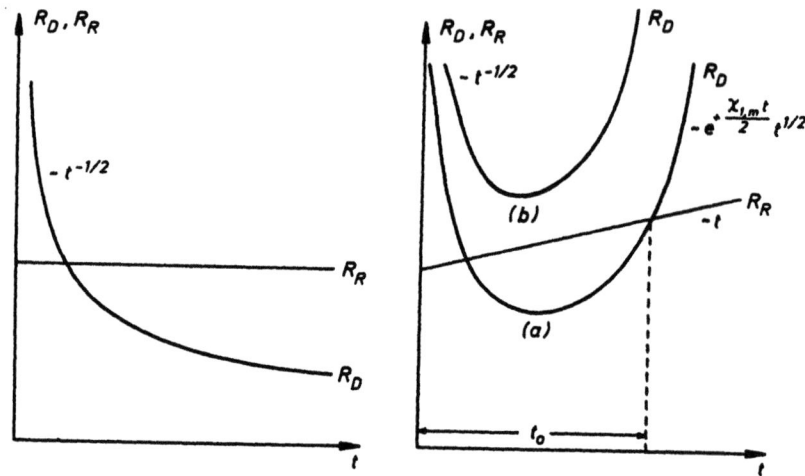

Fig. 1. Static solid solution effect

Fig. 2. Dynamic solid solution effect

The velocity of the corrosion front during a tensile test may finally be obtained from (7), (8), (15) and (18).

4. DISCUSSION

For both the static [eqs.(5), (6)] and the dynamic [eqs.(3), (15)] solid solution effect the rate R [eq.(7)] has a maximum for $c_A = c_B = 50\%$, in full agreement with the experiments [1, 5]. It is of interest that for $c_A = 0$ no static effect is predicted, whereas a dynamic effect is possible. Indeed, unalloyed copper shows a slight intergranular corrosion in ammonia under load [6, 7].

The static solid solution effect is reaction-controlled at the very first moment, but becomes and remains diffusion-controlled at later times (Fig.1). Since even at early times the experimental data follow an $(R \sim t^{-1/2})$-law, R_R must be fairly high.

The dynamic solid solution effect (Fig.2) is at early times reaction-controlled, too. Depending on the values of the parameters entering eqs. (15), at intermediate times it may either become diffusion-controlled (curve (a)) or remain reaction-controlled (curve (b)). In any case after a transition period t_o stress corrosion cracking is reaction-controlled, and the velocity of the corrosion front increases linearly with t, thus leading to catastrophic rupture.

It should be emphasized that the above model does neither take into account the stress concentration at the notches formed at the corrosion front nor the stress increase due to the corrosion-induced reduction of the cross-section of the specimens. Both effects tend to accelerate stress corrosion cracking even more.

ACKNOWLEDGEMENT

The authors are very grateful to Professor Dr. A. Seeger for critically reading the manuscript.

REFERENCES

1. Graf, L., in: The Theory of Stress Corrosion Cracking in Alloys, Scully, J.C., Ed., NATO Scientific Affairs Division, Brussels, 1971, p. 399.
2. Frank, W., and Graf, L., Z.Metallkde., Sept. 1975, in press.
3. Grienauer, H.S., and Mayer, K.R., in: Lattice Defects in Semiconductors 1974 (Inst.Phys.Conf.Ser.No. 23), Huntley, F.A., Ed., The Institute of Physics, London and Bristol, 1975, p. 550.
4. Kronmüller, H., in: Moderne Probleme der Metallphysik, Seeger, A., Ed., Springer, Berlin-Heidelberg-New York, 1965, Vol. 1, p. 126.
5. Graf, L., and Lacour, H.R., Z.Metallkde, 51, 152, 1960.
6. Graf, L., and Wittich, W., Werkstoffe und Korrosion, 17, 595, 1966.
7. Pugh, E.N., Craig, J.V., and Sedriks, A.J., in: Fundamental Aspects of Stress Corrosion Cracking, Staehle, R.W., Forty, A.J., and van Rooyen, D., Eds., NACE, Houston, 1969, p. 118.

DISCUSSION

Comment by H.-J. Engell:

I wonder about the following details of the theory presented:

1) Does dissolution along a grain boundary (which is a highly disturbed crystal area) require surface steps?

2) Can the active ion of the electrolyte (concentration C) diffuse along the surface steps at the grain boundary by some type of solid state diffusion and why should it diffuse inwards?

3) A crack propagation time law like $v \sim e^t$ or $v \sim t$ (all other conditions kept constant) has not been observed up to now as much as I know. On the contrary, v=const. for fixed K_I, T, C, E...is more familiar.

Reply:

1) The experimental evidence that the grain boundaries of alloys show a high reactivity with aqueous electrolytes or liquid metals, fairly independent of both the chemical nature of the alloys and the agents, implies that this arises from a special structure of the grain boundaries of alloys. It is reasonable to assume that due to the difference in size and hardness between the A- and B-type atoms, the grain boundaries of alloys show a roughness on a fine (e.g., atomistic) scale which is not present in unalloyed metals.

2) We did not state that the diffusion of the electrolyte along the grid of slip steps on the grain boundaries is like that in a solid, though it might be similar to pipe diffusion along dislocation lines. In any case it must be thermally activated, as implied by experiments. This appears to be reasonable if one imagines that the ions of the electrolyte have to squeeze through the pipes formed by the slip steps.

3) If the reduction of the cross-section during stress corrosion cracking and if the stress concentration at the corrosion front are not taken into account, the motion of the corrosion front should follow a $v \sim t$ law after a small transition period provided that the strain rate is kept constant and that stress depends linearly on strain. These prerequisites are usually not fulfilled simultaneously, so that deviations from this law must be expected.

Comment by R. W. Staehle:

The theoretical development does not acount for well known electrochemical influences. For example, the scc of the noble alloys is very sensitive to the oxidizing parameters. Secondly, the model does not consider the effects of compositional changes of the grain boundary. This alternative probably offers the most obvious interpretation. Thirdly, this work does not account for the morphological or compositional changes investigated by Swann and by Pickering.

Reply:

The present treatment preferentially aims at an understanding of the <u>physical</u> aspects of intergranular corrosion, stress corrosion cracking, etc., e.g., at explaining <u>why</u> the reactivity at the grain boundaries of solid solutions is extraordinarily high. Nevertheless (electro) chemical influences naturally enter the model, namely through the constants of reaction (μ_1, μ_m) or the diffusivities (D_1, D_m). It is true that in its present version the theory does not take into account morphological or compositional changes like those known to occur in the vicinity of grain boundaries of various alloys. An extension of the theory in this direction is straight forward and underway. In the case of alloys of noble metals, however, such changes play no decisive role. This follows from the observation that in these systems the "solid solution effects" are most pronounced at a <u>spatial average</u> solid solution concentration of $\sim$ 50 at %.

Comment by R. Bullough:

Concerning your mechanism for forming dislocations near the affected grain boundary in binary alloys, I certainly agree that the strain energy of a dislocation is much lower near a "soft" surface. However, if you wish to say the dislocation energy is dominated by its core energy then it is difficult to understand how you can simultaneously argue that it has a large negative interaction energy with solute atoms. Such large interaction energies arise from the large stress field of the dislocation which you are reducing by the presence of the nearby surface. Thus the interactions with such point defects will be correspondingly lower, and it is thus difficult to see how this interaction energy can ever exceed the self energy of the dislocation.

Reply:

Perhaps it is misleading to call these defects on the grain boundaries of alloys "dislocations" at all. The only point I wanted to make is that due to the difference in size and hardness between the A- and B- type atoms of a binary alloy the grain boundaries must be expected to show more roughness than those of pure metals. One can imagine that the steps providing this roughness on an atomistic scale are sites at which the reactivity with aqueous electrolytes or liquid metals is strongly enhanced. One may describe these steps as dislocations which have penetrated into the grains so deep that the produced dislocation self-energy is smaller than the release of elastic energy around the solute atoms. This description is advantageous, since the application of an external stress makes dislocations enter the grains right at these step-like nuclei.

CLEAVAGE vs SHEAR AT CRACK TIPS IN METAL CRYSTALS

W. R. Tyson

Physical Metallurgy Research Laboratories,
Canada Centre for Minerals and Energy Technology,
Department of Energy, Mines and Resources
Booth Street, Ottawa, Canada.

ABSTRACT. A simple criterion is developed to predict whether failure at the tip of a crack in a material will be accompanied by dislocation nucleation or will occur by fully brittle cleavage. The criterion is tested by application to those metals for which relevant data is available.

1. INTRODUCTION

Perhaps the single most important property required of a metal for use as a structural material is the ability to absorb energy at the tip of a crack. This can be achieved by the motion and multiplication of pre-existing dislocations in the stress field of the crack, or by generation of dislocations at the crack tip itself. Clearly, if conditions at the crack tip are such that dislocations must be nucleated there at stresses large enough to break bonds and propagate the crack, then some energy in excess of that required to form new surfaces must be absorbed in crack extension. This condition allows a simple criterion to be developed to predict whether failure at the tip of a crack in a material will be accompanied by dislocation nucleation or will occur by fully brittle cleavage. It is the purpose of this contribution to develop this criterion for the brittle failure of metals and to test it using the most recent available experimental data.

2. CRITERION FOR BRITTLE FAILURE

The possibility of dislocation nucleation at crack tips has been recognized for some time. Armstrong [1] was perhaps the first to treat this problem quantitatively, by comparing the stresses required to propagate a penny-shaped shear crack with those required to emit dislocations from the crack tip. Equality of these stresses is controlled by the value of $\gamma/\mu b$ where γ is the surface energy, μ is the shear modulus, and b is the lattice parameter; Armstrong predicted ductile or brittle behaviour according to whether this parameter is larger or smaller than a value determined to be ~0.084. For the case of a tensile crack, Rice and Thomson [2] considered in some detail the conditions under which dislocations would be generated, and showed that their results could be represented approximately by the criterion that fully brittle behaviour is expected if the parameter $\mu b/10\gamma$ is greater than 2 for fcc or 2/3 for bcc and hcp crystals ($\gamma/\mu b$ <0.05 or 0.15 respectively).

Both of these treatments are continuum theories, in the sense that the material in the vicinity of the crack tip is treated as an elastic continuum in deriving expressions for the energy of a dislocation in the stress field of the crack. However, dislocation generation is synonymous with local shear failure of the crystal lattice at the crack tip itself, and must involve local shear strain of interatomic bonds beyond the maximum strain sustainable by a perfect crystal during formation of the dislocation core. This condition was incorporated in the treatment by Kelly, Tyson and Cottrell [3] who proposed that fully brittle cleavage is possible only if the ratio of the ideal (or "theoretical") tensile strength σ_{th} to the ideal shear strength τ_{th} is smaller than the ratio R of the largest tensile stress σ to the largest shear stress τ at the crack tip, so that the lattice fails in tension before shear.

The major uncertainties in testing this hypothesis derive from the difficulty of reliably estimating σ_{th}, τ_{th}, and R. As recognized earlier [3], the problem is essentially an atomistic one; not only is the stress state at the crack tip complicated by hydrostatic components which makes definition of a suitable τ_{th} difficult, but atomic movements during shear relaxation must alter the stress state from its continuum (linear elastic) value. The problem can only be fully resolved by atomistic calculations using potentials that adequately describe the properties of real materials, a problem that is beyond the grasp of present theoretical methods except, perhaps, for the simplest solids such as the alkali metals.

Nevertheless, the heart of the method can be displayed, in the spirit of a first approximation, by combining some simple expressions. The ideal strengths in tension and shear may be roughly estimated by the Orowan-Frenkel relations

$$\sigma_{th} = \sqrt{\frac{E\gamma}{d}} \quad (1)$$

and

$$\tau_{th} = \frac{\mu}{2\pi} \cdot \frac{b}{d} \quad (2)$$

where E is Young's modulus, b is the lattice vector in the shear direction (the Burgers vector), and d is the interplanar spacing of the cleavage or shear planes. Using a sinusoidal approximation for the relationship between tensile stress and displacement, it is easy to show that

$$\gamma = \frac{E}{d} \left(\frac{a}{\pi}\right)^2 \qquad (3)$$

where a is the range of the interplanar forces. Combining Eqs. (1) to (3) and taking $a \approx d \approx b$, we have

$$\frac{\sigma_{th}}{\tau_{th}} = 2\pi^2 \frac{\gamma}{\mu b} \approx 20 \frac{\gamma}{\mu b} \qquad (4)$$

The value of R was estimated earlier [3] on the assumption that the crack tip region is in a stress state of biaxial tension and plane strain, implying that the largest shear stresses occur on planes bisecting the crack front and the normal to the cleavage plane. However, large shear stresses also exist on planes parallel to the crack front. According to Sneddon and Lowengrub ([4], p. 31 and 32) the maximum values of such shear stresses in the immediate vicinity of a tensile crack tip in plane strain are given by

$$\tau = \sigma_{app} \sqrt{\frac{c}{8y}} \quad \text{on } x = 0$$

while

$$\sigma = \sigma_{app} \sqrt{\frac{c}{2x}} \quad \text{on } y = 0$$

where σ_{app} is the applied stress, c is the half-length of the crack, and the coordinates (x,y) are as shown in Fig. 1. The

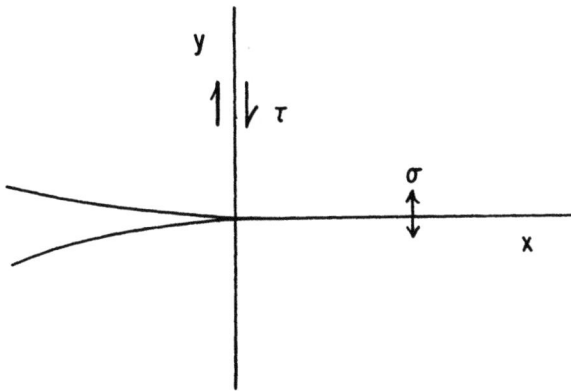

Fig. 1. Coordinate system at crack tip, showing the nature of the largest shear and tensile stresses at a given distance from the crack front in the approximation of linear elasticity.

approximation of linear elasticity inherent in these expressions breaks down within the distance x~y~b of the crack tip, where the stresses attain their highest values. Assuming that the nonlinear behaviour near the crack tip alters both σ and τ by approximately the same factor, we may deduce

$$R = \frac{\sigma}{\tau} \sim \sqrt{\frac{8}{2}} = 2.$$

The criterion for fully brittle failure may thus be stated in the form

$$20 \frac{\gamma}{\mu b} \leq 2, \text{ or } \frac{\gamma}{\mu b} \leq 0.1.$$

3. CORRELATION WITH EXPERIMENTAL DATA

While the physical properties μ and b may be determined experimentally with a high degree of accuracy, the surface energy γ is very difficult to measure. Furthermore, while experimental methods are mostly applicable near the melting point, for the purposes of the present discussion it is the value at absolute zero that is required.

Methods for experimental and theoretical determination of γ have been reviewed recently by the present author [5], and "best values" of surface energy and entropy have been suggested for the metallic elements. This allows a more thorough test of the ductile/brittle criteria proposed above than has been possible previously. Values of γ appropriate to 0°K from that report [5] are given in Table 1 for the fcc, bcc, and hcp metals along with

TABLE 1. Surface energy γ at 0°K, Burgers vector b, shear modulus μ, and parameter 20γ/μb for metals. Values of γ in brackets are calculated using data on heat of formation. (Note that $1 \text{ J/m}^2 = 10^1 \text{ d/cm}$, $1 \text{ N/m}^2 = 10 \text{ d/cm}^2$).

Metal	Structure	γ (N/m)	b (10^{-10} m)	μ (10^{10} N/m²)	20γ/μb
Ag	fcc	1.34	2.88	2.20	4.23
Al		1.20	2.86	2.73	3.07
Au		1.56	2.88	2.04	5.31
Cu		1.79	2.55	3.33	4.22
Ir		(2.95)	2.71	19.1	1.14
Ni		2.27	2.49	6.85	2.65
Pb		0.61	3.49	0.672	5.19
Pd		(1.63)	2.74	3.61	3.30
Pt		2.59	2.77	5.86	3.20
αTh		(1.44)	3.59	1.93	4.16
Cs	bcc	(0.088)	5.27	0.030	11.0
Cr		2.32	2.49	13.07	1.43
αFe		2.37	2.48	6.48	2.94
K		(0.130)	4.53	0.053	10.8
Li		(0.530)	3.03	0.16	22.1
Mo		2.28	2.72	13.37	1.25
Na		(0.240)	3.66	0.098	13.3
Nb		2.57	2.85	4.56	3.95
Rb		(0.100)	4.84	0.038	11.1
Ta		2.90	2.85	6.19	3.30
V		(2.28)	2.62	5.25	3.30
W		3.07	2.74	16.36	1.37
Be	hcp	(2.05)	2.28	16.62	1.08
Cd		(0.37)	2.97	2.42	1.03
αCo		2.21	2.50	8.24	2.15
αHf		(1.98)	3.19	6.00	2.07
Mg		(0.47)	3.20	1.84	1.60
Re		(3.36)	2.76	16.85	1.44
Ru		(2.89)	2.70	18.91	1.13
Zn		0.92	2.66	4.60	1.50

values for b (taken from lattice parameter data compiled by Pearson [6]) and μ (derived from elastic moduli collated by Simmons and Wang [7] using the most recent data at the lowest temperatures); the relationships used to obtain b and μ are listed in Table 2[†]. Values of γ enclosed in brackets are calculated from a semi-empirical correlation between γ and the heat of formation [5].

TABLE 2. Relation for b and μ for the three crystal structures; a_o is the lattice parameter and c_{ij} the single crystal elastic constants.

Structure	b	Slip System		μ
fcc	$a_o \frac{\sqrt{2}}{2}$	[110]	(1$\bar{1}$1)	$\frac{3c_{44}(c_{11}-c_{12})}{4c_{44}+c_{11}-c_{12}}$
bcc	$a_o \frac{\sqrt{3}}{2}$	[111]	(1$\bar{1}$0)	"
hcp	a_o	[11$\bar{2}$0]	(0001)	c_{44}

The last column of Table 2 should, according to the discussion in Sec. 2, be a measure of the ratio of ideal strengths σ_{th}/τ_{th}. If this value is less than about 2, crystals of the metal should be fully brittle at 0°K. Fig. 2 displays this relation graphically, where γ and μb are compared for the metals listed in Table 1; the line γ/μb = 0.1 should separate the plastic from the fully brittle metals at 0°K.

[†]It is recognized that, for logical consistency with the derivation of the ductile/brittle criterion given above, the slip direction should be parallel to the cleavage plane normal and the slip plane normal parallel to the cleavage direction, as shown in Fig. 1. This is particularly violated for the hcp slip system assumed in Table 2, since in hcp metals the cleavage plane is (0001). However, it is felt that, as a first approximation, the values of Table 2 are representative of the relevant shear moduli and Burgers vectors.

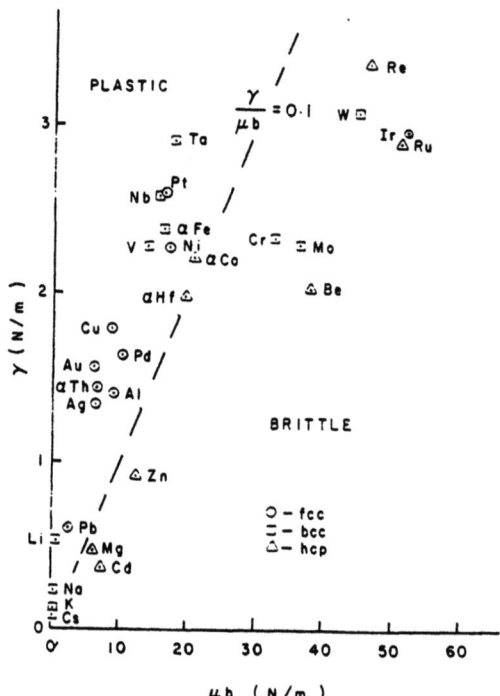

Fig. 2. Surface energy γ compared with the product of shear modulus μ and Burgers vector b for metallic crystals. The broken line indicates the criterion proposed to separate brittle from plastic behaviour at 0°K.

4. DISCUSSION

The behaviour indicated in Fig. 2 is in broad agreement with the fracture behaviour of the metallic crystals. The alkali metals are all predicted to be quite ductile, as expected. The hcp metals as a group tend toward brittle behaviour, the fcc metals toward ductility, and the bcc transition metals are split between the two types of behaviour.

One striking exception to the ductile behaviour in fcc metals is predicted in Fig. 2 to be Iridium. This is in satisfying agreement with experience, Ir being known to be brittle [8]. The malleability of the platinum metals as a group is explained well by Fig. 2, with Pt and Pd being fully ductile and the brittleness of the remaining metals known experimentally to increase in the order Rh, Ir, Ru, Os. Unfortunately, c_{ij} data is not available for Rh and Os which prevents their inclusion in Fig. 2.

Correlation of γ/μb with the brittle or ductile behaviour of metals is quite successful, and is predicted by all three of the treatments discussed above ([1] to [3]) as well as by the present simplified approximation. In spite of the somewhat different approaches taken in these treatments to the formulation of the problem, all lead to remarkably similar numerical criteria, namely $\gamma/\mu b \leq 0.1$ for brittle behaviour. Detailed discrimination between them is of doubtful value, as they are in essential agreement regarding the nature of the problem; further progress must come from fully atomistic treatments. It is the atomic movements involved in the lattice breakdown in tension or shear at the crack tip that is at issue.

Some suggestive results from lattice simulation studies along these lines are already available. Tyson and Alfred [9] demonstrated dislocation generation at the tip of a crack in a material for which $\sigma_{th}/\tau_{th} = 3.4$. Since this is greater than 2, ductile behaviour is expected as observed. Weiner and Pear [10] have investigated crack propagation in a material of variable σ_{th}/τ_{th}; they find fully ductile behaviour when this parameter is equal to 2.04, and brittle behaviour when it equals 1.0 (although dislocations were emitted when the crack attained high velocities). Further work along these lines, particularly on the effects of temperature, is needed.

5. CONCLUSIONS

A simple treatment has been presented of the way in which material properties interact to control brittleness or ductility, viewed as lattice failure in tension or in shear at the tip of a crack. The criterion derived in this straightforward fashion is in good agreement with experiment and with other treatments of the problem, the common limitation to all of these theoretical approaches being the assumption of linear elasticity and continuum mechanics. Further progress in the description and eventual control of material failure at a crack tip depends on the development of reliable atomistic models and lattice simulation of crack propagation.

REFERENCES

1. Armstrong, R. Mat. Sci. Eng. 1, 251 (1966).

2. Rice, J. R. and Thomson, R. Phil. Mag. 29, 73 (1974).

3. Kelly, A., Tyson, W. R. and Cottrell, A. H. Phil. Mag. 15, 567 (1967).

4. Sneddon, I. N. and Lowengrub, M. "Crack Problems in the Classical Theory of Elasticity", Wiley, N.Y. (1969).

5. Tyson, W. R. "Surface Energies of Solid Metals", PMRL report ERP/PMRL-75-3(J), accepted for publication in Can. Met. Quart.

6. Pearson, W. B. "A Handbook of Lattice Spacings and Structure of Metals and Alloys", Pergamon Press (1964).

7. Simmons, G. and Wang, H. "Single Crystal Elastic Constants and Calculated Aggregate Properties: A Handbook", MIT Press, Cambridge, Mass., 2nd edition (1971).

8. Reid, C. N. and Routbort, J. L. Met. Trans. $\underline{3}$, 2257 (1972).

9. Tyson, W. R. and Alfred, L.C.R. "Crack Propagation on an Atomic Scale", in "Corrosion Fatigue, NACE-2", p 281, Houston (1972).

10. Weiner, J. H. and Pear, M. J.A.P. $\underline{46}$, 2398 (1975).

NONDESTRUCTIVE, NEAR-SURFACE PLASTICITY DETERMINATION BY ELECTRON CHANNELING

David L. Davidson

Southwest Research Institute, P. O. Drawer 28510, San Antonio, Texas 78284, USA

ABSTRACT. The phenomenon of electron channeling may be used to assess the crystallography and defect structure of the near surface (100 nanometers) region of bulk crystals. Thus, the relation between surface (environmental) effects and crystal plasticity may be determined. A short literature review is included to indicate the breadth of use of the technique. Examples from work in our laboratory are given of surface related experiments.

1. INTRODUCTION

The crystallography and defect structure of a crystalline solid near-surface may be determined non-destructively using the electron channeling technique. This technique may conveniently be divided into two methods utilizing (1) selected-area electron channeling patterns (SAECP) and (2) diffraction contrast, also called channeling contrast. Both of these electron diffraction techniques utilize the same basic inelastic electron backscattering mechanism and differ only in instrument configuration. The purpose of this paper is to concentrate on the applications of this technique to materials problems, some of which are surface science related; the reader is left to a review of readily available literature for details on the physics of the electron beam-lattice interaction which causes electron backscattering to yield an electron channeling pattern.

2. TECHNIQUE AND INSTRUMENT CONDITIONS

General reviews of the physics and technique of electron channeling may be found in books,[1,2,3] and journal articles.[4,5,6] The mechanism of electron channeling is inelastic scattering of electrons, causing part of the impinging electron beam to be selectively backscattered, and part to be asborbed (or channeled) by the various crystallographic planes of the solid. This channeling of the electrons is manifested in large single crystals, provided the proper electron optical conditions exist, by the superposition of a pattern onto the normal topographic SEM image.[1] This is how the phenomenon was discovered by Coates.[7] When observing polycrystalline metals, individual grains show up as various levels on a gray scale, depending on their orientation, giving rise to what is termed diffraction contrast or channeling contrast. Selected area electron channeling patterns may be made by changing the normal scanning mode of the microscope so that the beam is rocked about a selected spot on the specimen surface. One of several instrumental configurations commercially available for generating SAECP is shown in Fig. 1b. In this case the electron channeling pattern (ECP) is being detected by a backscattered electron detector (solid state), but specimen current imaging may also be used. Spot size as small as 1μm has been reported,[8] although spot sizes of 10μm or greater are used more often (it is desirable to be able to adjust spot size to a desired value).

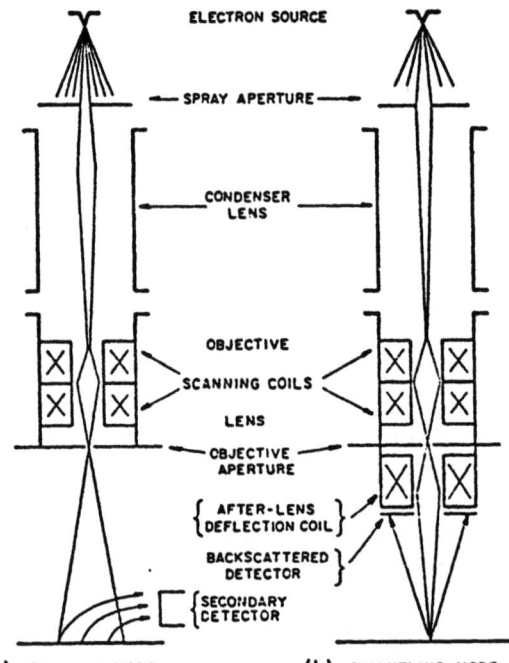

Figure 1. Electron optical conditions for normal and selected area electron channeling modes.

Instrumentation geometry for very small spot sizes usually necessitates use of specimen current imaging.

Specimen preparation is perhaps the limiting factor in practical use of the channeling technique. A specimen with initial conditions of relatively low dislocation density, and with no oxide layer, or at most a thin and uniform oxide layer is necessary, thus specimens must be very carefully machined and heat treated, followed by electropolishing or chemical polishing. Many of the same polishing problems encountered in thin foil preparation for transmission electron microscopy may also be encountered in preparation for channeling.

3. APPLICATIONS (REFERENCED)

3.1 Crystallographic orientation

Electron channeling patterns may be used to determine crystallographic directions in the grain being observed. The widths and angular relationahips between bands and lines in the pattern and the symmetry of crystal structure manifest in the patterns (as it is in Kikuchi patterns) are all used in orientation determination. The simplest, and often most convenient, way of determining orientation from an ECP is by comparison with a channeling map. Complete stereographic triangle maps exist for fcc, [9] bcc, [10] and diamond cubic [11] structures, and partial maps for hcp titanium [12] and silicon carbide [1] have been made, in some cases at several accelerating voltages. More formal methods of ECP orientation have also been discussed [2] and used. [13]

Applications of SAECP for crystallographic orientation have been many. Joy used SAECP to determine the orientation relationship between Fe-Ni martensite plates, between matrix and precipitates in Cu-Zn and Ag-Zn alloys, and between grains in a Pb-Sn alloy. [15] Booker [16] describes crystallographic texturing studies done on 316 stainless steels, and a Pb-1.5% Sn alloy. Ruff [17] has done similar work on titanium. Davidson [10] has oriented faceted fractures on single crystals, and Newbury, et al., [18] have used the technique in the same way on polycrystalling iron.

3.2 Deformation

Whereas the determination of crystallographic orientation is straightforward, assessment of crystalline perfection, or defect density, is complex. Two types of observation are possible: changes in SAECP line acuity and contrast and changes in channeling contrast. Stickler, et al.,[19] have shown that changes in ECP line acuity are related to dislocation density. Davidson[20] has shown that ion implantation causes degradation in ECP line acuity. Subgrain formation due to deformation may be observed using channeling contrast,[21] as may substructure information due to cyclic deformation.[22,23] An investigation of the effects of multiaxial loading in 304 SS using SAECP and channeling contrast has been reported.[24]

4. APPLICATIONS (ILLUSTRATED):

The uses of electron channeling discussed in this section have been taken from work done at SwRI and are selected because they are (1) applicable to surface versus bulk determination of plasticity, or (2) related to surface effects directly. Perhaps the work most directly applicable to surface effects was done in determining plastic zone size for a propagating stress corrosion crack in 304 stainless steel loaded in 10 N H_2SO_4 + 0.5 N NaCl solution at room temperature. Figure 2a is a backscattered electron (BSE) micrograph of the crack tip, with the plastic strain distribution shown in Fig. 2b. The details of the method used to quantitatively determine strains used in this and other crack tip plasticity work have been discussed,[25,26] along with the problems and limitations of the method. On the other hand, Fig. 2c shows both channeling contrast and SAECP of a 304 SS sodium sulfate dryer service failure caused by chloride induced stress corrosion cracking. Comparison of these ECPs with those from sensitized and unsensitized 304 SS indicates that the material probably was not sensitized. Examination of the BSE micrograph and ECPs shows that no discernible plastic deformation is associated with the crack, in contrast to Fig. 2b, and that there appears to be no effect of crystallography. The two observations shown in Fig. 2 are apparently inconsistent, leaving the role of plasticity in the stress corrosion of 304 SS to be the subject of further investigation.

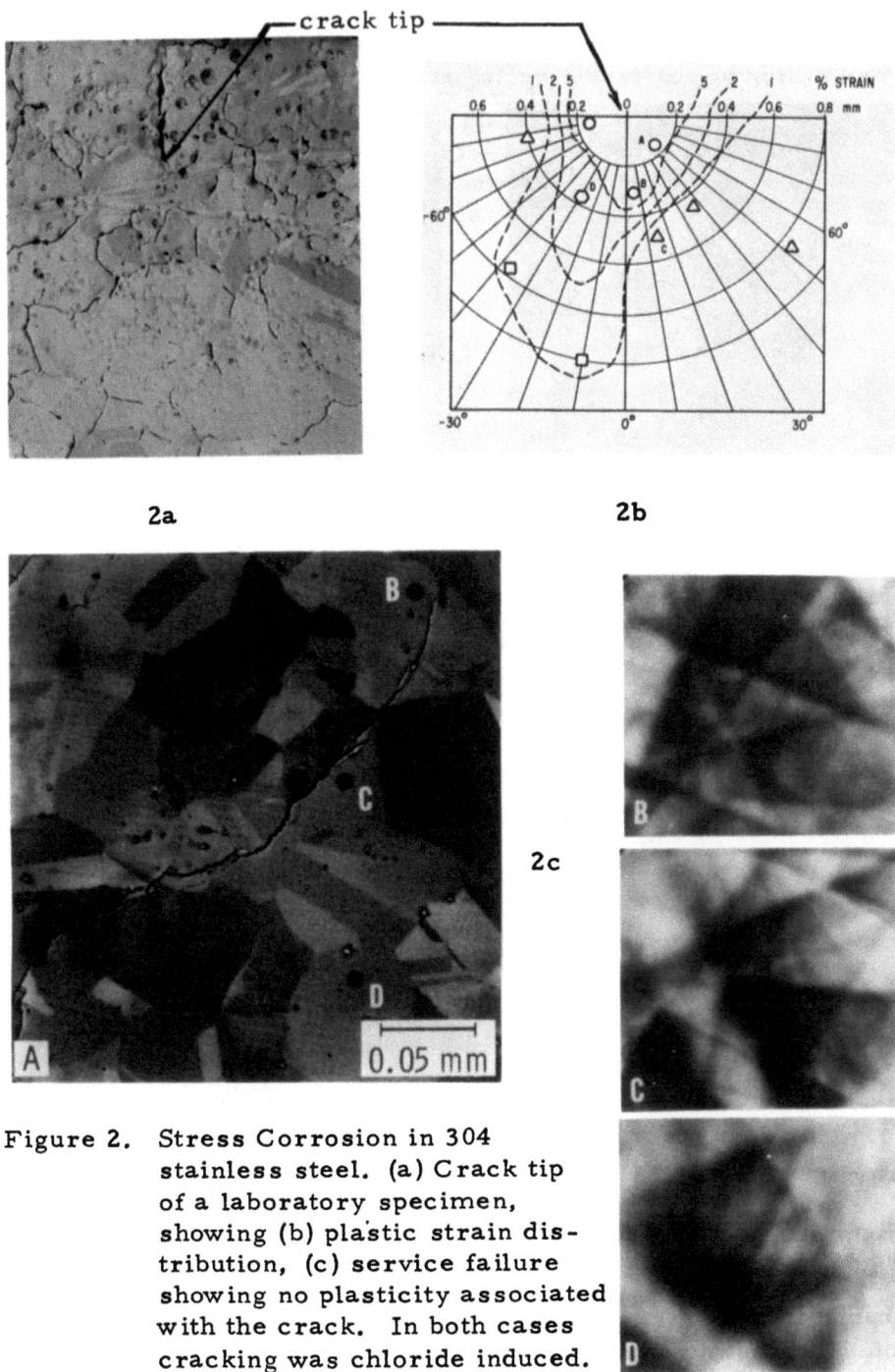

Figure 2. Stress Corrosion in 304 stainless steel. (a) Crack tip of a laboratory specimen, showing (b) plastic strain distribution, (c) service failure showing no plasticity associated with the crack. In both cases cracking was chloride induced.

Examples of the revelation of subgrains formed by cyclic deformation in low carbon steel are shown in Figures 3 and 4 for two different cyclic stress intensity factors.

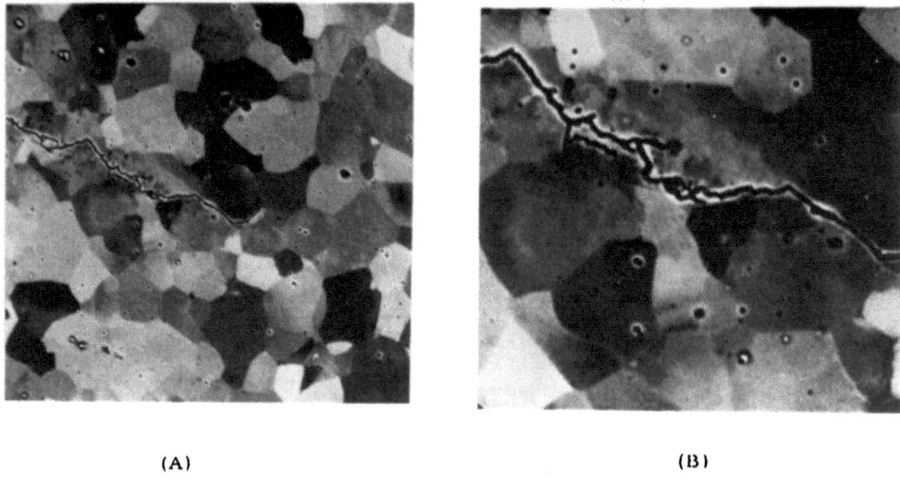

(A) (B)

Figure 3. Low carbon steel fatigue crack. $\Delta K = 5.9$ MN/m$^{3/2}$
(A) 400X (B) 1000X showing little subgrain formation.

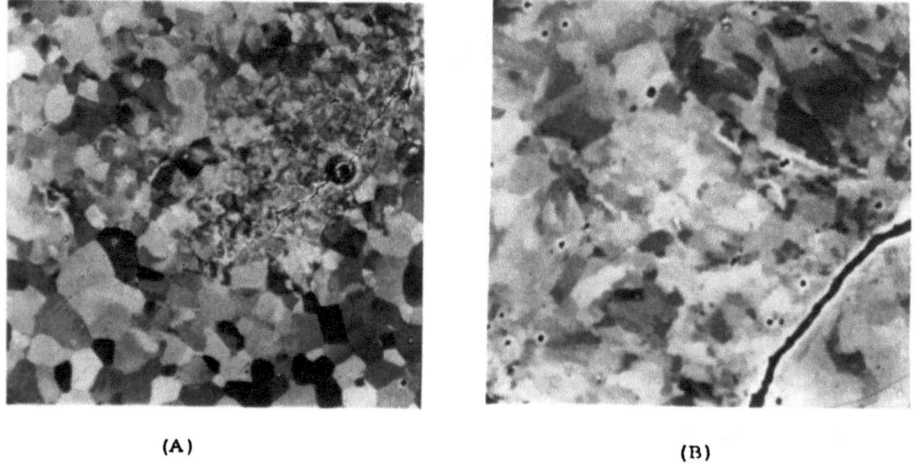

(A) (B)

Figure 4. Low carbon steel fatigue crack. $\Delta K = 26$ MN/m$^{3/2}$
(A) 400X (B) 1000X showing well developed subgrains.

Channeling contrast, in this case, shows up what SAECP could not. The plastic zone boundary must in both cases be determined using SAECP with the importance of using this technique being dependent on the formation of subgrains by the material under the test conditions imposed. Plastic zone boundary and distribution of strains within the plastic zone have been found for 6061-T6 aluminum alloy, Fe3Si and 304 SS using SAECP. (27) The effect of an aggressive environment on fatigue crack plastic zone size is currently under investigation in our laboratory.

A final example of plasticity assessment directly applicable to surface effects, is shown in Fig. 5.

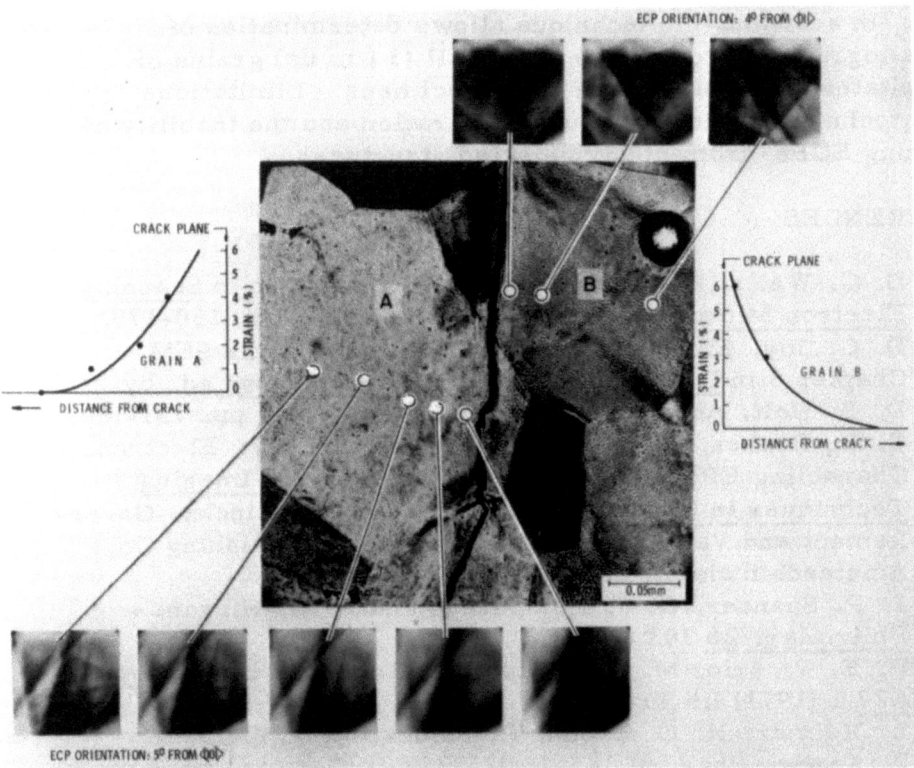

Figure 5. Fast fracture in Fe-3Si showing strain profiles perpendicular to crack plane for two grains.

This is a section perpendicular to the plane of fast fracture in Fe-3Si showing the depth profile of deformation as determined from SAECP's. Both depth and magnitude of deformation are expected to be affected by changes in surface energy, or other environmental factors. Of course, both temperature and rate of crack propagation will also be important.

5. SUMMARY

Selected area electron channeling patterns and diffraction contrast can add significantly to an understanding of the metal plasticity-environment interface because the inelastic backscattering of electrons is a near surface phenomenon. Assessment of surface related near-surface crystal plasticity may therefore be done nondestructively. Electron channeling seems particularly well suited to the study of crack tip plasticity. In addition, the technique allows determination of crystallography of both large and small (5 µm up) grains or precipitates at the surface of bulk specimens. Limitations of the technique include surface preparation and the inability of obtaining ECPs from highly defected structures.

REFERENCES

1. D. C. Wells, Electron Channeling, Chapter 6 in Scanning Electron Microscopy, McGraw-Hill, 1974, pp. 160-179.
2. D. C. Joy, Electron Channeling Patterns in the SEM, Chapter 6 in Quantitative Scanning Microscopy, ed. by D. B. Holt, Academic Press, London, 1974, pp. 131-182.
3. G. R. Booker, Scanning Electron Microscopy: Electron Channeling Effects in Modern Diffraction and Imaging Techniques in Material Science, ed. by Amelinckx, Gevers, Remaut and Van Landuyt, North-Holland Publishing Co., Amsterdam and London, 1970, pp. 613-653.
4. J. P. Spencer, C. J. Humphreys and P. B. Hirsch, Phil. Mag. 26 193 (1972).
5. P. E. Vicario, M. Pitaval and G. Fontaine, Acta Cryst. A27 1 (1971) (in French).
6. L. Reimer, H. G. Badde, H. Seidel and W. Buhring, Z. angew. Phys. 31 145-151 (1971) (in German).
7. D. G. Coates, Phil. Mag. 16 1179 (1967).
8. D.C. Joy and D. E. Newbury, J. Materials Sci. 7 714 (1972).
9. C. G. van Essen, E. M. Schulson and R. H. Donaghay, J. Matls. Sci. 6 213 (1971).
10. D. L. Davidson, J. Matls. Sci. 9 1091 (1974).
11. D. G. Coates, SEM/1969, p. 27.*

12. A. W. Ruff, Met. Trans. 5 601 (1974).
13. J. D. Ayers and D. C. Joy, Acta Met. 20 1371 (1972).
14. D. C. Joy, G. R. Booker, E. O. Fearon and M. Bevis, SEM/1971, p. 499.*
15. D. C. Joy and D. E. Newbury, SEM/1971, p. 113.*
16. G. R. Booker, SEM/1971, p. 465.*
17. A. W. Ruff, private communication; abstract published in Proceedings Electron Microscope Society of America, 1973.
18. D. E. Newbury, B. W. Christ and D. C. Joy, Met. Trans. 5 1505 (1974).
19. R. Stickler, C. W. Hughes and G. R. Booker, SEM/1971, p. 473.*
20. S. M. Davidson, Nature 227 487 (1 August 1970).
21. D. C. Joy, D. E. Newbury and P. M. Hazzledine, SEM/1972, p. 97.*
22. D. C. Joy and D. E. Pease, SEM Study of Dislocation Sub-cells in Deformed Copper and Stainless Steel, submitted to Metallurgical Transactions, 1974.
23. D. L. Davidson, J. Lankford, T. Yokobori and K. Satō, Fatigue Crack Tip Plastic Zones in Low Carbon Steel, submitted to International Journal of Fracture, 1975.
24. J. A. Cornie, D. L. Harrod and C. W. Hughes, Scientific Paper 73-ID4-HTRFC-P1, Westinghouse Research Labs., Pittsburgh, Pennsylvania 15235, USA, 1973.
25. D. L. Davidson and F. F. Lyle, Corrosion 31 135 (1975).
26. D. L. Davidson, SEM/1974, p. 927.*
27. D. L. Davidson and J. Lankford, Jr., Trans. ASME (H), Jnl. Eng. Matls. and Tech., in press.

---

*SEM/Year refers to Proceedings of the Annual Scanning Electron Microscopy Symposium/Year, ed. by Om Johari, et al., IIT Research Institute, Chicago, Illinois, 60616, USA.

CORRELATION BETWEEN PITTING AND STRESS CORROSION CRACKING OF
STAINLESS STEELS IN CHLORIDE SOLUTIONS

Z. Szklarska-Smialowska

Institute of Physical Chemistry of the Polish
Academy of Science, Warsaw, Poland

It is known from the industrial practice that in austenitic stainless steels the transgranular stress corrosion cracks often originate from pits or crevices [1]. The view is also well established that inside of the pits the metal is in the active state. It is postulated, on the other hand, that during SCC the anodic dissolution is restricted to the crack tip area and that the surface of the surrounding metal is covered by a protective film and acts as cathode. These two viewpoints seem to be contradictory.

Up to now, the composition of the solution within the pit is unknown, and the conditions under which the pits can develop into cracks are not defined well enough. Therefore, experimental studies have been assumed in our laboratory with the aim to determine the concentration of chloride ions within the pits, to establish potential ranges and other conditions that favor the occurrence of either SCC or pitting, and to look into the nature of surface films grown on steel under various conditions.

I should like to mention here some of our results and conclusions. The accumulation of Cl^- in pits grown in specimens of 18Cr-12Ni-12Mo-1Ti austenitic stainless steel was studied at 20°C in 0.5 N NaCl + 0.1 N H_2SO_4.

Fig. 1 shows the concentration of chloride ions accumulating in the pits growing at constant potential of 850 mV after 5, 15, 30, 60, 120, 180, 240 and 360 mins as a function of the diameter of the pits from which the solutions were collected for analysis [2]. There are two curves in Fig. 1 Curve 1 is for slowly growing pits, and curve 2 for quickly growing pits. Different rates of growth of pits are probably due to different tightness of the films covering them. In order to present the time dependence of Cl^- concentration the experimental data were divided according to the pit growth rate into two groups consisting of less rapidly growing pits, which after 120 min. had diameters of 0.48-0.65 mm, and more rapidly growing ones, which after the same time had diameters of 0.65-1.4 mm. As can be seen, the concentration within the pits growing for a longer time is due to decreasing tightness of the film that covers the pits and to increasing convection and diffusion.

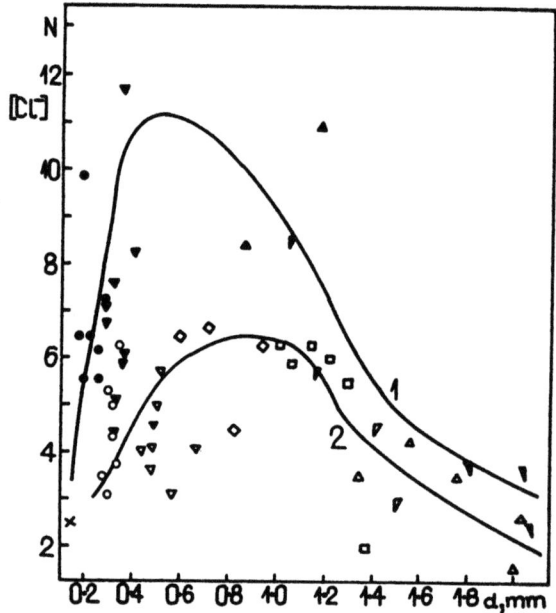

Fig. 1 Chloride ion concentration within the pit as a function of its diameter, (1) slowly growing pits, (2) quickly growing pits, ✕ 5 min., ○ 15 min., ▽ 30 min., ◊ 60 min., □ 120 min., △ 180 min., ▷ 240 min., ◁ 360 min.

It is generally believed that the concentration of the solution contained in cracks formed during stress corrosion is analogous to that occurring in corrosion pits. Hence, a similar accumulation of Cl^- should be also expected, but one may think that owing to the narrowness of the crack, the Cl^- concentration at its tip after reaching the maximum enrichment will not change with time very much.

The reason for the low pH values found by different workers in artifical pits [3], crevices [4], and cracks [5] grown in type 304 stainless steel are obscure. The hydrolysis cannot produce such a high acidity as experimentally observed, and attempts to explain this phenomenon on the basis of other assumptions were unsuccessful [6].

It has been acknowledged long ago [7,8] that in unbuffered water solutions, such as sulfuric or hydrochloric acids, the activity of hydronium ions considerably increases, when chlorides are added. The effect depends upon the cation present. Unfortunately; nothing is known about the influence of $FeCl_2$, $NiCl_2$ and $CrCl_3$. In order to assess that effect, pH measurements were performed at various concentrations of $FeCl_2$ and $CrCl_3$ in HCl. The results are shown in Fig. 2. Fig. 3 shows the pH of separate $CrCl_3$ and $FeCl_2$ solutions as a function of their concentration. As can be seen, the higher concentration of Cl^-, the lower pH of the solution.

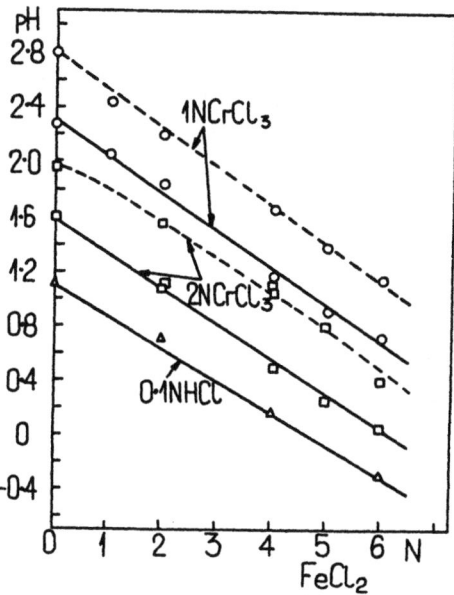

Fig. 2 Effect of $FeCl_2$ addition on pH of 1N $CrCl_3$ and 0.1 N HCl solutions.

-- O -- or --- ☐ -- 30 min. after the preparation of the solution.

— O — or —— ☐ — 24 h after the preparation of the solution.

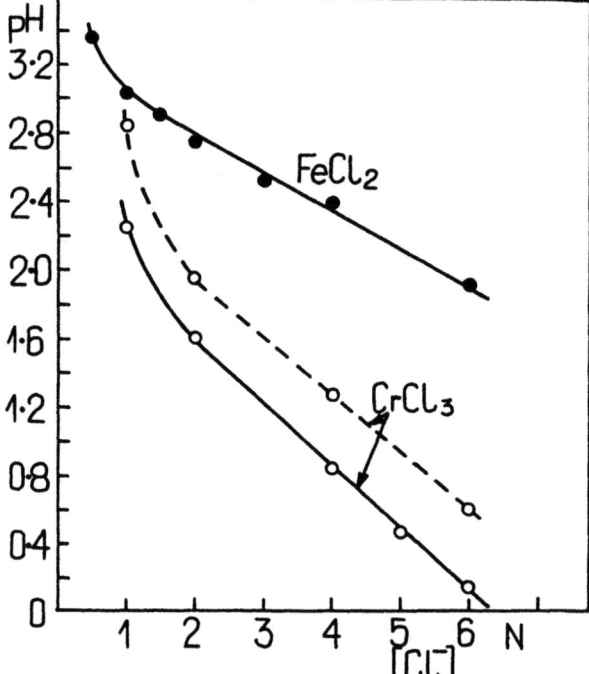

Fig. 3 Effect of $FeCl_2 \cdot 4\,H_2O$ and $CrCl_3 \cdot 6\,H_2O$ concentrations of pH of the solutions.

---- O ---- 30 min. after the preparation of the solution

——— O ——— 24 hrs. after the preparation of the solution

It follows from these results that the low pH of the solution within the pit (and also in cracks) is a complex effect of both the hydrolysis of corrosion products and high concentration of chloride ions. In the presence of high $FeCl_2$ and $CrCl_3$ concentrations is the H^+ activity greatly increased.

In another series of experiments we tried to check if, under laboratory conditions, a distinct correlation could be found between pitting and stress corrosion cracking [9].

Flat specimens of low-carbon 18Cr-9Ni austenitic stainless steel were submitted to axial tensile stress and corrosion in either 5% or 35% $MgCl_2$ solution, pH 4.6 at 60°C and 90°C and at various potentials. A constant load equal to 64% of the ultimate tensile strength (T.S.) was applied. Measurements of the elongation (Δl) of specimens as a function of corrosion time (t) enabled us to determine the induction time of SCC (from the deflection of the Δl vs. t curve). After corrosion the specimens were examined microscopically.

In separate experiments the range of potentials was established within which there occurs pitting and the induction time for pitting was determined from the deflection of the current density- time curve.

Fig. 4 illustrates the dependence between the induction time for stress corrosion cracking (t_{scc}) and induction time for pit formation in unstressed specimens (t_p) in 5% and 35% $MgCl_2$ at 60°C and 90°C at various potentials. It is seen that in 5% $MgCl_2$ at 60°C and 90°C, and in 35% $MgCl_2$ at 60°C, there occurs a linear relationship between log t_p and log t_{scc}, but in 35% $MgCl_2$ at 90°C another linear relationship with a different slope is observed.

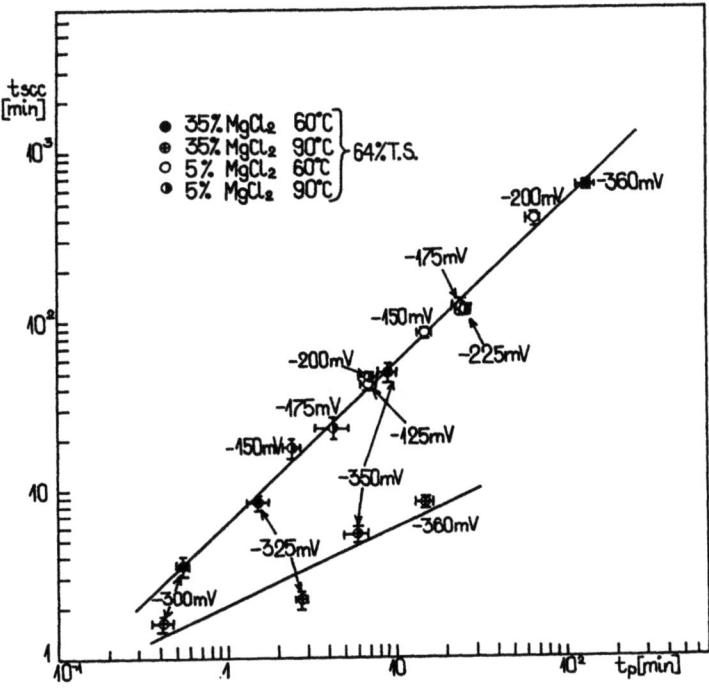

Fig. 4 Dependence between induction times for stress corrosion cracking and pitting.

Microscopic observation revealed significant differences in the mode of crack nucleation related to the slope of the log t_{scc} vs log t_p curves. In 5% $MgCl_2$ at 60°C and 90°C and also in 35% $MgCl_2$ at 60°C, in the range of potentials given in Fig. 4, the nucleation of cracks chiefly occurred on the bottom of the pits, but in 35% $MgCl_2$ at 90°C the majority of cracks originated from the supposedly smooth alloy surface.

Within the studied range of potentials there could occur both pitting corrosion and SCC. In 5% $MgCl_2$ at 60°C and 90°C, and also in 35% $MgCl_2$ but at lower temperature (60°C), the pitting was preferred ($t_p < t_{scc}$). On the contrary, in 35% $MgCl_2$ at 90°C the stress corrosion cracking occurred more rapidly ($t_{scc} < t_p$).

The following conclusions can be drawn from the above results:

The composition of the solution within the pits (and also in cracks) suggests that their bottoms are in the active state. They are covered with a salt layer [10] which cannot be considered as highly protective. If, however, under the effect of stress and glide a small portion of the metal surface becomes denuded from the salt layer, corrosion speeds up there and a crack nucleates. Hence, the nucleation of cracks can occur on the active metal surface where the concentration of Cl^- and H^+ is high.

The assumption can be made that during stress corrosion tests made in boiling concentrated $MgCl_2$ solutions the steel surface is active, though covered by a semi-protective salt layer. In the case of our experiments, such a state during which there occurred a slow general corrosion process was observed in 35% $MgCl_2$ at 90°C.

The results of ellipsometric investigations seem to confirm the above conclusion. Thus, as shown by preliminary measurements [11], in 35% $MgCl_2$ at 40°C and 60°C the steel surface is coated with a film exhibiting quite different optical properties in comparison with that present at 90°C. In the first case a complex refraction index of $2.30(1-0.13i)$ is observed. This presumably corresponds to an oxide film, while at 90°C the low refraction index of $1.62(1-0.10i)$ seems to suggest the presence of a salt layer. Initially, such a layer can probably contain a small amount of oxides. Under these conditions a general unhomogeneous pit-to-pit corrosion is observed. Further studies concerning that problem are in progress.

REFERENCES

1. H. Spaehn, G. H. Wagner and U. Steinhoff, Proc. Inter. Conf. Stress Corrosion Cracking and Hydrogen Embrittlement of Iron Base Alloys, Uniex-Firminy, France, 10-16 June 1973 (Houston NACE).
2. J. Mankowski and Z. Szklarska-Smialowska, Corr. Sci., 15, 493 (1975).
3. T. Suzuki, M. Yamabe and Y. Kitamura, Corrosion, 29, 18 (1973).
4. M. H. Peterson, T. J. Lennox Jr. and R. E. Groover, Materials Protection, 9, 1,23 (1970).
5. M. Marek and R. F. Hochman, Corrosion, 26, 5 (1970).

6. P. R. Rhodes, *Corrosion*, <u>25</u>, 462 (1969).
7. A. W. Thomas and M. E. Baldwin, *J. Am. Soc.*, <u>41</u>, 1981 (1919).
8. K. Schwabe, *Electrochim. Acta*, <u>12</u>, 67 (1967).
9. Z. Szklarska-Smialowska and J. Gust., to be published.
10. H. H. Strehblow, K. J. Vetter and A. Willgallis, *Ber. Bunsenges. physik. Chem.*, <u>75</u>, 822 (1971).
11. Z. Szklarska-Smialowska and N. Lukomski, to be published.

WHAT I KNOW AND WOULD LIKE TO KNOW CONCERNING ENVIRONMENTAL EFFECTS
ON THE MECHANICAL PROPERTIES OF METALS

R. W. Staehle

Corrosion Center, The Ohio State University, Columbus,
Ohio 43210, U.S.A.

1.0   Objectives

1.1   Prevent or mitigate damaging effects of environments,
      especially through stress corrosion cracking and corrosion
      fatigue, in all environments including hydrogen and liquid
      metals.

1.2   Apply SCC and CF effects to productive ends such as machining,
      comminution, biodegradation, preparation of raw materials.

1.3   Develop critical criteria for applying atomic theory (e.g.,
      surface and bulk) to prediction of metal- environment -
      stress interaction.

2.0   Known

2.1   Mode of Crack Propagation

      Environmentally induced cracks propagate either intergran-
      ularly or transgranularly in virtually all alloy systems
      depending upon stress, environment, metallurgical structure.
      The mode shifts readily for only environmental causes.

2.2   Rate

      SCC in high strength steels may propagate as rapidly as
      10 cm/min. SCC in lower strength materials such as stain-
      less steels typically propagates at a maximum rate of 0.01
      cm/min. Rates vary greatly as a function of applied stress.

SCC often propagates at a rate independent of stress intensity over as much as 2/3 of range for which SCC occurs.

2.3 Dependance Upon Alloy Strength

For the high strength alloys (steel, aluminum, titanium), the tendency for cracks to propagate <u>increases</u> as the strength increases until a steel of 250,000 psi yield stress can tolerate defects of only 0.1 - 1.0 mm for reasonable mean stress.

In the lower strength, more ductile alloys such as carbon steel, stainless steel, and copper alloys, the tendency for SCC decreases as the strength increases.

2.4 Dependance Upon Stress

First, most SCC will initiate in the absence of preexisting flaws. The stress dependance must be separated into the case (I) of an initially smooth surface and of an initially flawed surface (II).

For the initially smooth surface (I):

2.4.1 SCC can initiate with no applied stress only from the growth of oxides in environmentally produced flaws.

2.4.2 SCC can initiate at applied stresses in fully annealed material as low as 5000 psi (copper alloys and stainless steels).

2.4.3 Some SCC, especially intergranular, tends to exhibit a threshhold at about 2/3 of yield for ductile materials.

2.4.4 The high strength materials have very low resistance in the presence of flaws. These materials can also self initiate from smooth surfaces.

2.5 Dependance Upon Major Alloy Composition

2.5.1 The addition of major alloying elements very often produces a parabolic dependance of cracking susceptibility upon the element added with the minimum varying greatly from 0.5% to 50%. This trend applies primarily to low strength alloys. The base upon which this parabolic effect occurs may be a binary (e.g., Fe-20Cr to which Ni is added).

2.5.2 Pure metals do not crack unless intergranularly owing to solute segregation.

2.6 Dependance Upon Minor Impurity Elements

2.6.1 As alloys are purified, their resistance to SCC increases.

2.6.2 The Fe-Cr-Ni alloys in their pure state are not at all susceptible to SCC in chloride environments. Adding about 0.01w/o nitrogen induces susceptibility; adding similar amounts of phosphorous or molybdenum produces the same effects.

2.6.3 The high strength steels are greatly improved also by purification.

2.7 Dependance Upon Environments

2.7.1 In general alloys which are stainless or very corrosion resistant in given metal-environment systems tend to be very sensitive to SCC.

2.7.2 High strength alloys are susceptible to SCC in very innocuous environments. Water vapor at RH > 1% will cause cracks to propagate in high strength steels and aluminum alloys. Dry hydrogen at pressures as low as 10 torr will propagate SCC in high strength steels.

2.7.3 Lower strength alloys tend to be susceptible under more specific circumstances, e.g.

1) Cu base in aqueous NH_3 environments

2) Stainless steel in aqueous chloride

3) Carbon steels in aqueous nitrate.

However, these are not the only specific environments for these alloys. They are not generally susceptible in RH or dry hydrogen at the range of ambient conditions.

2.7.4 Very low concentrations of environmental species can be causative, e.g.

1) 1.0 ppm chloride in single phase aqueous solution is sufficient to cause austenitic stainless steel to crack at 200°C.

2) 1 torr of H_2S will cause high strength steel to crack.

2.7.5 Aqueous environments are not always necessary, e.g.

1) Anhydrous NH_3 cracks carbon steel

2) Alcohols crack Ti base alloys

3) N_2O_4 cracks Ti base.

2.7.6 Hydrogen environments are not always necessary, e.g.

1) Dry Cl_2 cracks high strength steels

2) Fused halides crack Ti base

3) Liquid metals embrittle some solid metals

2.7.7 Oxidizing conditions exert a dominating effect upon many SCC processes in low strength alloys, e.g.

1) Raising oxygen from 0 to 1 ppm in aqueous solutions causes stainless steel to SCC in chloride environments.

2) The range of applied potentials produces in a single environment-alloy system the full range of surface morphologies from passivity to general attack for SCC.

2.7.8 Changing pH in aqueous solutions produces a dominating effect similar to the applied potential. Changing pH produces transitions from IG to TG and passivity to SCC.

2.8 Inhibitors

2.8.1 Adding 0.5 v/o of O_2 to H_2 stops SCC in high strength steels.

2.8.2 Adding chromate to aqueous caustic stops SCC of austenitic stainless steels.

2.9 Dependance Upon Temperature

2.9.1 Many metal-environment systems in liquid and gaseous phases crack readily at room temperatures. Some gas systems will crack as low at -40°C.

2.9.2 Raising the temperature tends to accelerate SCC of many low strength systems and tends to retard SCC of high strength steels in gaseous hydrogen.

2.10 Dependance Upon Alloy Structure

2.10.1 The grain size dependance follows generally the $1/d^n$ pattern as for fracture.

2.10.2 TG SCC follows generally a crystallographic pattern, e.g.

    1) (111) in stainless steels in chlorides

    2) 15° off the basal plane in Ti base alloys.

2.10.3 Substantial changes are achieved by varying coherence in quench and tempered material with more coherent transformation products producing higher susceptibilities.

2.11 Continuum Between Static and Cyclic Load Conditions

2.11.1 High cyclic frequencies (greater than 10 Hz) sustain little environmental effects.

2.11.2 da/dN accelerates as cyclic frequency is lowered, e.g.

    1 Hz → 0.001 Hz

2.11.3 Superposition of $\dot{a}$ as K from static SCC to cyclic da/dN as ΔK does not always happen.

3.0 Want to Know

3.1 Surface Energies

Need surface energies for aqueous, liquid metal, gaseous environments as influenced by alloy substrate.

3.2 Slip

3.2.1 At Surfaces

The relative coplanarity of slip in the crucial circumstances as noted below. It is influenced by alloy composition, alloy structure, applied stress, stressing rate, and surface properties.

Need definition of discrete and wavy slip as function of all key environment and substrate parameters in order to understand initiation in SCC and CF.

3.2.2 At Crack Tips

    1) Understand conditions for acuity and blunting of crack tips

    2) Understand breaking of films

    3) Understand reaction with environmental species

3.2.3 **At Precipitate Free Zones**

Apply to grain boundaries of precipitation hardening alloys.

3.2.4 **At Grain Boundaries With Solute Segregation**

Apply to IG SCC + IG CF.

3.3 **Significance of Metal Purity**

3.4 **Solute Segregation**

Need criteria and rates for segregation of major and minor solutes to grain boundaries and dislocations. Competitive effects and exclusion effects are important.

3.5 **Metal-Metal Bond Energies**

Need especially the influence of hydrogen and liquid metals. But also other species: N, O, P, C, S, Sn, Sb, As.

Need as function of concentration of species and also as influenced by matrix composition.

3.6 **Solute Drag by Dislocations**

Need criteria for drag of solutes with respect to accumulation at internal sites and surface sites. Need for H, C, O, N, S, P, Sn, Sb, As.

3.7 **Dislocation Contribution to Reactivity in Solutions**

  1) Static

  2) Dynamic

  3) Cyclic

3.8 **The Stressed Crack Tip**

  1) Need effect of crack tip stress and deformation upon solute segregation and rate thereof for H, C, O, N, S, P, Sn, Sb, As.

  2) Changes in reactivity due to elastic-plastic conditions.

  3) Changes in reactivity due to electron concentration.

3.9   Formation of Passive Films

Need rate of re-formation vs function of alloy-environment chemistry.

3.10   Chemical Conditions at Crack Tips

1) Environment alterations

2) Potential

3.11   Theory of Concentrated Electrolytes

1) In cracks there are concentrated solutions of metal cations

2) Formation of protective films in concentrated solutions

3.12   Epitaxy of Protective Films

1) Conditions for adherence

2) Local stresses

3.13   Significance of Cyclic and Static Stressing

1) Wave shape

2) Frequency effect

Workshop Summary: <u>Surface Effects in Dissolution-Related Embrittlement Phenomena</u>

Prepared by R. W. Staehle

Discussion Leaders: R. W. Staehle, D. J. Duquette, J. P. Hirth, H.G.F. Wilsdorf

Recorders: P. A. Clarkin, W. M. Mularie, B. Escaig, D. L. Davidson

1.0 Stress Corrosion Cracking Theory and Phenomenology (Emphasis upon hydrogen effects on crack propagation):

Initial consideration in the propagation of cracks as affected by environments requires some differentiation between processes which are probably controlled by anodic processes and those controlled by hydrogen related processes. Somewhat more refined models incorporate considerations of surface energy reductions in the effects of environments on the breaking of bonds of the crack tip.

Discussions in these workshop sessions were aimed at considering the role of hydrogen in crack advance. Thus, the role of hydrogen was considered with respect to processes of surface energy lowering, bond breaking, void formation, associated dissolution, and formation of surface films.

Hydrogen reduces the ductility of metals by other means: notably and most importantly by the formation of hydrides. This is common both in body-centered cubic materials such as tantalum and niobium and in hexagonal close-packed materials such as titanium, zirconium, and yttrium. Here the embrittlement process seems to involve the formation of a brittle hydride. This brittle phase is either broken directly or serves as a pile-up site.

The participants agreed that there are few if any significant new ideas or breakthroughs which would rationalize existing data or provide <u>a priori</u> predictive capacity.

Nonetheless, several important observations were mentioned for serving as boundary conditions to be applied to any forthcoming explanations. For example, the data from Williams and Nelson (1) determined for a high strength steel in gaseous hydrogen and showing the effect of temperature on the crack velocity is a crucial observation. Here, slightly below room temperature the crack velocity increases with increasing temperatures; above room temperature the crack velocity decreases again to negligible values.

These data suggest that the crack propagation process consists of at least two fundamental and competing processes.

Secondly, there is the well known effect first observed by Johnson (2) that a small amount, notably 0.5 volume per cent, of oxygen is sufficient to stop cracking. This is particularly interesting in view of the fact that hydrogen from water molecules causes cracking relatively easily despite the fact that the oxygen is available from the dissociation. Certainly, the role of oxygen in this case is a subtle one. Finally, work using other hydrogen-containing molecules such as H_2S, HCl, HBr show that each of these molecules substantially accelerates the crack velocity relative to pure hydrogen (3). There is no clear explanation for these effects except for the qualitative correlation that hydrogen sulfide in aqueous solutions substantially accelerates the hydrogen entry rate.

Among the other important observations which must be considered in developing a model, the most important of these is the effect of strength. The hydrogen phenomena seem not to operate in lower strength steels until the yield strength exceeds approximately 150,000 psi. Above this value the susceptibility to hydrogen cracking phenomena increases progressively with higher plateau velocities and lower values of the stress intensity for the onset of stress corrosion cracking ($K_{I_{SCC}}$).

2.0 Model and Atomistic Interpretation of Hydrogen Effects

The concensus for approaching the problem of modeling the crack tip should include use of:

1. Continuum models
2. Atomic models with a characteristic set of atomic potentials
3. Dislocation motion model

Central to the development of any model, one must be concerned with the location of hydrogen. Whether the hydrogen is located only at the surface or penetrates appreciably into the material is a central question. There appears to be no question that the hydrogen can easily propagate into the material for crack velocities lower than about 10^{-3} cm/sec. The location of hydrogen in the crack tip, of course, if difficult to measure owing to experimental problems. Owing to the relatively small size of the hydrogen it was the general consensus that hydrogen would be absorbed into the material rather than simply being adsorbed. Work with tritium has shown that hydrogen reacts with dislocations and is adsorbed thereto. Hydrogen appears to diffuse easily to regions of high triaxial tensile stresses.

It was also considered that hydrogen could exert its influence in cracking through nonlocal models where the hydrogen may participate by:

1. Activating dislocation sources near the crack tip
2. Promoting decohesion and void formation ahead of the crack tip
3. Altering the dislocation mobility
4. Producing elastic interaction stresses

There is evidence that in the pinning of dislocations, hydrogen may exert a critical role in embrittlement; thus, when the dislocations tend to be pinned by hydrogen, a brittle appearance may occur at relatively higher crack velocities with a ductile appearance at lower velocities. Evidence was cited from work on silicon-iron for a transition from plastic to brittle behavior depending on the crack velocity.

Work on the role of hydrogen affecting void formation suggested that hydrogen may aid in stabilizing voids. Wilsdorf, using silver specimens 1 μm thick, has shown that cracks pop-in at a point considerably ahead of the main crack. Further straining produced holes with [110] sides. However, it was not clear as to whether the hydrogen could be shown to play a significant role in the process.

With respect to the question of dislocation locking, it seems that hydrogen may participate with other species such as carbon or silicon. They may act together in locking dislocations. If locking processes are not possible, cracking may not occur. The dislocation locking process may correlate well with the fact that relatively high strength steels are susceptible to hydrogen cracking.

The model for hydrogen effects proposed by Oriani received considerable attention. This model suggests that hydrogen plays a role in promoting decohesion by reducing the maximum force required to separate atoms. While there is considerable interest in this model, the theoretical problem of predicting the force-distance curves is a major difficulty. There appeared to be no fundamental basis for determining the effects of hydrogen on the force-distance laws. Further, it is not clear how such a model explains in such a detail the role of oxygen in inhibiting the cracking process.

In further discussion of the importance of atomic forces versus the more macroscopic view, Lothe noted that the Griffith criterion always gives the minimum necessary force for crack advance provided that no highly irreversible step occurs at the crack tip. The actual bond breaking at the tip is not the relevant process controlling the rate but the net work is the more important criterion. This involves the surface energy. The role of hydrogen in reducing surface energy could be a more important factor. However, if there were large Peierl's forces for crack advance, the matter of irreversibility would enter the problem and Oriani's model, now different from an Rebinder model, could apply with a maximum force identified as a Peierl's force in Figure 1.

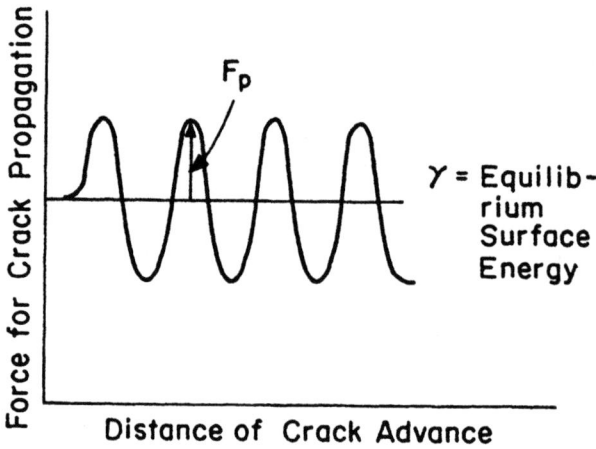

Figure 1   Force for crack propagation in a Peierls-type model with Fp the Peierls force.

3.0   Suggested Experimental Efforts

The problem of rationalizing a particular model for the advance of cracks is made difficult owing to the inaccessibility of the crack tip on an experimental basis. Thus, most of the experiments suggested were related to those which could be performed on accessible flat surfaces and then applied to the crack tip. Experiments suggested in this regard are the following:

1. Compare adsorption of hydrogen, sulfur, and oxygen as well as their interplay

2. Study the structural atomic changes of surfaces accompanying hydrogen adsorption using the following techniques:

   a. magnetic susceptibility relaxation in the KHz range, sensitive to a 10μm near-surface region

   b. x-ray or electron diffraction studies of small particles 100 to 200 nm in size to measure lattice parameter changes

c. Study influence of hydrogen on metal-metal bonding at the surface by measuring changes in the vibrational spectra of the latter by low energy electron loss spectroscopy.

3. Study adhesion of surfaces as a function of hydrogen adsorption, perhaps by methods such as those described at this Conference by Cuthrell.

4. Repeat the experiments of Johnson (2) where gaseous hydrogen causes cracking, the entry of oxygen in the metal system blocks cracking and its removal permitted cracking to proceed again. Focus on temperature and time dependence of the change in crack velocity at points such as A and B in Figure 2 to deduce information concerning the oxygen blocking effect.

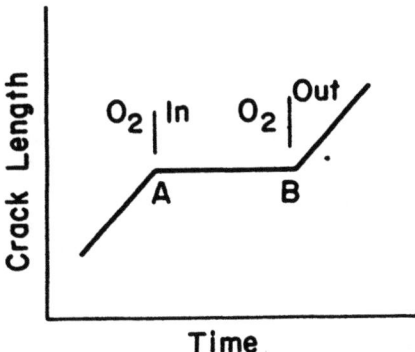

Figure 2 Crack growth of steel in a hydrogen atmosphere with and without oxygen present.

5. Perform Gorsky effect measurements on metal platelets which have both large surface to volume ratios and can withstand large (1 to 2 per cent) elastic strains. These should provide information on the kinetics and amounts of hydrogen redistribution in both the bulk and surface but additionally indicate the influence of elastic strain in these processes.

There is a substantial interest presently in computer modeling of crack tips where the type of iron bond is affected by the insertion of hydrogen. These have shown interesting possibilities and should be pursued.

Experiments should be performed to measure the surface diffusivity of hydrogen in the presence of adsorbates such as oxygen.

Unfortunately, the scale of many of these surface experiments is different from that of the crack tip and care must be taken in interpretation. It was noted also that there are experimental difficulties in correlating the behavior of surface films in relatively high vacuum environments and those, for example, in more technological environments such as aqueous solutions.

It was suggested that a scattering experiment might be done in which iron atoms would be scattered from iron surfaces with the incoming and exit energies measured. This would provide information on the force versus distance relationship. However, there were objections suggesting that condensation would occur. This experiment might be conducted in connection with substrates having oxygen or hydrogen or both adsorbed.

Studying the interaction of straining materials with tritium might be useful. However, the precise techniques here are not clear.

Finally, the influence of hydrogen on iron might be approached also by taking a more systematic look at the effect of hydrogen in the various elements of the periodic table. Thus, by appropriate extrapolations to the iron and deductions from electronic structure additional useful information might be obtained.

4.0 References

1. D. P. Williams and H. G. Nelson, Met. Trans., 1, 63 (1970).

2. G. G. Hancock and H. H. Johnson, Trans. Met. Soc. AIME, 236, 513 (1966).

3. G. E. Kerns and R. W. Staehle, Scripta Metallurgica, 6, 1189 (1972).

APPLICATIONS OF SURFACE EFFECTS IN CRYSTAL PLASTICITY TO TECHNOLOGY

Chairman: A. R. C. Westwood

APPLICATION OF CHEMOMECHANICAL EFFECTS TO
FRACTURE-DEPENDENT INDUSTRIAL PROCESSES

A. R. C. Westwood and J. J. Mills

Martin Marietta Laboratories, Baltimore,
Maryland, U. S. A.

ABSTRACT. The creation of new surface area by fracturing is fundamental to much of materials processing technology. Yet relatively little scientific attention has been given to improving the efficiency of such fracture-dependent operations as comminution, grinding, machining, drilling and tunneling by developing means of facilitating the fracture processes involved. This paper considers the possibility of applying chemomechanical effects -- the effects of adsorbed surface-active species on near-surface flow and fracture properties -- to this end. Estimates indicate that if such effects could be successfully applied to produce a 10% reduction in the cost of the operations mentioned above, and this seems a not too ambitious goal, then the savings would approach those resulting from the elimination of corrosion-induced failures entirely.

1. INTRODUCTION

That the physical and chemical state of the surficial layers of a component can markedly affect its strength and reliability has been known for about a hundred years [1, 2]. And, for the past forty years or so, much scientific attention has been devoted to minimizing the well-known detrimental influences of adsorbed or absorbed species and corrosive environments on mechanical properties. Essentially, the goal of this work has been prevention of premature failure, usually by the selection of some thermal treatment or choice of operating environment which reduces the solid's ability to fracture.

On the other hand, the creation of new surface area by various fracturing processes is fundamental to much of materials processing technology. Comminution, for example, is used in both the initial stages of extracting a metal from its ore, and in the final stages of producing useful non-metallic materials such as Portland cements. The industrially important shaping operations of machining, grinding, drilling, and polishing also involve the creation of new surfaces by fracture. It is somewhat surprising to note, therefore, that relatively little scientific attention has been devoted to improving the efficiency of these operations by developing means of facilitating the fracture processes involved. Yet the return on a successful research investment with this objective could be significant. For example, the cost of machining the billions of components produced in the U.S. in 1975 will be of order $75 billion*. And the anticipated annual cost of tunneling or boring through various kinds of rock for mineral exploitation, or for the development of rapid transit systems and utilities, etc., is about $8 billion [4]. Thus, if it were possible to reduce the cost of just these two types of operations by some 10%... which will be shown to be a not unreasonable goal ... then the savings to the U.S. would be about $8 billion.

Now, the annual cost of corrosion-induced failures to society is well recognized as being intolerably large, perhaps amounting to as much as $10 billion p.a. in the U.S. [5]. And recognition of this cost has inspired significant R&D expenditures oriented towards the control and prevention of corrosion. Yet, it can be seen that R&D oriented towards the production of relatively modest reductions in the costs of machining, drilling or comminuting operations by facilitating the fracture processes involved could lead to savings approaching those produced by eliminating corrosion-induced failures entirely.

Facilitation of the fracture processes involved in industrial operations can in theory be achieved in two ways, (i) via improvements in cutting tool or comminutor materials, design or fracturing action, and (ii) by changing the mechanical properties of the solid to be fractured such that crack initiation or propagation is made easier. This paper will be concerned only with the latter option. We will endeavor to show that the significant effects of adsorbed surface active species on the near-surface flow and flow-dependent fracture behavior of solids, known as chemomechanical effects, exhibit a potential for profitable application which merits the attention of both scientists and technologists. For reasons of convenience and the limitations

---

* Approx. 5% of the Gross National Product [3].

of space, examples selected to support this proposition have been taken primarily from the results of studies conducted by the authors and their colleagues at Martin Marietta Laboratories. But, of course, noteworthy contributions to this field have been made by workers at many institutions, the pioneering work of the group led by the late Academician Rebinder at the Institute for Physical Chemistry, Academy of Sciences of the U.S.S.R., in Moscow, being pre-eminent in this regard [6-8].

2. SOME FACTORS IN THE APPLICATION OF CHEMOMECHANICAL EFFECTS TO FRACTURE-DEPENDENT INDUSTRIAL OPERATIONS

The results of recent investigations [9-13] indicate that (i) the specific fracturing action of a cutting tool or comminutor, (ii) the deformation characteristics of the workpiece, and (iii) the influence of the surface active environment on the near-surface mechanical properties of the workpiece, must be considered in toto before any real understanding of the observed effects of the environment on the cutting or comminuting operation can be expected and, accordingly, before any reliable recommendation can be made for improving its efficiency. Other factors also to be considered, the relative importance of which depends on the specific solid-environment-tool (SET) "system" under consideration include (i) the viscosity, coolant, and particle-dispersing characteristics of the environment, (ii) the effective rate of delivery of the surface active species to freshly created surfaces, which will depend on the concentration of the surfactant, the flow rate of its carrier medium, adsorption kinetics, and the rate of production of new surface area, (iii) possible effects of the cutting environment on the mechanical behavior of the tool itself, and (iv) the load on the tool and its rate of application to the workpiece.

To the authors' knowledge, no comprehensive study of each of these variables has yet been undertaken for even one SET system. Moreover, there may be other important factors, as yet unidentified, ignorance or incomplete understanding of which could lead to apparently contradictory results between workers. To illustrate how such a problem could arise, consider the drilling behavior of calcite monocrystals in buffered water environments, Fig. 1 [12]. Entirely opposite effects of pH on drilling efficiency are obtained when different types of bit are used. It is now appreciated that such variations are related to the differences in the cutting action of the bits chosen. But if these two sets of data had been produced and published by different workers, perhaps working in different establishments, and with each being unaware of the significant differences between the cutting actions of the bits they used, then a certain scepticism

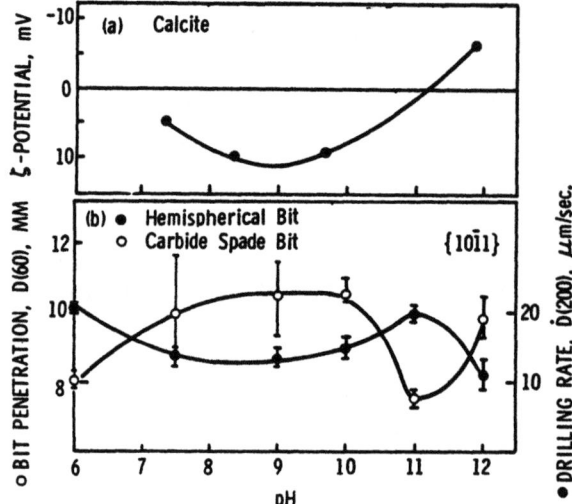

Fig. 1. Variation of (a) ζ-potential, and (b) penetration by a carbide spade bit in 60 sec, or rate of penetration by a diamond-studded bit after 200 sec, for calcite monocrystals as a function of pH in buffered water environments [12].

on the part of the scientific community as to the validity of the environmental effects reported would not be unjustified.

Such a problem has indeed arisen in the past, and only careful and correlated work in the future will overcome the "credibility gap" which can occur in consequence.

Surface active environments can influence fracture behavior in a variety of ways, but the most important of these as far as practical application is concerned arises because of their influence on the mobility of near-surface dislocations. Such effects are readily revealed as changes in microhardness [9, 14-16]. In this instance, the relative ease with which dislocations move and interact to form cracks is altered and, consequently, so is fracture behavior. Surface active environments can also influence surface cohesion, and such effects can sometimes be revealed directly via double cantilever beam determinations of the cleavage fracture energy of the solid [17].

Adsorption-induced changes in the hardness of non-metallic solids were first reported by Rebinder in 1928 [6], and are often interpreted in terms of adsorption-induced reductions in the surface free energy of the solid. Considerations of the nature of adsorption on insulator solids, and of the factors which determine the hardness of such solids, however, have led to the suggestion that such effects are more likely to be caused primarily by chemisorption-induced changes in the near-surface electronic structure of the solid. These, in turn, influence the ease of formation and/or motion of kinks on near-surface dislocations and, subseqnently, the nature of the interactions between dislocations, and between dislocations and near-surface point defects [12, 15, 18].

Studies have also revealed an interesting and useful correlation between hardness and the environment-sensitive ζ-potential* of a solid, namely that hardness is a maximum when $\zeta \simeq 0$ [12, 15, 16, 19, 20]. Such a correlation has now been established for alumina, calcite, feldspar, fluospar, magnesium oxide, quartz (see Fig. 2), silver chloride, soda-lime glass, and zinc in a variety of aqueous and non-aqueous solutions. It appears, therefore, that such a relationship may be generic to most inorganic materials, whether crystalline or non-crystalline.

The origin of the "ζ-correlation" is not yet clear. One speculative possibility [21] is that when the surface of the solid is at its iso-electric point ($\zeta = 0$), its electronic structure most closely approaches that of the bulk. Then, since the electronic structure of a solid can markedly influence or even determine its mechanical behavior, at $\zeta = 0$ the near-surface mechanical

---

* When a non-metallic solid is immersed in an electrolyte it acquires a surface potential, ψ_o, as a consequence of either (i) the adsorption of specific ions from the environment, or (ii) a change in the concentration or distribution of mobile point defects in the near-surface regions of the solid, or (iii) the establishment of equilibrium between ions making up the surface of the solid and the same constituent ions in solution, the concentrations of which are determined by the solubility product of the solid in the environment. A charge-balancing distribution of counter-ions is then established in the liquid phase, the concentration of the counter-ions decreasing exponentially with distance from the solid surface. Under conditions where there is relative motion between the solid and the electrolyte, it is possible to shear off the outer, more diffuse portion of the double layer from the inner, more strongly bound portion. The potential at the plane at which shearing occurs is defined as the ζ-potential and, in an ideal system, this potential is related in sign and magnitude to ψ_o, the surface potential. Most methods for determining ζ-potential make use of the electrokinetic phenomena resulting from the impression of either a potential gradient or a pressure gradient across the system of interest. For a comprehensive review of this topic, see Ref. [40].

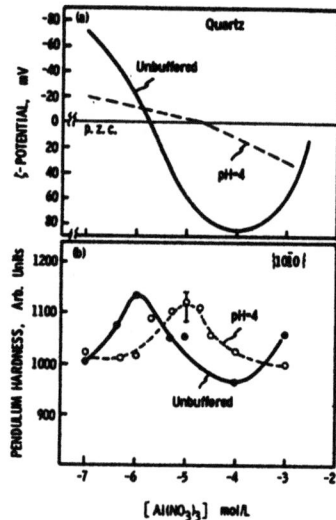

Fig. 2. Variation in (a) ζ-potential and (b) pendulum hardness of {10$\bar{1}$0} quartz in Al(NO$_3$)$_3$ solutions [11].

properties of the solid also closely approach its intrinsic, bulk properties... which presumably reflect the properties of the solid in its most stably bonded state.

When chemisorption occurs, however, the solid either acquires electrons from, or donates electrons to, the adsorbate. Now, if the electrons involved are donated from surface bonding orbitals, or accepted into surface antibonding orbitals, then some weakening of interatomic bonds at the surface will result. In other words, the acquisition of either a plus or a minus surface charge will weaken surface bonding somewhat, presumably making the generation and motion of kinks on dislocations easier within a region at most a few Angstroms deep at the surface. Such an effect would of necessity be truly a surface phenomenon, since the number of electrons involved would be far too small to have any significant influence on cohesion for more than an atomic layer or so into the solid. Presumably, similar reductions in surface cohesion can occur in metals also, despite the screening influence of the free electron cloud, vide liquid metal embrittlement [17].

The importance of matching the cutting action of the tool to the deformation characteristics of the solid to be fractured can be appreciated by recalling that a twist bit designed for use with soft metals will be ineffective in drilling granite, and an impact-bit designed for use with granite will not be very successful in penetrating soft aluminum. Recognizing this, some fracture-inducing devices or tools are designed to cut most efficiently through solids which exhibit a significant amount of plasticity. Their cutting action usually depends on a ploughing and shearing of the material. Such tools include twist and spade bits and non-cascading ball mills. Other tools, on the other hand, are designed for use with more brittle solids, and their effectiveness depends on their ability to initiate and propagate brittle cracks in the material. Such tools include pneumatic hammers, diamond-loaded bits, cascading ball mills and, possible, autogenous or fluid energy mills.

Knowing all this, one can sense intuitively how a given environment could produce opposing effects on the cutting efficiency of a given solid. Consider the situation illustrated in Fig. 3. Suppose that the environment selected softens the surface of the solid by 20% or so, which is not unreasonable [14]. As a consequence, the cutting action of the ploughing type of tool shown on the left will be facilitated, while that of the impact tool on the right will be hindered, because relatively more of its (impact) energy for fracturing will be dissipated by plastic flow processes, and less will be available to initiate and propagate brittle cracks.

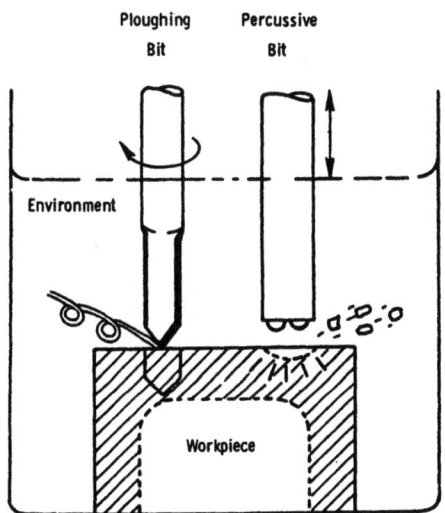

Fig. 3. Illustrating how a given environment can cause opposite effects on drilling efficiency depending on the cutting mode of the bit.

Given this simplified characterization of all fracture-producing processes as either predominantly plastic flow (PF)- or brittle fracture (BF)-dependent, and given an awareness of the relationship between ζ-potential and hardness, one may conclude that the fracturing efficiency of PF-dependent tools will be facilitated by chemical environments which produce large surface potentials on the solid (i.e., $\zeta \gtrless 0$). Conversely, the efficiency of BF-dependent tools will be enhanced by chemical environments which produce a zero surface potential on the surface of the workpiece. Evidence in support of such an hypothesis will be presented in the next section.

Of course, as mentioned above, there are other important environmental factors also. For example, in abrasive grinding operations, the hydrodynamic factors which determine asperity to asperity contact, and which are markedly influenced by the viscosity of the working medium, can sometimes overwhelm any effects related to the chemistry of the environment. And sometimes, in the machining of metals, minimizing frictional forces between the workpiece and the tool can have a sufficiently beneficial effect on tool life for it to represent a very practical way of reducing machining costs. For example, electrolytic environments which rapidly produce a thin, friction-reducing, oxide film on the freshly created surfaces of the metal can be effective cutting media [22]. In such circumstances a knowledge of the polarization diagrams for various metal-electrolyte systems can be of much value.

3. APPLICATIONS OF CHEMOMECHANICAL EFFECTS - EXPERIMENTAL OBSERVATIONS

3.1 Drilling and Machining

Nonmetallic Solids. Most nonmetallic, inorganic solids are notch brittle, and for crystalline ionic and ceramic materials fracture normally occurs as a consequence of dislocations piling up at some stable obstacle after only a limited amount of flow has occurred. The cracks thus nucleated then propagate rapidly to cause catastrophic failure. For such solids, therefore, the fracture stress is determined by, and is approximately equal to, the flow stress. Consequently, some correlation between the effects of a given environment on the near-surface flow behavior of such materials, i.e., on near-surface dislocation mobility, and on their fracture behavior might be anticipated. And such a correlation has indeed been observed in studies with MgO [23]. The measure of dislocation mobility adopted in these experiments was related to the extent to which edge dislocations (revealed by etch pitting) moved away from a standard microhardness impression in 1000 sec., defined as $\Delta L(1000)$. The measure of fracture behavior, albeit a somewhat complicated one, was the extent to which a carbide spade bit (a ploughing type of cutter) penetrated the crystal under given conditions of thrust, rate of rotation and time. Figure 4 shows the direct correlation noted between $\Delta L(1000)$ and depth of bit penetration in 600 sec., D(600), for MgO in dimethyl formamide-dimethyl sulfoxide (DMF-DMSO) solutions. Spade bit penetration is greatest for those environments which maximize dislocation mobility and minimize hardness. A similar correlation has been noted also for CaF_2 [23]. Thus a simple relationship exists between environment-sensitive dislocation behavior and drilling efficiency in these ideal cases.

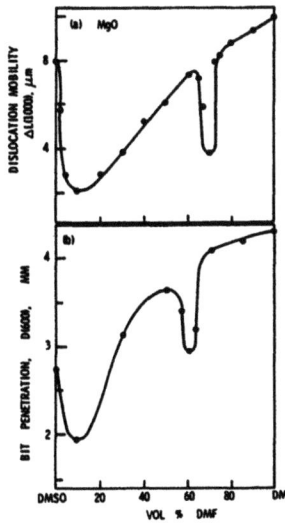

Fig. 4. Variation in (a) edge dislocation mobility and (b) penetration by a carbide spade bit in 600 sec for MgO monocrystals in DMSO-DMF solutions [23].

For harder and more brittle solids than MgO or CaF_2, however, various types of diamond-loaded bits would usually be employed to minimize wear. For this type of cutter, an opposite correlation between hardness and drilling efficiency is found [9], namely, that drilling rate is greatest when hardness is a maximum, i.e., when $\zeta \simeq 0$ [15]. Figure 5(c) shows that for a hemispherical-headed diamond-studded bit drilling into s.l. glass, D(200) -- defined as the average rate of penetration during the interval from 150-250 sec after drilling is commenced -- is greatest in heptyl alcohol, the environment which maximizes the hardness of this material, Fig. 5(b). And, as with the maximum in hardness, the maximum in drilling rate can also be reproduced in appropriate binary solutions of n-alcohols, one of which has $N_C^* < 7$ and the other $N_C > 7$. (See inserts to Fig. 5.)** Of practical interest is the observation that $\dot{D}(200)$ for this particular s.l. glass in heptyl alcohol is some twenty times greater than that obtained when drilling in water.

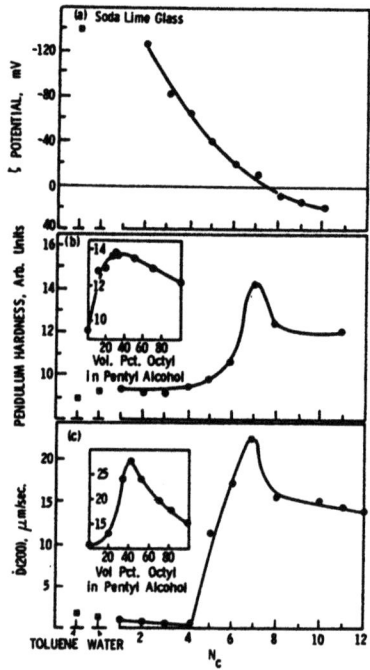

Fig. 5. Variation in (a) ζ-potential, (b) pendulum hardness and (c) rate of penetration with a diamond-studded bit of s.l. glass in toluene, water and n-alcohol environments [20]. Note influence of binary alcohol solutions illustrated in inserts to (b) and (c).

---

*N_C is the number of carbon atoms in the n-alcohol molecule.

** The mechanistic significance of this is discussed in Ref. 2.

A similar relationship between ζ-potential, hardness and drilling efficiency is shown for alumina in Fig. 6 [10]. Note particularly the values of D(200) for pentyl and octyl alcohols, which are some eight to ten times greater than values obtained when water is the cutting fluid.

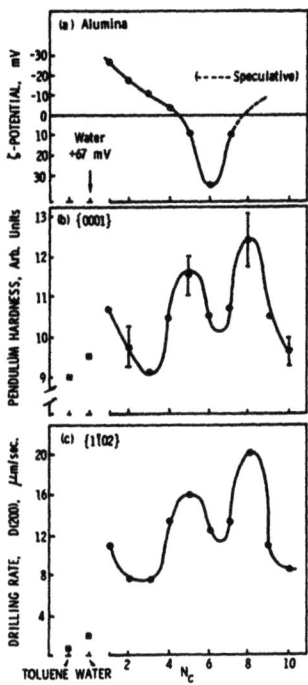

Fig. 6. (a) ζ-potential, (b) pendulum hardness and (c) rate of core bit drilling for alumina in toluene, water and n-alcohol environments [10].

An example of the opposite effects of a given environment on the rates of penetration of a given solid (calcite) by different types of bits was presented earlier in Fig. 1. These data have significance beyond establishing that cutting mode is important. That such opposing effects can occur in the same environment, in the presence of the same impurities, and with the same cooling, lubricating and dissolution properties, demonstrates that none of the latter factors exerts the controlling influence on drilling rate in this system. Yet "cutting" fluids are conventionally formulated primarily on the basis of their lubrication and coolant properties!

Consider next the technological implications of the observation that the maximum penetration rate obtained in heptyl alcohol when drilling s.l. glass with a diamond bit may be reproduced in appropriate binary solutions of other alcohols, Fig. 5. This implies that any liquid which imparts the same ζ-potential (i.e., the same surface hardness) to the solid to be drilled is likely to provide essentially the same drilling rate (other factors being equal). If this is so, then a wide choice of cutting fluids should be available for each substance, and it should be possible to find or formulate one that is cost-effective as well as non-toxic and non-polluting. Bear in mind, however, that a given cutting fluid will normally be optimum only for the particular combination of solid and cutting tool for which it was developed. Any change of tool which alters the mechanism of chip formation will require a different cutting fluid to achieve optimum performance.

These concepts are now being evaluated as part of a program intended to develop cutting fluids and techniques capable of significantly reducing the high cost of hard rock drilling. The results to date appear quite promising. For example, increasing the hardness of the quartz phase in a granite by immersion in appropriate n-alcohols can double the rate of penetration of the granite by diamond core bits, Fig. 7 [11]. Of course, the n-alcohols are not realistic drilling aids for large scale use, because they are noxious and costly. However, the negative ζ-potential exhibited by most silicates in water can be increased to zero by addition to the water of appropriate concentrations of various surfactants.

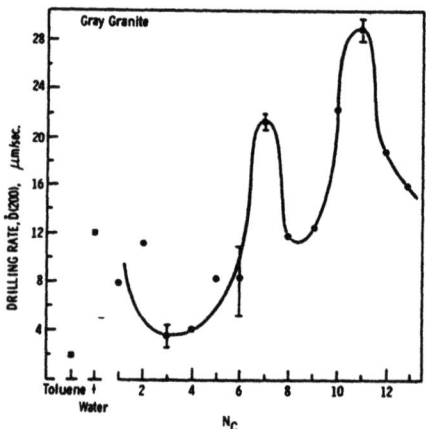

Fig. 7. Rate of drilling of gray granite with hemispherical-ended diamond bit in toluene, water and n-alcohol environments [11].

For quartz and Westerly granite, for example, streaming potential studies have revealed that the addition of 10^{-3} - 10^{-4} moles/l of the cationic flotation agent DTAB (dodecyl trimethyl ammonium bromide) to water produces this result [24]. Thus, aqueous DTAB solutions might be expected to significantly influence the rotary diamond drilling of quartz and Westerly granite, and such is the case, Fig. 8 [24]. Note the eight-fold increase in drilling rate in DTAB solutions of concentration close to that producing $\zeta \simeq 0$. Other studies showed that the drilling rate of s.l. glass could also be increased three-fold in appropriate DTAB solutions.

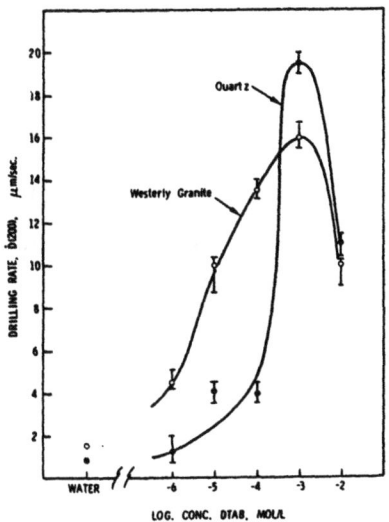

Fig. 8. Rate of core bit drilling (6 mm o.d.) of quartz monocrystals and Westerly granite as a function of concentration of aq. DTAB environments. Bit rotation speed 2200 r.p.m. [25].

Small scale tests using a dental drill showed that six-fold increases can be obtained even for bit rotation speeds of 25,000 r.p.m. And three-fold increases have also been obtained in work on the percussive drilling of Westerly granite using 18 mm dia. carbide wedge bits, Fig. 9 [25].

Work now in progress is intended to determine whether such environment-produced improvements in penetration efficiency are likely to be obtained when tunneling with large scale mole borers. To simulate under static conditions the fracturing action of the roller wedge type of cutters employed by such

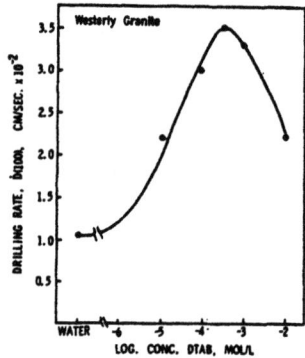

Fig. 9. Variation in rates of percussive drilling of Westerly granite using 18 mm diam. carbide wedge bits in aq. DTAB environments [25].

equipment, indentation tests are being conducted with both sharp and blunt carbide wedges of radius of curvature 4 cm and 75° included angle. Estimates of the specific chip generation energy (SCGE) involved are derived by dividing the energy absorbed in producing the crater (determined from measurements of the area under the force-displacement curves) by the volume of chips produced (measured by filling the crater with a low surface tension liquid via a calibrated microsyringe). The results are given in Table I [26]. The values shown are the means of 7-12 tests, and the standard errors of the means.

Table I

				Environment		
				aq. DTAB 2.5x		
		Air	Water	10^{-3}M	10^{-4}M	10^{-5}M
Blunt Wedge	SCGE (MJ/m^3)	235 $\pm$12	268 $\pm$18	230 $\pm$26	196 $\pm$19	221 $\pm$14
Sharp Wedge	SCGE (MJ/m^3)	145 $\pm$10	196 $\pm$12	133 $\pm$11	134 $\pm$8	128 $\pm$12

It can be seen that the energy to create chips under an aq. 2.5×10^{-4}M. DTAB environment (in which ζ-quartz $\simeq 0$ [24] is $\sim 27\%$ less than under water when the blunt wedge is used, and some 32% less when the sharp wedge geometry is used. It was also found that the load at which crushing was first evident under the blunt wedge was $\sim 35\%$ less in aq. 2.5×10^{-4}M. DTAB than in water. (Note that air is not usually an acceptable cutting medium because of dusting problems.)

Of course, the critical issue in transferring any development to practice is the promise of economic benefits, and analysis of such a possibility in this case must take into account other factors besides increased penetration rate, e.g., environment-sensitive tool life. Earlier work by Strebig et al. [27] showed that additions of 8 v/o of an anionic detergent to water increased the working life of 3 cm. diam. diamond bits when drilling quartzite up to six-fold, as well as increasing penetration rates by $\sim 70\%$. The economic impact of these improvements was a 30% reduction in the cost of drilling a 300 m hole. Subsequent work by Unger et al. [28], involving the diamond drilling of basalt in various aqueous surfactant solutions, indicated 28-35% increases in rates of penetration, and 22-26% reductions in cost. In neither of these two investigations, however, was any attempt made to optimize the concentration of the surface active agent used to produce maximum impact on penetration rate. On the other hand, preliminary economic analyses based on data similar to that presented in Fig. 9, which did not take into account the possibility of enhanced tool life but which did assume that the maximum observed increases in rates of drilling in lab scale tests could also be obtained in field operations, infer savings of similar magnitude. For example, for the percussive drilling of granite with 15 cm. diam. star bits, savings of up to 35% over costs in water are theoretically possible. Assuming that improvements similar to those shown in Fig. 8 can be obtained when diamond core bit diameter is increased from ~ 0.6 cm. to 4 cm., then savings of order 70% over costs in water are theoretically possible.

Not a great deal is known about the influence of surface-active environments on the wear mechanisms of cutting tools. But experiments using carbide spade bits and a variety of simple solids, such as MgO [23], LiF, and soda-lime glass indicate that water is a poor cutting environment as far as carbide tools are concerned. For example, it was noted that the addition of 10 v/o of water to heptyl alcohol produces a marked reduction in the cutting rate of s.l. glass due both to chemomechanical effects and to increased bit wear, Fig. 10. Conversely, however, calcite drills substantially better in water than in butyl alcohol, and experiments revealed that the addition of 10 v/o of water to butyl alcohol increases the rate of penetration of a carbide spade bit into this substance, Fig. 10 [26].

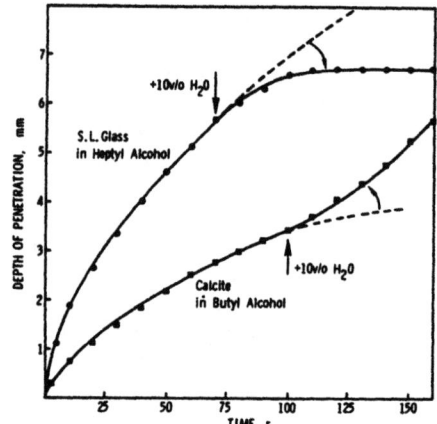

Fig. 10. Penetration vs. time curves for (i) soda-lime glass in heptyl alcohol to which 10% of water is added, and (ii) calcite in butyl alcohol to which 10% of water is added. Carbide spade bits. Note opposite effects on drilling rate of water additions.

In view of these apparently opposing influences of water on cutting behavior, therefore, carbide spade bits have been used to drill calcite in either butyl alcohol or water for extended periods. It has been found that, although the bits drilled faster initially under water, penetration ceased after ~ 10 min. due to bit wear, Fig. 11. Bits cutting more slowly under butyl alcohol, on the other

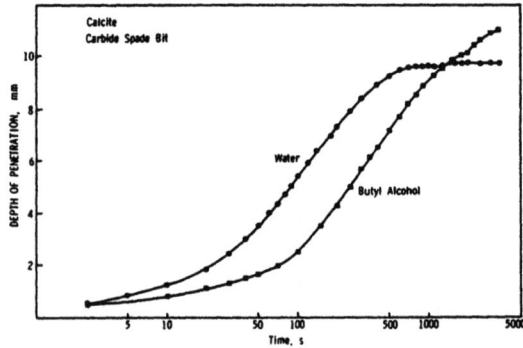

Fig. 11. Effect of environment on wear behavior of carbide spade bits when drilling calcite. Water initially facilitates cutting, but rate of bit wear is greater.

hand, continued to penetrate, matching the penetration achieved under water after 20 min., and continuing to cut at an appreciable rate well beyond 30 min. Thus, it appears that water is indeed an intrinsically bad environment for carbide cutters, and that even the occasional beneficial chemomechanical effect of water may be overcome in time by the detrimental influence of this environment on bit durability.

Metals. The significant effects of liquid metal and electrolyte environments on the fracture behavior of metals can be used to markedly increase the rate and efficiency of metal cutting operations. Shchukin et al. [29], for example, have observed that the rate of penetration of a bit through a stainless steel can be increased eight-fold by drilling under a Sn-Zn eutectic environment at 470°K, and two to three-fold in an In-Ga-Sn eutectic at room temperature, over that in air, Fig. 12. Likewise, Wood's metal increases the rates of drilling of aluminum by two orders of magnitude, and the In-Ga-Sn eutectic decreases the energy required to cut aluminum with a spade bit by a similar factor. It is also reported [29] that drilling a tempered U8 steel (Rockwell C Hardness ~45) is almost impossible in air using an alloy bit, but that quite satisfactory cutting rates ($\sim 10^{-2}$ cm/sec) can be achieved under the In-Ga-Sn eutectic. Copper can be drilled 30 times faster under a Pb-Sn eutectic than in air.

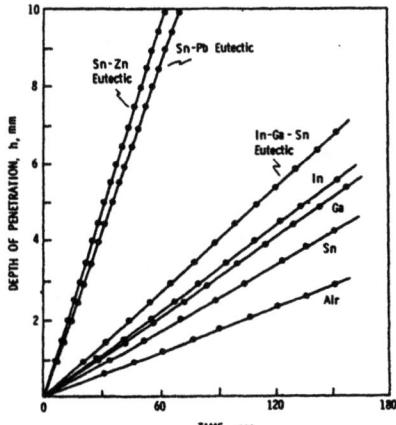

Fig. 12. Penetration as function of time when drilling a stainless steel in various liquid metals [29].

Shchukin et al. [29] suggest that facilitation of drilling can be achieved by liquid metals which do not affect the mechanical properties of the solid metal in conventional tensile tests.

They cite, for example, that gallium does not reduce the tensile strength of steel, but does accelerate its drilling. However, the question of whether or not the gallium actually penetrates the oxide film and contacts the underlying metal during conventional tensile testing conditions may be raised.

No systematic study of such effects has yet been reported. But, if reliable ways can be found for removing the active liquid metal species from the workpiece following the cutting operation, this novel approach to metal cutting might be well worth pursuing, especially for some of the otherwise hard to machine metals such as nickel superalloys or titanium alloys. And if removing the active liquid metal species proves to be troublesome, the less active but still extremely effective cutting aid oleic acid [29] might be tried. This substance reportedly also permits 2-10 fold increases in the rates of drilling of aluminum, copper, and ferrous alloys, and so potentially could be very useful in meeting a 10% cost-savings goal.

Studies of the effects of electrolytes on the mechanical behavior of metals have led also to the development of a new type of metal cutting procedure called electromechanical machining (EMM) [22]. In conventional metal machining operations, the removal of material is markedly dependent on the coefficient of friction between the tool and the workpiece, and the hardness and near-surface work hardening characteristics of the workpiece. Significantly, both of these factors can be affected by electrolytes [30, 31]. To examine the influence of such effects, a 18 cm. shaper was modified to allow nickel and superalloy workpieces to be totally submerged in Na_2SO_4 electrolytes. It was found that, for polycrystalline nickel at + 1600 mV, the work required to remove a given amount of metal was less than when the best commercially available cutting fluid was used. And, more importantly, the surface finish was much improved. For example, Fig. 13(a) shows that at the rest potential (i.e., no applied potential) surface quality is poor, with evidence for seizing and galling being readily apparent. As the potential was made more positive, however, a steady improvement in surface quality resulted, and at + 1600 mV -- a potential at which oxygen is being evolved from an oxide coated surface -- surface quality is excellent, Fig. 13 (b).

Subsequent studies have shown that electromechanical effects can also facilitate drilling [32]. For example, the rate of penetration of a nickel workpiece by a carbide spade bit is twice as fast at + 1800 mV than at -600 mV, Fig. 13(c). And the effects of switching from one potential to the other as the drill penetrates the workpiece are shown in curve C of Fig. 13(c), illustrating clearly the predictable and reversible influences of applied potential. Drilling experiments with CG27, a nickel-base superalloy, indicate behavior similar to that of nickel.

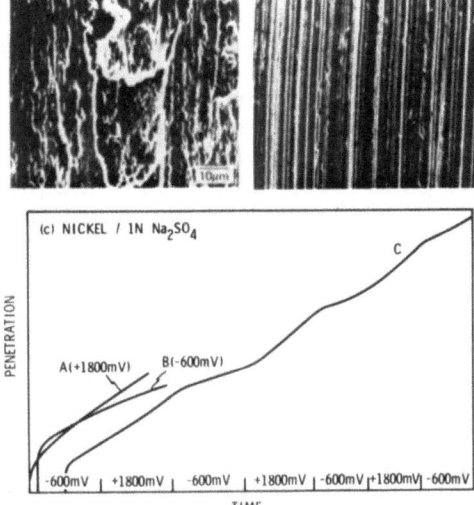

Fig. 13. SE micrographs of nickel surfaces produced (a) with no applied potential, (b) at +1600 mV potential [22], (c) carbide spade bit drilling of polycrystalline nickel as function of applied voltage in 1N. Na_2SO_4 electrolyte [32].

Nickel-base materials are difficult to machine because they are characteristically "gummy" and work harden rapidly. Intuitively, therefore, one would expect that machining behavior would be improved if the effect of the electrolyte was to harden the surface of the workpiece -- which an oxide film would do. Conversely, one might anticipate that the best cutting potential for a hardened steel, such as 4140, would be one at which the electrolyte softens the surface. This would be expected to occur in the active dissolution region (-400 to -600 mV in 1N. Na_2SO_4), and this is exactly what is observed.

Other advantages of EMM are increased tool life, improved tolerances, and smaller swarf particles -- leading to cleaner working conditions. A preliminary cost benefit analysis of the application of EMM to the drilling of a nickel superalloy indicates that savings of order 20% might be expected over more conventional procedures.

As with chemomechanical effects, the selection of an electrolyte and cutting potential combination appropriate for each material-tool system will be important, for cutting potentials which are effective for one material will not necessarily be suitable for another. However, the availability of numerous Pourbaix diagrams should facilitate the choice of both electrolytes and potentials.

Incidentally, EMM should not be confused with electrochemical machining (ECM) or electrical discharge machining (EDM). In these techniques, metal removal occurs predominantly by some procedure other than via direct contact between the tool and the workpiece. In EMM, however, metal removal is achieved by conventional means, the only difference being that the workpiece is electrochemically polarized. Thus, in principle, and in practice, any standard drill press, lathe or milling machine can be modified to use EMM.

3.2 Comminution, Grinding and Abrasive Material Removal

<u>Nonmetallic Solids.</u> Comminution and grinding are extremely inefficient processes, only about 1% of the energy consumed being used to generate new surface area. Sometimes they also constitute a significant cost factor in production. For example, the energy required to grind first the raw materials and later the clinker costs about 10% of the total cost to produce a barrel of cement [33]. Clearly, if comminuting and grinding efficiencies could be increased from their current 1% to perhaps 2% by the use of appropriate grinding aids, the economic impact would be significant. Naturally, much effort has been applied towards this laudable goal, and the literature is voluminous. However, as Somasundaran and Lin [34] comment in their recent comprehensive review of such work, the results of most previous studies are of little use for scientific interpretation because hardly any of the important system variables -- such as the concentration of the active species, surface potential of the particles, thermal treatment or composition of the solid, pH, flotation parameters, etc., -- have been varied in a logical manner, or in some instances even measured or controlled in any manner. Thus, no scientific basis yet exists upon which logical development work leading to optimum utilization of environmental effects can proceed.

In principle, there are two ways in which surface active species can influence comminution and grinding behavior, (i) by facilitating fracture processes, and (ii) by preventing agglomeration of the freshly created particles. Work by Locher and Seebach [33] and others has established beyond question the value of the latter approach, especially for fine grinding applications. However, there is still doubt as to whether grinding aids actually facilitate fracture in ball milling operations. Resolution of this issue is difficult because attrition in ball milling is usually either via (i) massive impact by cascading balls, leading to rapid propagation of brittle cracks through what are typically porous, inhomogeneous solids (such as cement clinkers or naturally occurring minerals). Such a process is certainly unlikely to be responsive to the relatively modest effects of adsorbed species on near-surface flow and fracture properties, or (ii) abrasive wear, which appears to be the predominant mechanism for ball milling in the absence of cascading. In the latter case, the softer the surface of the solid, the greater is likely to be the rate of comminution. Thus an environment which produces a relatively large surface potential on the solid should facilitate this type of milling, as is evident from the data presented in Fig. 14 [15]. However, these results can also be interpreted in terms of the influence of surface potential on agglomeration. If the grinding environment is such that the zeta potential of the solid approaches zero, then coagulation and agglomeration will be maximized and, as a consequence, grinding efficiency will be reduced [35], as is observed. Clearly, critical experiments designed to discriminate between such alternate explanations are now required.

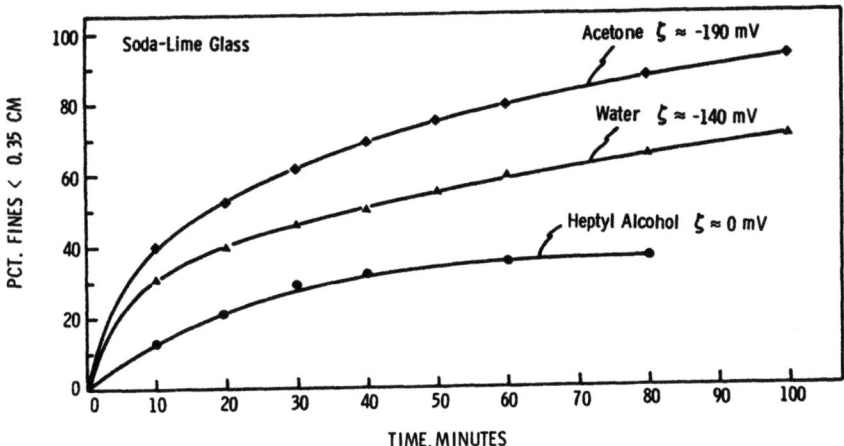

Fig. 14. Data from ball milling soda lime glass in a non-cascading mode.

In this regard, studies of the environment-sensitive, abrasive grinding behavior of alumina have produced some relevent observations [10, 13]. The stimulus for this work was the fact that the cost of producing the high quality surface finish required on alumina substrates for the microelectronics industry can amount to 60% of the total cost of the component. Initial experiments, involving measurements of weight loss as a function of environment for polycrystalline alumina samples ground on diamond-impregnated metallographic wheels, revealed two distinct environmental influences -- a viscosity dependence and a ζ-potential dependence [13]. Experiments conducted under n-alcohol environments revealed no dependence on ζ-potential such as might have been anticipated in the light of the data of Fig. 6, but rather a monotonic decrease in rate of material removal with increasing N_C. A similar relationship between abrasive grinding rate and N_C had been found earlier for s.l. glass by Weiderhorn and Johnson [36]. Evidently, hydrodynamic lubrication effects are involved when abrading under the higher alcohols (the variation in viscosity being a factor of thirty between $N_C = 1$ and $N_C = 12$), and viscosity becomes the dominant parameter controlling material removal rates in these environments. (In contrast, of course, the complex shape of the $\dot{D}(200)$ vs N_C plot obtained when drilling alumina with diamond core bits, Fig. 6(c), implies that viscosity is <u>not</u> an important environmental parameter in this case.)

Next, abrasive grinding rate and diamond core bit drilling experiments were conducted with alumina in dilute aqueous solutions of NaOH, lauric acid and oleic acid, for which viscosity is not a significant composition-dependent variable. In this case, Fig. 15, chemomechanical effects markedly influenced material removal rate, and abrasion was least when $\zeta \simeq 0$, i.e., when the alumina surface was at maximum hardness. Moreover, for these environments, cf. Figs. 15(b) and (c), an inverse relationship was noted between the rate of drilling alumina with diamond-loaded bits and the rate of grinding it with diamond-impregnated metallographic wheels.

SEM studies of ground alumina surfaces also confirmed Koepke's [37] earlier observation that, even for solids as hard and brittle as alumina, material is removed in grinding primarily by plastic flow and flow-dependent fracture processes. Thus, since the least amount of near-surface plastic flow occurs at the p.z.c., the observed minimum rate of material removal by abrasion at $\zeta \simeq 0$ is to be expected. On the other hand, since material removal in rotary diamond bit drilling is primarily dependent on brittle fracture processes, the opposite environmental dependence of rates of drilling and abrasive wear is what one would expect.

No new and markedly superior grinding fluid for alumina has yet been developed, but it has been established that surface potential and environment-sensitive fracture behavior can play dominant roles in abrasive grinding situations where agglomeration is not an important and simultaneous factor. And, in proprietary studies of the abrasive grinding of zirconia, production rate improvements of more than 30% have been obtained by modest reformulation of the cutting fluid along lines suggested by this work.

<u>Metals</u>. Considerably less work has been conducted on the environment-sensitive comminution of metals than of non-metals.

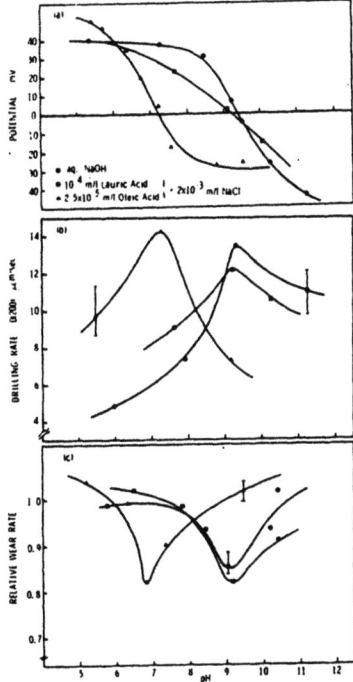

Fig. 15. Effect of pH on (a) ζ-potential, (b) drilling rate and (c) abrasive grinding rate relative to that in distilled water for polycrystalline alumina. Drilling rates pass through a maximum and grinding rates through a minimum at the p.z.c.'s appropriate to each environment [13].

However, some experiments conducted by Pertsov et al. [38] indicate the extraordinary potential for improvements in this field. Ordinarily, it is quite difficult to pulverize a relatively ductile metal such as zinc in a conventional vibratory ball mill. But the addition of 1% of gallium was found to increase its rate of grinding by 1-2 orders of magnitude, and to decrease the particle size achieved after 30 min. by about a factor of 10, Fig. 16. Gallium also exerts a marked influence on the grinding behavior of tin, cadmium, aluminum and bismuth. Appreciable improvements have also been reported using various organic media, such as isopentanol and glycerol for iron powders [39], and aqueous surfactant solutions for other metals [8].

4. IN CONCLUSION

In this brief review it has been possible to mention only a few of the many interesting and sometimes important examples of the

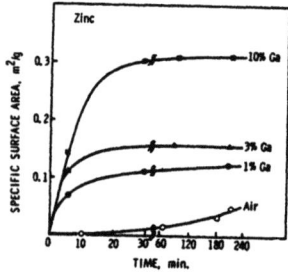

Fig. 16. Dependence of specific surface area on time of grinding for pure zinc in air and following introduction of various concentrations of gallium. Ref. [38].

influence of chemically active environments on the fracture behavior of solids. For example, their important effects on the properties of organic crystals and polymeric materials have been mentioned not at all. Clearly, however, the systematic application of chemomechanical phenomena -- on the basis of improved understanding -- to the facilitation of comminution, machining, drilling and finishing operations is only just beginning. And our ability to apply the effects of appropriately selected environments to such fracture-dependent processes will depend, in part, on our level of mechanistic and conceptual understanding of such effects. At present, such understanding is extremely limited. This is a consequence of our inadequate comprehension of some of the fundamental processes involved, e.g., of the specific electronic interactions which occur between adsorbates and surfaces during chemisorption, and of the influence, if any of lattice strain on such interactions. Another obstacle to progress is the interdisciplinary nature of this field of study, involving as it does the complicated interplay of variables arising from the chemistry of the environment, the physics of the near-surface layers of the solid, and the materials science and engineering of the bulk. At present, the different disciplines are using different "languages" to discuss similar or related phenomena. Continued and improved dialogue between workers in these fields, begun at this meeting, will be invaluable.

In the meantime, however, awareness of the empirical relationship between ζ-potential and near-surface hardness, and recognition of the need to take into account the predominant mode of fracture probably involved (i.e., brittle vs. ductile), should begin to permit significant improvements over current practice in the processing of non-metallic solids, either in terms of increased production rates, reduced tool wear, or lower energy consumption. In the light of the results presented above, a 10% cost reduction goal does not seem too ambitious. Awareness of the ζ-correlation will also permit considerable flexibility in the selection of fracture-aiding environments. Thus, it should be possible in most cases to formulate inexpensive but efficient water-base fluids which are neither toxic, corrosive nor polluting.

In summary, then, at this stage of its development, the field of environment-sensitive fracture behavior presents both an intriguing challenge to research scientists and a golden opportunity for development engineers [2].

ACKNOWLEDGEMENTS

The preparation of this review, and much of the work reported therein, was supported in part by Grants APR73-07787 and DMR75-05443 from the National Science Foundation. Their continued support for this field of study is greatly appreciated.

REFERENCES

1. Reynolds, O., Manchester Lit. Phil. Soc., 13, 93, 1874.
2. Westwood, A. R. C., J. Matl. Sci., 9, 1871, 1974.
3. Private communication, Merchant, M. E., Cincinnati Milacron Inc., May 1975.
4. Armstrong, E. L., Paper presented at Symp. on The Underground Environment, Geol. Soc. of America, Washington, D. C., Nov. 1971.
5. Liechtenstein, S., Technical News STR-3454, NBS, Washington, D. C., 1966.
6. Rebinder, P. A., Proc. Sixth Physics Conf., Moscow, 29, 1928.
7. Rebinder, P. A., Schreiner, L. A. and Zhigach, K. F., Hardness Reducers in Rock Drilling, Acad. of Sciences, Moscow, 1944, Transl. C. S. I. R. O., Melbourne, 1948.
8. Rebinder, P. A., and Shchukin, E. D., Prog. in Surface Sci., 3, 97, 1972.
9. Westwood, A. R. C. and Latanision, R. M., Science of Ceramic Machining and Surface Finishing, NBS, Washington, D. C., Spec. Pub. 348, 141, 1972.
10. Westwood, A. R. C., Macmillan, N. H. and Kalyoncu, R. S., J. Am. Ceram. Soc., 56, 258, 1973.
11. Westwood, A. R. C., Macmillan, N. H., and Kalyoncu, R. S., SME Trans., 256, 106, 1974.
12. Macmillan, N. H. and Westwood, A. R. C., Surfaces and Interfaces of Glass and Ceramics, Plenum, New York, 493, 1974.
13. Swain, M. V., Latanision, R. M. and Westwood, A. R. C., J. Am. Ceram. Soc., in press, 1975.

14. Westbrook, J.H. and Jorgensen, P.J., *Trans. AIME*, 233, 425, 1965.
15. Westwood, A.R.C. and Macmillan, N.H., *Science of Hardness Testing*, Am. Soc. Metals, Metals Park, Ohio, 377, 1972.
16. Latanision, R.M., Opperhauser, J. and Westwood, A.R.C., loc. cit., 432.
17. Westwood, A.R.C. and Kamdar, M.H., *Phil. Mag.*, 8, 787, 1963.
18. Macmillan, N.H., Huntington, R.D., and Westwood, A.R.C., *Phil. Mag.*, 28, 923, 1973.
19. Heins, R.W. and Street, N., *Soc. Pet. Engrs. J.*, 5, 177, 1965.
20. Westwood, A.R.C. and Huntington, R.D., *Mechanical Behavior of Materials*, Soc. Matls, Sci., Tokyo, 383, 1972.
21. Mills, J.J. and Westwood, A.R.C., Unpublished work, Martin Marietta Labs., 1975.
22. Latanision, R.M., Neilsen, K.C. and Kirschbaum, R., *Modern Machine Shop*, Feb., 1974.
23. Westwood, A.R.C. and Goldheim, D.L., *J. Am. Ceram. Soc.*, 53, 142, 1970.
24. Macmillan, N.H., Jackson, R.E., Mularie, W.M. and Westwood, A.R.C., *SME Trans.*, in press, 1975.
25. Jackson, R.E., Huntington, R.D., Westwood, A.R.C., Rept. to NSF on Grant GI-38114, Martin Marietta Labs., June, 1974.
26. Swain, M.V., and Westwood, A.R.C., Unpublished work, Martin Marietta Labs., 1975.
27. Strebig, K.C., Schultz, C.W., and Selim, A.A., *Mining Eng.*, Oct. 1969, 73.
28. Unger, H.F., Snowden, B.S. and Engelman, W.H., Preprint SPE-4236, Soc. Pet. Eng. AIME, Jan. 1973.
29. Shchukin, E.D., Polukarova, Z.M., Yushenko, V.S., Bruyukhanova, L.S., and Rebinder, P.A., *Sov. Phys. Doklady*, 17, 699, 1973.
30. Bockris, J.M., Argade, S.D. and Gileadi, E., *Electrochim. Acta*, 14, 1259, 1969.
31. Latanision, R.M. and Staehle, R.W., *Acta Met.*, 17, 307, 1969.
32. Latanision, R.M. and Neilsen, K.C., Unpublished Work, Martin Marietta Labs, 1973.
33. Locher, F.W. and Seebach, H.M., *Ind. Eng. Chem. - Proc. Prod. Dev.*, 11, 190, 1972.
34. Somasundaran, P. and Lin, J.J., *Ind. Eng. Chem. - Proc. Prod. Dev.*, 11, 321, 1972.
35. Ghigi, G. and Rabottino, L., *Dechema-Monogr.*, 57, 427, 1967.

36. Weiderhorn, S. M. and Roberts, D. E., <u>Wear,</u> <u>32</u>, 51, 1975.
37. Koepke, B. G., Ref. 9, p. 317.
38. Pertsov, A. V., Goryunov, U. V., Pertsov, N. V., Shchukin, E. D., and Rebinder, P. A., <u>Dokl. Acad. Nauk.</u>, <u>172</u>, 1137, 1967. See Ref. 8, p. 171.
39. Lowrison, G. C., <u>Crushing and Grinding,</u> CRC Press, Cleveland, 109, 1974.
40. Sennett, P., and Oliver, J. P., <u>Chemistry and Physics of Interfaces,</u> Am. Chem. Soc., Washington, D. C., 75, 1965.

DISCUSSION

Comment by I. R. Kramer:

It was noted that the data presented for drilling was restricted to rather short times (∼2 min.) and shallow depths. Are the data available for deep holes and longer times. It would have been expected that the power expended during the drilling and machining would have provided a great measure of the overall efficiency of the tool wear and the machining process.

Reply:

All the data presented to date has been from laboratory scale experiments. During the next year, however, we will be conducting a full-scale (4-6" diam. bits) evaluation of the possibility of chemically enhancing the drilling of granite using appropriately instrumented equipment capable of measuring power utilization, etc.

Comment by A. S. Argon:

1) I believe that the dramatic increases in the efficiency of machining of soft metals is probably not correctly explained as a result of reduction of flow stress alone. Machining consists of deformation and removal of metal chips for which the total work can be given as the product of the flow stress and the strain to fracture. This must indicate that the result comes about from a reduction of both flow stress and perhaps more significantly the strain to fracture.

2) In some high temperature super alloys machining produces local martensite transformation. It would seem to me that an improvement in the machinability of these materials may require suppression of martensite transformations during cutting.

Reply:

1) No doubt you are correct. We have only just begun to measure the specific chip generation energies associated with chemomechanical effects in rocks. For Westerly Granite at $\zeta = 0$, the load to produce chips, the fracture path, and the volume of chips produced are all altered, and the energy per cc of rock produced is reduced by $\sim$ 35%. Workers at the U.S. Bureau of Mines had previously observed such reductions also.

2) We have no relevent observations to offer on this possibility. At this time, I think the effects reported are caused primarily by oxide (or hydroxide) film-induced changes in the sliding interactions between the tool and the workpiece.

Comment by R. Bullough:

The only failure you described to correlate with the zeta potential explanation was explained by variations in viscosity of the fluids. Because we do not really understand why there should be a correlation between hardness and the zeta potential, it is important to hear whether you have had any inexplainable failure to correlate.

Reply:

Yes. If the fracture behavior of the solid is not markedly dependent on flow processes, then the effects I have described are not relevant. For example, in certain granites, fracture behavior is determined by preexisting or readily formed cleavage cracks in the microcline phase. In studies with this material, no evident influence of cutting environment was noted.

ENVIRONMENTALLY-ASSISTED HARD
MATERIALS MACHINING

N. V. Pertsov

Moscow Lomonosov State University, Chemistry Department
Moscow, U.S.S.R.

1.   INTRODUCTION

Destruction and plastic deformation of a wide variety of solids are the bases of the overwhelming majority of modern machining and processing methods. These methods include practically all kinds of metals machining; cutting of glass, ceramics and construction stone; comminution of minerals to extract ores, and of cement clinker and compact catalisors; rock drilling; and many others. Almost all these processes employ liquid media which may be affected by complicated physical, physicochemical, and chemical phenomena.

Modern industry, which is highly efficient and automated, requires special and efficient liquid media designed for application in definite technological processes. For metals cutting there should exist a wide choice of coolant-lubricants. Pressing requires special lubricants, liquid and solid. Active aqueous compositions that assist comminution and drilling also are necessary.

Growing demands of industry prompt the science to continuously search for new, efficient cooling-lubricating media. A large branch of the oil and chemical industry deals with the investigation and manufacture of assorted coolant-lubricants necessary for metals processing. Newly composed coolant-lubricants are usually created with joint efforts of specialists in organic chemistry and metals treatment technology. This long and experimentally complex work is complicated by the absence of precise and detailed ideas on mechanisms of effect, data on quantitative relations among different functions, and accurate scientific criteria for choosing components of coolant-lubricants.

The main functions of coolant-lubricants are usually considered as follows: (1) a lubricant between the workpiece and the tool, (2) to facilitate deformation and destruction of the machined material, (3) to clean the instrument from the products of dispersing, (4) to cool the instrument and the surface being machined, and (5) to stabilize the products of destruction. At present most investigators concentrate on the lubricating effect as the main function of coolant-lubricants; the destructive or "cutting" effect is generally paid far less attention to, probably because the mechanism of this phenomenon is more complicated and less studied than that of boundary lubrication.

However, critical analyses of early data and the results of recent investigations have made us revise the present point of view concerning the mechanism of coolant-lubricants, and pay more attention to special physicochemical influences of the active environment on deformation and destruction of materials in the course of mechanical shaping. It is essential to engage P. A. Rebinder's and his followers' ideas on adsorption-induced strength reduction in this field of science and technology.

2. THE DEVELOPMENT OF POSSIBLE APPLICATIONS OF THE REBINDER EFFECT IN THE MECHANICAL SHAPING OF SOLIDS

Soon after P. A. Rebinder in 1928 had discovered the effect of strength reduction of solids due to lowering of their surface energy, wide investigations commenced into various applications of this phenomenon in processes of mechanical treatment of materials. According to the general definition of the thermodynamic nature of the active environment action, given by Rebinder in the very first article [1], detailed investigations were conducted on the influence of wetting and adsorption of surfactants on processes of destruction, scratching, grinding, drilling, etc. of different solids [2-4]. The results showed that adsorption of organic surfactants from corresponding solutions reduces considerably the hardness of brittle solids with ionic bonds (as the specific work of new surface formation decreases). The concentration dependence of the effect of the active substance turned out to be important for understanding the nature of the phenomenon. It has a form analogous to that of the Shishkovsky surface-tension isotherm, and the efficiency of solutions (e.g., in the case of dispersing marble in aqueous solutions of amines and aliphatic acids) grows with the length of the hydrocarbon chain of the surface-active molecule in accordance with the Duclos-Traube rule. Hardness of various minerals was also found to be reduced by aqueous solutions of certain salts. The basic difference between the effects of organic surfactants and surface-active electrolytes is that the efficiency of the former continuously grows with increasing concentration, while the influence of electrolytes possesses a strongly pronounced concentrational maximum, its position depending on the valency of the ion being adsorbed and to a smaller extent on chemical composition of the mineral under destruction. The hardness of granite and quartzite rocks appeared to be very efficiently reduced by, one one hand, aqueous solutions of aluminum and magnesium chlorides of 0.1 - 0.5% concentration, and, on the other hand, by solutions of a mixture of naphthic acid and sodium salts (naphtha soap).

Investigations of the drilling of hard minerals led to most important practical results -- i.e., special drilling fluids were worked out. Solutions of aluminum chloride (0.1% to 1%) which allow increased drilling speeds of 40-50% and reduction of wear of the cutting instrument by 25-35% (Fig. 1) had widest application. Loamy drilling fluids in oil-well drilling showed increased efficiency if cellulose sulphite liquor (a waste of the paper industry) or certain wastes of the petro-chemical industry were added to them.

Synthetic surfactants also turned out to be highly efficient activators of drilling fluids. Solutions of ionogenic surfactants [5] and cellulose sulphate [6] increase the speed of well drilling by a factor of 2, solutions of OP-10 alkyl phenol by a factor of 3, and surfactants like ditalan, sulfanol and prevatzell by factors of 1.2 to 1.4, the rocks being of great hardness.

According to Rebinder, the mechanism of hardness reducers consists of accelerated development of the pre-destruction zone (i.e., the near-surface part of the solid). The pre-destruction zone is considered as a structure with a developed network of microcracks filled with the surrounding liquid medium. The environment causes the growth of microcrackness, crumbling and weakening of the material, and also prevents the cracks from being healed in periods between subsequent impacts of the instrument. To explain the influence of salts it is presumed that the decrease in strength occurs due to lowering of the surface energy of the crystal as a result of wetting with water, the role of electrolytes being to create optimum hydration of the surface when the wide diffused part of the double electric layer is being formed.

The Rebinder effect is also widely used in comminution of various materials. Thus, for instance, glycols, amines, silicons and acetates are successfully used in cement comminution [9-10]; certain electrolytes noticeably intensify various grinding processes [11]. Detailed investigations by G. Khodakov established that active environments do not just intensify the process of fine comminution, but also prevent the particles from sticking together again and intergrowth [12,13]. It was found that a number of media different in chemical nature (water, benzene, acetone, triethanolamine) cause substantial facilitation of quartz and other minerals dispersing. In order to explain the effect of these liquids (which was approximately the same), it was proposed that molecules of the active substance adsorbed on juvenile surfaces of the dissociated solid so that the dissociation products -- the most mobile atoms or radicals -- penetrate the defects of the structure.

Soon after a new form of the Rebinder effect (i.e., adsorption plastification [14]) had been discovered in 1936, thorough studies began of the regularities and mechanisms of the effect of coolant-lubricants on pressing and cutting of metals. It was found that additives of surfactants (polar organic compounds with a developed hydrocarbon radical) to inactive non-polar organic liquids may cause three different effects:

(i) Lowering of friction at the boundary between two contacting surfaces (e.g., those of the workpiece and the tool, or in friction joints). The mechanism and regularities of this phenomenon, which is of great importance for machining processes, were studied in detail [15,16], but its description is beyond the limits of our survey.

(ii) Adsorption plastification, manifested as lowering of the yield point and as increase in creep rate of metal single crystals [17-19]. In the case of polycrystalline metals, adsorption of organic surfactants from non-polar solvents leads to plastification of a thin (about 1 μm) surface layer which serves as a highly efficient non-extrudable lubricant between the detail and the instrument during pressing and cutting [20-22] (Fig. 2).

(iii) Direct facilitation of destruction of the material being machined, decrease in work of forming or specific work of cutting [2,23] (Fig. 3). The greatest influence of surfactants is displayed in processes of cutting both very brittle and very plastic materials. At the same time it turned out that many organic liquids (and water in some cases) may be effective "cutting" media without being surface-active in the common sense of this term (after Gibbs). G. Yepiphanov offered an idea that the molecules of the liquid may undergo catalytic disintegration at new surfaces in the destruction zone. Products of thermodestruction (atoms) diffuse into the lattice and strengthen it, hence embrittling [23,24]. Though this point of view was not further developed, the concept of catalytic disintegration during destruction may be very fruitful for the analysis of coolant-lubricants' mechanism of effect.

The basic result of experiments with the Rebinder effect was a theoretical concept about the existence of special "cutting" properties of liquid media, so that the newly worked out coolant-lubricants should contain hardness reducers -- surface-active compounds in Rebinder's terms -- i.e., the ones leading to strength reduction as a result of physicochemical processes occurring at the surface of the solid and decreasing its surface energy. The main practical result was creation of a number of coolant-lubricants with "cutting" properties for metals machining. Among them were aqueous compositions which allowed replacement of kerosene during World War II and securing of enterprises against air raids. However, it should be pointed out that within the latest decades Rebinder's concepts have not been used widely enough by engineers and specialists on coolant-lubricants.

At the same time, at the end of the 1930s, a theory of chemical polishing was offered by I. V. Grebenschchikov [25]. According to this theory chemically active components of coolant-lubricants and various pastes react with the solid to form a thin layer of a softer chemical compound on its surface. This layer can be easily removed mechanically. Later this theory failed to be developed in detail and is used now only for explanation of environmental influence on polishing (mainly of glass) and electro-chemical grinding of metals.

3. USE OF SUBSTANCES CAUSING GREAT DECREASE IN SURFACE ENERGY AT INTERFACIAL BOUNDARIES FOR MACHINING OF SOLIDS

Broad investigations of the regularities and mechanism of destruction of solid materials in the presence of melts similar to them in molecular nature, carried out in Rebinder's, Rostoker's, Westwood's and other laboratories, resulted in discovery of such aspects of this phenomenon, which can be successfully used in processes of mechanical treatment of various materials. These are: a possibility of great decrease in surface energy and, respectively, in strength (up to 10 times and more); high rate of development of the fracture facilitation process; physicochemical reversibility of the effect of environment, consisting in complete rehabilitation of mechanical properties of the material after the active melt is removed; intensification of the effect as the initial strength of the solid increases; extremely small quantities of the active substance necessary; and so on.

The basic condition of successful use of melts akin to the solid in chemical composition and structure is their ability to compensate the bonds ruptured in the destruction process; that is, to decrease greatly the free surface energy of the solid. Along with that, the condition of great reduction of the surface energy is necessary but not sufficient for noticeable reduction of strength. The degree of influence of the environment is connected with many circumstances, including the actual defect structure and the initial strength of the given solid, temperature, and character of propagation of the melt to the destruction zone. An important role is played by the whole of external conditions, first of all, by character and intensity of the stressed state. The surface-active medium does not cause destruction by itself, it only helps the mechanical stresses to break the bonds; that is why the effect of the environment manifests most clearly in rigid stressed states with predomination of extending components.

Many kinds of mechanical treatment of solids -- for instance, various sorts of cutting -- are characterized with a favorable complex of conditions for manifestation of the Rebinder effect: high local stresses, cyclic loading, small depths of penetration of the cutting part of the instrument, assisting the transportation of the active substance to the destruction zone. These conditions make it possible to use the considerable strength reduction and the emerging brittleness of the solid under treatment right in course of the machining process [26, 27].

Thus, for example, E. Shchukin, L. Bryukhanova et al. managed to demonstrate high efficiency of drilling hard steels in the presence of metal melts [28, 29]. A special device added to a drilling machine, which was used in these experiments, allowed registration of the time of drilling and the value of the torque under conditions of constant axial pressure of the drill at the sample and constant depth of penetration. This device made it possible to calculate the rate and the specific work of cutting.

Efficiency of drilling of various steels (mild, stainless, special construction, carbon tempered, etc.) appeared to be increased best by the tin-zinc eutectics. When used for drilling of stainless steel 1H18N9T (Fig. 4), it is two times more efficient than oleic acid usually used; the specific work of cutting decreases here 8 times, while in oleic acid (the effect of which is very high due to chemosorption processes occurring), it decreases only two times. Especially demonstrative are the results for U8 steel. Drilling of this tempered steel in air, in emulsol and even in some metal melts is practically impossible, the tungsten carbide drill being used. Use of the liquid Sn-Zn eutectics sharply accelerates and facilitates the process of drilling of this steel (Fig. 4b).

Besides facilitation and acceleration of the drilling process, the surface-active metal introduced into the cutting zone causes considerable dispersing of shavings, which are no longer continuous but look like needle-like particles about 1 mm thick. Considerable reduction of the energy parameters of the drilling process in a metal melt leads to a sharp increase in endurance of drills, sometimes 10 times and more. An essential fact here is that the instrument (made of tungsten carbide) and the material under treatment (steel) are different in their chemical properties and in strength, as well as in sensitivity to the metal melt chosen. Under conditions of drilling, the material machined is brought to the limiting state and is destroyed, while the instrument is far from such a state. On the contrary, use of drills made of high-speed steel showed that their endurance was much lower than in the case of drilling with no medium. Under the given conditions of drilling the eutectic melt influenced the instrument more than the material machined.

In a number of cases it may be necessary to remove the residue of the surface-active melt from the surface under treatment so that the strength would not be reduced in course of exploitation of the material. Special experiments had shown that the danger of subsequent manifestation of the effect of the active substance is small, since the substance apparently diffuses completely into the lattice of the solid and does not cause further reduction in strength.

Together with studies of the influence of metal melts on intensity of industrial steels drilling, it was necessary to carry out the comparison of the degree of strength reduction of hard materials due to the action of liquid surface-active media reducing the interfacial energy, with acceleration of mechanical treatment in such systems. For this purpose, model systems were used which were studied in detail in early investigations of embrittlement by metal melts [18, 30, 31]. The results of the tests established quite definitely that the media causing large strength reduction are more efficient in providing mechanical treatment and accelerating it.

Thus, for instance, mercury and gallium influence cadmium and brass drilling in different ways. Gallium causes quite considerable increase in rate of cadmium drilling (145%), compared to mercury on brass (45%). At the same time, mercury influences the rate of cadmium drilling much less (28% increase) as does gallium on brass (27% increase). This corresponds with the strong influence of gallium on the strength of cadmium and mercury on brass, and the absence of considerable strength reduction in mercury-cadmium and brass-gallium systems.

Mercury and gallium cause approximately the same increase in zinc drilling rate (about 2.2 times); at the same time both mercury and gallium decrease the strength of zinc single crystals two-fold. In addition, gallium reduces the strength of polycrystalline zinc about one and a half times more than mercury does. This fact indicates that in drilling processes destruction takes place across the grains.

Correlation between the degree of strength reduction and

the increase in efficiency of mechanical treatment is a result important for understanding the mechanism of coolant-lubricants effect in cutting, showing the paramount importance of the Rebinder effect among other functions of coolant-lubricants. This conclusion was confirmed in the case of non-metals with investigations on influence of alcohols on the rate of rock salt drilling (Fig. 6) and of their influence on fused quartz grinding (Table 1). Traskin and Skvortsova [32] had found that water is the most efficient substance for reducing the strength of ionic crystals, and the efficiency of alcohols falls monotonously in their homologous series. These results agree well with those of mechanical treatment tests. Investigations of grinding demonstrate the influence of surface energy reduction on parameters of the machining process especially distinctly. The most active media (water, lower alcohols) provide the lowest work of quartz dispersing, despite the fact that higher alcohols possess better lubricating qualities.

High efficiency of fusible metals as reducers of strength of metals and non-metals allowed the use of them for facilitation of grinding. F. Danilova had worked out diamond grinding wheels with organic bundles containing optimum quantities of fine-disperse fusible metal powder as an active filler [33, 34]. As the wheel is warmed up in the process of grinding, the surface-active metal fuses, thus providing continuous transportation of the active substance to the destruction and considerable facilitation of dispersing. Use of such activized wheels increases the grinding rates several times, improves the resistance of the instrument to wear and also decreases the level of residual stresses in the superficial layer.

In order to identify the mechanism of fusible fillers' effect, grinding of tungsten carbide samples with a diamond wheel was carried out directly in a bath with metal melt. This offered the possibility of proving that the most active melts can reduce the specific work of dispersing 10 times, the effect of metals amounting to facilitation of destruction of tungsten carbide grains.

4. INFLUENCE OF CHEMICAL PROCESSES ON DESTRUCTION AND MECHANICAL TREATMENT OF SOLIDS

Analysis of thermodynamic conditions of the Rebinder effect shows that the greatest strength reduction takes place in cases when the energy of interaction at the boundary of the solid with the environment is close to the one of interaction between particles of the solid [27, 35]. It occurs, in particular, in case of contact of a solid with a liquid akin to it in molecular nature. At the same time this scheme of analysis may be extended on account of considering some more processes to be reasons for strength reduction in solids. These may be any superficial processes whose intensity is close to the cohesion energy of the solid, in the first place chemical and chemosorption phenomena developing at the boundary of the solid with the chemically active environment.

Apparently one should expect chemical processes to influence destruction facilitation only in case these processes are intense enough and manage to occur directly in the destruction zone. According to the first condition (which is in essence a thermodynamic one), chemically active media may influence the process of destruction of a solid if the surface energy of the solid is close to the energy of interaction with the active component of the environment. The second condition is purely kinetic: the active component is to reach the destruction zone in time, hence assisting the facilitation of destruction and the reconstruction of bonds.

In order to check this hypothesis there were carried out investigations of the influence of chemically aggressive media on processes of mechanical destruction of nickel, quartz, and some other materials (scratching, drilling, grinding). Since the destruction took place at the given rate, which was sometimes extremely high, the corrosion processes, such as dissolving or chemical changing of the material at any substantial depth, were excluded or took place to a very small extent. The rates used and the relatively deep propagation of destruction made sure that the chemical processes could involve just the superficial monomolecular layer of the solid in the destruction zone. As any topochemical reaction with a solid phase participating at its first stage is chemosorption of the active component at the solid surface, this approach to the influence of chemical processes on destruction of solids is completely analogous to the concepts developed by Rebinder for the case of physical adsorption and wetting influence on mechanical properties.

The results presented in Tables 2 and 3 confirm in general the concept that destruction facilitation is caused by media whose intensity of chemical interaction with the solid (evaluated as heat of chemical reaction per bond) is not very great and is close to the value of the surface energy of the material under treatment. Thus for nickel, the surface energy of which is about 16 kcal/mol [36], the greatest influence was found as aqueous solutions of mercury and copper chlorides were used, as well as iodine and sulfur organic solutions. Destruction of quartz is greatly influenced by alkali and hydrogen fluoride solutions; of brass, by solutions of copper-ammonia complexes; and of aluminum oxide, by some inorganic salts.

In case visible chemical interaction is absent or weak, environments, as a rule, do not display any substantial influence.

At the same time interactions of high intensity (that is, with great heats of reaction) also do not cause facilitation of destruction. This conclusion is in accordance with the well-known data concerning stress corrosion of metals: corrosion cracking is not, as a rule, observed in environments manifesting intense frontal corrosion, but only in environments which react with the stressed metal relatively weakly [37].

We should, however, emphasize here that the comparison of results of investigations of mechanical treatment in the presence of chemically active media with energies of chemical interaction has sometimes but qualitative character. First, we do not know how all the stages of chemical reactions proceed, and we do not know exactly what stage takes part in the process of rupture of bonds in the solid -- only more or less basic assumptions may be proposed here. Second, calculation of reaction heats is based on reference to thermochemical data, the accuracy of which is not always sufficient, the more that we are interested in small heats of reactions, obtained as a result of subtraction of great heats of formation of the compounds participating in the reaction. Thus the accuracy of these calculations is often commensurable with the value we look for; the reactions between elements may be an exclusion.

An essential detail of our approach is the circumstance that under conditions of supplying the solid with mechanical energy, chemical reactions may take place without activation, since the elastic energy accumulated in the solid (first of all in places where the stresses are localized) is often present in amounts sufficient to provide a relatively high rate of mechanochemical reactions [38].

5. MECHANICAL AND PHYSICOCHEMICAL CONDITIONS OF MANIFESTATION OF THE LIQUID MEDIA EFFECT WHILE MACHINING

The very possibility and the degree of strength reduction of solids under the action of surface-active media are determined to a considerable extent with their initial mechanical properties and the type of stressed state. The stronger the material and the more rigid the loading conditions, the more intense is the effect of environment [30, 31]. This regularity was found in investigations of mechanical treatment in the presence of surface-active media: it is most expedient to use metal melts for drilling hard tempered steels; the efficiency of the environment grows also as the cutting instrument is more heavily loaded (Fig. 7). Apparently this growth has a certain limit, when factors begin to manifest which hinder the transportation of the active medium to the cutting zone. As the cutting instrument gets blunted, the effect of the environment considerably decreases.

In contrast to the regularities of effect of surface-active substances (ions and organic surfactants) soluted in inactive solvents, when optimum concentrations causing maximum strength reduction were some parts of a percent, the effect of chemically active substances becomes appreciable at concentrations somewhat greater. Thus, for instance, nickel drilling is influenced at most with saturated (17%) iodine solution alcohol; but the 2% solution turns out to be only just slightly less efficient (Fig. 8).

Increase in the viscosity of the environment may cause considerable weakening of its effect in processes like grinding. For instance, the efficiency of glycerine aqueous solutions in aluminum oxide ceramic diamond grinding falls sharply as the concentration of glycerine reaches 50% to 60%, though both glycerine itself and its aqueous solutions cause quite appreciable increases in the width of scratches in sclerometry tests where the speed of the instrument movement is low.

6. THE ROLE OF KINETIC FACTORS IN THE DEGREE OF INFLUENCE OF ACTIVE MEDIA ON PROCESSES OF MECHANICAL TREATMENT

Active liquid environments may facilitate the treatment of solids only in case their transportation to the destruction zone is provided in time. Respectively, the critical rate of destruction of the material at which the effect of the active medium would yet manifest, may be determined with the rate of transportation of the active substance. Development of macroscopic destruction cracks in the presence of a surface-active melt is determined basically by viscous flow of the melt along the crack (or its walls) as a capillary [18, 30, 39, 40]. Thus in mechanical treatment processes, viscous flow may determine the supply of the active liquid to the cutting surface of the instrument as well. In most cases, however, much more important is the process of movement of the active substance directly in the cutting zone, where the material under treatment contacts intimately with the instrument and where relative rates of formation of the new surface are high. The sole mechanism which can provide high rates of the active substance supplying to the destruction zone is surface diffusion. The rates of surface diffusion may theoretically reach those of gas molecules [41]. Usually this is not observed, since the speeds of movement are not investigated, but the Brownian movement or the resultant movement of the adsorbed atoms frontier. But if the source moves continuously, following the cutting tool, and atoms are to cover a short distance to reach the place where the bonds in the solid are being broken, high speeds may be acquired due to two-dimensional pressure and may approach the molecular ones.

The maximum speed V_m of surface migration of adsorbed

atoms may be presented as a product of frequency of atoms transitions ν and a single step -- i.e., the lattice parameter "b" of the adsorbent:

$$V_m = \nu b = \nu_o b\, e^{-U/RT}$$

where ν_0 is the frequency of atomic oscillations, U is the activation energy of surface diffusion, R is the gas constant, and T is temperature. In cases of adsorption and chemosorption U is approximately 0.5 of the binding energy of the adsorbed atom with the lattice [42]. The greater is U -- i.e., the higher the binding energy--the slower is the movement of the adsorbed atom along the surface. It is this circumstance that determines the absence of strength reduction (or, anyway, sharp deceleration of its manifestation) when chemical interaction of the solid with the environment is intense. If U is known, it is possible to determine the maximum rate of destruction at which the medium would yet manage to get to the destruction zone in time. Vice versa, we can determine this critical rate and calculate from it the parameters of the interaction of the solid with the medium.

In order to check these ideas, the second way was chosen. The degree of influence of active media depending on the speed of movement of the cutting instrument was studied. Drilling with a tungsten carbide spiral drill 5.5 mm in diameter and rotating at various speeds were the investigation methods. Some of the results are presented in Fig. 9. For all the liquid media causing strength reduction and, respectively, intensification of machining processes, a certain threshhold was observed after which the effect of the medium considerably decreases and then disappears completely. This decrease in liquid active media influence can be connected with the kinetics of supplying the destruction zone with the active substance; the medium does not manage to follow the cutting facet of the instrument as the latter moves faster.

The data obtained for brass and cadmium result in an important conclusion: i.e., the more intense the interaction of the solid with the environment (and, respectively, the greater the relative strength reduction and facilitation of treatment under comparable conditions), the lower are the rates of cutting at which the medium ceases to influence the destruction process. This conclusion means that in gaining efficiency we always loose in rate; so when choosing liquid media for efficient facilitation of treatment it is necessary to keep under careful control the speeds of movement of the instrument so that they stay within the permissible limits.

Water and alcohols cease to influence the drilling of rock salt crystals in approximately the same speed interval, though the degree of the influence is sharply different. It is connected with the fact that the critical speed is determined with the energy of interaction of the solid with the adsorbate molecules (basically with the hydroxylic group), which is practically the same for all alcohols and water. As for the decrease in efficiency of alcohols as the length of the hydrocarbon chain grows, this decrease is caused mainly by general decreases in hydroxylic groups concentrations, as found by V. Yu. Traskin [27].

The interval in which weakening of the effect of the media occurs is relatively wide. This interval is determined apparently by a number of factors: different speeds of molecules and irregularity of their movement; the non-isothermic character of the process (heat emission is greater at greater drilling rates); and distribution of linear speeds of the cutting facet movement along the diameter of the drill. Nevertheless the obtained data allow evaluation of a certain critical rate of treatment at which the kinetic character of the Rebinder effect manifests quite noticeably.

If a speed at which the effect of the environment decreases twofold is assumed to be such a critical value, one may evaluate with the help of the correlation given above the activation energy of surface diffusion and compare it with existing data. Results of such calculation are presented in Table 4. Let us consider some of them. The binding energy Ni-I is about a half of ΔH^o_{298} NiI, or about 10 kcal/mol; if we assume, as is usually done, the activation energy of surface diffusion to be equal to half of the binding energy, we shall get a value close to that obtained in the experiments. The heat of water adsorption on sodium chloride is about 11 kcal/mol [43]. This value also corresponds with the double value of activation energy found.

Thus our experimental data well confirm the proposed scheme of penetration of the active substance into the zone of cutting. These ideas may also be useful for the analysis of the lubricating effect of coolant-lubricants in cutting treatment, since in this case the problem of transportation of the lubricant to the cutting zone is also important. It should be pointed out, however, that this scheme is, of course, simplified and is not yet developed enough to pretend to explain the variety of physicochemical processes accompanying mechanical treatment of solids.

7. CONCLUSIONS

New ways and methods of application of the Rebinder effect in machining and comminution processes are still one of the main objects of investigations in the field of physicochemical mechanics. At present there are two basic problems to solve. The first one is to find definite technological methods of using highly surface-active media, in the first place liquid fusible metals, in cutting and grinding hard metals and metalloceramics. The solution of this problem is connected with working out optimum fusible compositions, special instruments and devices. The second problem is to determine the relative role of adsorption-induced strength reduction among other kinds of physicochemical influences of environment on the material machined and to investigate in detail the mechanism of cutting effect of coolant-lubricants.

REFERENCES

1. Rebinder, P.A., Proceedings of the VI Congress of Physicists, Moscow, 1928.

2. Rebinder, P.A., Kalinovskaya, N.A., Zh.F.Kh., 2, 776, 1931; Zh.F.Kh., 5, 332, 1934.

3. Rebinder, P.A., Vestnik AN SSSR, 8-9, 5, 1940.

4. Rebinder, P.A., Shreiner, L.A., Zhigatch, K.F., Hardness Reducers in Rock-Drilling, Moscow, AN SSSR, 1944.

5. Babalyan, G.A., in coll. "Application of Surfactants in Oil Industry", Moscow, 1966.

6. Akmulin, M.Sh., ibid.

7. Babalyan, G.A., Dyushushe, M.Zh., Sentsova, E.P., Neftynoye Khozyaystvo, N3, 1967.

8. Epstein, E.F., Corchagin, L.V. et al., Izv. Visshikh Uchebnikh Zaved., Geologiya i razvedka, 12, 143, 1969.

9. Deshko, Yu.I., Creymer, M.B., Krykhtin, G.S., Comminution of Materials in Cement Industry, Moscow, Stroyizdat, 1966.

10. Shaw, P.H., Industry and Economics, 62 (1), 36, 1970.

11. Somasudaran, P., Lin, J., Effects of the Nature of Environment on Comminution, N.-Y., 1972.

12. Khodakov, G.S., Rebinder, P.A., DAN SSSR, 127, 1070, 1959.

13. Khodakov, G.S., Physics of Comminution, Moscow, Nauka, 1972.

14. Rebinder, P.A., Venstrem E.K., Izv. OMEI AN SSR,

ser. fiz., 4-5, 531, 1947.

15. Akhmatov, A. S., Molecular Physics of Boundary Friction, Moscow, Fizmatgiz, 1963.

16. Bowden, F. P., Taylor, D., Lubrication and Friction, Moscow, Mashinostroyeniye, 1969 (translated from English).

17. Likhtman, V. I., Rebinder, P. A., Karpenko, G. V., Influence of Surface-Active Environment on Deformation of Metals, Moscow, AN SSSR, 1954.

18. Likhtman, V. I., Shchukin, E. D., Rebinder, P. A., Physicochemical Mechanics of Metals, Moscow, AN SSSR, 1962.

19. Rebinder, P., Likhtman, V., Proc. of the II Intern. Congress on Surface Activity, L., 1957, p. 563.

20. Veyler, S. Ya., Likhtman, V. I., Influence of Lubricants on Pressing Treatment, Moscow, AN SSSR, 1960.

21. Fux, G. I., Gantsevitch, I. B., Kolloidn. zh., 29, 304, 1967.

22. Kanayev, A. A., L'vov, V. N., Veyler, S. Ya., Rebinder, P. A., DAN SSSR, 187, 314, 1969.

23. Rebinder, P. A., Yepifanov, G. I., in coll. "Development of Friction and Wear Theory", Moscow, AN SSSR, 1957.

24. Epifanov, G. I., Pletnyova, N. A., Rebinder, P. A., DAN SSSR, 97, 277, 1954.

25. Grebenshchikov, I. V., Role of Chemistry in Polishing Processes, Moscow, Sots. Revolutsiya i nauka, v. 2, 1935.

26. Pertsov, N. V., Shchukin, E. D., Fiz. i khim. obr. mat., 2, 60, 1970.

27. Rebinder, P. A., Shchukin, E. D., UFN, 108, 1, 1972.

28. Shchukin, E. D., Polucarova, Z. M., Yushchenko, V. S., Bryukhanova, L. S., DAN SSSR, 205, 86, 1972.

29. Shchukin, E. D., Polucarova, Z. M., Bryukhanova, L. S. et al., Fiz. i khim. obr. mat., 3, 90, 1973.

30. Rostocker, U., McCogie, G., Marcus, G., Embrittlement by Liquid Metals, Moscow, IL, 1960 (translated from

English).

31. Westwood, A., in coll. Environment-Sensitive Mechanical Behavior, Moscow, Mir, 1969 (translated from English).

32. Skvortsova, Z. N., Traskin, V. Yu., Fizico-khimicheskaya mekhanika materialov, 4, 38, 1974.

33. Pertsov, N. V., Shchukin, E. D., Goryunov, Yu. V., Danilova, F. B., Storchak, E. A., coll. Almazy, 5, 42, 1970.

34. Pertsov, N. V., Danilova, F. B., et al., coll. Almazy, 10, 12, 1972.

35. Shchukin, E. D., Yushchenko, V. S., Fiziko-khimicheskaya mechanika materialov, 2, 143, 1966.

36. Taylor, J. W., Metallurgiya, 50, 161, 1954.

37. Logan, H. L., The Stress Corrosion of Metals, N.-Y., 1967.

38. Butyagin, P. Yu., Uspekhi Khimii, 40, 1935, 1971.

39. Shchukin, E. D., Summ, B. D., coll. Surface Diffusion and Spreading, AN SSSR, Moscow, 1969.

40. Summ, B. D., Mukhamed, Ya., Pertsov, N. V., et al., Fizikokhimicheskaya mekhanika materialov, 2, 33, 1972.

41. De Boer, J. H., The Dynamical Character of Adsorption, Oxford, 1953.

42. Hayward, D. O., Trapnell, D. M. W., Chemosorption, London, 1962.

43. Hugher, M., Compt. Rend., 264 (B), 243, 1967.

TABLE 1.

Results of fused quartz grinding in water and aliphatic alcohols. Diamond wheel M1, grinding pressure 16 kg/sm^2, speed of the grinding wheel 260 rot/min (0.4 m/sec).

Environment	Q	F	W	S	A·10^{-5}
H_2O	11.2	2.1	8.3	5	1.4
C_2H_5OH	13.5	3.0	12.1	6	1.5
C_3H_7OH	12.5	2.8	11.3	5	1.7
C_4H_9OH	12.4	2.7	10.7	5	1.8
$C_7H_{15}OH$	11.9	2.3	9.2	3	2.7
$C_9H_{19}OH$	10.5	2.2	8.8	2	3.4
$C_{10}H_{21}OH$	10.0	2.0	8.4	2	3.7

Q = grinding rate, mg/sec
F = friction force, kg
W = power spent, wt
S = specific surface, M^2/g
A = work of dispersing, erg/sm^2

TABLE 2.

Results of investigations of the influence of various media on nickel and quartz scratching process. C = concentration, mol/l; d/d_0 = relative change in the width of the scratch, %; $H^0{}_{298}$ = calculated heat of a presumed chemical reaction.

Material	Environment	C	d/d_0	Products of the Presumed Reaction	$H^0{}_{298}$
Ni (indentor load 15 g)	H_2O	-	10	-	-
	aqueous solutions				
	$HgCl_2$	0.2	28	$NiCl_2 + Hg$	-22
	$HgSO_4$	0.05	3	$NiSO_4 + Hg$	-47
	$Hg(NO_3)_2$	0.5	-5	$Ni(NO_3)_2 + Hg$	-44
	HCl	1	7	$NiCl_2 + H_2$	+7
	H_2SO_4	10	10	$NiSO_4 + H_2$	-2
	HNO_3	1	0	$Ni(NO_3)_2 + NO_2$	-130
	$FeCl_3$	3	3	$NiCl_2 + FeCl_2$	-35
	$CuCl_2$	1	18	$NiCl_2 + Cu$	-18
	$AgNO_3$	1	9	$Ni(NO_3)_2 + Ag$	-54
	KOH	10	7	-	-
	I_2 in KI solution	1	11	NiI_2	-20
	ethanol, acetone, benzene		9	-	-
	I_2 ethanol solution	1	20	NiI_2	-21
	S ethanol, acetone, benzene solutions	satur.	13	NiS	-20
quartz (indentor load 200 g)	H_2O	-	16	$H_2Si_2O_5$	-7
	aqueous solutions of				
	Na_2CO_3	1.6	60	Na_2SiO_3	-8
	NaOH	10	190	$Na_2SiO_3 + H_2O$	-11
	HF	20	340	$SiF_4 + H_2O$	-19

TABLE 3.

Influence of various media on drilling of nickel. C = concentration, mol/l; v_o and v are the rates of drilling in air and in the medium, mm/sec; v/v_o is the relative change in the drilling rate, %; H^o_{298} is the calculated heat of the presumed reaction, kcal/mol.

Environment	C	v_o	v	v/v_o	H^o_{298}
H_2O	-	0.18	0.21	16	-
ethanol	-	0.18	0.20	11	-
I_2 ethanol solution	1.0	0.16	0.53	230	-21
aqueous solutions					
$Hg(NO_3)_2$	0.5	0.18	0.17	-5	-44
$HgCl_2$	0.2	0.20	0.75	275	-22
HCl	1.0	0.12	0.13	8	+7

TABLE 4.

Calculated values of activation energy U (kCal/mol) of the active media surface migration.

Material	Environment	U
Nickel	J_2 solution in C_2H_5OH	5.1
Brass	Hg	7.0
Brass	Ga	5.6
Zn	Hg	5.3
Zn	Ga	4.7
Cd	Ga	5.5
NaCl single crystals	H_2O	5.5
" " "	C_2H_5OH	5.4
" " "	$C_6H_{13}OH$	5.3
" " "	$C_{10}H_{21}OH$	5.2

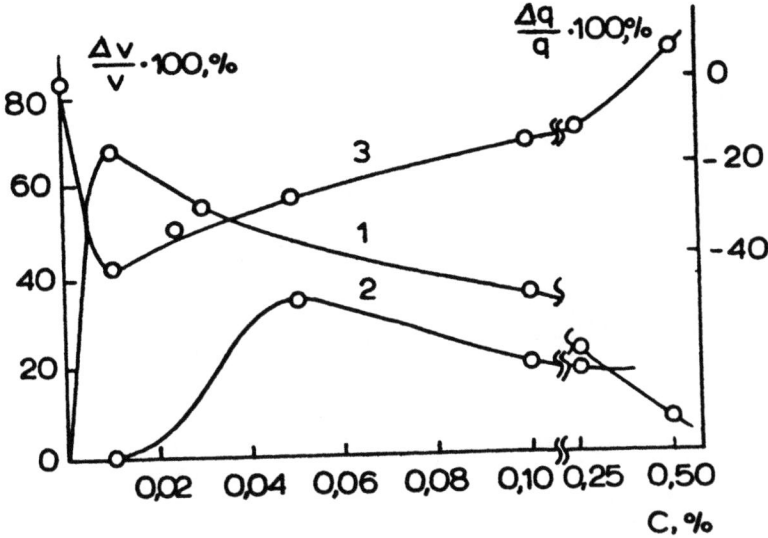

Fig. 1. Dependence of relative increase in quartzite (1) and granite (2) drilling rate $\Delta v/v_0$ and relative decrease $\Delta q/q_0$ in the wear of the bits (3) on concentration "c" of aluminum chloride aqueous solution [4].

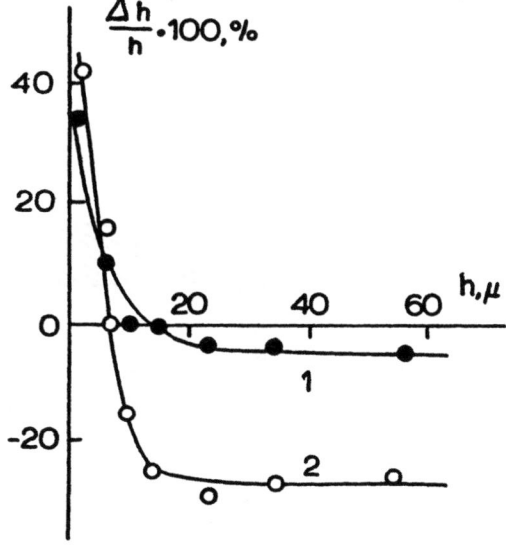

Fig. 2. Dependence of relative change in the depth of the scratch h/h_0 on the depth "h", scratching being carried out in aliphatic alcohols (1) and oleic acid (2) [22].

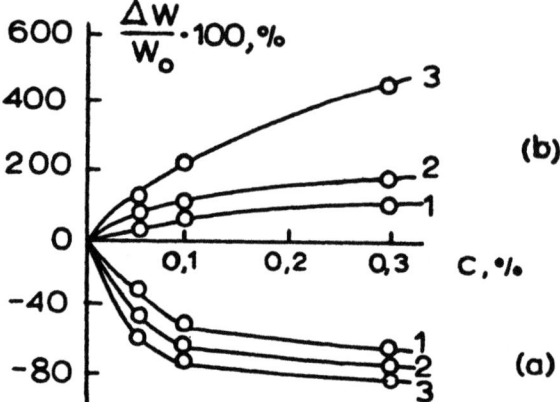

Fig. 3. Relative change in the specific work of cutting $\Delta W/W_o$ of aluminum (1), steel (2) and copper (3) as depending on concentration of palmic acid in paraffin oil, lubricating effect (a) and cutting effect (b) predominating [19].

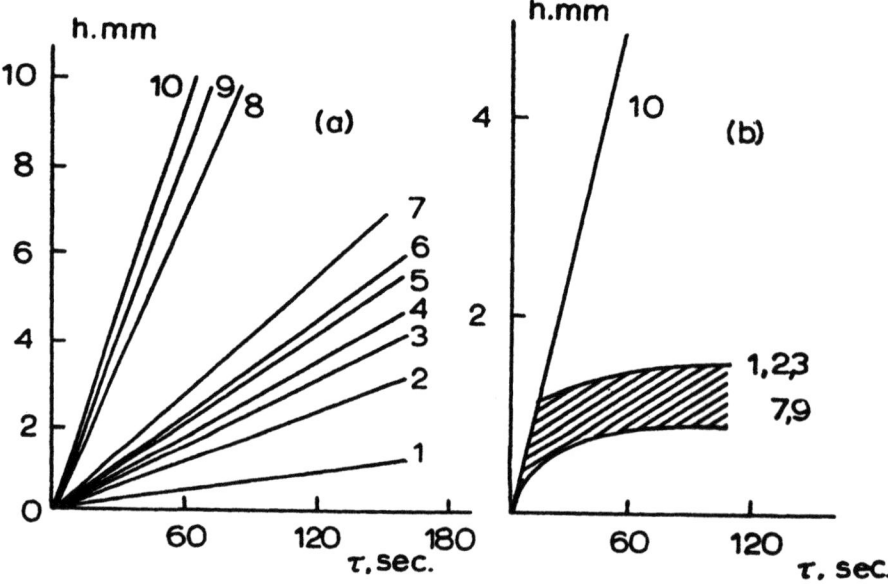

Fig. 4. Dependence of depth of drilling "h" on time of drilling τ for the 1H18N9T steel machining in paraffin oil and emulsion (1), in air (2), in the Wood alloy (3), in oleic acid (8), in melts of tin (4), gallium (5), indium (6), and of eutectics: indium-gallium-tin (7), tin-lead (9), tin-zinc (10). Axial force is 58 kG, $\omega = 450$ rotations per minute, the drill "a" is a flat drill 4.5 mm in diameter, made of R18 steel; the drill "b" of the same shape is 5 mm in diameter and is made of VK6 alloy [28].

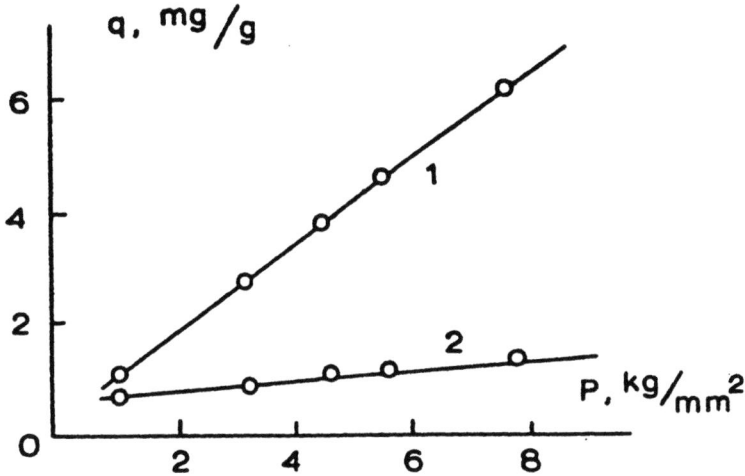

Fig. 5. Dependence of specific expenditure of diamond "q" on the axial force P_d pressing the samples of T15K6 alloy to the grinding wheel in air (1) and in liquid gallium (2) [33].

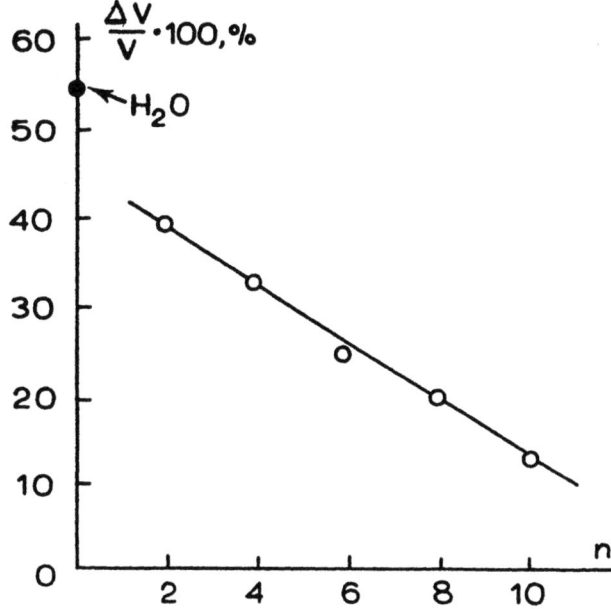

Fig. 6. Dependence of the relative increase in the drilling rate $\Delta v/v_0$ of NaCl single crystals on the number "n" of carbon atoms in the molecule of the alcohol. $\omega = 120$ rotations per minute, $P_d = 2.3$ kG, diameter 5.5 mm.

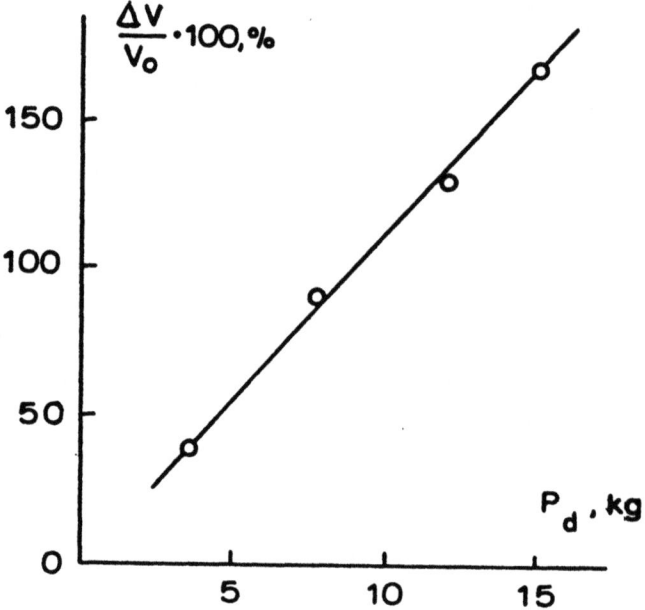

Fig. 7. Dependence of the relative increase $\Delta v/v_o$ in rate of drilling on axial force applied at the drill P_d, nickel being drilled in iodine alcohol solution (c = 1 mol/l, ω = 1600 rotations per minute, diameter 5.5 mm).

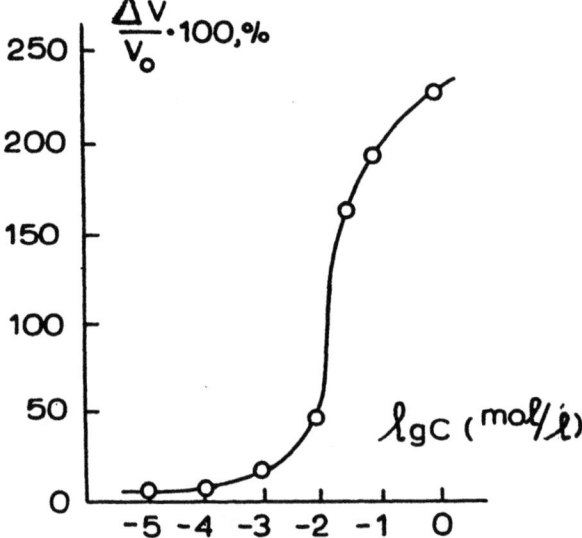

Fig. 8. Relative increase in drilling rate $\Delta v/v_o$ of nickel as depending on concentration of iodine in alcohol (ω = 1600 rotations per minute, P_d = 15 kG, diameter 5.5 mm).

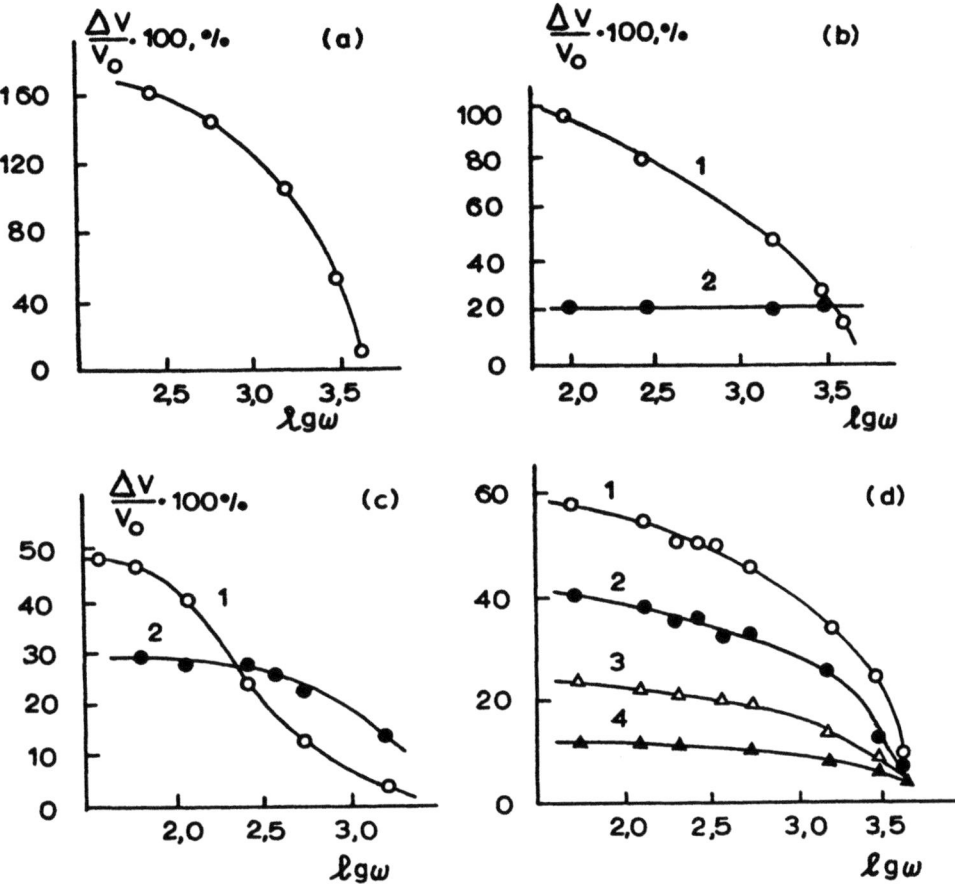

Fig. 9. Dependence of the relative increase in drilling rate $\Delta v/v_0$ on speed of rotation of the drill ω for: a) nickel in iodine alcohol solution; $c = 1$ mol/l, $P_d = 15$ kG; b) cadmium in gallium (1) and mercury (2), $P_d = 2, 3$ kG; c) brass in mercury (1) and gallium (2), $P_d = 10$ kG; d) NaCl single crystals in water (1) and alcohols: ethyl (2), hexyl (3), decyl (4) alcohol, $P_d = 2.3$ kG.

EROSION OF METALS AND ALLOYS

C. M. Preece

Department of Material Science
State University of New York at Stony Brook
Stony Brook, N. Y. 11794, U.S.A.

1. INTRODUCTION

The surfaces of solids can be eroded by a variety of means including cavitating liquids, liquid particle impact and solid particle impact. Erosion is becoming an increasingly prevalent mode of failure as the velocities of moving systems (eg. turbines, surface effect vehicles, etc.) are increased in erosive environments and as the environments themselves (eg. in coal gasification plants, pipelines, etc.) are moved at higher velocities. Erosion also contributes significantly to the irretrievable loss of many materials of limited supply which are considered essential to current and future technology. On the other hand, erosion is the basis of many beneficial processes including sand blasting, shot peening, jet cutting, ultrasonic cleaning, sputtering and ion implantation.

Despite the numerous detrimental and beneficial aspects of erosion, the mechanisms by which the impacted material is damaged and its surface is eroded have received relatively little scientific attention from the materials community. The purpose of this paper is to review that research which has been done and, hopefully, stimulate more interest in this technologically challenging area of surface effects.

2. SOLID PARTICLE EROSION

Erosion by solid particle impact is the simplest form of erosion to analyze since the size, velocity and number of particles can be accurately determined. Detailed theoretical analyses, however, have been largely restricted to brittle

materials, which, to a first approximation, can be considered as
an elastic half-space and to which the Hertzian theory [1] of
impact can be applied. Nevertheless, even for brittle materials,
their properties under dynamic loading conditions and the role of
microstructure in determining crack nucleation sites and paths are
not well understood [2]. Moreover, the analyses to date [2-7]
have been of single impact. The reality of multiple impact is
considerably more complex. For ductile materials, the analyses
have been limited to a prediction of the influence of properties
of the impacting particle (velocity, mass, diameter, temperature
and angle of incidence); the response of the target material has
not been studied analytically.

Experimentally, two types of response to solid particle
impact have been observed, one being characteristic of ductile
materials, mostly metals, and the second being characteristic of
brittle materials such as glasses and ceramics. The difference
in response is exemplified by the effect of impact angle of the
eroding particle on the mass, or volume, of material removed.
For brittle materials, this is a maximum at normal incidence
whereas, for ductile materials, erosion is greatest at angles
$\sim 30°$, Fig. 1 [8]. While this division is convenient, it is, of
course, oversimplified: many materials undergo a transition from
ductile to brittle behavior with temperature, strain rate, etc.
[9] and brittle materials can show a response to particle impact
similar to that of ductile materials within certain particle size
and velocity limits [10].

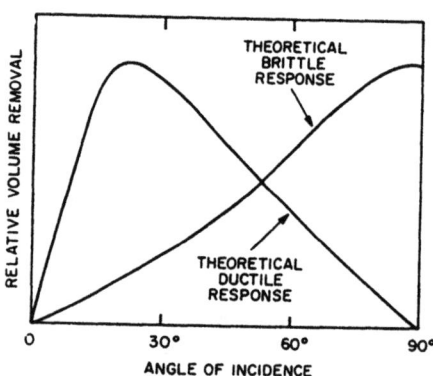

Fig. 1   Theoretical variation in erosion with angle of
incidence of impacting solid particles (Ref. 8).

The maximum in erosion rate at low angles of impact, has
been attributed to various factors. Finnie [11, 12] has suggested

that angular particles cut into the metal surface effectively removing material by micromachining, whereas Hutchings and Winter [15] have shown that metal flows upwards around the edge of the impacting particle and assume that this rim is eventually removed by ductile shear. Tilly and Sage [13, 14] have observed similar raised lips but suggest that these are removed by fragments of the impacting particle, which shatters on impact. On the other hand, Smeltzer et al. [8] consider the surface to be melted during impact and the material removed in the form of molten drops.

An attempt [16] has been made to differentiate between the mechanisms of "cutting", as described by Finnie [11], and "ploughing" in which the material is removed by plastic deformation, Fig. 2. For a constant impact angle, the "rake angle", as defined in Fig. 3, of an angular particle was found to be significant. For an impact angle of 25° and rake angles of +20°, +5°, -25° and -45°, cutting deformation was observed whereas, at a rake angle of -70°, ploughing occurred.

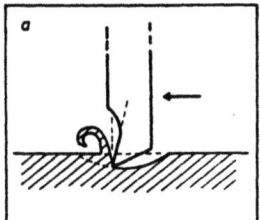 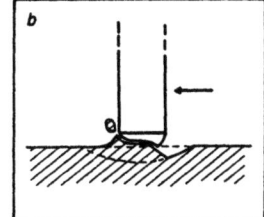

Fig. 2 (a) "Cutting" mechanism: with positive rake angle, metal is removed by micromachining. (b) "Ploughing" mechanism: with negative rake angle of tool, metal is removed by plastic deformation (Ref. 15).

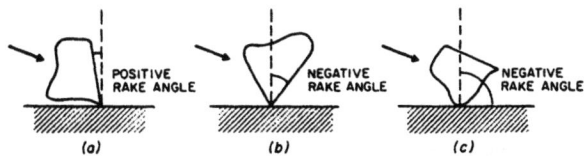

Fig. 3 Definition of rake angle (Ref. 16).

The discrepancy observed at high impact angles between the theoretical and experimental erosion rates, Fig. 4, probably results from a combination of some of the factors described above. For example, fragmentation of a particle impacting at 90° would give rise to a much lower angle of impingement by the fragments. Alternatively, impact of a ductile metal at 90° would produce an identation, possibly surrounded by a raised lip. Subsequent impacts on either the identation or lip would effectively be at

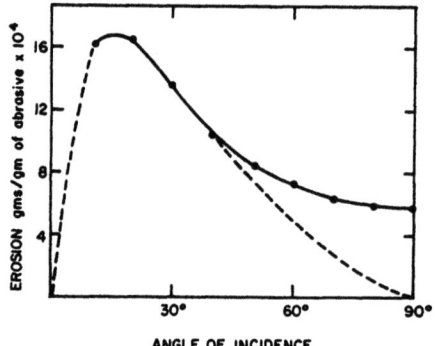

Fig. 4 Predicted and observed values of erosion of commercially pure aluminum (Ref. 10).

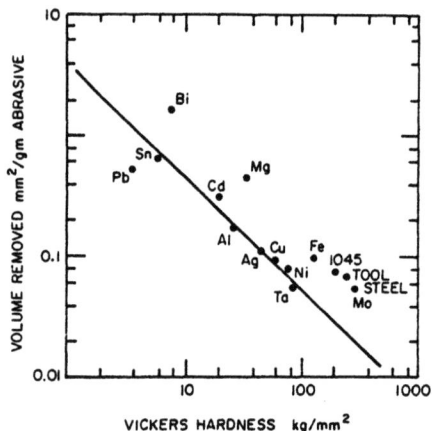

Fig. 5 Erosion of various metals plotted versus their hardness values (Ref. 12).

angles < 90°.

The only property of the target material included in the analytical models [11, 13, 17-21] of erosion of ductile materials is flow stress. Flow stress is, of course, related to hardness and Fig. 5 shows that a correlation can, in fact, be made between erosion and hardness. To investigate this relationship further, Smeltzer et al. [8] studied the erosion of 2024Al, Ti-6Al-4V, 410SS and 17-7PH steel in both the annealed and fully heat-treated conditions. The models predict that the volume of material removed is inversely proportional to the flow stress. Thus, the annealed samples, with the lower flow stress, should erode ∿ 3.0 - 4.5 times more readily than the heat treated samples. In fact, the differences in weight losses between annealed and heat-treated samples were less than 5% in all cases. It was concluded from this data that "flow strength, hardness and metallurgical structure have little bearing on the inherent erosion resistance of a specific target material" [8]. It is obvious, however, that this conclusion cannot be maintained for erosion of all materials under all conditions. For example, it is unlikely that localized melting will play a major part in the erosion of refractory metals by relatively low velocity particles, yet erosion does occur. In the above experiment, localized heating may have annealed the surface layers of the sample so that the microstructure and properties of the initially heat treated sample were very similar to those of the annealed sample. Alternatively, strain aging of the latter may have produced a structure similar to that of the heat treated sample.

Other metallurgical factors which have been investigated, albeit over a very limited range, are pre-strain [14, 22], and grain size [22, 23]. Prior cold work was found in one investigation [22] to have little influence on the erosion. This observation is attributed to the fact that the surface is rapidly work hardened by many "non-cutting" particles and that the strains produced by the "cutting" particles will themselves be very high. On the other hand, cold work was found to decrease the resistance of copper to erosion [14] and it was suggested that, in the work hardened metal, the deformation is concentrated in the surface layers whereas, in the annealed metal, the impact energy is spread through a large volume.

The sand erosion of two grain sizes each of pure aluminum and pure iron has been measured [23]. For the aluminum, the grain size had no appreciable influence on the erosion whereas the small grain sized iron showed an increasingly greater degree of erosion with increasing particle impact velocity than did the larger grained samples.

An interesting observation of Smeltzer et al. [8] was that, for each test condition, all of the target materials tested lost the same amount of weight (within ±10%) although, of course, the volume losses varied appreciably. The authors infer from these results and those of Finnie et al. [24], which are illustrated in Fig. 6, that the energy of impact is sufficient to produce localized melting and, therefore, the most significant material parameter influencing erosion is the target melting temperature. By this reasoning, however, cobalt and nickel alloys, which typically have similar melting temperatures, should exhibit similar erosion resistances. Yet the data of Tilly & Sage [13] indicate that a cobalt alloy suffers more than 300% the erosion of a nickel alloy and 10% more than an aluminum alloy, the melting temperature of which is only about 0.4 that of the cobalt alloy. The conclusion of the investigators [13] is "that the data for the erosiveness of different materials show that there is little scope for any substantial improvement over the engineering alloys in current usage, i.e. steel, nickel and titanium alloys. The cobalt alloy is very disappointing because it is one of the best materials under rain erosion as well as being generally good in other

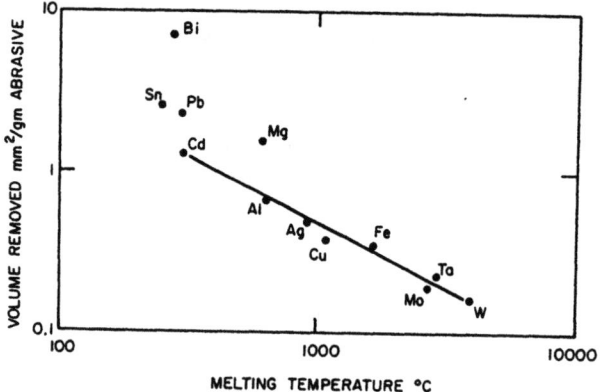

Fig. 6 The data from Fig. 5 plotted against melting temperature of the metals (Ref. 8).

situations involving wear". However, it is apparent that the micro-structural parameters controlling (i) the absorption of impact energy, (ii) the localized deformation modes and (iii) the mechanism of material removal, must be thoroughly investigated before any appraisal can be made of the potential of future materials for use in erosive environments.

3. LIQUID PARTICLE EROSION

Erosion by liquid particle impact is a maximum at 90° for all types of target materials [25], Fig. 7, and is caused by (i) the direct compressive force of the drop as it impacts the solid and (ii) the shear force on the surface produced by the outward radial flow of the liquid, Fig. 8. There is some controversy in the literature regarding the temporal magnitude and distribution of these forces [26] but theoretical calculations have been made of the pressure distribution due to a liquid drop striking a smooth, rigid surface. These are based on the water hammer equation for a rigid solid or, for an elastically deformable solid, a modified form of this equation:

$$P = \frac{\rho_o c_o v_o}{1+(\rho_o c_o / \rho_1 c_1)}$$

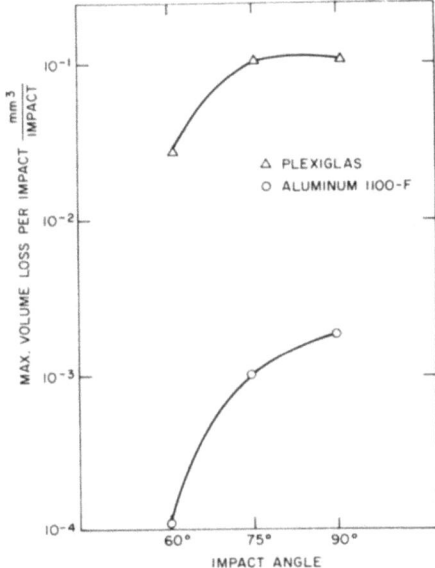

Fig. 7 Experimental variation of erosion rate as a function of angle of impingement of water drops (Ref. 25).

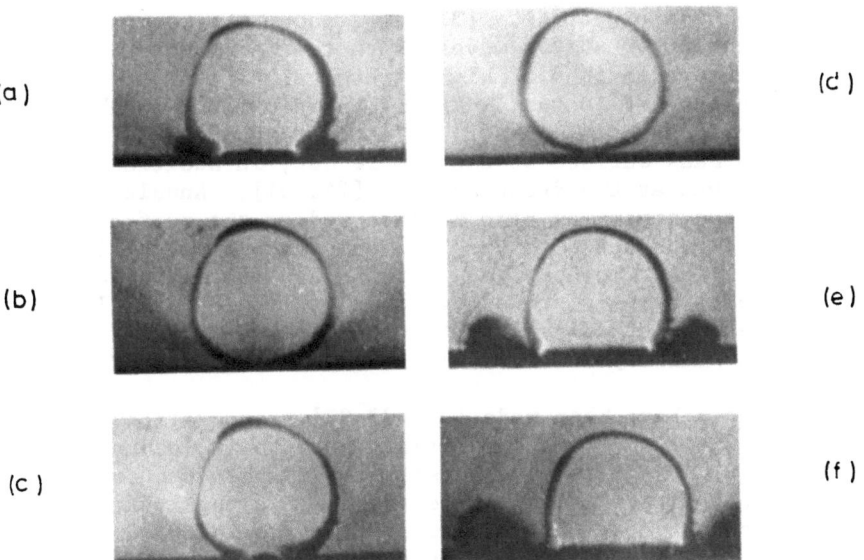

Fig. 8 Impact of a solid against a 5.0 mm diam. water drop at 100 m/s. The time interval between frames is 2.7 μs (Ref. 29).

where P is the pressure produced in the solid by a liquid drop impacting at a velocity, V_o, and ρ_o, ρ_1 and C_o, C_1 are the densities and acoustic velocities of the liquid drop and solid target respectively. This pressure is maintained only until the pressure wave in the liquid, emitted from the point of impact, intersects a free surface of the drop, at which time a pressure release wave is initiated and travels inwards. This wave relieves the impact pressure over the contact area and allows the high velocity lateral flow out of the compressed region. Thereafter, the contact pressure is reduced to $1/2 \, \rho_o V_o^2$ [27]. Since the initial contact perimeter between the drop and the solid moves radially outward at a velocity greater than the pressure wave in the drop, the solid is initially subjected to a ring of surface stress which increases in magnitude and diameter until the rate of increase of contact area falls below the pressure wave velocity and lateral outflow occurs. The maximum pressure is predicted to be greater than the water hammer pressure and to occur at about the radius at which outflow begins [28]. This has been confirmed experimentally by Rochester and Brunton [29], who used a transducer technique to determine the distribution of both the impact and shear stresses on the surface of a solid as it impacts a disc shaped water drop. The peak pressure for a 5mm diameter water disc impacted at 100 m/s was ~ 275 MN/m^2, Fig. 9, which is approximately 1.8 $\rho_o C_o V_o$. The shear stress for a smooth surface is relatively small, Fig. 10, but it should be noted that for a roughened surface, the shear stress can play a major role in the erosion process [30]. Using a similar impact technique and high speed photography, Field et al. [31] have shown the spatial and temporal development of stress waves in photoelastic materials and also confirm the predictions of high stress values at the expanding edge of the area of contact. As in other forms of shock deformation, failure is sometimes observed to occur by cracking and spalling at the back surface of the sample but, in addition, ring cracks are observed at the front surface [26, 31]. Annular rings of deformation can sometimes also be observed on the surface of ductile materials subjected to single liquid drop impacts [32]. However, for multiple impacts, such details are readily obscured and the surface of samples eroded by solid or liquid particles or by cavitation are practically indistinguishable. This can pose a problem in failure analysis if metallography is the only available tool.

Many attempts have been made [27, 33-37] to correlate the erosion resistance of different materials with their mechanical properties. While there is some consistency in the data for erosion and hardness or strain energy for specific alloy systems, there is no apparent correlation between alloys of different base metals. In the absence of any direct relationship between erosion resistance and any other mechanical property, erosion has generally been attributed to fatigue because of the repetitive stressing mode [38].

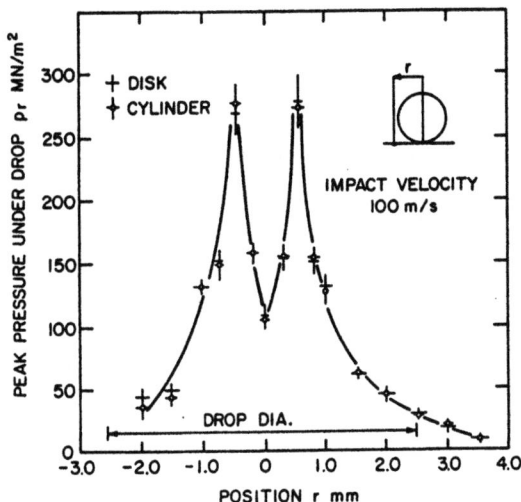

Fig. 9   Peak pressure under a water drop, P_r, versus position, r. The drop diameter was 5.0 mm and the impact velocity 100 m/s. (Ref. 29).

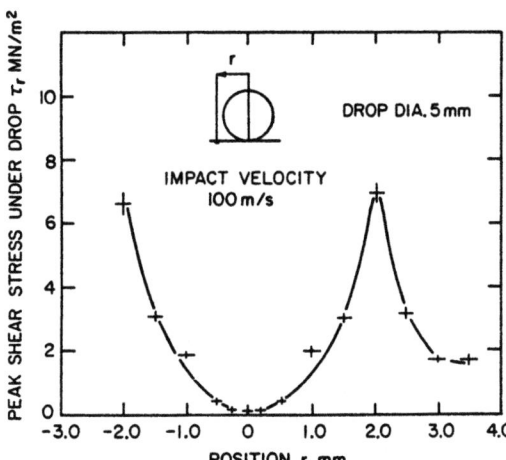

Fig. 10   Peak shear stress under a water drop, τ_r, versus position, r. The drop diameter was 5.0 mm and the impact velocity 100 m/s. (Ref. 29).

4. CAVITATION EROSION

Cavitation occurs in any fluid system in which there is a sufficiently high pressure differential (spatially or temporally) to cause the repeated growth and collapse of cavities or bubbles in the liquid. It may occur in a flow system in which the bubbles are formed in a localized region of low pressure and travel downstream to a higher pressure region where they collapse, or in a vibrating system such as a lubricating fluid in a vibrating machine. Any solid in the vicinity of the bubbles as they collapse will be susceptible to erosion. This erosion has been variously attributed to accelerated corrosion, localized melting, work hardening leading to brittle fracture, shock deformation and fatigue. It has been shown [39] that the imposition of a cathodic potential on the solid can reduce the rates of weight loss by cavitation, indicating that corrosion may provide a significant contribution to the erosion. However, other investigators [40, 41] have suggested that the hydrogen produced at the cathodic potentials in these experiments would cushion the collapse of the bubbles, thereby reducing their damaging effect.

Consequently, some form of mechanical stressing is now regarded as the primary cause of erosion. Fatigue has gained the greatest acceptance over the last decade, but it has recently been shown [42, 43] that the damage produced in a metal and the mechanism of material loss is more characteristic of that produced by shock or explosive deformation than by fatigue.

It has been shown theoretically that a collapsing bubble emits a shock wave into the surrounding liquid which, depending on the pressure differential and properties of the liquid is $\sim 10^2$-10^3 MN/m^2 [44-48]. However, the calculations indicate that the stress wave is attenuated within a distance comparable to its initial radius. Thus, it has been concluded that only those bubbles within this distance from the solid surface are able to produce any damage in the solid [46, 49, 50]. Moreover, bubbles adjacent to a solid surface do not collapse spherically but, because of geometric constraints, become involuted and form a jet of liquid which impacts the surface [51-53]. This observation has led many investigators to attribute the damage produced by cavitation almost exclusively to these jets and to assume a direct correlation between liquid impact and cavitation erosion. On the basis of this correlation, it has been assumed [54, 55] that the mechanism held responsible for liquid impact erosion, namely fatigue, is also responsible for cavitation erosion. In practice, however, a very large number of bubbles are produced and the theoretical calculations do not take into account the additive effect of their stress waves. Moreover, it is difficult to attribute the surface damage produced by cavitation, Figs. 11 and 12 [42, 56] to individual bubbles acting independently on the surface.

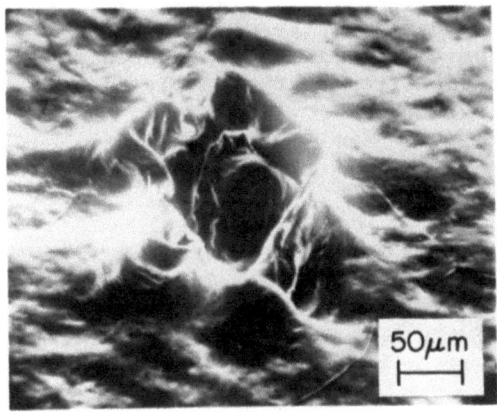

Fig. 11   Scanning electron micrograph of aluminum subjected to cavitation for 10 sec in a vibratory system (Ref. 38).

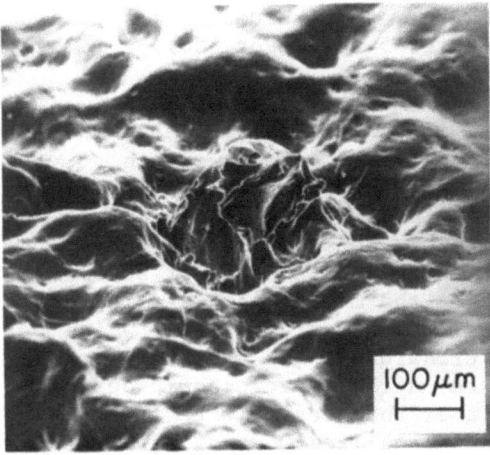

Fig. 12   Scanning electron micrograph of aluminum subjected to cavitation for 45 sec in a flow system (Ref. 47).

It has been proposed [42], therefore, that the small, approximately few μm dia., pits result from individual bubble collapse and that the grain boundary delineation, long slip lines and large undulations are caused by the concerted collapse of a large number, or "cloud", of bubbles. The shock pulses emitted by the repeated collapse of this cloud in a vibratory system, or the collapse of successive clouds in a flow system, deepen the undulations into craters, Figs. 11 and 12, from the edges of which material is eventually removed by ductile rupture. Transmission electron microscopy of aluminum which has been subjected to cavitation shows a dislocation structure, Fig. 13, which is typical of those produced by shock [57]. Transmission microscopy of nickel and stainless steel [58] also indicates that cavitation produces an entirely different mode of deformation from that normally associated with fatigue, tension, rolling, etc.

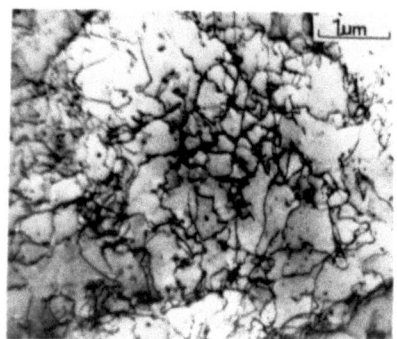

Fig. 13 Transmission electron micrograph of aluminum subjected to cavitation in a vibratory system for 5 sec. (Ref. 38).

As for liquid particle impact, many attempts [59] have been made to correlate erosion rates of different materials with various bulk mechanical properties, for example, hardness, yield strength, UTS and strain energy, but these have been singularly unsuccessful. Because of the nature of the shock pulse, the deformation produced by cavitation extends to a depth of several millimeters [43, 60]. In the bulk of the metal, the deformation is microscopic; simply an increase in dislocation density which is manifested as an increase in microhardness and residual stress, Fig. 14 and 15. Only on the surface is there macroscopic deformation. It is likely, therefore, that the erosion rates are controlled by the microstructure and localized mode of deformation, rather than the bulk mechanical properties. The research in the author's laboratory has, therefore, been aimed at understanding these factors. Firstly, the actual stress applied to the material by cavitation and the changes produced by this stress in the structure and properties of the material are being measured. Secondly, an investigation is being made of the influence of microstructure on (i) the length of the incubation period, t_o, i.e. time of exposure to cavitation during which no detectable weight loss occurs, (ii) the mechanism of damage at exposure times $< t_o$ and (iii) mode of material removal for times $> t_o$.

Results to date indicate that dimensional parameters such as grain size or second phase particle size are only significant when they are of the same order or smaller than the average bubble

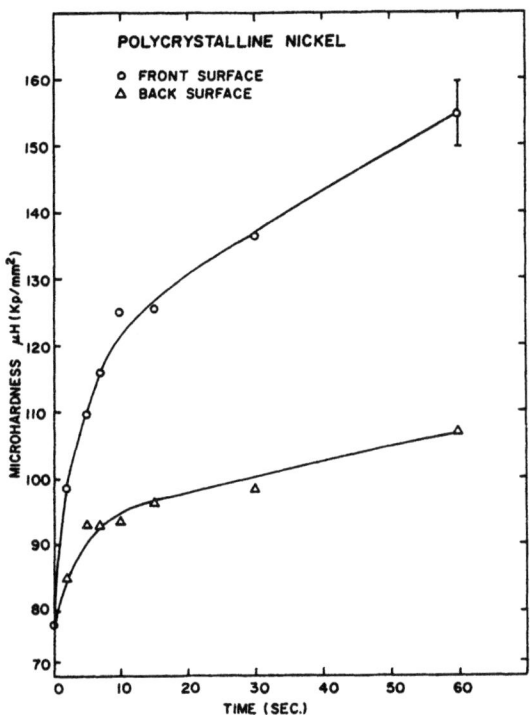

Fig. 14 The change in microhardness of nickel due to exposure to cavitation in a 20 KHz vibratory system. Data are shown for the exposed (front) surface and the unexposed (back) surface (Ref. 50).

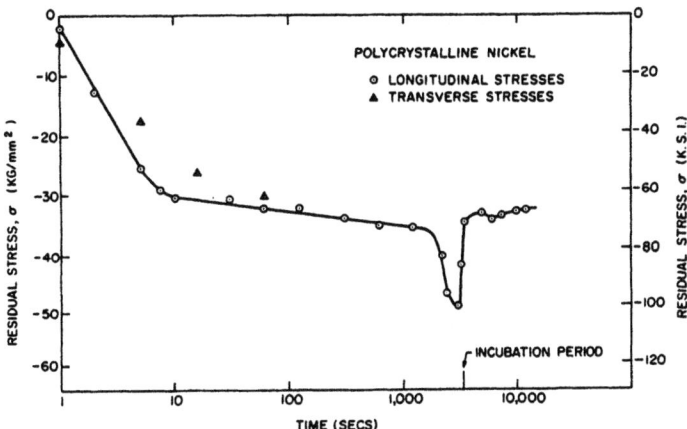

Fig. 15 The change in residual stress (determined by X-ray diffraction) of the surface of nickel due to exposure to cavitation in a vibratory system (Ref. 39).

size [61]. The effects of alloying are not general but depend on such factors as the change in slip mode and, in heat treatable alloys, the degree of metastability of the structure [61, 62].

The magnitude, σ_0, of the stress produced by cavitation generated in water by a 20KHz vibratory probe, and measured by a transducer technique [60], is found to be $\sim 10^2$–10^3 MN/m^2 depending on the power input to the vibratory probe and the distance in the water between the probe tip and the measuring transducer. This stress exceeds the bulk yield stress of most metals and alloys. Since it (a) is repeated every 20 μs and (b) is effectively a shock wave passing through the metal which can nucleate slip in each grain individually, rather than propagating macroscopic deformation from one grain to the next, it is not surprising that cavitation produces such extensive damage.

The incubation period, t_0, was found to be inversely proportional to both σ_0 and to the change in microhardness, $d\mu H/dt$, of nickel, Fig. 16. Thus, there is a direct correspondence between the applied stress, σ_0, and the change in microhardness at $t < t_0$.

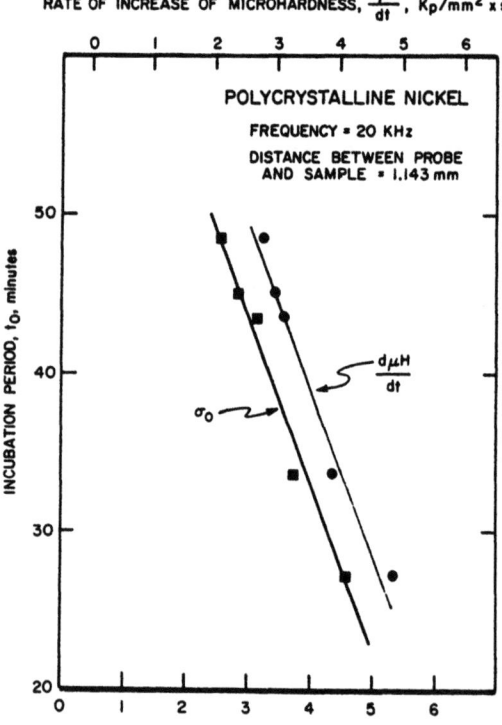

Fig. 16 The relationship between rate of increase in microhardness, incubation period and the magnitude of the applied cavitation stress (Ref. 50).

All the studies of cavitation produced by a vibratory probe indicate that erosion is primarily a result of mechanical stressing due to the concerted collapse of a cloud of bubbles in the liquid adjacent to the solid. While it is easy to visualize such a concerted collapse in a vibratory system in which the pressure is applied simultaneously to all the bubbles, it is not as readily apparent in a flow system in which bubbles are continuously transported from a low- to a high-pressure region. Nevertheless, high speed photography has shown that, even in a flow system, bubbles do not travel independently but are created, travel and collapse in a cloud [63]. Since there is no detectable difference in the surface topography of samples subjected to flow or vibratory cavitation, Figs. 11 and 12, it may be reasonably assumed that the mechanism of erosion is the same for both.

5. EPILOGUE

While division of the erosion process into three categories is convenient, it can be artificial since two, or even all three, processes may occur simultaneously. For example, in a high temperature system such as a coal gasification plant, both solid and liquid particles may be impacting together. In a turbulent flow system, liquid particle impact and cavitation may both occur and, once erosion has started, the liquid will contain the eroded solid particles which may then contribute to further erosion.

Moreover, although the mechanical aspects of erosion are emphasized in this paper, the corrosion aspects should not be ignored. For erosion by particle impact this factor is probably minimal, but the repeated growth and spalling of a surface film could contribute significantly to weight losses especially at elevated temperatures. In some cavitation systems, corrosion may play a major role. It has been shown that cavitation can produce an anodic shift in corrosion potential [64] which could give rise to a transition from the passive to the transpassive region and result in rapid corrosion. Cracking of the passive film in a corrosive media could also produce very high localized anodic regions. Localized dissolution could create pits which then act as further sites for mechanical erosion. Such factors have received some attention [65-67] but a more thorough investigation is needed to provide a complete understanding of the electrochemical contribution to erosion.

Finally, it is apparent that, for all types of erosion, the interaction between the complex stress mode and the microstructure of the impacted material must play a major role in determining the erosion rates. Moreover, this interaction may bear little, or no, relationship to the bulk mechanical properties of the solid. A thorough understanding of the erosion process will require an experimental determination of the mechanism of material loss, in conjunction with realistic theoretical modelling of these processes, which should include the appropriate dynamic materials properties and structural parameters.

ACKNOWLEDGEMENT

This work was supported in part by the National Science Foundation under Grant No. DMR 72-03227 A 0 1.

REFERENCES

[1] H. Hertz, Miscellaneous Papers, MacMillan and Co. London, (1896).
[2] W.F. Adler, Bell Aerospace Rep. No. AFM6-TR-74-210 Sept. (1974).
[3] F.C. Frank and B.R. Lawn, Proc. Roy. Soc. (London) A299 291 1967.
[4] T.R. Wilshaw and B.R. Lawn.
[5] A. Evanc, Rockwell International Technical Rep. No.
[6] B. Hamilton and H. Rawson, G. Mech. Phys. Solids 18 127 (1970)
[7] I. Finnie and S. Vaidyanathan, Fracture Mechanics of Ceramics 1, ed. by R.C. Bradt et al., Plenum Press, New York (1974).
[8] C.E. Smeltzer, M.E. Gulden and W.A. Compton, Trans. ASME 92D 639 (1970).
[9] A. Kelly, W.R. Tyson and A.M. Cottrell, Phil. Mag. 15 567 (1967)
[10] G. L. Sheldon and I. Finnie, Trans. ASME 88B 398 (1966)
[11] I. Finnie, Wear 3 87 (1960).
[12] I. Finnie, Wear 19 81 (1972).
[13] G.P. Tilly and W. Sage, Wear 16 447 (1970).
[14] G.P. Tilly, Wear 23 87 (1973).
[15] G.M. Hutchings and R.E. Winter, Wear 27 121 (1974).
[16] R. E. Winter and I.M. Hutchings, Wear 29 181 (1974).
[17] W.G. Mead and M.E. Harr, Wear 15 1 (1970)
[18] J.G.A. Bitter, Wear 6 5 (1963).
[19] J.G.A. Bitter, Wear 6 169 (1963).
[20] J.M. Neilson and A. Gilchrist, Wear 11 111 (1968).
[21] M.B. Moore, Tech. Rep. No. NYO-3477-12, Rutgers - The State University, New Jersey, (1968).
[22] W. J. Mead, L.D. Linebach and C.R. Manning, Wear 23 291 (1973).
[23] A. Behrendt, Proc. 3rd. Int. Conf. on Rain Erosion and Allied Phenomena, ed. by A.A. Fyall and R.B. King, Royal Aircraft Establishment UK (1970).
[24] I. Finnie, G. Wolak and Y. Kabie, J. Mat. 2 682 (1967).
[25] F.G. Hammitt, J.B. Hwang, L. Do, F.G. Mammitt Jr., Y.C. Huang, E.E. Timm and R.D. Hughes, Proc. 4th Int. Conf. on Rain Erosion and Allied Phenomena, ed. by A.A. Fyall and R.B. King, Royal Aircraft Establishment, UK (1974).
[26] W.F. Adler, Tech. Rep. No. D9244-953001 Bell Aerospace Company, September (1974).
[27] F.J. Neymann, Eng. Rep. No. E-1460, Westinghouse Electric Corp. March (1968).
[28] F. J. Neymann, J. Appl. Phys. 40 5113 (1969).
[29] M.C. Rochester and J.G. Brunton, Proc. 4th Int. Conf. on Rain Erosion and Allied Phenomena, ed. by A.A. Fyall and R.B. King, Royal Aircraft Establishment, UK (1974).
[30] G.P. Thomas and J.G. Brunton, Proc. Roy. Soc. (London) A314 549 (1970).

[31] J.E. Field, J.J. Camus, D.A. Gorham and D.G. Rickerby, Proc. 4th Int. Conf. on Rain Erosion and Allied Phenomena, ed. by A.A. Fyall and R.B. King, Royal Aircraft Establishment, UK (1974).
[32] E.F. Beutin, F. Erdmann-Jesnitzer and H. Louis, ibid.
[33] F.G. Hammitt, Y.C. Huang, C.L. Kling, T.M. Mitchell, Jr., and L.P. Solomon, ASTM STP474 288 (1970).
[34] B.C.S. Rao, N.S.L. Rao and K. Seetharamiah, Trans. ASME 92B 573 (1970).
[35] G.S. Springer and C.B. Baxi, ASTM STP 567 106 (1974).
[36] W.C. Brunton and J.M. Hobbs, NEL Rep. No. 479, National Engineering Laboratory, U.K. (1971).
[37] M.O. Speidel, Proc. 4th Int. Conf. on Rain Erosion and Allied Phenomena, ed. by A.A. Fyall and R.B. King, Royal Aircraft Establishment, U.K. (1974).
[38] N. L. Hancox and J.H. Bennton, Phil. Trans. Royal Soc. (Lon) A, 260, 121, 196, 1966.
[39] W.H. Wheeler, Cavitation in Hydrodynamics, HM SO London 1956.
[40] W.C. Leith and A.L. Thompson, Trans. ASME, J. Basic Eng., 820, 795-807, 1960.
[41] M.S. Plesset, Trnas. ASME, J. Basic Eng., 82D, 808-820, 1960.
[42] B. Vyas and C.M. Preece, ASTM STP 567 p. 77 (1974).
[43] B. Vyas and C.M. Preece, Proc. 4th Int. Conf. on Rain Erosion and Allied Phenomena, ed. by A.A. Fyall and R.B. King, Royal Aircraft Establishment, U.K. (1974).
[44] Lord Rayleigh, Phil. Mag. VI 34 94 (1917).
[45] T.B. Benjamin and A.T. Ellis, Phil. Trans. Roy. Soc. 206A 221 (1966).
[46] J.H. Brunton, Proc. 3rd Int. Conf. on Rain Erosion and Allied Phenomena, ed. by A.A. Fyall and R.B. King, Royal Aircraft Establishment, U.K. (1974).
[47] R. Schulmeister, Proc. 1st Int. Conf. on Rain Erosion and Allied Phenomena, ed. by A.A. Fyall and R.B. King, Royal Aircraft Establishment, U.K. (1974).
[48] R. Mickling and M.S. Plesset, Physics of Fluids 7 7 (1964).
(49) R.T. Knapp, Trans. ASME 77 1045 (1955).
[50] R.T. Knapp, Trans. ASME 80 91 (1958).
[51] M. Kornfield and L. Suvorov, J. App. Phy., 15, 495, 1944.
[52] C.F. Nande and A.T. Ellis, Trans. ASME, J. Basic Eng., 83, 648, 1901.
[53] T.B. Benjamin and A.T. Ellis, Phil. Trans., A, 260, 221-240, 1967.
[54] C.P. Thomas, 2nd Conf. on Rain Erosion held in Meersburg, Germany, 150-165, 1967.
[55] A. Thiruvengadam, ASTM, STP 408, 22-41, 1967.
[56] J. Erdmann-Jesnitzer and H. Louis, ASTM STP 567 171 (1974).
[57] M. F. Rose and T. L. Berger, Phil. Mag. 17 (150) 1121 (1968).
[58] J. W. Tichler and H. B. Zeedyk, 4th Int. Conf. on Rain Erosion and Allied Phenomena, Meersburg, 1974.

[59] J. M. Hobbs and D. McCloy, Metals and Materials 6 (1) 27 (1972).
[60] B. Vyas and C. M. Preece, to be published.
[61] S. Dakshinamoorthy and C. M. Preece, to be published.
[62] S. Dakshinamoorthy, S. Prasad and C. M. Preece, to be published.
[63] A. F. Conn, private communication.
[64] C. M. Preece and B. Vyas, Proc. 4th Int. Conf. on Rain Erosion and Allied Phenomena, ed. by A. A. Fyall and R. B. King, Royal Aircraft Establishment, U. K. (1974).
[65] W. H. Wheeler, Cavitation in Hydrodynamics, HMSO, 1956.
[66] P. Eisenberg, Trans. Soc. Naval Arch. and Marine Eng., 73, 241, 1965.
[67] H. Wiengand and H. H. Piltz, Werkstoffe und Korrosion, 15, 212-220, 1964.

DISCUSSION

Comment by H. Mughrabi:

I am inclined to say that your specimens undergo some sort of fatigue, since, after all, you apply vibrations. Also, the surface damage you presented reminds me in a distant manner of persistent slip bands. This leads me to the following remarks:

1) Cyclic deformation is known to introduce a high concentration of vacancies and their agglomerates which, I suspect, are also produced in your specimens.

2) The dislocations you showed remind me of the dislocations that form as a result of the production of a high density of point defects in a high voltage electron microscope. This could be another indication of the role of point defects in your specimens.

3) To check the effect of point defects it could be useful if you performed your experiments over a certain range of temperatures, allowing vacancies to migrate or remain immobile at will.

Reply:

At early stages of deformation by cavitation, some slip steps do appear to be developing into intrusions and extrusions similar to those observed in fatigue samples. However, they do not develop and I have not found any trace of crack initiation and propagation in our own or any other studies of the phenomenon. Moreover, the dislocation structures we observe are very different from those produced by fatigue, even at ultrasonic frequencies. I agree that the mode of stressing is likely to introduce large numbers of vacancies and, in fact, a large number of voids is produced throughout the thickness of the sample (∼3mm), with the largest concentration in a region just below the bottom of the surface "crater".

With regard to the change in temperature experiments, it is not possible to maintain constant bubble dynamics over a range of temperatures since the vapour pressure, viscosity, gas and impurity content, etc. all vary. Because of this the cavitation pressure varies significantly over a temperature range of ∼20°C.

Comment by S. Weissmann:

My colleague Professor V. Greenhut is studying in our laboratory the effects of shockwaves on metals and alloys. To show the massive vacancy production which is obtained by the interaction of jogged dislocations at short ranges, aluminum alloys were studied in the precipitation stages (metastable stages) which are then transformed into the next stage of aging in the aging sequence. These metastable stages functioned, therefore, as internal markers for the massive vacancy production. May I suggest that a similar approach may prove useful in erosion studies if the bubbles, as Dr. Preece suggests, have a shockwave-like impact on the material.

Reply:

We are currently studying the effect of heat treatment of Al-Cu alloys on their resistance to cavitation, and it would certainly be very helpful to correlate our results with those of Professor Greenhut.

Comment by R. Bullough:

I am curious to know whether the concept of vacancy waves mentioned by Professor Weissmann is accepted in such rapid deformation conditions that prevail during cavitation.

Reply:

It is known that the passage of a shock wave through a metal does introduce a large concentration of point and line defects. Also interactions between incident and reflected waves are found to produce internal voids and sometimes spalling of the metal. Optical metallography of cross sections of metals exposed to cavitation indicates that internal voids form by this process also and suggests that the two processes may be similar.

Comment by A. S. Argon:

In your transducer induced cavitation experiments where bubbles are collapsing at nearly the speed of sound, the local indentation strain rates on the surface may be as much as 10^{10} higher than the conventional machine strain rates. This is supported by your observation of the dislocation structure in aluminium which resembles more the structure at 77°K than at room temperature. Based on this observation it may be appropriate to compare the behavior of different f.c.c. metals by normalizing the transducer wave pressures that were used by the yield stress of these materials at low temperature, say for example the 0°K yield stress. This may make the material removal rates in these materials more comparable.

Reply:

I agree that this might prove to be appropriate. However, since there is a linear relationship between incubation period and applied pressure, the correlation should be even simpler: $t_o/\sigma_{F_{4°K}}$ = const. for a particular class of metals, e.g., high SFE, FCC metals. It appears that deformation mode is important in determining erosion rates and so $t_o/\sigma_{F_{4°K}}$ may not be constant for different classes of metals.

Comment by R. W. Staehle:

The state of the surface in cavitation can be changed in the following important ways with respect to your experiments.

1) Film free to film covered by going from active to passive conditions.

2) Film density and thickness by using various oxidants combined with changing the potential.

3) Changing the rate of hydrogen entry by adding poisons such as arsenic or sulfides.

4) Using halide or sulfide ions which affect the stability of the film.

I suggest that you control the potential and measure the current. It will give more meaningful and useful information.

Reply:

We have, in fact, controlled the potential and attempted to measure the current, but only in the passive region. The results, however, were too erratic to be meaningful because cavitation caused cracking of the passive film and resulted in random, extreme fluctuations in current. Nevertheless, this is obviously the more appropriate procedure and should produce interpretable data in the active region.

ENHANCEMENT OF FATIGUE, CREEP, AND STRESS-CORROSION RESISTANCE BY SURFACE TREATMENTS

I. R. Kramer

Naval Ship Research and Development Center,
Annapolis, Maryland, USA

ABSTRACT. By considering the nature of the surface layer formed as a result of plastic deformation, substantial improvements may be made in the fatigue, creep, and stress-corrosion cracking resistance. When the surface layer stress is decreased, the fatigue and stress-corrosion cracking resistance is increased. When its relaxation rate is decreased, the creep rate is also decreased.

It has been found that the surface layer can strongly influence the crack initiation and propagation behavior of metals. It will also be shown that, for some metals, the creep rate at elevated temperatures can be materially decreased by the elimination of the surface layer.

The evidence that the surface layer could influence crack initiation and propagation was first noted [1,2] from the decrease in the ductile-brittle transition temperature of molybdenum when the surface layer was removed during deformation. Polycrystalline specimens of molybdenum were strained axially in an electrolytic bath containing sulphuric acid and metal was removed at a constant rate. The strain rate was 5×10^{-5} sec^{-1} and the temperature increased only 0.5° C during the testing. The results of these tests, Fig. 1, showed that the ductile-brittle transition temperature decreased 15° C when the surface layer was removed at a rate of 15 μm/min. In comparison, a hundredfold decrease in the strain rate without removing the surface layer decreased the transition temperature by 25° C.

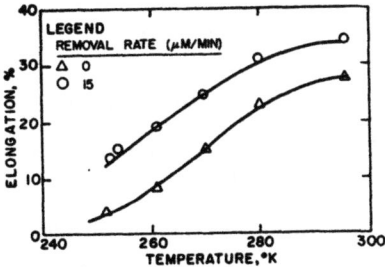

Fig. 1. Effect of removing surface layer on the ductile-brittle transition temperature of molybdenum.

Others have also reported that the surface can affect the brittle behavior of metals. Gilman [3] observed that an electrodeposited layer of copper approximately 3000 Å thick on zinc monocrystals made them much more brittle at room temperature. The plating reduced the stress for fracture. In contrast, the work of Greenough and Ryder [4] on copper plated zinc crystals indicated for orientations between $20° < \chi < 70°$, the cleavage stress was approximately twice that for unplated crystals of zinc. With $\chi > 75°$, the uncoated specimens were stronger than the coated specimens. Weiner and Gensamer [5] also reported that the normal fracture stress of zinc was increased some 40 percent upon applying coatings of copper, gold, zinc oxide, or tin thicker than about 500 Å. It was also reported [6] that for zinc crystals oriented $\chi = 83°$, the surface films act as a barrier to dislocations and decrease the fracture stress, making it independent of temperature.

In a recent investigation [7] of the fatigue behavior of alloys of aluminum (2014-T6), titanium (6Al-4V), and a 4130 steel, it was stated that the surface layer hardened during fatigue cycling. A propagating crack was initiated when the surface layer stress reached a critical value. This value was reported to be independent of the stress amplitude and the environment. Recent unpublished work has also shown it to be independent of the prior fatigue history. An example of the increase in the surface layer stress as a function of the number of cycles (R = -1) at various stress amplitude is given in Fig. 2 for 2014-T6 aluminum. The asterisks in the figure represent the values of the surface layer stress at N_o, where N_o is the number of cycles to form a propagating crack. It was also shown that removal of the surface layer before this critical value of the surface layer stress was reached prolonged the fatigue life until the specimen was too thin for further use. The extension of the fatigue life by the removal of the surface material for the metals investigated (2014-T6, Ti(6Al-4V)) was not thought to be associated with the removal of persistent slip bands or cracks [8]. An examination of the surface by means of the scanning electron microscope after the specimens had been cycled showed that only very light slip bands were present. In addition to these two observations, there is also an apparent correlation between the fatigue life of metals in various environments and the surface layer stress. In vacuum, the fatigue life is increased [9-11] and the surface layer stress is decreased, while in solutions containing CH_3OH-HCl for titanium and NH_4OH-$Cu(NO_3)_2$ for copper, the fatigue life decreased and the surface layer stresses increased [10]. In keeping with the above observations, it should be clear that a process that inhibits or prevents the formation of the surface layer would enhance the fatigue resistance. This may be done by mechanical processing and by surface treatments.

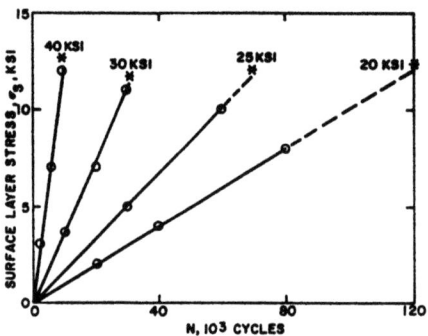

Fig. 2. Increase in surface-layer stress in 2014-T6 aluminum with number of fatigue cycles and stress amplitude, R = -1.

1. IMPROVEMENT IN FATIGUE RESISTANCE BY MECHANICAL PROCESSING

The method used to improve the fatigue life of metals by mechanical processes will be referred to as the surface layer elimination (SLE) process. In this method, the metal is stress, usually up to the proportional limit, and the surface layer formed is eliminated. The SLE may be accomplished by two methods. It may be eliminated by relaxation in those cases where the dislocations are not too strongly pinned (for example, in Ti, steels, Cu) or by chemical or electrochemical dissolution for strongly pinned dislocations (for example, precipitation hardened aluminum alloys). The relaxation method is preferred because of the decrease in fatigue life and hydrogen embrittlement often associated with chemmilling.

The changes on the structure that occur when the metal is stressed to the proportional limit has been described previously [12,13]. In brief, a surface layer and the interior are work hardened. When the surface layer is eliminated, the material is uniformly work hardened throughout its cross section.

An example of the effect of surface layer removal on crack propagation is given in Figs. 3 and 4 for a 4130 steel. The data in Fig. 3 show the increase in the crack length as a function of the number of cycles needed to propagate the crack, N_p. Compared to the untreated material, when the prestress for the SLE treatment was 60 ksi (413.7 MPa), the crack growth was very slow. When the prestress was increased to 80 ksi (551.6 MPa), the crack propagation resistance decreased. The cracking rate, however, was still lower than that of the untreated specimen.

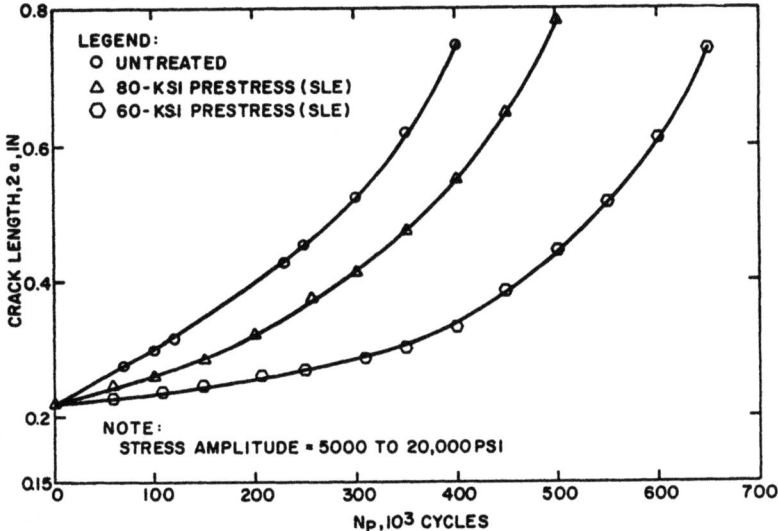

Fig. 3. Crack growth in 4130 steel, center-notched specimens, 0.06 in. thick, YS = 100 ksi.

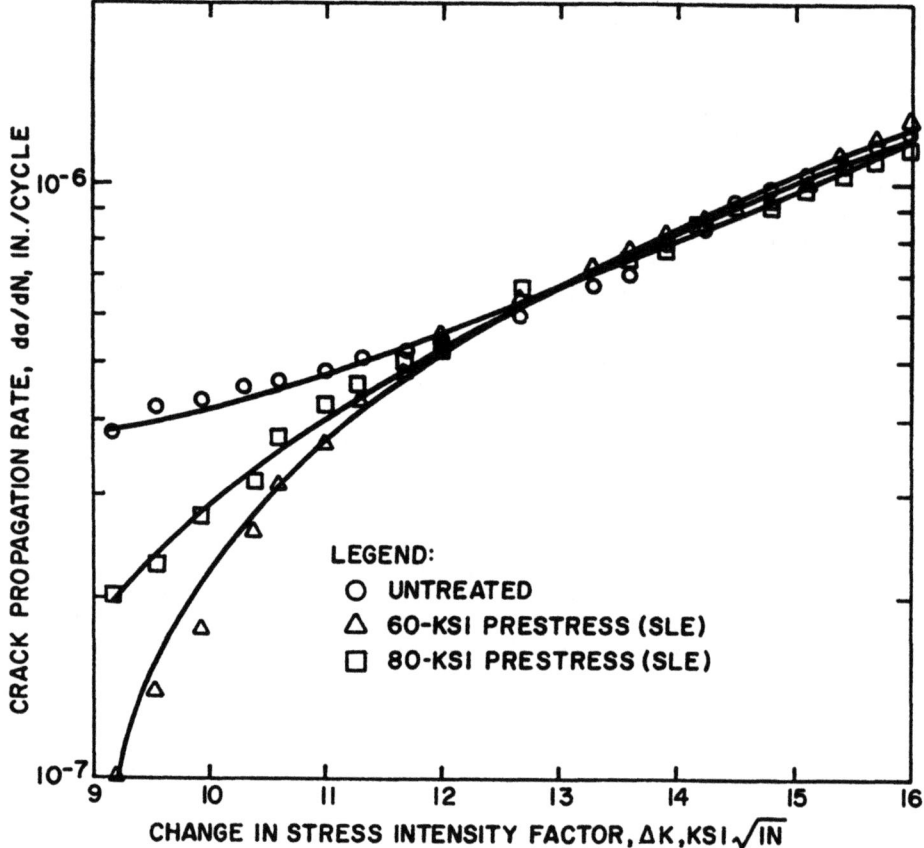

Fig. 4. Crack propagation rate in 4130 steel under plane stress conditions, YS = 100 ksi.

Fig. 4 shows the crack propagation rate, da/dN, as a function of the change in the stress intensity factor, ΔK. Note that at the lower ΔK values, the crack velocity of the SLE-treated specimens is about 4 times lower than that of the untreated specimens. This improvement tends to decrease as the stress in front of the crack approaches that of the prestress. At about 13 ksi $\sqrt{\text{in}}$. (14.26 MPa $\sqrt{\text{m}}$), the curves merge.

The data for the crack propagation rate under plane strain conditions for the 4130 steel are presented in Fig. 5. This figure also includes points obtained from duplicate specimens.

Again, these curves show the SLE treatment improves crack propagation resistance. At a ΔK of about 16 ksi $\sqrt{\text{in}}$. (17.44 MPa $\sqrt{\text{m}}$), the SLE-treated specimens have a crack velocity 4 times smaller than that of the untreated specimens. As ΔK increases, the degree of improvement decreases, and at values in excess of

18 ksi $\sqrt{\text{in.}}$ (19.75 MPa $\sqrt{\text{m}}$), the improvement is 40 percent. This degree of improvement remains constant at least to a ΔK of 22 ksi $\sqrt{\text{in.}}$ (24.13 MPa $\sqrt{\text{m}}$).

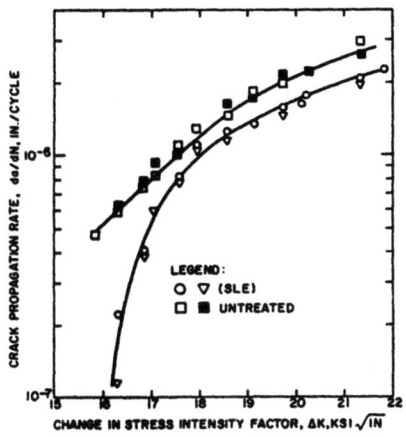

Fig. 5. Crack propagation behavior of 4130 steel, compact-tension specimens 0.625 in. thick under plane strain conditions, YS = 180 ksi.

The rates of crack growth in the center of the welds in 4130 steel as a function of ΔK are given in Fig. 6. Like the other data obtained for the 4130 steel, the SLE treatment decreased the crack propagation rate. For specimens prestressed to 70 ksi (482.6 MPa) before removing the surface layer, the crack velocity decreased by a factor of about 3 at a ΔK of about 9 ksi $\sqrt{\text{in.}}$ (9.87 MPa $\sqrt{\text{m}}$). The curves tend to merge as ΔK increases, and at about 15 ksi $\sqrt{\text{in.}}$ (16.45 MPa $\sqrt{\text{m}}$), they coincide.

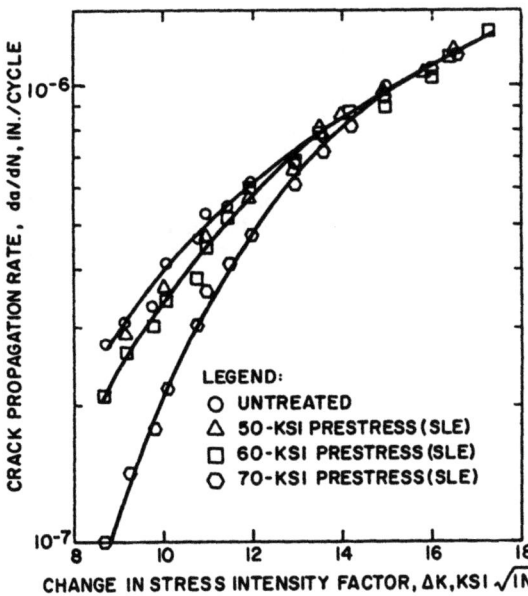

Fig. 6. Crack propagation rate in center of weld of 4130 steel center-notched specimens, 0.06 in. thick, YS = 100 ksi.

The crack growth rates in the heat-affected zone under a cyclic stress amplitude between 5 and 20 ksi (34.47 and 137.9 MPa) is given in Fig. 7. These data show that, in addition to improving the crack propagation resistance within the weld, the resistance in the heat-affected zone was also increased. A comparison of the crack propagation rates in the heat-affected zone and in the weld at a ΔK value of 9 ksi $\sqrt{\text{in}}$. (9.87 MPa $\sqrt{\text{m}}$) indicates the amount of improvement imparted by the SLE treatment was the same: the crack velocity was decreased by a factor of about 3.

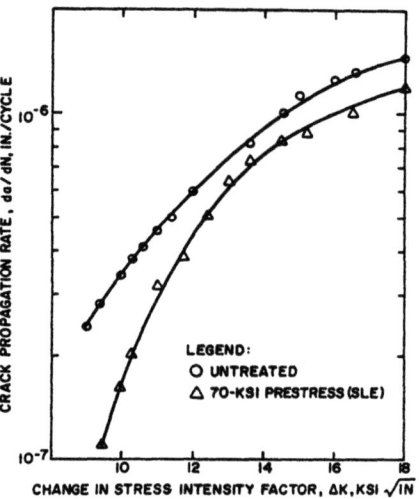

Fig. 7. Crack propagation rate in heat-affected zone of 4130 steel, center-notched specimens, 0.06 in. thick, YS = 100 ksi.

The effect of prestressing and removal of the surface layer on specimens of 7075-T6 aluminum tested in tension-compression is shown in Fig. 8. The optimum improvement in fatigue life was obtained by prestressing at 50 ksi (344.7 MPa). Prestressing in the plastic flow region caused the fatigue life to be less than that of the unstressed material. As shown, the SLE treatment increased the endurance limit at 10^7 cycles from 23 to 34 ksi (158.6 to 234.4 MPa), an increase of about 48 percent. At stresses above 40 ksi (275.8 MPa), the two curves coincide. From observations of the widths of the hysteresis loops during the fatigue tests, it was apparent that some plastic flow occurred at stresses above 34 ksi (234.4 MPa). The width of the loop increased with increasing stress, and it follows that the surface layer was partially formed.

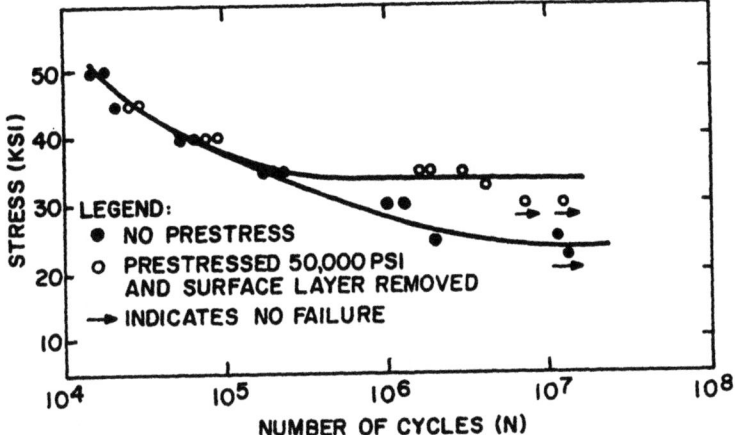

Fig. 8. Effect of removal of surface layer on the fatigue life of 7075-T6 aluminum tested in tension-compression.

2. IMPROVEMENT IN FATIGUE RESISTANCE BY CHEMICAL TREATMENT

It has been shown that the fatigue life as well as corrosion and stress-corrosion cracking resistance may be improved by a suitable modification of the surface of aluminum alloys. When aluminum alloys are anodized in a sulphuric acid bath to produce an anodized coating in the order of 0.005 in. (0.013 mm) thick and the open pores are infiltrated with a long chain fatty acid, amine, or an alcohol, the fatigue life is improved [13].

The improvement in fatigue life of an aluminum alloy (7075-T6) obtained by impregnating the porous anodized coating with polar organic molecules may be seen in Figs. 9 and 10.

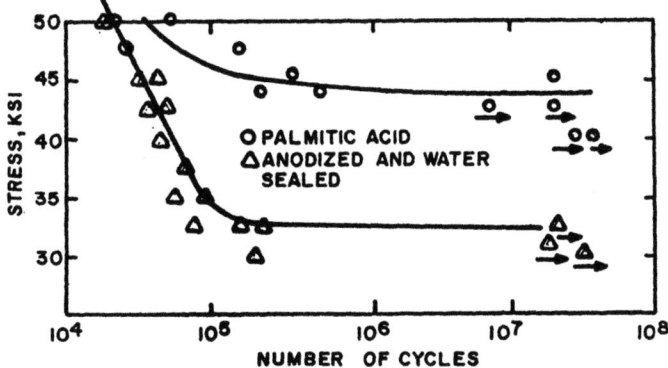

Fig. 9. Tension fatigue test of 7075-T6 aluminum alloy sheet, notch factor $K_t = 1$.

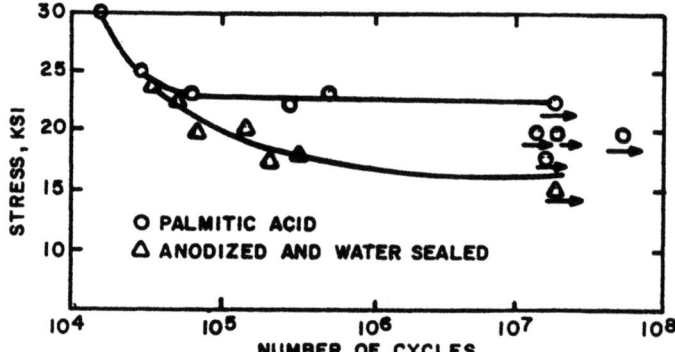

Fig. 10. Tension fatigue test of 7075-T6 aluminum alloy sheet, notch factor K_t = 2.37.

Axial-tension fatigue curves obtained from specimens which had been anodized and treated with palmitic acid show that the fatigue strength, taken at 10^7 cycles, was increased from 32.5 to 44 ksi (224 to 303 MPa) when a notch factor of 1 was used (Fig. 9), and from 16.5 to 23 ksi (113 to 158 MPa) when a notch factor of 2.37 was used (Fig. 10). The fatigue strength of specimens tested in flexural fatigue was increased 45 percent when the specimens were treated with stearic acid. The fatigue strength was increased from 19 to 27.5 ksi (131 to 189 MPa). It is of interest to note that the increase of the fatigue strength for the three types of tests was between 38 and 45 percent. In these tests, no difference between the fatigue life of bare specimens and anodized specimens was detected. The same percentage of improvement in the fatigue strength was indicated when aluminum alloys of 2014-T6, 2014-T6 Alclad, 5456-343, and 6061-T6 were tested with the organic acids, amines, and alcohols.

The effect of various surface-active agents on the fatigue life of the 7075-T6 aluminum was studied. The data show that a wide variety of polar organic compounds are effective in improving fatigue life. While the impregnation of the porous anodic oxide films with long chain acids, alcohols, and amines are effective in improving the fatigue life, essentially non-polar molecules impart only a small improvement in the fatigue life.

It is of interest to note that surface coatings of the organic coatings without the impregnation into the pores of the anodic coatings are ineffective. Specimens of the 7075-T6 alloy with unanodized surfaces treated in a palmitic acid bath at 130° C for 20 minutes did not show any improvement in flexural fatigue life at a stress level of 25 ksi (172 MPa). For comparison, when tested at 26 ksi (179 MPa), anodized specimens impregnated with palmitic acid required 30,000,000 cycles to failure.

3. IMPROVEMENT OF CREEP BEHAVIOR BY MECHANICAL PROCESSING

In a preceding portion of this paper, it was shown that, by prestressing a metal and removing the surface layer, the fatigue resistance was improved. This SLE process has also been found to improve the creep resistance of some metals.

At high temperatures, it would be expected that, in addition to the relaxation of dislocation obstacles in the interior of the metal, the rate of relaxation of obstacles in the surface layer could also contribute to the creep rate [14]. In those cases where the surface layer relaxes at a faster rate than obstacles in the interior, the creep rate is expected to be decreased by eliminating or decreasing the surface layer. When the SLE process is applied to such metals as titanium alloys (6Al-4V), Haynes 188, and 321 steel, the creep rate was reduced [15]. Previous work has shown that the surface layer in 2014-T6 and 7075-T6 does not relax at a rate faster than the dislocation obstacles in the interior at 150° C. Therefore, the SLE process was not expected to improve the creep resistance. Experimental results confirmed this expectation. The creep curves of the untreated and SLE-treated specimens tested at 150° C were the same.

In the following, the specimens were stressed to the proportional limit at room temperature and the surface layer was eliminated by allowing it to relax at the test temperature.

As a result of the SLE treatment, both the primary and steady-state creep rates were decreased. The enhancement in the creep resistance may be described [16] in terms of Eq. 1

$$\dot{\epsilon} = C\sigma^n \exp(-U/kT) \qquad (1)$$

where U is the activation energy, k the Boltzmann constant, T the temperature, $\dot{\epsilon}$ the steady-state creep rate, n the creep rate applied stress exponent, σ the applied stress, and C is a constant related to the rate of stress relaxation. In the steady-state region where C is constant with time (strain), a linear relationship exists between $\ln \dot{\epsilon}_s$ and $\ln \sigma$. The slope of the curve is n. The curves given in Fig. 11 are typical of the change of σ as a function of the applied stress for Haynes 188 at 760° C (1400° F), 321 steel at 760° C (1400° F), and titanium (6Al-4V) at 343° C (650° F). Note that the improvement increases rapidly as the applied stress is decreased. At the proportional limit at the temperature used in the tests, the creep rates for the untreated and SLE-treated specimens are the same. At this stress, the surface layer is reformed. The data presented in Table 1 are the values for n, U, and C for the Haynes 188, 321 steel, and titanium (6Al-4V) as a function of temperature. Note that when the specimens were treated to eliminate the fast relaxing surface layer, the coefficient C decreased markedly (for example, from 10^{-30} to 10^{-50} for the Haynes 188. In addition, the stress dependent exponent n increased and the activation energy also increased.

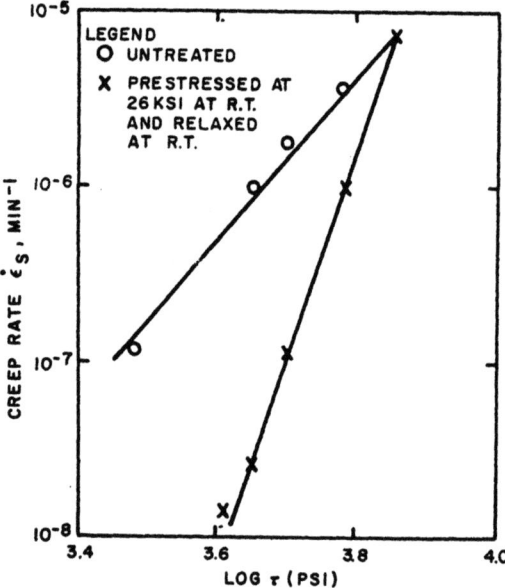

Fig. 11. Stress dependence of secondary creep rate of 321 steel at 760° C.

Table 1
Parameters of the Creep Eq. (1)

Material	Condition	Temp. °C	n	U koals/mole	C, min^{-1}/(dyne/cm^2)^n
Haynes 188	Untreated	870	4.0	62	7.45 × 10^{-30}
		760	4.0	62	7.45 × 10^{-30}
	SLE	870	6.6	75	1.7 × 10^{-50}
		760	6.6	75	1.7 × 10^{-50}
321 Steel	Untreated	650	4.0	88	2.9 × 10^{-22}
		760	4.0	88	2.9 × 10^{-22}
	SLE	650	8.5	137	1.7 × 10^{-34}
		760	8.5	137	1.7 × 10^{-34}
Titanium (6Al-4V)	Untreated	290	2.0	28	3.4 × 10^{-17}
		315	2.0	28	3.4 × 10^{-17}
		343	2.0	28	3.4 × 10^{-17}
	SLE	290	4.0	49	3 × 10^{-29}
		315	4.0	49	3 × 10^{-29}
		343	4.0	49	3 × 10^{-29}

4. IMPROVEMENT IN STRESS-CORROSION CRACKING AND CORROSION

The method of anodizing and impregnation of the pores with long chain polar organic molecules is applicable to the improvement of the corrosion and stress-corrosion cracking of aluminum alloys. Since these organic materials are both oleophobic and hydrophobic, the metal is not wetted and corrosive conditions do not readily attain. The following affords a comparison between the aluminum alloys impregnated with docosanoic acid, anodized, and bare aluminum.

For convenience, the term Mar-M sealed will be used to designate treatments that incorporate the organic polar molecules into the anodized layer.

An example of the corrosion resistance of Mar-M sealed panels, as rated by the CASS test, is shown in Fig. 12 for aluminum 7075-T6. The CASS test is a copper accelerated salt spray test. It measures the tendency toward pitting. As seen from the data, pitting started within 15 hours for the anodized specimens, whereas for the Mar-M sealed material, the pitting is delayed substantially. At the end of about 40-60 hours, the surfaces of the anodized metals were badly pitted, while for the Mar-M sealed materials, less than about 0.1 percent of the surface area was affected.

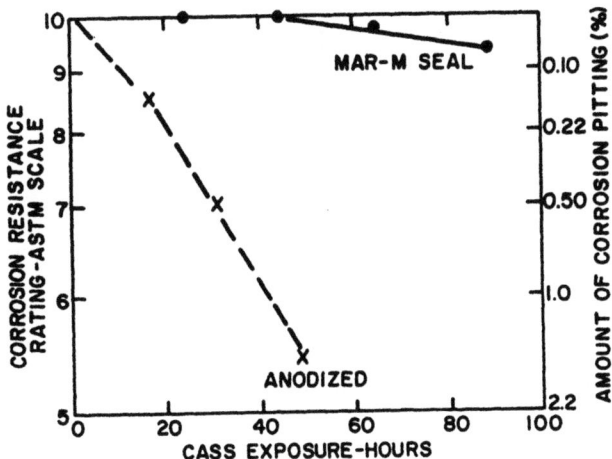

Fig. 12. CASS corrosion test of 7075-T6.

The resistance to hot salt water corrosion of the Mar-M sealed panels may be seen in Fig. 13. Note that the treated panel is completely unaffected by the salt water after an exposure of 6552 hours. The anodized and water sealed treated panel is badly corroded. When tests were conducted over 18,000 hours (2 years), no corrosion was detected.

Control—Hot Water Sealed	Mar-M Seal

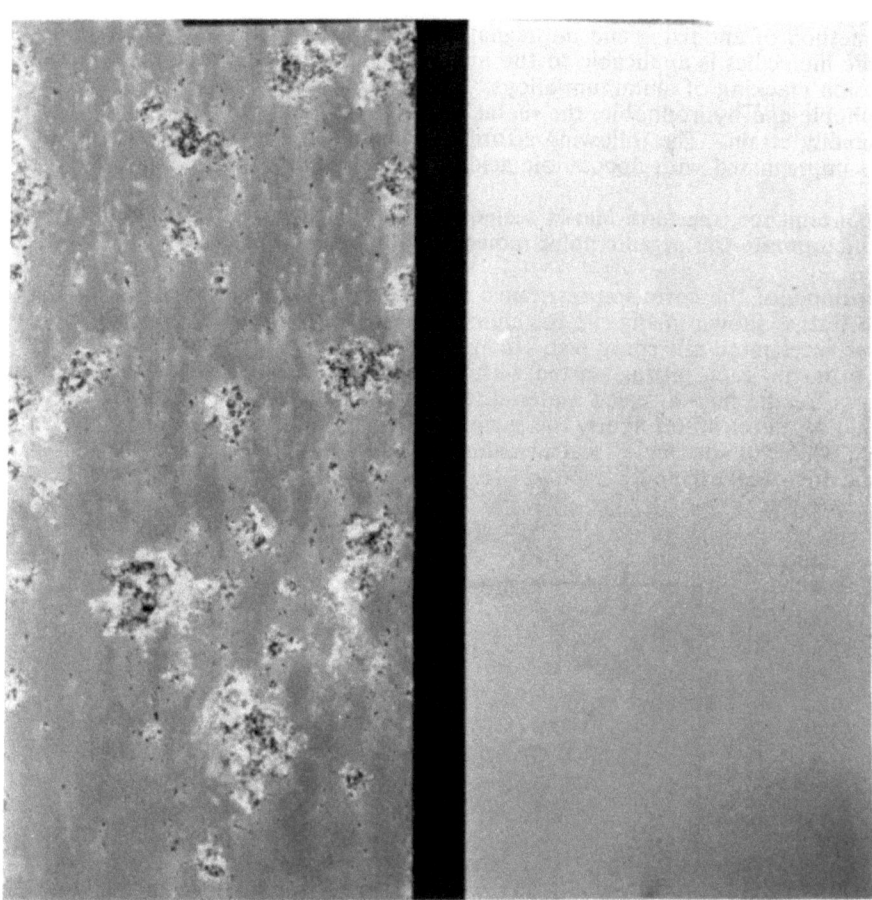

Fig. 13. 2014-T6 aluminum alloy corrosion panels after salt spray exposure of 6552 hours.

The data in Table 2 show the improvement in the stress-corrosion cracking behavior imparted by the Mar-M seal treatment. When tested in the long transverse and short transverse directions in various salt water solutions, no failure occurred. The addition of a chromate salt caused failure but in a much longer time than that obtained by other treatments.

Table 2
Stress-Corrosion Cracking of 7075-T6

Treatment	Stress % YS	Failure Time days
Test in Long Transverse Direction		
Solution: 3.5% NaCL + 1% H_2O_2		
Bare	50	14
Mar-M	50	N.F.
Mar-M	75	N.F.
Solution: Kure Beach (Natural Seawater)		
Bare	75	84
Mar-M	75	N.F. (366)
Solution: Salt Chromate		
Anodized and Chromate Sealed	85	5
Anodized and H_2O Sealed	85	4
Mar-M	85	51
Solution: 3.5% NaCL		
Bare	90	3
7075-T73	90	100
Anodized and H_2O Sealed	90	8
Anodized and Oil Sealed	90	20
Shot Peened	90	41
Mar-M	90	N.F. (255)
Test in Short Transverse Direction		
Solution: Salt Chromate		
Anodized and H_2O Sealed	80	0.1 (hours)
Anodized and Dischromate Sealed	80	0.2 (hours)
Furnane Resin	80	0.4 (hours)
Mar-M	80	3.7 (hours)

REFERENCES

1. I. R. Kramer, AFML Report TR68/65 (MRC-67-421), Jan. 1968
2. I. R. Kramer, Trans. Met. Soc., (239), 529, 1967.
3. J. J. Gilman, Trans. Met. Soc., (191), 1148, 1951.
4. G. B. Greenough and D. A. Ryder, J. Inst. Metals, (84), 467, 1955.
5. L. C. Weiner and M. Gensamer, J. Inst. Metals, (85), 441, 1956-57.

6. L. C. Weiner, Trans. Met Soc., (212), 342, 1958.
7. I. R. Kramer, Trans. Met. Soc., (5), 1735, 1974.
8. N. T. Thompson, N. J. Wadsworth, and N. Louat, Phil. Mag., (1), 113, 1965.
9. H. J. Gough and D. G. Sopwith, J. Inst. Met., (49), 93, 1932.
10. K. O. Snowden and J. N. Greenwood, Trans. Met. Soc., (212), 626, 1958.
11. N. J. Wadsworth, Proc. of the Symposium on Internal Stresses and Fatigue Metals, Elsevier Publ. Co., 1959.
12. I. R. Kramer and N. Balasubramanian, Acta. Met., (21), 695, 1973.
13. I. R. Kramer, Proc. Tenth Sagamore Army Mat. Conf., (110), 245, 1964.
14. I. R. Kramer, Trans. ASM, (60), 311, 1967.
15. I. R. Kramer, Trans. Met. Soc., (4), 431, 1973; 315, 1964.
16. J. Friedel, Dislocations, Pergamon Press, N. Y., 315, 1964.

MICROCUTTING: A METHOD TO STUDY SURFACE PLASTICITY AT VERY HIGH STRAIN RATES.

J.F. Prins

Department of Physics, University of Pretoria,
Pretoria, South Africa.

ABSTRACT. An experimental procedure is discussed where a diamond point excecutes single cuts of depth in the micrometer range on the surface of a material. Simultaneously the forces involved in the cutting process are measured. If viewed as a novel way to study surface plasticity, information may be gained on the dislocation mechanisms near a free surface under extreme conditions of deformation.

1. INTRODUCTION. The majority of products produced in our present technological society undergo some sort of machining process during their manufacturing cycle. With the development of harder and more specialized materials, more and more attention has been focussed on the process of grinding where material is removed from the workpiece in the form of very thin chips. This process is effected by using small abrasive grits bonded onto the periphery of a wheel by means of a suitable bonding material. New grit materials, like synthetic diamonds and cubic boron nitride, have contributed to a booming multimillion dollar abrasives industry.

Although the mechanics of machining has been studied by numerous investigators; very little attention has been paid to the fundamental aspects of this process from a materials point of view. The primary object of most of the research done to date, has been to develop better abrasive materials and to reduce the cost of material removal.

Grinding is a cooperative process where a number of grit particles penetrate the surface of the workpiece, deform it plastically and shear off small chips with dimensions in the micrometer range. No basic model of this process can be developed without a thorough knowledge of the interaction between a single grit particle and the workpiece surface. Various studies have been published in an attempt to elucidate this interaction [1,2,3,4,5,6, 7,8,9,10]. This type of investigation becomes more exciting when it is not approached purely from the machining point of view, but as a novel way to study surface properties under extreme conditions of deformation. The rate of deformation can to a certain extent be controlled by changing the wheel speed but is generally of a very high order of magnitude (> 10^5 sec^{-1}). If a shaped diamond is used as a cutting tool, the experimental procedure may be viewed as a dynamic hardness testing experiment.

2. THE GEOMETRY AND MECHANICS OF THE CHIP FORMING PROCESS

Figure 1 is a schematic representation of the chip forming process during cutting of a material which deforms plastically. Plastic deformation and shearing occurs along the plane AB which connect the tip of the diamond point and the free surface [11] and which is called the primary deformation zone. The resultant chip flows upwards against the leading diamond face AC, where it is further deformed by frictional forces (secondary deformation zone). Two angles are of importance for the process: the rake angle $-\alpha$ between the vertical and the leading face of the cutting tool, and the shear angle ϕ between the primary deformation plane AB and the horizontal plane. As shown, the rake angle is considered negative. In macro-machining operations, for example lathe turning, the rake angle is usually positive which limits the amount of plastic deformation to lower values in the secondary deformation zone.

Both points A and B are surface points, and the cutting process should be affected by the dislocation interactions with the newly created surface at A and the existing surface at B. It has been suggested [12] that the dislocations are produced by stress

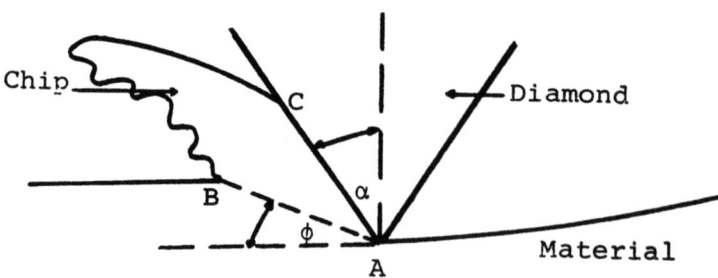

Figure 1: The chip forming process.

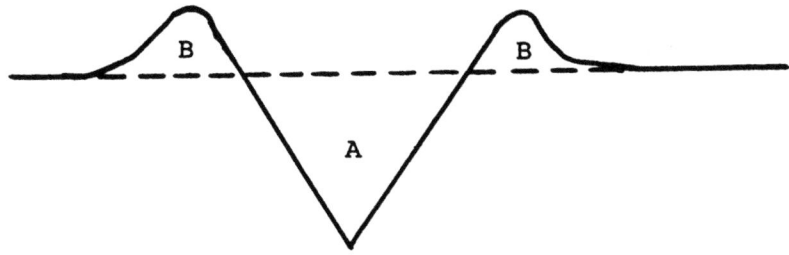

Figure 2: Tallysurf trace across cut made by pyramidal diamond point.

concentrations at A and B. The role of surfactants or foreign thin films at B should be of great interest. By increasing the wheel speed, less time is available for heat generated to dissipate and the process can be considered as adiabatic. It is even possible to reach the melting temperature of the sample at suitably high wheel speeds.

When the diamond tip penetrates the sample, plastic flow occurs and material flows up against the lateral sides of the diamond. Figure 2 is a typical example of a tallysurf trace obtained across such a cut. The ratio of the areas B to area A (B/A) is a measure of the amount of plastic deformation while (A-B)/A describes the effectiveness of the material removal process.

F_T and F_N are the tangential and normal forces which are measured during the short contact times (50-150 μ secs). For ϕ very small, F_T is mainly responsible for the chip removal process while F_N determines the ease of penetration.

3. SOME RESULTS ON STEEL

The sharp point in the [110] direction of a cubo-octahedral diamond was used as the cutting tool. A rake angle $\alpha = 54.74°$ was obtained by choosing the direction of cutting relative to the point as the [001] direction. Cuts of depth 10 μm where executed on samples of Pitho steel (containing 0,9 - 1,0% Carbon, 1,10 - 1,30% Mn, 0,45 - 0,65% Cr, 0,45 - 0,80% W, 0,15% max Vanadium) at different wheel speeds (6, 7, 16, 24 and 28 ms^{-1}).

The normal forces F_N measured at the maximum depth of cut (10 μm) showed a larger scatter in values than the corresponding tangential forces, but did not change within this scatter, with the speed of cutting. A definite decrease in the tangential forces were however observed with an increase in cutting speed. (See Figure 3).

Pitho steel specimens of original hardness 210 Vickers, were quench hardened to 900 Vickers and then tempered to give a range

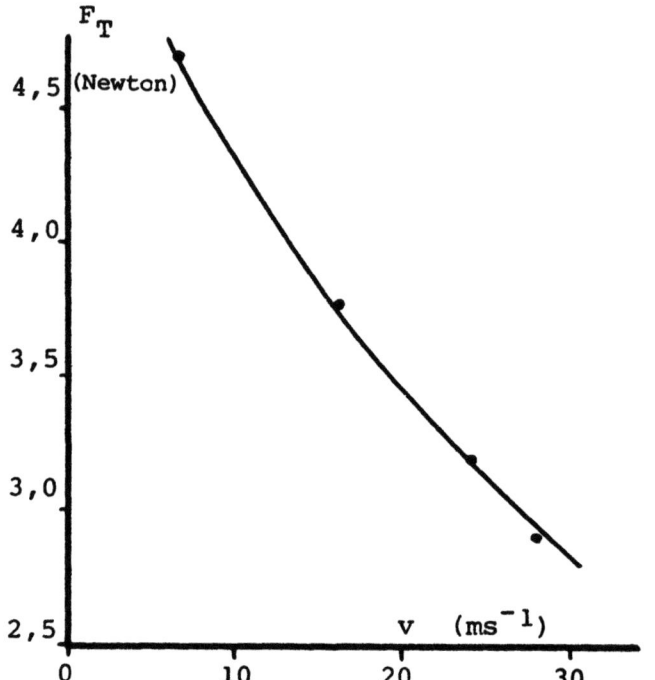
Figure 3: Tangential force as a function of cutting speed.

of hardnesses. Within the relatively wide scatter of the normal forces F_N, a linear relationship was found between F_N and the measured Vickers hardness of the specimen. The tangential forces F_T remained fairly independent of the hardness except for a slight tendency to decrease towards higher hardness values. An exception was found for the specimen of hardness 900 Vickers, which showed a minimum in F_T at about 25 ms^{-1} cutting speed. F_T then increases with higher cutting speeds. Recent measurements on softer materials also show a minimum which, however, occurs at a higher cutting speed [13]. More accurate measurements of the normal force indicate a possible weak dependence of F_N on the wheel speed [13].

Scanning electron micrographs of the chips formed during cutting show a wrinkled or "bamboo" structure on the leading face of each chip. At higher speeds, characterised by lower tangential forces, the wrinkles are less numerous per unit length indicating a decrease in the amount of plastic deformation. Tallysurf traces across the cuts also show a correlation between the factor B/A (See Figure 2) and the change in tangential force [13]. The decrease in F_T with cutting speed is thus accompanied by a corresponding decrease in plastic deformation and accordingly an increase in the effectiveness of chip removal.

4. DISCUSSION

When the diamond point penetrates the workpiece material, plastic deformation will occur. Simultaneously some kind of start-stop fracture mechanism along the primary deformation zone AB (in Figure 1) would be required for chip removal. Fracture would again require a large enough stress concentration along the planes AB to produce a microcrack at B or between A and B which either enlarges from or spreads toward B. The unexpected fact being observed is that for increasing cutting speed, less plastic flow is needed to obtain this stress concentration; notwithstanding an increase in temperature at the cutting point. An effective increase in brittle behaviour is thus observed up to some critical temperature after which plastic deformation starts to increase again.

The dislocation mechanism responsible for this apparent anomaly is unclear at present. If it is assumed that fresh dislocations which are created at A and B in Figure 1, move towards each other on slip planes along AB, an increase in ϕ with speed will shorten the effective slip planes and the component in the horizontal direction of the force along AB will become less. Studies on single crystals where the active slip planes cut the free surface at different angles, should provide interesting information.

REFERENCES

1. Takenaka, N., Annals of the C.I.R.P., 13, 183, 1966.
2. Eiss, N.S. Jr., Jour. Eng. Ind. Trans. of ASME, p. 463, Aug. 1967.
3. Sagarda, A.A., Sintetitcheskie Almazy, 2, 9, 1969.
4. Lindenbeck, D.A., Maschinemarkt, 76, 1892, 1970.
5. Gielisse, P.J. and Stanislao, J., The Science of Ceramic Machining and Surface Finishing, Nat. Bur. of Standards Special pub. 348, 1972.
6. Tanaka, Y, Ikawa, N. and Konishi, M., Techn. Reports of the Osaka University 20, 787, 1970.
7. Prins, J.F., Ind. Diamond Rev. p. 364, Sept. 1971.
8. Prins, J.F., Ind. Diamond Rev. p. 497, Dec. 1971.
9. Brecker, J.N., Komanduri, R. and Shaw, M.C. Annals of the C.I.R.P., 22, 219, 1973.
10. Komanduri, R. and Shaw, M.C., Nature, 248, 582, 1974.
11. For a review see Bailey, J.A., Wear, 31, 243, 1975.
12. Von Turkovich, B.F., Jour. of Eng. for Industry, p. 151, Feb. 1970.
13. Bredell, L.J. and Prins. J.F. To be published.

SUMMARY SESSION

Chairman: R. M. Latanision

SURFACE EFFECTS IN CRYSTAL PLASTICITY: CONCLUDING SUMMARY LECTURE

F. C. Frank

H. H. Wills Physics Laboratory
University of Bristol
Bristol, U.K.

It will be granted to me, I think, that a week's work cannot be summarized within an hour without omissions: so I apologize in advance to anyone whose favourite topic, or whose brilliant contributions to that topic, I fail to mention. I'm just not trying to mention everything worthy of attention that has been said.

What we were told about the investigative methods of surface science can, I think, be summarized because there was one clear consistent message--namely, that each of a variety of techniques is optimally suited to answering one kind of question about the surface. To answer that question you need only invest about 50,000 of some kind of units of account in the appropriate piece of ultra-high vacuum hardware, and maybe three years trouble-shooting on your particular problem, and you have the answer. If you then find you asked the wrong question, that's just too bad. It's well to consider your questions carefully first.

But it was with malice aforethought that I asked Charlie Duke a question about ice. That is an exceptional substance whose mechanical properties are of importance very close to and precisely *at* the melting point, and its vapour pressure is rather high. Its surface is rather important. None of those vacuum techniques are of any use for such a material. It seems that NMR and neutron scattering are the proper techniques to apply there, and those are things we didn't go into.

Several times during the meeting I found myself groaning inwardly "Willard, thou shouldst be living at this hour". Josiah Willard Gibbs, undoubtedly the greatest American scientist that ever lived, was mentioned on and off in connection with a formula, but seldom in connection with a principle, unless one takes a remark from one of our delegates that "only Gibbs and God never made mistakes" as a principle. Gibbs knew very well that it is sometimes necessary to represent the surface of a material as a mathematical surface, and considered carefully how that should be done. He never fell into the trap of supposing that it is a mathematical surface, or a plane, which can be so subtly misleading.

The geometrical pinning of dislocations by even very small roughness in a clean surface was something I didn't hear mentioned.* Bullough I think gave us a figure of 7eV per atomic

---

* Professor Nabarro tells me this was discussed in his workshop group. In a subsequent letter he told me that after the conference he had further discussion on this at Braunschweig with Professor Schwink and Dr. H. Neuhäuser. He writes: "Neuhäuser came up with an idea on the ζ-potential which seems to me to be very hopeful. Continuing your (and my working party's) idea that the surface of a crystal will usually have many ledges, and particularly when the crystal is undergoing dissolution, he suggested that the density of these ledges during a steady-state dissolution process would surely be greater the lower their surface energy. While it is far from obvious, it is at least reasonably plausible to say that we expect the energy of the ledge to have a term proportional to $+\zeta^2$, hence the density of ledges to be greatest at $\zeta = 0$, and the dislocation mobility to be least at this point. (This reverses the Masing-Raffelsieper argument that as a function of externally applied potential the enthalpy of a step will be greatest when it is uncharged.) It worries me a little that the same argument should apply in the case of a metal. We know that metals are different from non-metals in phenomena like indentation creep and the photomechanical effect, but I have not had time to check whether the ζ-potential effect is absent in metals or has merely not been studied in them. It seems to me that this idea does more than has been done before to resolve the Westwood-Rebinder controversy.

P.S. I sat on this for over a week wondering if I could really believe the argument. I am gradually coming to the conclusion that I can. Supposing I have an ambient solution which can contain either of two adsorbates. One of these does not reduce the surface energy as much as it might because it gives rise to a surface double layer which has a certain energy. The other gives rise to a double layer of the opposite sign. If I mix these, I can get the same adsorption energy without the surface double layer, and here I will expect to find most steps during dissolution."

These ideas of Neuhäuser and Nabarro bear thinking about.

unit of length as the elastic energy of a dislocation, and mentioned Cottrell's remark that in a crystal of 1 cm dimensions more than 1/2 of that energy is more than a micrometre away from the dislocation: but by the same token more than a fourteenth of that energy, 1/2 an eV, is closer than half a nanometre, so that an edge dislocation intersecting even a monostep increases its energy rather abruptly as it increases its length by the thickness of a monolayer. Add a comparable energy from one exposed lattice site, and we have a pinning energy of an eV or so, which should have considerable consequence. I do suppose that the removal of such geometrical pinning of dislocation ends plays a part in promoting ductility in the Joffe effect. Of course, when the experiments were done, the Griffith theory of fracture was in existence, and the dislocation theory of ductile yield was not, so people thought about little cracks in their interpretation, and I think have gone on doing so.

Really surface steps are very like dislocations. They exist in continuity with dislocations: they have not dissimilar energy. They interact at short range with dislocations in much the same way as dislocations do. Only, they do not have long-range elastic interactions, either with dislocations or each other. When we cover them with an impenetrable overlayer, then they being to have actions at a distance, with a range of the order of the overlayer thickness.

Gibbs, I think would have appreciated the point which Lothe made in discussion about the non-unique and non-local nature of crystal strength. As he said, a layer near the surface may differ from an interior layer either in dislocation density or in dislocation behaviour. In fact, there are excellent reasons to expect either larger or smaller dislocation densities near a surface, and doubtless both cases occur: but also, even at the same density dislocations are more manoeuverable near a surface where on one side they interact only with sympathetic images which never move out of turn. This difference is there as soon as we have or make a surface. If we now forget those surfaces, as we easily may if our terminology fails to distinguish "surface" from "near-surface layer", and interpret results in terms of regions of different flow stress acting in parallel we can arrive at distinctly contradictory conclusions.

If, with uniform dislocation density, freer dislocation motion near the surface makes the effective load-bearing thickness of a specimen of thickness D_1 only D_1-2d, then slicing the specimen will give us specimens of equal strength, though the sum of their yield loads will be less than that of the unsliced specimen. But, removal of layers from the surface, reducing the overall thickness from D_1 to D_2 will make a <u>more than proportional</u> reduction in yield load: so that talking in terms of local flow stress we shall think we have removed a harder layer. Fourie and Kramer could have the same condition in a specimen, and, making their different tests, describe it in opposite terms.

Two parts of one crystal are _not_ equivalent to those two parts made independent and loaded in parallel: they send dislocations into each other. Nor do I know of any proof that if one part is acting as pace-maker for another, it will necessarily be found either softer or harder when made independent.

I suppose that Fourie and Kramer's results are both correct, and not imcompatible. I think it is Kramer who takes the greater risks of using hypothesis-related terminology which may mislead.

We have heard that Rebinder only allowed his name to be attached to surface effects attributable to reduction in surface free energy. However, that is not a large renunciation. We know from Gibbs that there is a one-to-one relation between reduction in surface energy and adsorption. So Rebinder effects are attributable to adsorption: they may or may not be directly caused by the surface-energy reduction, but they must be inevitably correlated with it. That puts us back where we were.

MacMillan, discussing causes of Rebinder effects in his excellent paper, says "The remaining possibilities can conveniently be grouped under four headings:
   (i) dislocation egress effects
  (ii) surface drag effects
 (iii) surface anchoring effects
  (iv) electronic effects"

Why does he say "electronic"? Electric, yes, and in electronic semiconductors, clearly electronic, but in ionic crystals I must seek evidence to distinguish whether my charge carriers are electrons or ions.

MacMillan, Huntington and Westwood did successive polishing and etching experiments with MgO to convince themselves that pH at the surface influences dislocation mobility as deep into the crystal as 10 or 20 μm (for screws and edges respectively). That is such an important result that I feel that its 14-line presentation in the Phil. Mag. paper is hardly enough. I want to see the details of that dislocation configuration, and satisfy myself that there was no alternative way in which, say, the dislocations themselves could carry their message to the depths.

Why do we have this precise correlation with the ζ-potential? Correlation with some linear function of the ζ-potential, not necessarily having its zeros when ζ does, would be so much easier to understand. The ζ-potential is a measure of the amount and stand-off distance of space-charge in the mobile liquid. We think we are concerned with the amount and stand-off distance of space-charge in the adjacent solid. Is there some principle I don't know about which says that a charge layer at the surface must partition its balancing space-charges fairly between the two media that surface separates? If so, how is that rule evaded in a vacuum? Or is it an illusion to suppose there are charge layers at the surface at all: do we only have space charge layers of opposite sign balancing each other, one in each medium? Or is it merely that the cases known so far are selected ones, discovered when the rules work?

One nice unity of ideas in disparate application has appeared. It is charge transfer causing charge repulsion, as the leaves of a gold-leaf electroscope fly apart. In Haasen's interpretation of doping effects on the mobility of dislocations in germanium, it is charge transfer to the dislocation, charge repulsion, irrespective of the sign of the charge, tending to destabilize the straight configuration of the charged line, and thus promoting the formation of double kinks. In embrittlement phenomena, whether by hydrogen or by liquid metals, it is proposed, I do not know by whom first, that charge transfer to metal atoms at the tip of a crack makes them repel and come apart. And Johnson remarked in our discussion that that is just what happens in catalytic cracking of a hydrogen molecule.

Someone has rightly commented on the unreliability in arguments that X cannot control Y because there is so much more energy in Y than in X. The usual analogy is to a trigger or to a switch. Rather more relevant sometimes, I think, is Boys' jet. Sir C. V. Boys, F.R.S. gave a variety of instructive demonstrations with jets of water (described in "Soap-Bubbles, Their Colours, and the Forces Which Mould Them": London, Society for Promoting Christian Knowledge, 1912. Chapter on "Liquid Cylinders and Jets"). A fountain jet, issuing as a smooth cylinder from its orifice, soon breaks up into drops, which spatter widely: Boys received them on a large sheet of paper to show this happening and make it audible. A liquid cylinder is unstable under surface tension if its length exceeds its periphery; waists any further than this apart grow continually deeper until the cylinder breaks into drops. The instability increases with the wave-lenth of perturbation, but the longer the wave-length the further water has to flow in the development of a perturbation. Waists grow fastest when about $4\frac{1}{2}$ diameters apart, but nearly as fast for rather longer or shorter separations. Left to itself, the jet breaks into drops of irregular size, corresponding to about this separation of waists. Being of irregular size, they have irregular velocities, collide and scatter. With a tiny initial perturbation produced by a musical tone - one way of producing it was to blow across the hole in the end of a key - he could make all drops have the same size. Then they didn't collide, and all hit the paper target at the same spot. Here we have an example of a very weak regulator synchronizing some motions, preventing an interference which would otherwise occur, with large consequential effects. The possibility of such phenomena in dislocation motions seems real to me: in particular that surface sources might act as regulators in this sense when interior sources exhibit more complex mutual interference.

In this complicated subject, where there is usually more than one possible explanation for any result, I have been looking throughout the conference for simple clean experiments which prove one point. There weren't many of these. I thought Lohne's result, showing a dramatic breach of Schmidt's law in predicting secondary slip modes, and that a satisfactory prediction was obtained through surface-promoted cross-slip as a multiplying mechanism was one of these.

I am left with a few little questions. In metal turning, with continuous chip formation, the whole thickness of the chip must deform. I see difficulties about how the surfactant reaches the inner side of the chip, but none about it reaching the outer side. Why does the surfactant benefit go to zero at a certain limiting rate? Why don't I retain <u>half</u> that benefit at any speed?

When tools <u>rub</u> over a work-piece, in an electrolyte, they shear the space-charge layer, the seat of the ζ-potential, and should therefore produce kino-electric potentials. Has anyone thought about consequences of that?

Overall, I fear this is an inescapably complicated subject. One man may favour one interpretation of a phenomonon, another man an alternative one: almost inevitably he will prefer an explanation which he was the first to think of. But <u>both</u> may be right. The basic principle of statistical mechanics applies, that anything which <u>can</u> happen will. Any chemomechanical mechanism you can devise either contains its own logical inconsistency, or it is a real mechanism, likely to happen in <u>some</u> case.

I finish on the keynote quotation from Dr. Fischer:

   THE REALITY IS SOMEWHAT
    MORE COMPLICATED!

LIST OF PARTICIPANTS

NAME	ADDRESS
Argon, A. S.	Department of Mechanical Engineering, Massachusetts Institute of Technology, Cambridge, Massachusetts 02139, U.S.A.
Basinski, Z. S.	Divison of Physics, National Research Council of Canada, Ottawa, Canada
Bauer, E.	Physikalisches Institut, Technische Universität Clausthal, 3392 Clausthal-Zellerfeld, Leibnizstrasse 4, F.R.G.
Bonzel, H. P.	Institut für Grenzflächenforschung und Vakuumphysik, Kernforschungsanlage Jülich, 517 Jülich, F.R.G.
Brookes, C. A.	Department of Engineering Science, University of Exeter, Exeter EX4 4QF, United Kingdom
Bullough, R.	Theoretical Physics Division, AERE Harwell, Didcot, Berkshire, United Kingdom
Castaing, J.	Laboratory of Physical Metallurgy, C.N.R.S.-Bellevue, 92190-Meudon, France
Cetincelik, M.	Association of Engineers and Architects in Turkey, Chemical Engineering Department, P.O. Box 400-Kizilay, Ankara, Turkey
Champier, G.	Institut National Polytechnique de Nancy, Laboratoire de Physique de Solide, Ecole des Mines, 54042 Nancy Cedex, France
Clarkin, P. A.	Metallurgy Program, Office of Naval Research, Department of the Navy, Arlington, Virginia 22217, U.S.A.
Cuthrell, R. E.	Surface Metallurgy Division-5834, Sandia Laboratories, Albuquerque, New Mexico 87115, U.S.A.
Davidson, D. L.	Southwest Research Institute, P. O. Drawer 28510, San Antonio, Texas 78284, U.S.A.

Delamare, F.	Ecole Nationale Superieure des Mines, 60, Boulevard Saint-Michel, 75272 Paris Cedex 06, France
Duke, C. B.	Webster Research Center, Xerox Corporation, Webster, New York 14580, U.S.A.
Duquette, D. J.	Materials Division, Rensselaer Polytechnic Institute, Troy, New York 12181, U.S.A.
Engell, H.-J.	Max-Planck-Institut für Eisenforschung, 4 Düsseldorf, Max-Planck-Strasse 1, F.R.G.
Escaig, B.	Lab. Struct. Prop. de l'Etat Solide, Universite des Sciences et Techniques de Lille, B.P. 36, 59560 Villeneuve d'Asco, France
Firestone, R. F.*	Department of Metallurgical and Materials Engineering, School of Engineering, University of Pittsburgh, Pittsburgh, Pennsylvania 15213, U.S.A.
Fischer, T. E.	Exxon Corporate Research Laboratories, P.O. Box 45, Linden, New Jersey 07036, U.S.A.
Fourie, J. T.	National Physical Research Laboratories, P. O. Box 395, Pretoria, South Africa
Frank, F. C.	University of Bristol, H. H. Wills Physics Laboratory, Royal Fort, Tyndall Avenue, Bristol BS8 1TL, United Kingdom
Frank, W.	Max-Planck-Institut für Metallforschung, Institut für Physik, 7 Stuttgart, Seestrasse 75/EWB, F.R.G.
Gatos, H. C.	Department of Materials Science and Engineering, Massachusetts Institute of Technology, Cambridge, Massachusetts 02139, U.S.A.
George, A.	Institut National Polytechnique de Nancy, Laboratoire de Physique de Solide, Ecole des Mines, 54042 Nancy, France
Gibala, R.	Department of Metallurgy and Materials Science, Case Western Reserve University, Cleveland, Ohio 44106, U.S.A.

Graf, L.	Max-Planck-Institut für Metallforschung, Institut für Metallkunde, 7000 Stuttgart 1, F.R.G.
Greenfield, I. G.	School of Engineering, University of Delaware, Newark, Delaware 19711, U.S.A.
van Groenou, B.A.	Philips Research Laboratories, Eindhoven, The Netherlands
Hirth, J. P.	Department of Metallurgical Engineering, Ohio State University, Columbus, Ohio 43210, U.S.A.
Johnson, K. H.	Department of Materials Science and Engineering, Massachusetts Institute of Technology, Cambridge, Massachusetts 02139, U.S.A.
Johnson, R. M.	Mechanical Engineering Department, University of Texas at Arlington, Arlington, Texas 76019, U.S.A.
Kitajima, S.	Kyushu University, Faculty of Engineering, Department of Nuclear Engineering, Fukuoka, 812, Japan
Kramer, I. R.	Naval Ship Research and Development Center, Materials Department, Code 2802, Annapolis, Maryland 21402, U.S.A.
Kröner, E.	Max-Planck-Institut für Metallforschung, Institut für Physik, 7 Stuttgart 1, Azenbergstrasse 12, F.R.G.
Latanision, R. M.	Department of Materials Science and Engineering, Massachusetts Institute of Technology, Cambridge, Massachusetts 02139, U.S.A.
Lohne, O.	Universitet I Trondheim, Institutt for Fysikalsk Metallurgi, Norges Tekniske Hogskole, 7034 Trondheim, Norway
Lothe, J.	Institut of Physics, University of Oslo, Olso, Norway
Macmillan, N. H.	Materials Research Laboratory, The Pennsylvania State University, University Park, Pennsylvania 16802, U.S.A.

Maugis, D.	Equipe de Recherche de Mechanique des Surfaces du C.N.R.S.-Bellevue, 1, Place A. Briand, 92190 Meudon, France
Meister, G.	Physikalisches Institut, Technische Universität, 3992 Clausthal-Zellerfeld, Leibnitzstrasse 4, F.R.G.
van der Merwe, J.H.	University of Pretoria, Department of Physics, Pretoria, South Africa
Mills, J. J.	Martin Marietta Laboratories, 1450 South Rolling Road, Baltimore, Maryland 21227, U.S.A.
Moskovitz, J.*	Department of Materials Science and Engineering, Massachusetts Institute of Technology, Cambridge, Massachusetts 02139, U.S.A.
Mughrabi, H.	Max-Planck-Institut für Metallforschung, Institut für Physik, 7 Stuttgart 1, Azenbergstrasse 12, F.R.G.
Mularie, W. M.	3M Company, IEP Building 207, 3M Center, St. Paul, Minnesota 55101, U.S.A.
Müller, H.	DECHEMA, 6 Frankfurt/Main, Theodor-Heuss-Allee 25, F.R.G.
Nabarro, F.R.N.	Department of Physics, University of the Witwatersrand, Johannesburg, South Africa
Neumann, P.	Max-Planck-Institut für Eisenforschung, 4 Düsseldorf, Max-Planck-Str. 1, F.R.G.
Niehus, N.	Physikalisches Institut, Technische Universität, 3392 Clausthal-Zellerfeld, Leibnizstrasse 4, F.R.G.
Oliveira, A.M.C.	Chemistry Department, Faculty of Science and Technology, University of Coimbra, Coimbra, Portugal
Oriani, R. A.	United States Steel Corporation, Research Laboratory, 125 Jamison Lane, Monroeville, Pennsylvania 15146, U.S.A.
Pertsov, N. V.	Chair of Colloid Chemistry, Moscow State University, Moscow, U.S.S.R.

Preece, C. M.	Department of Materials Science, State University of New York, Stony Brook, New York 11790, U.S.A.
Prins, J. F.	University of Pretoria, Department of Physics, Pretoria, South Africa
Schröter, W.	IV. Physikalisches Institut, Universität Göttingen, 34 Göttingen, Bunsenstrasse 11-15, F.R.G.
Schwink, Ch.	Institut A für Physik, Technische Universität Braunschweig, 33 Braunschweig, Mendelssohn-strasse 1, F.R.G.
Seeger, A.	Max-Planck-Institut für Metallforschung, Institut für Physik, 7 Stuttgart 1, Azenbergstrasse 12, F.R.G.
Shchukin, E. D.	Chair of Colloid Chemistry, Moscow State University, Moscow, U.S.S.R.
Smialowski, M.	Institut of Physical Chemistry, Polish Academy of Sciences, ul. Kasprzaka 44/52, Warsaw, Poland
Sparnaay, M. J.	Philips Research Laboratories, Eindhoven, The Netherlands
Staehle, R. W.	Department of Metallurgical Engineering, Ohio State University, Columbus, Ohio 43210, U.S.A.
St. John, C. F.	Centre des Materiaux, Ecole des Mines de Paris, B.P. 114, 91102 Corbeil-Essonnes, France
Swain, M. V.	Cavendish Laboratory, Department of Physics, University of Cambridge, Cambridge CB3 OHE, England
Szklarska-Smialowska, Z.	Institut of Physical Chemistry, Polish Academy of Science, ul. Kasprzaka 44/52, Warsaw 42, Poland
Tyson, W. R.	Department of Energy, Mines, and Resources PMRL, CANMET, 568 Booth Street, Ottawa, Ontario, Canada K1A 0G1

Viefhaus, H.	Max-Planck-Institut für Eisenforschung, 4 Düsseldorf, Max-Planck-Str. 1, F.R.G.
Weinberg, W. H.	Department of Chemical Engineering, California Institute of Technology, Pasadena, California 91109, U.S.A.
Weissmann, S.	Department of Mechanics and Materials Science, College of Engineering, Rutgers University, New Brunswick, New Jersey 08903, U.S.A.
Wessel, A.	Institut für Werkstoffkunde (B), Technische Universität, 3 Hannover, Appelstrasse 24A, F.R.G.
Westwood, A.R.C.	Martin Marietta Laboratories, 1450 South Rolling Road, Baltimore, Maryland 21227, U.S.A.
Wilsdorf, H.G.F.	Department of Materials Science, University of Virginia, Thornton Hall, Charlottesville, Virginia 22901, U.S.A.

---

* Travel support provided through an NSF NATO Travel Grant.

If you have any concerns about our products,
you can contact us on
ProductSafety@springernature.com

In case Publisher is established outside the EU,
the EU authorized representative is:
**Springer Nature Customer Service Center GmbH
Europaplatz 3, 69115 Heidelberg, Germany**

Printed by Libri Plureos GmbH
in Hamburg, Germany